Weather Radar and Hydrology

Edited by:

ROBERT J. MOORE
Centre for Ecology & Hydrology, Wallingford OX10 8BB, UK

STEVEN J. COLE
Centre for Ecology & Hydrology, Wallingford OX10 8BB, UK

ANTHONY J. ILLINGWORTH
Department of Meteorology, University of Reading, Reading RG6 6BB, UK

Proceedings of a symposium held in Exeter, UK, April 2011

IAHS Publication 351
in the IAHS Series of Proceedings and Reports

Published by the International Association of Hydrological Sciences 2012

IAHS Publication 351

ISBN 978-1-907161-26-1

British Library Cataloguing-in-Publication Data.
A catalogue record for this book is available from the British Library.

The papers included in this volume have been reviewed and some were extensively revised by the Editors, in collaboration with the authors, prior to publication.

IAHS is indebted to the employers of the Editors for the invaluable support and services provided that enabled them to carry out their task effectively and efficiently.

Publications in the series of Proceedings and Reports are available from:
IAHS Press, Centre for Ecology and Hydrology, Wallingford, Oxfordshire OX10 8BB, UK
tel.: +44 1491 692442; fax: +44 1491 692448; e-mail: jilly@iahs.demon.co.uk

Printed by Berforts Information Press

Cover illustration: The Cobbacombe weather radar, 1-km river flows from the Grid-to-Grid hydrological model over southwest England, and aftermath of the Lynmouth 1952 flood disaster.

nd Hydrology

Recent IAHS Publications

GRACE, Remote Sensing and Ground-based Methods in Multi-Scale Hydrology

Editor Mohsin Hafeez
Co-editors
    Nick Van De Giesen, Earl Bardsley,
    Frederique Seyler, Roland Pail &Makoto Taniguchi

Publ. 343 (2011) ISBN 978-1-907161-18-6, 196 + x pp. £50.00

Recent advances in measuring hydrological variability by means of the Gravity Recovery and Climate Experiment (GRACE) mission, and other remote sensing platforms (TRMM, Landsat and MODIS), offer great potential for estimating spatio-temporal surface water balances, spatially-averaged water budgets, hydrodynamics, hydrological processes, and characterization of groundwater systems in gauged and ungauged basins, at regional and global scales. In parallel, advances in ground-based measurement techniques, such as distributed temperature sensing and geological-weighing lysimeters, are being incorporated into research and practice for determining hydrological parameters. Collectively, the 30 peer-reviewed papers provide an overview of these techniques and their use with hydrological models for understanding multi-scale hydrological processes.

Remote Sensing and Hydrology

Editors: Christopher M. U. Neale & Michael H. Cosh
Publ. 352 (August 2012) ISBN 978-1-907161-27-8

Remote sensing continues to expand the ability of scientists to study hydrological processes. With each new technological development, more of the hydrological cycle is revealed. This impacts both the scientific understanding of hydrological processes and the models used for forecasting, and so the ability to improve decision-making processes and other applications is increasing. This compendium of >100 papers from the latest ICRS International Symposium on Remote Sensing and Hydrology, reviews the status of technologies and highlights new directions and opportunities for hydrological remote sensing.

Revisiting Experimental Catchment Studies in Forest Hydrology

Editors A. A. Webb, M. Bonell, L. Bren, P. N. J. Lane,
    D. McGuire, D. G. Neary, J. Nettles, D. F. Scott,
    J. D. Stednick & Y. Wang

Publ. 353 (April 2012) ISBN 978-1-907161-31-5, 240 + viii pp. £56.00

Most of what we know about the hydrological role of forests is based on paired catchment experiments whereby two neighbouring forested catchments are jointly monitored during a calibration period of several years, after which one of the catchments is kept untouched as a reference (control), while the second is submitted to a forest treatment (impact). This volume, generated from a workshop that gathered forest hydrologists from around the world, with the aim of revisiting results and promoting a renewal of international collaboration on this topic, is divided into four sections:

1   Addressing new questions using historical data sets
2   Impacts of fires
3   Water quality and sediment loads
4   Ecosystem services

BENCHMARK PAPERS IN HYDROLOGY

The IAHS Series that collects together in themed volumes the papers that provided the scientific foundations for hydrology.

RIPARIAN ZONE HYDROLOGY AND GEOCHEMISTRY

by *T. P. Burt, G. Pinay & S. Sabater*
BM5 ISBN 978-1-907161-09-4 (2010) A4 format, hardback, 490 pp, £65.00

Study specifically of riparian zones is relatively new in hydrology, and while the oldest of the 36 benchmark papers selected for this volume dates to 1936, others were published in the 1970s and 1980s. They are grouped under the topics: Landscape ecology; Hydrology of the riparian zone; Linking riparian zone hydrology to solute transport; Biogeochemical processes and methods; Riparian buffering of surface and subsurface flows; and In-stream processes. Together, the papers and the editors' commentaries map the breakthroughs in the development of this important subdiscipline.

HYDRO–GEOMORPHOLOGY, EROSION AND SEDIMENTATION

by *Michael J. Kirkby*
BM6 ISBN 978-1-907161-14-8 (2011) A4 format, hardback, 640 pp, £70.00

A systematic analysis of the relationships between hydrology and geomorphology with commentaries on the papers that have been most influential in the development of research at the hydrology/geomorphology interface. Thirty-seven papers are reprinted in full or in part, the majority published pre-1970, including early contributions by Fisher (1866), Davison (1889) and Gilbert (1909), and seminal papers by Hack, Strahler, Wolman & Miller, and Melton, among others.

FOREST HYDROLOGY

by *David R. DeWalle*
BM7 ISBN 978-1-907161-17-9 (2011) A4 format, hardback, 474 pp, £65.00

The papers selected include the early review of forest and water by Zon (1927) and the Wagon Wheel Gap paired watershed study (Bates & Henry, 1928) report, but covers all aspects from evapotranspiration to water yields and quality to soil erosion.

 SAHRA   Sponsored by SAHRA, University of Arizona

IAHS Publications can be ordered from the online bookshop at
www.iahsmembers.info/shop.php

or by contacting:
IAHS Press, CEH Wallingford, Oxfordshire OX10 8BB, UK
email: jilly@iahs.demon.co.uk
tel: +44 (0) 1491 692442   fax: +44 (0) 1491 692448

Full details of publications available at www.iahs.info

Preface

The topic of "Weather Radar and Hydrology" brings together important science and technology challenges concerning the monitoring and forecasting of rainfall over space and time and how the pattern of rainfall is transformed by a varied landscape into surface water runoff and river flow across a city, region or country. It has significant practical application across a range of water resource functions, including flood forecasting and warning, flood design, urban drainage management, water supply and environmental services. The subject concerns developments in weather radar technology in combination with advances in hydrological application, and thus is of relevance to researchers in these fields, practitioners in the water industry and suppliers of weather radar systems.

These Proceedings bring together over 100 peer-reviewed papers presented at the International Symposium on "Weather Radar and Hydrology" (WRaH 2011), convened from 18 to 21 April 2011 at the University of Exeter, UK: see www.WRaH2011.org for details. The symposium was the 8th in a series that began in 1989 at the University of Salford (UK) under the title "Hydrological Applications of Weather Radar". Subsequent symposia have been convened in Germany, Brazil, USA, Japan, Australia and France. WRaH 2011 marked a return to the UK after 20 years of successful symposia across the world. More than 250 people attended from a range of organisations – governments, academia, research bodies, national hydrometeorological services and consultancies – and travelled from countries spanning four continents. WRaH 2011 provided a forum for the exchange of experiences and ideas on the use of weather radar in hydrology with a particular emphasis on user applications for flood forecasting and water management. These Proceedings serve as a valuable record of this activity.

The set of papers are arranged in the Proceedings under seven themes as follows.

(1) **Weather radar theory, technology and systems** including the topics: radar network compositing; correcting for attenuation, clutter, bright band and vertical profile of reflectivity (VPR) effects; radar reflectivity *versus* rain-rate (Z-R) relations; polarimetric radars at X-, C- and S-band; dual frequency, microwave and adaptive phase array radar technology; rain microphysics; and long-term diagnostic monitoring.

(2) **Rainfall estimation and quality control** including the topics: multi-sensor precipitation estimation; polarimetric precipitation estimation; VPR and orographic corrected precipitation estimation; data quality-control; blended radar and raingauge rainfall estimation; performance evaluation of precipitation estimations; and space–time variability of rainfall estimates.

(3) **Rainfall forecasting (nowcasting and numerical weather prediction)** including the topics: precipitation field advection estimation; blended radar rainfall advection and numerical weather prediction (NWP) model forecasts; nowcasting of orographic rain; probabilistic forecasting using ensembles; radar data assimilation for NWP; radar quality monitoring using NWP; and convective cell identification.

(4) **Uncertainty estimation** including the topics: precipitation estimation error models; bias in radar calibration; quality indices for radar data; probabilistic rainfall warning; and impact of rainfall uncertainty on flow forecasts.

(5) **Hydrological impact and design studies** including assessing the impact of summer thunderstorms and hailstorms, and a multifractal study of storm dynamics, using weather radar.

(6) **Hydrological modelling and flood forecasting** including the topics: operational perspectives on flood forecasting; rainfall estimation for flood forecasting including use of

X-band and polarimetric radars, and raingauge and radar data in combination; use of rainfall forecasts in deterministic and ensemble form for flood forecasting; influence of rainfall spatial variability and storm motion on modelled flood response; distributed hydrological models using gridded rainfall estimates for catchment, region and countrywide flood warning; data-based flood forecasting; and hydrological modelling using radar rainfall for hydropower generation, water quality and environmental management.

(7) **Urban hydrology and water management applications** including the topics: review of radar for urban hydrology; radar resolution requirements for urban applications; precipitation forecasting for urban surface runoff and flow prediction; rainfall depth-duration-frequency analysis and use with radar for monitoring urban drainage compliance; Z-R relations developed for urban and water management applications; and use of radar in predicting bathing water quality.

These seven themes serve to provide structure to the contents of the Proceedings, although in practice it is common for papers to overlap more than one theme.

The "Inter-Agency Committee on the Hydrological use of Weather Radar" (www.iac.rl.ac.uk) initiated and coordinated the WRaH 2011 Symposium, with the Royal Meteorological Society and the British Hydrological Society serving as joint convenors. The committee and the society convenors are thanked for their significant support. Members of the WRaH 2011 scientific committee served as reviewers of the papers published here: many thanks are due for their hard work and constructive suggestions that commonly led to a paper of improved quality. The editors of these proceedings served on behalf of the Inter-Agency Committee and as members of it.

Publication of these Proceedings by IAHS Press was managed by Cate Gardner with Penny Perrins responsible for its production: many thanks are due to their help and encouragement.

Some papers from these Proceedings have been developed further for publication in a Special Issue of the *Hydrological Sciences Journal* on "Weather Radar and Hydrology".

ROBERT J. MOORE
Centre for Ecology & Hydrology, Wallingford OX10 8BB, UK

STEVEN J. COLE
Centre for Ecology & Hydrology, Wallingford OX10 8BB, UK

ANTHONY J. ILLINGWORTH
Department of Meteorology, University of Reading, Reading RG6 6BB, UK

ACKNOWLEDGEMENTS

The International Symposium on "Weather Radar and Hydrology" (WRaH 2011) was jointly convened by the Royal Meteorological Society and the British Hydrological Society.

The following government and private bodies are thanked for their support of WRaH 2011:

Environment Agency

Scottish Environment Protection Agency

Met Office

Centre for Ecology & Hydrology (Natural Environment Research Council)

Baron Services

Enterprise Electronics Corporation

Gematronik

Halcrow

Hydrologic (The Netherlands)

Hydrometeo

Vaisala

Contents

1   Weather radar theory, technology and systems

2   Rainfall estimation and quality control

3 Rainfall forecasting (nowcasting and numerical weather prediction)

4   Uncertainty Estimation

5   Hydrological impact and design studies

6 Hydrological modelling and flood forecasting

7   Urban hydrology and water management applications

1 Weather radar theory, technology and systems

Weather Radar and Hydrology
(Proceedings of a symposium held in Exeter, UK, April 2011) (IAHS Publ. 351, 2012).

3

Weather radar for hydrology – the UK experience and prospects for the future

MALCOLM KITCHEN

Met Office, Fitzroy Rd, Exeter EX1 3PB, UK
malcolm.kitchen@metoffice.gov.uk

Abstract The national weather radar network in the UK has now been operational for a quarter of a century. It was established by a consortium of agencies to provide a real-time rainfall monitoring capability. Today those same agencies, and their successors, are still involved in the maintenance and development of this national infrastructure on behalf of the wider stakeholders. An attempt is made here to identify some lessons that have been learnt along the way, and suggest how the benefit to hydrology can be increased in the next 25 years.

Key words weather radar; rainfall; refractivity

USE OF RADAR IN HYDROLOGY

The Lynmouth flood of 1952 was associated with the greatest loss of life in the UK from a fluvial flood in historical times. At the time of this flood, there were no warning systems in place and indeed very little possibility of warnings based on real-time observation of rainfall over a catchment. Weather radars in the 1950s looked rather different to those today (see Fig. 1), and there were technological limitations which prevented any sort of routine operation or regular data supply. On the other hand, the physical principles underlying the measurement of rainfall by radar remain exactly the same today. Many of the limitations and uncertainties in the radar rainfall technique were understood at that time: the bright band, the assumptions involved in transforming from reflectivity to rainfall, and the problems with measurements at longer ranges. Although considerable progress has been made with the understanding and reduction of these uncertainties, it has been hard-won and frustratingly slow.

The technical limitations of the early radars, which prevented any operational exploitation, have been largely overcome. In the UK, the weather radar networks can deliver high-resolution data to users within minutes and with availabilities in the high 90s%. Whereas instabilities in the systems commonly used to cause serious systematic biases (>3 dB) between different radars in the network, the radars' hardware is now much more stable in performance (<1 dB). It is to be

(a)   (b)

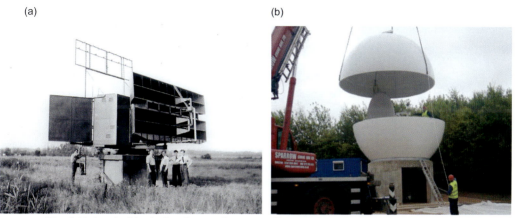

Fig. 1 (a) A radar identical to the very first Met Office weather radar – installed at East Hill, Dunstable, UK in 1946. (b) Prototype operational dual-polarisation radar being assembled at Exeter, UK, 2009.

expected that technological development will lead to gains in the accuracy of rainfall measurement; but technology does not offer complete answers by itself, and optimal solutions normally involve complex combinations of techniques.

The struggle for accuracy

In a project to pilot the radars to be deployed in the UK operational network, Harrold *et al.* (1975) reported an RMS error in measured storm rainfall accumulations as low as 13%. Whilst the operational radar network generally provided excellent qualitative data to assist flood and weather forecasters, it proved very difficult to routinely replicate the level of accuracy achieved in the closely monitored and expertly supervised R&D environment. Furthermore, during the 1980s, most of the available radar expertise was focused on the physical expansion of the network, and the limited resources then available to work on data quality were committed to the direct treatment of the symptoms, rather than the causes. Real-time adjustment of the radar data using gauges was the main weapon for trying to deal with some worryingly large errors in the basic data. This period, when the radar network was rapidly expanding to cover almost the entire UK land area (Fig. 2), was then also a time when new potential users in the water industry experienced some disappointment with the accuracy of the rainfall products.

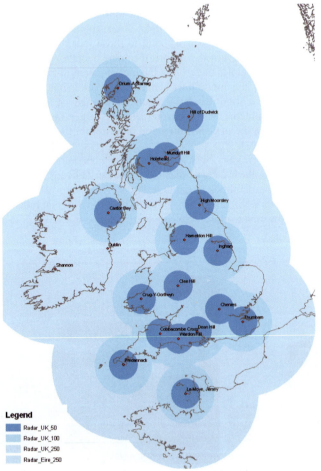

Legend

- Radar_UK_50
- Radar_UK_100
- Radar_UK_250
- Radar_Eire_250

Fig 2 The UK weather radar network in 2011. Range rings are at 50, 100 and 250 km. The locations of radars in Eire and the States of Jersey are also marked.

Fortunately, once the radar network reached more or less its present extent in the early 1990s, more resources became available to work on quality improvement. More computer power and communications bandwidth also facilitated the development of more sophisticated data correction algorithms. These worked alongside gauge adjustment to correct the relatively large, but localised, errors due to, e.g. the bright band, and orographic growth of rainfall below the radar beam. Following a review commissioned by the Environment Agency of England and Wales, a fundamental change was made in the data processing, whereby the quality control and processing of the data was relocated from the radar sites to a central processing system.

The last 15 years has seen a slow improvement in the accuracy of rainfall estimates from radar, although accuracy is still some way short of some key hydrological requirements (see Fig. 3). The current position is perhaps typified by the Ottery St Mary storm of 30 October 2008, exactly the sort of extreme intensity, small-scale event that represents the *raison d'être* for the radar network (Fig. 4). Extensive post-event analysis has demonstrated that the radar products have been shown to represent very good quality information. However, the uncertainty associated with the radar estimates was too large to produce flood forecasts with an acceptable level of confidence. In this case, as in many others, the areal extent of the storm was too small for there to be corroborating gauge data available in real-time.

THE NEXT 25 YEARS - SOME OPPORTUNITIES AND A THREAT

Prospects for improved accuracy in hydrological applications

In the UK, as in many other countries presently, the current generation of weather radars is being replaced with dual-polarisation systems (see Fig. 1(b)). Dual-polarisation offers good prospects for

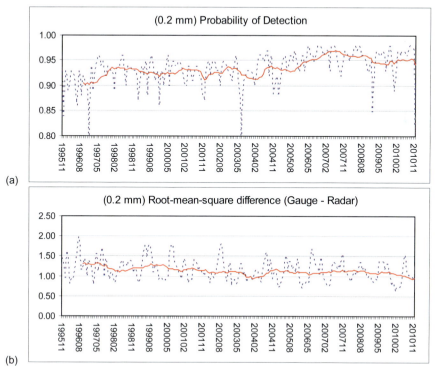

Fig. 3 (a) A 15-year time series of the monthly probability of the radar network detecting hourly gauge rainfall above a threshold of 0.2 mm. (b) 15-year time series of the monthly RMS differences between radar and gauge hourly accumulations above a threshold of 0.2 mm.

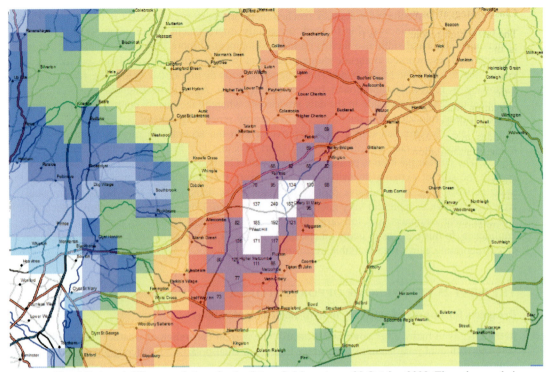

Fig. 4 Radar rainfall accumulations for the Ottery St Mary storm, 30 October 2008. The colour scale is logarithmic with the white area denoting accumulations of greater than 128 mm in 3 h. The rainfall squares are of dimension 1 km.

improved quality control and correction. It helps remove some of the remaining uncertainty in the transformation of reflectivity measurements into rainfall estimates. A reasonable expectation should be continued incremental improvement, rather than a series of dramatic step changes.

In the short-term, a very profitable area of development is also in the real-time estimation of uncertainty, and its communication to downstream users. There is a wealth of information available as to the likely accuracy of radar rainfall estimates. For example, the level of agreement with gauges in the recent past, the magnitude of corrections that have been applied to the data, and the tractability of the errors known to be affecting the radar data at a particular location in space and time (often a function of the meteorological situation). If we accept that uncertainty is one of the main limitations on the exploitation of radar data in operational hydrology, then skilful estimates of uncertainty should enable the maximum benefit to be derived from the data. It will permit the data to be used optimally in objective decision-making and risk management processes – just as probabilistic rainfall forecasts are an effective response to the limitations on the predictability of the future atmospheric state.

Whilst computer power and communications bandwidth has allowed the spatial resolution of radar rainfall to steadily improve (see Harrison *et al.*, 2009), the sampling interval has remained the same throughout. A mismatch has built up between the best available spatial resolution radar data (1 km), and the sampling interval (5 min), given that the typical advection speeds of storms is of the order of 1 km per minute. Even for slow-moving convective storms with typical lifetimes of an hour, a 5 min sampling is probably not adequate to resolve temporal variations in the surface rainfall. Some mitigation is achieved using interpolation techniques, but further thought is being given as to whether there needs to be some improvement in the radar sampling characteristics.

Closer integration with Numerical Weather Prediction (NWP)

The latest high-resolution NWP model for the UK has a grid spacing of just 1.5 km. Weather radar is the only mainstream observing system capable of delivering data on this horizontal scale. Although some surface rainfall data derived from radar has been assimilated into European scale NWP models for the last decade, work is now in hand to assimilate a much higher volume of reflectivity data into the models, at the height at which it is measured, and also to use the highest resolution radar winds retrieved by the Doppler technique.

As well as the trend towards higher spatial resolution in NWP, the development of NWP includes more explicit modelling of the microphysical and precipitation processes, and more frequent assimilation of observational data. Given these trends, it seems probable that at some point in the next 25 years (for example), the best estimate of surface rainfall at any time (including observation time) will come from within these NWP models, rather than by "external" physical or statistical correction of radar data. Reconciliation and combination of different rainfall measurements from radar and gauges could also take place within the NWP model, as part of the assimilation process. In turn, the radar data will have a much greater influence on the rainfall forecasts, and there is the prospect of achieving higher reliability and/or longer lead times in hydrological forecasts.

Measurement of another component of the water cycle

Fabry (1997) suggested that weather radars should be able to measure changes in the refractive index of the air by detecting small changes in the apparent range of fixed ground clutter. The refractive index of the air is a function of temperature, pressure and water vapour content, but at mid-latitudes, changes over timescales of a few hours tend to be dominated by changes in humidity. In principle then, a weather radar can produce maps of a quantity strongly related to the humidity of the air near the ground, but only in those areas where there is potentially ground clutter contamination of the radar signal (about 25% of the land area of the UK). Research in several countries is now pointing to the feasibility of making operational measurements. The development is significant because convergence of moisture at low levels in the atmosphere can be the trigger for severe convective storms, and there is currently no other means of detecting this convergence on horizontal scales less than about 50 km. The hope is that if these humidity data can be assimilated into the new high-resolution NWP models, they will improve the forecasting of the onset of severe convective storms, and increase the lead time available for flood warnings.

Currently in the research domain is another idea put forward by Fabry (2001) – that weather radars can be used to detect the microwave emissions from storms. The radar receivers in the UK network are now sensitive enough to make these passive radiometric measurements, which are completely independent of the radar's own microwave transmissions. The measurement is partly a bulk estimate of the rain in the radar field of view, but with significant contributions (contamination) from other sources such as the protective radome and the ground surface. If it proves feasible to isolate the contribution from storms, this could either be assimilated directly into NWP models as a separate quantity, or else used as a strong constraint to correct the radar reflectivity measurements for the effects of attenuation.

The greatest threat to radar hydrology?

Frequencies currently utilised by weather radars are also particularly suitable for use in wireless communications. Arrangements for sharing the radar band between the different uses are in place in most parts of the world, but sharing relies upon effective governance and regulation in order to avoid problems with interference. Despite best efforts, there have now been numerous instances of disruptive interference to weather radar operations in the UK. It is essential that meteorological and hydrological users of radar work together with the regulator to maximise the level of protection afforded to the radar frequencies. Otherwise, there is a risk that operations could be compromised completely at some point in the future.

SUMMARY

In the 25 years since inception, the UK weather radar network has moved from a R&D demonstration to become established as an essential component of national infrastructure. The rainfall images and data have become familiar to thousands of professional users in the water industry and millions of the general public. Pooling of investment by the meteorological and hydrological communities has produced a relatively dense (by world standards) network of medium-sized C-band radars – well suited to the requirement and an excellent platform for further progress. The requirement for *observations* of heavy rainfall to support warning services remains the main driver behind the operation and development of the network. High-resolution NWP is the key to improving *forecast* lead time, and weather radar is poised to make a significant contribution in this field as well, providing rainfall, wind and humidity data at the required resolution. The radars themselves are proving to be remarkably versatile instruments, able to perform well beyond their initial design specification.

REFERENCES

Fabry, F., Frush, C., Zawadzki, I. & Kilambi, A. (1997) On the extraction of near-surface index of refraction using radar phase measurements form ground targets. *J. Atmos. Ocean Technol.* 14, 978–987.

Fabry, F. (2001) Using radars as radiometers: promises and pitfalls. In: *Preprints, 30th Int. Conf. on Radar Meteorology* (Munich, Germany, American Meteorol. Soc.)

Harrison, D. L., Scovell, R. W. & Kitchen, M. (2009) High resolution precipitation estimates for hydrological uses. *Water Management* 162(2), 125–135.

Harrold, T. W., Nicholas, C. A. & Collier, C. G. (1975) The measurement of heavy rainfall over small catchments using radar. *Hydrol. Sci. Bull.* 20(1–3), 69–76.

Weather Radar and Hydrology
(Proceedings of a symposium held in Exeter, UK, April 2011) (IAHS Publ. 351, 2012).

9

EUMETNET OPERA Radar Data Centre: providing operational, homogeneous European radar rainfall composites

STUART MATTHEWS[1], PASCALE DUPUY[2], ROBERT SCOVELL[1],
ANTOINE KERGOMARD[2], BERNARD URBAN[2], ASKO HUUSKONEN[3],
ALISON SMITH[1] & NICOLAS GAUSSIAT[1]

1 *Met Office, FitzRoy Road, Exeter EX1 3PB, UK*
stuart.matthews@metoffice.gov.uk
2 *Météo France, 42, avenue Corriolis, 31057 Toulouse, France*
3 *Finnish Meteorological Institute, PO Box 503, 00101-Helsinki, Finland*

Abstract The main objective of the third EIG EUMENET OPERA Programme is the development and operational running of a European Radar Data Centre (Odyssey). Odyssey, which went live in January 2011, has the ability to ingest raw polar volume radar products from almost 200 operational weather radars operated by European National Meteorological Services. Composites of rain-rate, maximum reflectivity and hourly accumulations are produced every 15 min at 2 km resolution. Odyssey's algorithms have been designed to process data in a consistent way, allowing Odyssey to generate homogenous radar products for the whole European domain. Shared operational capability across two centres, Météo France and the Met Office, provides high levels of operational resilience. It is anticipated that these new Odyssey products should improve flood forecasting capability (especially for large river basins, e.g. the Danube, Elbe and Rhine) in their own right or by helping produce more accurate short range NWP forecast products.

Key words operational; composite; multinational catchments; homogenous; quality index

ODYSSEY PROJECT

The main objective of the third EUMETNET OPERA programme is the development and operational running of a European Radar Data Centre (Huuskonen *et al.*, 2012). The development of Odyssey builds on previous OPERA work to develop a Pilot radar data hub. The Pilot radar data hub was developed and hosted by the Met Office and has been running since 2006 (Harrison *et al.*, 2006). Having verified that the concept of centralised processing was viable, EUMETNET gave the go-ahead for the Odyssey project. The project started in 2007–2008 by gathering user requirements (including JRC) and producing a technical specification for the outputs (Chèze *et al.*, 2009). In spring 2009, EUMETNET approved a joint Météo France–Met Office bid to develop Odyssey and development started in autumn 2009. The Odyssey development project, see Dupuy *et al.* (2010), was split into six functional modules each being led by Météo France or the Met Office. The first operational products were disseminated in January 2011 and the system is now in its early life support phase with full operational sign-off and acceptance expected at the spring 2011 OPERA meeting.

INCOMING DATA

One of the key aims of Odyssey was to receive and ingest raw polar volume radar observations from all operational radars within Europe. By processing "raw" radar scans (i.e. polar reflectivity), Odyssey can apply one set of radar processing algorithms across the whole European domain.

By February 2011 Odyssey was receiving data from 10 National Meteorological Services (UK, France, Iceland, Slovakia, Czech Republic, Poland, the Netherlands, Sweden, Finland and Estonia), potentially supplying data from 77 operational radars. More countries will be added during 2011 and Odyssey was expected to have almost full coverage over the European domain by the end of 2011.

Challenges

A major set of challenges revolve around producing a standardised set of raw radar data. Firstly, the OPERA programme needed to define exactly what was meant by "raw" observation and to clearly define these. OPERA decided that the only treatment that should be applied to these incoming data was the removal of ground clutter. OPERA also had to describe the output format of the individual scans sent to Odyssey. Thanks to the good work done for many years by the OPERA group towards exchanging and harmonizing radar data processing in Europe, these hurdles were overcome by the adoption of a common information model, named ODIM (OPERA Data Information Model). This information model was defined and accepted before the Odyssey development project started. During their 2009 autumn meeting, the OPERA members agreed on two input formats (HDF5 and BUFR) compliant with the ODIM. These two formats are the only formats that are accepted by Odyssey and the next step was for the supplier National Meteorological Services (NMS) to provide their operational raw radar data in one of these two formats according to the ODIM specifications. As several radar manufacturers are also involved in designing the radar production chains, converging towards two formats is not a trivial activity. To facilitate the production of compliant input files, OPERA offers assistance to data providers.

A second major constraint is the volume of raw data that Météo France and the Met Office receive; nearly 300 000 individual scans per day will be delivered to both centres when all OPERA countries supply data to Odyssey. This is made more challenging by the demanding user requirements for timeliness and availability.

Finally, there is the conundrum of how to accurately record different types of precipitation across the whole continent. For example, northern European countries want greater resolution of snowfall measurements, whilst southern European nations will be more interested in convective storms. Meeting these conflicting demands may require an increase in data resolution and therefore file sizes.

Pre-processing

To produce a homogeneous product for Europe a set of pre-processing algorithms and quality measures needs to be agreed and then applied to the incoming data. At this time, pre-processing is in its infancy on Odyssey and we expect to see significant improvements over the coming years.

Current pre-processing

Currently, only a very basic pre-processing is applied on Odyssey. These are currently limited to:

- A consistency check of the received scan against the details stored in the Odyssey database. If the scan is not found in the database, the details of the new scan are added to the database and some pre-calculations are performed to allow the composite algorithm to be more efficient when subsequent scans are received.
- A simple estimate of the quality of the radar observation is produced, for each range gate, which is based on the height of the radar beam. The quality value is given by $u(r) = \exp(-\text{alpha} * (h(r) - h\_0))$ where $h(r)$ is the height as a function of range r from the radar, alpha is a constant chosen such that $u\,(h = 1000.0\ m) = 0.5\,u\,(h = h\_0)$ and where $h\_0 = 200.0\ m$ is a threshold height above the radar, below which the radar observations are rejected.
- A file containing the detection count for the received scan is updated. These files are kept on a rolling monthly cycle. It is the intention to use the detection counts to identify and exclude permanent clutter from the received data. Further work is required to determine a suitable rejection threshold (percentage of the maximum number of counts) in order to apply the filter. Once this algorithm is implemented, it will be possible to exclude cluttered pixels from the composites.

Future pre-processing

From a Numerical Weather Prediction (NWP) perspective, both for data assimilation and verification, there is now a vital requirement to build on the good foundation provided by the development of the Odyssey if European weather forecasting is to profit. It is essential that quality control procedures, namely: corrections for beam blockage, residual clutter removal, Vertical

Profile of Reflectivity identification, gauge adjustment, attenuation correction, enriched quality index, and advection correction, will be implemented in the near future. Fortunately, within OPERA and beyond there is a lot of expertise on how to process raw radar data to create high quality composite products that can meet the expectations of a wide range of users, including hydrologists.

It is envisaged that the current estimate of the quality, based on the beam height, will be replaced with a more sophisticated scheme (see Norman *et al.*, 2010), once more quality information becomes available. However, this will be complicated by the probability that countries will provide differing amounts of information.

Compositing

For the first time, processed raw polar data will be used to create a composite for the whole of Europe (the Pilot Data Hub uses processed Cartesian products, Harrison, *et al.*, 2006). One of the key elements of Odyssey software is successfully mapping polar pixels of the input data onto the Cartesian pixels of the composite products. The challenges are both on the quality of the method, in terms of the number and pertinence of raw data pixels used to fill a Cartesian pixel, as well as on its performances, in terms of CPU and memory used. The requirement is that raw radar data from across Europe is decoded, processed, projected and composited in less than 5 min. Considering the number of European radars (almost 200) and the numbers of scans performed by each radar during 15 min (15 on average), the challenge is real. One of the first activities of the Odyssey project was to define and test different mapping methods. The method adopted by the Odyssey service is described in detail by Dupuy *et al.* (2010) and Scovell *et al.* (2010).

The method relies on the pre-calculation of the mapping of polar cells to composite grid cells. For each composite grid cell, a list of polar cell identifiers are stored for polar cells which are known to be near to the composite cell centre. A polar cell identifier will contain a unique index identifying the scan (this is a database key in the database of scans mentioned earlier) and an index to the radar observation in the polar grid (the ray number and bin number).

The pre-calculated mapping data is generated using a technique which we refer to as "reverse mapping". It works by looping over all radar observation points (polar cells) from all available radars. For each radar observation it converts the coordinates of the centroid of the polar cell to coordinates on the surface of the 2-D composite map. It then establishes which composite cells lie within a specified range of the point in the composite map, e.g. ±2.5 km (see Fig. 1). The polar cell identifier is then added to an array containing a list of polar cell identifiers for each composite cell within range. Along with the polar cell identifier, the range between the composite cell centre and the centroid of the polar cell is stored. This value allows a range-weighting technique to be used for the purposes of deriving the composite data value.

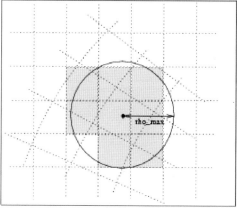

Fig. 1 Reverse mapping polar cells to composite Cartesian grid.

IT INFRASTUCTURE AND COMMUNICATIONS

Operational environment

The operational Odyssey service runs in parallel at the Météo France and Met Office operational centres in Toulouse (France) and Exeter (UK), respectively. Both centres receive and process the same raw data using the same algorithms to create identical composites. The only difference between the two centres is that only one is disseminating products at any given time. This cross organisational support gives the Odyssey service a very high level of resilience.

Availability and timeliness monitoring of incoming data and the Odyssey composites takes place at the EIG EUMETNET EUCOS Quality Management Portal (QMP) hosted at Deutscher Wetterdienst (DWD). This means:

- Measurements of Odyssey availability and timeliness available from the same source and consistent with other operational EUMETNET observations programmes.
- There is web access to Odyssey data performance.
- Information is available in near real-time.
- Monitoring is independent of Odyssey, therefore indicating what the customer actually received and when.

The OPERA community has agreed the thresholds for availability and timeliness of the input data and of the output products of the Odyssey.

IT communications

Moving all the raw radar input data and composite products around Europe, and also between Toulouse and Exeter, requires a robust IT communications network. The WMO Global Telecommunications System (GTS) within Europe (RMDCN or Regional Meteorological Data Communication Network) delivers such a network as it provides a very high level of guaranteed service. Therefore, all data traffic between Toulouse and Exeter goes via RMDCN (see Fig. 2). National Meteorological Services have also been encouraged to supply raw data products to Odyssey using RMDCN. Unfortunately there are bandwidth constraints within RMDCN and most NMS have therefore chosen to supply data via the internet. As National Meteorological Services upgrade their bandwidth on RMDCN we hope to see an increased use of RMDCN and therefore an increased level of resilience in the data supplied to Odyssey.

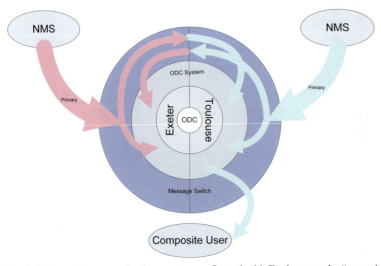

Fig. 2 Odyssey IT communications routes configured with Toulouse as the "operational" node.

PRODUCTS

Composites

Currently Odyssey produces three composite products:

- Surface rain rate: weighted average of the lowest valid pixels of each contributing radar, with a weight inversely proportional to the beam altitude.
- Max reflectivity: maximum of all the available local pixels values.
- Hourly rainfall accumulation: sum of the four 15 min previous rainfall intensity composites.

Future versions of Odyssey will see the resolution of rain-rate, accumulation, and maximum reflectivity composites improve both temporally and spatially, including the provision of 3D composites. These developments will depend on the availability of input data of sufficient resolution and also on the capability of the IT communications networks in Europe to distribute these extra data volumes.

Table 1 Odyssey composite characteristics.

Domain	Whole of Europe (see Fig. 3)
Projection	Lambert Equal Area
Update frequency	Every 15 min
Spatial resolution	2 km
Issue time	Approximately 15 min after data time
Format	HDF5 or BUFR

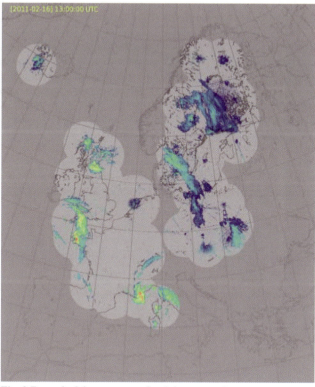

Fig. 3 Example Odyssey max reflectivity composite.

Raw reflectivity data

One consistent demand from NWP Modellers has been the requirement for the re-distribution of the raw radar scans to assimilate volume radar observations directly into their models. OPERA is now working on how to meet these needs.

Data policy

Another challenge for OPERA and EUMETNET as a whole is the development of a data policy for Odyssey products. Until a data policy is in place Odyssey products can only be made available to OPERA members for their core activities, excluding commercial exploitation. A task team has been established to solve this problem and a report with recommendations was presented to EUMETNET Assembly for ratification in May 2011. This agreement will see the way forward for the release of Odyssey products to a wider user base, including hydrologists.

Archive

The three composite types are routinely archived by Météo France in BUFR format. These will be made available, subject to the Data Policy, via a web interface to the Météo France archive.

CONCLUSION, PERSPECTIVES AND THE FUTURE

The main objective of OPERA-3 has been achieved. OPERA owns a real-time data centre, initially gathering data from 10 countries, but potentially covering the whole of Europe by the end of 2011. Having developed a solid element of the European infrastructure, the OPERA community will now have a place to test and implement new treatments and algorithms.

Improving and measuring the quality of the data in the Odyssey composites is the next big challenge to be addressed. To facilitate this, improvements have been included in the proposals for the next phase of OPERA (2013–2020), currently under development. The Numerical Weather Prediction community is building high expectations that the assimilation of the new products could improve regional analysis of precipitation and allow more accurate short range forecasts of severe weather events to be produced.

In Europe there is a fantastic level of knowledge on how to measure and improve the quality of radar data. Many processing algorithms exist at a national and regional (including the BALTRAD project, http://www.baltrad.eu/) level. The next challenge is to agree on what pre-processing algorithms should be deployed on Odyssey and should these algorithms be universally or regionally applied.

The development and launch of Odyssey is a major step forward in the production of operationally supported, high resolution, homogenous rainfall accumulations across the European domain.

REFERENCES

Chèze, J-L., Hafner, S., Holleman, I., Matthews, S. & Michelson, D. (2009) Specification of the EUMETNET operational Weather Radar Data Center. *OPERA deliverable OPERA_2008_02.*

Dupuy, P., Matthews, S., Gaussiat, N., Scovell, R. & Kergomard, A. (2010) Developing a European Radar Data Centre. In: *Proceedings 6th European Conference on Radar in Meteorology and Hydrology,* Sibiu.

Harrison, D. L., Scovell, R. W., Lewis, H. W. & Matthews, S. J. (2006) The development of the EUMETNET OPERA radar data hub. In: *Proceedings of the 4th European Conference on Radar in Meteorology and Hydrology,* Barcelona, Spain.

Huuskonen, A., Delobbe, L. & Urban, B. (2012) The European Weather Radar Network (OPERA): New opportunities for hydrology! In: *Weather Radar and Hydrology* (Proc. Symp. held in Exeter, UK, April 2011). IAHS Publ. 351. IAHS Press, Wallingford, UK (this issue).

Norman, K., Gaussiat, N., Harrison, D., Scovell, R. & Boscacci, M. (2010) A quality index scheme to support the exchange of volume radar reflectivity in Europe. In: *Proc. 6th European Conference on Radar in Meteorology and Hydrology,* Sibiu.

Michelson, D., Lewandowski, R., Szewczykowski, M. & Beekhuis, H. (2008) EUMETNET OPERA weather radar information model for implementation with the HDF5 file format. *OPERA deliverable OPERA_2008_03.*

Scovell R., Urban B. & Kergomard A. (2010) ODC composites definitions and compositing methods. *OPERA deliverable OPERA_2010_01* (not issued yet).

Weather Radar and Hydrology
(Proceedings of a symposium held in Exeter, UK, April 2011) (IAHS Publ. 351, 2012).

15

Tri-agency radar networks in Korea: where are we heading?

GYUWON LEE[1], SUNG-HWA JUNG[1], JUNG-HOON LEE[1], YO-HAN CHO[1], KWANG-DEUK AHN[1], BOK-HAENG HEO[2] & CHOONG-KE LEE[3]

1 *Dept. of Astronomy and Atmospheric Sciences, Kyungpook Natl' University, Deagu, Korea*
  gyuwon@knu.ac.kr
2 *Weather Radar Center, Korea Meteorological Administration, Seoul, Korea*
3 *Han Flood Control Office, Ministry of Land, Transport and Maritime Affairs, Seoul, Korea*

Abstract Korea has three radar networks operated by three agencies: Korea Meteorological Administration (KMA), Korean Air Force and Flood Control Office. A recent tri-agency agreement opens a new era for common use of data, similar maintenance procedures and operations, and possibly unification of radar types. KMA built the Weather Radar Center (WRC) to facilitate this agreement and expects WRC to be a focal point for the Korean radar networks. We will discuss the current status and future plans for the radar networks. Some advantages of using tri-agency networks are demonstrated through a simulation study. Issues in the current networks and ongoing research to resolve them are discussed in terms of data quality control, radar calibration, etc. The fuzzy logic based algorithm of quality control is developed. The radar reflectivity calibration is performed by intercomparison and with a disdrometer. A new scanning strategy is proposed to optimize the ground rain estimation and wind retrieval in space.

Key words radar networks; Korea; Weather Radar Center; quality control; calibration

INTRODUCTION

Korea owns three radar networks operated by three agencies: Korea Meteorological Administration (KMA), Korean Air Force (KAF) and Ministry of Land, Transport and Maritime affairs (MLTM). Each agency runs their radar network based on specific purposes. Recently, there was a tri-agency agreement for common use of data and, furthermore, the unification of maintenance procedures, operations and radar specification. In addition, KMA built the Weather Radar Center (WRC) to facilitate this agreement by benchmarking the Radar Operations Center in the USA, and plans to utilize WRC as a focal point for the Korean radar networks. In this paper, we present the current status and future plans for the radar networks. Some advantages of the use of tri-agency networks are demonstrated through a simulation study. Issues in the networks and ongoing research to resolve them are discussed, in terms of data quality control, radar calibration, etc.

Current and future radar networks

The current radar networks are composed of 11 S-band, 14 C-band and 1 X-band radars (see Fig. 1) operated by KMA, KAF, MLTM, US Air Force (USAF) and Korea Aerospace Research Institute (KARI). Currently, there are two dual-polarimetric radars: one S-band (RBSL) and one X-band (NIMR-X). MLTM is in the middle of building its dual-polarimetric radar network that is composed of six S- and five X-band radars. KMA also plans to replace its existing network by 10 S-band dual-polarimetric radars (Fig. 1) and one for a test-bed. As seen in Fig. 1, KMA radars are mostly located in coastal areas to extend coverage and monitor weather systems that approach from the west and south, whereas MLTM radars are being installed at inland high mountain locations to measure rainfall over the land. In addition, KAF radars are located at inland airbases to monitor weather around these bases.

Observational capability with current networks

Radar networks from KAF and MLTM can significantly improve coverage and accuracy of radar measurements over inland areas. An advantage of using different radar networks *versus* KMA alone is shown in Fig. 2. The KMA network shows good coverage at lower heights around coastal areas but with poor coverage over mountainous areas. Furthermore, the lowest observable heights are higher over southern coastal areas. However, the use of different radar networks significantly improves observational coverage at lower heights: see the right panel in Fig. 2. The networks can provide data below 1 km for most of the inland and coastal areas. This excellent observational capability at lower heights should significantly improve the accuracy of surface rainfall estimates. In addition, the use of different networks increases the number of data at a given position and thus improves the accuracy of multiple Doppler analyses (not shown here).

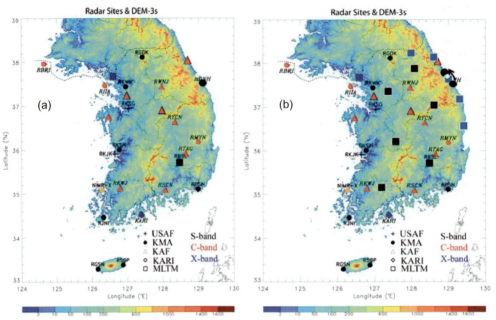

Fig. 1 (a) Current and (b) future radar networks in Korea from different agencies. The colour indicates the height of topography.

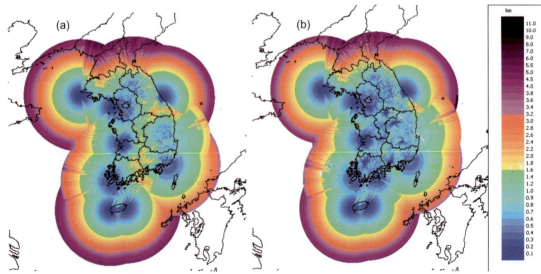

Fig. 2 Lowest observable heights (km) from (a) current KMA radar network and (b) from combined KMA, KAF, MLT, and KARI radar networks.

ISSUES

Although Korea owns extremely dense radar networks and use of the different networks is beneficial, there are several issues that need to be addressed before applying retrieval algorithms: for example reflectivity calibration, quality control and scanning strategy.

Radar reflectivity calibration biases are derived by inter-comparison and disdrometer-based calibration. In this method, the reference radar is first calibrated by comparing radar reflectivity Z with disdrometer measurements and then by inter-comparison with other radars. The derived calibration biases from KMA radars vary from –3 to –9 dB (Park & Lee, 2010). By correcting these calibration biases, the normalized error in rain estimation from Z improved from 59% to 37% (shown in Fig. 3). This technique has been adapted for real-time use and derives individual calibration biases every 10 minutes. Radar calibration with these biases depends on the weather situation and is determined by the radar operators.

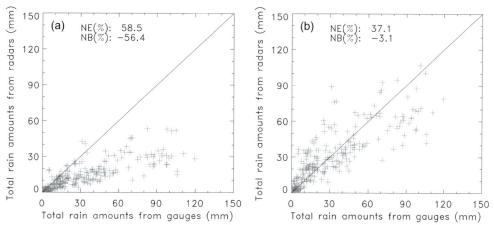

Fig. 3 Comparisons of total rainfall amounts between gauge measurements and radar estimates: (a) the radar estimates are from reflectivities before calibration and (b) after calibration (Park & Lee, 2010).

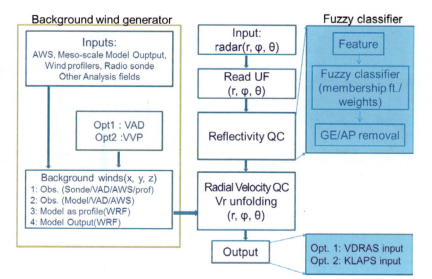

Fig. 4 Flowchart of quality control of radar reflectivity and unfolding of Doppler velocity.

A fuzzy-based algorithm has been developed to remove the ground echoes and anomalous propagation (Cho *et al.*, 2006). The algorithm is modified to use the characteristics of radar reflectivity only and is adapted for KMA radar data. Due to use of an infinite impulse response filter (IIR), Doppler radial velocity and spectral width are filtered so as to not contain returns from stationary objects. The modified algorithm (shown in Fig. 4) works successfully and shows different characteristics of fuzzy membership functions and weightings for different radars.

The Doppler radial velocity is unfolded by the procedure described by Lim *et al.* (2010) that was originally developed for the Variational Doppler Radar Analysis (VDRAS). This unfolding algorithm works on data in polar coordinates and utilizes the local and global unfolding with the background winds from numerical weather prediction models and observations (surface observations, soundings, volume velocity processing, wind profilers, etc.), as shown in Fig. 4. The gradient of radial velocity is also checked in horizontal and vertical directions.

The scanning strategies of KMA radars had been evaluated by examining wide areas of pseudo-CAPPI, uniformity of data in space, the number of elevation angles within a volume scan, etc. We have developed new scanning strategies to optimize rain measurements and to obtain uniform information in space. The lowest elevation angle is determined by simulating the beam propagation and optimal lowest coverage. The highest elevation angle is derived from the detectable storm top at a fixed range. The number of elevation angles is determined by the number of samples and antenna rotation speed. Finally, the specific elevation angles are assigned in between the lowest and highest angles with a method that optimizes angles for providing uniform data at 5 km CAPPI. This procedure provides good coverage near the surface. Results show that the new scanning strategy provides better lowest coverage and more uniform information throughout space than the current KMA scanning strategy. The effect of optimization is examined by the number of radars that can provide Doppler radial velocity at a given height (a.m.s.l.), as shown in Fig. 5. The optimized scan shows broader and uniform areas of multiple Doppler analyses. A thorough investigation of the scanning strategy for multiple Doppler analyses is underway.

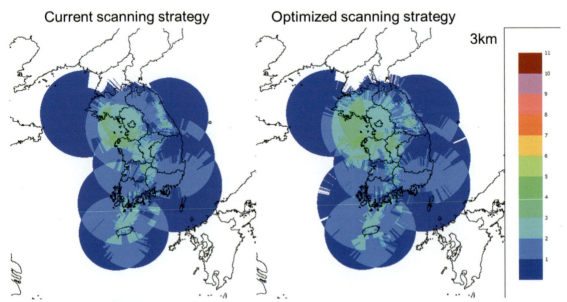

Fig. 5 Number of radars that can provide Doppler radial velocity at a given location and height (a.m.s.l.) with current and optimized scanning strategy.

SUMMARY AND FUTURE PLANS

Korea is one of the countries investing most in the construction of radar networks due to the needs of various users. However, the relevant research and development for optimal use of radar data are relatively under-invested. This fact was recognized by the community and led to a multi-agency agreement for common use of radar data and further coordinated planning. In addition, KMA organized the Weather Radar Center to facilitate coherent planning and research/development.

A new development is underway to improve radar data quality by optimizing operations and data processing. Currently, a new radar pre-processing system is being developed to resolve non-meteorological echoes, velocity folding, mis-calibration, etc. Furthermore, a comprehensive data processing system is under development to improve the quality of information and to satisfy potential users of radar data.

Acknowledgement This study was financially supported by the Construction Technology Innovation Program (08-Tech-Inovation-F01) through the "Research Center of Flood Defence Technology for Next Generation" in the Korea Institute of Construction & Transportation Technology Evaluation and Planning (KICTEP) of the Ministry of Land, Transport and Maritime Affairs (MLTM).

REFERENCES

Cho, Y. H., Lee, G., Kim, K. E. & I. Zawadzki (2006) Identification and removal of ground echoes and anomalous propagation using the characteristics of radar echoes. *J. Atmos. Ocean. Technol.* 23, 1206–1222.

Lim, E. & Sun, J. (2010) A velocity dealiasing technique using rapidly updated analysis from a four-dimensional variational Doppler radar data assimilation system. *J. Atmos. Ocean. Technol.* 27, 1140–1152.

Park, S-G. & Lee, G. W. (2010) Calibration of radar reflectivity measurements from the KMA operational radar network. *Asia-Pacific J. Atmos. Sci.* 46, 243–259.

Sun, J. & Crook, N. A. (1997) Dynamical and microphysical retrieval from Doppler radar observations using a cloud model and its adjoint. Part I: Model development and simulated data experiments. *J. Atmos. Sci.* 54, 1642–1661.

Compositing international radar data using a weight-based scheme

THOMAS EINFALT[1] & ARNOLD LOBBRECHT[2]

1 *hydro&meteo GmbH & Co. KG, Breite Str. 6-8, D-23552 Luebeck, Germany*
info@hydrometeo.de
2 *HydroLogic BV, Stadsring 57, 3811 HN Amersfoort, The Netherlands*

Abstract In the northeastern part of the Netherlands, the Dutch radars of De Bilt and Den Helder have only limited coverage, while the German Emden radar is just opposite the border. Therefore, hydro&meteo and HydroLogic developed a new radar composite for this part of the Netherlands, starting from the basic polar radar products of both national weather services. The composite should be available in near-real time. The paper presents a case study of an interesting rainfall event, using various filtering and correction algorithms. The result shows very good results when compared with independent raingauges. The independent verification demonstrates that the new composite is similar to the one of the Dutch weather service on average for the Netherlands, and in addition it is much better in the northeastern part of the country, due to the Emden radar data. The algorithms are now ready for use in operational water management.

Key words precipitation; radar; rainfall; composite; raingauge; flood

INTRODUCTION

Weather radar networks are generally optimized for national monitoring requirements. Sometimes the coverage is incomplete and in other cases radar measuring goes beyond state boundaries. Compositing radar information from bordering countries can be used to fill in coverage gaps and extend the usage of cross-border measurements. The situation of incomplete coverage by Dutch radars happens along the Dutch–German border. In particular, the measurement coverage in the northeastern part of the Netherlands, the province of Groningen, is not optimal. This has already been addressed by water boards in the region for many years.

A European-wide approach to establish cross border composite data is being undertaken by the EU funded EUMETNET group of the European weather services in their OPERA programme where they are working on a European composite (Odyssey project).

Since the weather services are currently not intending to create an operational cross-border coverage, the challenge was taken up to develop a Dutch-German weight-based compositing algorithm which is able to produce precipitation information with high resolution in time and space (5 minutes 1 × 1 km), using online ground measurements for radar adjustment.

The Dutch weather service (KNMI) is providing an adjusted composite for water management purposes with 3 h time resolution, which is available ~1.5 h after compositing. The calibration is done with an average correction for the entire country, based on hourly monitoring data. In addition there is a 24-h adjusted product which is available after ~36 h. This composite is spatially adjusted on the basis of daily monitoring data. Furthermore, an unadjusted 5 min composite exists (available after ~7 minutes). All these composites are provided in 1 × 1 km grids. To enable a good merging of the Dutch and German radars, research datasets were made available by KNMI and the German weather service (DWD), which includes the original 5 min polar information from two Dutch radars: De Bilt and Den Helder, and from one German radar: Emden.

A research project was performed by the companies hydro&meteo, Luebeck, and HydroLogic, Amersfoort, to assess and demonstrate whether a composite for the whole of the Netherlands with similar performance as the one of KNMI could be created out of the original Dutch polar data and whether an improvement in measurement could be obtained for the northeastern part of the country when the Emden radar is added.

Fig. 1 Overview of the raingauges and the ranges of the radars. The red stars indicate the radar sites, the red dots indicate the 32 Synop KNMI raingauge sites (hourly data), and the small black dots indicate the 325 independent raingauges (daily data).

AVAILABLE RESEARCH DATA

Raingauges

Continuous gauges For a hydrologically relevant rainfall event of 11 and 12 May 2010, hourly raingauge data from 32 automatic weather stations are available on the KNMI website (http://www.knmi.nl/klimatologie/uurgegevens/). These were used for the radar-raingauge adjustment performed by the SCOUT software (hydro&meteo, 2009). In the remainder of this article we denote these raingauges by "Synops KNMI".

Daily gauges The KNMI also operates a network of 325 raingauges. The measurements are performed each day at 08:00 UTC; after a validation by the KNMI, the data are published on their website (http://www.knmi.nl/klimatologie/monv/reeksen/). The data from these raingauges are used as independent gauges in the analysis for verification of the compositing result throughout this paper. After excluding the raingauges that are not in the range of the composite, 295 raingauges are left for analysis. Figure 1 presents an overview of the raingauges and the radar ranges.

Radar products

Reflectivity measurements from three radars located in De Bilt, Den Helder, and Emden were available from 11 May 2010 00:00 UTC till 12 May 2010 24:00 UTC.

KNMI polar radar data The KNMI radar data were available for De Bilt and Den Helder as a volume scan with 14 elevations, performed every 5 min. For the composite, the first choice was to use the same elevation angle for all three radars, 0.8°. Later, the optimal choice proved to be 0.4° for Den Helder. The data resolution was 1° × 1 km for a range of 240 km. The data from KNMI have been pre-processed for clutter and anaprop for which no additional information has been available.

German DX radar product The DX product of the German Weather service DWD is one elevation measurement at 0.8° elevation angle, and has a range of 128 km with a spatial resolution of 1° × 1 km. The measurement takes place every 5 min. The data are pre-processed for clutter using a statistical clutter filter. This filter is the standard pre-processing for the only currently available PPI product DX.

DATA PROCESSING CHAIN

Data pre-processing

Speckle filter The speckle filter eliminates items up to a defined size. Here, a limit of 16 (polar) pixels has been used so that ships could be deleted from the radar image.

Gabella filter The texture based filter (Gabella & Notarpietro, 2002) is used to smooth extreme peaks in the image, e.g. due to ground clutter within a rain field.

Bright band filter The bright band filter used in this research is a further development of the one presented in Golz *et al.* (2006). It uses image processing techniques, combined with a temperature bandwidth to allow the elimination of bright band and the correction of measurements beyond the bright band using a reflectivity–intensity relationship for snow.

DATA COMPOSITING

For compositing of the reflectivity data from the three radars, several approaches were employed: basic methods applying the maximum method and the weight-based method, explained below. Furthermore, the Dutch radars having a radius of 240 km could either be used with their full range or be reduced to a hydrologically useful range of 120 km. Finally, a selection on the optimal elevation angle was made for the Dutch radar data.

Maximum *versus* weight-based

The maximum composite method is using the maximum value of all radars for a pixel. This creates sharp edges at the boundaries of the radar range. An alternative, but more complex method is a weight-based method. Here, each radar is associated with a weight matrix. This matrix is then used to determine the influence of the values from each radar in the composite. For this, the composite was created based on the weight and reflectivity value of each radar *i*:

$$C = \sum_{i=1}^{k} w_i R_i \qquad (1)$$

C is composite image, w_i is relative weight matrix for radar *i*, R_i is measurement of radar *i*.
The weights were calculated by:

$$w_i = \frac{Q_i}{Q_0} \qquad (2)$$

Q_i is weight matrix for each radar *i*, Q_0 is cumulated weight matrix where Q_0 is defined as:

$$Q_0 = \sum_{i=1}^{k} Q_i \qquad (3)$$

and the individual weight factors in the Q_i simple inverse distances from the radar. This method produced composite images with smooth edges and the resulting adjustment had better results in the verification.

120 km *vs* 240 km

Although data beyond approx. 120 km from the radar are barely useful for quantitative applications, the results for the 240 km range application were better than the ones of the 120 km range application.

Elevation angle

The fixed elevation angle of Emden radar is 0.8°. Better results for the Dutch radars could be obtained using an elevation of 0.4° for Den Helder radar and 0.8° for De Bilt radar, the difference probably due to a less urbanised area around Den Helder radar. These have been applied.

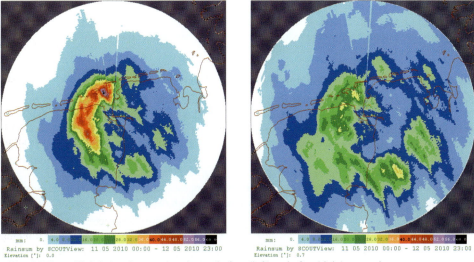

Fig. 2 Effect of bright band correction: image before (left) and after (right) correction.

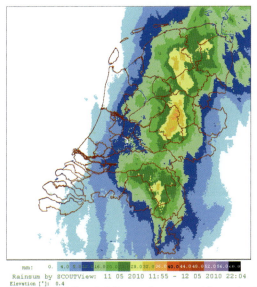

Fig. 3 Final daily rainfall sum after compositing and adjustment.

Data adjustment

Using SCOUT, one correction factor field is calculated and used for the whole event. The adjustment with the correction field values is based on inverse distance interpolation of the rainfall of the nearest four raingauges (Wilson & Brandes, 1979).

Raingauges and radar pixels with a measured precipitation amount <0.3 mm are automatically excluded from the correction factor computation. Raingauges with gaps in the data have also been excluded by SCOUT, in which case other stations were taken instead. One correction factor field per day is calculated, and the maximum distance for interpolation is set to 200 km.

The raingauges used for this computation were the gauges from the online network of KNMI (Synops stations of Fig. 1), to use the same station density as the KNMI online version.

RESULT VERIFICATION

The correction procedures were able to substantially improve the radar data as compared to the raw incoming data. Figure 2 shows the reduction of the bright band effect for the Emden radar, and Fig. 3 presents the final result after compositing and adjustment.

Comparison with KNMI station data for the area covered by Emden radar

Results were first compared with only the data of the Emden radar. Evaluated were event sums of the values obtained by the adjustment of the quality corrected radar data to the 32 online available Synop raingauges and 98 independent daily measuring raingauges within a range of 128 km from Emden radar. The daily raingauges were taken as the reference to assess the quality of the obtained composite by SCOUT. The comparison took place with the standard 3-h running adjustment (KNMI 3 h), the ones of the 24 h daily adjustment of KNMI (KNMI 24 h), and the daily raingauges (RG).

Table 1 shows that the use of Emden radar clearly improves the information for the area covered by this radar: the SCOUT results are better than the KNMI 3 h results. The correlation between radar and raingauges is higher, the absolute mean error (or mean bias) is less as well as the RMSE (root mean square error). The 24 h data are fully corrected by KNMI to meet the daily raingauges and therefore the result is always very close to the 24 h raingauge result (RG).

Another parameter is the number of stations with a deviation of radar derived rainfall from the gauge values larger than 5 mm or 10 mm: KNMI 3 h has a clear weakness in performance in the northeastern part of the Netherlands, close to the Emden radar (Table 2).

Comparison with KNMI station data for the complete area of the Netherlands

For the whole area of the Netherlands, the adjustment produced on the basis of 32 online available stations was verified with 319 independent raingauges (Fig. 1). The SCOUT results were slightly better than the ones by KNMI 3 h (Table 3). The SCOUT mean error ranges between KNMI 3 h and KNMI 24 h with an average underestimate of less than 1 mm. Also, the RMSE for SCOUT is between KNMI 3 h and KNMI 24 h, whereas the correlation factor of the SCOUT results outperforms both KNMI approaches for this event.

Table 1 Verification of adjustment results for SCOUT and KNMI methods for the region close to Emden radar.

Only Emden	KNMI 3 h	KNMI 24 h	SCOUT	RG
Correlation RG/radar	0.54	0.70	0.65	–
Mean error	−1.82	−0.15	1.08	–
RMSE	5.57	3.04	4.13	–
Mean rainfall	18.6	20.3	21.5	20.4

Table 2 Number of deviations between radar rainfall and gauge rainfall larger than 5 mm and 10 mm.

Only Emden	KNMI 3 h	KNMI 24 h	SCOUT
Number with D > 10 mm	10	2	1
Number with D > 5 mm	40	7	16

Table 3 Verification of adjustment results for SCOUT and KNMI methods.

	KNMI 3 h	KNMI 24 h	SCOUT	RG
Correlation RG/radar	0.75	0.82	0.83	–
Mean error	−1.64	−0.29	−0.76	–
RMSE	4.36	2.71	3.66	–
Mean rainfall	15.5	16.3	15.9	16.7

Table 4 Deviation of radar values from raingauge measurements: number of large and medium size deviations.

	KNMI 3 h	KNMI 24 h	SCOUT
Number with D > 10 mm	11	3	4
Number with D > 5 mm	78	12	46

Table 5 Deviation of radar from raingauge values without Emden radar coverage.

Not Emden	KNMI 3 h	KNMI 24 h	SCOUT
Number with D > 10 mm	1	1	3
Number with D > 5 mm	38	5	30

The number of differences of radar to raingauge values larger than 5 mm or 10 mm showed that the results from SCOUT are nearly as good as the ones from KNMI 24 h for deviations larger than 10 mm, and in between KNMI 3 h and KNMI 24 h for deviations larger than 5 mm (Table 4).

Influence of the Emden radar

Comparison of Tables 2 and 4 shows that measurements based on the Emden radar are an improvement to the Dutch composite. Table 5 is the result for those areas not covered by the Emden radar and indicates that the performance of SCOUT is close to the one of KNMI 3 h. In other words: the better performance of SCOUT compared to KNMI 3 h in Tables 2 and 4 is mainly resulting from the additional availability of Emden radar.

CONCLUSIONS

The presented example shows how the German Emden radar can help to obtain a better quality of radar composite data for the Groningen province in the Netherlands. Based on data from a bright band case, measurements of the two Dutch radars, De Bilt and Den Helder, have been composited with Emden radar and adjusted to 32 Dutch raingauges. The preliminary reflections on suitable compositing strategies as well as on the corrections to apply to the data and the elevation to choose, as well as data driven limitations, have been discussed.

The verification of the outcome was performed based on 319 independent gauges and compared to current standard practice for operational water management in the Netherlands: the online adjustment based on a 3 h composite which is available after ~1.5 h. The weight-based composite method was able to improve the adjusted data in terms of mean bias as well as in terms of regression to the independent gauges and is able to incorporate different data quality factors.

The result is now ready for use in near-real time hydrological applications in urban hydrology and in fluvial hydrology. This is of particular interest to water boards and municipalities in the border region.

Acknowledgements We gratefully appreciate that the German Weather Service and KNMI made available the test dataset.

REFERENCES

Gabella, M. & Notarpietro, R. (2002) Ground clutter characterization and elimination in mountainous terrain. In: *2nd Eur. Conf. On Radar Meteorology*, 305–311. Delft, Netherlands.

Golz, C., Einfalt, T. & Galli., G. (2006) Radar data quality control methods in VOLTAIRE. *Meteorologische Zeitschrift*, 15(5), 497–504.

hydro & meteo (2009) *The SCOUT Documentation* version 3.30. Lübeck, 69 pp.

Wilson, J. W. & Brandes E. A. (1979) Radar measurement of rainfall – A summary, *American Meteorological Society* 60, 1048–1058.

Weather Radar and Hydrology
(Proceedings of a symposium held in Exeter, UK, April 2011) (IAHS Publ. 351, 2012).

Estimating weather radar coverage over complex terrain

EDWIN CAMPOS

Argonne National Laboratory, Environmental Science Division, Argonne, Illinois, USA
ecampos@anl.gov

Abstract Minimizing terrain blockage is a basic consideration when assessing the efficacy of weather radar sites. A numerical model for simulating surveillance coverage of weather radars in mountain terrains is presented. As input, the simulation uses a high-resolution terrain digital model; weather radar parameters; and radiosonde observations of the vertical profile of temperature, pressure, and vapour mixing ratio. The coverage model is validated using observations from Environment Canada's C-band weather radar located at Mt Sicker (British Columbia, Canada).

Key words beam propagation; terrain blockage; surveillance area

INTRODUCTION

The challenge of deciding the optimal location of a weather radar site is particularly demanding over complex terrain, since the aim is to maximize the surveyed area of interest by minimizing mountain blockages of the radar beams. Placing the radar at a higher altitude or on a summit may reduce the beam blockage, but it also makes the radar beam miss the lower levels of precipitation. This dilemma is solved by finding the lowest altitude of the radar site which can survey the maximum area of operational application. Therefore, an objective estimation of the beam blockage is required, which involves numerical simulations for expected propagations of the radar beam.

Numerical simulations of this kind generally involve an electromagnetic beam propagating simply under atmospheric standard conditions (e.g. Doviak & Zrnic, 2006; Rinehart, 2006) or a terrain model that is considered only roughly (e.g. Wood *et al.*, 2003). In addition, simulations of this kind often lack objective validation (e.g. Brown *et al.*, 2002), or present validations over relatively flat topography (e.g. Kucera *et al.*, 2004). Under these limitations, the height h of the radar beam above the antenna level is generally computed by using this simplified equation:

$$h = \sqrt{r^2 + a_e^2 + 2 r a_e \sin\theta} - a_e \tag{1}$$

where r is the range (distance between the antenna and the target), θ is the antenna elevation angle), and a_e is the equivalent Earth radius:

$$a_e = \frac{R}{1 + R\,(dn/dh)} \tag{2}$$

where R is the Earth radius and dn/dh is the vertical gradient of atmospheric refractive index, n (defined later in this section). Using the equivalent Earth radius in equation (1) makes the geometry of the problem easier, by making the radar beam look like a straight line within a layer of constant refractivity. A common approximation (generally valid for heights below 2 km a.s.l.) is

$$\frac{dn}{dh} = -39 \times 10^{-6}\ \text{km}^{-1} \tag{3}$$

This is also known as standard refraction condition (i.e. refractive index conditions that would make the radar beam propagate with a curvature equal to one-fourth of the Earth's sea-surface curvature). Notice that standard refraction is also known as the "4R/3 approximation" because it leads to:

$$a_e \approx 4R/3 \tag{4}$$

Equation (1) is valid for conditions where the vertical gradient of refractive index is approximately constant in the height interval of interest. However, real refractivity gradients are not constant in the atmosphere. To overcome this limitation, actual changes in refractive index were considered by discretizing its vertical profile in small, finite intervals (e.g. according to the radiosonde vertical resolution). In this new approach, dn/dh is approximated to a constant only within finite intervals (piece wise). In the following sections, numerical simulation of radar coverage over complex terrain is presented.

METHODS

Refractive index

The atmospheric refractive index, n, is defined as the ratio between the speed of the electromagnetic beam moving in a vacuum and the speed of this electromagnetic beam moving in the atmosphere. For the electromagnetic beams used by weather radars (wavelengths in the X-, C-, and S-bands), n is a function mainly of the air temperature, pressure, and humidity. Notice that the refractive index can be expressed as refractivity N, using:

$$N = (n-1) \times 10^6 \tag{5}$$

Therefore, the refractivity for a radar beam propagating at a particular atmospheric point can be computed using:

$$N = \frac{77.6}{T} \left(p + 4810 \frac{e}{T} \right) \tag{6}$$

where the air temperature T is given in Kelvins, the air pressure p is given in hPa, and the air vapour pressure e is given in hPa.

In practice, n was computed using vertical profiles of the vapour mixing ratio (e is obtained from the pressure and mixing ratio), pressure (p), and temperature (T), which are typically available from radiosonde observations. Then it is assumed that the dn/dh profile measured during a particular radiosonde launch is representative of the entire sampling volume of the weather radar. This is a limitation due to the observations available. However, horizontal variability in the dn/dh profiles can be easily introduced when more thermodynamic profiles are available within the radar domain.

Computing the beam propagation

Let us consider the radar electromagnetic beam, which propagates along a height–horizon direction, crossing points 0, 1, and 2. When this electromagnetic beam moves from a denser to a less dense medium, the beam can be refracted or reflected in the incident medium. This is explained by the Snell-Descartes law for an electromagnetic beam propagating from medium 0 to medium 1:

$$\frac{\sin \alpha_0}{\sin \beta_1} = \frac{n_1}{n_0} \tag{7}$$

where α_0 is the angle of incidence, β_1 is the angle of refraction, and n_i is the refractive index of material/medium/environment i.

Now, let us assume that the refractive index is constant within discrete height steps, 0–1 and 1–2. Therefore, the change in beam height can be obtained from:

$$\Delta h_i = \frac{\Delta r_i \sin(\pi/2 + \theta_i)}{\sin \varphi_i} - a_i \tag{8}$$

where Δr_i is the change in radar range for a beam propagating from point i to point $i + 1$, θ_i corresponds to the beam elevation angle at point i, and a_i corresponds to the distance from the Earth's centre to point i (Earth radius at point i). The distance between the Earth's centre and the radar antenna is given by:

$$a_0 = R + H \tag{9}$$

where H is the altitude of the antenna (from sea level) and R is the distance from the Earth's centre to sea level at the radar latitude (and longitude). To compute this use:

$$R = \sqrt{\frac{(a_{max}^2 \cos \Phi)^2 + (a_{min}^2 \sin \Phi)^2}{(a_{max} \cos \Phi)^2 + (a_{min} \sin \Phi)^2}} \tag{10}$$

where Φ is the latitude of the radar site, $a_{max} = 6378.1370$ km is the distance from the Earth's centre to the Equator, and $a_{min} = 6356.7523$ km is the distance from the Earth's centre to the North (or South) Pole. The previous distance values are in accordance to the Global Positioning System, GPS, which uses the standards of the World Geodetic System 1984 (NIMA, 1997). Note also that the same Earth radius was used in the entire radar sampling volume.

Note that:
$$a_{i+1}=a_i+\Delta h_i \tag{11}$$
The angle connecting point i, the Earth's centre, and point $i+1$ is given by:
$$\phi_i = \arctan\left[\sin\left(\frac{\pi}{2}-\theta_i\right)\Big/\left(\frac{a_i}{\Delta r_i}+\cos\left(\frac{\pi}{2}-\theta_i\right)\right)\right] \tag{12}$$

The distance from a point directly below the radar antenna to a point directly below the radar target, all along a line at sea level, is called the Earth surface distance (or horizontal distance, S). Then, the segment of the surface distance corresponding to the beam segment from point i to $i+1$ is given by:
$$\Delta S_i = R\phi_i \tag{13}$$
To obtain the beam elevation angle at the point where the beam meets the new refractive index isoline, one of the following four conditions was selected:

(1) If $\theta_i \geq 0$, then:
$$\theta_{i+1}=\frac{\pi}{2}-\arcsin\left[\frac{n_i}{n_{i+1}}\sin\left(\frac{\pi}{2}-\theta_i-\phi_i\right)\right] \tag{14}$$

However, if the argument in the arcsin function in equation (14) is greater than one, then there are reflection (anomalous propagation) conditions. Therefore, β_{i+1} will correspond to an angle of reflection and will be equal to the angle of incidence α_i. In that case:
$$\theta_{i+1}=-|\theta_i+\phi_i| \tag{15}$$

(2) If $\theta_i<0$ and $\Delta h_i<0$, then:
$$\theta_{i+1}=-\left|\frac{\pi}{2}-\arcsin\left[\frac{n_i}{n_{i+1}}\sin\left(\frac{\pi}{2}-|\theta_i|-\phi_i\right)\right]\right| \tag{16}$$

However, if the argument in the arcsin function in equation (16) is greater than one, then there are reflection (anomalous propagation) conditions. In that case:
$$\theta_{i+1}=|\theta_i|-\phi_i \tag{17}$$

(3) If $\theta_i<0$ and $\Delta h_i \approx 0$, then:
$$\theta_{i+1}=\frac{\pi}{2}-\arcsin\left[\frac{n_{i-1}}{n_i}\sin\left(\frac{\pi}{2}-|\theta_i|\right)\right] \tag{18}$$

However, if the argument in the arcsin function in equation (18) is greater than one, then there are reflection (anomalous propagation) conditions. In that case:
$$\theta_{i+1}=\theta_i \tag{19}$$

(4) If $\theta_i<0$ and $\Delta h_i>0$, then:

this case breaks the main assumption that the refractive index is constant within the range interval Δr_i. To avoid this, the Δr_i is made smaller until $\Delta h_i \leq 0$. Then the beam position (i.e., ΔS_i and θ_{i+1}) is recomputed accordingly.

Projecting radar pixels into a terrain map

This procedure also requires projecting the radar gates (pixels) into the terrain map. For this, consider first that the most common display for weather radar information is the plan position indicator (PPI). The PPI maps radar observations, for a particular elevation angle, on polar coordinates (azimuth *versus* range) in plan view centred at the radar site. However, to project the radar observations in a terrain map, the range is not needed but the Earth surface distance, S, which is obtained using the following expression in conjunction with equation (13):
$$S_i = \sum_0^i \Delta S_i \tag{20}$$

The proper way to project radar observations in a terrain map requires multiple conversions between the radar coordinates (range, azimuth and elevation angle), beam heights, polar coordinates (azimuth angle and surface distance), rectangular coordinates (X, Y, and beam height), and geographic coordinates (latitude, longitude and altitude).

For the rectangular-geographic conversions the following were used

$$1° \ \Delta\Phi = 111 \ km \tag{21}$$
$$1° \ \Delta\varphi = 111 \ km \times \cos\Phi \tag{22}$$

where Φ is the latitude, $\Delta\Phi$ is the difference in latitude, φ is the longitude, and $\Delta\varphi$ is the difference in longitude.

Computing the beam blockage

Whether a given radar pixel is experiencing total or partial blockage can be evaluated by computing the percentage of the beam that is being blocked by the mountains. We did this computation as follows.

For a given radar gate (pixel), we first computed the height at the centre and at the lower part of the radar beam. This height was computed by using equation (8) and:

$$h_i = \sum_0^i \Delta h_i \tag{23}$$

Second, a matrix h_T was obtained containing the terrain altitudes (in metres above the mean sea level) inside the ground projection of the radar pixel (as in the previous section).

Using these terrain altitudes and the heights of the radar beam, the blockage fraction was then computed as follows:

$$BF = n_b / N_t \tag{24}$$

where BF is the blockage fraction for a particular radar pixel, N_t is the total number of terrain points inside the ground projection of a radar pixel, and n_b is the number of these terrain points that have a height above the radar beam centre. Equation (13) was used to project each radar pixel in the Earth's surface. (The azimuth limits for the radar pixel projected at ground are the same as for elevation. That is, $\alpha_z \pm \Delta\theta/2$, where α_z is the azimuth angle of the radar pixel.)

The total blockage fraction, BF_T, was then defined as the blockage fraction when n_b corresponds to the number of terrain points with heights above the centre of the radar beam ($h_T > z_{mid}$). Similarly, the partial blockage fraction, BF_P, was defined as the blockage fraction when n_b corresponds to the number of terrain points with heights between the lower part and the centre of the radar beam ($z_{down} < h_T \leq z_{mid}$).

The regions of total blockage were identified as those radar pixels where >75% of the terrain pixels have an altitude higher than the centre of the radar beam (i.e. $BF_T > 0.75$). Those radar pixels at the same azimuth but at farther range than a pixel already identified with total blockage were also categorized as regions with total blockage. The conditions for ground clutter over regions without total blockage were then evaluated. In those regions, a search was carried out for radar pixels where >75% of the terrain pixels have an altitude higher than the lower part of the beam (i.e. $BF_P > 0.75$). These were then categorised as regions of ground clutter.

RESULTS

Beam propagating over flat terrain

Figure 1(a) presents the computations for the beam propagation over a flat terrain, according to the atmospheric conditions over Quillayute station at 00:00 UTC, 11 March 2007. The simulations (solid lines) correspond to a radar antenna placed at 0.7 km above the sea level (i.e. the altitude at Mount Sicker), and pointing at elevation angles of –0.5, 0.0, 0.5, 1.0, 2.0, 3.0, 5.0, and 10.0°. As a reference, Fig. 1 also plots (in dashed lines) the corresponding values for beams propagating under standard refraction conditions (i.e. the $4R/3$ approximation).

Beam propagating over complex terrain

The conditions for the weather radar located at Mt. Sicker were then simulated, with its antenna pointing at 0.3° in elevation angle, having a resolution of 1 km in range and 1° in azimuth, for ranges smaller than 150 km. (In fact, this radar operates at 5 cm wavelengths, C-band, and has a beamwidth of 0.66°.) The equivalent Earth radius was computed by using the radiosonde observations at Quillayute station, on 11 March 2007 at 00:00 UTC.

For the terrain elevations, the GLOBE 1.0 dataset (Hastings & Dunbar, 1999) was used. This is a global digital elevation model on a nominal 1 km grid (i.e. a latitude–longitude grid spacing of 30 arc-seconds, or 0.008333°), which has an absolute vertical accuracy of approximately plus or minus 30 m at 90% confidence.

During the radar coverage simulations, the combination of resolutions in the radar system and in the terrain dataset is such that no terrain elevations were found within radar pixels at the closest ranges. Therefore, we increased the pixel size in the azimuth direction (to 1.5, 3.0 or 4.5 times the radar-azimuth resolution) for those radar pixels with <4 terrain points. Notice that this increase is only for blockage fraction computations using equation (24). After this change, numerous radar pixels at ranges smaller than 2 km were still free from terrain points. At ranges smaller than 21 km, however, it was possible to find about 1–3 terrain points per radar pixel. For the farther ranges, from 4 to 9 terrain points were used per radar pixel.

It was realized that the previous approach is an *ad hoc* solution to a fundamental problem: the beam blockage is very sensitive to details of local topography. It is expected that this solution will widen the areas of computed blockage at ranges smaller than about 20 km.

With these considerations, the beam blockage estimates from equation (24) were obtained. The results are given in Fig. 1(b). In this figure, the radar pixels are projected into the terrain map using equation (20).

Validation

The numerical simulations were validated by using observations from the weather radar located at Mt Sicker, in British Columbia, Canada (located at 48.86° latitude north and 123.76° longitude west). The site elevation is about 716 m a.s.l., although the terrain digital model locates this site at 638 m a.s.l. This radar operates at C-band (5 cm wavelength), and its antenna is located at near 31 m a.g.l. (27.1 m to tower top–radome base–plus 3.5 m to dish centre). Radar reflectivity observations taken at an elevation angle of 0.3° were first analysed (with a resolution of 1° in azimuth and 1 km in range) during widespread precipitation. These observations were then compared to numerical simulation of the beam blockage for this event (i.e. Fig. 1(b)).

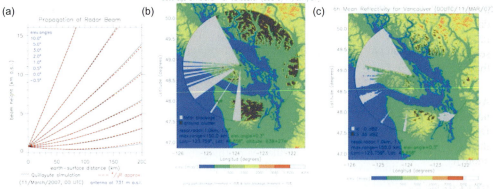

Fig. 1 (a) Simulation of the propagation of radar beams at eight different elevation angles (from –0.5° to 10°) over flat terrain. (b) Numerical simulation of radar coverage for the weather radar at Mt Sicker. The external black circle corresponds to a 150 km range at 0.3° elevation from the radar site. (c) Reflectivity observations for the weather radar at Mt Sicker. The light grey area in the southeast direction is due to precipitation absence.

Regarding the radar observations, the average of equivalent reflectivity factors (in mm^6/m^3) obtained at each radar pixel were computed during a 6 h period, 37 radar scans, from 21:00 UTC on 10 March 2007 to 03:00 UTC on 11 March 2007. (For each radar pixel and particular time, these reflectivities are calculated from the total received power, including precipitation and ground clutter signals, and assuming radar backscattering cross-sections of pure water.) This is an event of widespread precipitation associated with a warm front. Under these conditions, it was expected that (on average) very high reflectivities over the terrain that blocks the radar beam (also known as ground clutter) and very low reflectivities behind the terrain that blocks the beam (i.e. regions that cannot be reached by the radar beam because of total beam blockage by the mountains), would occur.

Figure 1(c) presents the radar observations (6 h mean reflectivities), where the light grey colours correspond to regions with average reflectivities <0 dBZ, and the dark grey regions correspond to radar pixels with average reflectivities >35 dBZ. Notice that the light grey colours are associated to echo-free regions, and the dark grey areas correspond ground clutter. The echo-free regions may be due to total beam blockage or to absence of precipitation during the analysis period.

The results in Fig. 1(b) and (c) are in general agreement, although the simulated ground-clutter areas are larger than the corresponding radar observations. Differences in the blockage patterns are associated, it is believed, to horizontal variations in the thermodynamic fields of the atmosphere (temperature, pressure, and moisture). In addition, the ground clutter classification in Fig. 1(c) is crude, and the simulation of the mountain returns in Fig. 1(b) is more basic than other clutter models available in the literature (e.g. Delrieu *et al.* 1995; Archibald 2000). In any case, effort is focused on the beam propagation simulation, which proved to be robust enough for supporting the selection of weather radar sites in the region.

CONCLUSIONS

A numerical model for simulating the surveillance coverage of weather radars is presented here. It is based on the Snell-Descartes law, considers the Earth curvature, is valid for positive and negative elevation angles, and includes reflection (anomalous propagation) conditions. The simulations use as input a digital model of the terrain, as well as radiosonde observations for the vertical profile of temperature, of pressure, and of vapour mixing ratio. Proper validations of these simulations required knowing the latitude and longitude of the radar site with a precision on the order of 10 arc seconds. The model was validated using observations from Environment Canada's weather radar located at Mt. Sicker (in British Columbia).

Future work includes the improvement of the digital terrain model used in these simulations. The terrain elevation used in this study has an accuracy of 30 m, which seems too large for mountain clutter modelling, especially at short ranges. In contrast, the widely-used SRTM terrain dataset (Farr *et al.*, 2007) has an absolute vertical accuracy better than 9 m. The quantification of radar returns from mountains can also be improved, for example, by implementing a radar cross-section for mountains (e.g. Delrieu *et al.*, 1995) and by including a treatment for the sidelobes of the antenna radiation pattern (e.g. Rico-Ramirez *et al.*, 2009).

Acknowledgements Funding for this work was provided in part by Argonne National Laboratory under US Dept. of Energy Contract DE-AC02-06CH11357. The author is indebted to Dr Norman Donaldson (from Environment Canada at Toronto) for his valuable discussion and for providing the dataset presented in Fig. 1(c). Prof. Guy Delrieu (from LTHE at Grenoble, France) and another anonymous reviewer provided useful comments to improve this work.

REFERENCES

Archibald, E. (2000) Enhanced clutter processing for the U.K. weather radar network. *Physics and Chemistry of the Earth, Part B: Hydrology*, 25: 823–828.
Brown, R. A., Wood, V. T. & Barker, T. W. (2002) Improved detection using negative elevation angles for mountaintop WSR-88Ds: Simulation of KMSX near Missoula, Montana. *Weather and Forecasting*, 17(2), 223–237.

Delrieu, G., Creutin, J. D. & Andrieu, H. (1995) Simulation of radar mountain returns using a digitized terrain model. *Journal of Atmospheric and Oceanic Technology*, 12, 1038–1049.

Doviak, R. J. & Zrnic, D. S. (2006) *Doppler Radar and Weather Observations*, 2nd ed. Dover Publications, Inc., Mineola, NY, USA. 562 pp.

Farr, T. G., Rosen, P. A., Caro, E., Crippen, R., Duren, R., Hensley, S., Kobrick, M., Paller, M., Rodriguez, E., Roth, L., Seal, D., Shaffer, S., Shimada, J., Umland, J., Werner, M., Oskin, M., Burbank, D. & Alsdorf, D. (2007) The Shuttle Radar Topography Mission. *Reviews of Geophysics*, 45 (RG2004): 1–33.

Hastings, D. A. & Dunbar, P. K. (1999) Global Land One-kilometer Base Elevation (GLOBE) Digital Elevation Model, Documentation, Volume 1.0. In: *Key to Geophysical Records Documentation (KGRD) 34*. National Oceanic and Atmospheric Administration, National Geophysical Data Center, Boulder, Colorado, U.S.A. Available on the Internet at http://www.ngdc.noaa.gov/mgg/topo/report/index.html.

Kucera, P. A., Krajewski, W. F. & Young, C. B. (2004) Radar beam occultation studies using GIS and DEM technology: an example study of Guam. *J. Atmos. Oceanic Technol.* 21(7), 995–1006.

NIMA (1997) Department of Defense World Geodetic System 1984, Its Definition and Relationships With Local Geodetic Systems. Third Edition, 4 July 1997. *Technical Report TR8350.2*. National Imagery and Mapping Agency, Bethesda, MD, USA. Available on the Internet at http://earth-info.nga.mil/GandG/publications/tr8350.2/tr8350_2.html.

Rico-Ramirez, M. A., Gonzalez-Ramirez, E., Cluckie, I. D. & Han, D. (2009) Real-time monitoring of weather radar antenna pointing using digital terrain elevation and a Bayes clutter classifier. *Meteorological Applications*, 16, 227–236.

Rinehart, R. E. (2006) *Radar for Meteorologist*, 4th ed. Rinehart Publications, Columbia, USA. 482 pp.

Wood, V. T., Brown, R. A. & Vasiloff, S. V. (2003) Improved detection using negative elevation angles for mountaintop WSR-88Ds. Part II: Simulations of the Three Radars Covering Utah. *Weather and Forecasting* 18(3), 393–403.

Weather Radar and Hydrology
(Proceedings of a symposium held in Exeter, UK, April 2011) (IAHS Publ. 351, 2012).

33

Evaluation and improvement of C-band radar attenuation correction for operational flash flood forecasting

STEPHAN JACOBI[1], MAIK HEISTERMANN[1] & THOMAS PFAFF[2]

1 *Institute for Earth and Environmental Science, University of Potsdam, Karl-Liebknecht-Strasse 24–25, 14476 Potsdam, Germany*
stjacobi@uni-potsdam.de

2 *Institute of Hydraulic Engineering, University of Stuttgart, Pfaffenwaldring 61, 70569 Stuttgart, Germany*

Abstract Signal attenuation is, even for C-band radars, an important reason for underestimating precipitation rates during heavy convective rainfall events. Gate-by-gate simulation of specific attenuation based on the conventional power law relation with fixed parameters is prone to instability with increasing distance from the radar location. Hence Krämer (2008) developed an attenuation correction algorithm which optimizes attenuation parameters iteratively for each beam and time step, dependent on the stability of corrected reflectivity. In cases of very high path-integrated attenuation (PIA) this stability criterion is not sufficient for the rainfall events examined; thus a second criterion based on PIA is introduced and the specific attenuation in cases of low reflectivity is limited. With the objective of verifying operational robustness, the correction approaches are compared with uncorrected radar data for several rainfall events, including the severe flash flood event on the River Starzel, Germany.

Key words QPE; signal attenuation; attenuation correction; attenuation threshold; flash flood; Starzel, Germany

INTRODUCTION

As a consequence of their sudden occurrence, flash floods can often cause damage. Hence an accurate and near-real-time estimation of precipitation is indispensable for reliable hydrological forecasts. Conventional raingauge networks cannot adequately detect convective precipitation events. In contrast, radar-based quantitative precipitation estimation (QPE) provides many advantages in terms of temporal and spatial resolution. But radar-based QPE is also limited by serious errors. Especially for cases of severe rainfall, C-band radars are prone to rainfall-induced signal attenuation. The magnitude of attenuation at any point in range is approximately proportional to the rain-rate, but its effects are cumulative with range. Under extreme conditions the entire radar signal may be lost (Harrison *et al.*, 2000) or at least be subject to severe attenuation losses. Therefore, it is important to identify operationally viable algorithms in order to correct for the underestimation of rainfall as induced by attenuation effects.

Hitschfeld & Bordan (1954) proposed a gate-to-gate correction strategy using the estimated rain-rate in one range bin to correct for the following one. However, this procedure turned out to be stable only for short distances of up to some 10 km (Bringi & Chandrasekar, 2001). For this reason Harrison *et al.* (2000) capped the attenuation correction at a factor of two increase in rain-rate, which equates to an attenuation of approximately 4.5 dB, dependent on the Z-R relationship. But convective precipitation events especially can induce wide-ranging attenuation effects of 10 dB and more (Kraemer, 2008).

Based on reference measurements of attenuation by microwave links, Kraemer (2008) analysed the integral attenuation coefficients along single beams and modified the original gate-to-gate approach. The algorithm adjusts the event-based attenuation coefficients to values which allow for a stable correction.

In the present study the basic concept of the methodology proposed by Kraemer (2008) was adopted. The aim of this study is to verify the benefit of this attenuation correction procedure for the catastrophic flash flood events caused by the River Starzel in June 2008 and River Glems in July 2009 and 2010, in southwest Germany. For these events, the radar-based rainfall estimation without attenuation correction led, as well as the interpolation of raingauge measurements, to a severe underestimation of event totals (Bronstert *et al.*, 2012).

METHODS

Study area and data

The study was conducted on the basis of radar data from the C-band radars Feldberg and Tuerkheim in the southwest of Germany. Both stations are part of a radar network operated by the German Weather Service (DWD). The catchment of the River Starzel is located in the overlapping range of both radar stations, which enhances the capability to verify the attenuation effects and is helpful for the visual interpretation of the rainfall event. The radar stations provide scans at the lowest possible elevation angle at intervals of 5 minutes. In order to verify the attenuation correction procedure, we used 184 raingauges available in the federal state Baden-Wuerttemberg (see Fig. 1).

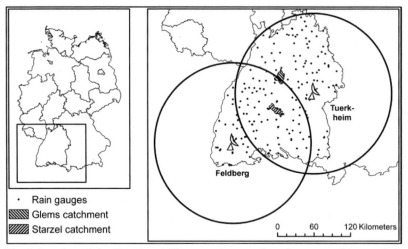

Fig. 1 Location of the Starzel and Glems catchments, the raingauge positions and the radar stations Feldberg and Tuerkheim.

Attenuation correction

The radar equation used in operational mode does not account for propagation effects by the precipitation itself. However, for wavelengths less than 10 cm and high rainfall rates, these effects are not negligible and should be corrected for in order to derive quantitative precipitation estimates. The specific attenuation is difficult to quantify, because it depends on the actual rain properties at each gate. Although there is no direct physical relation between specific attenuation and radar reflectivity, both quantities can be derived by means of the event-specific drop size distribution and the working radar wavelength. Some methodologies (Hitschfeld & Bordan, 1953; Marzoug & Amayenc, 1991; Kraemer, 2008) relate the specific attenuation k (dB/km) to the radar reflectivity Z (mm^6/m^3) by a power law relation:

$$k = a \cdot Z^b \tag{1}$$

In equation (1), a and b are the attenuation coefficients which are dependent on the operating radar wavelength and the drop size distribution model. The correction is then applied from gate to gate. First, the measured reflectivity at the second gate (Z_{2mea}) is multiplicatively corrected by the specific attenuation computed from the measured reflectivity of the first gate (Z_{1mea}). The corrected reflectivity at the second gate (Z_{2cor}) is used to correct the reflectivity on the third gate and so on.

Due to the cumulative growth of attenuation the algorithm tends to instability by exponential overestimation. In order to investigate this problem, Kraemer (2008) analysed the attenuation parameters a and b for different precipitation events. Hildebrand (1978) pointed out that when the radar calibration error is larger than the maximum attenuation (dB/km), attempts to correct for attenuation degrade the rainfall estimate in comparison with the attenuated rainfall estimate. But for cases of very intense storms with maxima up to 60 dBZ and calibration errors of ±2 dB maximum, his algorithm improved the estimation values. To avoid perturbations due to these errors, Kraemer used only data where the attenuation was greater than the expected calibration errors. Based on reference observations by a microwave link, they found that the path-integrated attenuation (PIA) could be adequately represented if the exponent b is kept constant and only the linear coefficient a is adapted to the respective rainfall situation. As a consequence, they propose to reduce the linear coefficient a in case the correction procedure turns out to be unstable for too many adjacent radar beams. The attenuation correction is considered as unstable if the corrected reflectivity exceeds a value of $Z_{max} = 59$ dBZ. For each beam, Kraemer (2008) proposed an iterative procedure in which the linear coefficient a is reduced until the correction becomes stable. If there are just a few adjacent unstable beams, the PIA for the last gate is interpolated between two stable beams as reference attenuation. Based on the interpolated reference attenuation, Kraemer applies a backward algorithm to calculate suitable attenuation parameters. The basic concept of Kraemer was adopted, except in the case where there are just a few adjacent unstable beams the PIA is interpolated and the parameters are backward calculated. The exponential attenuation parameter b is set to 0.7 and the linear attenuation parameter starts with a_{max} at 1.67×10^{-4}. In case of instability, a is iteratively reduced towards $a_{min} = 2.33 \times 10^{-5}$. If a decreases below a_{min} the beam will not be corrected.

Obviously, the validity of the procedure depends on its ability to detect instability. From an operational perspective, the rather arbitrary choice of the threshold Z_{max} is legitimate. However, it might not be sufficient to detect instabilities in case very low reflectivity values are corrected by very high and unstable PIA-factors. In order to enhance the robustness of the procedure, another threshold PIA_{max} was introduced in order to detect instability. According to this threshold, the attenuation correction within one radar beam is considered as unstable if the PIA exceeds values of 20 dB. In contrast to the approach of Harrison *et al.* (2000), this implies that we do not simply truncate PIA values higher than PIA_{max}, but instead reduce the linear coefficient a iteratively until they fall below PIA_{max}. In that way consistent PIA values along the entire radar beam are guaranteed. A second modification is then introduced to prevent instability. The basic idea is that radar gates with low to moderate reflectivity values are not expected to significantly contribute to the "true" PIA. However, gates with low to moderate reflectivity can significantly contribute to the instability of a gate-to-gate correction procedure. This is because they account for a large proportion of the entire radar image (both in convective and advective situations) and thus enable long-range propagation and accumulation effects which can finally lead to instability. The gate-to-gate correction was thus modified in a way that the current specific attenuation k_i is only allowed for summation to the PIA if the attenuation-corrected reflectivity $Z_{i,cor}$ of gate i is greater than a threshold $Z_{min} = 34$ dBZ (corresponding to a moderate rain-rate of about 5 mm/h).

It might be argued that this procedure inherently prevents the full retrieval of the true PIA and thus might – depending on the choice of Z_{min} – be susceptible to underestimation of the true rain-rate. This means a main source of instability is removed at the cost of losing the ability to correct for extreme losses of the radar signal. This is because very low reflectivity values could also be due to the almost entire loss of the radar signal behind extreme convective cells. However, in case the radar signal is almost entirely lost, it should be admitted that the recovery of the true rain-rate by multiplicative procedures is nearly impossible or at least associated with massive uncertainties. Thus the implementation of the Z_{min} criterion is a compromise of the stabilization of the corrected reflectivities and the possibility of underestimating the rain-rate. For the performance requirements of operational flash flood forecasting, a fast and pragmatic approach is needed. Despite the fact that proper attenuation-correction of non-coherent and single polarised C-band radar data is unfeasible without reference PIA measurements for nearly every radar beam concerned, the probability of rain-rate underestimation is less than with data without attenuation-correction.

Controlling the stability of the correction results gives no certainty of estimating the true PIA. It just sets an upper limit for the estimation. Since the linear attenuation parameter starts with a value for the maximum plausible attenuation, the algorithm can tend to overestimate the rain-rate. But for flash flood generating rain-rates it is assumed that the attenuation usually reaches maximum values temporarily, and for the remaining periods the possible overestimation is limited by the Z_{min}- and PIA_{max}-thresholds.

Furthermore attenuation effects due to radome wetting are not considered. Collier (1989) shows radome wetting can cause serious two-way attenuation up to 8.3 dB. But the high variability of wet radome attenuation makes an acceptably quantitative consideration during real-time QPE difficult. Without consideration of the radome attenuation an underestimation of reflectivity in situations while it is raining over the radar location and for a short period after that is expected as well as before the attenuation correction. In further studies it would be interesting to introduce a wet radome attenuation factor controlled by rain intensity data measured with a raingauge nearby the radar station.

Finally, clutter can be a considerable source of instability for the correction procedure. Thus, it was indispensable to carefully eliminate clutter from the radar images. For this purpose, a static clutter map was generated by accumulating the radar-based precipitation amounts over one year.

Verification

A verification of the attenuation corrections is carried out by comparing the precipitation sums derived from the corrected and uncorrected radar data with event-based raingauge measurements. As quality criteria, the root mean square error (RMSE) and the mean error (ME) were used.

RESULTS AND DISCUSSION

The results of the attenuation correction are examined for four reference periods: three extreme convective rainfall events and one stratiform event. The River Starzel was affected by a flash flood on 2 June 2008, with a peak discharge having a return period of more than 100 years.

Figure 2 shows the accumulated precipitation amount between 14:40 and 20:40 CEST for the radar stations Feldberg and Tuerkheim. The Starzel catchment is marked by the red outline. The configuration is quite unique as the convective precipitation cell over the catchment was flanked by two heavy rainfall cells in the northeast and the southwest. Due to the beam directions of the radar stations, both flanking precipitation cells led to a severe attenuation of the "Starzel cell" for

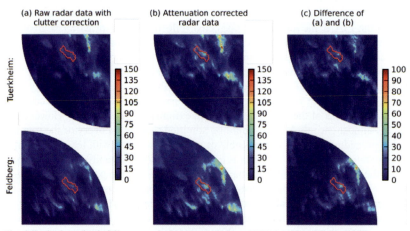

Fig. 2 Radar based rainfall accumulation (mm): 2 June 2008 14:40 to 20:40 CEST.

both radar stations. Due to the modified correction, the precipitation sums for the six hours were increased by up to 67 mm (Fig. 2(c)). The mean precipitation sum over the entire coverage of the radar station Feldberg rises from 23 mm to 36 mm after modified attenuation correction.

The maximum PIA for the Starzel event calculated with the Kraemer-algorithm adds up to 44 dB for the radar station Tuerkheim on the azimuth of 246° at 17:30 CEST. The dashed line in Fig. 3 illustrates the PIA without the additional thresholds, where the three convective cells are visible. This figure reveals the difficulties with instability, too. Due to the extreme PIA in the outer ranges, the rainfall rate at gate 113 is corrected from 0.56 to 175 mm/h (white bars in Fig. 3). Although the algorithm was not aborted by exceeding the Z_{max}-threshold of 59 dB for the corrected reflectivity, which is equivalent to a rain-rate of 177.6 mm/h using the Z-R relationship of Marshall & Palmer (1948), the corrected rain-rate seems to be definitely overestimated. The spatially corresponding gate of the radar station Feldberg shows in the spatio-temporal environment (±1 gate, ±1 azimuth, ±5 minutes) only a maximum of 7.7 mm/h for the attenuation corrected data. For that reason the modified attenuation correction was applied, with the effect that for the ranges between 30 and 45 km as well as between 75 and 101 km the modified PIA is not increasing. Correspondingly, the rainfall rate, especially for the outer convective cell tends, not to be overestimated (grey bars in Fig. 3).

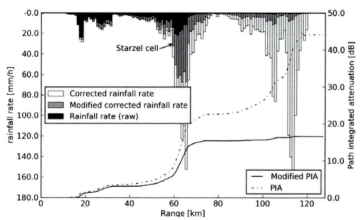

Fig. 3 Path-integrated attenuation along a beam (radar station Tuerkheim, azimuth 246°, 17:30 CEST).

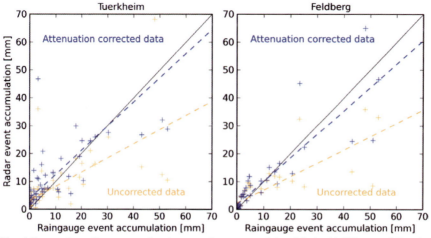

Fig. 4 Scatterplots of uncorrected and attenuation corrected (modified with PIA$_{max}$ and Z_{min} thresholds) precipitation sums for the Starzel event. Linear regressions are displayed by the dashed lines.

The scatterplots in Fig. 4 show the comparison of radar and raingauge precipitation sums during the Starzel event. The gradient of the regression line of attenuation-corrected data shows an improvement compared with the uncorrected data. However, the high mean variation of values of the radar station Tuerkheim limits the regression line as a quality criterion.

As indicated in Table 1, the ME and the RMSE of the attenuation-corrected data show no consistent improvement. In most cases for radar station Tuerkheim, the quality criteria seem to be worse compared with the uncorrected data and tend to overestimation. Only the modified attenuation correction algorithm provides considerable benefits compared to the conventional attenuation correction procedure. The quality criteria for the radar station Feldberg are an exception. The attenuation-corrected radar data with the two new thresholds seem to enhance the quality of the uncorrected data. Possibly the superior location of the radar station Feldberg compared to radar station Tuerkheim, with absolute altitudes of 1517 m and 765 m, respectively, may account for the better correction results.

Table 1 Verification results for all four events with the uncorrected (raw) and the attenuation-corrected (corr.) results and the attenuation-corrected results with modified algorithm (mod. corr.).

		ME			RMSE		
		raw	corr.	mod. corr.	raw	corr.	mod. corr.
2.6.2008 (convective)	Feldberg	−1.93	1.81	0.44	8.79	8.93	6.55
	Tuerkheim	−1.08	4.92	2.69	10.45	14.61	11.29
3.7.2009 (convective)	Tuerkheim	−0.78	6.71	5.43	3.94	11.59	9.12
4.7.2010 (convective)	Tuerkheim	−0.17	3.72	2.54	5.54	10.63	8.88
1.4.2008 (stratiform)	Tuerkheim	−3.58	3.38	3,52	4.78	4.61	4.72

CONCLUSION

The cases of severe convective storms examined here emphasise the specific problem of signal attenuation due to precipitation itself. The effects of convective storms on the discharge can be very serious, especially for small catchments of up to about 1000 km^2.

The iterative attenuation correction method of Kraemer (2009) tends to significant overestimation. The additional implementation of a threshold for a maximum PIA and the constraint of adding the specific attenuation to the PIA only on gates with a reflectivity greater than 34 dBZ, enhances the conventional correction approach of Kraemer for the periods and radar stations examined here.

Since the verification of the attenuation methods shows only good results for the Feldberg radar station, further investigations should clarify if this results from insufficient clutter elimination for the less exposed position of the Tuerkheim radar station.

REFERENCES

Bringi, V. N. & Chandrasekar, V. (2001) *Polarimetric Doppler Weather Radar.* Cambridge University Press, Cambridge, UK.

Bronstert, A., Heistermann, M. & Kneis, D. (2012) Hochwasservorhersage und –warnung – Umgang mit Unsicherheiten. In: *Management von Hochwasserrisiken* (ed. by B. Merz, U. Grünwald, K. Piroth, R. Bittner). Schweizerbart Verlag (in press).

Collier, C. G. (1989) *Applications of Weather Radar Systems.* Ellis Horwood, Chichester, UK.

Harrison, D. L., Driscoll, S. J. & Kitchen, M. (2000) Improving precipitation estimates from weather radar using quality control and correction techniques. *Met. Appl.* 6, 135–144.

Hildebrand, P. H. (1978) Iterative correction for attenuation of 5 cm radar in rain. *J. Appl. Met.* 17, 508–514.

Hitschfeld, W. & Bordan, J. (1954) Errors inherent in the radar measurement of rainfall at attenuating wavelengths. *J. Met.* 11, 58–67.

Kraemer, S. (2008) Quantitative Radardatenaufbereitung für die Niederschlagsvorhersage und die Siedlungsentwässerung. Dissertation, Gottfried Wilhelm Leibniz University, Hannover, Germany.

Marshall, J. S. & Palmer, W. M. (1948) The distribution of raindrops with size. *J. Met.* 9, 327–332.

Marzoug, M. & Amayenc, P. (1991) Improved range-profiling algorithm of rainfall rate from a spaceborne radar with path-integrated attenuation constraint. *Trans. Geosci. Remote Sens.* 29(4), 584–592.

Weather Radar and Hydrology
(Proceedings of a symposium held in Exeter, UK, April 2011) (IAHS Publ. 351, 2012).

Emission: a simple new technique to correct rainfall estimates from attenuation due to both the radome and heavy rainfall

ROBERT THOMPSON[1], ANTHONY ILLINGWORTH[1] & JAMES OVENS[2]

1 *Dept of Meteorology, University of Reading, Reading RG6 6BB, UK*
a.j.illingworth@reading.ac.uk
2 *Meteorological Office, Fitzroy Rd, Exeter EX1 3PB, UK*

Abstract We present a new technique for correcting errors in radar estimates of rainfall due to attenuation which is based on the fact that any attenuating target will itself emit, and that this emission can be detected by the increased noise level in the radar receiver. The technique is being installed on the UK operational network, and for the first time, allows radome attenuation to be monitored using the increased noise at the higher beam elevations. This attenuation has a large azimuthal dependence but for an old radome can be up to 4 dB for rainfall rates of just 2–4 mm/h. This effect has been neglected in the past, but may be responsible for significant errors in rainfall estimates and in radar calibrations using gauges. The extra noise at low radar elevations provides an estimate of the total path integrated attenuation of nearby storms; this total attenuation can then be used as a constraint for gate-by-gate or polarimetric correction algorithms.

Key words attenuation; emission; radome; rainfall estimation; weather radar

INTRODUCTION

Attenuation is a severe problem for C-band radars, especially in the very storms where the rain is heaviest and accurate rainfall rates are needed for improved flood predictions. Gate-by-gate correction schemes are notoriously unstable (Hildebrand, 1978) because any small initial calibration error at the first gate is increasingly magnified as subsequent gates are corrected. Polarisation techniques using, for example, the differential phase shift between the horizontally and vertically polarised returns are very powerful, but the coefficient linking the phase shift to the attenuation is uncertain and variable (Bringi & Chandrasekar, 2001). Phase shifts of up to 300° were observed in the storms producing the floods in London on 20 July 2007, but correction was difficult due to the uncertainty of up to a factor of two in this coefficient (Tabary *et al.*, 2008). Recent experiments with artificial rain on radomes indicate an attenuation of 3 dB can occur in rainfall of 15mm/hr (e.g. Kurri & Huuskonen, 2008). However, operationally there has been no way of monitoring this radome attenuation so it has always been ignored. In this paper we will revive the suggestion, made by Fabry (2001) in a short note, that attenuation due to storms can be corrected using the microwave emissions from the attenuating targets. The operational system has been functioning since November 2010. We report the first measurements made with an operational radar of attenuation by a wet radome. Attenuation from storms is rare in the winter, but we will describe one case where path attenuation of 2 dB was inferred for a heavy shower.

THE EMISSION TECHNIQUE

Attenuation can be detected by the increased noise in the radar receiver as a consequence of the emission from any attenuating target. The effect of an attenuating storm with a physical temperature, T_p, and a one-way optical thickness at radar frequencies of τ, will be an increase in the brightness temperature, ΔT_b, detected by the receiver, given by:

$$\Delta T_b = (T_p - T_g)(1 - \exp(-\tau)) \tag{1}$$

where T_g is the brightness temperature due to gaseous emission from the atmosphere in the absence of any attenuating radar targets. Suppose the target has a physical temperature, T_p, of 280K and the gaseous emission has a T_g of 30 K, so that $T_p - T_g$ is 250 K; the equation is approximately linear for two-way attenuations of up to 6 dB, or 75% loss of signal, with a gradient of 1 dB two-way loss for each ΔT_b of 20 K. If we wish, initially, to estimate the two-way attenuation to 10% (in dB), then this implies T_p–T_g must be known to 25 K. This is easily achieved because the temperature of the attenuating water on the radome or in the storm (remembering that the beam elevation will only be at a height of 1–2 km) is usually known to about 5 K; and the gaseous attenuation can be estimated to this accuracy from sonde ascents or the forecast model. In the mid-latitudes 75% of the gaseous attenuation is due to oxygen, which depends on the pressure, only about 25% is due to the variable water vapour. For much larger attenuations the emission will saturate as τ increases, so for an attenuation of 16 dB (97.5% loss of signal) ΔT_b must be known to within ±5 K if the attenuation is to be estimated to within 1.3 dB.

A large number of samples are needed to detect the increased noise level when emission is occurring. For example, the equivalent T_b for the receivers in the UK network is about 1000 K. If this is to be known to 10 K so that the attenuation can be estimated to 0.5 dB, then 10 000 independent estimates of the noise are needed. The weather radars in the UK use a low prf of about 300 Hz, giving them a maximum range of nearly 500 km. Twenty one pulses are transmitted for each one degree of azimuth of the higher elevation beams. Beyond a range of about 350 km the noise from 1453 gates, extending over a distance of 108 km, is sampled, giving 30 513 independent estimates of the noise, or a temperature accuracy of about 6 K if the receiver noise temperature is around 1000 K. The two lower elevation beams scan more slowly with 44 pulses per degree, so the temperature should be known to about 4 K. Remembering that a ΔT of 10 K is equivalent to 0.5 dB two-way attenuation, this accuracy should be acceptable. By sampling the noise for gates beyond 350 km problems with spurious returns from precipitation or ground echoes should be minimised.

DATA QUALITY AND CALIBRATION

Discrete targets such as aircraft appear in a single gate and are efficiently removed by a speckle filter. Sunrise and sunset are predictable and give much larger signals than are possible from attenuating targets and thus can be recognised and removed. To recognise any anomalous rays remaining where a few gates are still affected by targets, the mean noise signal averaged for the 1453 gates for each degree of azimuth (a "ray") is recorded along with the standard deviation of these estimates; occasional rays with a higher standard deviation can then be identified and rejected.

Absolute calibration is achieved by injecting a known (1165 K) noise source for five gates just before the 1453 empty distant gates. Calibration from day to day is found to be remarkably stable, providing dry days are used. The calibration of neighbouring radars can differ by almost a factor of two. The reason for this is unclear, so a cross-check of the relative values of the calibration of neighbouring radars has been carried out on dry days. By comparing the values of T_g as the elevation angles increases and the path through the atmosphere is less, the relative calibrations of the radars were found to be consistent with those derived from the noise source.

WET RADOME ATTENUATION

The analysis has concentrated on two radars. Predannack, on the Cornish coast close to the sea where a new "orange-peel" radome was installed in the summer of 2010, and Cobbacombe, Devon, with a 15-year-old "faceted" radome. Both radars have a five-minute scan sequence in which time five PPIs are executed at four elevation angles. Tests on dry days at Cobbacombe with the upper two beams at 4° and 2° confirmed that the changes of noise with azimuth were very small and equivalent to less that 3 K change in T_b, or less than 0.15 dB change in two way attenuation.

Figure 1 shows the two-way attenuation at Cobbacombe inferred from the increase in noise level as a function of the azimuth angle for the upper two elevation angles of 4° and 2° for the 576 scans during a two-day period starting at midnight on 16 November 2010. The difference in the attenuation between these two elevations is plotted in the third panel and is effectively zero.

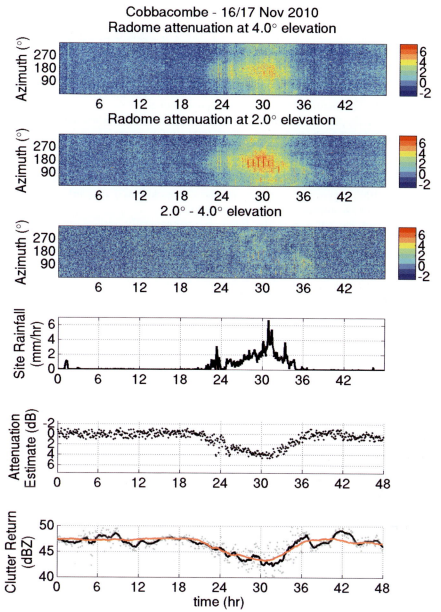

Fig. 1 The two-way attenuation inferred from the increased noise as a function of azimuth for the 48 h starting at 00:00 h on 16 November 2010; the upper two panels, for the 4° and 2° beams, and (third panel) the difference between these two beams. The fourth panel is the rainfall rate at the radar site. The fifth panel is the attenuation from the emission for each scan averaged over nine neighbouring azimuths centred on the 130° azimuth (standard deviation about the running mean 0.6 dB). Sixth panel: grey points – return from the clutter target on the 130° azimuth for each scan, black line – clutter return averaged over 1 h; red line – attenuation from the emission averaged over 1 h.

This confirms the suggestion that the upper beams are detecting the same radome attenuation. The fourth panel shows the rainfall intensity at the radar site. Because no automatic gauge is available, the rainfall is derived by averaging the composite radar rainrates over a 5 × 5 km² centred on the radar. The attenuation along the 130° azimuth averaged over nine neighbouring azimuths inferred from the emission is displayed in the fifth pane, and then in the bottom panel is compared with the changes in apparent reflectivity of a strong ground clutter target on the same 130° azimuth using the lowest (0°) elevation radar beam. The other elevations are 0.5° and 1°. The following features are evident from Fig. 1:

– Attenuation coincides with the rainfall over the radar site. Studies of many days for the upper two beams at the two radars during November 2010 and January 2011 confirm that when it rains at the radar site there is always an increase in the noise.
– At Cobbacombe the radome attenuation exceeds 4 dB for periods when the rain rate is only 2 mm/h. We find such high attenuation with all light rainfall events.
– The difference in attenuation at the two beams is small, suggesting that the layer of water on the radome is uniform for small vertical displacements just above the equator.
– The radome attenuation is highly dependent upon the azimuth; the 4 dB maximum value extends over 100°, whereas at other angles it can be as low as 1 or 2 dB. As a result it is difficult to predict the attenuation from the rainfall rate.
– The structure of the patterns in the figure is predominantly vertical, indicating that, at Cobbacombe, the whole radome gets wet at the same time.
– The short vertical red lines for the 2° elevation data, indicate enhanced emission for individual scans extending over about 100° in azimuth. These events can be identified by the anomalously high standard deviation of the noise from the gates and so could be removed. These spurious signals are only found when it rains; we believe there must be some subtle intermittent coupling of radio frequency interference via the wet radome or perhaps they are caused by streams of water on the radome.
– The change in the apparent reflectivity of the strong clutter echo on azimuth 130° tracks the attenuation inferred from the emission almost perfectly, thus providing independent validation of the technique.

EMISSION FROM DISTANT ATTENUATING STORMS

In theory, the technique for detecting radome emissions from the upper beams can be applied to the lower beams, where any additional signal should be due to emission from distant storms. We have not had the opportunity to fully evaluate this approach, because attenuating storms are rare in the UK during the winter time. We do find that lower beams are affected by "glowing" clutter which is detected via the sidelobes and increases the noise at all gates. This "glowing clutter" can be recognised by azimuthal variations in brightness temperature of up to 60 K, but on dry days the pattern of the azimuthal variation is very constant and reproducible to within 6 K. This suggests that it should be possible to identify the emission from storms by a positive excursion of the noise over a few degrees of azimuth. However, observations of the noise in the lower beams on wet days, when no attenuation by heavy rain is expected, are less encouraging; the clear signal from the upper beams in Fig. 1 is lost, and superposed is a complex and irregular increase in noise at many azimuths. This warrants further examination, but we think it may be due to water gathering in drops at the equator of the radome where it then forms drips which fall to the ground.

Figure 2 is an example from the Predannack radar where emission can be identified from the upper two beams. The figure reveals an additional emission signal for the 2.4° elevation beam between 12:00 h and 13:30 h which is not apparent in the 4° elevation beam. A series of radar images from the Predannack radar of the movement of an intense radar echo to the E and NE of the radar is also displayed in the figure.

In summary the figure shows:
– The attenuation for this new radome is less than at Cobbacombe, but it is still significant; at 13:30 h a two way attenuation of 2 dB coincides with 2–4 mm/h rainfall.
– There is also evidence of an azimuthal wetting which varies in time; at 09:00 h the side of the radome between 120° and 250° azimuth appears to attenuate before the rest of the radome.

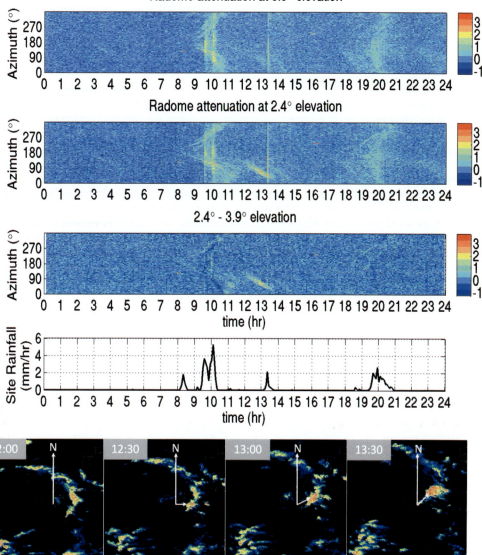

Fig. 2 Attenuation inferred from the Predannack radar on 19 November 2010. The upper four panels are as for Fig. 1, showing the two-way attenuation in dB as a function of azimuth, firstly for the 4° beam, then the 2° beam then the difference between the two beams, and finally the rainfall rate at the radar site. Note the streak of attenuation between 12:00 h and 13:30 h in the third panel during which the azimuth direction changes from E to NE. The final row of images are the radar composites between 12:00 h and 13:30 h over a region of about 250 × 250 km with the arrows indicating the azimuth of the intensifying cell to the E and then the NE of the radar site. The position of this cell is consistent with the extra emission in the 2° beam, indicating that the total path attenuation through this cell is reaching 2 dB (two-way).

This feature is common at Predannack and in heavier rain is much clearer than this example. We suspect the increased wind at the exposed radar site is responsible for differential wetting of the radome.

– Emission from a distant storm is evident between 12:00 h and 13:30 h, starting on an easterly azimuth and gradually increasing to be equivalent to an attenuation of about 2 dB (two-way) as it moves to the northeast. The sequence of composite images in the figure confirms this interpretation; an intensifying echo can be clearly seen moving in the direction inferred from the emission observations.

CONCLUSION AND DISCUSSION

An operational system for monitoring attenuation both due to the wet radome and to more distant storms has been installed on the operational radar network in the UK. First results are encouraging and indicate that the technique is viable. The two-way radome attenuation can be inferred by monitoring the increased brightness temperature for beams at 2° and 4° with an accuracy of 0.25 dB (5 K). Increased emission can be detected even in the lightest rain. For the 15-year-old Cobbacombe radome the attenuation is 3–4 dB for rainfall rates of only 1–2 mm/h; these values have been independently confirmed by monitoring changes in the reflectivity of a strong and constant clutter target. For the new Predannack radome, attenuation is less, but can be significant at 1–2 dB for 2 mm/h. The attenuation varies markedly with azimuth, so that it is difficult to predict from a knowledge of the rain rate. If such changes are at all typical then they may be affecting radar calibrations using close-in gauges, and they may be the source of the rather disappointing reports of disagreement of ±60% when hourly or daily gauge totals are compared with those inferred from radar in Switzerland (Germann *et al.*, 2006) and France (Tabary *et al.*, 2008).

We report one case where emission was higher on the 2° elevation beam than the 4° beam, equivalent to an additional two-way attenuation of 2 dB, and show that the azimuths over which this occurs track a developing intense radar echo. More observations are needed during the summer time to evaluate the accuracy with which the total path integrated attenuation through attenuating radar echoes can be derived from the increased noise level of the lower elevation beams. The advantage of this technique is that it works equally well for heavy rain and wet hail and no assumption is made about the microphysics of the attenuating storm. Total path attenuation of up to nearly 20 dB can, in theory, be accurately estimated, and then used as a constraint for the total attenuation computed using gate-by-gate or polarimetric correction algorithms.

REFERENCES

Bringi. V. N. & Chandrasekar, V. (2001) *Polarimetric Doppler Weather Radar*. Cambridge University Press, 636 pp.

Fabry, F. (2001) Using radars as radiometers: Promises and pitfalls. In: *Proc 30th Int. Conf. on Radar Meteorology* (19–24 July 2001, Munich), 197–198, Amer. Meteorol. Soc, Boston, USA.

Germann, U., Gasllli, G., Boscacci, K. & Bolliger, M. (2006) Radar precipitation measurement in mountainous regions. *Quart. J. Roy. Met. Soc.* 132, 1169–1692.

Hildebrad, P. H. (1978) Iterative correction for attenuation of 5cm radar in rain. *J. Appl. Met.* 17, 508–514.

Kurri, M. & Huuskonen, A. (2008) Measurement of the transmission loss of a radome at different rain intensities. *J. Atmos. Oceanic Technol.* 25, 1590–1598.

Tabary, P., Desplats, J., Do Khac, K., Eideliman, F., Gueguen, C. & Heinrich, J.-C. (2007) The new French operational radar rainfall product. Part II. Validation. *Weather and Forecasting* 22, 409–427.

Tabary, P., Vulpiani, G., Gourley, J. J., Illingworth, A. J. & Bousquet, O. (2008) Unusually large attenuation at C-band in Europe: How often does it happen? What is the origin? Can we correct for it? In: *5th European Conf. on Radar in Meteorology and Hydrology* (Helsinki, Finland, 30 June to 4 July 2008).

Weather Radar and Hydrology
(Proceedings of a symposium held in Exeter, UK, April 2011) (IAHS Publ. 351, 2012).

45

Techniques for improving ground clutter identification

J. C. NICOL[1], A. J. ILLINGWORTH[1], T. DARLINGTON[2] & J. SUGIER[2]

1 *University of Reading, Reading, UK*
  j.c.nicol@reading.ac.uk
2 *Met Office, Exeter, UK*

Abstract Several radar parameters quantifying signal variability in single-polarisation radar measurements (Power Ratio, PR; Clutter Phase Alignment, CPA; and Absolute Power Difference, APD) are evaluated using Bayes' theorem in terms of the separation between the returns from ground clutter and precipitation. As these parameters are not independent, the intention is to identify the parameter providing the best separation. It is shown that either PR or CPA, in combination with a radial measure of texture of reflectivity (in dBZ), provides excellent separation of ground clutter and precipitation returns on a gate-by-gate basis. The demonstrated skill in clutter identification is comparable to that only previously reported using dual-polarisation measurements. This approach is well-suited for anomalous propagation as clutter maps are not used. The findings suggest that ground clutter identification is likely to benefit from measurements of PR or CPA, even when dual-polarisation parameters are available.
Key words weather radar; ground clutter; precipitation; Bayes classifier

INTRODUCTION

Ground clutter returns are a well-known problem affecting quantitative weather radar applications. Radar data quality control is increasingly important as the assimilation of this data in NWP models becomes more common. The UK Met Office has recently developed its own in-house digital radar receivers, allowing processing of the in-phase and quadrature samples. The collection and evaluation of two new parameters capturing pulse-to-pulse signal variability has recently begun. This work is a preliminary investigation of the identification of ground clutter returns for operational, single-polarisation weather radar. A Bayes classifier is used to estimate the class of scatterer (clutter or precipitation) on a gate-by-gate basis, using likelihood functions of signal variability and reflectivity texture parameters established for clutter and precipitation from calibration data sets. Independent data sets are used for evaluation. The approach employed here only uses real-time measurements and does not rely on previously established clutter maps. This approach provides greater potential for clutter identification during anomalous propagation (AP) conditions, when atypical patterns of ground clutter coverage occur. However, clutter identification in standard and AP conditions have yet to be determined separately due to the limited number of AP cases currently available. Future improvements in clutter identification may come through spectral techniques (e.g. Warde & Torres, 2009), though these have yet to be demonstrated and evaluated on operational weather radars.

RADAR SPECIFICATIONS

This work involves the evaluation of ground clutter identification for radars of the UK operational weather radar network. Recent modifications allow the collection of extra radar parameters capturing the pulse-to-pulse variability of the returned signal. The network radars operate at C-band (5 cm) wavelengths and are generally only capable of single-polarisation measurements. These systems are to be upgraded for dual-polarisation measurements, so any clutter identification algorithm needs to easily accommodate dual-polarisation data when available. The data considered here have been collected by two of the operational radars (Cobbacombe and Chenies) in southern England since 13 January 2011. Data have been available for the lowest elevation PPIs at these sites (0° and 0.4°, respectively), which are repeated every 5 minutes. The radar specifications for these PPIs are; pulse duration = 2 μs, beamwidth = 1°, PRF = 300 Hz, scan rate = 1.2 rpm and max. range = 255 km. The integration and spacing of data is 1° in azimuth and 600 m in range.

GROUND CLUTTER IDENTIFICATION

In this work, a Bayes classifier is used to determine the likelihood that a given radar measurement is due to ground clutter rather than precipitation. Parameters that are used for this purpose should provide good separation between ground clutter and precipitation echoes. Ground clutter echoes may be characterized as having zero-velocity and low spectral widths compared to weather echoes (Doviak & Zrnic, 1993). However, precipitation echoes may also have near-zero radial velocity and low spectral widths. Spectral width estimates are also not very accurate, particularly at low spectral widths and low signal-to-noise ratios. Another distinguishing characteristic of ground clutter echoes is that the reflectivity fields are characteristically much more textured than they are for precipitation. We therefore consider these two characteristics for the separation of clutter and precipitation; signal variability and texture of reflectivity.

Signal variability parameters

Three signal variability parameters are considered for the identification of clutter echoes. These parameters are independent of echo intensity and hence the overall calibration of the radar system. First-of-all, PR is defined as "power of the mean complex signal divided by the mean complex signal power". In terms of power spectra, PR is the ratio of the power at zero-frequency to the total power (over all frequencies). PR can only take values between 0 and 1 inclusive. PR = 1 corresponds to a completely constant signal in both power and phase. Variability in either power or phase will reduce this value. The expected value of a completely incoherent signal is equal to $1/N$, where N is the number of pulses. Secondly, CPA, from Hubbert et al. (2009a), is equivalent to PR except that it is calculated in terms of signal amplitude (A) rather than power ($P = A^2$). They found that CPA outperformed PR when tested on NEXRAD radars in the USA and it is now part of the operational ground clutter mitigation scheme. As CPA is amplitude- rather than power-weighted, the improvement in performance was attributed to the fact that CPA is less affected by dramatic changes in signal strength as the radar beam scans past strongly backscattering ground clutter targets. Like PR, CPA can only take values between 0 and 1 inclusive. Finally, APD is the mean absolute power difference (in dBZ) using a specified lag between pulses. For the operational weather radars in the UK, a 2-pulse lag is employed (delay of 6.66 ms). This measure is currently used in the operational ground clutter identification scheme in the UK (Sugier et al., 2002), referred to therein as the Clutter indicator (Ci). Values of APD vary from 0 dB (signal has constant power) to about 6 dB (incoherent signal). APD has also been shown to provide accurate estimates of low spectral widths (Melnikov et al., 2002), which suggests it may provide good clutter identification. While the first two parameters are influenced by the mean radial velocity of the target, APD is completely independent of velocity though closely related to spectral width.

Texture parameters

Two parameters capturing the spatial variability of reflectivity are considered. Firstly, Texture (TEX) is simply defined as the standard deviation of reflectivity (in dBZ) over an $n \times n$ window (i.e. n rays by n gates). Texture calculated in this manner has been widely used in clutter schemes (e.g. Gourley et al., 2007; Hubbert et al., 2009b). In contrast, Radial Texture (RTX) is the mean squared difference of reflectivity (in dBZ) only between adjacent gates in range, within an $n \times n$ window. For example, a 3×3 window consists of six pairs of adjacent range-gates, hence RTX would be calculated from these six differences at a given location. RTX is expected to be a better indicator of the texture of the underlying field given that the radar beam typically smoothes much more in azimuth than in range relative to the typical sampling and averaging of the received signal.

Bayes classifier

The use of a Bayes classifier has previously been proposed for the identification of ground clutter (e.g. Moskowicz et al., 1994; Rico-Ramirez & Cluckie, 2008). This approach is now briefly summarised. The posterior probability $P(C|x)$ represents the probability of the occurrence of ground clutter (C) given the observation of parameter x. It is assumed here that all observations are either due to ground clutter or precipitation (R), i.e. $P(C) + P(R) = 1$. The posterior probability or Probability of Clutter (POC) may be expressed using Bayes' theorem as:

$$P(C \mid x) = \frac{P(x \mid C) P(C)}{P(x \mid C) P(C) + P(x \mid R) P(R)} = \frac{P(x \mid C) P(C)}{P(x \mid C) P(C) + P(x \mid R)(1 - P(C))} \tag{1}$$

POC depends on the likelihood functions for ground clutter $P(x|C)$ and precipitation $P(x|R)$ which may be empirically determined, making the separation between clutter and precipitation using calibration data sets. The prior probabilities $(P(C)$ and $P(R))$ represent the expected probabilities of C and R. We assume that these quantities are unknown for any given observation, so we are equally likely to expect clutter or precipitation, e.g. $P(C) = P(R) = 0.5$. This is known as a naïve Bayes classifier and may be written as:

$$P(C \mid x) = \frac{P(x \mid C)}{P(x \mid C) + P(x \mid R)} \tag{2}$$

for a single input parameter, or extended to two parameters, x and y, (e.g. PR and RTX) as:

$$P(C \mid x, y) = \frac{P(x \mid C) P(y \mid C)}{P(x \mid C) P(y \mid C) + P(x \mid R) P(y \mid R)} \tag{3}$$

EVALUATION OF GROUND CLUTTER IDENTIFICATION

The first step in the application of a Bayes classifier for ground clutter detection is to define both "cluttered" and "clutter-free" regions, which are to be used for calibration. The following steps were taken to define these regions. Initially, a dry period was objectively identified using a combination of radar and satellite observations. The frequency of occurrence (FOD) for significant radar returns (>15 dBZ) was then calculated during this 24-h period (19 January 2011) at both radar sites, each based on 288 PPIs. For our purposes, returns have been broadly categorized as "clutter-free" (FOD < 1%), "intermittent" ($1\% \leq$ FOD $\leq 90\%$) and "cluttered" (FOD > 90%). Thus, the clutter-free regions exhibited no more than one significant return (>15 dBZ) from the 288 PPIs considered. Cluttered regions covered an area of about 6% and 3% of the entire radar domain at Cobbacombe and Chenies, respectively. Intermittent echoes covered about 3% and 2% of the entire domain at these two sites. However, due to the fact that these intermittent echoes occur infrequently, returns from cluttered regions comprised >80% of all significant returns during the calibration period. It is not necessary to include all possible observations in the calibration set to accurately characterize clutter returns. Sensitivity tests have shown that the FOD threshold for cluttered regions can vary between 50% and 99% without significantly affecting the results presented later in this work.

The next step is to establish the likelihood functions for clutter and precipitation. The likelihood functions for ground clutter, $P(x|C)$, have been estimated from returns from the previously defined cluttered regions during the same dry 24-h period (19 January 2011). Likelihood functions have been derived for the three signal variability parameters (PR, CPA and APD) and the two texture parameters (TEX and RTX). The texture parameters have been calculated using a 3×3 window. To assess the benefits of smoothing the signal variability parameters, likelihood functions have been determined for these parameters averaged over a 3×3 window (denoted PR3, CPA3 and APD3), in addition to the raw measurements. The likelihood functions for precipitation, $P(x|R)$, have been established during a 24-h period (13 January 2011) from the clutter-free regions for each radar. This day was characterized by widespread and prolonged rain across southern England, exhibiting a mixture of rainfall from frontal and convective origins. It is implicitly assumed using the approach employed here, that the characteristics of returns from ground clutter mixed with precipitation will tend towards those of the dominant component.

These likelihood functions (given the occurrence of ground clutter (C) or precipitation (R)) have been used to derive the POC as a function of each of the parameters considered using Bayes' theorem. The likelihood functions for precipitation (blue) and clutter (red) are shown in Fig. 1(a)–(c) for PR, CPA and APD, respectively, all smoothed using a 3×3 window. One may observe the excellent separation and small overlap between these two distributions for PR and CPA. The corresponding cumulative likelihood functions along with the POCs (green) are shown in Fig. 1(d)–(f), respectively. When POC = 0.5, returns have an equal likelihood of originating from clutter as they do from precipitation. This determines the threshold above which returns are more likely to be due to ground clutter than they are to be from precipitation. If returns with POC $\geq$ 0.5 are identified as clutter, both the fraction of clutter which is successfully identified (probability of detection, POD) and the fraction of precipitation echoes which is falsely identified (false alarm rate, FAR) may be determined. The Critical Success Index, CSI = POD/(1+FAR), combines the performance in clutter and precipitation as a single quantity.

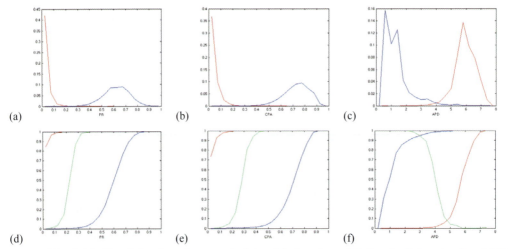

Fig. 1 Likelihood functions for ground clutter (blue) and precipitation (red) for signal variability parameters; (a) PR3 (PR averaged using 3×3 window), (b) CPA3 and (c) APD3. Cumulative likelihood functions (blue and red) and posterior probability or Probability of Clutter (green) for signal variability parameters; (d) PR3, (e) CPA3 and (f) APD3. (minimum reflectivity threshold = 15 dBZ).

To assess and validate the identification of ground clutter, "dry" and "wet" validation data sets have been developed. The "dry" data sets correspond to 48-h periods with no precipitation over the entire radar domain at Cobbacombe (30–31 January 2011) and Chenies (28–29 January 2011). The "wet" data sets correspond to the same 4-day period (14–17 January 2011) with widespread and prolonged rain over the majority of southern England, again exhibiting a mixture of frontal and convective rainfall. Evaluation statistics have been determined from these data sets as follows. POD (clutter identified as clutter) is calculated over the entire radar domain during the "dry" period, essentially assuming that all returns are due to clutter. FAR (precipitation identified as clutter) is calculated only from the "clutter-free" regions throughout the "wet" period. We assume that any significant returns in regions previously deemed to be free of clutter in the FOD maps are due to precipitation. The results (POD, FAR and CSI) from this preliminary analysis are shown in Fig. 2(a) and (b) for Cobbacombe and Chenies, respectively, using various minimum thresholds (15, 25 and 35 dBZ) which broadly represent the lower bounds for light, moderate and heavy rainfall (0.3, 1.3 and 5.6 mm/h).

In addition to the unsmoothed (PR, CPA and APD) and smoothed (PR3, CPA3 and APD3) signal variability parameters and the texture parameters (TEX and RTX), statistics have also been derived for the combination of parameters (equation (3)), PR and RTX, both with (PR3+) and without (PR+) smoothing. In terms of the signal variability parameters, improved separation between clutter and precipitation on a gate-by-gate basis is achieved by smoothing using a 3 × 3 window. Inferior results (not presented here) were obtained using smoothing and texture calculations with windows larger than 3 × 3, indicating the trade-off between accuracy and resolution. Overall, the performance of PR and CPA is practically identical and both of these parameters outperform APD. CSI values from PR3 and CPA3 were typically higher than those using APD by about 0.06 (6%). Low values of APD, typical of clutter, correspond to narrow spectral widths. High values of PR and CPA, also typical of clutter, require both near-zero velocities and narrow spectral widths. It is believed that this explains the improved separation of clutter and precipitation using either of these parameters relative to APD. Regarding the texture parameters, the radial texture (RTX) provides significantly better separation than the isotropic texture (TEX). In terms of CSI, RTX is consistently about 0.05 (5%) higher than TEX at both radar sites. It is believed that the effective smoothing of the beam in azimuth leads to a higher degree of correlation between adjacent rays than adjacent range-gates. This reduces the variability in intensity of clutter returns in azimuth. Hence, RTX is better than TEX at capturing the "point" target-like nature of the underlying clutter field.

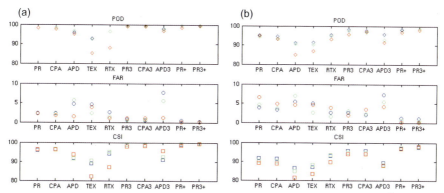

Fig. 2 (a) Clutter identification POD, FAR and CSI for the radar at Cobbacombe for various signal variability and texture parameters. Minimum reflectivity thresholds (15, 25 and 35 dBZ), which broadly represent the lower bounds for light (blue), moderate (green) and heavy (red) rainfall, respectively, have been employed. (b) Same for the radar at Chenies.

The overall performance using PR3 and RTX at Cobbacombe is exceptional. CSI values of 0.993, 0.996 and 0.993 were obtained for the minimum thresholds of 15, 25 and 35 dBZ respectively. These correspond to PODs greater than 99.2% with FARs less than 0.3% in all three classes. The performance using PR3 and RTX at Chenies was also very good. CSI values of 0.977, 0.978 and 0.982 for the same thresholds were obtained with PODs close to 99% and FARs less than 1%. The poorer performance at Chenies may be related to other forms of interference and clutter due to the proximity to London and Heathrow airport (~30 km from radar) in particular. These results compare favourably with those previously reported using dual-polarisation measurements from a radar at Thurnham in the UK (Rico-Ramirez & Cluckie, 2008). They found CSI values of around 96% using single-polarisation measurements and the highest CSI value using an optimally-weighted combination of available dual-polarisation observations was 98.2%. Their scheme also used clutter frequency maps (based on the prior probability of occurrence) and textures were calculated using a 5 × 5 window. Similar performance has been obtained here for the radar at Chenies and even better performance at Cobbacombe using just two single-polarisation input parameters, PR and the radial texture of reflectivity.

EXAMPLES

Here, we provide two examples of clutter identification in special circumstances; AP and concurrent returns from clutter and precipitation. A dry-weather AP episode is shown in Fig. 3(a) (reflectivity) for the radar at Cobbacombe at 2100 UTC, 31 January 2011. The echoes beyond 100 km (~15–25 dBZ) are due to AP. Figures 3(b)–(d) show POC based on APD3, PR3 and PR3+, respectively. POC ≥ 0.5 indicates ground clutter. While APD3 is only partially successful at identifying the AP returns, improved identification is found using PR3 and excellent (PR3+) detection achieved using PR3+. The second example, shown in Fig. 4(a) (reflectivity), shows a widespread region of precipitation around the radar at Chenies. POC using PR3+ is shown in Fig. 4(b). The detected regions correspond very well with the standard day-weather clutter map even in the presence of rain. Bright band effects may be observed to the west of the radar, indicating the presence of rain and ice within the radar domain. It may be inferred that partial beam-filling and beam blocking effects do not seem to degrade the separation of clutter and precipitation.

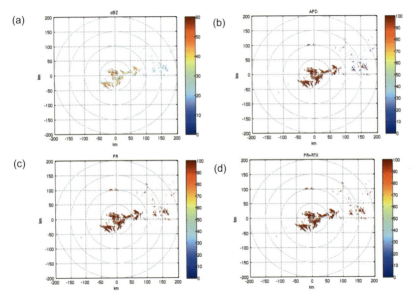

Fig. 3 Images of (a) reflectivity (in dBZ) and the Probability of Clutter (POC) using (b) APD3, (c) PR3 and (d) PR3+ during an AP episode (21:00 h 31 January 2011) for the radar at Cobbacombe. Clutter is detected when POC ≥ 0.5. The echoes beyond about 100 km (~15–25 dBZ) are due to AP.

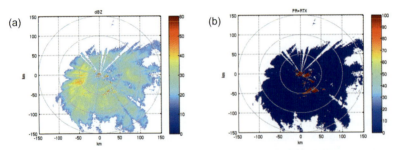

Fig. 4 Images of (a) reflectivity (in dBZ) and (b) Probability of Clutter for the radar at Chenies using PR3+ at 06:45 h 17 January 2011, demonstrating ground clutter identification in the presence of rain.

CONCLUSIONS

We have presented some preliminary findings on ground clutter identification using signal variability parameters on two radars of the UK operational weather radar network. This has involved the use of a naïve Bayes classifier on various parameters capturing signal variability and the texture of reflectivity. The two new parameters considered (PR and CPA) provide a significant improvement over that currently used in the operational ground clutter scheme (APD). No significant improvement was found using CPA over PR, in contrast to the findings in Hubbert *et al*. (2009a) using NEXRAD radars in the USA. It is believed that the difference is due to the slower operational scan rates employed at low elevations in the UK (7.2°/s). Radial texture was also found to significantly outperform an isotropic measure of the texture of reflectivity. In combination, PR3 (PR averaged using a 3 × 3 window) and RTX seem to provide excellent discrimination between clutter and precipitation with results comparable to those obtained from dual-polarisation measurements. Inferior results were obtained using larger windows (e.g. 5 × 5, …). Examples illustrating good performance for dry-weather AP and when clutter and precipitation co-exist have been presented. The findings suggest that PR or CPA would likely improve ground clutter identification even when dual-polarisation measurements are available.

REFERENCES

Doviak, R. & Zrnic, D. (1993) *Doppler Radar and Weather Observations*. Academic Press, UK, 592 pp.

Gourley, J., Tabary, P. & Parent du Chatelet, J. (2007) A fuzzy logic algorithm for the separation of precipitating from non-precipitating echoes using polarimetric radar observations. *J. Atmos. Oceanic Technol.* 24, 1439–1451.

Hubbert, J., Dixon, M., Ellis, S. & Meymaris, G. (2009a) Weather radar ground clutter. Part I: Identification, modelling and simulation. *J. Atmos. Oceanic Technol.* 26, 1165–1180.

Hubbert, J., Dixon, M. & Ellis, S. (2009b) Weather radar ground clutter. Part II: Real-time identification and filtering. *J. Atmos. Oceanic Technol.* 26, 1181–1196.

Moszkowicz, S., Ciach, G. & Krajewski, W. (1994) Statistical detection of anomalous propagation in radar reflectivity patterns. *J. Atmos. Oceanic Technol.* 11(4), 1026–1034.

Rico-Ramirez, M. A. & Cluckie, I. (2008) Classification of ground clutter and anomalous propagation using dual-polarisation weather radar. *IEEE Trans. Geosci. Remote Sens.* 46(7), 1892–1904.

Sugier, J., Parent du Chatelet, J., Roquain, P. & Smith, A. (2002) Detection and removal of clutter and anaprop in radar data using a statistical scheme based on echo fluctuation. In: *Proceedings of ERAD*, 17–24.

Warde, D. A. & Torres, S. M. (2009) Automatic detection and removal of ground clutter contamination on weather radars. *Proc. 34th Conf. Radar Meteorology*, Williamsburg, VA, USA, AMS, P10.11.

A probability-based sea clutter suppression method for polarimetric weather radar systems

RONALD HANNESEN & ANDRÉ WEIPERT

Selex-SI GmbH, Gematronik Weather Radar Systems, Raiffeisenstr. 10, 41470 Neuss, Germany
r.hannesen@gematronik.com

Abstract Beyond calibration, the mitigation and suppression of clutter signals is still a challenge in radar remote sensing. The weather radar market trend (for aviation and hydrological/meteorological applications) shows explicitly that the decision for polarimetric radar systems is continuously increasing, since the potential capabilities and benefits of dual-polarization radar systems are well known. This publication presents an automatic discrimination method between weather and sea clutter based on multi-parameter polar datasets (Doppler and polarimetric) as well as the generation of a sea clutter probability index. Additionally, a new polar-based clutter type map will be introduced and the suppression of sea clutter signals outlined.

Key words sea clutter; clutter suppression; dual polarization

INTRODUCTION

In the past decade, an increasing amount of operational weather radar systems have been equipped with polarimetric measurement capabilities. Polarimetric radar data allow for better attenuation correction and precipitation estimation, compared to single-polarization radar systems (e.g. Schuur *et al.*, 2003; Rhyzhkov *et al.*, 2005; Selex-Gematronik, 2007).

A particular focus of polarimetric radar data application is on the identification and removal of non-weather radar signals. While echoes from stationary targets can usually be removed very efficiently by application of Doppler filters in the radar signal processor, echoes from moving non-weather targets often pass signal processor thresholds and may eventually affect radar data processing algorithms. Typical errors of such data not being filtered are non-real precipitation accumulation data or false velocity profile estimations. Also, false wind shear alerts may result from transition zones between meteorological and not filtered non-weather signals.

One type of radar target which usually cannot be filtered effectively by Doppler filters is sea clutter: waves on the sea surface cause a signal which is typically as strong as that of weak to moderate precipitation and is usually not removed by Doppler filters due to the waves' motion of typically several metres per second.

To identify and eventually to remove sea clutter signals, an automatic sea clutter detection and suppression algorithm has been developed. This probability-based fuzzy-logic algorithm can be applied to both polarimetric and single-polarization radar data, where the use of polarimetric data provides much better results. The algorithm is described next, followed by some results and a summary of the study.

ALGORITHM DESCRIPTION

Previous observations of sea-clutter have shown that radial velocity values are rather small and smoothly distributed (due to the rather homogeneous velocity of sea surface waves), differential reflectivity (ZDR) and differential phase shift (Φ_{dp}) are very variable, and the polarimetric correlation (ρ_{hv}) is rather low (Rhyzhkov *et al.*, 2002; Sugier & Tabary, 2006; Rico-Ramirez & Cluckie, 2008). Furthermore, sea clutter usually appears only in the lowest elevation slices of a radar scan and exhibits reflectivity strongly decreasing with height. Table 1 summarizes typical parameters and values of sea clutter and of weather echoes, and indicates the usefulness of each parameter to distinguish between sea clutter and weather echoes.

The first seven parameters of those listed in Table 1 are available for every range gate of a radar scan. The corresponding measurements are converted into conditional probabilities p_{ij}

Table 1 Parameters to distinguish between sea clutter and weather signals.

i	Parameter	Typical range for sea clutter	Typical range for weather echoes	Usefulness
1	Reflectivity (Z)	low to moderate	low to high	–
2	Texture of Z	moderate to high	low to moderate	o
3	Vertical gradient of Z	negative	variable	o
4	Radial velocity	variable	low	o
5	Texture of ZDR	moderate to high	low	+
6	Texture of Φ_{dp}	high	low	+
7	ρ_{hv}	low to moderate	high	+
8	Spatial area	over sea; mostly low-level	anywhere	o

Usefulness: +, very useful, o, less useful, –, not useful

$(0 \leq p_{ij} \leq 1)$ for the corresponding parameters (i = 1, …, 7) and for one of the assumed echo types (j = 1 for sea clutter; j = 2 for weather echoes). The values of p_{ij} are one for the typical ranges given in Table 1, and zero outside these ranges (with a linear change in a small transition interval). These typical ranges are configurable for fine-tuning, with initial settings according to the above references.

Following the sum method in the inference step of a fuzzy logic classifier, the conditional probabilities of all seven parameters are combined to conditional probabilities P_j by:

$$P_j = \sum_i w_i p_{ij} / \sum_i w_i \qquad (1)$$

where the weight factors w_i $(0 \leq w_i \leq 1)$ are higher for the useful parameters than for the others. It was decided to use the weighted sum rather than the weighted product of the conditional probabilities p_{ij} in order to avoid combined probabilities P_j becoming zero in the case that just one p_{ij} is zero, e.g. due to random errors of the measured parameter. For the same reason, a Bayes classifier of the form:

$$P_j = P\left(C_j | x_1, ..., x_7\right) = \frac{P(C_j) \prod_i p_{ij}}{P(x_1, ..., x_7)}$$

was not used.

To reduce the amount of false alarms, i.e. echoes being identified as sea clutter which in reality are from meteorological targets, the combined conditional probability P_j for sea clutter is multiplied with an *a priori* probability of sea clutter $P_{SC,a}$ which is derived as follows:

$$P_{SC,a} = P_R P_E P_{Sf} \qquad (2)$$

where P_R is an *a priori* probability between 0 and 1 depending on the range to the radar site, P_E is an *a priori* probability between 0 and 1 depending on the elevation angle, and P_{Sf} is an *a priori* probability (0 or 1), depending on the Earth surface (i.e. water or no water).

The combined conditional probability P_j for weather echoes is not modified.

A radar measurement is finally classified as sea clutter if the modified combined conditional probability P_j for sea clutter is larger than the combined conditional probability P_j for weather echoes; otherwise a measurement is classified as meteorological echo. If, however, both combined conditional probabilities are small, the measurement is not classified.

The algorithm can be applied to single-polarization weather radar data. In this case, instead of seven parameters, only the first four parameters of Table 1 are used. The treatment of the *a priori* probability of sea clutter $P_{SC,a}$ is performed in the same way as in the dual-polarization case.

RESULTS

Polarimetric radar data

Figure 1 shows 0.5° PPIs of various data types scanned with a METEOR 1500 CDP C-band radar of the Republic of China Air Force (ROCAF), located on Makung Island west of Taiwan, during the passage of cyclone "Fanapi". All data are Doppler filtered. Areas of sea-clutter are indicated by blue ellipses. The sea clutter with reflectivity up to above 30 dBZ, corresponding to >2 mm/h of rainfall, would significantly affect precipitation accumulation algorithms unless corrected. Furthermore, the transition from sea clutter to weather echoes causes sharp gradients in the radial velocity data, which is a potential source of wind shear false alarms.

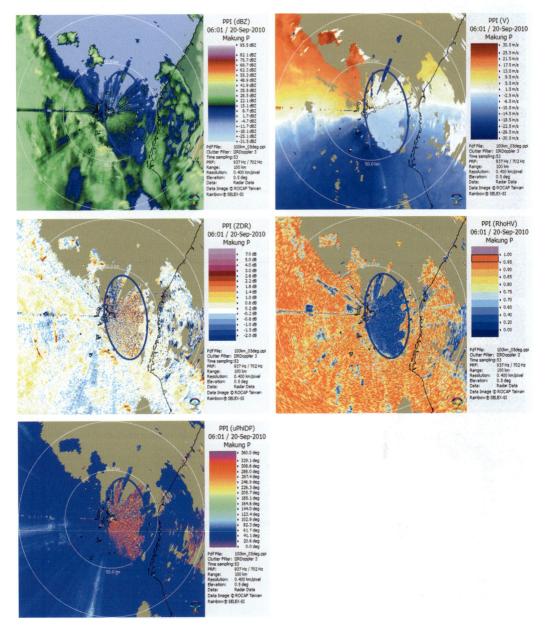

Fig. 1 0.5° PPIs scanned with a METEOR 1500 CDP C-band radar west off Taiwan. Shown are reflectivity (top left), radial velocity (top right), differential reflectivity (middle left), polarimetric correlation coefficient (middle right), and differential phase shift (bottom). Data Images copyright © of ROCAF, Taiwan.

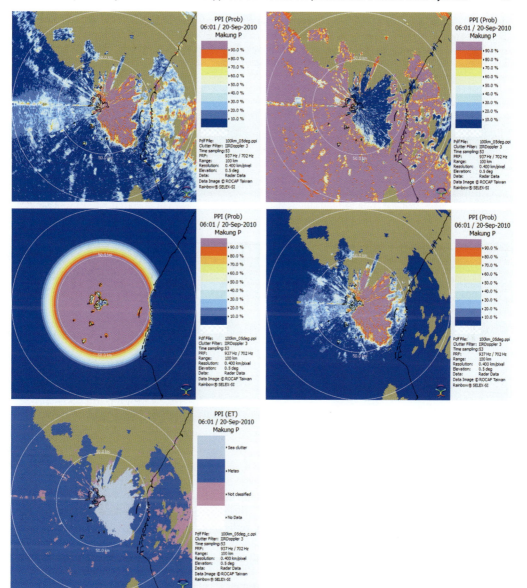

Fig. 2 Conditional probabilities and classification results derived from the data of Fig. 1. Shown are the combined conditional probabilities for sea clutter (top left) and for weather echoes (top right), the *a priori* probability of sea clutter (middle left), the modified combined conditional probability for sea clutter (middle right), and the classification result (bottom). Data Images copyright © of ROCAF, Taiwan.

Figure 2 shows the combined conditional probabilities for sea clutter and for weather echoes calculated from the given data according to equation (1), the *a priori* conditional probability of sea clutter according to equation (2), the modified combined conditional probability for sea clutter and the resulting classification.

Figure 3 shows the original reflectivity data (repeated from Fig. 1) and the corrected reflectivity data, i.e. data classified as sea-clutter being suppressed. The example, demonstrates that the sea-clutter echoes are suppressed very effectively and almost all meteorological echoes are kept.

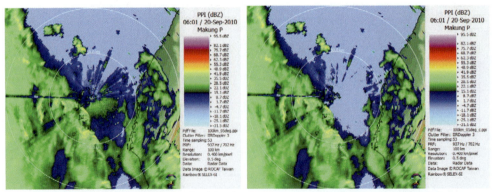

Fig. 3 Uncorrected and corrected reflectivity data. Data images copyright © of ROCAF, Taiwan.

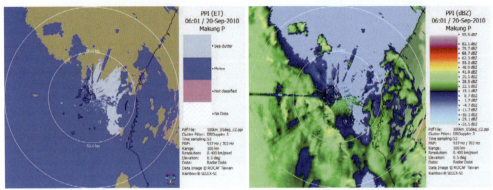

Fig. 4 Classification results and corrected reflectivity data using single-polarization data only. Data Images copyright © of ROCAF, Taiwan.

Single-polarization data

If polarimetric data are not used or not available, the algorithm provides results for single-polarization data. Figure 4 shows the resulting classification and corrected reflectivity data for the same data as in the above sub-section, where only reflectivity, radial velocity and the reflectivity derivatives (as discussed under Algorithm Description) have been used for classification. A large part of the sea-clutter has been removed successfully, and most of the weather echoes remain, but it is clearly evident from comparison with Fig. 3 that the algorithm based on single-polarization data provides worse results than that based on polarimetric radar data.

Removal of ship echoes

The algorithm can also be applied to remove echoes from ships and vessels. Similar to wave echoes, echoes of moving ships are not removed by a conventional Doppler filter. Figure 5 shows 0.3° reflectivity PPIs of a METEOR 1600 SDP S-band radar of the National Environment Agency, Singapore, located at Changi International Airport in Singapore. As can be seen from the images, the Doppler filter very effectively removes the ground clutter signals from the city of Singapore area, whereas echoes from ships (indicated by red ellipses) are mostly kept. Intensities peak up to about 80 dBZ (corresponding to rainfall intensities above 1000 mm/h), and the signals appear sometimes as arch-like signatures due to side-lobe returns.

Figure 6 shows the classification result (based on polarimetric data which are not shown) and the final reflectivity data after suppression of sea clutter signals. The image clearly shows the improvement compared to application of a conventional Doppler filter only, as in Fig. 5 (left).

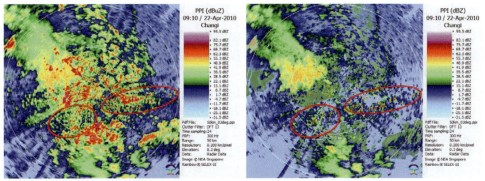

Fig. 5 0.3° reflectivity PPIs scanned with a METEOR 1600 SDP S-band radar in Singapore. Shown are uncorrected reflectivity (left), and reflectivity processed with a DFT Doppler filter (right). Images copyright © of the Meteorological Services, National Environment Agency, Singapore.

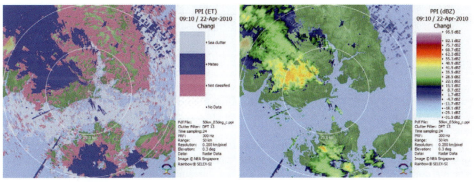

Fig. 6 Classification results (left) and corrected reflectivity data with sea clutter being suppressed (right). Images copyright © of the Meteorological Services, National Environment Agency, Singapore.

SUMMARY

Radar returns from sea surface, which are usually not removed by a conventional Doppler filter in the radar signal processor, can effectively be suppressed with the probability-based algorithm described in this study. When being applied to polarimetric radar data, the algorithm provides the best results; it can, however, also be applied to single-polarization data. When applied to polarimetric data, the algorithm also effectively removes ship echoes.

Acknowledgements The authors are grateful to Lt. Col. Chun-Chieh Chao and Lt. Col. Dai of the Republic of China Air Force, Taiwan, and to Ang Chieng Hai of the National Environment Agency, Singapore, for providing the radar data.

REFERENCES

Rico-Ramirez, M. A. & Cluckie, I. D. (2008) Classification of ground clutter and anomalous propagation using dual-polarization weather radar. *IEEE Trans. Geosci. Rem. Sensing* 46(7), 1892–1904.

Rhyzhkov, A., Zhang, P. Doviak, R. & Kessinger, C. (2002) Discrimination between weather and sea clutter using Doppler and dual-polarization weather radars. *XXVII General Assembly of the International Union of Radio Science, Maastricht, The Netherlands, paper #1383.*

Rhyzhkov, A., Giangrande, S. E. & Schuur, T. J. (2005) Rainfall estimation with a polarimetric prototype of WSR-88D. *J Appl. Met.* 44(4), 502–515.

Schuur, T., Ryzhkov, A. & Heinselman, P. (2003) Observations and classification of echoes with the polarimetric WSR-88D radar. *NOAA National Severe Storms Laboratory Tech Report, Norman, Oklahoma, USA.*

Selex-Gematronik (2007) *Dual-Polarization Weather Radar Handbook. An Overview of Dual-Polarization Weather Radar: Theory and Applications (2nd edn)* (ed. by V.N. Bringi, M. Thurai & R. Hannesen). Selex-SI Gematronik, Neuss, Germany, 163 pp.

Sugier J. & Tabary, P. (2006) Evaluation of dual-polarisation technology at C-band for operational weather radar network. OPERA-2 work packages 1.4 &1.5, deliverable b; http://www.knmi.nl/opera/deliverables2.html.

58

Weather Radar and Hydrology
(Proceedings of a symposium held in Exeter, UK, April 2011) (IAHS Publ. 351, 2012).

Design of a clutter modelling algorithm based on SRTM DEM data and adaptive signal processing methods

E. GONZALEZ-RAMIREZ[1], M. A. RICO-RAMIREZ[2], I. CLUCKIE[3], J. I. DE LA ROSA VARGAS[1] & D. ALANIZ-LUMBRERAS[1]

1 *Autonomous University of Zacatecas, Zacatecas, Mexico*
gonzalez_efren@hotmail.com
2 *University of Bristol, Bristol, UK*
3 *Swansea University, Swansea, UK*

Abstract This paper presents an algorithm for clutter modelling based on the radar equation, radar characteristics and digital elevation data. Optimization methods from adaptive signal processing theory were used to calculate the weights of an adaptive linear combiner representing the radar system for clutter modelling. Modelled clutter showed an acceptable precision demanded by applications in meteorology and hydrology for radar rainfall estimation.

Key words weather radar; modelling; SRTM; adaptive linear combiner

INTRODUCTION

Weather radar is an important system for qualitative and quantitative rainfall estimation. The radar antenna is oriented to scan close to the ground to allow the best estimation of rainfall reaching the ground. This technique is inconvenient since at these grazing angles of the antenna the radar beam hits the ground as well as meteorological targets.

The signal arriving at the radar antenna contains rainfall information embedded with ground echoes. These echoes are known as ground clutter and they have to be filtered to minimize their effects on rainfall estimates. Several techniques are currently used to remove clutter, e.g. the clutter map is a technique consisting of measuring the intensity of echoes when rainfall is not present. These echoes are usually flagged and removed when the radar system is used for rainfall estimation. This technique is effective in removing most of the clutter under standard atmospheric conditions (Harrison *et al.*, 2000). Other techniques that employ radar data from single- and dual-polarisation weather radar are used for automatic classification and removal of ground echoes as well as anomalous propagation (AP), as described in Berenguer *et al.* (2006), Cho *et al.* (2006), Gourley *et al.* (2007) and Rico-Ramirez & Cluckie (2008). The presence of ground clutter not only has a detrimental effect on rainfall estimation, but can also be used to sense the performance of the weather radar system such as the calibration of weather radars at attenuating wavelength (Serrar *et al.*, 2000) and to control radar signal stability (Sempere-Torres *et al.*, 2001, 2003). When filtering a signal it is of great importance to know its characteristics in order to recover it when such a signal is mixed with other signals and noise. High-resolution Digital Terrain Model (DTM) data allow the modelling of clutter in a simple manner. Examples of such clutter models are described in Delrieu *et al.* (1995), Andrieu *et al.* (1997), Archibald (2000) and Gonzalez-Ramirez (2005).

Data

The radar data used to validate the results were obtained from the Thurnham radar located in the southeast of England (latitude 51.2942°, longitude 0.6059° and altitude 219 m a.m.s.l.). The radar system is a multi-parameter C-band radar with simultaneous transmission and reception of horizontal and vertical polarized waves. This radar system has the power divider and receiver in the pedestal to maximise data quality. The nominal beam width of the radar is 0.95°, with an antenna diameter of 4.2 m and with typical gate resolutions of 125, 250 and 500 m. The transmitting peak power is 250 kW. Every PPI scan has one ray per degree.

The DTM data were obtained from the Shuttle Radar Topography Mission (SRTM), which are available for the entire world with a resolution of 3 arc second (approximately 90 m resolution at the Equator). There are also data available with 1 arc second resolution, but this dataset is not available for all countries. Farr *et al.* (2007) describes in detail the instrumentation and data processing to generate the SRTM DTM data. For the purpose of this research, 3 arc second resolution SRTM DTM data were employed. The SRTM DTM data are available for download as 5° × 5° tiles in the geographic coordinate system WGS84. The SRTM DTM data were processed and mapped to horizontal 50 m × 50 m squares in the United Kingdom National Grid (UKNG) system to be employed by the clutter predictor software. The transformation of the SRTM geographic grid to UKNG was carried out by transforming the geographic coordinates of SRTM data cell vertices based on the WGS84 datum to geographic coordinates based on the Airy datum, converting these geographic coordinates to UKNG easting and northing coordinates and approximating the SRTM curved grid to a grid with four non-parallel linear sides. Elevation values for each 50 m × 50 m square were computed as a weighted average of the elevation value of the overlapping 3 arc sec × 3 arc sec SRTM grids. The weight for each overlapping 3 arc sec × 3 arc sec SRTM grid used in the average was taken as the ratio between its overlapped area and the total area of the 50 m × 50 m square.

Modelling of ground clutter echoes

The averaged reflected power from meteorological echoes is given by (Battan, 1973; Skolnik, 1980):

$$\overline{P}_r = \frac{C_0 |K|^2 Z}{r^2} \tag{1}$$

where C_0 is a constant that depends on the characteristics of the radar system (see Battan, 1973), $|K|^2$ is the dielectric constant of the scattering particles (assumed to be equal to 1), Z is the radar reflectivity factor and r is the range from the radar to the meteorological echoes (i.e. the target). In a similar way, equation (1) can also be employed to compute the reflectivity measured by the radar from ground clutter echoes if P_r is known.

The received backscattered power at the radar antenna from a 50 m-square-shaped portion of the ground r metres away from the radar site can be calculated by use of the radar equation for any kind of target (this includes hydrometeors and portions of the ground) as given in Battan (1973):

$$P_r = (P_T G^2 \lambda^2 \sigma_t)/(4^3 \pi^3 r^4) \tag{2}$$

where the parameters P_T, and λ are constant; σ_t is based on the land type according to the results shown by Billingsley (2002) and demonstrated in Gonzalez-Ramirez (2005). The value of G depends on the angular offset of the target with respect to the radar main beam axis according to the antenna pattern. The backscattered power from all targets within the resolution volume can be computed by using equation (3) (see also Delrieu *et al.*, 1995) considering a symmetric beam shape with respect to the main beam axis (i.e. depending only on r and θ) and assuming no attenuation by atmospheric gases; that is:

$$\overline{P}_r = \frac{P_T \lambda^2}{(4\pi)^3} \iint_S \frac{G^2(r,\theta,\phi)\sigma_t(r,\theta,\phi)I_t(r,\theta,\phi)}{r^4} \, dS \tag{3}$$

The refractive index gradient plays an important role in clutter returns. It has a profound influence on the energy arriving at each point of the ground. Under normal atmospheric conditions the effective Earth radius is usually taken as 4/3 of its actual value. This effective Earth radius may be different under normal conditions and even more under anomalous atmospheric conditions. The effect of the refractive index gradient is considered in the value of $G(r, \theta, \phi)$ for a point with spherical coordinates (r, θ, ϕ) (the shape of $G(r, \theta, \phi)$ is taken according to the beam propagation).

The quantity I_t takes the value of one when the target is illuminated and zero otherwise. Clutter reflectivity from a beam volume unit can be formulated as an adaptive linear combiner in which the received power is equal to the summation of all received powers from ground targets within the beam volume and using equation (1) to transform it to reflectivity. The energy from ground targets can be grouped according to their angle from the radar main axis θ. The total received power from a ground target located at range r_i away from the radar antenna can be approximated as:

$$k_{i1}f(\theta_1)+k_{i2}f(\theta_2)+\ldots+k_{iN}f(\theta_N)=Z_{r_i} \qquad (4)$$

where the values k_{ij} (j = 1, 2, ..., N) are computed based on DTM data, radar parameters and considering results from Billingsley (2002) for the computation of σ_t of ground targets. $f(\theta_j)$ is a function which depends on the angle θ_j of the ground targets measured from the main beam axis, and Z_r is the reflectivity measured by the radar.

The system equation (4) can be solved by the adaptive Least Mean Square (LMS) algorithm described by Widrow (1985).

RESULTS

Most of the clutter at zero degrees of elevation is concentrated in the first 50 km and shows low contamination of backscattered energy from sidelobes, which is more visible at higher elevation angles. Therefore, data from the lowest elevation angle were used to validate modelled clutter (i.e. zero degrees). The clutter reflectivity data were available in a 230 × 360 matrix where the rows represent the gates with a resolution of 250 m and the columns represent the rays with one-degree resolution. Data were filtered row by row using the Wiener filter to minimize detected noise, assumed to be a time-invariant additive Gaussian white noise with 0.9355 of standard deviation.

The ground clutter was modelled using the functions $f(\theta_j)$ found in the LMS algorithm considering normal atmospheric conditions and were compared with measured data from the Thurnham radar prior to conversion to reflectivity. Measured and modelled clutter are shown in Fig. 1(a) and (b), respectively. The correlation between measured and modelled ground clutter echoes was 0.74. 51.4% of clutter present in the measurements was also shown by the clutter modelling algorithm. From the total number of clutter pixels detected by the Thurnham radar, 48.3% were located on the sea, where the clutter algorithm shows no significant echoes. In this region, measured reflectivity data showed some geometric textures probably formed by echoes from ships and man-made structures on the coast and in the sea which were not included in the DTM dataset. Other types of clutter are also shown with low intensity echoes. This can be observed in Fig. 1(a) located approximately 49.5 km from the radar location at 149° in azimuth. A

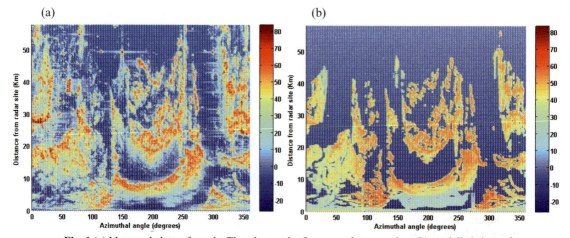

Fig. 1 (a) Measured clutter from the Thurnham radar for a complete scanning, (b) modelled clutter from the Thurnham radar for a complete scanning.

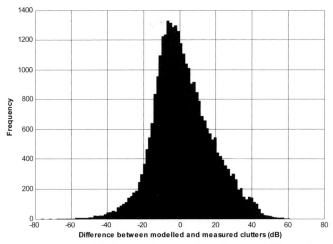

Fig. 2 Histogram of differences between clutter from modelling and measurements.

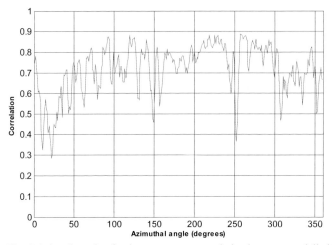

Fig. 3 Azimuth angle of radar rays *versus* correlation between modelled and measured clutter.

trace formed to the right and left of this set of clutter was probably caused by the backscattered energy from sidelobes. Small distortion is also observed being more evident in the azimuth range from 300 to 360°. This distortion is produced by the radar system and is more noticeable as the elevation angle increases.

There were 0.3% of clutter echoes present at places marked as obscured ground in the modelling algorithm, representing a small percentage overall. Discrete objects (those high objects with high capacity of reflection, but very narrow so unnoticeable in the DTM dataset) may cause these type of clutter since they are visible from the radar antenna. Deviation of the real effective Earth radius to the one considered in this work may also contribute to the presence of this kind of clutter.

Figure 2 shows the distribution of the differences between modelled and measured clutter. The measured intensity of the clutter echoes from the Thurnham radar was subtracted from the modelled clutter using the algorithm proposed in this paper. The results show that differences have a mean of 1.77 dB and a standard deviation of 16.12 dB. The skew observed in this histogram is probably caused by the effect of contamination of noise created in the receiver of the radar system, by echoes from sidelobes and by echoes from ground points reached when anomalous propagation conditions are present. Echoes from sidelobes and from anomalous propagation were not taken into account in the algorithm.

Figure 3 shows the correlation between modelled and measured clutter. The minimum correlation between these two sets is present at 23° with 0.295 and the maximum correlation is at 257° with a correlation of 0.886. The average correlation was 0.714. The worst performance of clutter modelling was found between the azimuth angles of 9 and 43°. This region is mainly covered by the sea and this explains the poor correlation. Also, it is likely that ships and man-made structures in this region are also affecting the results.

CONCLUSIONS

A ground clutter algorithm based on DTM data, the radar equation taking normal atmospheric conditions and results from Billingsley (2002) has been developed. The algorithm allows the computation of the position and approximate intensity of most of the clutter present in any weather radar. The main advantage of calculating ground clutter echoes from DTM data is its applicability for reducing the effect of this unwanted signal over echoes with hydrological and meteorological interest. Another important advantage is the availability of DTM data from the Shuttle Radar Topography Mission (SRTM), which are freely available on the web.

There were measured ground clutter echoes not shown by the ground clutter model. These echoes are mainly from discrete objects on the land such as towers and buildings, and on the sea such as ships and man-made structures which are not included in the DTM data. For future work, the refractive index gradient must be considered to estimate the beam propagation path in order to reduce the difference between measured and modelled clutter.

Acknowledgements The authors would like to acknowledge the financial support provided by CONACYT-MEXICO, the EPSRC Flood Risk Management Research Consortium (GR/S76304/01). We also thank the Met Office for providing the radar data. The SRTM data were downloaded from http://srtm.csi.cgiar.org/.

REFERENCES

Andrieu, H., Creutin, J. D., Delrieu, G. & Faure, D. (1997) Use of weather radar for the hydrology of a mountainous area. Part I: radar measurement interpretation. *J. Hydrol.* 193, 1–125.

Archibald, E. (2000) Enhanced clutter processing for the UK weather radar network. *Physics and Chemistry of the Earth, Part B: Hydrology* 25, 823–828.

Battan, L. J. (1973) *Radar Observation of the Atmosphere.* The University of Chicago Press, USA.

Berenguer, M., Sempere-Torres, D., Corral, C. & Sanchez-Diezma, R. (2006) A fuzzy logic technique for identifying nonprecipitating echoes in radar scans. *J. Atmos. and Oceanic Technol.* 23, 1157–1180.

Billingsley, J. B. (2002) *Low-Angle Radar Land Clutter: Measurements and Empirical Models.* William Andrew Publishing.

Cho, Y., Lee, G., Kim, K. & Zawadzki, I. (2006) Identification and removal of ground echoes and anomalous propagation using the characteristics of radar echoes. *J. Atmos. Ocean. Technol.* 23, 1206–1222.

Delrieu, G., Creutin, J. D. & Andrieu, H. (1995) Simulation of radar mountain returns using a digitized terrain model. *J. Atmos. Ocean. Technol.* 12, 1038–1049.

Farr, T. G., Rosen, P. A., Caro, E., Crippen, R., Duren, R., Hensley, S., Kobrick, M., Paller, M., Rodriguez, E., Roth, L., Seal, D., Shaffer, S., Shimada, J., Umland, J., Werner, M., Oskin, M., Burbank, D. & Alsdorf, D. (2007) The Shuttle Radar Topography Mission. *Rev. Geophys.* 45, RG2004, 1–33.

Gonzalez-Ramirez, E. (2005) Weather radar data analysis oriented to improve the quality of rainfall estimation. PhD Thesis, University of Bristol, UK.

Gourley, J. J., Tabary, P. & du Chatelet, J. P. (2007) A fuzzy logic algorithm for the separation of precipitation from non-precipitation echoes using polarimetric radar observations. *J. Atmos. Ocean. Technol.* 24, 1439–1451.

Harrison, D. L., Driscoll, S. J. & Kitchen, M. (2000) Improving precipitation estimates from weather radar using quality control and correction techniques. *Met. Appl.* 7, 135–144.

Rico-Ramirez, M. A. & Cluckie, I. D. (2008) Classification of ground clutter and anomalous propagation using dual-polarization weather radar. *IEEE Trans. Geoscience and Remote Sensing* 46(7), doi:10.1109/TGRS.2008.916979.

Sempere-Torres, D., Sanchez-Diezma, R., Berenguer, M., Pascual, R. & Zawadzki, I. (2003) Improving radar rainfall measurement stability using mountain returns in real time. In: *31st Int. Conf. on Radar Meteorology* (Seattle, WA), AMS).

Sempere-Torres, D., Sanchez-Diezma, R. & Cordoba, M. A., Pascual, R. & Zawadzki, I. (2001) An operational methodology to control radar measurements stability from mountain returns. In: *30th Int. Conf. on Radar Meteorology* (Munich, Germany), AMS.

Serrar, S., Delrieu, G., Creutin, J. D. & Uijlenhoet, R. (2000) Mountain reference technique: Use of mountain returns to calibrate weather radars operating at attenuating wavelengths. *J. Geophys. Res.* 15, 2281–2290.

Skolnik, M. (1980) *Introduction to Radar Systems*. McGraw-Hill.

Widrow, B. & Stearns, S. D. (1985) *Adaptive Signal Processing.* Prentice-Hall, Inc.

Radar bright band correction using the linear depolarisation ratio

ANTHONY ILLINGWORTH & ROBERT THOMPSON

Dept. of Meteorology, University of Reading, Reading RG6 6BB, UK
a.j.illingworth@reading.ac.uk

Abstract The enhanced radar return associated with melting snow, "the bright band", can lead to large overestimates of rain-rates. Most correction schemes rely on fitting the radar observations to a vertical profile of reflectivity (VPR) which includes the bright band enhancement. Observations show that the VPR is very variable in space and time; large enhancements occur for melting snow, but none for the melting graupel in embedded convection. Applying a bright band VPR correction to a region of embedded convection will lead to a severe underestimate of rainfall. We revive an earlier suggestion that high values of the linear depolarisation ratio (LDR) are an excellent means of detecting when bright band contamination is occurring and that the value of LDR may be used to correct the value of reflectivity in the bright band.

Key words bright band; rainfall estimation; weather radar

INTRODUCTION

Errors in precipitation intensity of up to a factor of five can result from the enhanced radar return due to melting snow (the bright band) if left uncorrected (Joss & Waldwogel, 1990). The challenge is first to identify when bright band contamination is occurring so that any rain-rates derived from reflectivity, Z, can be flagged as error prone. The second challenge is to derive an accurate rainfall at the ground when the beam is sampling the bright band. The approach generally adopted is to compute a mean vertical profile of reflectivity (VPR) by analysing the operational radar data themselves and then use the shape of this mean VPR to predict the rain at the ground. For example, Tabary *et al.* (2007) in France use a correction scheme based on a mean "daily" VPR with four adjustable parameters that is derived from all observations within 250 km over a 24-h period. The scheme in the UK (Kitchen & Davies, 1994) uses a more constrained "standard" high-resolution VPR; the height of the top of the bright band is fixed to the level of the 0°C isotherm in the operational forecast model, the bright band depth is fixed at 700 m, and the enhancement is a prescribed function of the rain-rate. The VPR has just one degree of freedom, which is the scaling factor in Z, and this is chosen iteratively so that when the VPR is multiplied by the appropriate beamwidth of the operational radar, it agrees with the Z observed by the radar. Mittermaier & Illingworth (2003) have shown that the average error in the height of the top of the bright band predicted from the forecast model is <150 m. The advantage of this scheme is that a local VPR is computed for each radar gate. The performance of this approach is analysed in Harrison *et al.* (2000).

One particular shortcoming of the "standard" or "mean" VPR approach is that embedded convection is quite common and has no bright band, so applying a VPR will lead to a drastic underestimate of rainfall at the ground. In practice VPRs are variable in space and time, so mean profiles are not representative. We propose to revive a scheme first proposed by Caylor *et al.* (1990) which uses the value of the linear depolarisation ratio (LDR) to identify the bright band; values of LDR between –14 to –18 dB result from the large oblate melting snowflakes which rock from side to side as they fall, and are associated with a value of Z in the bright band about 10 dB higher than in the rain below. In contrast to this, embedded convection has an LDR below –18 dB, because the melting graupel or soft hail pellets are much more spherical, and the value of Z at the freezing level is very similar to that in the rain below. Caylor *et al.* (1990) noted that the value of the co-polar correlation (ρ_{HV}) between the time series of the returns for the horizontally and vertically polarised transmitted pulses fell to about 0.9 in the bright band because of the variety of hydrometeor shapes present, whereas in rain ρ_{HV} is over 0.98. However, because the difference in LDR in the bright band and the rain is larger than the change in ρ_{HV}, they felt that the LDR method would be more reliable. Modern polarimetric radars transmit slant elliptical polarisation at 45°.

This means that LDR has fallen out of favour because dedicated scans are needed for its measurement. Matrosov *et al.* (2007) discuss the use of ρ_{HV} to identify the bright band in more detail; the comparative performance of the two techniques warrants further investigation.

THE CHILBOLTON DATASET

Examples of LDR data are displayed in Figs 1 and 2 which are vertical sections (RHIs) through stratiform rain and embedded convection, respectively. The data were obtained with the S-band Chilbolton radar which has a 300 m gate length and 0.25° beamwidth so that the vertical resolution at 57 km is just 250 m. These RHIs immediately reveal the vertical structure in a way which is not at all evident from the operational radars which scan in azimuth at a low elevation angle with a 1° beam. In Fig. 1, the rain is horizontally rather uniform; the bright band at 2.2 km altitude with a depth of about 700 m is very obvious, as is the large enhancement in Z, which we will call ΔZ, when compared with the rain below. Above the bright band in the ice, the Z values drop off very rapidly with height. The LDR signature of the bright band is very clear with values of about −16 dB which are 10 dB higher than in the rain below or the ice above. Note that clutter in the lowest elevation beam can be identified by its high values of LDR.

A very different picture is evident for the embedded convection in Fig. 2. The very high values of Z between 32 and 36 km range exceed 46 dBZ and reach an altitude of 7 km without any visible bright band. This region is surrounded by rain with quite a different VPR; Z values in the ice are much lower and there is a significant bright band. The LDR signatures clearly differentiate the embedded convection from the surrounding stratiform rain and bright bands. The bright band has an LDR of −14 dB which is only 700 m thick, whereas in the convective region there are some high values of LDR but generally at temperatures below freezing; these are associated with wet oblate hail which tumbles as it falls.

The extensive Chilbolton dataset of many RHIs can be used to examine how well LDR can identify the presence of bright bands, and if a correlation exists between the value of LDR close to the freezing level and the enhancement of reflectivity, ΔZ, in the bright band compared with the reflectivity in the rain below. The first analysis will involve the original high-resolution data, but the data can be degraded to mimic what an operational 1° beamwidth radar would measure. One

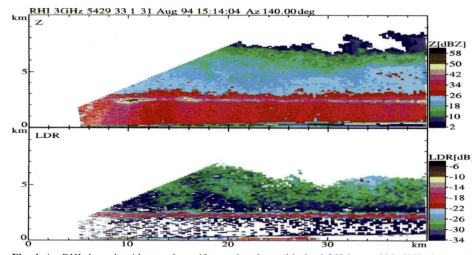

Fig. 1 An RHI through widespread stratiform rain taken with the 0.25° beamwidth Chilbolton radar, UK. Upper panel: Radar reflectivity, the bright band of enhanced reflectivity at 2.2 km height is clearly visible. Lower panel: Linear depolarisation ratio (LDR). The value is about −16 dB in the bright band but 10 dB lower in the ice above and the rain below.

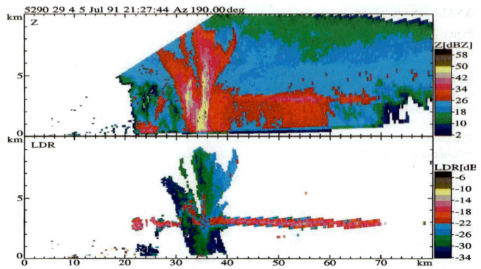

Fig. 2 As for Fig 1, but for a case with embedded convection. Note the high values of LDR coincident with the bright band in the stratiform areas. The LDR accompanying the high Z echo from 32 to 36 km range is generally lower, but some high values are found above the melting layer and are associated with melting hail.

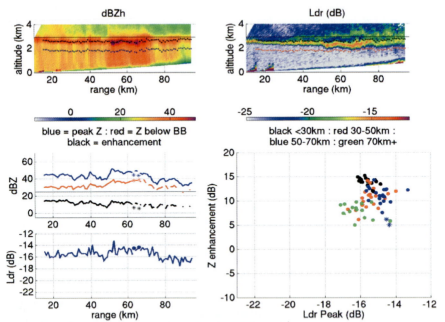

Fig. 3 Stratiform case; 13 August 2005 at 13:00 h. Left Hand Side: Upper plot – Straight black line – 0°C isotherm. The two black lines are the altitude where Z peaks and 750 m below. Middle plot: peak Z, Z 750 m below and the enhancement in Z in blue, red and black, respectively. Lowest plot: peak values of LDR. Right Hand Side: Upper plot: LDR values. Red line is the position of max Z. Lower plot: Peak LDR *versus* enhancement for various ranges.

issue of interest is the behaviour of LDR and ΔZ at larger ranges where both will be reduced because the bright band will not fill the radar beam.

ANALYSIS OF TWO CASE STUDIES

Figure 3 shows a uniform bright band case similar to that in Fig. 1. Superimposed on the two RHIs is a uniform horizontal black line which is the 0°C isotherm. The analysis is performed on vertical profiles using 0.9 km (three gate) running averages; this is to minimise wind shear effects which lead to sloping rather than vertical profiles. The first step, shown in the panels on the left hand side of the figure, is to locate the height of the maximum Z in each profile within ±750 m of the 0° isotherm, and if this maximum is within ±500 m of the isotherm and has a Z > 25 dBZ, find the value of Z 750 m below this maximum (shown as the two rows of black dots superposed on the RHI), and finally the difference in these two values of Z, which is ΔZ, the bright band enhancement. The ±750 m and ±500 m limits are to restrict the search to a region where a bright band is expected, and the 25 dBZ limit is introduced so that areas of rainfall less than about 1 mm/h are excluded. Such low echo regions are often associated with the ragged edges of clouds where the concept of a VPR is not valid. If no maximum Z is found within ±500 m, no bright band has been found, then the "enhancement" is computed from the Z value 250 m and 1 km below the 0°C isotherm. Such data points are plotted as a single "*" symbol; there are only three in this case, close to the maximum range of 90 km. The lower plot on the left is the maximum value of LDR which is found within ±750 m of the 0°C isotherm. On the right, the top plot is the LDR RHI, with the red dots indicating the position of the maximum Z in each profile. The plot on the lower right tests the hypothesis that the bright band reflectivity enhancement ΔZ is related to the maximum value of LDR; in the scatter plot the colour indicates the range of the profile. The points lie in a fairly tight cluster with enhancements between 5 and 15 dB and LDR ranging from –13 to –17 dB. The more distant green points at a range beyond 70 km have values of ΔZ and LDR which both average about 5 dB lower than for the points at shorter range. This is encouraging for any correction scheme; it suggests that the beam-filling effects at larger ranges affect LDR and ΔZ to the same extent.

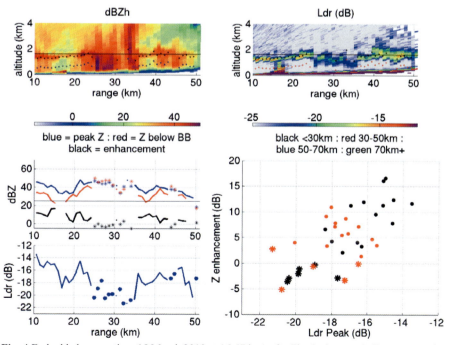

Fig. 4 Embedded convection: 15 March 2010 at 16:47 h. As for Fig. 3, the stars indicate gates where no maximum in Z was found, so a Z value 250 m below the 0°C isotherm was used.

Figure 4 shows a more challenging case of embedded convection. Between 25 and 35 km where Z values are highest, most of the profiles are indicated with a star indicating that no bright band was identified. These profiles have a maximum LDR of between −18 and −22 dB and generally a value of ΔZ which is slightly negative. The scatter plot of ΔZ against LDR for all profiles is very encouraging; Z enhancements range from −5 to +15 dB and are well correlated with LDR which increases from −21 to −13 dB. In this case there are no data beyond 50 km so beam-filling is not a problem.

FUTURE WORK AND OPERATIONAL IMPLEMENTATION

The correlations between ΔZ and LDR found in Figs 3 and 4 indicate that there may be some skill in a bright band correction scheme based on LDR. The next step is to examine the rest of the high-resolution dataset to provide more robust statistics. The final stage is to model the effects of a 1° beamwidth to see if the beam-filling affects LDR and ΔZ to the same extent so that the same correction algorithm can be used at all ranges. Currently, rainfall estimates are most accurate if the beam is sampling only the rain, so for a 1.7 km freezing height, the beam must not be higher than 1 km, leading to a 50 km maximum range for a 0° elevation scan. If rain-rates can also be derived from the bright band, the maximum height of the beam could be 2 km and the range could be doubled.

Most polarisation radars transmit V and H simultaneously and cannot detect the cross polar return, so a special LDR scan is needed when either H or V polarisation is transmitted. However, because the LDR signal is so large, the required accuracy for LDR is low and can be achieved with a scan rate three times faster than for the normal scan. A possible implementation would be to have a conventional Z scan at each elevation angle followed by a rapid LDR scan for the VPR correction algorithm.

REFERENCES

Caylor, I. J., Goddard, J. W. F., Hopper, S. E. & Illingworth, A. J. (1990) Bright band errors in radar estimates of rainfall: Identification and correction using polarization diversity. In: *Weather Radar Networking* (ed. by C. G. Collier & M. Chapuis), COST Project 73, EUR 12414 EN-FR, Document EUCO-COST 73/52/90, 294–303.

Harrison, D. L., Driscoll, S. J. & Kitchen, M. (2000) Improving precipitation estimates from weather radar using quality control and correction techniques. *Met. Appl.* 6, 135–144.

Joss, J. & Waldvogel, A. (1990) Precipitation measurements in hydrology. In: *Radar in Meteorology, Batten Memorial and 40th Radar Meteorology Conference* (ed. by D. Atlas), American Meteorological Society, 577–606.

Kitchen, M., Brown, R. & Davies, A.G. (1994) Real-time correction of weather radar data for the effects of bright band, range and orographic growth in widespread precipitation. *Quart. J. Roy. Met. Soc.* 120, 1231–1254.

Matrasov, S. Y., Clark, K. A. & Kingsmill, D. E. (2007) A polarimeteric radar approach to identify rain, melting-layer and snow regions for applying corrections to vertical profiles of reflectivity. *J. Appl. Met. & Clim.* 46, 154–166.

Mittermaier, M. P. & Illingworth, A. J. (2003) Comparison of model-derived and radar-observed freezing level heights: Implications for vertical reflectivity profile correction schemes. *Quart. J. Roy. Met. Soc.* 129, 83–96.

Tabary, P., Desplats, J., Do Khac, K., Eideliman, F., Gueguen, C. & Heinrich, J.-C. (2007) The new French operational radar rainfall product. Part II validation. *Weather and Forecasting* 22, 409–427.

Weather Radar and Hydrology
(Proceedings of a symposium held in Exeter, UK, April 2011) (IAHS Publ. 351, 2012).

Determining the vertical profile of reflectivity using radar observations at long range

KATE SNOW[1], ALAN SEED[2] & GEORGE TAKACS[1]

1 *University of Wollongong, Department of Engineering Physics, New South Wales 2522, Australia*
ks598@uowmail.edu.au
2 *Centre for Australian Weather and Climate Research, Bureau of Meteorology, GPO Box 1289, Melbourne,
Victoria 3001, Australia*

Abstract The Vertical Profile of Reflectivity (VPR) plays an important role when estimating the rain rate at the surface and has been the subject of radar meteorology research for many years. The VPR can either be sampled directly from observations that are close to the radar where the impact of the convolution with the beam pattern can be ignored, or the parameters for a theoretical form for the VPR are estimated using the available observations or climatology. In either case, a significant difficulty arises when a rain band approaches the radar and quantitative precipitation estimates are required before any detailed observations of the VPR at close range are possible. Long range in this context is the range where the height of the lowest elevation angle in the volume scan is greater than the wet bulb freezing level at that time, and therefore only limited information on the shape of the bright band is available. This paper uses a modified version of the VPR model proposed by Fabry (1997) and evaluates strategies to make optimum use of empirical observations, and how estimates for the model parameters could be updated in time. The technique is demonstrated using case studies of widespread rainfall over Sydney and Brisbane, Australia. Comparing the final technique to both the current short range and long range methods indicates that the parameterised VPR is able to provide similar VPR accuracies as the short range, with great improvement on the current long range method, making it suitable for rainfall corrections.

Key words vertical profile; parameterisation; long-range; beam convolution; Australia

INTRODUCTION

Accurate determinations of surface rainfall by radar are important in providing the spatial distribution of rainfall. This is significant in areas from aviation to flood forecasting. However, radar cannot measure surface rainfall directly, but relies on reflectivity measurements at various altitudes that vary according to the Vertical Profile of Reflectivity (VPR). One of the most significant effects is the increase of reflectivity at the 0°C isotherm, known as the bright band, occurring due to the melting of ice into liquid droplets. Considerable attention has been given to methods of determining the VPR over the past years (Andrieu & Creutin, 1995; Joss & Lee, 1995; Kitchen *et al.*, 1994; Vignal *et al.*, 2000, etc.).

The bright band is a significant increase in radar reflectivity over a relatively small change in altitude, so the profile is strongly dependent on the range of the observation due to the convolution of the Gaussian beam profile and the true vertical profile. The curvature of the Earth implies that the minimum observable height increases with range. These factors combine to limit the amount of information that can be inferred about the VPR from observations that are distant from the radar and illustrates why the VPR for many quantitative precipitation estimation systems is based on observations that are within 65 km of the radar. This paper assumes that the VPR is constant over the area under the radar and provides a method for inferring the VPR from observations at long ranges using a parameterised model. The basic idea that is presented in this paper is to fit VPR model parameters using data from a number of range intervals. Since some parameter estimates are more equal than others (as it were) a range weighted mean of the variable is then calculated.

Previous VPR parameterisations have been performed, in particular that by Fabry (1997):

$$z(h) = z_0 + 6*10^{-3} h_b - 6.5*10^{-3} h + 9 \exp\left(-\left|1.6(h-h_b)/d\right|^3\right)(h-h_b) \qquad h > h_b$$

$$z(h) = z_0 - 0.5*10^{-3} h + 9 \exp\left(-\left|1.6(h-h_b)/d\right|^3\right)(h-h_b) \qquad\qquad h < h_b \qquad (1)$$

where d is the bright band width, z_0 the ground reflectivity, h the height and h_b the bright band height. Fabry (1997) indicated that the VPR was not strongly dependant on rain rate, and such parameterised VPR's may be successful for rainfall correction purposes.

PARAMETERISATION

The parameterisation was initially defined as being similar to equation (1) using five parameters now with a, the slope in reflectivity below the bright band, b, the slope above the bright band and, c, the bright band intensity while an added parameter e was found to be required so as to account for the offset of reflectivity of rain above the bright band to ice below.

$$z(h) = z_0 + ah_b + b(h - h_b) + e + c\exp\left(-\left|(h - h_b)/d\right|^3\right) \qquad h > h_b$$

$$z(h) = z_0 + ah + c\exp\left(-\left|(h - h_b)/d\right|^3\right) \qquad\qquad h \le h_b$$

(2)

The VPR model can also be compared to other inverse methods such as that proposed by Kirstetter *et al.* (2010). This method however, requires the use of an *a priori* VPR within close range, making it unsuitable when only long-range data are available. It also has a time update of 1 h rather than 6 minutes, as in the VPR model proposed in this paper. However, unlike the model presented in this paper, which is restricted to long-range data, Kirstetter *et al.* (2010) allows for improved VPR determination and application over the entire radar domain.

ANALYSIS OF RAINFALL DATA METHOD

The two sets of rainfall data analysed include: (i) 122 six-minute profiles from Sydney (Terry Hills), New South Wales, Australia 25 May 2010, and (ii) 240 six-minute profiles from Brisbane (Mt Stapylton), Queensland, Australia 19 August 2010. The profiles were calculated from the volume scan data in 0.2 km intervals up to 9.8 km above the radar and 10 km range intervals up to 140 km from the radar. Typical or initial values for each parameter are taken as an overall estimate based on previous studies (Battan, 1973; Kitchen *et al.*, 1994; Fabry & Zawadzki, 1995; Matrosov *et al.*, 2007; Scovell *et al.*, 2008) and observed profiles from the case studies. The parameter estimates are then a = –0.85 dBZ km^{-1}, b = –3 dBZ km^{-1}, c = 7 dBZ, d = 0.35 km, e = –2.1 dBZ and h_b = 2 km.

With the initial parameters, the determined VPR is convolved with the beam profile so as to simulate the radar observations of the VPR at a particular range. The sensitivity of the VPR to each parameter is then determined as a function of range by calculating the RMSE variation between the initial model and the VPR obtained when one of the parameters is varied to indicate the relative error in that parameter (results for 10% error given in Fig. 1). Low sensitivity in the RMSE to perturbations in the value of the parameter implies a limited ability to determine the parameter from observations and allows for consideration of appropriate methods for determining each parameter inversely from the measured observations.

Once methods to determine the parameters are found, weighting with range must be applied to determine the final parameter values. The weighting is also performed so that parameter values may still be obtained even if data were only available at the furthest ranges, and these estimates are updated as data closer to the radar becomes available. Initially standard deviation is considered, however it is found for poorly measured profiles that outlier terms with low standard deviations occur, providing unsuitable weighting.

Slope of the VPR above the bright band, *b*

From Fig. 1(a), the RMSE for b is uniformly large for all ranges and therefore it should be possible to estimate the value of b reliably. Also, measurements above the bright band are well defined at long ranges, and a distinct value of b is attainable directly from the data using a linear regression of points measured 700 m above the wet bulb freezing level. This range is chosen as Scovell *et al.* (2008) found that values 500 m above the bright band peak allowed the measured slope to provide an accurate indication of the VPR slope regardless of range, and 700 m allows for errors in the estimate. Then, due to the uniform sensitivity, the mean *b* determined over the various ranges provides a robust estimate.

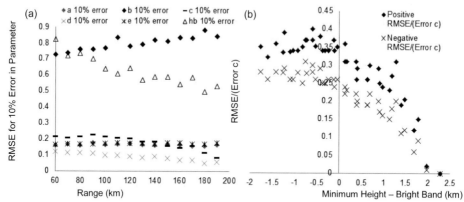

Fig. 1 (a) RMSE between the initial VPR convolved with the beam profile to the relevant range and the VPR when there is a 10% error in one parameter and (b) RMSE/(Error *c*) for positive and negative errors of *c* about 7 dBZ *versus* the minimum height at a particular range minus the bright band height.

Intensity of the bright band, *c*

While a fairly low RMSE occurs in Fig. 1(a) for *c*, the intensity of the bright band, changes in *c* only affect the VPR around the bright band meaning *c* may be determined from the measured data if the elevation of the base scan is below the height of the bright band. The relative error was assumed to be a function of the difference of the minimum observable height at a particular range and the bright band height as observed by the decreasing trend in Fig. 1(a). To determine this dependence *c* was varied about the "true" value of 7 dBZ in the positive direction up to 14 dBZ and downward to 0 dBZ in 0.05 dBZ increments using h_b = 1.5 km, 2 km and 2.5 km. The linear trend of RMSE *versus* error in *c* was then calculated with the corresponding RMSE/(Error *c*) values given in Fig. 1(b).

From Fig. 1(b), it is shown that uniformly accurate results occur for ranges with a minimum observable height 100 m above the bright band or less, and hence are provided equal weighting. For ranges beyond this, an exponentially decreasing weighting is applied:

$$w_i = 1 \qquad\qquad h_m - (fl - 0.5) < 0.1$$
$$w_i = -\exp(h_m - (fl - 0.5)) + 1.15 \qquad 0.1 < h_m - (fl - 0.5) < 2.1 \qquad (3)$$
$$w_i = 0 \qquad\qquad h_m - (fl - 0.5) > 2.1$$

where w_i is the weighting factor $0 < w_i < 1$, h_m is the minimum observable height at range *i* and *fl* is the wet bulb freezing level forecast from a Numerical Weather Prediction model (NWP). The final *c* is then calculated as the weighted mean using w_i^2.

Height of the bright band, h_b

In Fig. 1(a), the RMSE shows large sensitivity to perturbations in h_b, the height of the bright band. However, this ability decreases with range and it may be expedient to use the wet bulb freezing level as an estimate of h_b beyond some range. The combination of results should allow h_b to be known within a 200 m accuracy as required for a suitable VPR (Mittermaier & Illingworth, 2002).

The Nelder-Mead downhill minimisation method (Press, 2002) was used to estimate values of c, h_b and z_0 for each range interval. The function minimised was the absolute difference between the parameterised VPR, convolved to the relevant range, and the measured VPR at that range. Weighting for h_b is then applied where the wet bulb freezing level estimates provides the theoretical estimate of h_b. While standard deviation alone does not provide a suitable indication of accuracy, it may be compared to the RMSE of the wet bulb freezing level (RMS_{wbf}) and allow results to be weighted with the NWP estimate of the height of the bright band. In this case, let the bright band height determined from the data be h_{bm} with standard deviation ST_m. Then the weighting will depend on the ST_m compared to RMS_{wbf} in a linear fashion giving the final value of h_b for N available ranges as:

$$h_b = \frac{1}{N} \sum_{i=1}^{N} w_{hb} h_{bm} + (1 - w_{hb})(fl - 0.35) \quad \text{where} \quad w_{hb} = \frac{RMS_{wbf}}{ST_m + RMS_{wbf}} \tag{4}$$

Slope of the VPR below the bright band, *a*

From Fig. 1(a), a relatively small RMSE occurs for a. At long ranges changes in a effectively act to shift the VPR meaning changes are spread over the entire profile and VPRs with different a values will be hardly discernible. Taking into account ground clutter, it may then be most suitable to fix a based on climatological and orographic trends in the area. These values are -0.8 dBZ km^{-1} for Terry Hills and 0.2 dBZ km^{-1} for Mt Stapylton based on fitting parameters to the mean of the short range VPR's over the entire period and simply provide an idea of how a fixed value affects the results given some idea of trends of low level rainfall.

Rain–snow offset, *e*

Conclusions of fixing e are similarly determined since measuring e requires information below the bright band. Also, e is based on hydrometeor properties and has little dependence on precipitation intensity (Fabry & Zawadzki, 1995), hence it was initially fixed to -2 dBZ.

Initial VPR results, however, showed that it would be advantageous to measure e in some cases. For example, calculating the total mean profile over the period at Terry Hills 60 km from the radar and determining the parameterisation for fixed e, the results of the VPR below the bright band predict values near 20 dBZ while the measured values are closer to 22 dBZ. The deviation indicates that determining e would be advantageous, particularly in cases of higher bright bands. Yet as close as 90 km, the standard deviations in e means all results are within the limit -2 dBZ. Hence, determining e conditionally with range is most appropriate.

Width of the bright band, *d*

The d parameter provides the lowest overall sensitivity in Fig. 1, meaning a distinct value will be difficult to estimate. Since less information will be available at longer ranges and d is dependent on the bright band height, the most suitable approach is to fix d to 0.35 km, the standard profile bright band width used in studies such as Kitchen *et al.* (1994).

Updating the model parameters in time

The VPR parameters for the current time step are used to update the model parameters using:

$$s_t = (1 - r)x_{t-1} + rs_{t-1} \tag{5}$$

where s_t is the smoothed parameter at time t, x_t is the estimate of the parameter at time t, and r is the lag 1 autocorrelation of the x_t time series.

COMPARISON TO CURRENT METHOD

The current method of determining the VPR in Australia relies on data being within 65 km of the radar. Within this range, the VPR for each time step is calculated using the mean over range of the data at each height, no temporal smoothing is performed. This VPR is then convolved to ranges greater than 65 km so as to obtain the expected VPR at the longer ranges. If data are not available within 65 km, the VPR is assumed constant with height. The VPR from the parameterised method devised is compared to both these forms of VPR by calculating the RMSE between the determined VPR in each case and the measured data. This is performed for all 240 Mt Stapylton profiles at each range and the mean of the RMSE for each range taken and compared in Fig. 2(a) (similar results for Terry Hills).

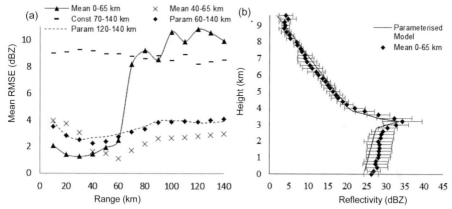

Fig. 2 (a) Comparing RMSE of VPR methods of mean 0–65 km, mean 40–65 km, constant long-range and parameterised profiles with minimum ranges of 60 km and 120 km for Mt Stapylton, and (b) VPR comparison of the parameterised 60–140 km and 0–65 km mean profile at Mt Stapylton 12:42 10 August 2010.

In Fig. 2(a), it is shown that the difference between the parameterised profiles, whether using data from 60–140 km or 120–140 km, is negligible. In addition, the magnitude of the RMSE of the parameterised profiles, when compared to the measured data at the short ranges, is of the same order of magnitude as the mean VPR indicating the parameterised VPR provides close approximations to the short range mean VPR. The RMSE for the parameterised VPR is far lower than the RMSE for the constant VPR, which is currently in use when there is no short range data available to calculate the mean VPR.

As another comparison between the 60–140 km parameterised VPR and the 0–65 km mean VPR, the RMSE between the profiles themselves at each 6-minute interval is calculated. Figure 2(b) shows the 60–140 km parameterised VPR for a typical time in the data and the 0–65 km mean VPR. The standard deviations between the mean VPR profiles were calculated and are shown as error bars around the mean VPR. Results show some variation below the bright band, however even with this slight deviation, it is indicated that a typical parameterised VPR result will be within the error range of the mean 0–65 km VPR and hence provides accuracies comparable to that of the 0–65 km mean VPR.

CONCLUSION

The sensitivity of each parameter was used to provide an indication of how accurately that parameter can be determined and an idea of suitable methods of measuring the parameters. Only the height of the bright band and the slope of the VPR above the bright band can be determined reliably using long-range data, and at some point even the height of the bright band becomes difficult to estimate. When there is little data below the bright band, NWP forecasts of wet bulb freezing levels and climatological values must be used, otherwise linear regression and minimisation techniques allow for suitable values to be obtained. Smoothing over time allows for the range weighted results to be updated with time and the final parameterised VPR obtained. These parameterized VPRs provide comparably accurate VPRs regardless of the minimum observable range and a typical profile comparison indicates that these profiles are within the error range of the mean 0–65 km profile. Hence, the method proposed is able to determine an approximate VPR from long-range data only and is suitable for rainfall corrections.

Acknowledgements The first author would like to acknowledge the Faculty of Engineering at the University of Wollongong for funding and support.

REFERENCES

Andrieu, H. & Creutin, J. D. (1995) Identification of vertical profiles of radar reflectivity for hydrological applications using an inverse method. Part 1: Formulation. *J. Appl. Met.* 34, 225–239.

Battan, L. J. (1973) *Radar Observations of the Atmosphere*. University of Chicago Press, Chicago, USA.

Fabry, F. (1997) Vertical profiles of reflectivity and precipitation intensity, In: *Weather Radar Technologies for Water Resources Management* (ed. by B. J. Braga & O. Massambani), 137–145. UNESCO Press.

Fabry, F. & Zawadzki, I. (1995) Long-term radar observations of the melting layer of precipitation and their interpretation. *J. Atmos. Sci.* 52, 838–851.

Joss, J. & Lee, R. (1995) The application of radar-gauge comparisons to operational precipitation profile corrections. *J. Appl. Met.* 34, 2612–2630.

Kirstetter, P. E., Andrieu, H., Delrieu, G. & Boudevillain, B. (2010) Identification of vertical profiles of reflectivity for correction of volumetric radar data using rainfall classification. *J. Appl. Met. & Climatol.* 49, 2167–2180.

Kitchen, M., Brown, R. & Davies, A. G. (1994) Real-time correction of weather radar data for the effects of bright band, range and orographic growth in widespread precipitation. *Quart. J. Roy. Met. Soc.* 120, 1231–1254.

Matrosov, S., Y., Clark, K. A. & Kingsmill, D. E. (2007) A polarimetric radar approach to identify rain, melting-layer, and snow regions for applying corrections to vertical profiles of reflectivity. *J. Appl. Met. & Climatol.* 46, 154–166.

Mittermaier, M. P. & Illingworth, A. J. (2002) Comparison of model-derived and radar-observed freezing-level height: Implications for vertical reflectivity profile correction schemes. *Quart. J. Roy. Met. Soc.* 123, 1–13.

Press, W. H., Teukolsky, S. A., Vetterling, W. T. & Flannery, B. P. (2002) *Numerical Recipes in C++: The Art of Scientific Computing,* Cambridge University Press, New York, USA.

Scovell, R., Lewis, H., Harrison, D. & Kitchen, M. (2008) Local vertical profile corrections using data from multiple scan elevations. Proceedings, *Fifth European Conference on Radar in Meteorology and Hydrology.* ERAD, Helsinki, Finland.

Vignal, B., Gianmario, G., Joss, J. & Germann, U. (2000) Three methods to determine profiles of reflectivity from volumetric radar data to correct precipitation estimates. *J. Appl. Met.* 39, 1715–1726.

Development of optimal functional forms of *Z-R* relationships

RAMESH S. V. TEEGAVARAPU[1] & CHANDRA PATHAK[2]

1 *Department of Civil, Environmental and Geomatics Engineering, Florida Atlantic University, Boca Raton,
Florida 33431, USA*
rteegava@fau.edu

2 *Operations and Hydro Data Management Division, South Florida Water Management District, 3301 Gun Club Road,
West Palm Beach, Florida, USA*

Abstract Use of appropriate functional reflectivity (*Z*)–rainfall rate (*R*) relationships is crucial for accurate estimation of precipitation amounts using radar. The spatial and temporal variability of several storm patterns combined with subjectivity in application of a specific functional *Z-R* relationship for a particular storm makes this task very difficult. This study evaluates the use of gradient and genetic algorithm-based optimization solvers for optimizing the traditional *Z-R* functional relationships with constants and coefficients for different storm types and seasons. The *Z-R* relationships will be evaluated for optimized coefficients and exponents based on training and test data. In order to evaluate the optimal relationships developed as a part of the study, reflectivity data and raingauge data were analysed for a region in south Florida, USA. Exhaustive evaluation of *Z-R* relationships and their utility in real-time improvement of precipitation estimates with optimization formulations were evaluated in this study.

Key words rainfall-rate-reflectivity relationships; optimization; genetic algorithms; precipitation; South Florida, USA

INTRODUCTION

Radar rainfall estimates derived from conversion of reflectivity are known to depend on the specific reflectivity (*Z*)–rainfall (*R*) relationship. Raghavan (2003) summarized over 50 different *Z-R* relationships (power functional forms) used in various regions of the world for different storm types for estimation of precipitation. *Z-R* relationships are based on assumed drop size distributions that may be typical for specific storm types. However, the evolution of the storm and the processes that produce precipitation may result in drop size distributions that differ from those assumed for the development of a specific *Z-R* relationship (Vieux, 2004). Radar-centric factors affecting the *Z-R* relationship include variations in antenna, radar transmitter, and receiver characteristics that diminish or increase signal strength. Derivation of a *Z-R* relationship that is specific to a radar and for a specific geographic region is expected to improve radar-based rainfall estimation. The work reported in this paper includes development of *Z-R* relationships and evaluating uncertainty inherent within that relationship aimed at improving the accuracy of NEXRAD (NEXt generation RADar)-based rainfall estimates. Details of the NEXRAD system used in the USA can be found in Vieux (2004). To assess improvements in radar rainfall estimation, *Z-R* relationships will be derived that are specific to each radar observation; appropriate to the season and type of precipitation; and representative of the climatology of the coastal and inland regions. These relationships can be derived through statistical characterization of gauge and radar observations and also be derived using functional optimization and inductive modelling. The derived *Z-R* relationships will be evaluated for an area where raingauge observations exist for comparative purposes. Data-withholding experiments will remove selected raingauges for cross validation. Improvements using the derived optimal *Z-R* relationships will be compared to using typical *Z-R* power relationships for tropical and convective storms.

Z-R RELATIONSHIPS

This section describes a few relevant studies that closely relate to the influence of *Z-R* relationships on radar-based precipitation estimates. In many studies typical power relationships were used and sensitivity of relationship parameters in relation to rainfall rate estimation was investigated. Observations from a 16-month field study using two vertically pointing radars and a disdrometer at Wallops Island were analysed by Tokay *et al.* (2008) to examine the consistency of the multi-instrument observations with respect to reflectivity and *Z-R* relations. The radar-based precipitation is in good agreement using reflectivity values at the collocated range gates and was superior to that between the disdrometer and either radar. Baldanado (1996) discussed an approach to determine the radar reflectivity–rainfall intensity (*Z–R*) relationship. Calculations of the simultaneous reflectivity factors and rain rates in a volume of air are performed based on the solution to the radar equations by infusion of the different radar parameters and the measurement of return power of the electromagnetic wave. Linear regressions are applied to the transformed data sets and constants and coefficients in the relations that take the form: $Z = aR^b$, were determined. A set of empirical *Z-R* relations for different locations in a river basin, as well as for particular storms, are obtained. A study by Malkapet (2007) focused on uncertainty caused by natural variability in radar reflectivity–rainfall relationships and how it impacts runoff predictions. Rainfall intensity, *R*, is estimated from one minute disdrometric reflectivity, *Z*, using different *Z-R* relationships. *Z-R* relationships are derived at different time scales: climatological, daily, event-based, and physical-process based. The effect of using different estimation methods, such as least square fitting, bias-corrected, fixed-exponent, and default operational relation, were also examined. *Z-R* relationships for physical processes were derived by classifying and separating rain records into convective, transitional and stratiform phases based on vertical profile of radar reflectivity, developed from volume-scan WSR-88D (Weather Surveillance Radar-88 Doppler) data. Lee *et al.* (2006) derived a *Z-R* relationship with the help of linear regression to convert the radar reflectivity *Z* into the rainfall rate *R*. The relationship was expressed in terms of an equation of the form as $Z = aR^b$.

Villarini *et al.* (2003) investigated the effects of random errors and used a procedure to convolve the radar-rainfall with a multiplicative random uncertainty factor, and discuss the effects of different magnitudes and spatial dependences. They indicate the basic systematic factors that can affect the results of such analyses that include the selection of a *Z-R* relationship, the rain/no-rain reflectivity threshold, and the distance from the radar. Ulbrich & Atlas (1977) described a method by which radar reflectivity and optical extinction are used to determine precipitation parameters, such as liquid water content and rainfall rate, and an exponential approximation to the raindrop size spectrum. The improvement in the accuracy was demonstrated by the authors using empirical relations and also by applying the method to a set of experimentally-observed raindrop size spectra. The authors examine these problems using high-resolution radar-rainfall maps generated with different parameters based on the Level II data from WSR-88 D radar in Kansas, USA. Trafalis *et al.* (2007) used artificial neural networks (ANNs), standard support vector regression (SVR), least-squares support vector regression (LS-SVR), and linear regression (LR) to relate reflectivity and rainfall rate (RR). They found that LS-SVR generalizes better than ANNs, linear regression and a rain rate formula in rainfall estimation and for rainfall detection, and SVR performs better than the other techniques. A simplified probability matching method is introduced by Rosenfield *et al.* (1994) that depends on matching the unconditional probabilities of *R* and Z_e, using data from a C-band radar and raingauge network near Darwin, Australia. This is achieved by matching raingauge intensities to radar reflectivities taken only from small windows centred about the gauges in time and space. The windows must be small enough for the gauge to represent the rainfall depth within the radar window, yet large enough to encompass the timing and geometrical errors inherent to such coincident observations. Amitai (2000) discussed spatial smoothing by the radar beam as well as post-detection integration that reduces the variability of the distribution of rainfall rate in space. It is shown that when raingauge data are smoothed in time there is an optimum smoothing time interval such that the random error and the bias are reduced to a negligible level. A method is suggested for the optimum comparison of radar and raingauge data and the possibility of determination of *Z-R* relationships from such comparisons is discussed. To provide a foundation for other radar studies in the Miami area, 50 comparisons were made by Woodley & Herndon (1969) between shower rainfall recorded by raingauges and observed with radar to evaluate the reflectivity *Z*, rainfall rate *R* relation, $Z = 300R^{1.4}$, referred to as the Miami *Z-R* relation.

ANALYSIS FRAMEWORK

The analysis framework consists of the development of optimization models based on reflectivity and raingauge data using a well-known and used simple power form of functional relationship,

$Z = aR^b$. Nonlinear optimization formulations were developed to obtain optimal exponents and coefficients of Z-R relationships for several candidate storm events selected. A simplified sample nonlinear formulation that is used in the current study is provided below. The temporal spread of reflectivity measurements is shown in Fig. 1, which forms the basis for optimization formulation.

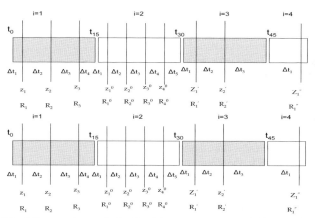

Fig. 1 Temporal spread of reflectivity measurements within a specific time interval.

Optimization model

$$\text{Minimize} \quad \sum_{i=1}^{no} (\hat{\phi}_i^m - \theta_i^m)^2 \tag{1}$$

Subject to:

$$R_n = \left(\frac{10^{\frac{dBZ_n}{10}}}{a} \right)^{\left(\frac{1}{b}\right)} \quad \forall n \tag{2}$$

$$\theta_i^m = \sum_{n=1}^{k} R_n \Delta t_n \tag{3}$$

$$T_i = \sum_{n=1}^{k} \Delta t_n \tag{4}$$

$$a_l \leq a \leq a_u \tag{5}$$

$$b_l \leq b \leq b_u \tag{6}$$

where θ_i^m is the estimated rainfall rate at a raingauge station m; no is the number of candidate events or all the events available in the database for which reflectivity and rainfall rate data are available; a and b are the coefficient and exponent in the Z-R relationship, a_u is the upper bound on the coefficient, b_u is the upper bound on the exponent, and Z_n is the reflectivity expressed in decibel units; and $\hat{\phi}_i^m$ is the observed rainfall rate at a gauge. The above formulation will be used for obtaining the optimal coefficients and exponents, a, b values, respectively. An objective function which is based on summation of absolute deviations (i.e. raingauge-radar estimate) or exponential function of deviations or squared summation of deviations can be used. The reduced gradient-based solver under the MATLAB environment and genetic algorithms are used for optimization.

CASE STUDY APPLICATION

The study evaluated the use of optimization models for optimizing the Z-R coefficients and exponents for different storm types and seasons. Precipitation data based on reflectivity data collected from National Climatic Data Center (NCDC) and raingauge data from southwest region of Florida, USA, over the same temporal resolutions were analysed using a Rain-Radar-Retrieval (R^3) system developed at Florida Atlantic University. Level II reflectivity data from the National Weather Service (NWS) or NCDC of NOAA for the Tampa NEXRAD radar station in south Florida (Fig. 1(a)) was collected for this study for analysis. The reflectivity data, which are available in approximately 5 minute intervals on a polar grid of $1° \times 1$ km resolution, are used to obtain radar-based rainfall estimates and to match with the resolution of gauge observations. Target area used for the current study includes the area covered between the Tampa radar station at the radius of 240 km and is shown in Fig. 2(a). Optimal *Z-R* relationships obtained using genetic algorithms applied for selected events are used to obtain precipitation estimates. Comparisons of absolute errors at 66 gauges in the region using these two relationships are shown in Fig. 2(b). Optimized *Z-R* relationships developed based on one set of events and applied to another set of events compared to standard *Z-R* relationships are experimented with. Results related to these experiments are provided in Table 1.

Optimized seasonal *Z-R* relationships developed based on one set of events within a given season are applied to another set of events in the same season. Results related to these experiments are provided in Table 2. Results suggest the use of optimal coefficient and exponent values improve the precipitation estimates as evidenced by performance measures.

Multiple *Z-R* relationships: moving temporal windows

A varying "temporal window" can be selected for optimizing different *Z-R* relationships over time as the storm event unfolds. Optimization of coefficient and exponent values is achieved by using any

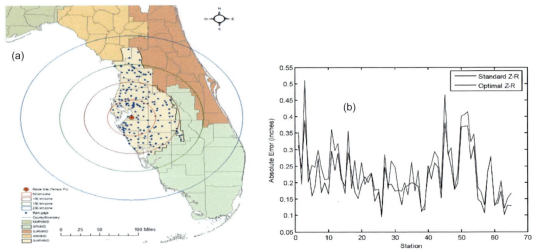

Fig. 2 (a) Location of radar site in Tampa, Florida, USA. (b) Absolute errors calculated based on standard and optimized *Z-R* relationships.

Table 1 Event based optimization of *Z-R* relationships.

Event	*a*	*b*	CC	AE	OCC	OAE
7 events	240	1.42				
9 events (test)			0.768	7.526	0.771	7.394
8 events	200	1.45				
8 events (test)			0.727	7.515	0.738	7.304

AE: absolute error; OAE: absolute error based on optimized *Z-R* relationship; CC: correlation coefficient; OCC: correlation coefficient based on optimized *Z-R* relationship.

mathematical programming formulation (optimization formulation). In this case as shown in Fig. 3(a) the temporal window size for optimization increases as the storm event unfolds, providing more information about the storm in optimizing the power relationship. Also a constant temporal window can be selected for optimizing the *Z-R* relationships as shown in Fig. 3(b). The near-time optimal *Z-R* relationship is more suitable for a storm type that is changing over time. A selected number of storm events can be selected for development of optimized *Z-R* relationships for a given season. The objective is to obtain a seasonal *Z-R* relationship based on select storm events.

Results related to a constant temporal window are presented in Table 3. Performance measures show improvements in the absolute errors and correlations as the optimization window moves towards the temporal duration in which the radar-based precipitation estimate is sought.

Insights and observations

The minimum and maximum values of coefficients and exponents need not be fixed *a priori* for optimization formulations. Improved estimates of precipitation have been realized when temporal windows preceding the window in which the optimal relationship of *Z-R* is applied are selected for optimal coefficients and exponents. In real-time, this information about the window will depend on the amount of information available about the elapsed time interval with observed precipitation at raingauges. Accurate continuous raingauge observations are critical for the success of the development of optimal *Z-R* relationships. Storm type information can be incorporated within the optimization or selection of *Z-R* relationships indirectly by analysing the raw reflectivity values. The number of raingauges used for developing optimal *Z-R* relationships depends on the performance criteria used for the selection. The constant and varying window approaches are applicable for optimization of *Z-R* relationships in real-time, provided an optimization solver is linked to a radar data processing tool and the execution time of the solver is reasonable. Average coefficient and exponent values for specific seasons can be recommended for further evaluation and applicability in real-time. The probability distribution of raw reflectivity values from the past storm events

Table 2 Optimized seasonal *Z-R* relationships.

Season	AE	OAE	CC	OCC	*a*	*b*
Dry	7.095	6.740	0.809	0.838	211	1.60
Transition	1.287	1.067	0.937	0.945	700	1.20
Wet	12.567	12.215	0.753	0.768	200	1.48
		Binary				
Dry	7.095	6.748	0.809	0.838	200	1.6
Transition	1.287	1.287	0.937	0.937	300	1.4
Wet	12.567	12.335	0.753	0.786	200	1.6
		Binary pairs				
Dry	7.095	6.748	0.809	0.838	200	1.6
Transition	1.287	1.287	0.937	0.937	300	1.4
Wet	12.567	12.335	0.753	0.786	200	1.6

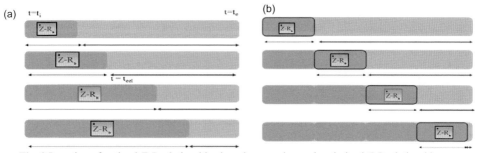

Fig. 3 Location of optimal *Z-R* relationships based on varying and optimized *Z-R* relationships.

Table 3 Performance measures based on moving window approach for estimation of optimal coefficients and exponents used in *Z-R* relationship.

Event*	Opt. Window	Test window	Stations	a	b	ϵ	ϵ'	ρ	ρ'
July	1–200	801–950	181	690	1.30	140.051	133.491	0.685	0.674
July	201–400	801–950	181	879	1.64	140.051	139.106	0.685	0.698
July	401–600	801–950	181	900	1.57	140.051	138.003	0.685	0.694
July	601–800	801–950	181	795	1.46	140.051	135.363	0.685	0.687
July	1–400	801–950	181	785	1.47	140.051	135.566	0.685	0.688
July	1–600	801–950	181	823	1.50	140.051	136.437	0.685	0.690
July	1–800	801–950	181	816	1.49	140.051	136.168	0.685	0.689

Event*: 10 day event in the month of July. Total number of windows: 960 (15 minute intervals)

can be used to obtain optimal coefficient and exponent values for similar events in the future. The seasonal average optimal coefficient and exponent values obtained from this study can be used as starting values for rainfall estimation before bias corrections are applied.

CONCLUSIONS

This paper reports the development, implementation and evaluation of optimization models for developing optimal reflectivity (*Z*)–rainfall-rate (*R*) relationships to obtain improved radar-based precipitation estimates. Results from a case study application of optimal *Z-R* relationships indicate that improved precipitation estimates are possible if standard *Z-R* relationships based on storm types for each event or a number of events in a season are optimized. The precipitation estimates are sensitive to the coefficient and exponent values used in *Z-R* relationships with the exponent values being the most sensitive. Once *Z-R* relationships are applied, bias corrections can be applied for improved rainfall estimates. Optimization of standard *Z-R* relationships based on data available from selected events suggests that real-time optimization of *Z-R* functional relationships is possible and is recommended for improvement of radar-based rainfall estimates.

REFERENCES

Amitai, E. (2000) Systematic variation of observed radar reflectivity-rainfall rate relations in the tropics. *J. Appl. Meteor.* 39, 2198–2208.

Baldonado, F. D. (1996) Experiments in determination of radar reflectivity – rainfall intensity (Z–R) relations in the Philippines. *Philippine Atmospheric, Geophysical and Astronomical Services Administration.* AGSSB Abstracts.

Hagen, M. & Yuter, S. E. (2003) Relations between radar reflectivity, liquid-water content, and rainfall rate during the MAP SOP. *Quart. J. Roy. Met. Soc.* 129, 477–493.

Lee, D. I., Jang, M., You, C. H., Kim, K. E. & Suh, A. S. (2002) Kuduck dwsr-88c radar rainfall estimation and Z-R relationships by poss during 2001 in Korea, *EGS XXVII General Assembly, Nice*, 21–26, abstract #2242.

Malakpet, C. G., Habib, E., Meselhe, E. A. & Tokay, A. (2007) Sensitivity analysis of variability in reflectivity-rainfall relationships on runoff prediction.. In: *87th AMS Annual Meeting* (San Antonio, Texas). Proc. 21st Conference on Hydrology.

Raghavan, S. (2003) *Radar Meteorology*. Atmospheric and Oceanographic Sciences Library, Springer Publication.

Rosenfeld, D., Wolff, D. B. & Amitai, E. (1994). The window probability matching method for rainfall measurements with radar. *J. Appl. Meteor.* 33(6), 689–693.

Tokay, A., Hartmann, P., Battaglia, A., Gauge, K. S., Clark, W. L. & Williams, C. R. (2008) A field study of reflectivity and Z-R relations using vertically pointing radars and disdrometers. *J. Atmospheric and Oceanic Technology*, doi: 10.1175/2008JTECHA1163.1.

Trafalis, T. B., Santosa, B. & Richman, M. B. (2005) Learning networks in rainfall estimation. *Computational Management Science*, 2, 3(7), 229–251.

Ulbrich, C. W. & Atlas, D. (1977) A method for measuring precipitation parameters using radar reflectivity and optical extinction. *Annals of Telecommunications* 32(11–12), 415–421.

Vieux, B. E. (2004) Precipitation measurement – distributed model input. In: *Distributed Hydrologic Modelling Using GIS,* 2nd edn, 149–176.

Villarini, G., Ciach, G. J., Krajewski, W. F., Nordstrom, K. M. & Gupta, V. K. (2007) Effects of systematic and random errors on the spatial scaling properties in radar-estimated rainfall. In: *Nonlinear Dynamics in Geosciences* (ed. by A. A. Tsonis). Springer.

Woodley, W. & Herndon, A. (1969) A raingauge evaluation of the Miami reflectivity-rainfall rate relation. *J. Appl. Met.* 9(2), 258–264.

Zrnic, D. S. & Balakrishnan, N. (1990) Dependence of reflectivity factor-rainfall rate relationship on polarization. *J. Atmospheric and Oceanic Technology* 7(5), 792–795.

Weather Radar and Hydrology
(Proceedings of a symposium held in Exeter, UK, April 2011) (IAHS Publ. 351, 2012).

81

Radar hydrology: new Z-R relationships over the Klang River Basin, Malaysia for monsoon season rainfall

SUZANA RAMLI & WARDAH TAHIR

Faculty of Civil Engineering, Universiti Teknologi MARA, 40450 Shah Alam, Selangor, Malaysia
suzana_ramli@yahoo.com

Abstract The use of Quantitative Precipitation Estimation (QPE) in radar–rainfall measurement for hydrological purposes is significantly important. For several decades radars have been deployed to monitor and estimate precipitation routinely in several countries. However, in Malaysia, radar application for QPE is still new and needs to be explored. This paper focuses on the Z-R derivation work of radar-rainfall estimation. The work develops new Z-R relationships for the Klang River in the Selangor area for the monsoon season; namely southwest monsoon rain, northeast monsoon rain and two inter-monsoon rains which distribute heavy rain (>30 mm/h). Looking at the high potential of Doppler radar for QPE, the newly formulated Z-R equations will be useful in improving the measurement of rainfall for any hydrological application, especially for flood forecasting.
Key words radar; Quantitative Precipitation Estimation; Z-R development; monsoon; flood forecasting

INTRODUCTION

The capability of weather radar to detect rainfall is a useful function in quantitative precipitation estimation (QPE) hydrological work. Radar has advantages in terms of spatial and temporal rainfall measurement since it is able to estimate rain over a larger area within short time intervals. However, accuracy is always an issue, especially during measurement of convective rain. Thus, many researchers have suggested combining weather radar and raingauge data as good practice in order to achieve better rainfall estimation. According to Einfalt *et al.* (2004) integration of weather radar and raingauge data will provide the best spatial estimate of rainfall as required for hydrological applications. In addition, several problems such as attenuation effects, ground clutter and Z-R equations should be assessed critically in order to eliminate errors.

Radar rainfall information shows potential in improving the ability to provide accurate flood predictions, although some uncertainties have been recognized as errors in radar-rainfall estimation (Malakpet *et al.*, 2005). To cope with biases, the joint use of radar data and raingauge data is recommended. Since weather radar cannot measure rainfall depth directly, the relationship between reflectivity, Z, and rainfall rate, R, is established empirically and this is referred to as a Z-R relationship. This equation is needed to convert reflectivity (dBZ) measured from radar to rainfall rate in mm/h. Normally, the process of converting the reflectivity into rain-rate uses one standard Z-R relationship: e.g. the Marshall & Palmer (1948) equation $Z = 200R^{1.6}$. Due to errors associated with Z-R variability, the use of different Z-R equations according to rain events and locations was introduced.

Various Z-R relationships are created according to the event types (convective, stratiform) and locations. Different equations are suitable for different atmospheric conditions. Thus, it is appropriate to ensure the radar conversion system uses the most suitable Z-R equations for radar-rainfall measurement purposes. Many studies have shown that with inappropriate use of Z-R relations, the rainfall estimates are proved to be inaccurate (Zogg, 2006). In order to minimize both underestimation and overestimation for light and heavy rainfall, respectively, newly developed Z-R relationships are needed in Malaysia to obtain better estimation and also rainfall prediction for flood forecasting purposes.

In Malaysia, the rainfall distribution is affected by local topography and seasonal wind flow patterns. Convective rain usually brings flood problems to low plain areas. Those floods were triggered by rain in the monsoon season, with four main types of monsoon rain: southwest monsoon, northeast monsoon and two inter-monsoons that carry heavy rain, especially during late evening. Rain in the southwest monsoon occurs between May and August, whilst in the northeast monsoon rain develops from November until February. The transitional season between the southwest and northeast monsoons is known as the inter-monsoon rain which happens twice, in March to April and in September to October. During the inter-monsoon, rain is usually convective and very intense for short time durations. Flash floods normally occur during the inter-monsoon season.

CASE STUDY AND DATA COLLECTION

The Klang River Basin in Selangor has experienced flooding for more than a decade. Since it is located in the midst of one of the busiest areas in Malaysia (between Selangor and Kuala Lumpur) it suffers from urbanization and a high population. The size of the Klang River Basin is 1288 km^2 in area, with a total stream length of approx. 120 km. Located at 3°17′N, 101°E to 2°40′N,101°17′E, it covers areas in Sepang, Kula Langat, Petaling Jaya, Klang, Gombak and Kuala Lumpur. In the case study, only the upper river basin is targeted within an area of 468 km^2 (Petaling Jaya, Klang, Gombak and Kuala Lumpur). Most of the flooding in Klang River occurred from soil erosion problems, and high rainfall intensity adds to serious degradation. Since 1998, more than RM20 million have been spent on flood mitigation of this river. It is essential to study the rainfall intensity effect for the Klang River Basin, and the combination of radar and raingauge will improve the rainfall estimation. Hence, it can be used for further hydrological work.

Radar reflectivity data were obtained from an S-band Terminal Doppler Radar, TDR, at Kuala Lumpa International Airport (KLIA), operated by MMD (Malaysia Meteorological Department) and located at an elevation of 37 m a.m.s.l. The conventional radar data are collected every 10 min up to the effective range of 230 km for three elevation scans (PPI) with elevation angles of 1.0°, 2.0° and 3.0°. Ground clutter is removed in the radar data calibration. The 1° × 2 km resolution maps were collected every 10 min and converted into rainfall intensity by means of the classical Marshall-Palmer relationship ($Z = 200R^{1.6}$) using the IRIS software program. The Doppler radar is situated in Bukit Tampoi, Dengkil, about 10 km north of KLIA, and was first introduced in 1998. The prime function of TDR is to detect and to alert KLIA of the wind shear problem and the microburst scenario. Both conventional and Doppler radars can detect rainfall intensity through its signal reflectivity.

Hydrological data such as rainfall, river discharge and water level were obtained from the Hydrology Division, DID, Malaysia. For this study 25 raingauge stations were selected that are located near the catchment area of the Klang River Basin, Selangor. Hourly rainfall data from DID were used, covering the period January–December 2009. There were >100 events recorded throughout these months. For Z-R development purposes, four categories of rainfall were included: southwest monsoon rain, northeast monsoon rain and two inter-monsoon rains which distribute heavy rain (>30 mm/h).

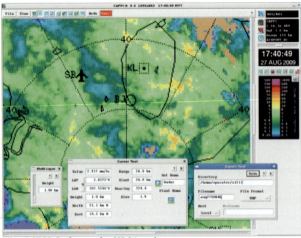

Fig. 1 IRIS Software for KLIA radar.

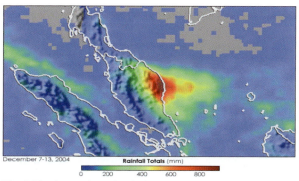

Fig. 2 Flood event in Malaysia in 2004.

METHODOLOGY: Z-R DERIVATION

The statistical method requires a combination of data from radar and raingauge stations. The Z-R relationship will then be obtained by measuring both data simultaneously. One of the techniques used for statistical estimation of the Z-R equation is an optimization method that relates measured values of radar reflectivity to rainfall rate. The approach is motivated by observation of the sampling properties and not driven by drop-size density control of Z-R relations. In the optimization-based approach, some measures of "closeness" of the radar rainfall products and the surface rainfall reference data obtained by raingauges is minimized. The optimization approach determines the relationship by using reflectivity data measured by radar and rainfall data from raingauge stations. Five minute time-interval data from radar-derived rainfall were obtained from CAPPI images and then accumulated to 1 h data. The Z-R relationship is deduced from radar reflectivity data and the surface rainfall rate from raingauges. The "best" values of a and b parameters in the relation $Z = aR^b$ are obtained by optimal curve fitting in the graph between reflectivity and rainfall rate. Issues with the method can arise from strong wind disturbances and partial evaporation of the falling rain. The values of the a and b parameters can also be determined from the literature with certain restrictions. The optimization approach implements an algorithm that uses the Z-R relationship as an empirical formula to obtain the unknown parameters of the Z-R relation. For a particular application, the products are optimized in an appropriate manner according to chosen criteria.

RESULTS AND DISCUSSION: NEW Z-R RELATIONSHIPS

At present, the KLIA weather radar has been using the classic Marshall-Palmer equation ($Z = 200R^{1.6}$) to convert reflectivity to rainfall rate in mm/h. The data shows that >80% of values obtained from the radar were overestimated when compared to raingauge observations. The need for new Z-R relations is crucial in order to improve the results. Applying the optimization approach, new Z-R relationships have been derived for four types of rainfall events, namely: southwest monsoon rain, northeast monsoon rain and two inter monsoon rains which distribute heavy rain (>30 mm/h). The classification is important to observe the rainfall rates in terms of error measurement.

For southwest monsoon rain, the original data can exceed >100 mm/h when converted using the Marshall-Palmer equation. Using the optimization method, the new Z-R relation obtained is $Z = 500R^{1.9}$. The results show reduction in the data values and improvement in the accuracy of hourly rainfall data. Figure 3 shows the plot between estimated radar rainfall using $Z = 200R^{1.6}$ and using the new $Z = 500R^{1.9}$ and raingauge data for January to December 2009. For northeast monsoon rain, the best Z-R obtained from optimization work is $Z = 166R^{1.9}$. Results produced are significantly better than radar rainfall with $Z = 200R^{1.6}$ when compared to raingauge data. It can be seen from Fig. 4 that the original radar rainfall gave a maximum value of 200 mm/h compared to 20 mm/h

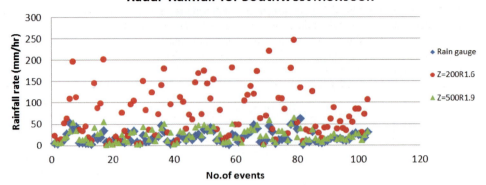

Fig. 3 Radar and raingauge estimates of rainfall for southwest Monsoon rain.

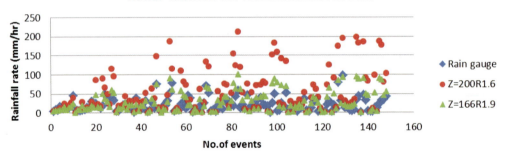

Fig. 4 Radar and raingauge estimates of rainfall for northeast Monsoon rain.

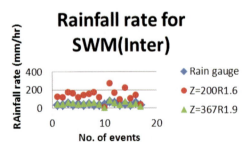

Fig. 5 Rainfall rate estimates for Inter SWM.

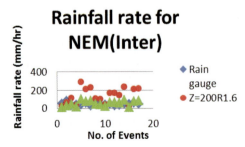

Fig. 6 Rainfall rate estimates for Inter NEM.

from the raingauge data. The new Z-R relation modified the results with better estimation. With relatively fewer events, new Z-R relations for both inter-monsoon rains were able to be derived. The data for inter-monsoon rain with rain events more than 30 mm/h in intensity gave interesting results. Most of the radar rainfall data were overestimated by more than 80% compared to raingauge values, with maximum data values reaching 300 mm/h. The new Z-Rs, which are $Z = 367R^{1.9}$ for inter SWM (southwest monsoon) and $Z = 260R^{1.9}$ for inter NEM (northeast monsoon), greatly reduce the disparity between radar rainfall and raingauge values. Obvious improvement can be observed from Figs 5 and 6.

The improvements obtained using the new Z-R relationships can be further described for the different types of rainfall events. The correlations increase between the radar reflectivity and the converted rainfall rates using the modified Z-R. The bigger the values of reflectivity the larger the difference in rain-rate obtained using the Marshall-Palmer equation and newly derived equations. For example, for the same reflectivity of 50 dbZ (northeast monsoon), Marshall-Palmer gives a rain rate of approximately 50 mm/h while the new modified Z-R estimates around 30 mm/h. Better results can also be estimated from southwest monsoon and inter-monsoon rain events. With correlations more than 0.7 (Figs 7 and 8), the newly developed Z-Rs have a huge potential in reducing errors between radar rainfall and raingauge data.

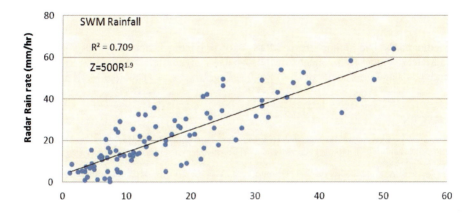

Fig. 7 Scatter plots of radar and raingauge values for $Z = 500R^{1.9}$.

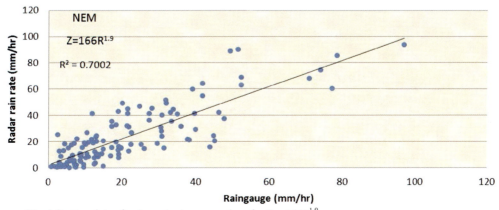

Fig. 8 Scatter plots of radar and raingauge values for $Z = 166R^{1.9}$.

Table 1 Statistical error of different Z-R equations for the Upper Klang River Basin.

Type of rain		New Z-R	Mean error	Mean absolute Error	RMSE	Bias
Southwest monsoon	All intensities, May–Aug.	$Z=500R^{1.9}$	−3.85	8.66	10.90	1.21
Northeast monsoon	All intensities, Nov.–Feb.	$Z=166R^{1.9}$	−6.98	13.03	19.68	1.33
Inter SWM	Convective > 30 mm/h, Sept–Oct.	$Z=367R^{1.9}$	−3.80	11.54	14.14	1.09
Inter NEM	Convective >30 mm/h, Mar.– Apr.	$Z=260R^{1.9}$	−10.13	32.04	36.49	1.22

CONCLUSIONS

Table 1 presents a summary of the statistical error of different Z-R equations. The rain has been classified into four categories depending on the rain monsoon seasons. It shows that for different rainfall seasons, the results also fluctuate. To improve the results, new Z-Rs have been derived by using an optimization approach. The modified equations show significant improvement in the rainfall rates obtained and the errors are also minimized. The rain regimes are a very important character since the Z-R relationships are hugely dependent on location and type of rain.

Acknowledgements The authors thank MOSTI for providing E-Science funds for this project, Malaysia Meteorological Department (MMD), Jabatan Pengaliran dan Saliran (JPS) for providing radar-rainfall data and the Faculty of Civil Engineering, Universiti Teknologi MARA Malaysia.

REFERENCES

Blanchard, D. C. (1953) Raindrop size distribution in Hawaiian rains. *J. Meteorol.* 10, 457-473.

Calheiros, R. V. & Zawadzki, I. (1987) Reflectivity rain rate relationships for radar hydrology in Brazil. *J. Clim. Appl. Meteorol.* 26, 118-132

Einfalt, T., Nielsen, K. A., Golz, C., Jensen, N. E., Quirmbach, M., Vaes, G. & Vieux, B. (2004) Towards a roadmap for use of radar rainfall data in urban drainage. *J. Hydrol.* 299, 186–202.

Krajewski, W. F. & Smith, J. A. (2002) Radar hydrology: rainfall estimation. *Adv.Water Resour.* 25, 1387–1394.

Marshall, J. S. & Palmer, W. (1948) The distribution of raindrops with size. *J. Meteorol.* 5,165–166.

Malakpet, C. G. (2005) Sensitivity analysis of variability in reflectivity-rainfall relationships on runoff prediction. Online article for NASA.

Rosenfeld, D., Wolff, D. B. & Amitai, E. (1994) The window probability matching method for rainfall measurements with radar. *J. Appl. Meteorol.* 33, 682–693.

Wilson, J. W. & Brandes, E. A, (1979) Radar measurement of rainfall – a summary. *Bull. Am. Meteorol. Soc.* 60, 1048–1058.

Zogg, J. (2006) How does radar estimate rainfall? National Weather Service Quad Cities IA/IL, Vol. 3, Issue 2.

Weather Radar and Hydrology
(Proceedings of a symposium held in Exeter, UK, April 2011) (IAHS Publ. 351, 2012).

87

Simultaneous measurements of precipitation using S-band and C-band polarimetric radars

ALEXANDER RYZHKOV[1,2,3], PENGFEI ZHANG[1,2], JOHN KRAUSE[1,2], TERRY SCHUUR[1,2], ROBERT PALMER[3] & DUSAN ZRNIC[2]

1 *Cooperative Institute for Mesoscale Meteorological Studies, University of Oklahoma, 120 David L. Boren Blvd, Norman, Oklahoma 73072, USA*
alexander.ryzhkov@noaa.gov

2 *National Severe Storms Laboratory, Norman, Oklahoma, USA*

3 *Atmospheric Radar Research Center, University of Oklahoma, Norman, Oklahoma, USA*

Abstract Simultaneous measurements of heavy tropical rain made by closely located S- and C-band polarimetric radars are examined. The performance of different algorithms for rainfall estimation is discussed. It is demonstrated that the polarimetric algorithm based on the combined use of specific differential phase and differential reflectivity yields the least biased estimate of rainfall at S-band. Similar estimation at C-band faces challenges.

Key words polarimetric radar; flash flood; tropical rain

INTRODUCTION

Beginning in May 2010, system and operational tests of the pre-prototype of the operational polarimetric WSR-88D radar (KOUN) have been conducted in central Oklahoma in preparation for the polarimetric upgrade to the US NEXRAD network. Since then, observations and rainfall estimates from the S-band KOUN polarimetric radar have been regularly computed and compared with simultaneous observations and rainfall estimates from the C-band OU-PRIME polarimetric radar, which is owned and operated by the University of Oklahoma. Because these closely located radars operated continuously, a large dataset simultaneously collected has been obtained in a relatively short period of time.

In this study, we focus on an extreme flash flood event that occurred on 13–14 June 2010, during which 100–200 mm of rain was recorded at several locations within the Oklahoma City metropolitan area. The storm was of a tropical precipitation type similar to the ones that produced the famous Fort Collins, Colorado, flash flood on 28 July 1997 (Petersen *et al.*, 1999) and the extensive central Oklahoma flooding associated with the remnants of tropical storm Erin on 19 August 2007 (Schuur *et al.*, 2008; Gourley *et al.*, 2010). Both the conventional R(Z) algorithm and the polarimetric algorithm accepted for initial WSR-88D deployment underestimated rain during these tropical events by about 40%. The primary goal of this study is to better understand the reason for this underestimation and to find a solution that might lead to algorithm improvement. We also compare S- and C-band radar measurements with a focus on optimizing the C-band rainfall estimation algorithms.

BACKGROUND

The initial version of the rainfall estimation algorithm for the polarimetrically upgraded WSR-88D is based on the results of hydrometeor classification. It stipulates the use of the R(Z,Z_{DR}) relation:

$$R(Z, Z_{DR}) = 1.42 10^{-2} Z^{0.770} Z_{dr}^{-1.67} \qquad (1)$$

for pure rain and the R(K_{DP}) relation:

$$R(K_{DP}) = 44.0 \, | \, K_{DP} \, |^{0.822} \, sign(K_{DP}) \qquad (2)$$

if the rain is mixed with hail, where Z represents radar reflectivity, Z_{DR} differential reflectivity, and K_{DP} specific differential phase (Giangrande & Ryzhkov, 2008). More recently, Schuur *et al.*

(2008) and Gourley *et al.* (2010) reported that the alternative $R(Z,Z_{DR})$ relation:

$$R(Z,Z_{DR}) = 6.70 10^{-3} Z^{0.927} Z_{dr}^{-3.43} \qquad (3)$$

which uses Z_{DR} much more aggressively than equation (1), performs better for tropical rain. Because of the large exponent in the Z_{DR} factor, however, it is very sensitive to possible errors in the Z_{DR} measurements.

The challenge is to find an algorithm that automatically takes drastic differences between drop size distributions (DSDs) in continental and tropical storms into account yet is also sufficiently robust with respect to the measurement errors of polarimetric radar variables. In order to evaluate the impact of DSD variability on the performance of different rainfall algorithms, we resort to the comprehensive disdrometer dataset collected over a 7-year period that contains 47 144 DSDs for multiple events representing a wide variety of seasons and precipitation regimes in central Oklahoma (Schuur *et al.*, 2005). The values of Z and Z_{DR} at S-band were simulated for all DSDs in the dataset and the Z–Z_{DR} pairs were segregated into five rain regimes using histograms of Z_{DR} for a given Z. Figure 1 (left panel) shows separation of Z_{DR} into 20% percentiles. At fixed Z, Z_{DR} decreases as continental rain type changes to tropical.

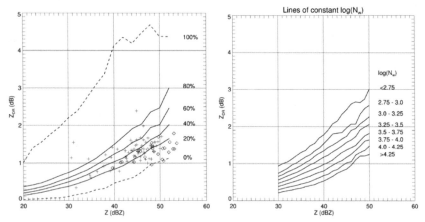

Fig. 1 Left panel. Partitioning of the Z–Z_{DR} plane for different rain regimes based on the histograms of Z_{DR} for given Z simulated at S-band from disdrometer data. The Z–Z_{DR} pairs from radar observations at locations of gauges with maximal tropical rain accumulations are denoted by crosses (14 June 2010 case) and diamonds (19 August 2007 case). Right panel. Z–Z_{DR} dependencies corresponding to constant median $\log(N_w)$ for eight intervals of the $\log(N_w)$ variability.

Different rain regimes can be distinguished using the normalized drop concentration N_w (Testud *et al.*, 2000). It is obvious that the lines of constant $\log(N_w)$ in the Z–Z_{DR} plane (Fig. 1, right panel) lie almost parallel to the lines of constant Z_{DR} percentiles (left panels). It can be also shown that the scatterplots of rain-rate R *versus* Z and K_{DP} become very narrow after these variables are normalized by N_w and that the $R(Z)$ and $R(K_{DP})$ relations can be parameterized by N_w as:

$$R(Z) = 1.57 10^{-3} N_w^{0.365} Z^{0.635} \qquad (4)$$

and

$$R(K_{DP}) = 5.73 N_w^{0.240} K_{DP}^{0.760} \qquad (5)$$

where N_w is measured in m^{-3} mm^{-1}. These relations are valid at S-band.

The fixed $R(Z)$ and $R(K_{DP})$ relations optimized for a whole disdrometer dataset inevitably overestimate extreme continental rain represented by the 80–100% percentile interval in Fig. 1 and underestimate very tropical rain in the lowest 0–20% percentile regime. The biases, however, are much larger for the $R(Z)$ relation. Obviously, utilization of the sole $R(K_{DP})$ relation (at least for higher rain-rates) is not sufficient to capture the full range of DSD variability in Oklahoma. The solution can be either the use of the $R(K_{DP},Z_{DR})$ relation optimized for the whole dataset:

$$R(K_{DP},Z_{DR}) = 95.9 K_{DP} Z_{dr}^{-1.89} \qquad (6)$$

or equation (5) provided that N_w is reliably estimated.

RESULTS OF OBSERVATIONS

Typical composite PPIs of Z, Z_{DR}, differential phase Φ_{DP}, and correlation coefficient ρ_{hv} measured at S- and C-bands for the flash flood case on 14 June 2010 are shown in Figs 2 and 3. A simple correction for attenuation at S- and C-bands was made using linear relations between Z and Z_{DR} biases and Φ_{DP} with the coefficients of proportionality equal to 0.02 and 0.07 dB/deg for Z and 0.002 and 0.02 dB/deg for Z_{DR} at S- and C-bands respectively. Locations of the three raingauges in Oklahoma City received largest amounts of rain (63, 45 and 58 mm) during the one hour period from 12 to 13 UTC are marked as diamonds. It is evident that the structure of the rain is very non-uniform. Indeed, at the moment illustrated in Figs 2 and 3, a very tropical rain with anomalously low Z_{DR} that barely exceeds 1 dB, combined with high Z, is observed at all three locations. However, an area of very high Z_{DR} is seen just southeast of the gauges at the periphery of the rain band. This area is marked by extremely high Z_{DR} combined with low ρ_{hv} in the C-band measurements. It is this "hotspot" that accounts for the high attenuation / differential attenuation at C-band that is not fully compensated by the simple attenuation correction scheme as Fig. 3 shows.

It is not surprising that the Z–Z_{DR} pairs measured by the WSR-88D radar at the gauge locations from 12 to 13 UTC (marked by crosses in the left panel of Fig. 1) fall into the tropical rain regime, as did the corresponding data for the Erin rain event on 17 August 2007, for the two gauges that recorded the highest rainfall accumulation (depicted by diamonds).

The corresponding fields of instantaneous rain-rates retrieved from simultaneous S- and C-band measurements are presented in Figs 4 and 5. The conventional WSR-88D $R(Z)$ relation $Z = 300 R^{1.4}$ and the polarimetric relation (1) provide low rainfall accumulations, with both under-estimating hourly rain total in the three gauges by a factor of about 2 (Fig. 4(a),(b)). The $R(K_{DP})$ relation combined with the $R(Z,Z_{DR})$ relation (3) for $Z < 40$ dBZ shows better performance but still falls short of the gauges (Fig. 4(c)). The best agreement with the gauges (with a slight negative bias of 4%) is achieved using the $R(K_{DP},Z_{DR})$ relation (6) combined with $R(Z,Z_{DR})$ from (3) for $Z < 40$ dBZ. This conclusion is consistent with that of Petersen *et al.* (1999), who found that the $R(K_{DP},Z_{DR})$ relation, although different from (6), yielded the least biased estimate for tropical rainfall during the Fort Collins flash flood.

The C-band $R(K_{DP})$ relation optimized for the Oklahoma disdrometer dataset:

$$R(K_{DP}) = 25.3 \, | K_{DP} |^{0.776} \, sign(K_{DP}) \qquad (7)$$

combined with $R(Z)$ for $Z < 40$ dBZ definitely outperforms the $R(Z)$ (Fig. 5) but still underestimates rain. Because the Z_{DR} data at C-band appear to be more erratic than at S-band due to uncompensated differential attenuation, resonance scattering, and larger statistical fluctuations due to lower ρ_{hv}, it is problematic to combine K_{DP} and Z_{DR} to improve the algorithm performance in the way it is done at longer wavelength. Sugier & Tabary (2006) also point to the fact that the quality of Z_{DR} can be impacted by the antenna radome which may further compromise the utility of Z_{DR} for QPE at C-band. Alternate techniques for rain estimation at C-band are being tested (Testud *et al.* 2000; Tabary *et al.*, 2010). They avoid direct use of the locally estimated K_{DP} and Z_{DR} and imply estimation of N_w either using the ZPHI method in the heavier rain areas or the Illingworth & Thompson (2005) routine based on estimating of spatially averaged Z_{DR} in the areas of light-to-moderate rain.

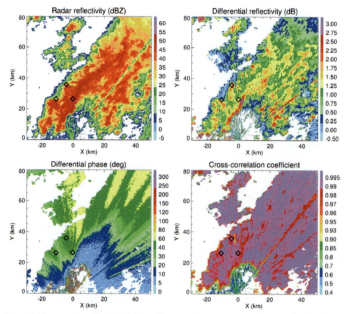

Fig. 2 Composite plot of Z, Z_{DR}, Φ_{DP} and ρ_{hv} measured in tropical rain at S-band at El = 1.4° at 1202 UTC 14 June 2010. Diamond signs indicate locations of gauges with largest hourly totals.

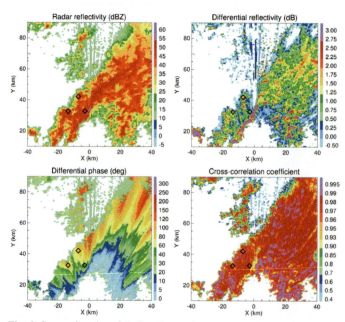

Fig. 3 Composite plot of Z, Z_{DR}, Φ_{DP} and ρ_{hv} measured in tropical rain at C-band at El = 1.5° at 1202 UTC 14 June 2010. Diamond signs indicate locations of gauges with largest hourly totals.

We used a routine similar to the one of Illingworth and Thompson to retrieve the fields of $\log(N_w)$ by finding the best matching curves in Fig. 1 (right panel) using the S-band Z and Z_{DR} data collected over 2 km × 2° regions (Fig. 6). The area of tropical rain with $\log(N_w)$ exceeding 4.25 is

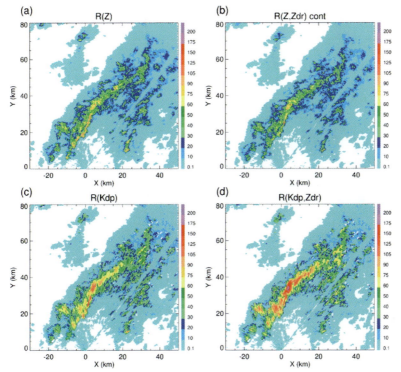

Fig. 4 The maps of rain rate estimated at S-band using (a) $R(Z)$, (b) $R(Z,Z_{DR})$ (equation (3)), (c) $R(K_{DP})$, and (d) $R(K_{DP},Z_{DR})$ at 1202 UTC 14 June 2010.

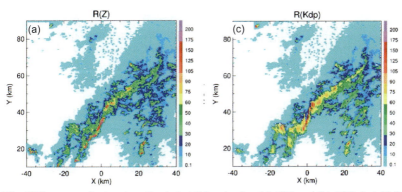

Fig. 5 The maps of rain-rate estimated at C-band using (a) $R(Z)$ and (b) $R(K_{DP})$ at 12:02 UTC 14 June 2010.

visible in the centre of the rain-band. In the regions of convective initiation to the southeast of the major rain-band, $\log(N_w)$ occasionally drops below 2.75, which is consistent with high Z_{DR} observed in that location. The N_w retrieval routine seems to work reliably at S-band but at the moment of writing this paper, we were not able to do robust retrieval of $\log(N_W)$ at C-band using the piecewise ZPHI technique described by Tabary *et al.* (2010). One of the possible reasons is the high variability of the A_h/K_{DP} ratio, which was assumed to be relatively stable in the Tabary *et al.* (2010) validation study that was performed in an area with a more temperate climate than that in Oklahoma.

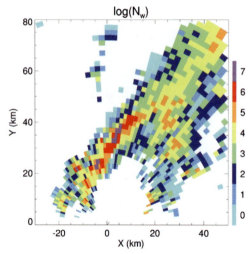

Fig. 6 The map of log(N_w) retrieved at S-band at 1202 UTC 14 June 2010. Colour scale: log(N_w) <2.75 ("0"), 2.75–3 ("1"), 3–3.25 ("2"), 3.25–3.5 ("3"), 3.5–3.75 ("4"), 3.75–4 ("5"), 4–4.25 ("6"), >4.25 ("7").

One of the advantages of simultaneous observations with S- and C-band polarimetric radars is that S-band estimates of polarimetric radar variables, rain-rates, parameter N_w, etc. can be utilized for optimization of the algorithms for attenuation correction and rainfall estimation at C-band. Since the results of N_w retrieval are very sensitive to possible Z_{DR} (and Z) artefacts, we found it useful to display the fields of log(N_w) to visualize the diversity of rain types in the radar echo and to control absolute Z_{DR} calibration and its correction for noise in the areas of weak echo.

CONCLUSIONS

The $R(K_{DP},Z_{DR})$ relation applied to heavy rain with $Z > 40$ dBZ performs the best for torrential tropical rain at S-band. However, the use of a similar relation at C-band may not be feasible due to the difficulties of ensuring high quality of Z_{DR} measurements. The use of the piecewise ZPHI technique and determination of N_w or the $R(K_{DP})$ relation combined with $R(Z)$ (for lighter rain) might be a preferable choice for C-band.

REFERENCES

Giangrande, S. & Ryzhkov, A. (2008) Estimation of rainfall based on the results of polarimetric echo classification. *J. Appl. Meteorol.* 47, 2445–2462.

Gourley, J., Giangrande, S., Hong, Y., Flamig, Z., Schuur, T. & Vrugt J. (2010) Impacts of polarimetric radar observations on hydrological simulation. *J. Hydrometeorology* 11, 781–796.

Illingworth, A. & Thompson, R. (2005) The estimation of moderate rain rates with operational polarization radar. In: *32nd Conf. Radar Meteorol.*, Albuquerque, NM, P9R.1.

Petersen, W., Carey, L., Rutledge, S., Knievel, J., Doesken, N., Johnson, R., McKee, T., Vonder Haar, T., & Weaver, J. (1999) Mesoscale and radar observations of the Fort Collins flash flood of 28 July 1997. *Bull. Amer, Meteor. Soc.* 80, 191–216.

Schuur, T., Ryzhkov, A. & Clabo, D. (2005) Climatological analysis of DSDs in Oklahoma as revealed by 2D-video disdrometer and polarimetric WSR-88D. Preprints. In: *32nd Conference on Radar Meteorology*, CD-ROM, 15R.4.

Schuur, T., Giangrande, S. & Ryzhkov A. (2008) Polarimetric WSR-88D reflectivity and differential reflectivity attenuation correction for tropical rainfall. In: *Int. Symp. Weather Radar and Hydrology* (Grenoble, France), CD-ROM P1-021.

Sugier, J. & Tabary P. (2006) Evaluation of dual-polarization technology at C-band for operational weather radar network. *Report of the EUMETNET Opera 2, Work packages 1.4 and 1.5, Deliverable b* [available at http://www.knmi.nl/opera].

Tabary, P., Boumahmoud, A., Fradon, B., Parent-du-Chatelet, J., Andrieu, H. & Illingworth, A. (2010) Evaluation of two integrated techniques to estimate the rainfall rates from polarimetric radar measurements and extensive monitoring of azimuthal-dependent Z_{DR} and Φ_{DP} biases. In: *ERAD 2010.* Sibiu, Romania, P13.22.

Testud, J., Le Bouar, E., Obligis, E. & Ali-Mehenni, M., (2000) The rain profiling algorithm applied to polarimetric weather radar. *J. Atmos. Oceanic Technol.* 17, 332–356.

Weather Radar and Hydrology
(Proceedings of a symposium held in Exeter, UK, April 2011) (IAHS Publ. 351, 2012).

93

French-Italian X- and C-band dual-polarized radar network for monitoring South Alps catchments

E. MOREAU[1], E. LE BOUAR[1], J. TESTUD[1] & R. CREMONINI[2]

1 *NOVIMET, 11 bd d'Alembert, 78280, Guyancourt, France*
emoreau@novimet.com
2 *ARPA Piemonte – Sistemi Previsionali, Torino, Italy*

Abstract In the framework of the CRISTAL (CRues par l'Integration des Systèmes Transfrontaliers Alpins) project, two dual-polarized X-band radars have been deployed for monitoring the catchment of the Roya, located in the south Alps at the French-Italian border. The French radar (Hydrix) has been installed in the Maritime Alps (Mt-Vial, 1500 m) and the Italian radar was installed at Col de Tende at 1800 m altitude during the summer of 2010. Two Italian C-band dual-polarized radars complete the network, ensuring a full monitoring of the Roya catchment. This paper focuses on the capability of the two operational X-band radars to complement each other when monitoring rain/flood events in a mountainous area. Also illustrated is their ability for gap-filling neighbouring C-band radars which are blinded by orography. The ZPHI® algorithm is applied to the whole set of radar data, correcting for signal attenuation and estimating drop-size distribution and surface rainfall without any use of raingauge information. A case study from summer 2010 is shown, by comparing various radar-derived rainfall mosaics.

Key words radar network; dual-polarized radar; catchment

INTRODUCTION

In Mediterranean Alpine regions flood risk is very high. Flash floods are caused by intense rainfall, catchments having high slopes and high vulnerability, particularly along coastal areas with intense urbanization. This implies the need for effective flood forecasting. The forecast implies a prior accurate estimate of rainfall over the catchments. Traditionally rainfall estimates are based on gauge data, but point measurement density can be particularly low in Alpine areas. More recently, weather radars have been used to provide spatial rainfall estimation. In particular, compact dual-polarisation X-band radars are especially well adapted for monitoring rainfall in such mountainous catchments. In the framework of the CRISTAL project, two dual-polarized X-band radars have been deployed for monitoring the catchment of the Roya, located in the south Alps at the French-Italian border. This paper first describes the instrumental configuration and the available dataset, and then focuses on the processing chain for calibrating radar, correcting for signal attenuation and estimating drop-size distribution and surface rainfall. Finally, a case study from summer 2010 is shown, by comparing various radar-derived rainfall mosaics.

Fig. 1 Location of the radars with range ring at 60 km. In red, the Roya catchment at the Italian-French border.

INSTRUMENTAL RADAR NETWORK

The radar network is composed of four dual-polarisation radars: a Hydrix® X-band radar installed in the French Maritime Alps (Mt-Vial, 1500 m altitude), a mobile Italian X-band radar installed at Col de Tende (1800 m altitude) during the field campaign, and two C-band radars from the operational Italian network located at Bric della Croce and Monte Settepani, respectively (Fig. 1). The main technical characteristics of the four radars are listed in Table 1. All radars have dual-polarisation capability, allowing the use of a similar processing chain.

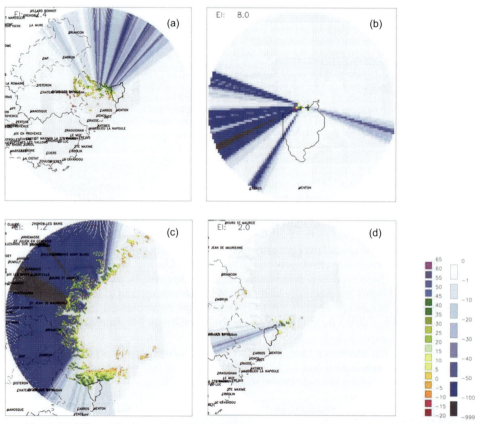

Fig. 2 Visibility of the four radars. Simulated partial beam blocking expressed in percent (blue scale) and simulated ground echo in dBZ, averaged at 1 km² resolution. For each radar, the lowest exploitable elevation is shown: (a) Mt-Vial at 2.4°, (b) Col de Tende at 8.0°, (c) Bric della Croce at 1.2°, and (d) Monte Settepani at 2.0°.

The complex orography leads to varying visibility of the Roya catchment for all four radars (Fig. 2). The radar visibility is simulated in terms of partial beam blocking and ground echo. A detailed description of the visibility simulation tool can be found in Moreau _et al._ (2010).

Despite the close range of the Italian X-band radar, the edge of the Col de Tende reduced the visibility toward the south leading to good visibility for the north part of the Roya catchment only. The Mt-Vial radar located at ~40 km range exhibits a relative good visibility except in the northern area of the Roya catchment. A mosaic of the two X-band radars is expected to offer a good visibility over the whole catchment. Despite the far range (~100 km) of the Bric della Croce radar, a low elevation (at 1.2°) means it exhibits only relatively small partial beam blocking (<10%). The

Table 1 Technical characteristics of the radar network.

Site	Mt-Vial, France	Bric della Croce, Italy	Monte Settepani, Italy	Col de Tende, Italy
Latitude	43.897°N	45.034°N	44.247°N	44.149°N
Longitude	7.149°E	7.733°E	8.199°E	7.560°E
Altitude	1500 m	773 m	1390 m	1858 m
Polarisation type	Linear H and V	Linear H and V	Linear H and V	Linear H and V
Beam width	1.5°	1.0° w/o radome	1.0° w/o radome	1.35° w/o radome
Antenna diameter	1.5 m	4.2 m	4.2 m	1.8 m
Radome	none	sandwich	sandwich	sandwich
Frequency	9.4 GHz	5.6 GHz	5.6 GHz	9.4 GHz

Monte Settepani radar offers good visibility at 2.0° except for the northern part of the Roya, with partial beam blocking reaching 30%.

PROCESSING FLOW

For each radar site, the data are processed following the same steps. First, reflectivity is corrected for partial beam masking due to orography. A first guess of the correction to apply is documented by a digital elevation model (DEM) for each elevation angle. Secondly, each sweep is processed, implying classification and estimation processing steps (see following). The outputs feed some statistics that are used, when sufficient, to estimate (potentially high) residues of Z and Z_{DR} directional biases due to effects not represented by the DEM data. The data are then re-processed, using the updated correction map of Z and Z_{DR}. Lastly, a surface rainfall rate map is provided by merging the rain-rate obtained at each elevation. The followings paragraphs describe in more detail some key steps involved in the processing flow synthesized aloft.

Classification

Taking advantage of dual-polarisation, a fuzzy logic classification is employed, adapted to attenuating frequencies. First, hydrometeor and non-hydrometeor targets are separated, essentially thanks to textures of measured variables. Then, hydrometeor types are subdivided into five classes: stratiform rain, convective rain, melting snow, dry snow and hail. For discriminating the melting snow, the height of the 0°C isotherm value is used, provided by a weather forecasting model.

ZPHI® processing

Z and Z_{DR} are corrected for attenuation by means of the rain profiling algorithm ZPHI® (Testud *et al.*, 2000), taking advantage of the dual-polarisation capability of the radar. In addition, the intercept parameter N_0^* is processed, allowing systematic tuning of the Z-R relationship needed because of natural rain variability. For light rain along which the differential phase shift is beyond a lower threshold of six degrees ($\Delta\Phi_{DP} < 6°$, i.e. PIA < 1.92 dB in X-band, PIA < 0.68 dB in C-band), an alternative $N_0^*(Z, Z_{DR})$ estimate is used, as described in Le Bouar *et al.* (2010).

Directional bias correction

Parameter N_0^* derived from ZPHI® is sensitive to calibration error or, more generally, any reflectivity bias δZ. A statistically stable average of N_0^* can be easily obtained (generally after gathering a couple of days of measurements), to reasonably associate its directional variations of Z bias, due to neighbouring obstacles or radome artefacts. Thus an azimuth-by-azimuth map of Z bias is built up for the three radar sites. Figure 3 illustrates the resulting map obtained for the Bric della Croce site. Z_{DR} correction maps are also processed, by comparing directional statistics of Z_{DR} for light rain, to be compared with a reference value of +0.35 dB.

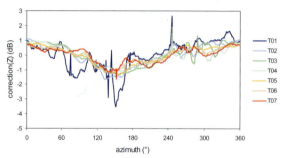

Fig. 3 Bric della Croce directional correction map for Z. Each curve refers to one PPI level, T01 (blue) being the lowest one (–0.1° elevation), and T07 (red) referring to elevation 5.8°.

Merged surface rain rate

First, the rainfall rate from each radar and from each elevation i, initially given in polar coordinates, is projected onto a Cartesian map with a 1 km^2 resolution pixel. Then, at each pixel a quality index q_i is processed as follows: $q_i = (1 - \sigma^2) - c_{pbb}$ where c_{pbb} is a coefficient that lowers by 0.4 the quality index when partial beam masking exceeds 20%. When the beam masking is greater than 50%, the whole quality index is set to zero. σ^2 is the relative variance (dimensionless) of the estimate, resulting from several contributions:

– representativeness error increasing linearly with altitude of measurement with respect to the ground. From a two year experimental campaign in France, the slope was found to be 0.06 km^{-1}.
– time sampling error proportional to the dwell time. The multiplicative coefficient, deduced from the same experimental campaign, is set to 0.01 min^{-1}.
– estimate error depending on the intrinsic relative error of algorithm used (10% for ZPHI®, 20% for Z-ZDR in light rain and 60% for Z-R in snow).

Finally, the rain rates processed from each elevation angle i are merged through a linear averaging weighted according to the quality index.

This ensures, for example, that low elevations are more strongly weighted than high ones, according to representativeness error. In contrast, the importance of low elevation tends to be reduced for the case of partial beam filling occurring near the ground. This also leads to filling gaps due to bright band or ground clutter.

A CASE STUDY

The first field campaign took place during summer 2010, when only a limited number of rainfall events occurred. The rainfall event of 19 June is shown here, characterized by multiple convective cells moving over the Roya catchment and eastward to Italy. Over the Roya catchment, instantaneous rainfall reached 20 to 30 mm/h, whereas more intense cells occurred in the Italian area, up to 100 mm/h. Unfortunately, the X-band radar located at Col de Tende was not available during this event, with only 3 of the 4 radars operating.

Figure 4 shows instantaneous rainfall (at 10 h UTC) observed by the X-band radar at Mt-Vial and by the two Italian C-band radars. Three cells (labelled A, B, and C) overpass the Roya. The northern cell (A) is well observed by the three radars with the same rainfall intensity and relative similar shape. The southern cells (B and C) are completely underestimated by the Bric della Croce radar compared to the two other radars. In contrast, cells (B) and (C) are seen similarly by both Mt-Vial and Monte Settepani radars, in terms of structure and intensity. The mosaic map of the three radars network is also shown in Fig. 4.

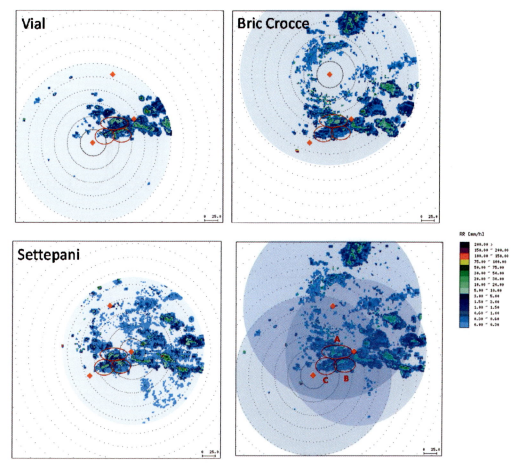

Fig. 4 Radar rainfall rate maps (in mm/h) measured on 19 June 2010 at 10 h UTC by each individual radar, and merged as a mosaic of the radar network. Red diamond symbols indicate the three available radars. Cells over the Roya are labelled A, B and, C (see text).

In terms of validation, two raingauge stations were available near the Roya catchment, located at Breil-sur-Roya and at Luceram. Figure 5 shows hourly rain accumulation estimated by the three individual radars and by the raingauges during this event. The best agreement is observed with the Mt Vial radar, with an overall bias of –15% at Luceram and –18% at Breil-sur-Roya when compared to ground measurements. At the Breil-sur-Roya station, we note an overestimation of the rainfall from the Settepani radar, in particular on 19 June from 12:00 to 16:00 h UTC, whereas the two others radars produce similar rainfall accumulations. At the Luceram station, we note a large variability in results during the two day event with the first peak (19 June at 14:00 h) being underestimated by the BricCroce radar. The second peak (19 June at 19:00 h) is well observed by the Mt Vial and BricCroce radars and is largely underestimated by the Settepani radar. The third peak (20 June at 06:00 h) is overestimated by the two Italian radars (+60%) and underestimated by the Mt Vial radar (–25%). Combining the three radar rainfall estimates together allows reducing the overall bias for this event.

A second field campaign is scheduled during summer 2011 in the framework of the CRISTAL project with the Italian X-band radar installed at Col de Tende during which we expect intense rainfall events over the Roya catchment.

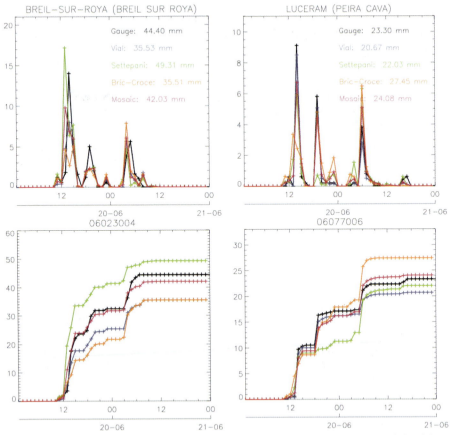

Fig. 5 Hourly rainfall rate in mm/h (top) and rainfall accumulation in mm (bottom) derived from the Mt-Vial (in blue), Settepani (in green), Bric-della-Croce (in orange), the radar mosaic (in pink) and measured by the raingauges stations (in black).

Acknowledgements This work has been co-funded by European Union - European funds for regional development – IT-F ALCOTRA Programme – 2007–2013 project CRISTAL no. 008.

REFERENCES

Le Bouar, E., Moreau, E. & Testud, J. (2010) Evaluation of light rain estimation applied to the X-band polarimetric radar HYDRIX data. In: *Proc Sixth European Conference on Radar Meteorology and Hydrology* (6–10 September 2010, Sibiu, Romania).

Moreau, E., Testud, J. & Le Bouar, E. (2010) Hydrix visibility in mountainous area. In: *Proc Sixth European Conference on Radar Meteorology and Hydrology* (6–10 September 2010, Sibiu, Romania).

Testud, J., Le Bouar, E., Obligis, E. & Ali-Meheni, M. (2000) The rain profiling algorithm applied to polarimetric weather radar. *J. Atmos. Oceanic Technol.* 17, 332–356.

Weather Radar and Hydrology
(Proceedings of a symposium held in Exeter, UK, April 2011) (IAHS Publ. 351, 2012).

99

Hail events observed by an X-band polarimetric radar along the French Mediterranean coast

ERWAN LE BOUAR, EMMANUEL MOREAU & JACQUES TESTUD

NOVIMET – 11, Bd d'Alembert, 78280 Guyancourt, France
elebouar@novimet.com

Abstract During summer 2010, in a mountainous and Mediterranean context, very strong reflectivities were observed by the Hydrix™ radar located at Mount Vial (1500 m height, near Nice, France), suggesting hail occurrences. However, the operational product processing failed to provide good results since no hail detection procedure was implemented. Thus, it expectedly produced very strong rainfall rates when gauge measurements showed very weak ones. A hail detection procedure taking advantage of the radar polarimetric capabilities has been tested in off-line processing, showing much better output performances, encouraging its operational implementation. This paper presents the obtained results, and describes the approach chosen for detecting the presence of hail.

Key words hail; X-band; dual-polarisation; classification

INTRODUCTION

Since 2006, an X-band dual-polarized weather radar (Hydrix™ radar) has been operating in the south of France, near Toulon from 2006, and then near Nice from late 2007. The present radar site is located at the top of Mont Vial (1500 m height above sea level), in a mountainous and Mediterranean environment, where flash flood monitoring is crucial. More recently, customers have taken advantage of services using data from this radar, for optimized rain alerts essentially based on rain accumulation criteria.

However, in such an environment favouring storm intensification, it is not uncommon to observe hail events, particularly in late summer and autumn, with their well known impacts in several sectors (e.g. damage to agriculture fields, aircraft structures, roofs), thus making hail detection an important issue in weather and hydrological radar monitoring.

Dual polarisation offers special benefits for identifying hail-bearing precipitation cells, thanks to additional variables providing extra clues on size, concentration and thus categories of observed precipitation. The most popular approach referenced for discriminating hydrometeor types is based on fuzzy logic, such as the one by Straka *et al.* (2000). However, whereas many studies have proven the efficiency of such an approach to identify hail precipitation at S-band (e.g. Heinselman & Ryzhkov, 2006), only a few refer to attenuation issues at C- or X-band frequencies (e.g. Vulpiani *et al.*, 2007, Tabary *et al.*, 2010).

In this paper, a case study is shown characterized by hail occurrence reported at the ground. A hail detection procedure and criteria are then described, to be implemented in the operational processing flow adapted to X-band radar data.

CLASSIFICATION AND ATTENUATION ISSUE

Classification

Basically, there are two purposes for performing a classification procedure. The first one is identifying the type of echo measured by the radar, leading to useful information on the microphysical processes involved in a storm cell, for example. The second one is applying the correct or best adapted estimation algorithm to particular range bins, that is often hydrometeor-type dependent. Practically, the basic estimated variable is the rainfall rate. For an attenuating frequency, the estimation algorithm can also refer to attenuation induced by the identified hydrometeor.

Typical classification procedures for polarimetric radar data are now based on a fuzzy logic approach, that combines for each hydrometeor category several weighted membership functions, selecting as output the hydrometeor class that provides the combination with the best score. Classification studies in X-band can be found in the literature (Dolan *et al.*, 2009), but hail-related categories have been omitted, as far as we know. Hereafter, membership function profiles in the Z_{DR}–Z and K_{DP}–Z spaces are shown for three categories: rain, rain/hail mixture, and hail (Fig. 1). Adjustment may be required because of the lack of experienced comparison with real measurements. The represented membership limits for rain and upper-limit for rain/hail mixture are derived from scattering rain model results, while the limits for hail and lower-limit for rain/hail mixture are derived from Straka *et al.* (2000), although originally set for S-band. For K_{DP}, the thresholds are actually extended by a factor of about 3 (S-/X-band wavelength ratio). These membership functions are delineated by considering attenuation corrected values of Z and Z_{DR}.

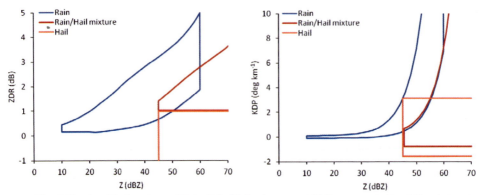

Fig. 1 Membership functions of Rain, Rain/Hail mixture, and Hail, represented by delineated regions in Z–Z_{DR} space (left) and Z–K_{DP} space (right).

It is noticeable that the upper limit for Z_{DR} in rain has been determined by considering a low intercept DSD parameter N_0^* (i.e. small drop concentration, here with $N_0^* = 10^5$ m^{-4}) for Z. The same low N_0^* value has produced the lower limit of K_{DP} in the Z–K_{DP} space. This is consistent with large drop characteristics, which exhibit quite high Z_{DR} and low K_{DP}. Inversely, a higher N_0^* value (i.e. $N_0^* = 10^8$ m^{-4}) has produced the lower limit in Z–Z_{DR} space, and the upper limit in Z–K_{DP} space. Thus, when rain/hail and pure-rain domains overlap each other in both Z–Z_{DR} and Z–K_{DP} spaces, i.e. around the lower limit in rain domain in both 2D-spaces, then the corresponding hydrometeors are more likely of rain/hail nature than of pure rain nature. This distinction is not apparent through standard fuzzy-logic hydrometeor classification using Z as the reference variable.

Attenuation correction

In X-band, occurring attenuation can complicate the classification decision tree, especially when this attenuation is quantitatively known *a priori*, adding more ambiguity or fuzziness. Rain-induced attenuation can be handled through the propagation differential phase Φ_{DP}. When attenuation originates from rain–hail mixed precipitation, Φ_{DP} is likely to reflect rain or liquid contribution only, since its half-derivative K_{DP} is rather sensitive to small particles in comparison with dry hail particle size, or because mixed phase particles (in particular a small ice core surrounded by a liquid torus) might appear as large water drops. Note that this assertion is widely admitted, and needs to be confirmed, e.g. by extending to X-band the study by Borowska *et al.* (2011). So, under this hypothesis, exploiting Φ_{DP} helps to determine the attenuation partially, and at least, reduces the ambiguity among the membership functions.

IMPLEMENTATION

In the current operational processing flow, a first classification step separates hydrometeor from non-hydrometeor echoes, using measured Z, co-polar correlation coefficient ρ_{HV}, and textures of the whole set of measured variables. The hydrometeor-typed pixels are then classified into three rough categories: rain, melting snow, and dry snow. Up to now, this basic classification has worked with a fairly satisfactory performance, by using measured and therefore attenuated Z and Z_{DR}, ρ_{HV}, and the 0°C isotherm documented by a NWP model. In the proposed implementation, the hail detection step follows the attenuation correction. For the latter, the rain profiling algorithm ZPHI® (Testud *et al.*, 2000) is used to correct both Z and Z_{DR}. However, this attenuation correction applies for rain only, and is not sufficient in the presence of hail.

Because ZPHI®-derived K_{DP} is valid only in rain and thus erroneous in hail, a standard K_{DP} estimate is preferred, obtained as a least square fit for the measured differential phase. Here, backscattering contribution is ignored, although not negligible at X-band, since Mie scattering applies at smaller sizes than at S- and C-band, in both rain and hail media. However, the large induced error in K_{DP} can be compensated for by reducing the weight of Z–K_{DP} membership functions in the aggregation stage of the classification.

A CASE STUDY: 5 AUGUST 2010 HAIL EVENT

This case has been registered by the French tornadoes and storms observatory KERAUNOS as an outstanding hail event (http://www.keraunos.org/grele-chutes-marquantes-grelons-orage-2010.htm). Hail cells moved easterly along the coast, over-passing Antibes and Cannes at about 12:00 UTC. During this event, some witnesses reported 2 cm-diameter hailstones, while KERAUNOS observatory mentions 5 cm-diameter hailstones.

These hail cells, as observed by the X-band radar at 12:10 UTC, are depicted in Fig. 2 (cells B and C), with maximum measured (i.e. not corrected for attenuation) reflectivity close to 60 dBZ, and characterized by relatively low differential phase shift enhancement. From the radar location, cells B and C are seen beyond a structure exhibiting raw reflectivity factors of around 50 dBZ (among which cell A is depicted), implying attenuation on cells B and C. Note, at the rear of these two cells, the region of textured Φ_{DP} and Z_{DR}, with weak ρ_{HV} and weak reflectivity. It turns out that this signature does not result from lowered SNR (since still >20 dB). Rather, slight radial extensions of Z behind the storm cores suggest that it might be associated with three body scattering, that could indicate the presence of hail in cell C and B.

Figure 3 shows the Z and Z_{DR} fields obtained after attenuation correction by the ZPHI® algorithm. Some negative Z_{DR} values persist (see also Fig. 4, among the black dots), indicating underestimated attenuation correction at the rear of hail-bearing cells. Indeed, ZPHI® fails in retrieving the whole attenuation, since it is adapted to rain only. At least, as mentioned earlier (see "Attenuation correction" subsection), Z and Z_{DR} are partially attenuation corrected. This limits the confidence of our proposed procedure to rain or hail cells with no hail cell located upstream.

Figure 4 shows scatter-plots of the PPI projected on both Z–Z_{DR} and Z–K_{DP} spaces, after attenuation correction by ZPHI®, with delineated membership functions superimposed. These Z–Z_{DR} and Z–K_{DP} plots are reproduced three times, highlighting one of the three cells A, B and C.

As shown in the upper row, cell A is characterized by high Z_{DR} and quite low K_{DP} (width of K_{DP} distribution is rather thick with negative values, but this might be an artefact of K_{DP} computation) indicating the presence of large drops exclusively, with coincident maxima of Z and Z_{DR} (53 dBZ and 5 dB, respectively). Such a large drop cell, with low K_{DP}, and therefore low concentration, is likely to produce weak attenuation beyond, in particular on cell B.

For cell B, the middle row in Fig. 4 clearly shows the presence of large drops (same diagnosis as for cell A), and rain/hail mixture, and hail. Although rain and mixed rain/hail ambiguity arises at first sight, the comment ending subsection "Classification" tends to exclude the presence of pure rain but large drops. Among these three categories, rain-hail produces the strongest reflectivity factors, reaching 65 dBZ or more.

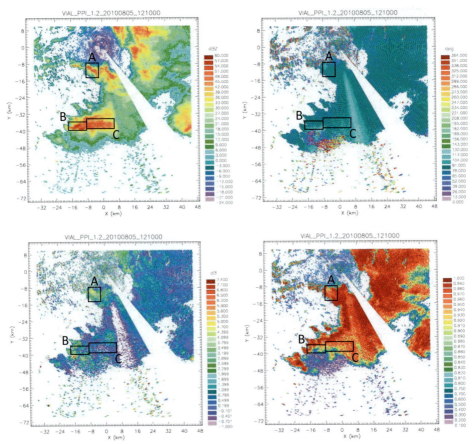

Fig. 2 Fields of attenuated Z (top left), Φ_{DP} (top right), attenuated Z_{DR} (bottom left) and ρ_{HV} (bottom right) measured by the Hydrix™ radar at 12.10 UTC 5 August 2010 at elevation 1.2°. The white wedge to the right of cells B and C in the Z_{DR} field results from off scale values below −1 dB, while the white wedge in all the panels is explained by the transmitter radiation off because of a pylon close to the radar.

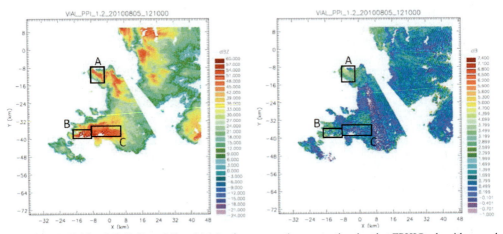

Fig. 3 Fields of Z (left) and Z_{DR} (right) after attenuation correction by the ZPHI® algorithm, and filtered from non-hydrometeor pixels. Light grey spots at the core of cells B and C in the Z field denote values higher than 60 dBZ.

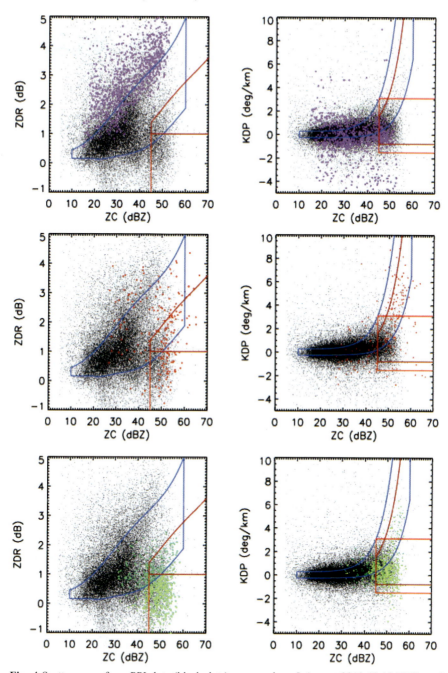

Fig. 4 Scattergrams from PPI data (black dots) measured on 5 August 2010 12:10 UTC at elevation 1.2°, projected onto Z–Z_{DR} (left column) and Z–K_{DP} (right column) spaces. Z and Z_{DR}(partially)ZPHI®. Superimposed coloured scatter plots highlight the cells depicted in the text: cell A (top), cell B (middle), and cell C (bottom).

Cell C (lower row in Fig. 4), which extends cell B to the east, appears composed of hail (light-green) and rain-hail mixture (turquoise blue), according to the same reasoning as for cell B. This is confirmed by depressions of co-polar correlation coefficient in Fig. 2, which result from mixed phase particles in the sampling volumes. One should be aware that observed negative values of Z_{DR} might be due to residual differential attenuation (at worse of 1 dB) not retrieved by ZPHI®. In this case, increasing Z_{DR} values by 1 dB would still produce hail-related classification output, but with more rain-hail mixture.

CONCLUSION

Other hail events have been submitted to the same analysis approach, encouraging the operational implementation of the above-described classification procedure. In the absence of a hailpad network providing ground truth, the resulting hail detection could be confirmed by three body scattering or at least by better accordance with raingauge measurements. Obviously, the hail detection procedure can still be improved (membership function or weight adjustment, K_{DP} processing after Φ_{DP} filtering, special attenuation correction in case of unexpectedly negative Z_{DR}, new variables or estimated parameter consistency like N_0^*).

REFERENCES

Borowska, L., Ryzhkov, A., Zrnić, D., Simmer, C. & Palmer, R. (2011) Attenuation and differential attenuation of 5-cm wavelength radiation in melting hail. *J. Appl. Meteor. Climatol.* 50, 59–76.

Dolan, B. & Rutledge, S. A. (2009) A theory-based hydrometeor identification algorithm for X-band polarimetric radars. *J. Atmos. Oceanic Technol.* 26, 2071–2088.

Heinselman, P. L. & Ryzhkov, A. V. (2006) Validation of polarimetric hail detection. *Weath. Forecasting* 21, 839–850.

Straka, J. M., Zrnić, D. S. & Ryzhkov, A. V. (2000) Bulk hydrometeor classification and quantification using polarimetric radar data: synthesis of relation. *J. Appl. Meteor.* 39, 1341–1372.

Tabary, P., Berthet, C., Dupuy, P., Figueras i Ventura, J., Fradon, B., Georgis, J.-F., Hogan, R., Kabeche, F. & Wasselin, J.-P. (2010) Hail detection and quantification with C-band polarimetric radars: results from a two-year objective comparison against hailpads in the south of France. In: *Proc. 6th European Conf. on Radar in Meteorology and Hydrology* (Sibiu, Romania), 7.46.

Testud, J., Le Bouar, E., Obligis, E. & Ali-Mehenni, M. (2000) The rain profiling algorithm applied to polarimetric weather radar. *J. Atmos. Oceanic Technol.* 17, 332–356.

Vulpiani, G., Tabary, P., Parent-du-Chatelet, J., Bousquet, O., Segond, M.-L. & Marzano, F. S. (2007) Hail detection using polarimetric algorithm at C-band; impact on attenuation correction. In: *Proc. 33rd Conf. on Radar Meteorology* (Cairns, Australia), Amer. Meteor. Soc., P10.3.

Weather Radar and Hydrology
(Proceedings of a symposium held in Exeter, UK, April 2011) (IAHS Publ. 351, 2012).

105

Getting higher resolution rainfall estimates: X-band radar technology and multifractal drop distribution

D. SCHERTZER[1], I. TCHIGUIRINSKAIA[1] & S. LOVEJOY[2]

1 *Université Paris-Est Ecole des Ponts ParisTech LEESU, 6-8 Av Blaise Pascal Cité Descartes, Marne-la-Vallée, 77455 Cx2, France*
daniel.schertzer@enpc.fr
2 *McGill University, Physics Dept., Montreal, PQ, Canada*

Abstract Hydrologists have been waiting for some time to have radar data with a resolution higher than the kilometre scale, especially for urban applications. This is now achievable with the help of polarimetric X-band radars, not only because of their higher frequency, but also because they are much more affordable and versatile. X-band radar networks are thus planned around megalopolises. However, to fully take advantage of the sophisticated polarimetric "self-calibration" requires further investigations of fundamental questions. For instance, ad-hoc homogeneity approximations and/or factorization of the drop distribution have led to the common practice to average several scans, and therefore to degrade the measurement resolution in an attempt to reduce the coherent backscattering due to heterogeneity of the drop distribution. With the help of high-resolution data from an infrared optical spectro-pluviometer, we come back to the question of the insights brought by multifractals on the corresponding statistical bias.

Key words X-band radar; urban hydrology; drop distribution; multifractals

INTRODUCTION

Weather radars remain the only measuring devices that provide space-time estimates of rainfall. However, their classical resolution scale of one kilometre does not meet the relevant scales of urban hydrology (e.g. Berne *et al.*, 2004), especially when there are increasing concerns about the sustainable development of large cities in the context of climate change. On the one hand, urbanisation sprawl induces considerable change in landscape and land-use, and requires more detailed observations and integrated predictions of the water balance. On the other hand, IPCC foresees an increase of hydrological extremes and heat waves (Solomon *et al.*, 2007). A key driver is the extreme variability of the rain field from planetary scales (Lovejoy *et al.*, 2008) to centimetre scales (Lilley *et al.*, 2006): the rain-rate, which is the classical precipitation observable, is strongly scale dependent. This feature corresponds to the fact that rain accumulation is a (mathematical) singular measure (Schertzer *et al.*, 2010). The latter property has many important and practical consequences, especially for small-scale observations. These consequences become even more important due to the recent breakthrough of dual polarimetry (Testud *et al.*, 2000; Le Bouar *et al.*, 2001; Anagnostou *et al.*, 2004; Maki *et al.*, 2005; Matrosov *et al.*, 2005; Berne & Uijlenhoet, 2006). Dual polarimetry provides a radar self-calibration with the help of an estimate of the drop size distribution (DSD) based on the differential reflectivity varying in response to drops flattening with their size. This self-calibration is rather indispensable to removing the rain attenuation bias, which increases with the radar frequency and which has been a long lasting problem (Hitschfeld & Bordan, 1954). X-band radars became thus able to measure, and not only to detect, rainfall. They have several attractive features: reduced transmitted power and antenna size, and reduced sensitivity to ground clutter. These features are particularly attractive in rugged topography and urban areas. For instance, X-net (Maki *et al.*, 2008) is a network around Tokyo of three polarimetric X-band radars run by INED and 4 X-band radars managed by other institutions. Similar networks are to be deployed around 10 other major towns in Japan (Maki *et al.*, 2010) and four polarimetric X-band radars of CASA-IP1 are deployed around Oklahoma City (Chandrasekar *et al.*, 2007). There are several other projects, including one in the Paris region, France, led by the Chair "Hydrology for Resilient Cities" of Ecole des Ponts ParisTech, sponsored by VEOLIA. All these projects are focused on getting higher resolution rainfall estimates and call attention to needed improvements to the present retrieval schemes of rain-rate from reflectivity and to consider more realistic assumptions than those that are currently used.

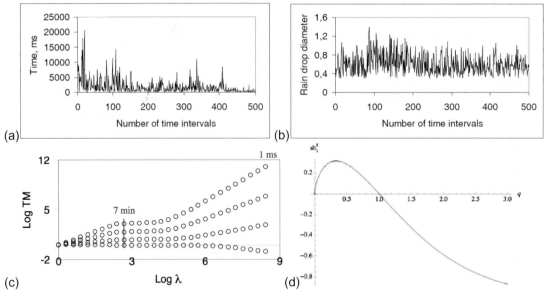

(a) (b) (c) (d)

Fig. 1 Example of an infrared OSP record: (a) series of the time interval durations between drops, (b) corresponding series of the drop diameters, (c) log–log plot of the statistical moments (of orders q = 0.5, 1.5, 2.5, 3.5, from bottom to top) of the drop volumes *vs* the time resolution λ (= $\Delta t/T$, Δt being the observation time scale, T the sample length (= 1 day)). The latter exhibit two clear scaling ranges (power laws, i.e. straight lines): from 1 ms to about 2 s and from 7 min up to 1 day (total sample time). Both ranges, separated by a transition plateau, yield the same universal parameter estimates (equation (19)). The latter are used to obtain (d) a semi-analytical estimate (equation (17)) of the corresponding relative speckle bias $sb_\lambda^{(q)}$ *vs* the statistical moment orders q's. One may note that this bias is already of 33% and 60% for q = 1.5, 2, respectively.

RADAR REFLECTIVIY AND RAINFALL VARIABILITY

Developments of polarimetric retrieval schemes have been mostly focused on getting observables less dependent on the DSD variability, and are based on the differential reflectivity Z_{DR} or the specific differential phase shift K_{DP} in order to get robust correction schemes for the absorption. The basic problem of the "speckle" effect or "drop rearrangement" has remained rather untouched. This effect results from the ubiquitous hypothesis of a homogeneous distribution of the drops that is mathematically convenient to factorize the drop distribution into a Poisson distribution of centres and a translation invariant DSD, although it is physically implausible. This hypothesis was rather compulsory when the computation means were rather limited and radars were mainly used for rain detection (Lawson & Uhlenbeck, 1950), whereas the turbulent wind as well as the coalescence processes tend to cluster drops, rather than to homogeneously distribute them in space and time. This clustering was empirically checked by Lovejoy & Schertzer (1990): the fractal distribution of raindrops on horizontal sections exhibits a fractal dimension of order 1.8, instead of 2 for a homogeneous distribution. This result was disputed by (Gabella *et al.*, 2001; Uijlenhoet *et al.*, 2009) and refined by (Desaulnier-Soucy *et al.*, 2001; Lilley *et al.*, 2006) with the help of 3D multifractal analyses of rain field volumes of about a cubic metre. The impact of the speckle effect was tentatively estimated by Lovejoy *et al.* (1996) with the help of a comparison of the effective reflectivity field Z_e, which is the Fourier transform of the backscatter distribution:

$$Z_{e,\lambda} = \left| \int_{B_\lambda} \sigma_\lambda(\boldsymbol{x}) e^{i k_r \cdot x} dx \right|^2 \tag{1}$$

and the traditional radar reflectivity factor Z_λ:

$$Z_\lambda = \int_{B_\lambda} \sigma_\Lambda(x)^2 \, dx \tag{2}$$

Both reflectivities are written with an apparent continuous integration and non-dimensional variables for convenience: $\lambda = L/l$ denotes the non-dimensional radar resolution, i.e. the scale ratio of a given outer scale L with respect to the radar pulse length l (e.g. $l \approx 100$ m for X-band radars) and B_λ is the corresponding radar pulse volume; the resolution $\Lambda = L/\eta$ corresponds to the inner scale η of the backscatters/rain field variability and the integration is in fact discrete; $\sigma_\Lambda(x) = v(x)/vol(B_\lambda)$ is the relative volume of the drop volume $v(x)$ centred at the non-dimensional location $x = r/L$; k_r is the non-dimensional radar pulse wave-number (i.e. also adimensionalized by L). The classical model corresponds to incoherent small-scale variability (homogeneous distribution of backscatters): the backscatter phases in equation (1) are independent identically distributed random variables and therefore for a large number of drops, the cross terms cancel leading to $Z_{e,\lambda} \approx Z_\lambda$. The deviation from unity of the ratio of the corresponding statistical moment of order q (<.> denotes the ensemble average):

$$sb_\lambda^{(q)} = < Z_{e,\lambda}{}^q > / < Z_\lambda{}^q > -1 \tag{3}$$

measures the speckle bias of order q. To take into account the attenuation, one has to distinguish the apparent (attenuated) radar reflectivity Z_a from the equivalent radar reflectivity Z_e through:

$$Z_a(x) = Z_e(x) \exp\left(-\int_0^x A(s) ds\right) \tag{4}$$

where A is the attenuation rate by unit distance. The classical DSD is defined by an exponential probability distribution (Marshall & Palmer, 1948):

$$N(D) = N_0 \exp(-D/D_m) \tag{5}$$

where N is the concentration by drop size diameter D, N_0 is the "intercept parameter", and D_m is the "median volume" diameter. The mathematical convenience of an exponential distribution is that its statistical moments of order q are all proportional to the corresponding power of the median volume diameter, as for a deterministic distribution (Γ denotes the Euler gamma function):

$$< D^q > = \int dD \ D^q N(D) = N_0 D_m^q \Gamma(m+1). \tag{6}$$

This convenient property may easily turn out to be a weak point, because it corresponds to a case of weak variability. Nevertheless, its practical success is that it easily yields a Z-R relationship for incoherent scattering, because Z corresponds to ($\sigma \propto D^3$):

$$Z \propto \int dD \ D^6 N(D) \tag{7}$$

whereas the rain-rate R corresponds to:

$$R \propto \int dD \ D^3 v(D) N(D) \tag{8}$$

where the terminal velocity is assumed to be scaling:

$$v(D) \propto D^\delta \tag{9}$$

The precise value of the exponent δ is a source of uncertainty: $\delta = 2$ corresponds to the Stokes law (in laminar flows), whereas $\delta = 1/2$ is commonly used in the high Reynolds number regime and according to Atlas & Ulbrich (1977) $\delta = 2/3$ is more accurate. This yields two types of Z-R relationships:

$$Z = aR^b; \ b = 6/(3+\delta), \ a = N_0^{1-b} \Gamma(7)/\Gamma(1+6/b)^b \tag{10}$$

$$Z = AR; \ A = D_m^{6(b-1)/b} \Gamma(7)/\Gamma(1+6/b) \tag{11}$$

that are parameterized by N_0 and D_m, respectively. Unfortunately, none of them is able to predict R

accurately from Z, because both N_0 and D_m have large random fluctuations. In contrast, due to the relation existing between Z_{DR} and D_m, the above Z-R relation parameterized by D_m (equation (11)) can be transformed into a two-parameter estimator $R(Z, Z_{DR})$, which can be less dependent on the DSD variability (Chandrasekar et al., 1990; Gorgucci et al., 1991). As suggested by (Jameson, 1991) and developed by Gorgucci & Scarchilli (1997), K_{DP} can be used as well. A relatively less dependence on the DSD is obtained with the help of slightly more general definitions (Testud et al., 2001) of the parameters N_0 and D_m (then called "mean volume diameter"):

$$D_m = 4 <D^4>/<D^3>; \quad N_0 = 4^4 LWC/(\pi \rho D_m{}^4) \tag{12}$$

that are defined for any DSD. However, they are only two among many possible parameters of a probability distribution and whose physical significance largely depends on the type of the actual DSD. Furthermore, their relative generality does not preclude any speckle effect. The latter has been indirectly estimated by (Le Bouar et al., 2001) for Z_a in the framework of the ZPHI algorithm (Testud et al., 2000) as of the order of ±25% for $\delta A/A$ and ±20% for $\delta R/R$ for 10 "independent" samples of the same segment. These estimates were used to estimate the standard error of Φ_{DP}, which was obtained by averaging over enough adjacent gates (11 in the precise case). This also corresponds to the general practice of averaging on a large enough number of samples to smooth out the phase dependency of Z_e and therefore to become closer to incoherent scattering.

SPECKLE AND MULTIFRACTALITY

Due to the scale dependence of various expressions derived above, the respective importance of various terms could be theoretically deduced from their scaling behaviour. This is particularly simple for the radar reflectivity Z_λ and the rain-rate R_λ because both correspond to an integration of a given power of the relative drop volume σ or of the corresponding diameter D. There are general reasons to believe that both are multifractal as the rain-rate is (Schertzer & Lovejoy, 1987; Lovejoy & Schertzer, 1995). This was analysed by Tchiguirinskaia et al. (2003) on a high-resolution time series of diameters and time intervals obtained by Salles et al. (1998) with the help of an infrared optical spectro-pluviometer (OSP), see Fig. 1(a)–(c). In particular, the statistical moments exhibit multiscaling/multifractality, i.e. for any (positive) order q:

$$\left\langle \sigma_\lambda^q \right\rangle = \lambda^{K_\sigma(q)} \left\langle \sigma_1^q \right\rangle \tag{13}$$

where the moment scaling function $K_\sigma(q)$, with $K_\sigma(0) = 0$ due to the probability normalisation, is convex and nonlinear, except for the exceptional case of mono-/uni-fractality. With the help of the technique of "normalized power densities" (Schertzer et al., 1997, 2002; Schertzer & Lovejoy, 2011) that yields:

$$\left\langle \left(\int_{B_\lambda} \sigma_\lambda{}^p dx \right)^q \right\rangle \propto \lambda^{K_\sigma(q,p)}; \quad K_\sigma(q,p) \equiv K_\sigma(q,p) - qK_\sigma(p) \tag{14}$$

one obtains from equations (2), (7)–(8), (13)–(14):

$$<Z_\lambda^q> \propto \lambda^{K_Z(q)}; \quad K_Z(q) = K_\sigma(q,2); \quad <R_\lambda^q> \propto \lambda^{K_R(q)}; \quad K_R(q) = K_\sigma(q,2/b). \tag{15}$$

On the other hand, due to the fact that the effective reflectivity field Z_e (equation (1)) is the energy spectrum component for the radar wave number k_r and is therefore the Fourier transform of the autocorrelation function of the backscatter field σ_Λ, one obtains, still with the help of the technique of "normalized power densities":

$$<Z_{e,\lambda}^q> \propto k^{K_Z(q)}; \quad K_Z(q) = K_\sigma(q,2) \tag{16}$$

As a consequence, the speckle bias scales like:

$$sb_\lambda^{(q)} = \left(k/\lambda\right)^{K_Z(q)} - 1 \tag{17}$$

To obtain an estimate of this bias, let us consider that the backscatter field σ_Λ is a universal multifractal field (Schertzer & Lovejoy, 1997) for which:

$$K(q,p) \equiv K(qp) - qK(p) = \frac{C_1}{\alpha-1} p^\alpha (q^\alpha - q); \quad K(q) \equiv K(q,1) \tag{18}$$

where the co-dimension of the mean field $C_1 \geq 0$ measures the mean intermittency ($C_1 = 0$ implies that the field is homogeneous) and the multifractality index $0 \leq \alpha \leq 2$ measures how the intermittency varies with various thresholds or statistical moment order ($\alpha = 0$ corresponds to a mono-/uni-fractal, whose intermittency of the extremes and the mean are the same). The following estimates (Tchiguirinskaia *et al.*, 2003) are obtained for the multifractal distribution of raindrop volumes in time:

$$C_{1,\sigma} \approx 0.35; \quad \alpha_\sigma \approx 0.82 \tag{19}$$

yield the speckle bias $sb_\lambda^{(q)}$ displayed in Fig. 1(d), for $k_r/\lambda \approx 3$ by considering the wavelength of an X-band radar $k_r^{-1} \approx 3$ cm and the inter-drop distance $\lambda^{-1} \approx 1$ cm. Using the Marshall-Palmer relationship (equation (10)), one can crudely estimate similar relative biases for the rain-rate, but for respective moment orders $q_R = qb$, with e.g. $b = 18/11 \approx 1.6$. It is worthwhile to note that a more detailed analysis is required for moment orders q_R's larger than the critical order q_D of the rain-rate statistical moment divergence, which is presumably of the order of 3 (Lovejoy & Schertzer, 1995). Therefore, more detailed analyses are required to assess the biases of higher order.

DISCUSSION AND CONCLUSION

Higher resolution rainfall data are now available with the help of polarimetric X-band radars. However, in spite of the increasing sophistication of the retrieval algorithms, the speckle problem remains rather unsolved. Using a multifractal approach, we first obtain a general expression of the speckle bias (equation (17)). Using empirical estimates of the multifractal parameters of the drop volume distribution obtained by Tchiguirinskaia *et al.* (2003) on a high-resolution OSP time series (Salles *et al.*, 1998), we obtain speckle bias estimates already of 33% and 60% for the statistical moments of order $q = 1.5$; 2, respectively, contrary to a previous, global estimate of 25% (Le Bouar *et al.*, 2001). We also point out similar biases for the rain-rate, as well as the necessity to pursue more detailed analyses. Nevertheless, the present results show that there is a necessity to work directly on the effective reflectivity rather than on the standard reflectivity.

Acknowledgements We greatly acknowledge the support of the Chair "Hydrology for Resilient Cities" of Ecole des Ponts ParisTech, sponsored by VEOLIA, as well as of the EU-FP7 project SMARTesT.

REFERENCES

Anagnostou, E. N., Anagnostou, M. N., Krajewski, W. F., Kruger, A. & Miriovsky, B. J. (2004) High-resolution rainfall estimation from X-band polarimetric radar measurements. *J. Hydrometeorology* 5, 110–128.

Atlas, D. & Ulbrich, C. W. (1977) Path- and area-integrated rainfall measurement by microwave attenuation in the 1–3 cm Band. *J. Appl. Meteorol.* 16, 1322–1331.

Berne, A., Delrieu, G., Creutin, J. D. & Obled, C. (2004) Temporal and spatial resolution of rainfall measurements required for urban hydrology. *J. Hydrol.* 299(3–4), 166–179.

Berne, A. & Uijlenhoet, R. (2006) Quantitative analysis of X-band weather radar attenuation correction accuracy. *Natural Hazards and Earth System Sci.* 6(3), 419–425.

Chandrasekar, V., Balakrishnan, N. & Zrnic, D. S. (1990) Error structure of multiparameter radar and surface measurements of rainfall. Part III: Specific differential phase. *J. Atmos. Oceanic Technol.* 7, 621–629.

Chandrasekar, V., Mclaughlin, D. V., Brotzge, J., Zink, M., Philips, B. & Wang, Y. (2007) Distributed collaborative adaptive radar network: The first results from the CASA IP-1 testbed. In: *Proc. International Symposium on X-band Weather Radar Network – Challenge the Severe Storms* (Tsukuba, Japan).

Desaulnier-Soucy, N., Lovejoy, S. & Schertzer, D. (2001) The continuum limit in rain and the HYDROP experiment. *J. Atm. Res.* 59–60, 163–197.

Gabella, M., Pavone, S. & Perona, G. (2001) Errors in the estimate of the fractal correlation dimension of raindrop spatial distribution. *J. Appl. Met.* 40, 664–667.

Gorgucci, E. & Scarchilli, G. (1997) Intercomparison of multiparameter radar algorithms for estimating rainfall rate. In: *28th Conf. on Radar Meteorology* (Am. Meteor. Soc., Austin, Texas), 55–56.

Gorgucci, E., Scarchilli, G. & Chandrasekar, V. (1991) Radar and surface measurements of rainfall during CaPE: 26 July 1991 case study. *J. Appl. Met.* 34, 1570–1577.

Hitschfeld, W. & Bordan, J. (1954) Errors inherent in the radar measurement of rainfall at attenuating wavelengths. *J. Meteor.* 11, 58–67.

Jameson, A. R. (1991) A comparison of microwave techniques for measuring rainfall. *J. Appl. Met.* 30, 32–54.

Lawson, J. L. & Uhlenbeck, G. E. (1950) *Threshold Signals.* McGraw-Hill, New York, USA.

Le Bouar, E., Testud, J. & Keenan, T. D. (2001) Validation of the rain profiling algorithm "ZPHI" from the C-band polarimetric weather radar in Darwin. *J. Atmos. Oceanic Tech.* 18, 1819–1837.

Lilley, M., Lovejoy, S., Desaulnier-Soucy, N. & Schertzer, D. (2006) Multifractal large number of drops limit in rain. *J. Hydrol.* 328, 20–37.

Lovejoy, S., Duncan, M. & Schertzer, D. (1996) Scalar multifractal radar observer's problem. *J. Geophys. Res.* 101(D21), 26479–26492.

Lovejoy, S. & Schertzer, D. (1990) Fractals, rain drops and resolution dependence of rain measurements. *J. Appl. Met.* 29, 1167–1170.

Lovejoy, S. & Schertzer, D. (1995) Multifractals and rain. In: *New Uncertainty Concepts in Hydrology and Water Resources* (ed. by Z. W. Kunzewicz), 62–103. Cambridge University Press, Cambridge, UK.

Lovejoy, S., Schertzer, D. & Allaire, V. (2008) The remarkable wide range spatial scaling of TRMM precipitation. *J. Atmos. Res.* 90, 10–32.

Maki, M., Iwanami, K., Misumi, R., Park S. G., Moriwaki, H., Maruyama, K., Watabe, I., Lee, D.-I., Jang, M., Kim, H.-K., Bringi, V. N., & Uyeda, H. (2005) Semi- operational rainfall observations with X-band multi-parameter radar. *Atmos. Sci. Lett.* 6, 12–8.

Maki, M., Maesaka, T., Kato, A., Shimizu, S., Kim, D.-S., Iwanami, K., Tsuchiya, S., Kato, T., Kikomuri, Y. & Kieda, K. (2010) X-band Polarimetric Radar Networks in Urban Areas. In: *Proc. Sixth European Conf. on Radar in Meteorology and Hydrology.*

Maki, M., Maesaka, T., Misumi, R., Iwanami, K., Suzuki, S., Kato, A., Shimizu, S., Kieda, K., Yamada, T., Hirano, H., Kobayashi, F., Masuda, A., Moriya, T., Suzuki, Y., Takahori, A., Lee, D.-I., Kim, D.-S., Chandrasekar, V. & Wan, Y. (2008) X-band polarimetric radar network in the Tokyo metropolitan area – X-NET. In: *Proc. Fifth European Conf. on Radar in Meteorology and Hydrology.*

Marshall, J. S. & Palmer, W. M. (1948) The distribution of raindrops with size. *J. Meteorology* 5, 165–166.

Matrosov, S. Y., Kingsmill, D. E., Martner, B. E. & Ralph, F. M. (2005) The utility of X-band polarimetric radar for quantitative estimates of rainfall parameters. *J. Hydromet.* 6, 248–262.

Salles, C., Creutin, J. D. & Sempere Torres, D. (1998) The optical spectropluviometer revisited. *J. Atmos. Oceanic Technol.* 15, 1215–1222.

Schertzer, D. & Lovejoy, S. (1987) Physical modeling and analysis of rain and clouds by anisotropic scaling multiplicative processes. *J. Geophys. Res. D* 8(8), 9693–9714.

Schertzer, D. & Lovejoy, S. (1997) Universal multifractals do exist! *J. Appl. Met.* 36, 1296–1303.

Schertzer, D. & Lovejoy, S. (2011) Multifractals, generalized scale invariance and complexity in geophysics. *J. Bifurcation and Chaos* 21(12) 3417–3456.

Schertzer, D., Lovejoy, S. & Hubert, P. (2002) An introduction to stochastic multifractal fields. In: *ISFMA Symp. on Environmental Science and Engineering with related Mathematical Problems* (ed. by A. Ern *et al.*), 106–179, 4, High Education Press, Beijing, China.

Schertzer, D., Lovejoy, S., Schmitt, F., Tchiguirinskaia, I. & Marsan, D. (1997) Multifractal cascade dynamics and turbulent intermittency. *Fractals* 5(3), 427–471.

Schertzer, D., Tchiguirinskaia, I., Lovejoy, S. & Hubert, P. (2010) No monsters, no miracles: in nonlinear sciences hydrology is not an outlier! *Hydrol. Sci. J.* 55(6), 965–979.

Solomon, S., Qin, D., Manning, M., Chen, Z., Marquis, M., Averyt, K. B., Tignor, M. & Miller, H. L. (eds) (2007) Contribution of Working Group I to the Fourth Assessment Report of the Intergovernmental Panel on Climate Change. Cambridge University Press, Cambridge, UK.

Tchiguirinskaia, I., Salles, C., Hubert, P., Schertzer, D., Lovejoy, S., Creutin, J. D. & Bendjoudi, H. (2003) Mulitfractal analysis of the OSP measured rain rates over time scales from millisecond to day. In: *IUGG2003, Sapporo.*

Testud, J., Le Bouar, E., Obligis, E. & Alimehenni, M. (2000) The rain profiling algorithm applied to polarimetric weather radar. *J. Atmos. Oceanic Tech.* 17, 332–356.

Testud, J., Oury, S., Black, R. A., Amayenc, P. & Dou, X. (2001) The concept of normalized distribution to describe raindrop spectra: a tool for cloud physics and cloud remote sensing. *J. Appl. Met.* 40, 118–140.

Uijlenhoet, R., Porra, J. N., Sempere Torres, D. & Creutin, J. D. (2009) Edge effect causes apparent fractal correlation dimension of uniform spatial raindrop distribution. *Nonlin. Processes Geophys.* 16, 287–297.

Weather Radar and Hydrology
(Proceedings of a symposium held in Exeter, UK, April 2011) (IAHS Publ. 351, 2012).

111

Improvement of the dual-frequency precipitation retrieval method for a global estimation of *Z-R* relations

SHINTA SETO[1] & TOSHIO IGUCHI[2]

1 *Institute of Industrial Science, the University of Tokyo, 4-6-1, Komaba, Meguro-ku, Tokyo 153-8505, Japan*
seto@rainbow.iis.u-tokyo.ac.jp
2 *National Institute of Information and Communications Technology, 4-2-1, Nukui-kita-machi, Koganei 184-8795, Japan*

Abstract *Z-R* relations between radar reflectivity factor *Z* and precipitation rate *R* have been used for operational radar measurements, but the relations are known to be highly variable in time and space and also to be dependent on precipitation types. The Dual-frequency Precipitation Radar (DPR), which will be carried on the core satellite of the Global Precipitation Measurement (GPM) mission hopefully from 2013, is expected to instantaneously estimate the 2-moment drop size distribution function and to finally derive global maps of the coefficients of *Z-R* relations. For this big goal, it is necessary to develop an accurate retrieval method for DPR. Mardiana *et al.* developed the iterative backward retrieval method (MA04) without the use of surface reference technique, which may cause significant errors over land. Some previous studies tested MA04 with simple settings of precipitation measurement, and found that MA04 cannot derive the true solution when the precipitation rate is relatively high. In the first part of this study, MA04 was tested with a simulated DPR measurement dataset, which is more realistic than those used in the previous studies. The retrieved surface precipitation rate is evaluated, and it is shown that MA04 has a negative bias which corresponds to 40% of the true precipitation rate. It is also shown that the estimated Z_e (equivalent radar reflectivity factor) by MA04 tends to be smaller at lower range bins, while the true Z_e does not change largely along the range. In the second part of this study, based on MA04, three modified retrieval methods are developed and they are tested with the same simulated DPR measurement dataset. To overcome the shortcomings of MA04, constraints to make vertically stable profiles of Z_e are introduced in the modified methods. In the best method, the bias is limited to 12% of the true precipitation rate.
Key words *Z-R* relation; drop size distribution; spaceborne radar; DPR; GPM

INTRODUCTION

Z-R relations between radar reflectivity factor *Z* and precipitation rate *R* have been used for operational radar measurements, but the relations are known to be highly variable in time and space and are also dependent on precipitation types. The Dual-frequency Precipitation Radar (DPR), which will be carried on the core satellite of the Global Precipitation Measurement (GPM) mission hopefully from 2013, consists of two radars: KuPR (at 13.6 GHz) and KaPR (at 35.5 GHz). As KuPR is similar to the Tropical Rainfall Measuring Mission (TRMM) Precipitation Radar (PR), which has been working for more than 13 years, we are expecting to have a combined long-term precipitation dataset. KaPR is sensitive to weak rainfall or solid precipitation, which are dominant in middle to high latitudes. Moreover, simultaneous measurement by KuPR and KaPR will enable us to instantaneously estimate the 2-moment drop size distribution function and to finally derive global maps of the coefficients of *Z-R* relations. For this big goal, it is necessary to develop an accurate retrieval method for DPR.

Mardiana *et al.* (2004) developed a new retrieval method (Ma04) for the GPM/DPR. Ma04 has the advantage that it does not rely on the surface reference technique (SRT), which is not very accurate, particularly over land. However, previous studies have suggested that Ma04 does not yield correct solutions for medium and heavy rainfall (e.g. Rose & Chandrasekar, 2005). In our previous study (Seto & Iguchi, 2011), we examined the conditions under which Ma04 can yield the correct solution (the conditions are referred to as "the applicability") under simplified precipitation conditions. The applicability of Ma04 should be examined for more realistic rainfall conditions with vertical variation in DSD. The first objective of this study was to evaluate Ma04 quantitatively with a simulation dataset for DPR. From the standard product of Tropical Rainfall Measuring Mission (TRMM) Precipitation Radar (PR), DSDs were derived and measured radar reflectivity factor (Z_m) at the two frequencies of the DPR were simulated. Ma04 and three modified methods were applied for the simulation dataset and the results were evaluated.

A SIMULATION DATASET FOR DPR

General settings

This section introduces the simulation dataset. Only liquid precipitation particles exist in a range bin, and the DSD follows the gamma distribution:

$$N(D) = N_0 D^\mu \exp\left[-(3.67 + \mu)D/D_0\right] \tag{1}$$

where D is particle size (in mm), $N(D)$ is the number density (in $mm^{-1} \cdot m^{-3}$), and N_0 (in $mm^{-1-\mu} \cdot m^{-3}$), D_0 (in mm), and μ are DSD parameters, with μ assumed to be known. Effective radar reflectivity factor Z_e (in $mm^6 \cdot m^{-3}$) (denoted by dBZ_e in decibels) and specific attenuation k (in decibels per km) can be expressed as functions of N_0 and D_0:

$$dBZ_e(N_0,D_0) \equiv 10 \log_{10} Z_e(N_0,D_0) = 10\log_{10}N_0 + F(D_0), \tag{2}$$

$$k(N_0,D_0) = N_0 \times G(D_0) \tag{3}$$

where F and G are functions of D_0, such that:

$$F(D_0) = 10\log_{10}\left\{C_z \int_{D=0}^{\infty} \sigma_b(D)D^\mu \exp[-(3.67+\mu)D/D_0]dD\right\} \tag{4}$$

$$G(D_0) = C_k \int_{D=0}^{\infty} \sigma_t(D)D^\mu \exp[-(3.67+\mu)D/D_0]dD \tag{5}$$

where σ_b is a backscattering cross-section (in mm^2), σ_t is the extinction cross-section (in mm^2), which are given by Mie theory for spherical particles; C_z depends on the refractivity and C_k is a constant for unit conversion.

As shown in Fig. 1, let us assume that DPR observes precipitation downward at the nadir. The number of range bins included in the dataset is denoted by N and the range bins are named range bin 1 to range bin N from top to bottom. The width of the range bin is denoted by L (in km), and L was fixed as 0.25 km throughout this study. The r axis is taken vertically downward so that r (in km) is the distance from the radar. At the top and bottom of range bin i, r is denoted by r_{i-1} and r_i, respectively. Z_m is expressed as dBZ_m in decibels, and its value at range bin i is denoted by $[dBZ_m]_i$, which is given by:

$$[dBZ_m]_i = [dBZ_e]_i - dBA(r_{i-1}) - \alpha \times 2 \times [k]_i \times L. \tag{6}$$

$[dBZ_e]_i$ and $[k]_i$ indicate the values of dBZ_e and k at range bin i, respectively. $dBA(r)$ is the attenuation in decibels which occurs on the two-way paths between the radar and the height of r. Therefore, $dBA(r_{i-1})$ shows two-way radar attenuation to the top of range bin i. $[dBZ_m]_i$ should be affected not only by $dBA(r_{i-1})$ but by attenuation by particles inside range bin i. This is called "internal attenuation" and is given as the third term of the right hand side of equation (6). α is a parameter for the internal attenuation, and theoretically α is larger than 0 and smaller than 1.

Conversion from TRMM/PR standard product to DPR simulation dataset

Range bins with liquid precipitation were extracted from the TRMM/PR standard product (2A25, Version 6; Iguchi et al., 2009). Generally, the target range bins were located between node-D (0°C water) to the clutter-free bottom (CFB). However, if a range bin was judged to be a noise range bin between node-D and CFB, the noise range bin and lower range bins of the same pixel were not used in the subsequent processes.

For each range bin, Z_e is given in the standard product. Then, k can be calculated as $k = \varepsilon \alpha_0 Z_e^\beta$, where α_0 and β depend on the type of precipitation and α_0 also depends on the height. ε is a correction factor of α_0 and may be affected by the SRT.

To retrieve DSD parameters from k and Z_e, taking the ratio of equations (2) and (3) gives:

$$k/Z_e = \frac{G(D_0)}{10^{F(D_0)/10}}. \tag{7}$$

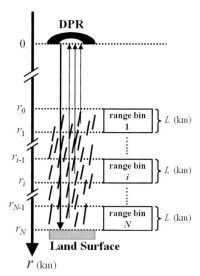

Fig. 1 Schematic representation of DPR measurement assumed in this study.

According to equation (7), k/Z_e is a function of D_0. For D_0 between 0.3 and 3.0 mm, k/Z_e is a monotonously decreasing function of D_0. The dependence of k/Z_e on the particle temperature T may not be negligible for D_0 smaller than 1.0 mm but is small for D_0 larger than 1.2 mm.

D_0 is retrieved from k and Z_e is given by the standard product and T; T is assumed to be 0°C at node-D and to increase by 1.5°C per range bin. Once D_0 is known, N_0 can be determined from equation (2) or (3). Then, [dBZ$_e$] and [k] are calculated by equation (2) and (3) for the frequency of KuPR (13.6 GHz) and KaPR (35.5 GHz), which compose the DPR.
[dBZ$_m$]$_i$ is calculated by equation (6). For this simulation dataset, only liquid precipitation particles were considered; attenuation by other particles was assumed to have already been corrected in [dBZ$_m$]$_i$. Therefore, dB$A(r_{i-1})$ is given as:

$$\mathrm{dB}A(r_{i-1}) = \sum_{j=1}^{i-1} 2 \times [k]_j \times L \qquad (8)$$

where j is a dummy parameter of range bin number. As described in previous studies (e.g. Mardiana *et al.*, 2004), α was set to 1. This setting is obviously an overestimation of internal attenuation, but is required for ease of retrieval. Then, equation (6) can be rewritten as:

$$[\mathrm{dBZ_m}]_i = [\mathrm{dBZ_e}]_i - \mathrm{dB}A\ (r_i) \qquad (9)$$

As DPR actually cannot distinguish signals from noise when [dBZ$_m$] is smaller than 12 dB, range bins with [dBZ$_m$] smaller than 12 dB are regarded as noise range bins and are excluded from the simulation dataset. In the same pixel, range bins located lower than the noise range bin are also excluded because the retrieval algorithms in this study were designed for connected range bins only. After this process, the number of range bins in a pixel is fixed to N. Generally, range bin N does not correspond to the actual land/ocean surface, but this study assumes that the bottom of range bin N is always at the land/ocean surface to define the path integrated attenuation (PIA) as dB$A(r_N)$.

Rain-rate is calculated from DSD with a function of falling velocity (Gunn & Kinzer, 1949):

$$v(D) = 4.854 \times D \times \exp(-0.195D) \qquad (10)$$

Hereafter, for simplicity, "rain-rate" presents that at range bin N.

In the processes discussed above, the simulation dataset is produced from all the orbits of TRMM/PR in July 2001 (490 orbits; the orbit numbers are 20675–21164).

METHODS

Ma04

Ma04 was applied to the simulation dataset. First, PIA was assumed to be 0 dB at the two frequencies. The backward retrieval method (BRM; Meneghini *et al.*, 1992) was applied to retrieve DSD sequentially from range bin N to range bin 1. $dBA(r_0)$ can be calculated from the retrieved DSDs and assumed PIA as:

$$dBA(r_0) = PIA - \sum_{i=1}^{N} 2 \times [k(N_0, D_0)]_i \times L \tag{11}$$

If $dBA(r_0)$ is judged to be 0 dB at the two frequencies, the retrieved DSD becomes the final estimate and the procedure is terminated. Otherwise, PIA is assumed to be $\sum_{i=1}^{N} 2[k(N_0, D_0)]_i L$, and the BRM is applied again; this process is iterated until $dBA(r_0)$ is judged to be 0 dB. Rain-rate estimates are calculated from the final estimates of DSD with equation (10).

SK

As the applicability of Ma04 cannot be applied for heavy precipitation, we developed three modified methods. The first modified method is a stepwise IBRM with constant-k assumption (abbreviated as SK). This method consists of N steps. In the first step, Ma04 is applied to the top range bin; it is assumed that the land/ocean surface is at the bottom of range bin 1 and PIA = $dBA(r_1)$. The first guess of $dBA(r_1)$ is 0 dB at the two frequencies. Estimated $dBA(r_1)$ is denoted by $dBA(r_1)_E$. In the ith step ($2 \leq i \leq N$), Ma04 is applied to the top i range bins. The first guess of $dBA(r_i)$ is given as $dBA(r_{i-1})_E \times i/(i-1)$ as shown in Fig. 2(a). To avoid selecting unnaturally small $[k]$ compared with those of the upper range bins, $[k]_i$ is assumed to be the average of $[k]$ of the upper range bins. Here, the constancy of $[k]$ is assumed, but it does not constrain the estimated $[k]$ after iterations. In the Nth step, all the range bins are targeted as in the original Ma04, but the first guess of PIA is $dBA(r_{N-1})_E \times N/(N-1)$, which is different from Ma04 (0dB).

SZ

The second modified method is a stepwise IBRM with constant-Z_e assumption (abbreviated as SZ). SZ is the same as SK except that the first guess of $dBA(r_i)$ is given by assuming the constancy of $[dBZ_e]$. In the first step, SZ is the same as SK. In the ith step ($2 \leq I \leq N$), SZ is the same as SK, but the first guess of $dBA(r_i)$ is given as $[dBZ_e]_{(i-1)E} - [dBZ_m]_i$ by assuming that $[dBZ_e]_i$ is equal to $[dBZ_e]_{i-1}$ as shown in Fig. 2(b), where $[dBZ_e]_{iE}$ is defined as $[dBZ_e]_{iE} = [dBZ_m]_i + dBA(r_i)_E$.

NKZ

The third modified method is a non-stepwise IBRM with constant-k and Z_e assumption (abbreviated as NKZ). NKZ is not a stepwise method, in contrast to SZ and SK. NKZ is the same as Ma04 except that the first guess of PIA is given by assuming the constancy of $[k]$ and $[dBZ_e]$. With $[dBZ_e]_1 = [dBZ_e]_N$ and $dBA(r_N) = dBA(r_1) \times N$, it follows that:

$$dBA(r_N) = [dBZ_e]_N - [dBZ_m]_N = [dBZ_e]_1 - [dBZ_m]_N = [dBZ_m]_1 + dBA(r_1) - [dBZ_m]_N$$
$$= [dBZ_m]_1 + dBA(r_N)/N - [dBZ_m]_N \tag{12}$$

With the exception of $N = 1$, it is derived from equation (12) that $dBA(r_N) = \{[dBZ_m]_1 - [dBZ_m]_N\} \times N/(N-1)$. This value is used as the first guess of PIA in NKZ. In the case of $N = 1$, the first guess of PIA is set as 0 dB.

In SK and SZ, the final estimates for all the range bins are determined in the Nth step. The results of $(N-1)$ steps are reflected in the first guess of PIA in the Nth step. The difference among the four methods lies only in the first guess of PIA.

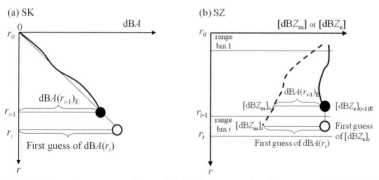

Fig. 2 Schematic representation explaining how to set the first guess of dB$A(r_i)$ in (a) SK and (b) SZ.

Table 1 Evaluation of the four methods with the simulation dataset for July 2001.

All the types	Bias (mm·h^{-1})	Bias (%)	RMSE (mm·h^{-1})
Ma04	−1.21	−40.38	4.66
SK	−0.70	−23.56	4.79
SZ	−0.37	−12.54	3.42
NKZ	−0.77	−25.70	3.92

RESULTS

Rain-rate estimates were evaluated by assuming that the true rain-rate (truth) is directly calculated from the DSD of the simulation dataset. The results for July 2001 are summarized in Table 1.

In Ma04, the bias was −1.21 mm h^{-1}, corresponding to almost 40% of the truth, and the RMSE was 4.66 mm h^{-1}. Thus, the degree of underestimation was severe. Figure 3(a) compares the estimates and the truths, given on the vertical and horizontal axes respectively. Both axes are on a logarithmic scale and range from 0.1 mm h^{-1} (−10dBR) to 100 mm h^{-1} (20dBR). Instead of scatter plots, the population of samples in each 1 dBR × 1 dBR grid is indicated by background grey colours: dark and light grey indicate larger and smaller populations, respectively. Moreover, the averages of the estimated values were calculated for each 1 dBR bin of the truth. The five lines are for the five categorizes of rainfall depth $N \times L = NL$ (km): 0–1, 1–2, 2–3, 3–4, 4–5 km. The estimate was larger than the true value in almost no pixels; instead, the estimates were mostly less than or equal to the true value. The average of the estimates was negatively biased when the true value was smaller than 1 mm h^{-1}, which occurred regardless of rainfall depth because of the multiple solution problem in BRM (Seto & Iguchi, 2011). If $NL = 0$–1 km and the true value is between 1 mm h^{-1} and 10 mm h^{-1}, the average of the estimates was very close to the true value, suggesting that the estimate was almost equal to the truth at instantaneous pixels as no overestimations were found in Ma04. For deeper rainfall, the estimates were close to the true value within a narrower range. If $NL = 4$–5 km, the average of the estimates was always smaller than the true value.

In SK, the bias was reduced to −0.70 mm h^{-1}, but the RMSE of 4.79 mm h^{-1} was larger than that of Ma04. While there was no overestimation in Ma04, severe overestimation (about 10 times larger than the true value) was found at some pixels in SK (Fig. 3(b)).

SZ gave the best performance among the methods, as the bias is −0.37 mm h^{-1} and the RMSE is 3.42 mm h^{-1}. When rainfall depth was smaller than 3 km and rain-rate was between 1 and 20 mm h^{-1}, the average of estimates was very close to the truth (Fig. 3(c)). When rainfall depth was 3–4 km, the range of rain-rates where SZ gave unbiased estimates was between 1 and 10 mm h^{-1}, and was between 1 and 5 mm h^{-1} when rainfall depth was 4–5 km. Heavy rainfall (> 10 mm h^{-1} is difficult to estimate accurately because lighter rainfall tends to be selected in IBRM (Seto & Iguchi, 2011).

Shinta Seto & Toshio Iguchi

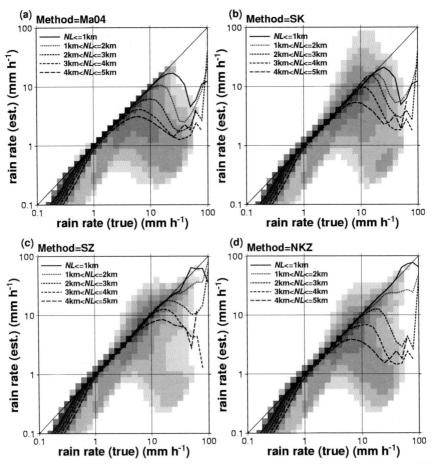

Fig. 3 Evaluation of rain-rate estimates in (a) Ma04, (b) SK, (c) SZ, and (d) NKZ. Five thick lines of different types show the averages of the estimates for different rainfall depth categories. Background grey colours indicate a two-dimensional histogram.

In NKZ, a negative bias of -0.77 mm h^{-1} remained, but the RMSE was 3.92 mm h^{-1}, which was better than those of SK and Ma04 but worse than that of SZ. When NL was smaller than 1 km, the average of the estimates was close to the truth even for heavy rainfall, but severe underestimation is found when NL is larger than 2 km (Fig. 3(d)). The results suggest that the assumption that $[dBZ_e]_1 = [dBZ_e]_N$ and $dBA(r_N) = dBA(r_1) \times N$ are not valid when NL is large.

REFERENCES

Gunn, R. & Kinzer, G. G. (1949) The terminal velocity of fall for water droplets in stagnant air. *J. Meteorol.* 6, 243–248.

Iguchi, T., Kozu, T., Kwiatkowski, J., Meneghini, R., Awaka, J. & Okamoto, K. (2009) Uncertainties in the rain profiling algorithm for the TRMM Precipitation Radar. *J. Meteor. Soc. Japan* 87A, 53–66.

Mardiana, R., Iguchi, T. & Takahashi, N. (2004) A dual-frequency rain profiling method without the use of a surface reference technique. *IEEE Trans. Geosci. Remote. Sens.* 42, 2214–2225.

Meneghini, R., Kumagai, H., Wang, J. R., Iguchi, T. & Kozu, T. (1997) Microphysical retrievals over stratiform rain using measurements from an airborne dual-wavelength radar-radiometer. *IEEE Trans. Geosci. Remote. Sens.* 35, 487–506.

Rose, C. R. & Chandrasekar, V. (2005) A systems approach to GPM dual-frequency retrieval. *IEEE Trans. Geosci. Remote. Sens.* 43, 1816–1826.

Seto, S. & Iguchi, T. (2011) Applicability of the iterative backward retrieval method for the GPM dual-frequency precipitation radar. *IEEE Trans. Geosci. Remote. Sens.* 47 (in print).

Weather Radar and Hydrology
(Proceedings of a symposium held in Exeter, UK, April 2011) (IAHS Publ. 351, 2012).

117

Dual-frequency measurement of rain using millimetre-wave radars: initial results

PETER SPEIRS & DUNCAN ROBERTSON

School of Physics and Astronomy, University of St Andrews, North Haugh, St Andrews, Fife KY16 9SS, UK
pjs27@st-andrews.ac.uk

Abstract This paper presents initial results from an investigation into the feasibility of measuring rain using a pair of horizontally-pointed FMCW radars operating at 38 and 94 GHz. Such a system could potentially offer data complementary to that provided by existing networks of meteorological radars. It may find application where higher resolutions are required, or where a small, portable, low-power system is desirable. The technique used is a variation on the well known dual-frequency extinction technique. Radar-measured rainfall rates and drop-size distributions are compared with data gathered from a disdrometer.

Key words millimetre-wave; radar; rainfall; dual frequency

INTRODUCTION

The need for making multi-parameter measurements to optimally determine rainfall rate has long been recognised, and to this end dual-polarization radars are now beginning to make appearances in national weather radar networks (Doviak *et al.*, 2000). The need for such multi-parameter measurements is even stronger at millimetre-wave frequencies, where Mie-scattering and high levels of attenuation result in far less pronounced differences in received power at different rainfall rates than are found at lower frequencies. Indeed, it has been suggested that rain could be used as a non-varying target for the calibration of 94 GHz radars (Hogan *et al.*, 2003).

Dual-polarization is perhaps the most common form of two parameter measurement, but other techniques are also available, including dual-frequency techniques such as the one used here. The small size of the radar hardware makes a dual-frequency system more practical to implement at millimetre-wave frequencies than it would be at lower frequencies. It has also been shown that the frequency pair chosen should be suitable for the measurement of rain (Meagher & Haddad, 2006).

At St Andrews, interest in the measurement of rain at millimetre-wave frequencies stems primarily from the All-weather Volcano Topography Imaging Sensor (AVTIS) project (Robertson & Macfarlane, 2004), which uses millimetre-wave radar to monitor changing volcanic topography. It has been suggested that some volcanic eruptions may be triggered by heavy rainfall (Matthews *et al.*, 2002), so there is interest in using the existing AVTIS instrument to monitor rainfall in addition to its other tasks.

There has also recently been some interest expressed in using denser networks of smaller, higher frequency radars than are currently used to monitor rain (McLauchlin *et al.*, 2009). At 10 GHz, the radars proposed are of a far lower frequency than those considered here, but it is not inconceiveable that at some stage even higher frequency radars may be of some interest.

The key advantage of such radars is that they can be small – indeed, the need to be portable by hand, car and helicopter is largely what dictated the 94 GHz frequency of AVTIS. Additonally, the choice of Frequency-Modulated Continuous Wave (FMCW) radars means that they can operate at much lower powers than their pulsed counterparts, making battery powered and autonomous remote operation possible. Such high frequency radars potentially also offer far higher resolution than their lower frequency counterparts, albeit over shorter ranges.

This paper considers one possible dual-frequency technique for making measurements at these high frequencies, outlined below. Sample radar data are compared with measurements of rainfall rate from a laser disdrometer. So far only preliminary data have been collected and so only preliminary conclusions can be drawn. However, it appears that while some additional refinement is needed, the technique does work for moderate rainfall rates.

THEORY

The method used here is based on that of (Goldhirsh & Katz, 1974). The objective is to use the difference in attenuation between the two frequencies to determine parameters in the rain Drop-Size Distribution (DSD). In order to do this, it is first necessary to measure the attenuation at the two frequencies. In general, the power return from a radar measuring some beam-filling volumetric target at some range R is given by:

$$P(R) = C \frac{\exp\left(-2 \int_0^R \kappa_{ext}(r)dr\right)}{R^2} \eta(R) \tag{1}$$

where C is a constant for a given radar system, $\eta(R)$ is the reflectivity of the target at R and $\kappa_{ext}(r)$ is the extinction coefficient of the media between the radar and the target. From a single measurement of a single range bin it is not possible to separate out variation in η from variation in κ_{ext}, but if it is assumed that both η and κ_{ext} are constant from some range R_1 to range R_2, then the ratio of the powers reflected from these two range bins will be given by:

$$
\frac{P(R_1)}{P(R_2)} = \frac{C \exp\left(-2\int_0^{R_1}\kappa_{ext}(r)dr\right)\eta R_1^{-2}}{C \exp\left(-2\int_0^{R_2}\kappa_{ext}(r)dr\right)\eta R_2^{-2}} = \frac{R_2^2}{R_1^2}\exp\left[2\int_{R_1}^{R_2}\kappa_{ext}(r)dr\right]
$$

$$
= \frac{R_2^2}{R_1^2}\exp\left[2\kappa_{ext}(R_1)(R_2 - R_1)\right] \tag{2}
$$

which is independent of η. Goldhirsh and Katz use this ratio of the powers at two ranges R_1 and R_2 to obtain an estimate for κ_{ext}. However, fuller use can be made of the data if equation (1) is rewritten as:

$$
\ln(P(R)R^2) = -2\kappa_{ext}(R_1)(R - R_1) - 2\int_0^{R_1}\kappa_{ext}(r)dr + \ln(C\eta(R))
$$

$$
= -2\kappa_{ext}(R_1)R + \left(2\kappa_{ext}(R_1)R_1 - 2\int_0^{R_1}\kappa_{ext}(r)dr + \ln(C\eta(R))\right) \qquad R_1 \le R \le R_2 \tag{3}
$$

where it is assumed that κ_{ext} is constant with value $\kappa_{ext}(R_1)$ from R_1 to R_2. It can be seen that this has the form $y = mx + c$. Thus, if a straight line is fitted to $\ln(PR^2)$ plotted against R from R_1 to R_2, the gradient term will be proportional to a constant approximating the local extinction coefficient. Note that for the y-intercept to be constant, η must also be constant over the range R_1 to R_2.

Assuming a uniform random spatial distribution of raindrops and that their diameters are distributed according to the normalised gamma distribution (Illingworth & Blackman, 2002), the theoretically calculated extinction due to rain can be written as:

$$
\kappa_{ext} = \frac{0.03 N_L (3.67 + \mu)^{\mu+4}}{\Gamma(\mu+4)} \int_0^\infty \left(\frac{D}{D_0}\right)^\mu \exp\left(-(3.67+\mu)D/D_0\right) Q_{ext}(D,f)dD \tag{4}
$$

with Q_{ext} the extinction coefficient for an individual raindrop of diameter D at frequency f, D_0 the median drop diameter for the particular DSD, μ the shape parameter and N_L the number density. Thus, the ratio of the extinction coefficients at the desired frequencies is given by:

$$
\frac{\kappa_{94}}{\kappa_{38}} = \frac{\int_0^\infty \left(\frac{D}{D_0}\right)^\mu \exp\left(-(3.67+\mu)D/D_0\right) Q_{ext}(D, 94\,\text{GHz})dD}{\int_0^\infty \left(\frac{D}{D_0}\right)^\mu \exp\left(-(3.67+\mu)D/D_0\right) Q_{ext}(D, 38\,\text{GHz})dD} \tag{5}
$$

which is dependent only on μ and D_0. Separating these two parameters would be challenging. However, for calculating rainfall rate it can be shown that the variation of μ is insignificant compared with the effect of the variation of D_0 and N_L on both the rainfall rate and the radar reflectivity. The value of μ is fixed at three for these calculations. A numerical determination of the RHS of equation (5) yields the curves shown in Fig. 1. A T-matrix model (Tsang *et al.*, 1985) is used to determine scattering from raindrops assuming the shapes given in (Beard & Chuang, 1987) and assuming the polarizations used by the radars described in "Experimental Methods".

It can be seen that there is some ambiguity in this curve at lower D_0's. At present this is avoided by excluding all D_0's at values less than that of the turning point from the relation used. Such D_0's are only likely to occur in light rain, so for heavier rain this technique should still be applicable. In the future, it may be possible to circumvent this ambiguity by determining if the rain is heavy or light based on the reflectivity, and using this to decide on which side of the turning point to operate.

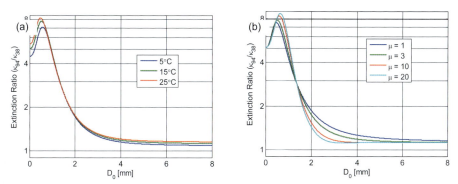

Fig. 1 The variation of the extinction ratio with (a) refractive index change as a function of temperature and (b) μ.

With D_0 determined, it can be substituted back into equation (4) and this used to determine a value for N_L. Two, generally slightly different, values for N_L are obtained for the two frequencies which are averaged to produce the actual value used. With the assumed value for μ the rainfall rate can then be determined by combining the DSD with a terminal fall velocity given by $v = 9.65 - 10.3 \exp(-6D)$ (m s^{-1}) (Atlas *et al.*, 1973).

It would also be possible to use a method similar to this, but dependent on reflectivities rather than attenuation. This would offer better range resolution and would eliminate the assumption that the reflectivity is constant over a long range. However, the precision of radar calibration required, the high levels of path attenuation and the strong dependence of attenuation on rainfall rate at these frequencies make such an approach difficult at present. The attenuation method eliminates any dependence on the radar parameters, as well as dependence on the path attenuation leading up to the volume being sampled. It is for these reasons that the attenuation method has been chosen here.

EXPERIMENTAL METHOD

Two radars were used for these measurements (Fig. 2). The first radar, "Bug-Eyes" (Robertson *et al.*, 2007) is a dual antenna, homodyne, 94 GHz FMCW radar using a frequency multiplied, low phase noise, linearly swept microwave oscillator. The transmit power is 7 dBm and the lens-horn antennas have gains of 40 dBi. The second radar, operating at 38 GHz, which has not previously been reported, is of the same architecture but uses direct FMCW modulation of a varactor-tuned Gunn oscillator. The transmit power is 21 dBm and the lens horn antennas have gains of 34 dBi. The polarizations of the radars were HH for 94 GHz and VV for 38 GHz, which were dictated by

Fig. 2 The 38 (top) and 94 (bottom) GHz radars used to collect the data presented.

available hardware. These polarizations must be taken into account in the post-processing but the actual choice of polarization is believed to be unimportant to the performance of the technique.

Initially two Stanford FFT Analysers sampling at 256 kHz were used to capture range profiles (500 averaged over 5 s). These have been replaced with two data acquisition cards (DAQs), typically sampling at 2048 kHz. Whilst this arrangement updates more slowly than the FFT Analysers (typically 400 range profiles per minute), it captures the raw voltage data, allowing far more flexibility in post-processing. FMCW measurements were made using the up-ramp of a triangle wave for modulation, with 5 ms chirp for the Stanfords and 4.65 ms for the DAQs. The modulation amplitudes, and hence transmit bandwidths, were chosen to give range bins of approximately 0.8–0.9 m in depth for both frequencies. This yielded instrumented maximum ranges of around 400 m and 3–4 km for the Stanford and DAQ cases, respectively. However, attenuation limited the practical maximum ranges in rain to around 800 m, depending on rainfall rate.

The radars were pointed out of an open ground-floor window overlooking a playing field, pointing slightly upwards to clear the tops of trees approximately 120 m away. A Thies laser disdrometer (www.thiesclima.com/disdrometer.html) was placed in the playing field approx. 70 m from the radar and directly underneath the radar beam. A weather station located approx. 600 m from the measurement site was used to provide coarse tipping bucket data to corroborate that from the disdrometer as well as to supply temperature data.

In post-processing, radar measurements from periods where the disdrometer reported no rain were averaged and the resultant "no-rain" range profile subtracted from the radar data to suppress clutter and interference. In a very small number of cases this resulted in negative values for the received power, which were excluded from further processing. Discrete interference lines and transient clutter were further suppressed by thresholding against a local average (similar to constant false-alarm rate techniques) before the data were processed using the method outlined above.

It was also found during processing that in a very small number of cases the interpolation from extinction ratio to D_0 values failed. These measurements were also excluded from the data as there is no apparent way to process them.

RESULTS

A set of sample rainfall rate measurements are shown in Fig. 3(a), and the combined results from all measurement periods to date are shown in Fig. 3(b). While the rainfall rate estimates at low rainfall rates are variable, it can be seen that the radar does successfully measure moderate rainfall rates with a better percentage accuracy. At lower rainfall rates the levels of attenuation that the radar is attempting to measure are very small, so noise and random fluctuations in η are very likely to dominate. It can be seen that the radar-measured rainfall rate does follow the broad trends of the disdrometer data, albeit with some disagreement in absolute values and some variation in time.

Note that in Fig. 3(b) results for disdrometer-determined rainfall rates of below 4 mm h^{-1} have been excluded for clarity, as noise and other fluctuations dominate. Additionally, all cases where

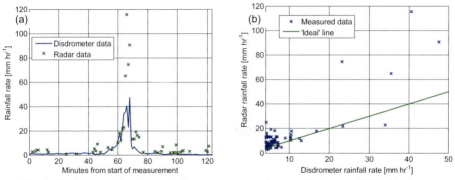

Fig. 3 (a) An example of a typical measurement, taken on 24 August 2010. (b) Comparison of the radar-measured rainfall rates with disdrometer rainfall rates for all measurements taken to date.

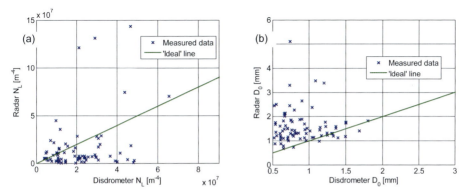

Fig. 4 (a) Comparison of N_L measured with the radar and the disdrometer for all data acquired to date. (b) Comparison of D_0 measured with the radar and the disdrometer for all data acquired to date.

the disdrometer measured a D_0 of <0.5 mm have been excluded as such values are omitted from the D_0/κ_{ext} interpolation curve used in the processing to avoid the ambiguity in the curve. It can be seen that, even with these exclusions, there is considerable variation in the measured data from the ideal line. Perfect conformity would not be expected since the radars and disdrometer are sampling different spatial volumes, but it seems unlikely that this explains all the variability.

For the highest rainfall rates it appears that the high sensitivity to variation in the extinction ratio for the associated larger D_0 values is a significant problem. Very slight errors in the extinction coefficients result in large variations in the calculated rainfall rate in these cases.

Figure 4 shows that the correlations between the radar measured and disdrometer measured D_0 and N_L values are skewed, with the radar data tending to yield under-estimates of N_L but over-estimates of D_0. The radar processing effectively attempts to fit a curve relating the measured extinction coefficient with the weighted integral of the individual sphere extinction coefficients via a specified functional form (equation (4)) to two measured data points. The radar calculations assume a μ of three, but the disdrometer-measured μ was found to have an average value of 1.9 (SD 0.9). In terms of rainfall rate this assumption leads to a maximum possible error of +1%/–5% for all D_0, N_L and μ value triplets with D_0 and N_L ranging from their minimum value to +2 SDs above their disdrometer-measured mean values (for all points not excluded for the reasons outlined above) and μ ranging from the minimum value to its mean +2 SDs for the same data. However, this fixing of μ can still be expected to skew the values of N_L and D_0 to fit as well as possible the imperfect functional form to the measured extinctions and hence also to the actual DSD. The resultant value pairs can still be expected to produce a reasonable (albeit somewhat less good) overall fit to the DSD, resulting in the small maximum errors in the rainfall rate values. Arguably, the presentation is not useful in determining the quality of the fit as the direct comparison of the parameters does not give a good measure of how well the curve produced by the value pair approximates the actual DSD curve. The combined effect of the value pair needs to be considered.

DISCUSSION

This algorithm is capable of making measurements of rain at far higher frequencies than would normally be considered appropriate for the task. More data are required to fully confirm the performance, but the initial results for moderate rainfall rates look promising. However, the technique is not without its limitations: it is not suitable for use at lower rainfall rates and struggles to estimate very high rainfall rates. For the lower rates, it may be appropriate to combine the technique presented here with a single-frequency reflectivity determination of rainfall rate. In low rainfall rates where attenuation is sufficiently low, this should perform acceptably, and certainly provide a better estimate of these low rainfall rates than is provided by the dual-frequency extinction technique. Better results at higher rainfall rates may be achievable by using more linear 38 GHz radar but the very nature of the curves in Fig. 1 means that the technique will never be as accurate for these very large D_0 values as it is for smaller particles.

As this is a two-parameter technique, it is not ideal for making measurements of the normalised gamma DSD, which strictly requires three parameters. However, whilst it may appear to be more applicable to two-parameter Marshall-Palmer type distributions of the form $N(D) = N_0 \exp(-\Lambda D)$, great care must be taken with such an approach as it will considerably over-estimate the number of very small drops present, which contribute significantly to reflectivity at these frequencies.

The results presented are a short range, test of principle demonstration of the millimetre-wave dual-frequency extinction technique. For practical deployment, greater ranges would be needed, requiring higher transmit power than used here, but which is currently commercially available. It would also be necessary to operate in a swept mode. With hardware and processing refinement this technique has the potential to find application in the hydrological science community.

REFERENCES

Atlas, D., Srivastava, R. C. & Sekhon, R. S. (1973) Doppler radar characteristics of precipitation at vertical incidence. *Rev. Geophys.* 11(1), 1–35.

Beard, K. V. & Chuang, C. (1987) A new model for the equilibrium shape of raindrops. *J. Atmos. Sci.* 44(11), 1509–1524

Doviak, R. J., Bringi, V., Ryzhkov, A., Zahrai, A. & Zrnić, D. (2000) Considerations for polarimetric upgrades to operational WSR-88D radars. *J. Atmos. Oceanic Tech.* 17, 257–278.

Goldhirsh, J. & Katz, I. (1974) Estimation of raindrop size distribution using multiple wavelength radar systems. *Radio Sci.* 9, 439–446.

Hogan, R. J., Bouniol, D., Ladd, D. N., O'Conner, E. J. & Illingworth, A. J. (2003) Absolute calibration of 94/95-GHz radars using rain. *J. Atmos. Oceanic Tech.* 20, 572–580.

Illingworth, A. J. & Blackman, T. M. (2002) The need to represent raindrop size spectra as normalized gamma distributions for the interpretation of polarization radar observations. *J. Appl. Met.* 41, 286–297.

Matthews, A. J., Barclay, J., Carn, S., Thompson, G., Alexander, J., Herd, R. & Williams, C. (2002) Rainfall-induced volcanic activity on Montserrat. *Geophys. Res. Lett.* 29(13), 22, doi:10.1029/2002GL014863.

McLauchlin, D., Pepyne, D., Chandrasekar, V., Philips, B., Kurose, J., Zink, M. *et al.* (2009) Short-wavelength technology and the potential for distributed networks of small radar systems. *Bull. Am. Met. Soc.* 12, 1797–1817.

Meagher, J. P. & Haddad, Z. S. (2006) To what extent can raindrop size be determined by a multiple-frequency radar? *J. Appl. Met. Climate* 45, 529–536.

Robertson, D. A. & Macfarlane, D. G. (2004) A 94-GHz dual-mode active/passive imager for remote sensing. In: *Passive Millimetre-Wave and Terahertz Imaging and Technology, Proc. SPIE* 5619, 70–81, Orlando.

Robertson, D. A., Middleton, R. J. & Macfarlane, D. G. (2007) A 94GHz FMCW instrumentation radar. In: *IRMMW-THz2007*, 919–921, Cardiff.

Tsang, L., Kong, J. A. & Shin, R. T. (1985) *Theory of Microwave Remote Sensing.* Wiley, New York, USA.

Weather Radar and Hydrology
(Proceedings of a symposium held in Exeter, UK, April 2011) (IAHS Publ. 351, 2012).

123

Adaptive phased array radar technology for urban hydrological forecasting

C. G. COLLIER[1], M. HOBBY[1], A. BLYTH[1], D. J. McLAUGHLIN[2], J. McGONIGAL[3] & C. P. McCARROLL[4]

1 *National Centre for Atmospheric Science, School of Earth & Environment, University of Leeds, Leeds LS2 9JT, UK*
c.g.collier@leeds.ac.uk
2 *University of Massachusetts, Amherst, Massachusetts, USA*
3 *Raytheon, UK*
4 *University of Massachusetts, Lowell, Massachusetts, USA*

Abstract Current Water Companies Asset Management Programmes (AMP5 2010–2015) will address: sewer flooding and the pollution which may arise from it. Such floods can result in considerable damage and pose health hazards. Forecasting these events, often the result of heavy convective rainfall, to enable preventive action to be taken, is a major challenge. Convective precipitation patterns may change rapidly within a few minutes and improvements to the scanning technology are needed. Phased arrays are used in many defence radars, and are a desirable technology because they do not require maintenance of moving parts and allow flexibility in beam steering. They are also more robust in respect of component failure. An important additional feature is that such antennas can potentially be mounted to the sides of towers and buildings. Phased tilt technology, including its associated signal processing, has not been explored in the context of quantitative precipitation estimation for urban flood forecasting. This is the subject of this paper. Phased arrays are well suited to the application of adaptive scanning which offers great potential for this application.

Key words radar; phase tilt; urban flooding; adaptive scanning; quantitative precipitation estimation

BACKGROUND

In urban areas runoff comes mainly from impervious areas. About 40% of flood damage in England and Wales is due to short duration rainfall (1–6 h), and 80% of recent floods were due to surface water drainage problems. Following the June 2007 floods in South Yorkshire and Gloucestershire, the Government Pitt Report recommended that a high priority area for future work was forecasting and dealing with urban flooding. Current Water Companies Asset Management Programmes (AMP5 2010–2015) will address: sewer flooding and the pollution which may arise from it. Such floods can result in considerable damage and pose health hazards as well as Environment Agency fines of at least £20k per event. Forecasting these events to enable preventive action to be taken is a major challenge.

RESEARCH NEEDS

The objective of this research is to apply adaptive phased array scanning technology to precipitation measurement by weather radar; to develop novel data processing to maximise the benefit derived from the technology; and to investigate how the data collected with such a system can be used to identify flood producing rainfall in urban areas. It is apposite to examine this technology as plans are now being developed and implemented in the Met Office for future radar developments.

Currently, operational weather radars complete a volumetric surveillance scan in about 5 minutes, scanning at a rate of 1–3 revolutions per minute to determine where precipitation is located. Operational radar systems such as in the UK employ an advection correction scheme to allow for changes in precipitation during the 5-min intervals between scans. There is some support for sub 5-min radar sampling, particularly for use in surface water and sewer flood forecasting. However, the quantitative evidence of benefit is currently lacking. Research underway addresses this need head-on, and will provide information on the strategy for radars in support of optimum scanning for hydrological applications. Significant development can occur very quickly in convective storms, which may be missed by the surveillance scans. It is critical therefore for forecasting the severity of events to perform rapid scanning on short time scales in-between the routine scans in order to capture storm advection, growth and decay. Scanning at intervals of (say) one minute may remove the need for advection correction schemes. Urban areas may generate ground clutter and interference which adversely impacts measurements at low altitudes.

Research is needed to: (a) determine the feasibility of rapid phased array scanning for precipitation measurements, and how this can best be implemented into operational forecasting; (b) develop a suitable software algorithm for implementing the rapid scanning; and (c) analyse the phased array radar data to identify their strengths and weaknesses. The participants in the USA's Engineering Research Center for Collaborative Adaptive Sensing of the Atmosphere, CASA) are developing a rapid scanning X-band dual polarisation phased array radar system.

PHASED ARRAY RADARS AND ADAPTIVE BEAM SCANNING

Phased arrays are used in many defence radars, and are a desirable technology because they do not require maintenance of moving parts and enhance flexibility in beam steering. They are also more robust in respect of component failure. An important additional feature is that such antennas can potentially be mounted to the sides of towers and buildings. McLaughlin *et al.* (2009) discuss the specifications of these systems as a work in progress, but this technology has now reached the stage of advanced testing.

An X-band phased array radar has been implemented for research studies of severe convective storms (Bluestein *et al.*, 2010). But so far the potential has not been explored in the context of urban flood forecasting, the subject of this research. Phased arrays are well suited to the application of adaptive scanning which offers great potential for this application.

A radical alternative radar technology and network architecture being developed by CASA is based upon the concept of Distributed Collaborative Adaptive Sensing (DCAS) which potentially involves thousands to tens of thousands of small-size, low-power, low-cost, short-range, densely overlapping weather radars that are electronically linked to amplify their individual capabilities (McLaughlin *et al.*, 2009). A novel feature of the system is that the sensing strategies can be dynamically changed to adapt to the needs of end users and the environment in which they are operating.

The CASA adaptive scanning technology has been developed and tested for forecasting tornadoes (McLaughlin *et al.*, 2009), and Quantitative Precipitation Estimation (QPE) using the mechanically scanning dual polarisation capability providing a single-parameter Kdp-based rainfall product (Wang & Chandrasekar, 2010). However, the use of phased tilt technology in a weather radar system for QPE over rapid response urban drainage areas as proposed here has not been deployed and demonstrated previously.

Rather than attempting to sense the whole atmosphere, the adaptive scanning system is "end-user driven" in that the radar beam is targeted on those locations that the end-users have indicated are most important to their information needs. This has the advantage of allowing the beam to dwell for longer on an area of interest, so increasing the accuracy of the measurements. Also, the time to return to the same feature is considerably reduced.

The new system must be designed to scan one complete surveillance scan at the lowest elevation as rapidly as possible. As an example appropriate to the current Met Office system, the lowest elevation scan takes approx. 20 seconds (3 revolutions per minute is quite possible with the current mechanically scanned radar). The next 40 seconds could be used for targeted sector scanning. This involves commanding the radar to scan a single sector, where a sector is a wedge in azimuth of a particular angular width in the range (60, 360) and a particular compass orientation in the range (0, 360) where 0 is due north. To scan a sector, the radar will sweep horizontally back and forth, starting at the lowest elevation tilt, and stepping up one discrete elevation angle with each sweep, for as many elevations as the radar can achieve in the 40 second time allocated for the scan. Hence, for a sector size of S degrees, a radar can scan $[40 \times (360/20)]/S = 720/S$ elevations.

The radar can therefore only do a 12 elevation volume scan if the sector size is limited to 60 degree, i.e. the radar can only scan about 1/7th of its surrounding volume in the 40 second scanning period. In an end-user responsive system, this limitation introduces both intra-user and inter-user resource conflicts, which have to be dealt with. These timings may be changed depending upon the radar rotational capabilities. However, given that some stability is required in the scanning procedure for comparison with conventional radar systems, it may be necessary to restrict changes in the procedure to periods of 5 minutes or greater. Although the CASA system has been developed using mechanical azimuthally and vertically scanning, the team have now developed a X-band radar which scans electronically in azimuth, and mechanically in elevation. This is the new aspect of the radar system known as a "phased-tilt radar", which is currently the subject of a research proposal being developed in the UK for urban hydrological forecasting.

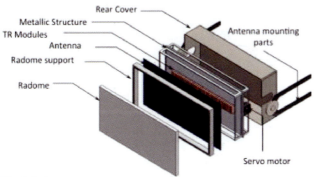

Fig. 1 Antenna.

THE CASA PHASE TILT RADAR

The characteristics of the Phase-Tilt antenna are shown in Fig. 1. The initial proposed Phase Tilt system architecture is shown in Fig. 2, and is summarized as follows:

Radar attributes

- 64 element dual polarization phased array
- 3.6 degree elevation beamwidth
- 1.8 degree beamwidth at broadside, 2.5 degree at ± 45 degree scan; this will need to be reduced in a truly operational system as the beamwidth will be detrimental to QPE correction procedures.
- ±45 degrees of electronic azimuth scanning over +20 degrees of elevation scanning
- Ability to mechanically reposition array over 360 degrees
- Single polarization or Alternating Dual polarization
- 30 km range, 4 kHz maximum PRF, 40 μS max pulse length
- Vaisala RVP900 Signal processor

Signal processing

- Multiple pulses (long, short) to cover blind ranges: Initial setup is scan long, scan short. Vaisala are implementing concatenated long and short pulses, and hope to have the capability by 2012
- Pulsing modes: blocks of single PRF, can be interleaved to implement dual PRF. *Implementing staggered PRF with time domain GMAP at CSU*
- I and Q Doppler processing
- Available data streams: raw I and Q, moment generation with attenuation correction (Z_h, Z_{dr}, velocity, spectral width, differential phase, ρ_{hv}), with or without clutter filter.

Control

– Manual scanning control mode via operator interface
– MC&C based control to mechanically reposition array, set elevation and electronic scan parameters.

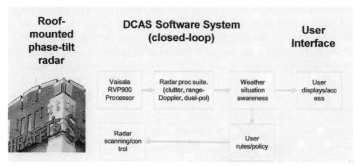

Fig. 2 Phased Tilt radar system architecture (DCAS – Distributed Collaborative Adaptive Sensing).

DEALING WITH END-USER CONFLICTS

Unlike most sensor networks that deliver the same data to all end-users, a radar using adaptive scanning must be able to modify data collection according to end-user requirements. For the UK a system for quantitative precipitation measurement for heavy rainfall and flood warnings in urban catchments using azimuthally-phased array scanning needs to be developed. The data requirements and preferences of the end-users must be tabulated in a set of rules. These rules tell the radar system what to scan, when to scan it, and how to scan it. The information contained will be gathered from a review of best practices, in-depth interviews as discussed in the next section of this paper, and demonstrations with simulated and actual system performance.

A "trigger" parameter will activate what scan sequence to use based upon weather type detections made by numerical weather detection algorithms. These are algorithms that process the weather echoes from previous scans, and their role is simply to identify regions where active weather is occurring. The radar detection will initially be based upon the following types of detection areas where low-level growth of precipitation is occurring, areas of elevated reflectivity not reaching the lowest elevation, areas with storm cells having reflectivity greater than (say) 20 dBZ, areas containing ground clutter and/or anomalous propagation echoes, areas affected by the radar bright-band (the region where snow melts to form rain, areas of possible snowfall, areas of hail detected using an algorithm based upon the reflectivity and the height of the storm, and areas with rotations (probably quite rare in the UK). When a specific detection from the specified algorithm occurs, the indicated rule is triggered to give the end-user's preference for how the region where the detection occurred should be scanned by the radar(s). A "revisit" parameter gives the sample rate, and relates to the horizontal movement of the weather system, its growth rate in size and its evolution. Finally a "sector" parameter gives preferences for how to scan the region that contains the detected weather phenomena.

The problem is to decide what sector to scan during the allocated time slot. A software module will assign two attributes to each weather feature, firstly how important the task is to the end-users, and secondly how well a particular set of sectors meets the scanning requirements of the rule(s). A function combines these two factors, and is optimized in order to define the preferential allocation of the radar system resources so that the optimal configuration of sectors is achieved. The form of the function which is maximised is:

$$J(s) = \sum U(t) I(Q(t,s) > 6Q_{max}(t))$$

where *s* is the sector to be scanned by the radar; *t* are the tasks with the summation being over all tasks; *U* is the function giving the importance of the task to the end-users; *Q* is the function giving the expected scan quality; Q_{max} is the maximum scan quality that could be achieved if task *t* were the only task; and *I* is the indicator function (1 = true or 0 otherwise) (McLauglin *et al.*, 2009).

This *scan quality function* will estimate the level to which the scan selected will satisfy the user requirements. This will involve deriving a function which accounts for how well the scan collects data over the azimuthal range required for the particular user. Then the function must take account of whether all the user elevations required are available in the time available, and finally the range of the user target area from the radar site which is related to the quality of the data.

THE LOCATION OF URBAN FLOODING AND THE SCAN QUALITY FUNCTION

Flooding may arise in several different ways: (i) through the inability of natural water courses to cope with excessive rainfall–fluvial flooding; (ii) through the inability of urban drainage systems (UDS) to cope with excessive rainfall–sewer surcharging; and (iii) direct runoff over land causing local flooding in areas not previously associated with natural or manmade water courses–pluvial flooding. Often these types of flooding occur together as a result of intense local rainfall, e.g. the floods of June 2007 in Sheffield and Tewkesbury. In urban areas the drainage systems may respond to rainfall very rapidly. Figure 3 shows the hydrograph for an urban river in Bolton, northwest England. A radar measuring the rainfall must be capable of doing so within minutes to enable timely hydrograph forecasts to be made in such cases.

Design criteria for urban drainage protection measures were based upon 1 in 10-year return period events; for river flooding 1 in 100-year events; and for coastal defences 1 in 200-year events. Water companies are aware of the parts of their systems likely to flood, and this information will be used to define the radar scanning requirements.

To forecast where intense rainfall will "pond" on the surface away from natural water courses requires detailed knowledge of the ground surface and the urban morphology. The Environment Agency (EA) airborne laser data set provides such information from which the EA has produced a map of the envelope of surface water flooding, assuming no absorption by the sewerage and drainage systems. This information, with other information gathered using a questionnaire approach similar to that shown in Table 1, will be used to help define the scan quality function mentioned in the previous section. The Environment Agency is now working on fine tuning this type of map by using improved models and by looking at the most appropriate basis for planning. The flooding locations are being pre-computed for use with the real-time rainfall forecasts.

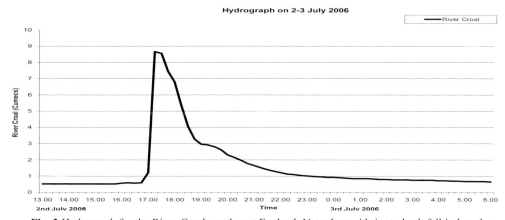

Fig. 3 Hydrograph for the River Croal, northwest England. Note the rapid rise to bank full in less than 30 minutes.

Table 1 Draft questionnaire to help specify the scan quality function for urban flooding.

Number	Question
1	What were the strengths or weaknesses of the radar data?
2	Would adaptive scanning help to identify severe weather?
3	Which scanning strategies are likely to be of most use near thunderstorms / heavy showers, far from thunderstorms / heavy showers, in snowfall and near active frontal systems?
4	What scanning characteristics which were not available would best suit your needs?
5	How did the present scanning strategy impact the spatial and temporal resolution of the data?
6	To what degree did faster update rates come at the expense of loss of useful information?
7	To what degree did the software follow the precipitation system evolution?
8	How did the performance of the software affect your capability to detect and analyse storm features and forecast the response of the urban drainage system?
9	How quickly does flooding occur; what is the time of concentration?
10	Are there situations not experienced during the trial which may be useful to operations?
11	If you issued a warning how did the data impact your warning decision making?
12	What are the radar scanning challenges?

PERFORMANCE ASSESSMENT

In order to assess the performance of the adaptive scanning system using the phased array, comparisons need to be made with data collected using the operational radar network. The Met Office has also offered to integrate the data from the phased-tilt radar, should it become available in a research project currently being developed, into rainfall products that mimic standard operational products supplied to the water industry. By this means, the added benefit derived from the new radar, and key project outcomes, will be readily assessed. The number of scanning tasks will be recorded in case studies representing different types of rainfall events. Analysis of the type and quality of the adaptive scanning data will be carried out.

REFERENCES

Bluestein, B., French, M. M., PopStefanija, I., Bluth, R. T. & Knorr, J. B. (2010) A mobile, phased-array Doppler radar for the study of severe convective storms. The MWR-05XP. *Bull. Am. Met. Soc.* 91, 579–600.

Bringi, V. N. & Chandrasekar, V. (2001) *Polarimetric Doppler Weather Radar Principle and Applications*. Cambridge University Press, USA.

McLauglin, D., Pepyne, D., Chandrasekar, V., Philips, B., Kurose, J., Zink, M., Droegemeier, K., Cruz-Pol, S., Junyent, F., Brotzge, J., Westbrook, D., Bharadwaj, N., Wang, Y., Lyons, E., Hondl, K., Liu, Y., Knapp, E., Xue, M., Hope, A., Kloesel, K., DeFonzo, A., Kollias, P., Brewster, K., Contreras, R., Dolan, B., Djaferis, T., Insanic, E. Frasier, S. & Carr, F. (2009) Short-wavelength technology and the potential for distributed networks of small radar systems. *Bull. Am. Met. Soc.* 90, 1797–1817.

Wang, Y. & Chandrasekar, V. (2010) Quantitative precipitation estimation in the CASA X-band dual-polarisation radar network. *J. Atmos. Ocean. Tech.* 27, 1665–1676.

Quantitative precipitation estimation using commercial microwave links

AART OVEREEM[1,2], HIDDE LEIJNSE[1] & REMKO UIJLENHOET[2]

1 *Royal Netherlands Meteorological Institute (KNMI), PO Box 201, 3730 AE De Bilt, the Netherlands*
overeem@knmi.nl
2 *Hydrology and Quantitative Water Management Group, Wageningen University, PO Box 47, 6700 AA, the Netherlands*

Abstract There is an urgent need for high-quality rainfall observations with high spatial and temporal resolutions in catchment hydrology, particularly in urban hydrology. Weather radars are in principle well-suited for that purpose, but often need adjustment. Usually, only a few raingauge measurements are available as input for hydrological models or to adjust the radar data in real-time. X-band radar data, specifically interesting for urban hydrology, are often not available. Previous studies have shown that (commercial) microwave link data are suitable to calculate path-averaged rainfall intensities and, therefore, are a potentially valuable source of additional rainfall information. This is further explored in this study using data from 321 links from a commercial cellular telephone network in the Netherlands. Some preliminary results are presented concerning the derivation of rainfall maps and the correction of radar data using microwave link data.

Key words microwave link; rainfall measurement; weather radar; the Netherlands

INTRODUCTION

The estimation of rainfall using microwave links from commercial cellular telephone networks is a new and potentially valuable source of information. Such networks cover large parts of the land surface of the earth, have a high density, and are widely used in mobile telecommunication. The data produced by the microwave links in such networks, received powers at one end of a link, is essentially a by-product of the communication between mobile telephones. Recent studies show that rainfall intensities can be estimated from microwave link data (e.g. Messer *et al.*, 2006; Leijnse *et al.*, 2007b). Rainfall attenuates the electromagnetic signals transmitted from one telephone tower to another (Fig. 1). By measuring the received power at one end of a microwave link as a function of time, the path-integrated attenuation due to rainfall can be calculated, from which the path-averaged rainfall intensity can be retrieved.

In this study, data from a commercial cellular telephone network are used to estimate rainfall intensities over the Netherlands (3.55×10^4 km^2). Overeem *et al.* (2011) show that data from this network are suitable for quantitative precipitation estimation in an urban area, the Rotterdam region. This study applies the rainfall retrieval algorithm of Overeem *et al.* (2011) to a network of 321 links covering the Netherlands. First, a description of the commercial microwave link and radar data is given. This is followed by a description of the rainfall retrieval algorithm. Subsequently, two case studies are presented based on country-wide daily rainfall depths from 7 July 2009 08.00 UTC–8 July 2009 08.00 UTC. The first case study considers the derivation of a link-based rainfall map, which is compared to rainfall maps derived from radar and raingauge data. In the second case study, the commercial link data are employed to adjust radar data. The link-adjusted and gauge-adjusted radar rainfall maps are compared. The paper ends with conclusions.

DATA

Commercial microwave link data

Received signal level data were obtained from a commercial microwave link network in the Netherlands. All links with at least 20 h of available data from 7 July 2009 08.00 UTC–8 July 2009 08.00 UTC were selected, resulting in 321 selected single-frequency links. Figure 2(a) displays the locations of the 321 commercial microwave links. The data set contains minimum and

Aart Overeem et al.

maximum received powers over 15-min intervals with a resolution of 0.1 dB, based on 10-Hz sampling. These links have an average data availability of 84.7% and only present part of the network of one provider. The frequencies of the microwave links range from approximately 13 to 39 GHz, and these links typically measure at a height of several tens of metres above the ground.

Fig. 1 Rainfall attenuates the electromagnetic signals transmitted from one telephone tower to another.

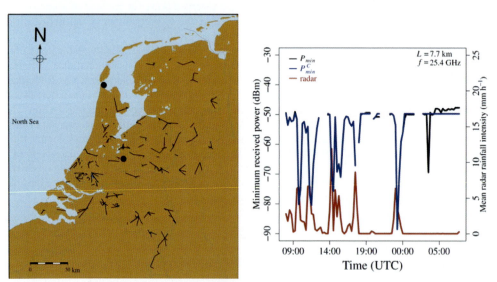

Fig. 2 (a) Map of the Netherlands with the locations of the 321 commercial microwave links (lines) and the two weather radars (dots), and (b) minimum received power (black), corrected power (blue) and mean gauge-adjusted radar rainfall intensity (red) of one link for 7 July 2009 08:00 UTC–8 July 2009 08.00 UTC, for a time step of 15 min. The corrected power has been obtained by using the radar approach to identify wet and dry spells. The selected link is plotted in blue on the map.

Radar data

A radar data set of path-averaged rainfall intensities over each link was constructed, to serve as a reference in the calibration of the rainfall retrieval algorithm. Gauge-adjusted radar data were obtained, following a similar procedure to that in Overeem *et al.* (2009a,b). Horizontal cross sections of the radar reflectivity factor at constant altitude (pseudo-CAPPI images) of 1500 m were utilized with a 5-min temporal and a 1-km spatial resolution. This composite was based on data from the two C-band Doppler weather radars in the Netherlands. After ground clutter removal, the reflectivity factors were converted to rainfall intensities using a fixed *Z-R* relationship. Raingauge networks were employed to adjust the radar data. A daily spatial adjustment, using daily raingauge rainfall depths (~1 station per 100 km^2), was combined with an hourly mean-field bias adjustment, using hourly raingauge rainfall depths (~1 station per 1000 km^2). On average, the adjusted radar rainfall depth is 60% larger than the raw radar rainfall depth for the chosen day. For 88% of the locations the adjusted radar rainfall depth is less than 100% larger than the raw radar rainfall depth. For a more elaborate description of the radar data and the employed adjustment methods, see Overeem *et al.* (2009a,b).

Subsequently, path-averaged rainfall intensities were derived for each link and each 5-min time step. The mean 15-min radar rainfall intensity over a link was calculated by averaging the three 5-min path-averaged radar rainfall intensities. Daily rainfall depths were calculated from these mean 15-min radar rainfall intensities. In a similar way, these mean 15-min rainfall intensities were also derived for the unadjusted radar data, where no raingauges were used to adjust the data (comparable to operational radar rainfall intensities). These unadjusted radar rainfall intensities were only used in the classification of wet and dry spells (the radar approach).

RAINFALL RETRIEVAL ALGORITHM

Several sources of error have been encountered in estimation of rainfall intensities using microwave links (see, e.g. Leijnse *et al.*, 2007a, 2010). Signal fluctuations not related to rainfall often occur and need to be removed. Overeem *et al.* (2011) show that a link approach, in which received signal levels from nearby links are combined, is successful to identify wet and dry spells, but may lead to a decrease in the number of selected links. The performance is similar to the radar approach, in which radar data are used to identify wet and dry spells, and which is adopted in this study. For each link and time step the path-averaged mean 15-min rainfall intensity along the link from unadjusted radar data is used. If this intensity is larger than 0.1 mm h^{-1}, the current and subsequent time step are classified as wet. The subsequent time step is included, because the radar measures at larger heights and it therefore takes some time for hydrometeors to reach the Earth's surface. Figure 2(b) shows the minimum received power (black and blue) over a 24-h period for a single-frequency link and the corresponding mean radar rainfall intensity (red), which can be seen to be strongly negatively correlated.

Another source of error is overestimation of the path-integrated attenuation because part of the decrease in microwave signal power is caused by water films on the antennas. This is taken into account in the rainfall retrieval algorithm, by subtracting a value A_a from the path-integrated attenuation. A low power resolution can deteriorate the accuracy of small rainfall intensities, but is hardly an issue in this study because of the power resolution of 0.1 dB. The temporal sampling strategy determines the number of available samples per unit of time and the applied methodology to obtain these samples. For the commercial data in this study the minimum and maximum received signal powers are available over 15-min intervals, which will lead to sampling errors, because the temporal variability of rainfall intensities can often be large. The rainfall retrieval algorithm incorporates this. The following model is used to compute the corrected path-averaged mean 15-min rainfall intensity (Overeem *et al.*, 2011):

$$\langle R \rangle = \alpha \cdot a \left(\frac{A_{\max} - A_a}{L} \right)^b H\left(A_{\max} - A_a\right) + \left(1 - \alpha\right) \cdot a \left(\frac{A_{\min} - A_a}{L} \right)^b H\left(A_{\min} - A_a\right) \qquad (1)$$

where H is the Heaviside function, L the link length (km), A_a the attenuation due to wet antennas (dB), and α a coefficient that determines the contribution of the minimum and maximum path-integrated attenuation (respectively, A_{min} and A_{max}) during a 15-min period, taking into account the temporal sampling strategy. The reference levels of received powers, P_{min}^{ref} and P_{max}^{ref}, are calculated for each link separately by taking the median of the received powers for all time steps classified as dry. The path-integrated attenuation (dB) is calculated for each time step and link using:

$$A_{min} = P_{max}^{ref} - P_{max}^{C} \tag{2}$$

$$A_{max} = P_{min}^{ref} - P_{min}^{C} \tag{3}$$

where P_{min}^{C} and P_{max}^{C} are the corrected minimum and maximum received powers (dBm), respectively, over 15-min intervals. The 15-min link-based rainfall intensities are accumulated to daily rainfall depths and compared to the corresponding gauge-adjusted radar rainfall depths to find optimal values of A_a and α for a low-frequency and a high-frequency class. These optimal values have been obtained from Overeem *et al.* (2011) and were found using a 18-d data set from June and July 2009, including 7 July 2009 08.00 UTC–8 July 2009 08.00 UTC, using on average 34 links from the Rotterdam region. For more details on the rainfall retrieval algorithm and its calibration, see Overeem *et al.* (2011).

CASE STUDY 1: LINK-BASED COUNTRY-WIDE RAINFALL MAP

The daily rainfall depth of 321 links for 7 July 2009 08:00 UTC–8 July 2009 08:00 UTC are used to obtain rainfall maps. The microwave links measure the average rainfall over a path. For simplicity, the link rainfall depths are assumed to represent a point measurement. The middle of the link is assigned the value of the daily rainfall depth. The daily rainfall depths from gauge and link are interpolated using curvature splines under tension (Smith & Wessel, 1990). The obtained rainfall maps for the Netherlands are shown in Fig. 3, together with the gauge-adjusted radar composite. A substantial part of the country received a rainfall depth of more than 15 mm, sometimes even exceeding 50 mm.

In general, the rainfall patterns of the different observational systems correspond fairly well. The radar reveals far more spatial detail in the daily rainfall depth. The link tends to underestimate rainfall in the southwestern and northern part of the country, which is related to the low link

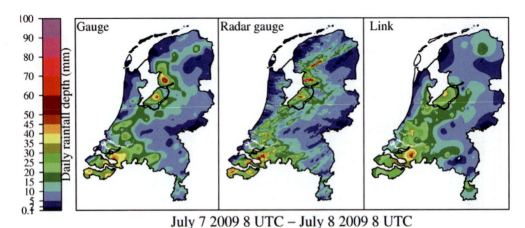

Fig. 3 Daily rainfall depth for 7 July 2009 08.00 UTC – 8 July 2009 08.00 UTC for manual raingauges (interpolated), spatially gauge-adjusted radar composite, and commercial links (interpolated).

density in these regions, as is apparent from Fig. 2(a). The links overestimate rainfall in the middle of the country. Note that this particular day has quite some influence in the calibration of the coefficients of the rainfall retrieval algorithm, but that only about 10% of the link data has been used in the calibration. Because of this, Fig. 3 can almost be considered an independent validation. Unadjusted radar data were only used to identify wet and dry spells for each link.

CASE STUDY 2: LINK-ADJUSTED COUNTRY-WIDE RADAR RAINFALL MAP

Now that the link rainfall map has shown to be in reasonable correspondence with the radar and raingauge rainfall maps, the link data are used to adjust the raw radar composite of daily rainfall depths. Again, the link rainfall depths are considered as point measurements. A daily spatial adjustment of raw radar composites is performed using a modified version of the Barnes adjustment, which is described in Overeem *et al.* (2009a). The spatial adjustment factor F_S^c follows from the ratio of a distance-weighted interpolation (Barnes, 1964) of the microwave link rainfall depths and the interpolation of the corresponding raw radar daily precipitation depths:

$$F_S^c(i,j) = \frac{\sum_{n=1}^{N} w_n(i,j) \times M(i_n, j_n)}{\sum_{n=1}^{N} w_n(i,j) \times R_{raw}^c(i_n, j_n)} \qquad (4)$$

where the subscript S denotes the spatial adjustment method, N is the number of radar-link pairs, and $w_n(i,j)$ is a weighting function, given by:

$$w_n(i,j) = \exp\left[-d_n^2(i,j)/\sigma^2\right] \qquad (5)$$

where σ determines the smoothness of the F_S^c field and $d_n(i,j)$ is the distance between the middle of link n and pixel (i,j). The daily composite of raw accumulated precipitation from the radars is adjusted for each day as:

$$R_S^c(i,j) = R_{raw}^c(i,j) \times F_S^c(i,j) \qquad (6)$$

The same methodology is applied also to the manual raingauge data. See Overeem *et al.* (2009a) for more information on this spatial adjustment method and the chosen parameter values.

Figure 4 reveals a clear underestimation of the daily rainfall depth for the raw radar composite compared to the gauge-adjusted radar composite, which can be considered as the ground-truth. The

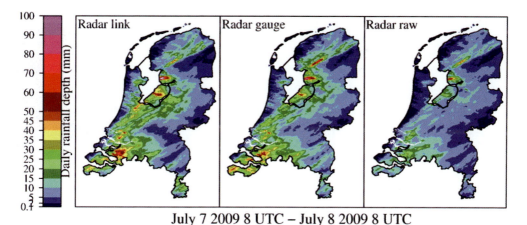

July 7 2009 8 UTC – July 8 2009 8 UTC

Fig. 4 Daily rainfall depth for 7 July 2009 08.00 UTC–8 July 2009 08.00 UTC for spatially link-adjusted radar composite, spatially gauge-adjusted radar composite, and the raw radar composite.

rainfall patterns in the link-adjusted radar composite correspond quite well to those from the gauge-adjusted radar composite. The link adjustment gives rise to overestimation in the middle of the country. It is apparent that some local extreme rainfall events, not that extreme in the raw radar composite, are now present in the link-adjusted radar composite, although their extent is sometimes quite large compared to the gauge-adjusted composite. In conclusion, this case study suggests that commercial microwave link data can in principle be used to adjust radar data.

CONCLUSIONS

Rainfall was estimated using a network of 321 commercial microwave links located in the Netherlands for one summer day with significant rainfall. These preliminary results of ongoing research show that the quality of link-based daily rainfall depths is reasonable and that they are potentially valuable for adjustment of radar data. This indicates that the several sources of error, of which wet antenna attenuation is one, are reasonably corrected for. The next step is to improve the rainfall retrieval algorithm for country-wide use and to verify the methodology using a longer data set. The calibration of the rainfall retrieval algorithm only provides a rough estimate of the wet antenna attenuation. To better estimate the wet antenna attenuation an experiment will be performed using the commercial microwave link data. Another important source of error is the temporal sampling strategy, which requires additional study. Specific attention will also be given to the generation of rainfall maps. The links have been considered as point measurements, which is a first-order approximation. Methods will be developed to optimally use the rainfall data from these path measurements to obtain high-quality rainfall maps. A similar development is needed to improve the adjustment of radar data using microwave link data. To conclude, microwave link data have the potential to improve radar rainfall accumulations, both in real-time as well as for climatological purposes.

Acknowledgements We thank Ronald Kloeg and Ralph Koppelaar from T-Mobile NL for providing the cellular telecommunication link data. We thank Technology Foundation STW for partially funding this research.

REFERENCES

Barnes, S. L. (1964) A technique for maximizing details in numerical weather map analysis. *J. Appl. Met.* 3, 396–409.

Leijnse, H., Uijlenhoet, R. & Stricker, J. N. M. (2007a) Hydrometeorological application of a microwave link: 2. precipitation. *Water Resour. Res.* 43, W04417, doi:10.1029/2006WR004989.

Leijnse, H., Uijlenhoet, R. & Stricker, J. N. M. (2007b) Rainfall measurement using radio links from cellular communication networks. *Water Resour. Res.* 43, W03201, doi:10.1029/2006WR005631.

Leijnse, H., Uijlenhoet, R. & Berne, A. (2010) Errors and uncertainties in microwave link rainfall estimation explored using drop size measurements and high-resolution radar data. *J. Hydrometeor.* 11, doi: 10.1175/2010JHM1243.1.

Messer, H. A., Zinevich, A. & Alpert, P. (2006) Environmental monitoring by wireless communication networks. *Science* 312, 713.

Overeem, A., Holleman, I. & Buishand, A. (2009a) Derivation of a 10-year radar-based climatology of rainfall. *J. Appl. Met. Clim.* 48, 1448, doi:10.1175/2009JAMC1954.1.

Overeem, A., Buishand, A. & Holleman, I. (2009b) Extreme rainfall analysis and estimation of depth-duration-frequency curves using weather radar. *Water Resour. Res.* 45, W10424, doi:10.1029/2009WR007869.

Overeem, A., Leijnse, H. & Uijlenhoet, R. (2011) Measuring urban rainfall using microwave links from commercial cellular communication networks. *Water Resour. Res.* 47, W12505, doi:10.1029/2010WR010350.

Smith, W. H. F. & Wessel, P. (1990) Gridding with continuous curvature splines in tension. *Geophysics* 55, 293–305.

Off-the-grid weather radar network for precipitation monitoring in western Puerto Rico

JORGE M. TRABAL[1,2], GIANNI A. PABLOS-VEGA[2], JOSÉ A. ORTIZ[2], JOSÉ G. COLOM-USTARIZ[2], SANDRA CRUZ-POL[2], DAVID J. MCLAUGHLIN[1], MICHAEL ZINK[1] & V. CHANDRASEKAR[3]

1 *University of Massachusetts, Amherst, Massachusetts, USA*
  jtrabal@ecs.umass.edu
2 *University of Puerto Rico, Mayagüez, Puerto Rico, USA*
3 *Colorado State University, Fort Collins, Colorado, USA*

Abstract Operational weather radars are challenged in providing low-altitude observations of rainfall due to the Earth's curvature and their deployment in "sparse" networks spaced hundreds of km apart. Given this limitation, work is underway to explore the feasibility of "dense" networks of small X-band radars. One approach uses low-cost networks of simple, single-polarization radars that are not dependent on existing infrastructure. This "Off-the-Grid" (OTG) concept is one that might provide a means to monitor rainfall and provide useful data where it is not feasible or cost-effective to deploy more costly and more accurate radars. This paper describes the OTG concept and design, and compares two data events from this network with measurements from an S-Band NEXRAD radar located in Puerto Rico, and rainfall data from a set of raingauges deployed in western Puerto Rico. Results show that OTG radar estimates were consistent with those from the S-band radar.

Key words X-band; radar network; off-the-grid; rainfall mapping and estimation

INTRODUCTION

The high acquisition and recurring cost of weather radars and the associated challenges of deploying radars into the infrastructure motivate today's deployment of "sparse" weather radar networks. Such networks are typically comprised of physically large, high power radars spaced hundreds of km apart. The curvature of the Earth between these radars results in the radars being blocked from observing rainfall and other weather in low altitudes (below ~2 km). This situation applies to the USA operational weather radar network and has been documented in the National Research Council (2002) report for weather technology beyond the present S-band network.

Today's approach to rainfall estimation using radar in the USA involves using observations taken aloft (e.g. 2 km or more above the ground) and extrapolating them toward the ground in order to estimate rainfall amounts close to the ground. This can result in large errors due to large vertical variations in rain-rate. Given the limitations of such techniques, work is underway to explore the feasibility of "dense" networks of small X-band radars (McLaughlin *et al.*, 2009). The concept is to place the radars close enough to one another (e.g. 30 km apart) to defeat Earth curvature blockage and measure rainfall amounts closer to the ground. These efforts show promise, but success will ultimately depend on being able to cost-effectively deploy networks comprised of hundreds, or even thousands, of small radars. The USA Center for Collaborative and Adaptive Sensing of the Atmosphere (CASA) project is presently concentrating on two radar network approaches. One approach uses a network of small dual-polarization radars installed in "Tornado Alley", Oklahoma. The second approach, which is the subject of this paper, uses a network of very low cost, single-polarization radars installed in Puerto Rico. This innovative design grew entirely out of a student-led research systems engineering educational project.

RADAR NETWORK INNOVATION CONCEPT

Much current effort on quantitative precipitation estimation (QPE) in the CASA project and elsewhere focuses on the use of dual-polarization waveforms and related retrieval algorithms (Wang & Chandrasekar, 2010). The innovation in this paper takes a different approach: we recognize that *some degree* of rainfall monitoring close to the ground is better than no monitoring (i.e. the present situation in some parts of the world) and potentially better than monitoring based on extrapolations from observations aloft. We recognize the challenge of designing low-cost radars that are capable of high quality QPE. We also recognize the challenge of deploying large numbers of radars in regions where adequate infrastructure (e.g. power, data backhaul) is either not available or too costly to utilize in a dense network. With these ideas in mind, CASA's student team have designed a radar network based on very low-cost (e.g. <$30 k [USD] in parts), single-polarization radars that are not dependent on any existing power-line, computation, or data communication infrastructure. This low infrastructure requirement, which we call an "Off-the-Grid" concept (Donovan *et al.*, 2006), potentially means the system can provide a means to monitor rainfall and provide useful data in a variety of deployments lacking power, communication, and computation infrastructure. Other groups, such as Pederson *et al.* (2008) and Turso *et al.* (2009), are also experimenting with very low cost X-band radar concepts; ours is differentiated by operating as an OTG network of overlapping radars.

We are exploring proof-of-concept of this design through deployment and experiments conducted in western Puerto Rico, where we operated a network of three such radars during the Central American and Caribbean Games (CAC) in summer 2010 in conjunction with the USA National Weather Service (NWS). Data from our test network has been used to identify small-scale rain cells with better temporal and spatial resolution, and with lower-level coverage, than the operational NEXRAD installation in Puerto Rico. The key to retrieving such useful data was to configure the radars with overlapping coverage, operate them at short range, and use image mosaics to improve the data quality above what would be achieved with a single radar.

OTG RADAR NETWORK DEVELOPMENT

OTG Radar System

The OTG radar has been outfitted with a photovoltaic system (solar power panel, batteries, and a maximum power point track solar charge control), a low power computer, and an A/D processor (see Fig. 1). The solar power subsystem was designed to provide 8 h of OTG radar operation with a full battery charge. A 802.11g wireless antenna was installed in each radar for control and communication between the radar and the server. A parabolic reflector "dish" antenna was installed to increase the radar sensitivity and to reduce ground clutter signal contamination. Due to the small size of the aperture and to reduce rainfall attenuation effects, no radome was installed on the antenna. Figure 1 (table) summarizes the radar specifications. Calibration and verification of the first radar prototype operation was successfully performed (Pablos *et al.*, 2010).

Network development and configuration

As a first attempt for the OTG radar network demonstration and user validation, the NWS asked the CASA team for the possibility to provide support with one or more radars during the CAC Games, to be hosted by the City of Mayagüez. This event provided the opportunity of deploying and demonstrating the new OTG radar technology and sharing weather products with the user community, which was comprised of forecast meteorologists as well as emergency managers in western Puerto Rico. The deployment, thus, served as the first step toward validating the use of this technology in user engagements centred around a demanding situation.

Our test-bed is being designed to provide radar coverage over an area of western Puerto Rico that presently has poor low-altitude (below 2 km) radar coverage provided by the single NEXRAD weather radar in Puerto Rico. The current locations for the three OTG radar nodes that compose the network are shown in the left image of Fig. 2. Each yellow dashed-line circle in Fig. 2 represents a 15 km OTG radar coverage ring, while the red dashed line emphasizes the 100 km coverage range for the NEXRAD radar. A tipping-bucket network of 28 raingauges (blue in Fig. 2) is located within the radars' coverage area and provides comparison data for verification of radar-derived rainfall precipitation estimates. The raingauges have a time resolution of 10 minutes, a minimum recorded rainfall amount of 0.1 mm, and a mean separation of 500 m (Harmsem *et al.*,

Radar Component	Parameter	Value
Transmitter	Operational Frequency	9.4 GHz
	Type	Magnetron
	Peak Power	4 kW or 25 kW
	PRF	600 Hz
	Pulse Length	800 ns
Receiver	Bandwidth	60 MHz
Antenna	Type	Parabolic Reflector
	3 dB Beamwidth	3.8 deg
	Gain	32.4 dB
	Polarization	Linear
	Dish Diameter	0.72 m
	1st Sidelobe Level	22 dB
	Rotation Rate	26 rpm
Data	Spatial/Temporal Res.	15 m/3 min

Fig. 1 CASA OTG radar, solar and wireless sub-systems, and radar processor (left). OTG radar summary list of specifications (right).

2008). Moreover, they are located at distances ranging from 1.2 km to 12 km from the OTG radars, and 108.3 km to 111.6 km from the NEXRAD radar.

The low-altitude sensing gap associated with the NEXRAD radar is illustrated in Fig. 2. The centre of that radar beam is ~2.5 km above the raingauges described above. In comparison, the height of the beam centres of our OTG radars ranges from 300 m to 600 m due to the shorter range. This illustrates why the CASA concept of dense networks of small radars is more capable of providing low-altitude coverage than a "sparse network" of large, long-range radars. As presently deployed, the OTG radar network contributes to filling this sensing gap in western Puerto Rico. In addition to providing improved low-altitude coverage, our OTG network also achieves better spatial resolution (median value of 647 m across the coverage domain, compared to 1846 m for the more distant NEXRAD radar) and better temporal (3 min compared to 5 min) resolution updates. Data integration is performed over 12 transmitted pulses and 10 PPI rotations with an equivalent reflectivity accuracy of about 20%.

DATA VERIFICATION AND RAINFALL ESTIMATION

OTG X-band radar network and S-band radar data evaluation

Data analysis and comparison between radars and raingauges was performed in a Cartesian grid map. The OTG radars were deployed and operated with overlapping coverage so that multiple simultaneous views of the rainfall were obtained with this network. A data merging algorithm was used to convert these multiple views into a single value. This algorithm is based on the selection of maximum reflectivity from the set of OTG radar views of a particular volume. These results were then used to create merged reflectivity images of the weather observed by the OTG network, which were then compared with reflectivity images from the NEXRAD radar and data from the raingauge network. Two convective rainfall events on 3 and 9 August 2010 with duration of 2.5 h and 1.5 h, respectively, were analysed. An example of data collected by each of the OTG radars and the resulting data merge is shown in Fig. 3 for the 9 August 2010 event.

Since the OTG radars are situated in a region of hilly terrain, some of the ray-paths are subject to clutter contamination and some are subject to blockage from elevated regions. A clutter map was used to eliminate the ground clutter contamination. The image of the OTG2 radar (Fig. 3 top centre) illustrates this: pixels lacking reflectivity data can be observed surrounded by pixels having high reflectivity, as a result of the ground clutter elimination algorithm. This effect is mitigated when the data from several OTG radars are combined, as shown in the merged radars image (Fig. 3 bottom left). This image shows how those pixels having "no data" from a particular radar tend to become filled with data from other radars in the network covering the same area. The merged radar

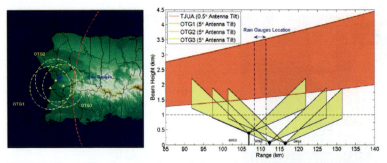

Fig. 2 CASA OTG radar network test bed experimental facilities. (Left) Top view of OTG X-band radars in yellow (15 km range circles), WSR-88D S-Band radar in red (100 km range circle) and raingauges in blue. (Right) Side view of NEXRAD radar gap fill by OTG radar network in Puerto Rico.

network data was obtained by mapping each radar data into a Cartesian map and selecting the data pixel with maximum reflectivity value for each pixel in the grid.

The lower right of Fig. 3 shows the NEXRAD radar image from 9 August 2010 side-by-side with the OTG radar network merged image. The increased spatial resolution of the OTG image is obvious when compared with the image from the larger, more distant radar. The rain cell in the bottom left between the OTG2 and OTG3 coverage circle limits shows an area with low reflectivity, surrounded by a ring of high intensity rain; this detailed structure is not resolved in the S-band radar image owing to the coarser resolution achieved with that operational radar.

Rainfall estimation and verification

In this section, we perform a comparison between rain-rates and accumulated rainfall amounts derived from OTG radar, NEXRAD radar, and raingauges. This is our first look at such a comparison, and we use relatively simple approaches to processing the radar data, such as the merged reflectivity described above, no attempt to compensate for rainfall attenuation, and clutter removal using a clutter map. Given this simplicity, we consider this as only a preliminary comparison. The rain-rate was estimated from the reflectivity measurements of both X- and S-band radars using the NWS Z-R relationship for convective type events $Z = 250R^{1.2}$, where Z is the reflectivity in non-logarithmic units and R is the rain-rate in mm. While the raingauge network is composed of 28 gauges, for the 3 and 9 August 2010 events, only five and four raingauges, respectively, were in operation and collected rainfall data.

The mean rainfall rate and accumulation were estimated over each raingauge, and the mean estimated values are plotted in Figs 4 and 5 for each of the two events. A line plot is included for each of the three X-band radars, the merged X-band data, the S-band radar and the raingauges estimate. For the X-band data, a 500 m area reflectivity mean was performed over each of the raingauges to obtain the rainfall estimates, while for the S-band radar a 3 km mean was selected. This will allow for minimization of any random errors within a single pixel of data.

From Fig. 4, it can be noticed that the OTG1 and OTG2 X-band and S-band radars all overestimate the instantaneous rainfall during the first part of the storm, and they all underestimate during the second part. Partial beam blockage in the direction of some of the raingauges may account for some of the underestimation of the rainfall for OTG2. Moreover, because a maximum reflectivity merging algorithm is used for the X-band network rainfall estimation, in addition to errors due to the difference in the radars' data collection times (>2 min) and calibration, an overestimation of the rainfall estimated from the merged data is expected. During the 9 August 2010 event (Fig. 5), all the radars overestimate the rainfall estimated by the raingauges.

The mean normalized bias and mean normalized standard deviation (normalized using the raingauges storm accumulation) for each of the individual radars and the merged network radar data were calculated for storm accumulations, using each individual raingauges accumulation as an iteration. Table 1 summarizes the results for the radar-gauge comparison for storm total

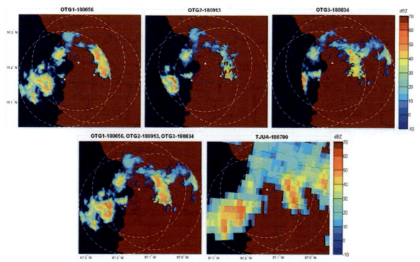

Fig. 3 OTG radar network data for the 9 August 2010 event. Top from left to right: OTG1, OTG2, and OTG3 radars. The bottom figures show a side-by-side data comparison between the OTG radar network merged reflectivity image (left) and the NWS S-band TJUA WSR-88D radar image (right).

accumulations. For the 3 August 2010 event, the best radar-gauge comparison error was obtained from the OTG1 radar with an estimated error of 26%. This result was attributed to the fact that the radar is located outside the city and area of mountain terrain clutter. Moreover, OTG2 resulted with the worst rainfall estimation performance, which we attributed to city clutter contamination and partial beam blockage from surroundings trees, buildings, and mountains in the raingauges' path. OTG3 comparison results show the lowest bias in rainfall estimation. The three X-band radars show a slight underestimation that can be related to some signal attenuation. Furthermore, the merged data estimates resulted in a slight overestimation of the rainfall. The positive bias is attributed to rain cell duplication due to the radars' PPI difference in data collection time (>2 min) and cross-calibration problems. These preliminary results from only two case studies show rainfall accumulation estimates with errors in the order of ~40% or better can be achieved with these low cost and minimal infrastructure OTG X-band radars.

These results need to be interpreted with care, and they should be considered as a preliminary result due to the limited amount of data analysed. Several random errors in the X-band radar data may be impacting these results. These errors include: radar calibration, attenuation, beam blockage, clutter, azimuth beam position calibration and the impacts of limited pulse integration time. We intend to focus on such errors as we go forward and put more work into this analysis.

SUMMARY

This paper provides a description of the OTG weather radar concept and design, and the network configuration deployed in western Puerto Rico. As an indication of the potential for this technology, a preliminary assessment of the use of OTG radar network data for precipitation estimation was performed by comparing single and composite radar data from the network with measurements from the S-band NEXRAD radar, along with rainfall data from a set of raingauges. The OTG network is configured with overlapping coverage and the use of image mosaics. Our results suggest this low cost radar network can provide rainfall estimation with an error of ~40% or better when errors in calibration bias and data quality are assessed. Preliminary results were obtained from two events involving less than two hours of heavy rain, and further events will be Rain-analysed. Moreover, further work on calibration, attenuation compensation, and other potential error sources is ongoing to further explore the capability of this technology.

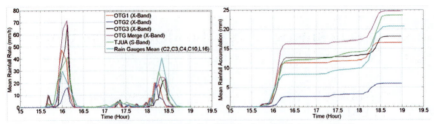

Fig. 4 Rain-rate (left) and event accumulation (right) plots for 3 August 2010.

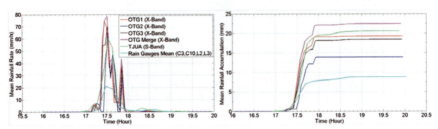

Fig. 5 Rain-rate (left) and event accumulation (right) plots for 9 August 2010.

Table 1 Radar-gauges normalized mean biases and standard deviations for the 3 August 2010 (5 raingauges) and 9 August 2010 (4 raingauges) events rainfall storm totals.

Event day	3 August 2010		9 August 2010	
Radar	Raingauges normalised mean bias	Raingauges normalised std deviation	Raingauges normalised mean bias	Raingauges normalised std deviation
OTG1 (X-band)	−0.19	0.26	1.04	1.34
OTG2 (X-band)	−0.47	0.61	0.84	1.71
OTG3 (X-band)	−0.07	0.48	0.94	1.23
OTG merge (X-band)	0.22	0.45	1.52	1.85
TJUA (S-band)	0.11	0.38	1.33	1.40

Acknowledgements This work was primarily supported by the Engineering Research Centers Program of the National Science Foundation under NSF Award Number 0313747.

REFERENCES

Donovan, B. C., Hopf, A., Trabal, J. M., Roberts, B. J., McLaughlin, D. J. & Kurose, J. (2006) Off-the-grid radar networks for quantitative precipitation estimation. In: *Proc. ERAD,* Barcelona, Spain.

Harmsen, E. W., Gomez-Mesa, S. E., Cabassa, E., Ramirez-Beltran, N. D., Cruz-Pol, S., Kuligowski, R. J. & Vasquez, R. (2008) Satellite sub-pixel rainfall variability. *Intl. J. Systems. App., Eng. & Dev.* 2(3), 91–100.

Pablos-Vega, G. A., Colom-Ustariz, J. G., Cruz-Pol, S., Trabal, J. M., Chandrasekar, V., George, J. & Junyent, F. (2010) Development of an off-the-grid X-band radar for weather applications. In: *Proc. IGARSS* (Hawaii, USA), 1077–1080.

National Research Council (2002) *Weather Technology Beyond NEXRAD*. National Academies Press.

McLaughlin, D. J., Pepyne, D., Chandrasekar, V., Philips, B., Kurose, J., Zink, M., Droegemeier, K., Cruz-Pol, S., Junyent, F., Brotzge, J., Westbrook, D., Bharadwaj, N., Wang, Y., Lyons, E., Hondl, K., Liu, Y., Knapp, E., Xue, M., Hopf, A., Kloesel, K., DeFonzo, A., Kollias, P., Brewster, K., Contreras, R., Dolan, B., Djaferis, T., Insanic, E., Frasier, S. & Carr, F. (2009) Short-wavelength technology and the potential for distributed networks of small radar systems. *Bull. Am. Met. Soc.* 90(12), 1797–1817.

Pedersen, L., Zawadzki, I., Jensen, N. E. & Madsen, H. (2008) Assessment of QPE results from 4 kW X-band local area weather radar (LAWR) evaluated with S-band radar data. In: *Proc. ERAD* (Helsinki, Finland).

Turso, S., Zambotto, M., Gabella, M., Orione, F., Notarpietro, R. & Perona, G. (2009) Microradarnet: An innovative high-resolution low-cost X-band weather radar network. In: *Proc. European Conference on Applications of Meteorology,* (Toulouse, France).

Wang, Y. & Chandrasekar, V. (2010) Quantitative precipitation estimation in the CASA X-band dual-polarization radar network. *J. Atmos. Oceanic Technol.* 27(10), 1665–1676.

Estimation of rain kinetic energy flux density from radar reflectivity factor and/or rain rate

NAN YU[1], GUY DELRIEU[1], BRICE BOUDEVILLAIN[1], PIETER HAZENBERG[2] & REMKO UIJLENHOET[2]

1 *UJF – Grenoble 1 / CNRS / G-INP / IRD, LTHE UMR 5564, Grenoble, France*
nan.yu@ujf-grenoble.fr
2 *Hydrology and Quantitative Water Management Group, Wageningen University, Wageningen, The Netherlands*

Abstract This study offers an approach to estimate the rainfall kinetic energy (KE) by rain intensity (I) and radar reflectivity factor (Z) separately, or jointly, based on the one- or two-moment scaled raindrop size distribution (DSD) formulation, which contains (a) I and/or Z observations, (b) dimensionless probability density function (*pdf*) and (c) some intrinsic parameters. The key point of this formulation is to explain all variability of the DSD by the evolution of observations, hence the *pdf* and intrinsic parameters are considered as constants. A robust method is proposed to estimate the climatic values for these parameters, and our 28-month DSD data are used to test this estimation process. The results show that three relationships (KE-I, KE-Z and KE-IZ) which link the observations (I and/or Z) to rainfall kinetic energy (KE) are well established based on the climatic scaled DSD formulation. In particular, the combination of I and Z yields a significant improvement of estimation of KE.

Key words rain intensity; radar reflectivity factor; raindrop size distribution; kinetic energy

INTRODUCTION

Soil erosion due to rain is a major issue in the agriculture, environment and water management fields. All studies on soil erosion have suggested that increased rainfall amounts and intensities will lead to greater rates of erosion (Parry *et al.*, 2007). In particular, rainfall kinetic energy has often been suggested as an indicator of rainfall erosivity (Fornis *et al.*, 2004; Van Dijk *et al.*, 2005). The kinetic energy flux density KE (J m^{-2} h^{-1}) can be expressed as:

$$KE = 3 \times 10^{-7} \rho \pi \int_0^\infty N(D) \, D^3 \, v_t^3(D) \, dD . \tag{1}$$

The KE describes the energy of raindrops falling on a unit surface during a unit time. It is nearly proportional to the 5th order moment of the raindrop size distribution (DSD).

The purpose of this study is to investigate KE-I, KE-Z and KE-IZ relationships, where I and Z are the rain intensity and the radar reflectivity factor, based on recent DSD scaling formulations which use one moment (Sempere Torres *et al.*, 1994) or two moments (Testud *et al.*, 2001; Illingworth *et al.*, 2002; Lee *et al.*, 2004) to normalize the DSD spectra. It is hoped that the rain intensity (I) associated with the radar reflectively factor (Z) jointly could improve the estimation of the KE. The DSD data used in this study were collected at Alès, situated in the south of France, during the activities of the Cévennes-Vivarais Mediterranean Hydrometeorological Observatory (Delrieu *et al.*, 2005). 28 months of disdrometer data (September 2006–December 2008) are selected as a climatic DSD database. And we integrated them into a 5-min interval dataset. All 5-min spectra with rain intensity <0.5 mm/h were removed. A double tipping-bucket raingauge next to the disdrometer (2 m) measured the rainfall simultaneously. The rainfall measured by the disdrometer is in good agreement with the raingauge observations.

GENERAL DSD SCALING FORMULATION AND RELATIONSHIPS OF MOMENTS

Considering the probability density function (*pdf*), a two-parameter gamma function is introduced to model the DSD scaled by the concentration (N_t) and characteristic diameter (D_c). Then, assuming power law relationships between the scaled variables (N_t and D_c) and available observations (I and/or Z), the one- and two-moment scaled DSD formulations are proposed. According to the 28 month disdrometer measurements, the parameters in the formulations were estimated using a robust method. Finally, three relationships (KE-I, KE-Z and KE-IZ) were established based upon the climatic DSD scaling formulations.

General scaling DSD formulation and moments' relationships

As pointed out by Sempere Torres *et al.* (1998) and Porrà *et al.* (1998), a raindrop-size distribution is a mixture of the concentration N_t (expressed in [m^{-3}]), and a dimensionless probability density function (*pdf*) $g(x)$ [–] of the scaled raindrop diameter x [–] as follows:

$$N(D) = \frac{N_t}{D_c} g(x) \text{ with } x = \frac{D}{D_c} \tag{2}$$

where D_c (mm) is the characteristic diameter of the raindrops. Due to classical observation problems with small raindrops and our interest in high order moments of the DSD ($k = 5$ for the rainfall kinetic energy flux density), the following characteristic diameter was chosen hereafter:

$$D_c = \frac{M_4}{M_3} \tag{3}$$

As was previously proposed by Lee *et al.* (2004), a two-parameter gamma function was selected to model the probability density function (*pdf*) $g(x)$, with:

$$g(x; \mu, \lambda) = \frac{\lambda^{\mu+1}}{\Gamma(\mu+1)} x^{\mu} \exp(-\lambda x) \tag{4}$$

Introducing the *pdf* model (4) into (2), yields a DSD scaling formulation:

$$N(D) = \frac{N_t}{D_c} \left[\frac{\lambda^{\mu+1}}{\Gamma(\mu+1)} x^{\mu} \exp(-\lambda x) \right] \text{ with } x = \frac{D}{D_c} \tag{5}$$

Equation (5) is a DSD formulation with respect to the concept of probability density function (*pdf*). However, the concentration N_t and characteristic diameter D_c can only be measured with a disdrometer at ground level. To extend the application of this DSD model, power law relationships are used to relate N_t and D_c to available DSD moment(s) (e.g. rain rate, radar reflectivity factor).

In case only one moment (M_i) is available, N_t and D_c may be expressed as:

$$N_t = C_i M_i^{\alpha_i} \tag{6}$$

$$D_c = K_i M_i^{\beta_i} . \tag{7}$$

In case two moments (M_i and M_j) are measured, N_t and D_c can be expressed as the following multiple power laws:

$$N_t = C_{ij} M_i^{\alpha_i} M_j^{\alpha_j} \tag{8}$$

$$D_c = K_{ij} M_i^{\beta_i} M_j^{\beta_j} \tag{9}$$

Introducing equations (6), (7) and (8), (9) into DSD formulation (5) yields the one- and two-moment scaled DSD equations (10) and (11), respectively:

$$N(D) = \frac{C_i M_i^{\alpha_i}}{K_i M_i^{\beta_i}} \frac{\lambda^{\mu+1}}{\Gamma(\mu+1)} \left(\frac{D}{K_i M_i^{\beta_i}} \right)^{\mu} \exp(-\lambda \frac{D}{K_i M_i^{\beta_i}}) \tag{10}$$

$$N(D) = \frac{C_{ij} M_i^{\alpha_i} M_j^{\alpha_j}}{K_{ij} M_i^{\beta_i} M_j^{\beta_j}} \frac{\lambda^{\mu+1}}{\Gamma(\mu+1)} \left(\frac{D}{K_{ij} M_i^{\beta_i} M_j^{\beta_j}} \right)^{\mu} \exp(-\lambda \frac{D}{K_{ij} M_i^{\beta_i} M_j^{\beta_j}}) . \tag{11}$$

Integrating these formulations with respect to D^k, any kth moment is expressed as:

$$M_k = \int N(D) D^k dD = \frac{\Gamma(\mu+k+1)}{\Gamma(\mu+1)} C_i K_i^k \frac{M_i^{\alpha_i+k\beta_i}}{\lambda^k} = A_{ik} M_i^{B_{ik}} \tag{12}$$

$$M_k = \int N(D) D^k dD = \frac{\Gamma(\mu+k+1)}{\Gamma(\mu+1)} C_{ij} K_{ij}^k \frac{M_i^{\alpha_i+k\beta_i} M_j^{\alpha_j+k\beta_j}}{\lambda^k} = A_{ijk} M_i^{B_{ik}} M_j^{B_{jk}} . \tag{13}$$

Considering $i = 3.67$ and/or $j = 6$ with $k = 5$, the raindrop kinetic energy flux density *KE* can be expressed by the rain rate I and/or radar reflectivity factor Z. All parameters (C, K, μ, λ, α and β) in these formulations are supposed to be independent of the observation moment(s). Hence, these parameters are considered to be constant in the present study. A robust parameter estimation process is proposed to yield their climatic values.

Estimation of parameters in the one- and two-moments DSD formulation

Despite the use of many parameters in the one- and two-moment DSD scaling formulations (10 and 11), some constraints exist among them. Specifically, for the one-moment scaled DSD formulation (10), by setting $k = i$ in equation (12), one obtains two self-consistency constraints as:

$$\alpha_i + i\beta_i = 1 \tag{14}$$

$$\frac{\Gamma(\mu+i+1)}{\Gamma(\mu+1)} C_i \left(\frac{K_i}{\lambda}\right)^i = 1 \tag{15}$$

Considering the definition of the characteristic diameter D_c and the moment expression (12), a third constraint can be derived as:

$$\frac{M_4}{M_3} = \frac{\mu+4}{\lambda} K_i M_i^{\beta_i} = \frac{\mu+4}{\lambda} D_c \tag{16}$$

which yields a self-consistency relationship between μ and λ:

$$\mu + 4 = \lambda \tag{17}$$

The self-consistency relationships in equations (14), (15) and (17) indicate that only three free parameters in the one-moment scaled DSD formulation (10) need to be estimated. Theoretically, this can be done based on the power law relationships (6) and (7). However, as the concentration N_t is always measured with a large uncertainty due to the limitation of the instrument to detect tiny raindrops, equation (11) cannot be applied in the real application. An alternative method called "ratio of coefficients", similar to the approach of Hazenberg *et al.* (2011), is proposed. The process of estimation for the one-moment formulation is described as follows:

(a) Establish the power law relationships $M_k = A_{ik}M_i^{B_{ik}}$ between all moments M_k ($k = 0$ to 6) and observed moment M_i ($i = 3.67$ or $i = 6$) to derive values of A_{ik} and B_{ik}.
(b) Estimate α and β by the traditional method (Sempere Torres *et al.*, 1994, Uijlenhoet *et al.*, 2003) based on the linear relationship between the exponent values B_{ik} ($k = 0$ to 6) and order k.
(c) Through self-consistency relationship (17), determine the values of K_i, λ and μ from a linear regression analysis on the ratios of consecutive coefficients A_{ik+1} and A_{ik} derived from the moment's relationship (12):

$$\theta_k = \frac{A_{ik+1}}{A_{ik}} = (u+1)\frac{K_i}{\lambda} + k\frac{K_i}{\lambda} \tag{18}$$

(d) Calculate the value of C_i by the self-consistency relationship in equation (15).

Three years of DSD data described in the previous section are applied to estimate the climatic parameters in the one-moment scaled formulation. Figure 1(a) and (b) display the values of B_{ik} as a function of the order of the moment k, for the scaled formulation based on rain-intensity ($i = 3.67$) and radar reflectivity factor ($i = 6$), respectively. In these two cases, a good linear relationship is found, except for the low moments ($k = 0$ and $k = 1$). This is probably explained by the uncertainty associated with the measurement of small raindrops, which is essential to determine the low moments. Hence the 0th and 1st moments are neglected in the estimation of β in

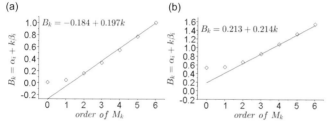

Fig. 1 Relationship between the exponent B_{ik} and the order of moment k, for Z- (a) and I- (b) scaled DSD formulations.

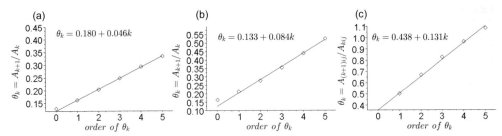

Fig. 2 Relationship between the ratio of consecutive coefficients ($\theta_k = A_{k+1}/A_k$), for Z- (a), I- (b) and ZI- (c) scaled DSD formulations.

the regression analysis. The values of α derived by the intercepts are slightly different from those calculated by the self-consistency relationship (14). This phenomenon has also been demonstrated in previous investigations (Sempere Torres *et al.*, 1994; Uijlenhoet *et al.*, 2003). Based on their experience, the self-consistency value, instead of the intercept, is retained to represent α.

After the estimation of α and β, the exponents B_{ik} in $M_k = A_{ik} M_i^{Bik}$ are slightly changed. By forcing the exponent equal to the new B_{ik}, we adapt the values of the coefficients A_{ik} to coincide with the shift of exponents. Figure 2(a) and (b) show the relationships between the ratios of consecutive coefficients ($\theta_k = A_{ik+1}/A_{ik}$) and the order k, for the scaled formulation based on the rain-intensity ($i = 3.67$) and radar reflectivity factor ($i = 6$), respectively. Again, a good linear relationship is used to estimate K_i, λ, and μ based on the linear equation (18). For the same reason of measurement uncertainty, the first ratio $\theta_0 = A_1/A_0$ is removed from the regression analysis. Finally, C_i is derived from the self-consistency relationship (15). All parameters in the one-moment scaled formulation are listed in Table 1. Introducing these values in the moment expression (12), one obtains two relationships which link the raindrop kinetic energy flux density KE to the rain intensity I and radar reflectivity factor Z, respectively:

$$KE = 10.3 I^{1.285} \tag{19}$$

$$KE = 0.124 Z^{0.803}. \tag{20}$$

Salles *et al.* (2002) derived a similar equation to (19) and pointed out the possible range of exponents is between 1 and 1.4. He suggested the variability of exponents is linked to the type of rain (convective or stratiform). Steiner *et al.* (2000) proposed a linear Z-KE relationship as $KE = 0.0536 Z^{0.91}$. However, as shown in Fig. 3(a),(b), a large uncertainty still exists in these estimations, especially for the *I-KE* relationship. In order to improve the quality of estimation, I and Z are applied jointly. Similar to the previous estimation process, by setting $k = i$ and $k = j$ in equation (13), we find six self-consistency relationships for two-moment scaled DSD formulation:

$$\alpha_i = -\frac{j}{i-j} \tag{21}$$

$$\alpha_j = \frac{i}{i-j} \tag{22}$$

$$\beta_i = \frac{1}{i-j} \tag{23}$$

$$\beta_j = -\frac{1}{i-j} \tag{24}$$

$$\frac{\Gamma(\mu+i+1)}{\Gamma(\mu+1)} C_{ij} \left(\frac{K_{ij}}{\lambda}\right)^i = 1 \tag{25}$$

$$\frac{\Gamma(\mu+j+1)}{\Gamma(\mu+1)} C_{ij} \left(\frac{K_{ij}}{\lambda}\right)^j = 1 \tag{26}$$

Table 1 Estimated parameters in different scaling DSD formulations.

	C	α_i	α_j	K	β_i	β_j	μ	Λ	C^*
I	634.0	0.213		1.116	0.214		0.578	4.578	
Z	1887	−0.184		0.318	0.197		2.923	6.923	
$I+Z$	2232292	2.575	−1.575	0.0997	−0.429	0.429	2.338	6.338	2215286

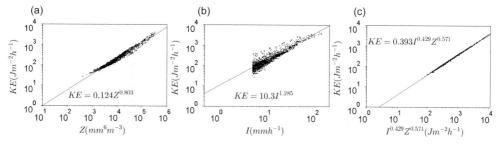

Fig. 3 Raindrop kinetic energy flux density as a function of available observations.

From equations (21) to (24), an interesting feature of the two-moment scaled formulation can be obtained: the exponents of the multiple power-law equations (8) and (9) only depend on the observed moment orders. In fact, the exponents of our two-moment scaling equation (11) are the same as the DSD formulation proposed by Lee *et al.* (2004). However, two supplementary parameters K_{ij} and C_{ij} are added in the formulation (11) to avoid the dependency between the observed moment orders (i, j) and the parameters (μ and λ). Expressing the ratio of the 4th moment to the 3rd moment by equation (13) yields the same self-consistency relationship as equation (17). Considering these self-consistency relationships, the following estimation process is proposed for the two-moment scaled formulation:

(a) Establish the multiple power law relationships $M_k=A_{ijk}M_i^{Bki}M_j^{Bkj}$, based on equation (13), between all moments M_k ($k = 0$ to 6) and observed moments M_i, M_j ($i = 3.67, j = 6$) by forcing the exponent B_{ki} and B_{kj} equal to appropriate values through equations (21)–(24).

(b) With the self-consistency relationship equation (17) and the moment equation (13), determine the values of K_{ij}, λ and μ from a linear regression analysis on the ratio of consecutive coefficients $A_{ij(k+1)}$ and A_{ijk}, as:

$$\theta_k = \frac{A_{(k+1)ij}}{A_{kij}} = (u+1)\frac{K_{ij}}{\lambda} + k\frac{K_{ij}}{\lambda} \tag{27}$$

(c) Calculate the last parameter C_{ij} either from the self-consistency relationship equation (25) or from equation (26). This step allows us to verify our theory by comparing C_{ij} values derived by different self-consistency relationships.

Figure 2(c) illustrates the ratio of consecutive coefficients θ_k as a function of moment order k, derived from the coefficient A_{ijk} in the multiple power law relationships $M_k=A_{ijk}I^{(6-k)/2.33}Z^{(k-3.67)/2.33}$. Again, due to the uncertainty associated with the measurement of small raindrops, the first ratio M_1/M_0 is removed from the linear regression analysis. Then the parameters (K_{ij}, λ, and μ) are estimated from the slope and intercept of the straight solid line. To compare with the one-moment DSD scaling formulation, the parameter values of the *IZ*-scaled DSD formulation are shown in Table 1. A climatological estimator of the raindrop kinetic energy flux density, based on moment expression (13), can be written as:

$$KE = 0.393I^{0.429}Z^{0.571} \tag{28}$$

Figure 3(c) illustrates a robust relationship between KE and $I^{(6-5)/2.33}Z^{(5-3.67)/2.33}$. One can immediately note the improvement in the estimation of KE when I and Z are applied jointly. The three equations (19), (20) and (28) derived from the DSD scaling formulations, represented by the solid lines in Fig. 3, agree well with the observations. The detailed performance of these estimated models are compared in the next section.

COMPARISON OF ESTIMATORS OF RAINFALL KINETIC ENERGY FLUX DENSITY

Four criteria [Bias (Bias), Root mean square error (RMSE), Correlation coefficient (ρ) and Nash-Sutcliffe efficiency (Nash)] between the estimated and observed kinetic energies KE are proposed to compare the performance of the estimation equations (19), (20), and (28). The results are listed in Table 2.

Table 2 Performance of three models for estimating the kinetic energy flux density KE.

Relationships of estimation KE	Bias	RMSE	ρ	Nash
I scaling DSD formulation (19)	1.02	79.24	0.956	0.910
Z scaling DSD formulation (20)	1.01	50.56	0.983	0.963
IZ scaling DSD formulation (28)	1.01	7.01	1.000	0.999

Generally, the four performance criteria exhibit the same tendency. I and Z jointly yield the best estimation of kinetic energy flux density. Concerning single-moment estimation, the radar reflectivity factor Z provides a better estimation than the rain intensity I. This is mainly because the kinetic energy flux density is proportional to the 5th order moment, which is closer to the radar reflectivity factor (6th order moment). The near perfect quality of the estimation from I and Z jointly shows the potential advantage of combining two moments in radar meteorological applications. However, future study is needed to investigate the improvement in operational applications which use real radar and raingauge data, taking all types of errors (instrument error, sampling error, theoretical error, etc.) into account. Such relationships will be aimed at spatializing the KE estimation over domains of typically 50×50 km^2 by using radar reflectivity factor data and/or raingauge data.

REFERENCES

Delrieu, G., *et al.* (2005) The catastrophic flash-flood event of 8–9 September 2002 in the Gard region, France: a first case study for the Cevennes-Vivarais Mediterranean hydrometeorological observatory. *J. Hydrometeor.* 6(1), 34–52.

Fornis, R. L., Vermeulen, H. R. & Nieuwenhuis, J. D. (2004) Kinetic energy-rainfall intensity relationship for Central Cebu, Philippines for soil erosion studies. *J. Hydrol.* 300(1–4), 20–32.

Hazenberg, P., Yu, N., Boudevillain, B., Delrieu, G. & Uijlenhoet, R. (2011) Scaling of raindrop size distributions and classification of radar reflectivity-rain rate relations in intense Mediterranean precipitation. *J. Hydrol.* 402(3-4). 179-192.

Illingworth, A. J. & Blackman, T. M. (2002) The need to represent raindrop size spectra as normalized gamma distributions for the interpretation of polarization radar observations. *J. Appl. Meteorol.* 41(3), 286–297.

Lee, G., Zawadzki, I., Szyrmer, W., Sempere Torres, D. & Uijlenhoet, R. (2004) A general approach to double-moment normalization of drop size distributions. *J. Appl. Meteorol.* 43(2), 264–281.

Parry, M. L., Canziani, O. F., Palutikof, J. P., van der Linden, P. J. & Hanson, C. E. (2007) Contribution of Working Group II to the Fourth Assessment Report of the IPCC, Cambridge University Press, Cambridge, United Kingdom and New York, 189.

Porrà, J. M., Sempere Torres, D. & Creutin, J. D. (1998) Modelling of drop size distribution and its applications to rainfall measurements from radar. In: *Stochastic Methods in Hydrology* (ed. by O. E. Bandorff-Nielsen *et al.*), 73–84. *Advanced Series on Statistics Science and Applied Probability*, Vol. 7, World Scientific Corporation,.

Salles, C., Poesen, J. & Sempere Torres, D. (2002) Kinetic energy of rain and its functional relationship with intensity. *J. Hydrol.* 257, 256–270.

Sempere Torres, D., Porrà, J. M. & Creutin, J. D. (1994) A general formulation for raindrop size distribution. *J. Appl. Meteorol.*, 33(12), 1494–1502.

Sempere Torres, D., Porrà, J. M. & Creutin, J. D. (1998) Experimental evidence of a general description for raindrop size distribution properties. *J. Geophys. Res.* 103, 1785–1797.

Steiner, M. & Smith, J. A. (2000) Reflectivity, rain rate, and kinetic energy flux relationships based on raindrop spectra. *J. Appl. Meteor.* 39, 1923–1940.

Testud, J., Oury, S., Black, R. Amayenc, P. & X. Dou (2001) The concept of "normalized" distribution to describe raindrop spectra: a tool for cloud physics and cloud remote sensing. *J. Appl. Meteorol.* 40(6), 1118–1140.

Uijlenhoet, R., Smith, J. & Steiner, M. (2003) The microphysical structure of extreme precipitation as inferred from ground-based raindrop spectra. *J. Atmos. Sci.* 60(10), 1220–1238.

Van Dijk, A. I J. M., Meesters, A. G. C. A., Schellekens, J. & Bruijnzeel, L. A. (2005) A two-parameter exponential rainfall depth-intensity distribution applied to runoff and erosion modelling. *J. Hydrol.* 300, 155–171.

Variability of rain microphysics using long-term disdrometer observations

TANVIR ISLAM, MIGUEL A. RICO-RAMIREZ & DAWEI HAN

Department of Civil Engineering, University of Bristol, Bristol, UK
tanvir.islam@bristol.ac.uk

Abstract This study explores variability of rain microphysics in terms of drop size distributions (DSD) using seven years of Joss-Waldvogel disdrometer data in a long-term perspective. Firstly, self-consistency evaluation of the disdrometer is performed against four raingauges. The result indicates that the disdrometer derived rain totals are in a good agreement to the raingauges with correlation coefficients ranging from 0.89 to 0.99. In addition, a total of 162 415 one-minute filtered raindrop spectra obtained from the disdrometer are fitted to the normalized gamma DSD model to understand DSD variability in different seasonal and atmospheric states. To characterize rain microphysics, four sets of DSDs are created from the entire raindrop spectra – two are based on seasonal "equinox" criteria and the other two are based on wet bulb temperature. It has been revealed that the normalized gamma DSD parameters, N_w, D_m, and μ vary from set to set because of seasonal and atmospheric variability. Finally, radar Z-R relations for the four DSD sets are developed and it is shown that coefficients differ meaningfully from state to state. In particular, the variability is found to be more substantial between those DSDs which have been separated using the wet bulb temperature.

Key words microphysics of precipitation; drop size distribution; radar remote sensing; reflectivity

INTRODUCTION

Since remote sensing instruments are not capable of observing rainfall directly, the microphysical properties of rain, represented by drop size distributions (DSD), play an important role in rainfall estimation. Commonly, rainfall measurements by radar, either ground-based or satellite-borne, are carried out in terms of a Z-R relation, where Z represents reflectivity factor and R represents rain rate. The Z parameter is a function of the 6th moment of the DSD, and inaccuracy in DSD parameterization contributes error to rain rate estimation. Numerous researchers contended that the Z-R relation needs to be justified based on DSD characteristics of the given region and sequence (Atlas *et al.*, 1999; Maki *et al.*, 2001). Petersen *et al.* (1999) showed that the National Weather Service's conventional $Z = 300\ R^{1.4}$ relation significantly underestimated rainfall during the Fort Collins flood events, whereas, $Z = 250\ R^{1.2}$ estimated rain rates were in a good agreement with gauge rain accumulations. Satellite-based rainfall estimation is also dependent upon DSD significance. Iguchi *et al.* (2009) stated that one of the main causes of rainfall retrieval uncertainty using the precipitation radar aboard TRMM is the misapprehension of DSD characteristics. In addition to the need to understand rain microphysics for algorithm development, calibration of remote sensing instruments also heavily depends on the DSD phenomenon. Attenuation correction at high frequency microwave signals also requires a good understanding of the DSDs during extreme events.

It is well documented that the DSDs vary with respect to time and space due to complex atmospheric and climatologic systems. In this study, an extensive seven years of disdrometer data are analysed to understand characteristics of DSDs in different seasonal and atmospheric scenarios of the mid-latitude UK region. In fact, this is the first attempt in understanding DSD characteristics of the UK region in a long-term perspective. In addition, variability of Z-R relationships in diverse seasonal and atmospheric conditions is also deliberated on this paper.

DATA SOURCES AND PROCESSING

Dataset

The disdrometer used in this study is an impact type Joss-Waldvogel (JW) RD-69 disdrometer, located in the Chilbolton Observatory, in the southern region of UK (51.1445°N, 1.437°W). The data are taken from April 2003 to March 2010; with the exception during mid-August 2004 to mid-

December 2004 and July 2005 to May 2006 when the instrument was returned to the manufacturer for maintenance. To evaluate self-consistency of the disdrometer, we have used four nearby raingauges from the disdrometer (within 10 m distance), located in the Chilbolton observatory. One of them is a tipping-bucket raingauge (TBR), and the remaining three are of drop counting types, denoted as RG2, RG1, RG3 and RG4, respectively. In addition, in this study, the wet bulb temperature has been used to partition DSDs distinguishing atmospheric conditions. The data are taken from the nearest Met Office – MIDAS Land Surface Observation Station located in Middle Wallop (51.1493°N, 1.56851°W). Any missing values are filled in from a nearby station located in the Southampton Oceanography Centre (50.8918°N, 1.3961°W).

DSD parameterization

In general, the concentrations of raindrops within a unit volume are expressed as DSD and measured from a disdrometer (Montopoli *et al.*, 2008):

$$N_m(D_i,t) = \frac{n_i(t)}{A.dt.v_i.dD_i} \tag{1}$$

where m indicates a measured quantity; D_i is the central drop diameter of the channel c_i in mm; $N_m(D_i,t)$ is the number of raindrops per unit of volume in the channel c_i at the discrete instant dt in units per millimetre per cubic metre; $n_i(t)$ is the number of drops counted in the ith channel at the instant t; A is the sensor area in square metre; dt is the time interval in seconds; v_i is the terminal velocities of rainfall in metres per second; dD_i is the ith bin width in millimetres. The terminal fall speed of raindrop is presumed as a function of the particle diameter by $v_i = 3.78\, D_i^{0.67}$ (Atlas & Ulbrich, 1977). The rain rate in mm/h is estimated by:

$$R = 0.6\pi \times 10^{-3} \int D^3 v(D) N(D) dD \tag{2}$$

For more accurate rainfall estimation, Marshall & Palmer (1948) expressed for the first time that DSDs can be represented by an exponential distribution using two parameters. Later, it was found that two parameters are not enough to describe the volumetric distribution of raindrops, and DSDs have been suggested in the form of a gamma or lognormal distribution. In this study, we have used the normalized gamma model described in Bringi *et al.* (2003):

$$N(D) = N_w f(\mu) \left(\frac{D}{D_m}\right)^\mu \exp\left[-(4+\mu)\frac{D}{D_m}\right]; \quad f(\mu) = \frac{6(4+\mu)^\mu}{4^4 \Gamma(\mu+4)} \tag{3}$$

The parameter D_m represents here a mass-weighted mean drop diameter, which is the ratio of the 4th to the 3rd DSD moments; N_w represents the normalized or scaled intercept parameter in mm^{-1} mm^{-3}; and $f(\mu)$ is given as a function of the shape parameter (μ). The normalized standard deviation of mass spectrum σ_M with respect to D_m can be related to μ as follows:

$$\frac{\sigma_M}{D_m} = \frac{1}{(4+\mu)^{1/2}} \tag{4}$$

DSD separation in atmospheric and seasonal context

Since we are interested in understanding the microphysical variation of raindrop spectra only, we have applied a special "rain partition" filter to retrieve raindrops separate from snow, using the wet bulb temperature (T_w). T_w is documented as a good measure in separating snow and rain. It is known that snow is probable only if T_w is below 2°C and some other atmospheric criteria are fulfilled. We have used $T_w = 3$°C as partition indicator, and removed all disdrometer data points below this temperature threshold. Also, the DSD with rain rates lower than 0.1 mm/h are considered as noise and also discarded from the database. Once filtered, a total of 162 415 one-minute raindrop spectra are obtained to fit in the normalized gamma DSD model.

To understand the variability of DSD in diverse seasonal and atmospheric conditions, we have separated the DSD dataset into two subdivisions using two criteria. The first separation is, based on climatic seasons – the DSDs taken from autumn and winter seasons (23 September–19 March) is defined as "cold" set DSD; and the DSDs taken from spring and summer seasons (20 March–22

Table 1 Division of rain classes and their range in rain rates (mm/h).

Class	1	2	3	4	5	6	7	8	9	10
Range	0.1–0.2	0.2–0.5	0.5–1	1–5	5–10	10–20	20–30	30–50	50–80	80–120

September) is defined as "warm" set DSD. The separation of the DSDs herein is built upon the concept of the "equinox" criterion. Another separation is based on wet bulb temperature. Wet bulb temperature is testified as a directly observable quantity that characterizes both surface temperature and surface humidity. Based on a study conducted by Eltahir & Pal (1996), it is revealed that there is a relationship between these surface conditions (temperature and humidity) and subsequent rainfall. They have established that storm rainfall is a linear function of wet bulb temperature and the likelihood of occurrence of convective storm accelerates as the wet bulb temperature rises. We have used $T_w = 12°C$ as the separator indicator and the DSDs obtained during $T_w \leq 12°C$ are denoted as "dry" set DSDs, whereas DSDs obtained during $T_w > 12°C$ are denoted as "wet" set DSDs. This threshold is selected based on the wet bulb temperature statistics of Chilbolton within our seven years dataset. The hypothesis of creating such two clusters is that the prevalence of generating rainfall from convective clouds is more in "high temperature" DSDs of "warm" and "wet" sets, whereas, stratiform type rain likelihood is applicable to low temperature DSDs of "cold" and "dry" sets. In addition, for a comprehensive investigation, rain rates are sub divided into 10 classes ranging from 0.1 mm/h to 120 mm/h, as shown in Table 1.

Self-consistency of disdrometer

To check the data quality of the disdrometer, we have compared disdrometer-derived hourly rain accumulations against four raingauges over a seven year period. We have used two statistical parameters as a performance indicator for the evaluation, the Pearson coefficient of correlation (r), and mean absolute error (MAE). Table 2 illustrates the statistical performance of the disdrometer and indicates that the disdrometer rain totals agree quite well with those measured by raingauges. The correlation is in the range of 0.89–0.99. The MAE is noted as between 0.10 mm and 0.46 mm. It is worth mentioning that rainfall estimation using raingauges is also subject to errors, in particular at low rain rate spectra. A large number of articles regarding raingauge-based rainfall estimation uncertainty have been published in the literature (Habib *et al.*, 2001; Wang *et al.*, 2008). The notable problems are associated with gauge calibration, sampling and systematic errors along with mechanical problems. While comparing between raingauge and disdrometer, natural and temporal variation of rainfall measurement by two instruments should also be accounted for. Generally, the uncertainty in the rainfall measurement is 10% for the drop counting raingauge and 25% for the tipping bucket raingauge. This uncertainty can be due to systematic error or random error. Hence, if we consider the uncertainties of rainfall estimation by the raingauge itself, it seems performance of our disdrometer is quite satisfactory. The statistical measures are also consistent throughout the whole study period. Furthermore, the results are in agreement with the past studies conducted by Wang *et al.* (2008) and Radhakrishna & Rao (2010).

RAINDROP SPECTRA CHARACTERIZATION

To understand the DSD variability, the DSDs are assessed in terms of the normalized gamma parameters. Figure 1 (left) displays the $<\sigma_M/D_m>$ distribution against $<\mu>$ values for four DSD sets. The symbol $<.>$ represents the average operator. Standard deviations of σ_M/D_m are also shown. The plot expresses that average μ values are considerably higher in "warm" and "wet" states than that of "cold" and "dry" states. During warm conditions, the $<\mu>$ value is found as 4.627 as compared to $<\mu> = 4.420$ in cold season. The variation is more prominent between the dry and wet DSD sets, showing a difference of 0.914. In this case, the average shape parameter is found as 5.136 in a wet atmosphere and 4.222 in a dry atmosphere. It is also worth mentioning that the median value of μ follows a similar trend, $<\mu>$ values are noted equal to 3.358 and 2.935 in wet and dry conditions,

Table 2 Statistical performance of JW disdrometer derived hourly rain accumulations taking four raingauges as the reference. The study period is 2003–2010. The MAE is given in mm.

	RG1		RG2		RG3		RG4	
Year	r	MAE	r	MAE	r	MAE	r	MAE
2003	0.95	0.24	0.89	0.28	0.92	0.46	0.93	0.33
2004	0.95	0.30	0.93	0.22	0.94	0.27	0.95	0.22
2005	0.98	0.25	0.97	0.16	0.97	0.18	0.98	0.13
2006	0.93	0.26	0.99	0.29	0.99	0.15	0.99	0.10
2007	0.89	0.28	0.94	0.26	0.98	0.15	0.99	0.11
2008	0.97	0.32	0.97	0.21	0.97	0.22	0.97	0.17
2009	0.97	0.35	0.97	0.18	0.97	0.30	0.97	0.24
2010	0.98	0.27	0.96	0.13	0.98	0.25	0.98	0.21

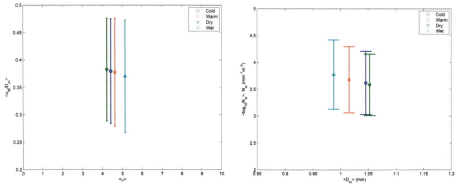

Fig. 1 $<\sigma_M/D_m>$ *vs* $<\mu>$ plot (left) and $<\log_{10} N_w>$ *vs* $<D_m>$ plot (right) for different DSD sets.

respectively. If we consider the probability of convective events is higher in "warm" and "wet" conditions than in "cold" and "dry" conditions, our results support the findings of Maki *et al.* (2001) and Testud *et al.* (2001) from airborne microphysical data during TOGA COARE.

Another way to observe DSD variations can be represented by plotting the averaged N_w–D_m relation. Figure 1 (right) shows the variation of the mean intercept parameter $\log_{10} N_w$ with respect to the mass-weighted mean drop diameter D_m for different DSD sets. The standard deviations of $\log_{10} N_w$ are also shown. These results indicate the dissimilarities of the DSD parameters in different conditions. For example, during the cold season, $<D_m>$ is found as 1.046 mm alongside $<\log_{10} N_w> = $ 3.616 with the std of 0.589. However, $<D_m>$ shows 1.016 mm during the warm season as opposed to $<\log_{10} N_w> = 3.674$ with the std of 0.616. As for the previous figure, the variations are more substantial between the dry and wet conditions. The average D_m and $\log_{10} N_w$ are found as 1.053 mm and 3.581 (std = 0.571), respectively, in dry conditions. In contrast, the mean drop diameter is much lower in wet conditions recording $<D_m> = 0.988$ and $<\log_{10} N_w> = 3.772$ with std = 0.644. The result implies that the proportion of smaller drops is higher at high temperatures (warm and wet), and the converse behaviour applies to low temperatures of cold and dry conditions. The results are in agreement with those of Moumouni *et al.* (2008), where they found a higher fraction of bigger drops for a given rain rate during stratiform rain events. This phenomenon could be due to the accretion incidence which takes place just above the melting layer and produces bulky ice particles. When these particles melt, the product particles then form into bigger drops (Maki *et al.*, 2001). Based on results presented here, it can also be perceived that the normalized intercept concentrations are slightly greater at higher temperatures than at lower temperatures. A thorough exploration can be accomplished if the $<N_w>$–$<D_m>$ relation is plotted for different rain classes (see Fig. 2). From Fig. 2 it can be clearly noticed that D_m increases with rain intensities, and this is true for all conditions.

Figure 3 shows the profiles of the normalized gamma DSD parameters at different rain intensities. The profiles are constructed by taking the mean values of the data for 10 rain classes as

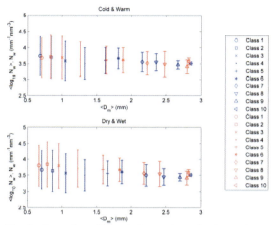

Fig. 2 $<D_m>$–$<\log_{10} N_w>$ relation against rain classes for "Cold and Warm" case (top) and "Dry and Wet" case (bottom). Blue bar denotes "cold" and "dry" DSDs, and red bar denotes "warm" and "wet" DSDs.

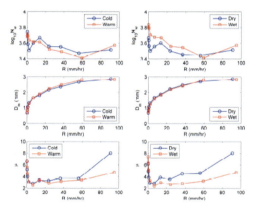

Fig. 3 Variation of $\log_{10} N_w$ (top), D_m (middle) and μ (bottom) as a function R.

established earlier. From Fig. 3, the difference in distributions of concentration parameter between the "cold" and "warm" DSD data sets is clearly apparent. The variation is also distinguishable in the cases of "dry" and "wet" set DSDs. Besides, the distribution of D_m is noticed as exponential against rain rates. Although, there is no significant variation in cold/warm and dry/wet conditions, it can be seen that the D_m value increases with rain intensity, similar to those experienced by other authors (Harikumar *et al.*, 2010). For high rain rates, D_m tends to follow the equilibrium-type distribution showing uniform trends. The most prominent variations are observed in the shape parameter. It is undoubtedly apparent that the distributions of μ values are significantly higher in the "cold" DSD data set as compared to the "warm" DSD data set. Indeed, the variation is proportional to rain intensities over 10 mm/h and as the intensity rises, the difference increases as well. The differences are much severe between "dry" and "wet" sets.

To investigate the variability of DSD characteristics in different seasonal and atmospheric states, four *Z-R* relations considering our four DSD sets were derived. Table 3 represents the *Z-R* relations acquired for four varied DSD sets by performing linear regression between the radar variable *Z* and disdrometer-derived rain rate *R* considering the Thurai *et al.* (2007) DSD model. It is clearly apparent that the coefficient *a* varies according to the classification (i.e. cold/warm or dry/wet), whereas *b* is in the range of 1.5 to 1.6. In particular, during the cold period, the coefficient *a* is found as 230, whereas during the warm period, the intercept obtained is much lower (*a* = 206). Similar to

Table 3 *Z-R* relationships for four seasonal and atmospheric DSD sets.

DSD set	*Z-R* relation	DSD set	*Z-R* relation
Cold	$Z = 230\ R^{1.582}$	Dry	$Z = 238\ R^{1.556}$
Warm	$Z = 206\ R^{1.528}$	Wet	$Z = 182\ R^{1.263}$

our previous findings in terms of DSD variation in "dry" and "wet" atmospheric weather, the coefficients difference in these two sets is also momentous. At this stage, the coefficient *a* is given as 238 in the "dry" DSD set as opposed to *a* = 182 in the "wet" DSD set.

SUMMARY AND CONCLUSIONS

In this study, 7 years of Joss-Waldvogel disdrometer data are analysed to understand the variability of the DSDs in diverse seasonal and atmospheric states. The disdrometer-derived hourly rain accumulations are compared with four raingauges to check the accuracy of the disdrometer, and the results have indicated good correlations. To understand the variability of DSDs, a total of 162 415 one-minute raindrop spectra obtained from the disdrometer are fitted to the normalized gamma model describing the DSDs using three parameters: the concentration parameter (N_w), the mass-weighted mean drop diameter (D_m) and the shape parameter (μ). It has been established that the DSD parameters are strongly dependent on the rain rate. Moreover, it has also been shown that the DSD parameters vary from set to set reflecting seasonal and atmospheric variability. This phenomenon has been further supported in terms of *Z-R* variability.

Acknowledgements The authors would like to thank the British Atmospheric Data Centre and the Radio Communications Research Unit at the STFC Rutherford Appleton Laboratory for providing the dataset. We also thank Prof. VN Bringi and Dr. Merhala Thurai of Colorado State University for providing the normalized gamma DSD model.

REFERENCES

Atlas, D., Ulbrich, C. W., Marks, F. D., Amitai, E. & Williams, C. R. (1999) Systematic variation of drop size and radar-rainfall relations. *J. Geophys. Res.-Atmos.* 104(D6), 6155–6169.

Bringi, V. N., Chandrasekar, V., Hubbert, J., Gorgucci, E., Randeu, W. L. & Schoenhuber, M. (2003) Raindrop size distribution in different climatic regimes from disdrometer and dual-polarized radar analysis. *J. Atmos. Sci.* 60(2), 354–365.

Eltahir, E. A. B. & Pal, J. S. (1996) Relationship between surface conditions and subsequent rainfall in convective storms. *J. Geophys. Res.-Atmos.* 101(D21), 26237–26245.

Habib, E., Krajewski, W. F. & Kruger, A. (2001) Sampling errors of tipping-bucket rain gauge measurements. *J. Hydrol. Eng.* 6(2), 159–166.

Harikumar, R., Sampath, S. & Kumar, V. S. (2010) Variation of rain drop size distribution with rain rate at a few coastal and high altitude stations in southern peninsular India. *Adv. Space Res.* 45(4), 576–586.

Iguchi, T., Kozu, T., Kwiatkowski, J., Meneghini, R., Awaka, J., and Okamoto, K. (2009) Uncertainties in the Rain Profiling Algorithm for the TRMM Precipitation Radar. *J. Meteorol. Soc. Jpn.* 87, 1–30.

Maki, M., Keenan, T. D., Sasaki, Y. & Nakamura, K. (2001) Characteristics of the raindrop size distribution in tropical continental squall lines observed in Darwin, Australia. *J. Appl. Meteorol.* 40(8), 1393–1412.

Marshall, J. S. & Palmer, W. M. (1948). The distribution of raindrops with size. *J. Meteorology* 5(4), 165–166.

Montopoli, M., Marzano, F. S. & Vulpiani, G. (2008) Analysis and synthesis of raindrop size distribution time series from disdrometer data. *IEEE Trans. Geosci. Remote Sensing* 46(2), 466–478.

Moumouni, S., Gosset, M. & Houngninou, E. (2008). Main features of rain drop size distributions observed in Benin, West Africa, with optical disdrometers. *Geophys. Res. Lett.* 35(23).

Petersen, W. A., Carey, L. D., Rutledge, S. A., Knievel, J. C., Doesken, N. J., Johnson, R. H., McKee, T. B., Vonder Haar, T. & Weaver, J. F. (1999). Mesoscale and radar observations of the Fort Collins flash flood of 28 July 1997. *Bull. Amer. Meteorol. Soc.* 80(2), 191–216.

Radhakrishna, B. & Rao, T. N. (2010). Differences in cyclonic raindrop size distribution from southwest to northeast monsoon season and from that of noncyclonic rain. *J. Geophys. Res.-Atmos.* 115.

Testud, J., Oury, S., Black, R. A., Amayenc, P. & Dou, X. K. (2001) The concept of "normalized" distribution to describe raindrop spectra: A tool for cloud physics and cloud remote sensing. *J. Appl. Meteorol.* 40(6), 1118–1140.

Thurai, M., Huang, G. J. & Bringi, V. N. (2007) Drop shapes, model comparisons, and calculations of polarimetiric radar parameters in rain. *J. Atmos. Ocean. Technol.* 24(6), 1019–1032.

Wang, J. X., Fisher, B. L. & Wolff, D. B. (2008) Estimating rain rates from tipping-bucket rain gauge measurements. *J. Atmos. Ocean. Technol.* 25(1), 43–56.

Long-term diagnostics of precipitation estimates and radar hardware monitoring: two contrasting components of a radar data quality management system

DAWN HARRISON & ADAM CURTIS
Met Office, FitzRoy Road, Exeter EX1 3PB, UK
dawn.harrison@metoffice.gov.uk

Abstract Quality is key to ensuring that the potential offered by weather radar networks is realised. To help ensure optimum quality, a comprehensive radar data quality management system, designed to monitor the end-to-end radar data processing chain and evaluate product quality, is being developed at the Met Office. Two contrasting elements of the system, monitoring of key radar hardware performance indicators and generation of long-term integrations of radar products, are described. Examples for January 2011 are presented and ways in which the information has been used to identify problems and formulate solutions are given.

Key words quality monitoring; precipitation estimation

INTRODUCTION

Quality and reliability issues currently restrict the use of radar-based quantitative precipitation estimates (QPE) in hydrological applications. Issues are many and varied in their source, ranging from problems with performance of radar hardware components to limitations of post-processing algorithms. In the case of radar QPEs, quality is often quantified using comparison with raingauge measurements. This can be very useful, but interpretation can be problematic due to the sampling differences. A comprehensive radar data quality management system (RDQMS) is being developed, which will deliver a range of monitoring information, with comparison of radar and gauge being just one component. It is envisaged that improved monitoring will ensure any radar or radar data problems are quickly identified, therefore facilitating faster resolution.

QUALITY MANAGEMENT IN THE CONTEXT OF RADAR DATA PROCESSING

Quality management (QM) can be defined as the process of ensuring that customers' quality expectations are met. Identification of quality metrics is key to the QM cycle (Fig. 1(a)) and evaluation of these metrics determines if action is required. This may be something simple such as a configuration parameter change, or complex, e.g. improving clutter identification algorithms. QM should be considered at every step in the data processing chain (Fig. 1(b)). A review of radar data quality management was carried out using information from several National Meteorological Services (Harrison *et al.*, 2010). As a result, several different RDQMS components were identified and initial priorities specified as follows:

(1) Establish a dedicated RDQMS (to avoid impact on operational radar data processing systems).
(2) Generate high resolution (polar form) daily and monthly QPE products. Devise analysis methods using these products to identify and quantify persistent problems.
(3) Monitor a range of radar hardware performance indicators (e.g. transmitter power, receiver noise, and antenna rotation rate) centrally.
(4) Use hourly radar and gauge comparisons (available for the last 10 years) to investigate trends in QPE quality metrics. Select preferred metric(s) and define targets based on this analysis.
(5) Set up routine comparison of radar measured and NWP model simulated reflectivity. (Although more typically used to verify NWP model analyses/forecasts, analysis of long-term differences between NWP model and radar reflectivity can also be useful for revealing persistent anomalies and relative calibration errors in the radar data (see Georgiou *et al.*, 2011)).
(6) Integrate noise monitoring into the RDQM, using data from extreme range.

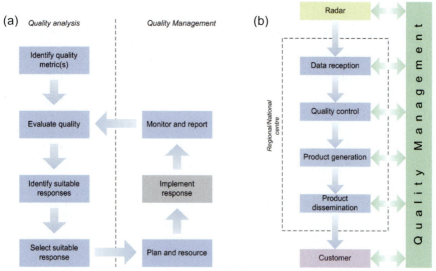

Fig. 1 (a) The quality management cycle, (b) Quality management and radar data processing.

The RDQMS will run in parallel to the Radarnet system at the Met Office in Exeter, UK. Radarnet performs quality control (QC) and product generation for data from the Met Office radar network (Harrison *et al.*, 2011) and will be developed incrementally with components added according to perceived importance. The first version, which was due in April 2011, consists of components 1–3, and 5 above. Initial outputs from components 2 and 3 are described here.

MONITORING RADAR SYSTEM HARDWARE PERFORMANCE

To ensure optimum radar product quality it is essential that radar system components are functioning as expected. The Met Office weather radars are un-manned so monitoring information needs to be available at a central location. Meta data contained in the reflectivity scan data files can be exploited on reception to provide real-time monitoring of the radar system. The parameters that have been initially chosen for constant monitoring are as follows. Significant deviation from expected values will trigger more in-depth investigation by engineers as appropriate.

Scan mean transmitter pulse power

The current transmitter incorporates a thyratron modulator driving a pulsed cavity magnetron. Peak power output is 220 kW minimum, with any significant deviation from this value indicating a problem with transmitter components including the power supply, modulator or magnetron unit.

Scan mean transmitter frequency (GHz)

Met Office radars operate at C-band (5.60 to 5.65 GHz). If the magnetron cavity volume is altered through thermal expansion, frequency drifts can occur. Therefore, temperature in the transmitter cabin needs to be kept constant. Frequency drift can suggest magnetron or air conditioning failures. Figure 2 shows an example of bi-modal transmitter frequency shifts at the Munduff Hill radar. Shifts of 100 kHz every 4 h were an indication of incorrect air conditioning system configuration. This resulted in different temperatures being set every four hours, resulting in two distinct transmission frequencies. Although the frequency shift is small and well within the electromagnetic spectrum bounds, avoidable thermal expansion could reduce magnetron longevity. Reconfiguration of the air conditioning was undertaken by engineers to solve the problem.

Scan mean receiver noise (dBc)

Ambient receiver noise is measured during "dead time" between radar pulses. Anomalies can indicate faults across the entire radar system. Noise and spurious emissions introduced into the radar from external sources (e.g. wireless LAN) can also be detected by this parameter. Within the "dead time", noise is injected into the waveguide and consequently to the receiver. This is used to derive the radar noise figure, a measure of the reduction of signal-to-noise ratio caused by internal noise generation in the radar receiver. Significant increases in the noise figure can indicate potential receiver faults. Figure 3 is a one-month time series for the High Moorsley radar. Large ambient noise spikes (over 60 dBc) occurred, which correlated in some instances with radar outages.. The cause is unclear and fault analysis is on-going. However the monitoring is aiding investigations and will help verify that systems are working correctly when the issue is resolved.

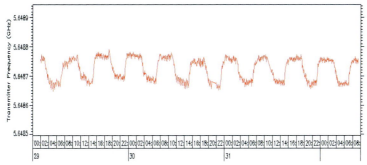

Fig. 2 Time series of transmitter frequency for a 3-day period, 29–31 January 2011, for the radar at Munduff Hill (Fife, Scotland).

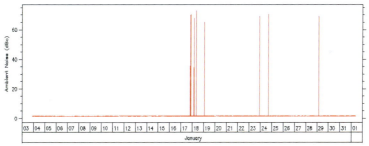

Fig. 3 Time series of ambient noise for a one month period, January 2011, for the radar at High Moorsley (Tyne and Wear, England).

Scan mean antenna rotation rate

To ensure consistent data quality it is desirable for the antenna rotation rate to be constant within a single scan. A constant rotation rate is demanded for each individual scan (typically 1.4 rpm for the lowest scans). The actual rotation rate is derived using scan start and stop time. Differences between demand and actual antenna rotation rates can indicate potential issues with the site processor, azimuth servo, azimuth DC torque motor or interconnections between the systems.

Scan elevation (degrees)

The elevation is encoded for each ray, therefore consistency within a single scan can be monitored. Deviation from the demand elevation can indicate potential issues with the site processor,

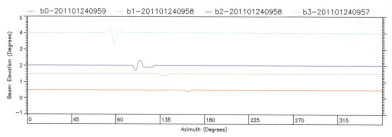

Fig. 4 Example of antenna elevation monitoring for a single volume (4 scans) from the Crug-y-gorllwyn radar, 24 January ~10:00 h.

elevation servo or DC torque motor or interconnections between the systems. Antenna elevation variation within a scan can degrade data quality, particularly at low elevations. Ground clutter intensities can fluctuate, which can compromise clutter identification. Antenna elevation monitoring has revealed that deviations up to 1° from the demand elevation occur over >20° azimuth at some sites. Figure 4 is an example from the Crug-y-Gorllwyn radar. Engineers rectified the elevation "nodding" by tuning the servo system to provide efficient damping in this case.

Monthly integrations of radar-based precipitation products

Long-term integrations of QPE products can help identify and quantify persistent anomalies, which may result from errors in the basic reflectivity measurement and/or limitations of any QC and correction algorithms applied (in this case those applied on the Radarnet system (Harrison *et al.*, 2011). When looking over periods of one month or more good quality products should, in general, have an appearance which reflects the climatological variation of precipitation, with variations mainly linked to significant orography and aspect. Precipitation probability, accumulation and average rate (conditional on precipitation rate >0.0 mm/h) have been produced from polar form radar products since October 2010. Figures 4–6 show examples for January 2011 for three UK radar sites, chosen to represent the range of quality seen: Chenies shows numerous sectors where poor correction for partial beam blockages is evident; Munduff Hill exhibits clutter breakthrough whereas Ingham shows consistent performance over virtually its entire domain.

In addition to quality monitoring, the potential for using these diagnostic products in real-time QC is also being investigated. For example, clutter from wind turbines is an increasing problem and existing clutter detection techniques do not generally work well as the target is often moving. Residual clutter can result in false flood or severe weather warnings where radar-based QPEs are used to drive meteorological and hydrological forecasting systems. Using a combination of long-term precipitation probability and average rate can be used to identify these small scale problematic features and exclude them from QPE products.

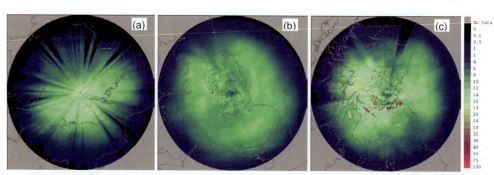

Fig. 4 Probability of precipitation (%), January 2011. (a) Chenies, (b) Ingham, (c) Munduff Hill.

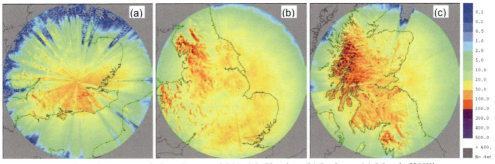

Fig. 5 Precipitation accumulation, January 2011. (a) Chenies, (b) Ingham, (c) Munduff Hill.

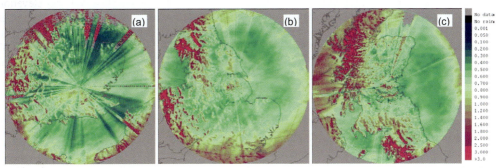

Fig. 6 Average precipitation rate (conditional that precipitation rate >0), January 2011. (a) Chenies, (b) Ingham, (c) Munduff Hill.

Probability of precipitation (PoP)

As part of the Radarnet QC, spurious (non-precipitation) echoes are flagged using a method largely based on pulse-to-pulse signal variability (Sugier *et al.*, 2002). The radar QPE uses reflectivity data from the lowest "usable" (unflagged) scan. Where there is no usable data in any scan the QPE is flagged as missing. Therefore, PoP (Fig. 4) across the radar domain should be relatively free from anomalies. Anomalously high PoP is indicative of clutter breakthrough. It is evident that there is significant clutter breakthrough to the south of Edinburgh from the Munduff Hill radar (Fig. 4(b)), with areas where PoP > 40% rather than around 15%, as detected at similar range. This suggests that the clutter identification processes are not working effectively for this site.

The other main feature of interest in these products is how PoP varies with range from the radar site. PoP should decrease with increasing range, as the lowest usable scan will begin to overshoot the top of shallow precipitation. In addition, the minimum detectable signal increases with range so very light precipitation will not be detected. Figure 7(a) shows average PoP *versus* range for January 2011. It is evident that some radar sites (e.g. Ingham) show relatively consistent PoP to beyond 120 km range, whereas with others (e.g. Chenies) PoP begins to steadily decline from as little as 50 km range. It also illustrates which sites have clutter breakthrough at short range.

The false alarm rate (FAR) in radar QPEs is often estimated using raingauge data as ground truth. Results have indicated that an FAR of 3–4% is typical. However, this approach is only of limited value as verification is limited to the gauge locations and relies on the gauge distribution to be representative. However, clutter is not random, with visible slopes and coasts being more clutter prone. These are often locations where gauges are less likely to be sited. An alternative approach is to use monthly or seasonal PoP data to estimate the FAR. This can be achieved by analysing the frequency distribution of PoP, filtering out points with anomalously high PoP and calculating a spatially averaged PoP without these outliers. The deviation of the PoP at a point to the spatially averaged PoP can provide an estimate of the FAR. This has the advantage of not requiring any other source of data as input and can therefore be used at most locations.

Precipitation accumulation (AccP)

Monthly or seasonal AccP, can be a useful product in its own right, but is also useful in illustrating radar data quality. Figure 5 shows AccP for January 2011 for the same sites as Fig. 4. It is immediately apparent that Chenies' domain has numerous sectors where precipitation is significantly less than elsewhere. This is further illustrated in Fig. 8(a) in contrast to Fig. 8(b) which show the variation in precipitation (averaged over all ranges) with azimuth. It is evident that there are large sectors within Chenies domain where precipitation is <50% of that elsewhere. The precipitation estimation process on Radarnet includes identification of unusable rays/sectors and a correction for partial beam blockage, but it is evident that this is not working effectively for Chenies. The reason is not entirely clear, but it is possible that changes in the radar horizon have not been captured or that the actual antenna elevation angle is not the same as that demanded/reported in the signal processing software.

The impact of the orographic enhancements, applied within the QPE process on Radarnet (Georgiou *et al.*, 2010), is evident in Fig. 5, where precipitation over significant orography is often several times that for lowland areas at similar range. These products proved useful in detecting an error with the corrections, introduced following a configuration change on Radarnet.

Average precipitation rate (CavP)

A further useful diagnostic is the conditional average precipitation rate (i.e. considering only instances where the QPE is >0.0 mm/h). Although AccP will decrease with range, CavP is expected to increase, since the likelihood of overshooting shallow precipitation and the minimum detectable signal will both increase. Figure 7(b) shows the variation CavP, averaged over all azimuths, with range. There are a number of characteristics to examine. Residual bright-band effects (the Radarnet QPE process includes a bright-band correction) would manifest in an increased rain rate in a distinct range band. In January this would be within 75 km range from the radar site. Most of the sites show relatively consistent performance between 15 and 150 km range, with only Predannack showing any clear evidence of an increase in CavP between 50 and 70 km range. At far range it is useful to compare radar performance. There are two radars with markedly different trends. Dean Hill shows lower CavP beyond around 100 km range, which could point to some sort of calibration error. Conversely, Hameldon Hill shows a rapid increase in rates beyond 200 km. It also has the lowest detection rate at far range. This could point to a greater than expected loss in sensitivity.

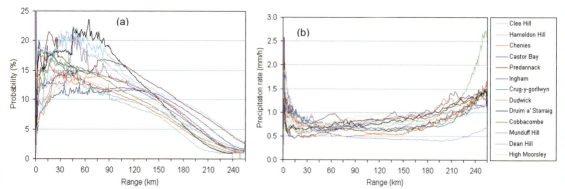

Fig. 7 Variation in monthly diagnostics with range from the radar site, January 2011. (a) Probability of precipitation, (b) conditional average precipitation rate.

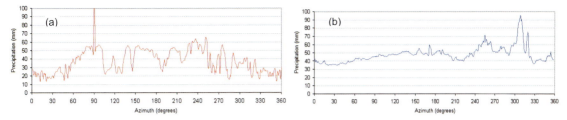

Fig. 8 Variation in average monthly precipitation with azimuth, January 2011. (a) Chenies, (b) Ingham.

SUMMARY, CONCLUSIONS AND FUTURE DEVELOPMENTS

Optimum quality, and therefore usability, of radar data and products is dependent on effective end-to-end quality management. The development of an RDQMS at the Met Office represents a pro-active approach to ensuring radar product quality. Contrasting types of monitoring, from either end of the processing chain, have been presented. Already these have been effective in identifying several faults with radar system components and post processing algorithms, which have, in turn, been resolved. The diagnostic products are also proving useful in their own right within the QC processes. Over the coming year work will focus on putting in place other elements of the RDQMS. These will include long-term diagnostics based on reflectivity volume data, comparison of Doppler and NWP model radial winds and additional quality metrics from radar and raingauge comparisons. It is anticipated that a complete RDQMS will lead to significant improvements in the quality and reliability of radar data and products, by quickly identifying any problems, thus enabling either short term solutions and/or long-term improvement strategies to be devised. It is also envisaged that the QM information will serve to inform customers about radar product quality issues thus increasing confidence in the use of the products or promoting realistic expectations.

REFERENCES

Georgiou, S., Ballard, S., Gaussiat, N. & Harrison, D. L. (2011) Quality monitoring of the UK network radars using synthesised observations from the Met Office Unified Model. In: *Proc. 8th Symp. on Weather Radar and Hydrology.* Exeter, UK.

Georgiou, S., Gaussiat, N. & Lewis, H. (2010) Dynamic modelling of the orographic enhancement of precipitation in the UK. In: *Proc. 6th European Conf. on Radar in Meteorology and Hydrology.* Sibiu, Romania

Harrison, D. L., Norman, K., Pierce, C. E. & Gaussiat, N. (2011) Radar products for hydrological applications in the UK. ICE *Water Management* (special publication on flood forecasting), vol. 164, issue WM1, 1–15 (in press).

Harrison, D. L., Hafner, S., Peura, M., Dupuy, P. & Boscacci, M. (2010) Radar data quality management in operational environments. *Proc. 6th European Conference on Radar in Meteorology and Hydrology.* Sibiu, Romania

Sugier J., Parent du Châtelet J., Roquain P. & Smith A. (2002) Detection and removal of clutter and anaprop in radar data using a statistical scheme based on echo fluctuation. In: *Proc. 2nd European Conf. on Radar in Meteorology and Hydrology*, Delft, 17–24.

2 Rainfall estimation and quality control

Weather Radar and Hydrology
(Proceedings of a symposium held in Exeter, UK, April 2011) (IAHS Publ. 351, 2012).

163

NMQ/Q2: National Mosaic and Multi-sensor QPE System

KENNETH HOWARD[1] JIAN ZHANG[1], CARRIE LANGSTON[2], STEVE VASILOFF[1], BRIAN KANEY[2] & AMI ARTHUR[2]

1 *NOAA/National Severe Storms Laboratory, National Weather Center, 120 David L. Boren Blvd., Norman, Oklahoma 73072, USA*
kenneth.howard@noaa.gov

2 *CIMMS/University of Oklahoma, National Weather Center, 120 David L. Boren Blvd., Norman, Oklahoma 73072, USA*

Abstract Accurate quantitative precipitation estimates (QPE) are critical for monitoring and prediction of water-related hazards and water resources. While tremendous progress has been made in the last quarter century in many areas of QPE, significant gaps continue to exist in both knowledge and capabilities that are necessary to produce accurate high-resolution precipitation estimates on a national scale for a wide spectrum of users. Toward this goal, a national Next-Generation QPE (NMQ/Q2) system has been developed at the National Oceanic and Atmospheric Administration's National Severe Storms Laboratory (NSSL). The NMQ/Q2 system has been running in real-time in the USA since June 2006. The system generates a suite of QPE products for the Conterminous United States at a 1-km horizontal resolution and 2.5 minute update cycle. The experimental products are disseminated in real-time to users and have been utilized in various meteorological and hydrological applications. In 2006, working with the United States National Weather Service's Office of Climate, Weather, and Water Services, NSSL began prototype testing of the high-resolution gridded NMQ/Q2 precipitation products as input into the Flash Flood Monitoring and Prediction program. Dissemination of Q2 products to selected River Forecast Centers (RFCs) began in 2007 with all RFCs currently having access through the Advanced Weather Interactive Processing System (AWIPS) Multi-sensor Precipitation Estimator (MPE).
Key words Multi-Sensor QPE System

INTRODUCTION

The National Mosaic and Multi-sensor QPE (Quantitative Precipitation Estimation), or "NMQ/Q2", system is a multi-radar multi-sensor system. The objective of NMQ/Q2 research and development is to assimilate different observational networks towards creating (1) high-resolution national multi-sensor QPEs for flash flood warnings and water resource management and (2) high-resolution national 3-D grid of radar reflectivity for data assimilation, numerical weather prediction model verification, and aviation product development. The NMQ/Q2 system is fully automated and has been running in real-time since June 2006. The system ingests base-level data from over 140 WSR-88D radars, 31 Canadian C-band weather radars, and Terminal Doppler Radars. The system generates high-resolution 3-D reflectivity mosaic grids (31 vertical levels) and a suite of severe weather and QPE products for CONUS at a 1-km horizontal resolution and 2.5 minute update cycle. Currently the system keeps a running three-year product archive online (http://nmq.ou.edu).

The experimental products are provided in real-time to users across government agencies, universities, research institutions, and the private sector. The products have been utilized in various meteorological, aviation, and hydrological applications and products. Further, the NMQ/Q2 system ingests a number of operational QPE products generated from different sensors (radar, gauge, satellite) and by human experts. A real-time verification system was established as part of the NMQ/Q2 system where operational and experimental QPE products can undergo systematic evaluations using independent gauge observations. This paper describes the major scientific components of the NMQ/Q2 system and provides examples of the real-time verification capability.

SYSTEM OVERVIEW

The NMQ/Q2 system uses a distributive computation architecture with four major modules. An overview flowchart of the NMQ/Q2 system is shown in Fig.1. Multiple data sources are used in all four major modules that constitute the NMQ/Q2 system: (1) single radar processing, 92) 3-D and 2-D radar mosaic, (3) the QPE (Q2, Vasiloff *et al.*, 2007), and (4) verification system. Data sources include the level-2 (base-level) data from the WSR-88D network, the Environment Canada weather radar network, the Federal Aviation Administration's Terminal Doppler Weather Radar (TDWR) network, numerical weather prediction model hourly analyses, and raingauge observations from the Hydrometeorological Automated Data System (HADS), a real-time data acquisition and data distribution system operated by the National Weather Service.

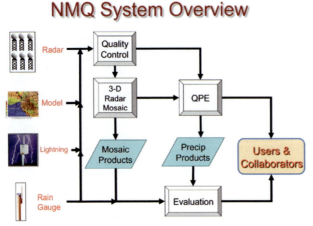

Fig. 1 An overview flowchart of the NMQ system.

Single radar processing

The base-level radar reflectivity data are quality-controlled (QC) to remove non-precipitation echoes, including those from clear air, biological targets (birds, bats, and insects), sun strobes, residual ground clutter, electronic interference, and anomalous propagation. The reflectivity QC module includes a pre-processing step, a neural network, and a post-processing step. The neural network approach is based on 3-D spatial characteristics of reflectivity (Lakshmanan *et al.*, 2007, 2010) such as intensity, gradients, texture, and depth of radar echoes. All the reflectivity bins with significant blockages or too close to the terrain (i.e. if the bottom of the bin is within 50 m of the ground) are removed in the pre-processing. Each volume scan of QC'ed reflectivity field is interpolated from the native polar coordinate system onto a 3-D Cartesian grid that is centred at the radar site. The Cartesian grid covers a 460 km range for coastal radars and 300 km for inland radars. It has a horizontal resolution of 0.01° (~1 km × 1 km), and 31 levels ranging from 500 m to 18 km above mean sea level. The analysis scheme includes a nearest-neighbour mapping on the range-azimuth plane and an exponential interpolation in the elevation direction (Zhang *et al.*, 2005; Langston *et al.*, 2007).

3-D and 2-D radar mosaic

Single radar reflectivity Cartesian (SRC) grids from multiple radars are combined into a 3-D reflectivity mosaic grid that covers CONUS and the southern part of Canada (Fig. 2). The mosaic domain spans from 130°W to 60°W in longitude and 20°N to 55°N in latitude. The grid is in the cylindrical equidistant map projection and has a resolution of 0.01° (longitude) × 0.01° (latitude). The resolution in the east–west direction is approximately 1.045 km at the southern bound of the domain and about 0.638 km at the northern bound of the domain. The resolution in the north–south direction is about 1.112 km everywhere. An exponential distance-weighting function is used when multiple radar observations cover a single grid cell (Zhang *et al.*, 2005). Figure 3 shows a horizontal and a vertical cross-section taken from the 3-D reflectivity mosaic grid for the Dallas hailstorm on 5 May 1995. The 3-D reflectivity grid depicts several convective cells at different stages of their life cycles. The 3-D reflectivity grid in conjunction with the environmental 3-D thermal field is also used to identify microphysical processes and to segregate precipitation regimes.

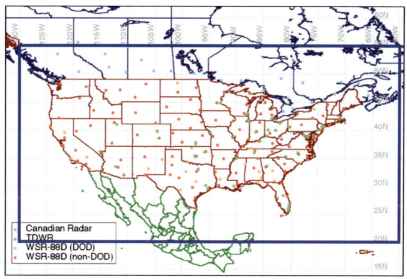

Fig. 2 The NMQ product domain (solid blue box). Dots of different colours represent different radar networks including WSR-88D (red and brown), Terminal Doppler Weather Radar (TDWR, green), and the operational Canadian weather radars (cyan).

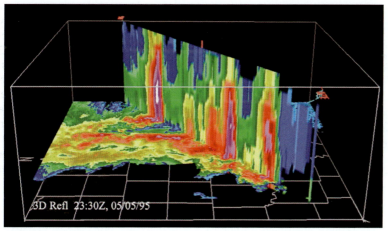

Fig. 3 Horizontal and vertical cross-sections from the 3-D reflectivity mosaic of the Dallas, Texas hail storm on 5 May 1995.

Q2 QPE

Q2 is a key component in the NMQ/Q2 system which performs automated precipitation classification and generates multi-sensor precipitation products ranging from instantaneous rates to monthly accumulations. The multi-sensor product suite includes radar-based QPE, local gauge bias-corrected radar QPE, gauge-only QPE, and a gauge, orography and precipitation climatology combined QPE. The Q2 radar-based QPE is computed from the radar hybrid scan reflectivity (HSR). Single radar HSR fields are mosaiced to produce a regional HSR field. This mosaic scheme yields QPE fields with better horizontal continuity than does a nearest-neighbour approach, since the latter can result in discontinuities in mosaiced data fields midway between neighbouring radars. The discontinuities are due to factors including different calibration among the radars and different sampling paths from the radars to the overlapping mosaic region.

The accuracy of radar QPE is largely dependent on the choice of a proper Z-R relationship for precipitation at a given location and time. In the NMQ/Q2 system, an automated precipitation classification is developed based on 3-D radar reflectivity structure and atmospheric environmental data. The NMQ/Q2 classification of precipitation regimes consists of a series of physically-based heuristic rules shown in Fig. 4. Each grid point is assigned a precipitation type based on 3-D reflectivity structure and the environmental thermal and moisture fields. Currently five precipitation types are identified: (1) stratiform rain, (2) convective rain, (3) warm rain, (4) hail, and (5) snow. Figure 5 shows mosaiced HSR, associated precipitation type (Fig. 5(b)), instantaneous rate (Fig. 5(c)) and 12-h accumulation (Fig. 5(d)) field for an event over the central USA on 28 February 2011. The event illustrates the complexity of precipitation systems that encompass all five classification schemes.

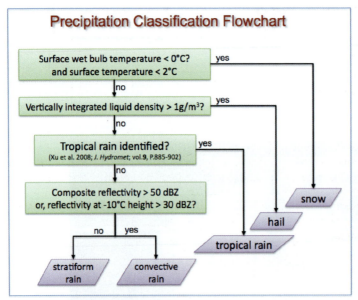

Fig. 4 The Q2 precipitation classification process.

A precipitation type field is computed at a high spatial (every 1 km) and temporal (every 2.5 min) resolution to capture small-scale variations of microphysical processes across space and in time. The NMQ/Q2 radar-only precipitation rate is obtained by applying Z-R relationships to the mosaiced HSR field pixel-by-pixel. Four Z-R relationships are used in association with the precipitation type field (e.g., Fig. 5(b)): convective (Fulton *et al.*, 1998) $Z = 300R^{1.4}$, stratiform (Marshall *et al.*, 1955) $Z = 200R^{1.6}$, warm rain (Rosenfeld *et al.*, 1993) $Z = 230R^{1.25}$, and snow (Radar Operations Center, 1999) $Z = 75R^{2.0}$. Here, Z represents the radar reflectivity in mm^6m^{-3}, and R represents rain-rate or snow water equivalent in $mm \cdot h^{-1}$. Hourly and three-hourly accumulations are computed every 5 minutes by aggregating the rate fields.

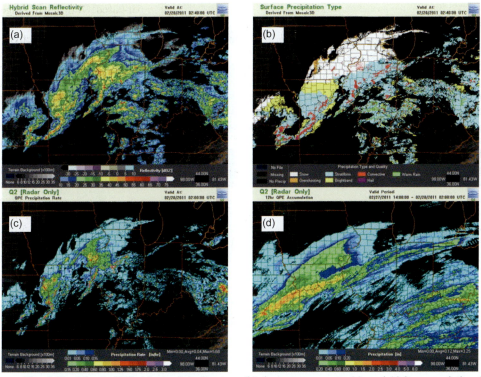

Fig. 5 The Mosaic HSR (a), precipitation type (b), instantaneous rate (c), and 12-h accumulation (d) fields at 02:00 UTC 28 February 2011.

The Q2 QPE future development will centre on the upgrade of the WSR-88D network to dual-polarization capability. Current research within the NMQ/Q2 system includes integration of polarimetric radar techniques for identification of non-meteorological echoes, hydrometeor classifications, and QPE. Additionally, research is focused on refining a Vertical Profile of Reflectivity (VPR) correction scheme for radar-based QPE because non-uniform VPR will remain a challenge for dual-polarized radar QPE. At the core of NMQ/Q2 is continued research and development on merging of multi-sensor QPEs, including those from satellite, especially for the mountainous west where radar coverage is limited.

VERIFICATION

A major component of the NMQ/Q2 system is a real-time verification system for QPE. A web-based system (nmq.ou.edu) was developed as an initial display system for quickly reviewing NMQ/Q2 products. The system was expanded to address the systematic evaluation of Q2 QPE product suites in comparison to operational and other experimental QPE techniques. The verification system was designed to ingest various gauge network data and other datasets useful for determining performance measures. A statistical engine was implemented that facilitates detailed event analysis of QPE performance for different regions and seasons. Several studies are currently underway to assess QPE/Q2 as well as operational QPE performances across multiple years. However, the verification system is open and allows researchers to assess the performance of various QPE techniques in real-time across varying geographical areas and time periods. Figure 6 contains several examples of the verification system analysis for an event in the central USA on 25 May 2011.

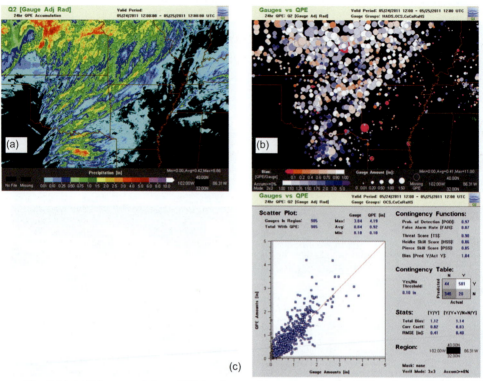

Fig. 6 (a) A 24-h accumulation of Q2 local gauge bias-corrected radar QPE ending at 12:00 UTC 25 May 2011, (b) a radar gauge bias bubble chart for three surface gauge networks for the same accumulation period as (a), and (c) a scattergram with performance measures (bias, RMSE, POD, FAR, Threat score, corr. coeff., etc.) in left column.

REFERENCES

Fulton, R., Breidenbach, J., Seo, D.-J., Miller, D. & O'Bannon, T. (1998) The WSR-88D rainfall algorithm. *Weath. Forecasting*, 13, 377–395.

Langston, C., Zhang, J. & Howard, K. (2007) Four-dimensional dynamic radar mosaic. *J. Atmos. Ocean. Tech.* 24, 776–790.

Lakshmanan, V., Zhang, J. & Howard, K. (2010) A technique to censor biological echoes in weather radar images. *J. Appl. Meteor. Clim.* 49, 453–462.

Lakshmanan, V., Fritz, A., Smith, T., Hondl, K. & Stumpf, G. J. (2007) An automated technique to quality control radar reflectivity data. *J. Appl. Meteor.* 46, 288–305.

Marshall, J. S., Hitschfeld, W. & Gunn, K. L. S. (1955) Advances in radar weather. *Adv. Geophys.* 2, 1–56.

Radar Operations Center (1999) Guidance on selecting Z-R relationships. http://www.roc.noaa.gov/ops/z2r_osf5.asp.

Rosenfeld, D., Wolff, D. B. & Atlas, D. (1993) General probability-matched relations between radar reflectivity and rain rate. *J. Appl. Meteor.* 32, 50–72.

Vasiloff, S., Seo, D.J., Howard, K., Zhang, J., Kitzmiller, D. H., Mullusky, M. G., Krajewski, W. F., Brandes, E. A., Rabin, R. M., Berkowitz, D. S., Brooks, H. E., McGinley, J. A., Kuligowski, R. J. & Brown, B. G. (2007) Q2: next generation QPE and very short-term QPF. *Bull. Amer. Met. Soc.* 88, 1899–1911.

Zhang, J., Howard, K. & Gourley, J. J. (2005) Constructing three-dimensional multiple radar reflectivity mosaics: examples of convective storms and stratiform rain echoes. *J. Atmos. Ocean. Tech.* 22, 30–42.

Weather Radar and Hydrology
(Proceedings of a symposium held in Exeter, UK, April 2011) (IAHS Publ. 351, 2012).

169

Quantitative precipitation estimate by complementary application of X-band polarimetric radar and C-band conventional radar

ATSUSHI KATO, MASAYUKI MAKI, KOYURU IWANAMI, RYOUHEI MISUMI & TAKESHI MAESAKA

National Research Institute for Earth Science and Disaster Prevention, 3-1, Tennodai, Tsukuba, Ibaraki 305-0006, Japan
maki@bosai.go.jp

Abstract In recent years, frequent flood damage and fatalities have occurred due to rising levels in urban rivers. Such floods are characterized by their very local nature and rapid development. To provide warnings about such floods, highly accurate Quantitative Precipitation Estimates (QPEs) at a high resolution and in real-time are required. Most QPE research involves the combination of data from raingauges and conventional radar. However, there are insufficient real-time data. The X-band polarimetric radar is useful for real-time QPE with high resolution. Compared with long wavelengths, X-band radars have the advantages of finer resolution, smaller-sized antennas, easier mobility (resulting from smaller antennas for the same beam widths), and lower cost. However, X-band radar has a relatively short observation range and is affected by strong signal attenuation during heavy rainfall. This study examines real-time quantitative rainfall estimation by complementary application of X-band polarimetric radar and C-band conventional radar. A comparison with ground raingauge data verifies that the proposed method is in good agreement with gauge data and is more accurate than conventional radar rainfall estimates.

Key words quantitative precipitation estimate; X-band polarimetric radar; urban flood; specific differential phase

INTRODUCTION

In recent years, urban flooding has occurred with a high frequency worldwide, with many reports of injuries and deaths due to localized torrential rainfall associated with the sudden development of cumulonimbus cells (e.g. Kato & Maki, 2009). Such floods are characterized by their very local nature and rapid development. To respond to such conditions, highly accurate quantitative rainfall information at a high resolution, in real-time, are required. In 2008, the water level of the Toga River in Kobe, Japan, rose by 1.5 m in just several minutes (Fujita, 2009). The river flow showed a very rapid response to heavy rain. To predict flooding, highly accurate rainfall information is required at a spatial scale of less than 1 km and a time scale of several minutes to 1 h. Given such a short time scale, the availability of real-time data is of great importance.

Most QPE research has been based on combined data from raingauges and from conventional radar. However, there remains a problem regarding the availability of real-time data. Multi-parameter radar has been proposed as a solution to the above problems, and has recently been the focus of an increasing number of studies (e.g. Bringi & Chandrasekar, 2001). Conventional radar estimates rainfall rate by receiving single polarized waves back-scattered from raindrops. Therefore, the radar reacts to raindrops of average size; i.e. it only uses power information. In contrast, multi-parameter radar simultaneously sends and receives horizontally and vertically polarized waves, compares them, and obtains various parameters with which to calculate rainfall rate. A parameter of particular importance is the specific differential phase (K_{DP}), which is the phase difference between the two polarized waves. K_{DP} is less sensitive to the raindrop size distribution (DSD) and hail mixtures, and it is resistant to radiowave attenuation, hardware calibration, and topographical blocking. The purpose of this research is to develop a new QPE method that uses X-band polarimetric radar to provide quantitative, real-time information on the current rainfall status that is necessary to predict the degree of inundation and damage resulting from flood events which have a rapid response time to localized heavy rain.

METHOD

Figure 1 shows the observation area of an MP-X (X-band Multi-parameter radar) operated at Ebina City in Kanagawa prefecture, Japan. The MP-X has been in operation since 2003, with the aim of providing improved, quantitative, polarimetric estimates of precipitation (Maki *et al.*, 2005).

A limited area of QPE analysis (within a 40 km radius) is employed to exclude errors in radar rainfall due to the influence of partial beam blockage and ground clutter, which are likely to be minimal within the selected 40 km radius because of the flat nature of the terrain. The influence of extension of the radar beam is also considered to be minor.

To estimate the rainfall distribution, volume-scan radar data are collected every 5 minutes from PPI (Plan Position Indicator) scans at five elevation angles (from 0.7° to 4.5°). The rainfall rate is calculated using the following composite method simplified from that of Park *et al.* (2005):

$$R = \begin{cases} 7.07 \times 10^{-3} Z_H^{0.819}, & (K_{DP} \leq 0.3° km^{-1}) \\ 19.6 K_{DP}^{0.823}, & (K_{DP} > 0.3° km^{-1}) \end{cases} \tag{1}$$

This approach is based on the R–K_{DP} and R–Z_H relationships, where R, K_{DP}, and Z_H are the rainfall rate, specific differential phase, and reflectivity factor, respectively. The threshold (K_{DP} = 0.3 km^{-1}) corresponds to a rainfall rate of 7.3 mm/h. The rainfall rate is interpolated from the radar coordinates (1° × 0.1 km) onto a Cartesian horizontal grid (0.5 × 0.5 km) using a Cressman weighting scheme (Cressman, 1959). The three-dimensional weighting function w is the product of horizontal and vertical weighting functions (w_h, w_v) using the Cressman function (i.e. $w = w_h \times w_v$); the functions are calculated using the radius and altitude of influence, respectively. The radius of influence is the linear function of the distance from the radar. The altitude of influence is the height of the melting band, which is located 1 km below 0°C altitude, as estimated by the operational mesoscale forecast (MSM) produced by the Japan Meteorological Agency (JMA). Herein, these data are termed R_{2p}.

The JMA operates a C-band single polarimetric radar network covering all of Japan; herein referred to as conventional radar. The locations of radars around the target area are shown in Fig. 1. JMA also operates the surface raingauge network (Automated Meteorological Data Acquisition System, AMeDAS), with an average spacing between stations of 17 km. Rainfall distribution information is provided in real-time and is generated using the C-band conventional radar network

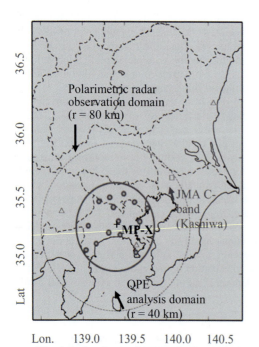

Radar type	MP-X
Photograph	
Frequency	9.375 GHz
Antenna type	2.13 mɸ parabola
Gain	41.6 dB
Beam width	1.3°
Transmitter	Magnetron 50kW
Pulse width	0.5 µs
PRF	≤ 1800 Hz
Receiver sensitivity	-111 dBm
Observation range	≥ 80 km
Parameters	T, Z, V, W, Z_{DR}, ρ_{hv}, Φ_{DP}, K_{DP}

Fig. 1 Radar observation area (radius 80 km), QPE analysis area (radius 40 km), radar locations, and radar specifications. (+) Polarimetric radar site (Ebina City). (○) AMeDAS surface raingauges in QPE target domain. (□) Radar operated by the Japan Meteorological Agency (JMA).

and AMeDAS (herein termed R_{JMA}). The production process is as follows (JMA, 1995). The constant altitude plan positioning indicator (CAPPI) at an altitude of around 2 km is generated for each elevation data. The reflection factor is converted to the rainfall rate based on the conventional Z–R relation ($Z = 200R^{1.6}$). Two correction procedures (overall correction and local correction) are performed using AMeDAS observation data. The grid point values (GPV) for a 1 km grid covering all of Japan are generated by combining the radar data corrected using the ground raingauge data.

X-band polarimetric radar outperforms the traditional technique, especially in the case of heavy rainfall (e.g. Kato & Maki, 2009). However, there are problems with using X-band polarimetric radar for hydrological applications, related to the radiowave extinction range and limited observation range; an example of these problems is provided in the Results section. This study proposes a simple technique for real-time compositing of R_{2p} and R_{JMA}.

The composited rainfall where $R_{3p\text{-}COR}$ is expressed by:

$$R_{3p\text{-}COR} = \begin{cases} R_{JMA\text{-}COR} & \text{(range of radiowave extinction or outside the range of radar)} \\ R_{2p} & \text{(the other)} \end{cases} \tag{2}$$

where $R_{JMA\text{-}COR}$ is corrected R_{JMA} and calculated by:

$$R_{JMA\text{-}COR} = \alpha R_{JMA} \tag{3}$$

The correction coefficient α is sequentially estimated in the area of QPE analysis where R_{2p} overlaps with R_{JMA}. For radar with a short wavelength influenced by attenuation, it is not possible to estimate the values of parameters required to estimate the rainfall rate within the radar area where the received signal is below the minimum detectable parameters of the radar receiver: this area is referred to as the radiowave extinction range.

A common approach to multi-sensor estimation is a weighted averaging of each estimate based on the relative error variances or distance from each radar. However, as shown in the following section, the accuracy of R_{2p} is higher than R_{JMP} so we use R_{2p} in the overlapping area. Of course, in the radiowave extinction area or outside the range of radar R_{2p} cannot be obtained, and we use R_{JMA} corrected by the method described in this section.

The verification was performed using raingauge data for events with an hourly rainfall of 5 mm or more in the target area for the rainy season in 2007–2008. The total number of sample data points (hourly rainfall data) is 19 024. The statistics used to verify the data are presented below.

The fractional bias FB and the root mean square error (rms error) S based on the regression line that passes through the zero point s is calculated by:

$$FB = \frac{\sum R_i}{\sum G_i} \tag{4a}$$

$$s = \sqrt{\frac{1}{n} \sum_{i}^{n} (R_i - aG_i)^2} \tag{4b}$$

where R_i, G_i, n and a are radar rainfall, surface raingauge data, the number of samples and slope of the regression line, respectively.

The fractional bias and rms error of class k of hourly rainfall range in steps of 5 mm, FB_k and s_k, are calculated by:

$$FB_k = \frac{\sum_{j}^{nk} R_j}{\sum_{j}^{nk} G_j} \tag{5a}$$

$$s_k = \sqrt{\frac{1}{n_k} \sum_{j}^{nk} (R_j - (a_k G_j + b_k))^2} \tag{5b}$$

where R_j, G_j and n_k are radar rainfall, surface raingauge data and the number of samples for class k, respectively. a_k and b_k are parameters of the regression line of class k.

The averaged fractional bias and the averaged rms error, $\overline{FB}$ and $\overline{s}$, are calculated by:

$$\overline{FB} = \frac{1}{N} \sum_{k}^{N} FB_k \tag{6a}$$

$$\bar{s} = \frac{1}{N} \sum_{k}^{N} s_k \tag{6b}$$

In addition, the normalized rms error $\tilde{s}_k$ is calculated by:

$$\tilde{s}_k = \frac{s_k}{\bar{G}_k} \tag{7}$$

where $\bar{G}_k = \frac{1}{n_k} \sum_{j}^{nk} G_j$.

RESULTS

Figure 2(a)–(c) show images of R_{2p}, R_{JMA} and $R_{3p\text{-}COR}$, respectively, at 19:50 UTC on 11 September 2007. R_{2p} detects the heavy rainfall areas A1, A2, A3, B3, and B5 in Fig. 2(a), which are not detected by R_{JMA}. The X-band polarimetric radar has the advantage of being able to detect heavy rainfall with high resolution and high accuracy (e.g. Kato & Maki, 2009). However, the area in black in Fig. 2(a) indicates the area of radiowave extinction range. This is a main disadvantage that presents itself when using single X-band polarimetric radar. Moreover, it is a problem that the observation range for single X-band polarimetric radars is limited for hydrological applications (e.g. total basin rainfall amount and input data for a hydrological model and precipitation nowcasting models). $R_{3p\text{-}COR}$ succeeds in detecting the whole precipitation system A1–A7 and B1–B6 in Fig. 2(c).

Figure 3 shows the verification results that were obtained comparing surface raingauges with R_{2p} and R_{JMA} for 19 024 samples of hourly rainfall from 2007 to 2008. The scatter plots in Fig. 3(a) and 3(b) show that R_{2p} is in good agreement with surface raingauges, although R_{JMA} is underestimated. Figure 3(c) and (d) show the mean values and the rms error $\bar{s}_k$ of each class of rainfall amount, as defined earlier. Figure 4(a) and 4(b) compare FB_k and s_k of R_{2p} and R_{JMA}, respectively. Figure 4(a) also shows $\overline{FB}$ for R_{2p} (horizontal solid line) and for R_{JMA} (horizontal dotted line). R_{2p} is overestimated for the class of 0–10 mm; however, it is in good agreement with observed data for the other classes, although slightly underestimated ($\overline{FB}$ of 0.89). In contrast, R_{JMA} is in good agreement with observed data during weak rain; however, it shows increasing underestimation with increasing rainfall amount. As a result, $\overline{FB}$ is 0.77. The rms error s_k for R_{2p} is smaller than that for R_{JMA}, in all classes. This decrease in estimation error can be attributed to the fact that $R(K_{DP})$ estimates are less sensitive to natural variations in DSD than are $R(Z_H)$ estimates (Ryzhkov & Zrnic, 1996). It should be noted that the number of samples of R_{2p} is decreasing with increasing rainfall amount (Fig. 5), because the area of radiowave extinction (see Fig. 2(a)) increases in heavy rainfall events. For example, the As ratio of number of samples of R_{2p} to R_{JMA} is about 50% at 40 mm h⁻¹.

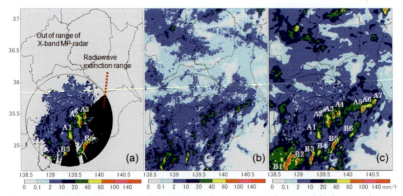

Fig. 2 Example of radar rainfall: (a) R_{2p}. (b) R_{JMA}, (c) $R_{3p\text{-}COR}$. Black area denotes the radiowave extinction range in which the radar signal is severely attenuated by heavy rainfall. The inner 40 km radius circle denotes the validation domain, and the outer 80 km radius circle denotes the observation domain of the MP radar.

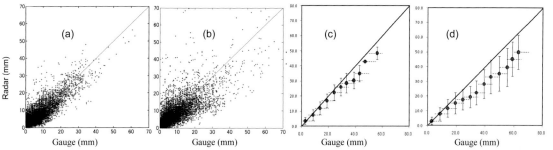

Fig. 3 Validation of R_{2p} and R_{JMA} using surface raingauge data: (a) scatterplot of R_{2p} *versus* surface raingauge data, (b) scatter plot of R_{JMA} *versus* surface raingauge data, (c)–(d) validation results based on classes of rainfall rate, corresponding to 3(a) and 3(b), respectively.

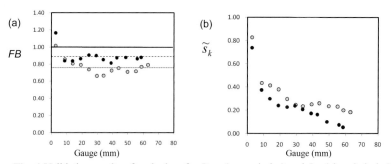

Fig. 4 Validation results of each class for R_{JMA} (open circles) and R_{2p} (closed circles): (a) is for fractional bias for each class based on gauge observations, (b) is for normalized rms error $\tilde{s}_k$ for each class.

Table 1 Results of QPE validation.

	R_{2p}	R_{JMA}	$R_{JMA\text{-}COR}$	R_{3p}	$R_{3p\text{-}COR}$
$FB\ [\ ]$	1.04	0.89	1.01	1.12	1.00
$S\ [mm]$	2.5	3.2	3.5	3.0	2.7
$\overline{FB}\ [\ ]$	0.89	0.77	0.87	0.89	0.86
$\overline{S}\ [mm]$	4.7	8.5	8.3	6.7	5.4

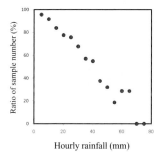

Fig. 5 Ratio of the number of samples of R_{2p} to R_{JMA} for each class.

Table 1 lists the statistics of five sets of radar rainfall. In terms of $R_{JMA\text{-}COR}$, the underestimation bias is reduced before the correction, with $\overline{FB}$ of 0.87. The results show that the dynamic correction

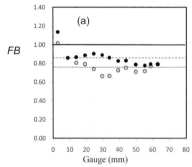
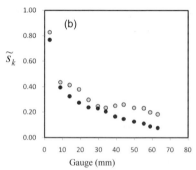

Fig. 6 Validation results of each class for R_{JMA} (open circles) and $R_{3p\text{-}COR}$ (closed circles): (a) is for normalized fractional bias FB_k of for each class based on gauge observations, (b) is for normalized rms error of each class.

procedure is effective. R_{3p} is somewhat underestimated, with $\overline{FB}$ of 0.89, and yields $\overline{s}_k$ of 6.7 mm. $R_{3p\text{-}COR}$ is somewhat underestimated ($\overline{FB}$ of 0.86), and yields smaller $\overline{s}_k$ than those for R_{3p} ($\overline{s}_k$ of 5.4 mm). Figure 5 is the same as Fig. 4, but $R_{JMA\text{-}COR}$ and R_{JMA}. $R_{3p\text{-}COR}$ is in good agreement with surface raingauges, and s_k for $R_{3p\text{-}COR}$ is smaller than that for R_{JMA}, in all classes.

SUMMARY AND DISCUSSION

The present paper compared R_{2p} (retrieved from an X-band polarimetric research radar operated by NIED) and R_{JMA} (retrieved from an operational C-band single polarized radar network together with surface raingauge network data operated by JMA). The error characteristics of each set of radar rainfall are also described. Furthermore, a technique for obtaining the real-time rainfall distribution with a high degree of accuracy is proposed, by using the two sets of radar rainfall in a complementary manner according to their characteristics, and verifying the reliability of this approach.

Verification by surface raingauge data revealed that R_{2p} was accurate in terms of the amount of hourly rainfall for amounts greater than 5 mm, although somewhat underestimated (average bias $\overline{FB} = 0.89$; average error $\overline{s}_k = 4.9$ mm), while R_{JMA} is not sufficiently accurate ($\overline{FB} = 0.77$, $\overline{s}_k = 8.5$ mm).

However, in terms of weak rain, the accuracy of R_{2p} was insufficient. This result may be influenced by the sensitivity limit to weak rain of differential phase data. In addition, for the ratio of the non-detection area to the rainfall field caused by radiowave extinction range, the number of samples of R_{JMA} is decreasing with increasing rainfall amount (e.g. about 50% of the sample for 40 mm h^{-1}). We also proposed a technique for obtaining highly accurate and seamless real-time rainfall distributions by making use of the advantages of each set of radar rainfall. We then verified the reliability of the technique. R_{3p}, as obtained by the proposed method, reproduced the entire precipitation system, which proved to be a difficult task for R_{2p} retrieved by the X-band polarimetric radar. The accuracy of R_{3p} ($\overline{FB} = 0.86$, $\overline{s}_k = 5.4$ mm) was superior to that of R_{JMA} ($\overline{FB} = 0.77$, $\overline{s}_k = 8.5$ mm).

Although the proposed technique was highly accurate in estimating the rainfall distribution, the accuracy is not high enough to enable its application in the field of hydrology, for which data on the catchment water balance are essential. Therefore, non-biased rainfall estimates are required. The proposed technique yielded slight underestimates of R_{3p}, related mainly to the underestimation of R_{2p}. Further analysis is required to understand the factors of R_{2p} and to improve the phase-based technique for quantitative rainfall estimation. Useful data in this regard would include drop-size distribution data retrieved by a disdrometer.

Acknowledgements We are grateful to Drs Suzuki, Shimizu, Wakatsuki, Hirano and Shusse for their constructive comments. We thank Mr Wan, Mr Yamada, Mr Kawada and Ms Endo for technical assistance.

REFERENCES

Bringi, V. N. & Chandrasekar, V. (2001) Polarimetric Doppler weather radar: principles and applications. Cambridge University Press, UK.

Cressman, G. P. (1959) An operational objective analysis system. *Monthly Weather Rev.* 87, 367–374.

Fujita, I. (2009) Water accident investigation in Toga river, Symposium on river disaster 2008, 1–8 (in Japanese).

Japan Meteorological Agency (JMA). (1995) Analytical method and accuracy of radar AMeDAS analyzed rainfall. *Sokko Jiho* (JMA's periodical) 62(6), 279–339 (in Japanese).

Kato, A. & Maki, M. (2009) Localized heavy rainfall near Zoshigaya, Tokyo, Japan on 5 August 2008 observed by X-band polarimetric radar—preliminary analysis. *SOLA* 5, 89–92.

Maki, M., Iwanami, K., Misumi, R., Park, S.-G., Moriwaki, H., Maruyama, K., Watabe, I., Lee, D-I., Jang, M., Kim, H.-K., Bringi, V. N. & Uyeda, H. (2005) Semi-operational rainfall observations with X-band multi-parameter radar. *Atmos. Sci. Letters* 6, 12–18.

Park, S. G., Maki, M., Iwanami, K., Bringi, V. N. & Chandrasekar, V. (2005) Correction of radar reflectivity and differential reflectivity for rain attenuation at X-band wavelength. Part II: Evaluation and application. *J. Atmos. Oceanic Technol.* 22, 1633–1655.

Ryzhkov, A. & Zrnic, D. S. (1996) Assessment of rainfall measurement that uses specific differential phase. *J. Appl. Meteor.* 35, 2080–2090.

176

Weather Radar and Hydrology
(Proceedings of a symposium held in Exeter, UK, April 2011) (IAHS Publ. 351, 2012).

X-band polarimetric quantitative precipitation estimation: the RHYTMME project

FADELA KABECHE, JORDI FIGUERAS I VENTURA, BÉATRICE FRADON & PIERRE TABARY

Météo France DSO-CMR, 42 Av. Coriolis, 31057 Toulouse Cedex, France Toulouse, France
fadela.kabeche@meteo.fr

Abstract This paper presents the current status of the radar data processing chain of the RHYTMME project, aimed at providing real-time quantitative precipitation estimations (QPE) at the local agents in order to minimize the economic and social impact of hazardous weather. The RHYTMME radar network will be composed of four X-band polarimetric radars that will feed data into a centralized processor which will process the data of each individual radar and produce a real-time composite QPE map, which will be transferred to the local operators. Currently there are two radars deployed and the X-band polarimetric processing chain is being finalized.

Key words X-band radar; precipitation; mountain; mask

INTRODUCTION

Sparsely distributed weather radar networks in mountainous areas usually have low efficiency because of problems such as increased ground clutter, partial beam blockage, presence of the bright band contamination in the measurements, etc. Météo France, in partnership with other French research institutes and administrations, is currently running a project named RHYTMME (Risques Hydro-météorologiques en Territoires de Montagnes et Méditerranéens) that aims to overcome this problem for a 300×300 km^2 mountainous region in southeast France. This region is prone to heavy and highly localised precipitation, causing torrential flooding, debris flows, landslides and, in the winter time, avalanches. In the frame of this project, a network of four polarimetric X-band radars, spaced approximately 60 km apart, will be deployed over the period 2010–2013. This network will provide local operators with the means to anticipate dangerous events and improve risk management in mountainous regions.

Currently, two different dual-polarization Doppler X-band weather radar datasets are available. The first dataset is from a Hydrix® radar located in Mont Vial, 30 km north of Nice, at an altitude of 1500 m. Data from this radar is available from January 2010. The second dataset is from a Selex radar located in Mont Maurel, about 50 km north of Mont Vial, at an altitude of 1778 m. This radar was deployed in October 2010 and data are available from February 2011. Both radars face challenges due to the orography: beam blockage by the Maritime Alps, north of

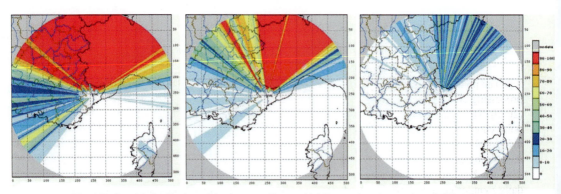

Fig. 1 Horizontal distribution of the amount of beam shielding at elevation angles of 0.4°, 1.2° and 2.4°.

the radars; sea clutter east of them, as well as the high altitude of their location. As an example of such challenging conditions, Fig. 1 shows the effects of topography on radar wave propagation of the Mont Vial Radar. The simulations have been performed using the approach proposed by Delrieu *et al.* (1995). The digital terrain model has a horizontal resolution of 250 m near Mont Vial. The maximum unambiguous range is approximately 150 km for Mont Vial and 250 km for Mont Maurel. The gate spacing is 300 m for Mont Vial and 240 m for Mont Maurel, while the azimuthal spacing is 0.5° for both radars.

RHYTMME RADAR DATA PROCESSING CHAIN

Figure 2 shows the flow diagram of the RHYTMME radar data processing chain. The polarimetric variables collected by the radars are horizontal reflectivity (Z_H), differential reflectivity (Z_{DR}), co-polar correlation coefficient (ρ_{HV}), and differential phase shift (Φ_{DP}) in polar coordinates. Mont Maurel also provides the standard deviation of the reflectivity from pulse to pulse (σ_Z) at 1 km^2 resolution. In addition, both radars provide mean Doppler velocity (v_D) and Doppler spectral width (σ_v). All these data are stored in a server and are available for off-line processing. Since the radars in the network are heterogeneous, the first step in the processing is, if necessary, a pre-treatment of the data to rend it in a uniform format so that it can be ingested into a polarimetric processing chain.

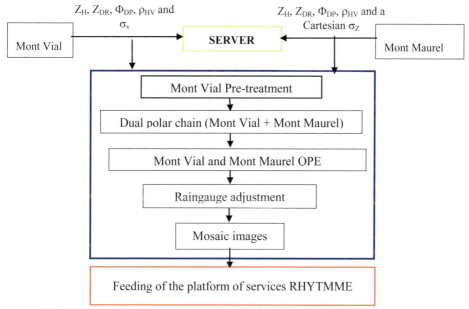

Fig. 2 Description of the real-time QPE evaluation.

The polarimetric processing chain has been adapted from an in-house developed S- and C-band processing chain. Modules of this processing chain are described in detail by Boumahmoud *et al.* (2010). The inputs of the dual-polarization chain are the polarimetric fields of Z_H, Z_{DR}, Φ_{DP}, ρ_{HV} in polar coordinates and a Cartesian field of pulse-to-pulse fluctuation of reflectivity obtained by each individual radar. It performs successively the following operations: calibration of Z_H and Z_{DR}, non-meteorological echo identification, ρ_{HV}-based bright band identification, Φ_{DP} offset removal and filtering, K_{DP} estimation, attenuation correction, hydrometeor classification and the computation of daily monitoring indicators (bias curves for Z_H and Z_{DR}, offset curves for Φ_{DP}, average ρ_{HV} in rain, etc.). Most of these modules are not frequency dependent. What is strongly dependent on the frequency, since different scattering modes are present, are the echo-type classification and the attenuation correction. Therefore, these modules have been adapted to the X-band.

The echo-type classification consists of two steps: a pre-classification in ground clutter, clear air and precipitation, and the classification of different types of precipitation. Gourley *et al.* (2007) developed the pre-classification algorithms at C-band implemented in the polarimetric chain. The scheme is based on a fuzzy logic algorithm. Note that the membership functions are obtained by empirical observations and objectively weighted according to the level of trust of each dataset for a particular radar site. It was observed that three parameters were the best discriminators between the different categories: ρ_{HV}, the texture of Z_{DR} and σ_Z.

The first step in the adaptation of the algorithm to the X-band has been therefore to empirically obtain the membership functions for these three parameters. The hydrometeor classification is also performed using fuzzy logic. If the use of a fuzzy logic precipitation classification at S- and C-band can be considered relatively mature, the same cannot be said at X-band due to the relatively recent interest of the community in this frequency band. Météo France is currently evaluating the performance of an algorithm developed by Snyder *et al.* (2010), which could be considered an extension of the algorithm by Park *et al.* (2009) to X-band.

Both the specific attenuation and the specific differential attenuation are considered proportional to Φ_{DP} with a constant of proportionality (γ_H for the specific attenuation and γ_{DP} for the specific differential attenuation, respectively) that is frequency dependent (Ryzhkov & Zrnic, 1995). The value of γ_H was experimentally estimated using the Mont Vial radar data from scatter plots of measured Z_H *versus* Φ_{DP} to be 0.28. The value of γ_{DP} was deduced from ratios of γ_H/γ_{DP} that appear in the literature (Bringi & Chandrasekar, 2001; Snyder *et al.*, 2010). We chose $\gamma_{DP} = 0.04$.

The polarimetric chain outputs are Z_{DR} and Z_H corrected for attenuation, Z_{DR} and Z_H not corrected for attenuation, co-polar correlation coefficient ρ_{HV}, K_{DP}, offset corrected Φ_{DP}, texture of Z_{DR}, the estimated horizontal attenuation and differential attenuation, σ_Z and the echo-type classification in polar coordinates. The corrected Z_H, the echo type and the Path Integrated Attenuation (PIA) are also provided in Cartesian coordinates. Météo France is currently at the latest stage of the evaluation of the performance of the X-band polarimetric processing chain.

After the polarimetric processing, the polarimetric information obtained, including the echo-type, will be combined to produce a composite image of precipitation intensity at ground-level. The composition will try to fully exploit the polarimetric data (attenuation level, echo-type, etc.) using polarimetric QPE algorithms and will combine data of all available radars in areas where there is overlapping. Finally, the resultant composite will be raingauge adjusted and the result fed into a service platform to be used by end-users.

EXAMPLE OF POLARIMETRIC VARIABLES

As an illustrative example of the challenges encountered using X-band data we will show here some data from an extreme precipitation event registered on 15 June 2010 in the Var region, approximately 70 km west of the Mont Vial radar. The event was a very localized phenomenon caused by a convective cell. More than 350 mm of precipitation was registered in less than 24 h. Examples of PPIs from some of the polarimetric parameters obtained at the 2.4° elevation at 15:00 UTC are shown in Figs 3 and 4.

The data shown here are still preliminary. However, the importance of a good attenuation can be easily noticed. The attenuation, estimated from offset corrected and filtered Φ_{DP} (See Fig. 4(b)) reaches values of more than 20 dB (see Fig. 3(e)) and the differential attenuation reaches values close to 4 dB (see Fig. 3(f)) in an area located northwest of the Vial radar. Therefore, if attenuation correction were not applied, this precipitation cell would be severely underestimated as it is qualitatively shown by comparing the non-corrected (see Fig. 3(a)) with the corrected reflectivity (see Fig. 3(b)).

Comparison of Fig. 4(a) with Fig. 4(b) highlights the necessity of filtering Φ_{DP}. The non-filtered Φ_{DP} in Fig. 4(a) is much noisier, which would both affect the quality of the K_{DP} estimation (shown in Fig. 4(d)) and the attenuation correction. The optimal filter length, which is a compromise between having sufficient range resolution to adapt to the natural variability of precipitation and the necessity to heavily filter phase noise, is still under investigation. At the moment the filter length for Mont Vial is 7.5 km.

As can be seen in Fig. 4(f), from the echoes placed north of the radar, the texture of Z_{DR} is a fairly good discriminator between ground clutter and precipitation. Ground clutter tends to have a high value, while precipitation has a much lower value. This is thus the most influencing parameter in the pre-classification to discriminate between ground clutter and precipitation. However, as can be seen from the final echo-type classification (in Fig. 4(c)) at the borders of the precipitation cells that are affected by noise, the texture may have large values and thus the pre-classification will erroneously classify these areas as ground clutter. An example of this miss-classification is seen in an area southwest of the radar.

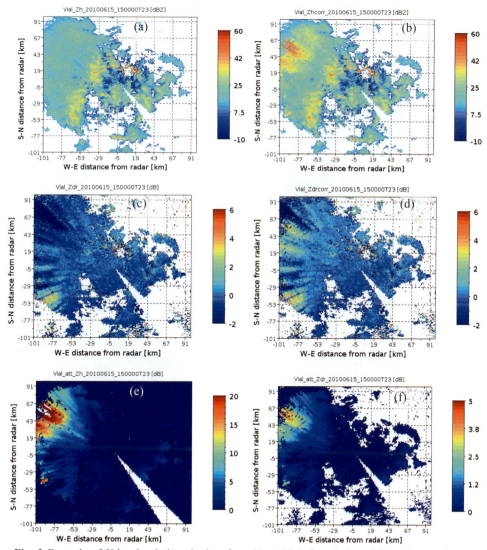

Fig. 3 Example of X-band polarimetric data from Mont Vial, from left to right and top to bottom: (a) ZH, (b) ZH corrected by attenuation, (c) ZDR (d) ZDR corrected by attenuation, (e) attenuation and

(f) differential attenuation.

The hydrometeor classification has qualitatively been evaluated to be satisfactory as for example the continuity between, the rain (in green), the wet snow (in orange) and the dry snow (in dark orange) in Fig. 4(c) shows.

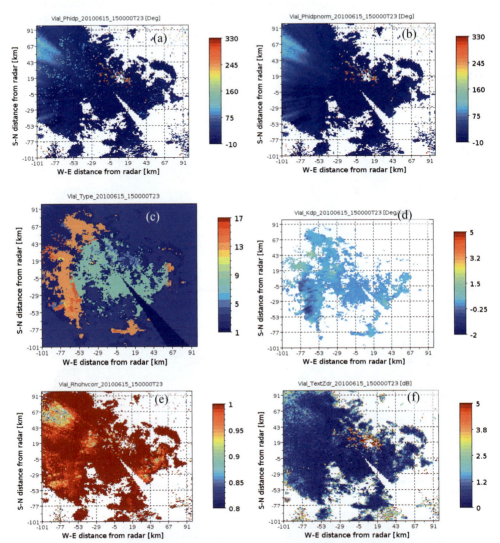

Fig. 4 Example of X-band polarimetric data from Mont Vial, from left to right and top to bottom: (a) ΦDP, (b) filtered ΦDP, (c) Echo-type classification: 1 = missing value of ZH, 2 = simple polarisation, 3 = noise, 4 = sea, 5 = ground clutter, 6 = insects, 7 = birds, 8 = low rain, 9 = medium rain, 10 = heavy rain, 11 = large drops, 12 = wet snow, 13 = dry snow, 14 = ice, 15 = graupel, 16 = hail, 17 = rain hail), (d) KDP, (e) ρHV and (f) the texture of ZDR.

CONCLUSION AND OUTLOOK

This paper has presented the current status of development of a polarimetric radar network data processing chain to provide accurate quantitative precipitation estimation to decision makers in a flooding-prone mountainous region. The polarimetric data processing of the individual radars is at the last stage of evaluation. The overall results are satisfactory, but issues have been observed on the discrimination between precipitation and ground clutter that will be addressed in the near future.

The next step will be to evaluate the quality of different polarimetric precipitation estimation algorithms, such as K_{DP}-based algorithms, conventional Z-R algorithms (with and without attenuation correction and with and without raingauge adjustment), Illingworth & Thompson (2005) integrated ZZDR technique, Hogan (2007) variational approach and Testud *et al.* (2000) ZPHI much in the same way as it is described in Figueras i Ventura *et al.* (2010); once the optimal algorithm has been determined a compositional strategy will be studied.

Acknowledgements Financial support for this study was provided by the European Union, the Provence-Alpes-Côte d'Azur Region, and the French Ministry of Ecology, Energy, Sustainable Development and Sea through the RHYTMME project.

REFERENCES

Boumahmoud A.-A., Fradon, B., Roquain, P., Perier, L. & Tabary P. (2010) French operational dual-polarization chain. In: *European Conference on Radar in Meteorology and Hydrology ERAD2010.*

Bringi V.N. & Chandrasekar, V. (2001) Polarimetric Doppler Weather Radar: Principles and Applications, Cambridge University Press.

Delrieu, G., Creutin, J. & Andrieu, H. (1995) Simulation of radar mountain returns using a digitized terrain model. *J. Atmos. Oceanic Technol.* 12, 1038–1049.

Figueras i Ventura, J., Kabeche, F., Fradon, B., Hogan, R., Boumahmoud, A-A., Illingworth, A. & Tabary, P. (2010) Extensive evaluation of Polarimetric Quantitative Precipitation Estimations (QPE) in ideal and less ideal conditions. In: *European Conference on Radar in Meteorology and Hydrology ERAD2010.*

Gourley, J. J., Tabary, P. & Parent-du-Chatelet, J. (2007) A fuzzy logic algorithm for the separation of precipitating from non-precipitating echoes using polarimetric radar. *J. Atmos. Oceanic Technol.* 24(8), 1439–1451.

Hogan, R. J. 2007 A variational scheme for retrieving rainfall rate and hail reflectivity fraction from polarization radar. *J. Appl. Meteor.* 46, 1544–1563.

Illingworth, A. J. & Thompson, R. J. (2005) The estimation of moderate rain rates with operational polarisation radar. In: *Proc. 32th Conference on Radar Meteorology.*

Park, H. S., Ryzhkov, A. V., Zrnic, D. S. & Kim, K. (2009) The Hydrometeor Classification Algorithm for the Polarimetric WSR-88D: Description and Application to an MCS. *Weather and Forecasting* 24, 730–748.

Ryzhkov, A. V. & Zrnic, D. S. (1995) Precipitation and attenuation measurements at 10-cm wavelength. *J. Appl. Met.* 34, 2121–2134.

Snyder, J. C., Bluestein, H. B., Zhang, G. & Frasier, S. J. (2010) Attenuation correction and hydrometeor classification of high-resolution, X-band, dual-polarized mobile radar measurements in severe convective storms. *J. Atmos. Oceanic Technol.* 27, 1979–2001.

Testud, J., Le Bouar, E., Obligis, E. & Ali-Mehenni, M. (2000) The rain profiling algorithm applied to polarimetric weather radar. *J. Atmos. Oceanic Technol.* 17, 332–356.

Evaluation of the performance of polarimetric quantitative precipitation estimators in an operational environment

JORDI FIGUERAS I VENTURA, BEATRICE FRADON, ABDEL-AMIN BOUMAHMOUD & PIERRE TABARY

Centre de Météorologie Radar, Direction de Systèmes d'Observation, Météo France, 42 Av. Coriolis, 31057 Toulouse Cedex, France
jordi.figueras@meteo.fr

Abstract This paper presents the evaluation of several polarimetric Quantitative Precipitation Estimation algorithms (Pol-QPE), candidates for operational implementation in the Météo France polarimetric weather radar network. The performance at C-band and in ideal conditions of three families of QPE algorithms have been studied: (1) algorithms based on simple Z-R relationships with and without attenuation correction using the differential phase Φ_{dp}, (2) algorithms based on reflectivity (Z_h) and differential reflectivity (Z_{dr}), and (3) algorithms based on specific differential phase (K_{dp}). The results confirm the superiority of polarimetric algorithms as reported repeatedly in literature. In particular, K_{dp}-based algorithms are shown to perform quite well at moderate and high rain rates. It is for this reason that at this stage Pol-QPE algorithms based on K_{dp} are preferred for operational use. To this end, a synthetic algorithm based on attenuation-corrected Z_h for low rain rates and K_{dp} for higher rain rates has been designed and tested successfully.

Key words polarimetry; polarimetric quantitative precipitation estimation

INTRODUCTION

In recent years polarimetry has reached a mature status and is currently being implemented in several operational radar networks in the world. The first Météo France polarimetric radar was installed in Trappes, near Paris, in 2004. In 2011, 10 out of the 24 radars composing the metropolitan French weather radar network are polarimetric (9 at C-band and 1 at S-band). Additionally, a network of 4 X-band radars (two of which are already installed) is being deployed to improve the coverage over the French southern Alps. Since the installation of the first polarimetric radar in Trappes, continuous effort has been laid on addressing the issues posed by the operational exploitation of polarimetry. Work has been done to design and implement robust and efficient procedures to monitor the data quality (Gourley *et al.*, 2006), correct for precipitation-induced attenuation (Gourley *et al.*, 2007), identify the bright band (Tabary *et al.*, 2006), classify the hydrometeors (Gourley *et al.*, 2007) and assess the effects of orography on the measurements (Friedrich *et al.*, 2009). All these efforts have led to the development of a first version of an operational polarimetric pre-processing chain that has been introduced in parallel with the legacy processing chain in July 2010 and is currently being evaluated (Boumahmoud *et al.*, 2010).

The extra information provided by polarimetry with respect to conventional radar can contribute to the mitigation of Quantitative Precipitation Estimation (QPE) errors since it facilitates the identification of the scatterers, the attenuation correction and the real-time retrieval of the Drop Size Distribution (DSD) parameters. A number of polarimetric QPE algorithms have been proposed in literature, but it is not yet clear which algorithm performs best in operational conditions for each (1) wavelength, (2) rainfall intensity regime, (3) noise level/biases on polarimetric variables, (4) partial beam blocking, and (5) the presence of hail. Polarimetric QPE algorithms are either based on horizontal reflectivity Z_h, differential reflectivity Z_{dr} or specific differential phase K_{dp} values or on a combination of two or more of these parameters. Numerous algorithms have been published in literature. A number of algorithms representative of the different families were selected for this study.

The first family of algorithms are the Z-R relationships of the type $Z = aR^b$. That is the algorithm that has been used in single polarization. The advantage brought by polarimetry to this estimator is that the estimation of and correction for attenuation is relatively straightforward using the differential phase Φ_{dp} (Gourley, 2006) and leads to better results than iterative approaches using A_h-Z_h

relationship (Hitschfeld & Bordan, 1954). (Hereafter the abbreviation "att" will be added to data where attenuation correction has been performed and "no att" will refer to non-corrected data). The parameters a and b are typically determined empirically using long series of disdrometer DSD data. The most widely used Z-R relationships, and the ones selected for evaluation in this study, are the Marshall-Palmer (1948) ($a = 200$, $b = 1.6$) (hereafter called MP) and the one used by the WSR-88D radars (a = 300, b = 1.4) (Fulton *et al.*, 1998) (hereafter called WSR88D).

Algorithms using Z_{dr} only are not very common due to the extreme sensitivity of the variable to radar interferences, partial beam blocking, etc. More common are estimators of the type $R = a_1 Z_h^{b1} Z_{dr}^{c1}$ whereby the values of Z_{dr} constrain the rainfall (see Gorgucci *et al.* (1994) for example). One of the problems of operational radars is that the quality of the Z_{dr} is limited due to the necessity of performing rapid scans and consequently having a small number of independent samples to average, which increases the variance of the estimation. Therefore, a spatial or temporal Z_{dr} average of some sort is typically performed. We selected for this evaluation a variant of an algorithm which was first developed by Illingworth & Thompson (2005), hereafter called Z-Z_{dr}. This algorithm attempts to adapt the a factor in the Z-R relation to the actual drop size distribution, which is proportional to the square root of the normalized drop concentration, N_w. b is considered constant with value 1.5. The algorithm attempts to derive the N_w of an area by obtaining the areal Z_{dr} that best fits into one of the *a priori* Z_h-Z_{dr} curves calculated from a particular N_w. The details of the algorithm implementation can be found in Figueras i Ventura *et al.* (2010).

The use of K_{dp} retrieval algorithms of the type $R = c(K_{dp}/f)^d$ where f is the frequency in GHz have been widely reported in literature (see for example Brandes *et al.*, 2001). The parameters c and d are derived from drop equilibrium shape distributions. Since at low precipitation rates Φ_{dp} is too noisy it is typically combined with Z-R relations below a particular precipitation rate. It should be noticed that unlike the other algorithms which are affected by attenuation, this algorithm performs equally at moderate and high rain rates. That makes this family of algorithms particularly suitable for use in more attenuating wavelengths such as C or X band. Moreover, compared to reflectivity-based algorithms, the K_{dp} retrieval is less sensitive to changes in the drop size distribution since it depends on the drop diameter by a factor 3 instead of the factor 6 of the reflectivity-based estimations (Ryzhkov & Zrnic, 1996). In this study we first evaluated two different K_{dp} only based algorithms: The Beard & Chuang (1987) ($c = 129$, $d = 0.85$) (hereafter called K_{dp} B-C) and Brandes *et al.* (2002) ($c = 132$, $d = 0.79$) (hereafter called K_{dp} Brandes). After the analysis of the results provided by each R-K_{dp} algorithm, a synthetic algorithm combining MP att and the K_{dp} B-C was analysed. Two thresholds, based exclusively on values of K_{dp}, have been tested: K_{dp} B-C used for data with $K_{dp} > 0.5°$ km^{-1} and with $K_{dp} > 1.0°$ km^{-1}, respectively.

SPECIFICATIONS OF THE QPE ALGORITHMS IMPLEMENTATION AND EVALUATION METHODOLOGY

A modular method has been developed to evaluate the pol QPE algorithms. The modules are the following: (1) Estimation of the instantaneous rainfall rate in regions that have been classified as rain, dry or wet hail, and big drops by the polarimetric chain using one of the various implemented algorithms. The other regions of the radar domain are assigned a No Data Available (NA) value. (2) Transformation of the outputs of the algorithm from polar to Cartesian coordinates using a Cressman analysis. (3) Oversampling of the rainfall rate maps from 5 min to 30 s using a 2-D advection field (Tuttle & Foote, 1990) and the accumulation of the 30 s rain fields to obtain a smooth 5 min rainfall accumulation map. (4) Computation of hourly rainfall accumulation by adding the twelve 5 min rainfall accumulation images. If one 5 min accumulation pixel is missing within the hour, then the hourly accumulation for this pixel is considered to be missing. That very strict criterion was introduced in order not to bias the evaluation results.

The hourly rainfall accumulation Cartesian maps obtained by the QPE algorithms can then be compared against hourly raingauges. The Météo France raingauge network consists of tipping bucket gauges with a bucket resolution of 0.2 mm, i.e. the minimum rainfall accumulation per hour that can be measured is 0.2 mm. All raingauge data are routinely quality-controlled. The radar–raingauge comparison is done by matching each raingauge with the corresponding radar pixel. In order to minimize the effects of partial beam blocking and possible Vertical Profile of Reflectivity (VPR) effects the comparison is restricted to ranges below 60 km from the radar. There are typically between 30 and 50 raingauges in the comparison area. No VPR correction is applied. Notice that VPR effects are reduced by the fact that the range is limited and the elevation angles are low.

Only one elevation angle is used to generate the radar QPE maps. The selection of the elevation angle is a compromise between minimizing beam height over the ground and minimizing partial beam blocking and ground clutter effects. The evaluation is carried out on a day-by-day basis. For practical reasons, the last hour of each day (from 23:00 to midnight UTC) has not been considered. The quality of the algorithms is evaluated using the normalized bias (NB) and the correlation (corr) between the raingauge and the radar. The present study has been focused on the warm period. A selection of about six events has been made in the July–August 2010 period. Table 1 summarizes the radar-events used in this study.

Table 1 Events selected for evaluation.

Radar	Elev [°]	Date
Avesnes	1.0	10, 12, 14, 23, 27 July 2010
Blaisy	1.0	1, 2, 14, 15, 16, 22, 23, 27 August 2010
Cherves	1.0	14, 21, 22, 23, 26, 27 August 2010
Montancy	1.2	2, 14, 15, 16, 23, 27 August 2010
Trappes	1.5	2, 3, 12, 14, 21, 23 July 2010

QPE EVALUATION RESULTS

Figure 1 summarizes the results obtained by the different algorithms on each radar and on all radars together. The base to create the rainfall rate products at Météo France is the MP no att and it is thus against that algorithm that the others should be compared. As can be seen in Fig. 1(a) MP no att largely underestimates the rainfall rate (NB = –0.30), particularly for rain rates >5 mm (NB = –0.48). The comparison between Fig. 1(a) and Fig. 1(c) (WSR88D no att) shows that the corr of the MP no att retrieval is slightly better than that of WSR88D no att (NB = 0.75 for MP against NB = 0.72 for WSR88D no att). In contrast, the NB for rain rates of >5 is significantly better for WSR88D no att (NB = –0.42 for WSR88D no att against NB = –0.48 for MP no att). Consequently the WSR88D no att seems to be better adapted for the type of events evaluated.

The application of the polarimetric attenuation correction significantly improves the results obtained by the Z-R relationship family of estimators. As it can be seen in Fig. 1(b) the corr of MP att increases from 0.75 to 0.81. The improvement is even greater at rain rate >5 mm, from 0.49 to 0.68. The NB at rain rate >5 mm is also greatly improved, from –0.48 to –0.28. It is remarkable that the improvement due to the attenuation correction of the WSR88D is even greater, due to the smaller exponent in the Z-R relationship.

The results of the Z-Z_{dr} no att algorithm shown in Fig. 1(e) exhibit a similar corr to that of the MP no att algorithm. However, in this case the NB is close to 0 for all the rain rates and has a value of –0.16 for the rain rates >5 mm. The effect of applying the attenuation correction to the Z-Z_{dr} algorithm results in an improvement of the corr of the same order of that obtained when applying it to MP algorithm (see Fig. 1(f)). The change of the NB is more marked than with the MP algorithm. It goes from 0.01 to 0.19 and the negative bias becomes a positive one at higher rain intensities.

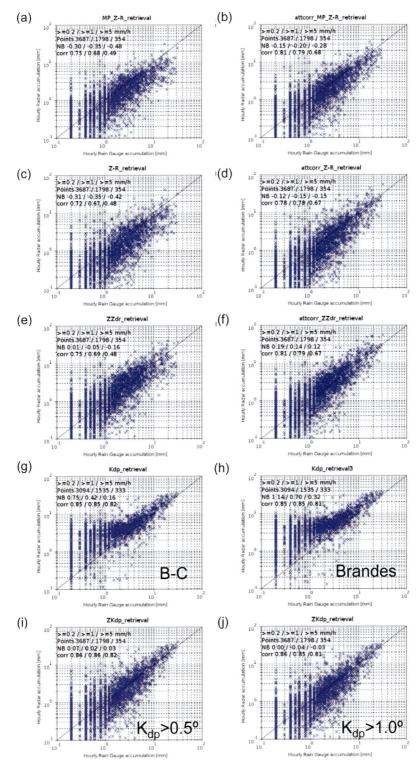

Fig. 1 QPE evaluation results.

The results of the family of K_{dp} retrievals (see Fig. 1(e),(g)) clearly show the effect of not considering the negative values of K_{dp}, which results in an overestimation of rain rate, particularly at lower rain rates. On the other hand, although the corr is much larger than with the MP att algorithm (0.85 compared with 0.75), particularly at rain rates >5 mm, where it reaches values of 0.82 for the K_{dp} B-C algorithm and 0.81 for the K_{dp} Brandes. Between the two K_{dp} algorithms studied, the K_{dp} B-C provides a smaller NB.

The fact that the K_{dp} algorithms gives the best results at moderate and high rain rates led to the creation of a synthetic algorithm that combines K_{dp} B-C at higher rain intensity with MP att at low rain rates. The resulting synthetic algorithm, the results of which can be seen in Fig. 1(i) and (j), provides the best results both in terms of corr and NB. The results show that the ZK_{dp} with threshold at $0.5°$ km^{-1} slightly overestimates the rainfall rate while with the threshold at $1.0°$ km^{-1} it obtains an optimal bias.

In Fig. 2 the results of each estimator are stratified according to rainfall intensity. The figure shows the inadequacy of the MP estimator to the type of precipitation predominant in those events. It tends to largely overestimate the weak rain and underestimate the intense precipitation. The WSR88D seems to be a better Z-R relationship for conventional radars. Both the overestimation for weak precipitation and the underestimation for large precipitation are smaller. In fact, when attenuation correction is applied the curve of the WSR88D becomes almost flat for precipitation larger than 1 mm h^{-1}, suggesting that the exponent of the Z-R relation is quite adequate and that is the *a* factor that should eventually be modified. The Z-Z_{dr} algorithm largely overestimates precipitation up to 5 mm h^{-1}. If attenuation correction is applied to the algorithm the precipitation is overestimated for all rain intensities. The K_{dp} algorithms have enormous bias (higher than 1) for weak precipitation due to the overestimation introduced by construction. This bias actually affects all the hourly accumulations of precipitation to some extent, hence the large overestimation for more intensive precipitation. The ZK_{dp} algorithm does not suffer as much from the bias due to phase noise, particularly at $1.0°$ km^{-1} threshold. However the bias is slightly higher than with the MP att for precipitation lower than 1 mm h^{-1}. For precipitation larger than that the curve is actually very close to 0, with the precipitation estimated by the $0.5°$ km^{-1} threshold slightly higher than that of $1.0°$ km^{-1}. In conclusion the ZK_{dp} algorithm seems to have the best behaviour in terms of bias at rain intensities higher than 1 mm h^{-1}.

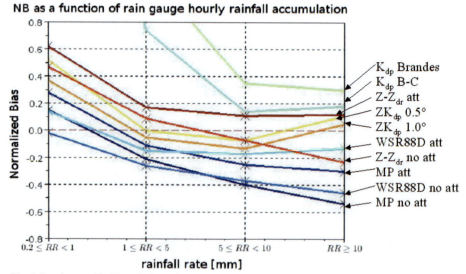

Fig. 2 Results stratified by rain intensity.

CONCLUSION

This article summarizes the results of a study aimed at determining the most suitable Pol-QPE algorithm in view of implementing it on the French Polarimetric Weather Radar network. Several Pol-QPE algorithms have been tested: algorithms based on simple Z-R relationships, an algorithm based on the relation between Z_h and Z_{dr} and algorithms based on K_{dp}. The results of the evaluation show, as it has been shown repeatedly in literature (see for example Bringi *et al.*, 2011 or Ryzhkov *et al.*, 2005), a superior performance of all the polarimetric algorithms with respect to the simple Z-R relationships. It has been shown that ZK_{dp} is a very good candidate because of various advantages, among them: insensitivity to calibration errors, partial beam blocking and the presence of hail and a smaller sensitivity to the drop size distribution of the precipitation. A synthetic Z-K_{dp} algorithm, which combines the superior performance of the R-K_{dp} relations for intensive precipitation and the relatively low bias of the Z-R relations corrected for attenuation in the weak rain, has been successfully tested rending it a very interesting option for a first version of an operational Météo France Polarimetric radar rainfall product.

Acknowledgements Dr Robert Thompson and Prof. Anthony Illingworth are thanked for their help in the implementation of the Z-Z_{dr} algorithm. We are grateful to Mr Laurent Perier for his explanations on the Castor processing of the radar data.

REFERENCES

Beard, K. V. & Chuang, C. (1987) A new model for the equilibrium shape of raindrops. *J. Atmos. Sci.* 44, 1509–1524.

Bringi, V. N., Rico-Ramirez, M. A. & Thurai, M. (2011) Rainfall estimation with an operational polarimetric C-band radar in the UK: comparison with a gage network and error analysis, *J. Hydromet.* (in press).

Boumahmoud, A.-A., Fradon, B., Roquain, P., Perier, L. & Tabary, P. (2010) French operational dual-polarization chain. In: *Proc. 6th European Conference on Radar in Meteorology and Hydrology, Sibiu, Romania, 6-10 September 2010.*

Brandes, E. A., Ryzhkov, A. V. & Zrnic, D. S. (2001) An evaluation of radar rainfall estimates from specific differential phase. *J. Atmos. Oceanic Technol.* 18, 363–375.

Brandes, E. A., Zhang, G. & Vivekanandan, J. (2002) Experiments in rainfall estimation with a polarimetric radar in a subtropical environment. *J. Appl. Meteor.* 41, 674–685.

Figueras i Ventura, J., Kabeche, F., Fradon, B., Hogan, R., Boumahmoud, A.-A., Illingworth, A. J. & Tabary, P. (2010) Extensive evaluation of Polarimetric Quantitative Precipitation Estimations (QPE) in ideal and less ideal conditions. In: *Proc. 6th European Conference on Radar in Meteorology and Hydrology, Sibiu, Romania, 6–10 September 2010.*

Fulton, R. A., Bredienbach, J. P., Seo, D.-J., Miller, D. A. & O'Bannon, T. (1998) The WSR-88 rainfall algorithm. *Wea. Forecasting.* 13, 377–395.

Hitschfeld, W. & Bordan, J. (1954) Errors inherent in the radar measurement of rainfall at attenuating wavelengths. *J. Meteor.* 11, 58–67.

Gorgucci, E., Scarchilli, G. & Chandrasekar, V. (1994) A robust estimator of rainfall rate using differential reflectivity. *J. Atmos. Oceanic Technol.* 11, 586–592.

Gourley, J. J., Tabary, P. & Parent du Chatelet, J. (2006) Data quality of the Meteo-France C-band polarimetric radar. *J. Atmos. Oceanic Technol.* 23, 1340–1356.

Gourley, J. J., Tabary, P. & Parent du Chatelet, J. (2007) Empirical estimation of attenuation from differerential propagation phase measurements at C band. *J. Appl. Meteor. Climatol.* 46, 306–317.

Gourley, J. J., Tabary, P. & Parent du Chatelet, J. (2007) A fuzzy logic algorithm for the separation of precipitating from non-precipitating echoes using polarimetric radar observations. *J. Atmos. Oceanic Technol.* 24, 1439–1451.

Illingworth, A. J. & Thompson, R. J. (2005) The estimation of moderate rain rates with operational polarisation radar. *Proc. 32nd Conference on Radar Meteorology, Albuquerque, NM, 24-29 Oct. 2005.* American Meteorological Society: Boston, Massachusetts, USA.

Marshall, J. S., Langille, R. C. & Palmer, W. K. (1947) Measurement of rainfall by radar. *J. Meteorology* 4, 186–192.

Ryzhkov, A. & Zrnic, D. (1996) Assessment of rainfall measurement that uses specific differential phase. *J. Appl. Meteor.* 35, 2080-2090.

Ryzhkov, A., Schuur, T. J., Burgess, D. W., Heinselman, P. L., Giangrande, S. E. & Zrnic, D. S. (2005) The Joint Polarization Experiment: Polarimetric Rainfall Measurements and Hydrometeor Classification. *Bull. American Meteorol.* Soc. 86, 809–824

Tabary, P., Le Henaff, G., Vulpiani, G., Parent-du-Châtelet, J. & Gourley, J. J. (2006) Melting layer characterization and identification with a C-band dual-polarization radar: a long-term analysis. In: *Pro. 4th European Conference on Radar in Meteorology and Hydrology, Barcelona, Spain, 18-22 September 2006.* Servei Meteorològic de Catalunya: Barcelona, Spain.

Tuttle, J. D. & Foote, G. B. (1990) Determination of boundary layer airflow from a single Doppler radar. *J. Atmos. Oceanic Technol.* 7, 218–232.

188

Weather Radar and Hydrology
(Proceedings of a symposium held in Exeter, UK, April 2011) (IAHS Publ. 351, 2012).

VPR corrections of cool season radar QPE errors in the mountainous area of northern California

YOUCUN QI[1,2], JIAN ZHANG[3], DAVID KINGSMILL[4] & JINZHONG MIN[1]

1 *College of Atmospheric Science, Nanjing University of Information Science and Technology, Nanjing 210044, China*
youcun.qi@noaa.gov
2 *CIMMS, University of Oklahoma, Norman, Oklahoma 73072, USA*
3 *National Severe Storms Laboratory, Norman, Oklahoma 73072, USA*
4 *CIRES, University of Colorado & NOAA/Earth System Research Laboratory, Boulder, Colorado, USA*

Abstract Non-uniformity of the vertical profile of reflectivity (VPR) is one of the major error sources for radar quantitative precipitation estimation (QPE) in the cool season, especially for mountainous areas. The error is due to two factors: one is that the radar beam samples too high above the ground and misses the microphysics at lower levels; the other is that the radar beam broadens with range and thus cannot resolve vertical variations of reflectivity structure. These errors have posed a major challenge for radar QPE in the complex terrain of northern California. The current study used precipitation profiler observations obtained in this mountainous area and developed a new VPR correction methodology for scanning radar QPE. The precipitation profiler data were used to determine slopes of a linear VPR model in the ice, bright band, and rain regions, and the slope parameters are derived for different geographical areas. The parameterized VPR is then used to correct for scanning-radar QPE. The new methodology was tested using a heavy rain case that occurred over the period 30 December 2005 to 1 January 2006 in northern California, and was found to provide significant improvements over the operational radar QPE.

Key words Vertical Profile of Reflectivity; VPR correction; radar QPE

INTRODUCTION

Non-uniformity of the vertical profile of reflectivity (VPR) is one of the major error sources for radar quantitative precipitation estimation (QPE) in the cool season, especially in mountainous areas. The error is due to two factors. One is that the radar beam samples is too high above the ground and misses the microphysics at lower levels. The other factor is that the radar beam broadens with range, and can not resolve vertical variations of reflectivity structure at far ranges. Many studies (e.g. Koistinen, 1991; Joss & Lee, 1995; Germann & Joss, 2002; Bellon *et al.*, 2005) have tried to correct for these errors based on radar data observed at close ranges, which had relatively high vertical resolution and were close to the surface. In mountainous areas, however, the VPR structure can change significantly in space due to the underlying topography. One mean VPR obtained from data close to the radar may not represent microphysical processes in areas far away from the radar. This has been a major challenge for obtaining accurate radar QPEs in the complex terrain of northern California. Andrieu & Creutin (1995) proposed a sophisticated inversion scheme to filter radar sampling effects (i.e. beam broadening as a function of range) and retrieved a mean VPR over the radar domain from two elevation angles. The scheme was later generalized by Vignal *et al.* (1999) to retrieve local VPRs over a small area of 20 km × 20 km using multiple elevation angles and was evaluated on Swiss (Vignal *et al.*, 2000) and USA (Vignal & Krajewski, 2001) radar data. Both evaluations showed that the local VPR approach provided more improvements in radar-derived QPE than the mean volume scan VPRs. However, the local VPR approach is relatively expensive computationally and is not easily implemented for operational applications.

The current study uses precipitation profiler observations from this mountainous area and develops a new VPR correction methodology that accounts for geographical variations of VPRs. A linear VPR model is used, which has different slopes in the ice, bright band (BB) and rain regions. The slope parameters of the linear VPR were derived from precipitation profiler data, while the BB layer heights were determined from precipitation profiler VPRs and atmospheric environment data. Further, the slope parameters are derived for different geographical areas to account for the horizontal variations of VPRs. The region-specific VPRs are then used to correct for scanning-radar QPEs in the corresponding areas.

The new VPR correction technique was tested using a heavy rain event that occurred over the period 30 December 2005 to 1 January 2006 in northern California, and results were compared with two other radar QPEs, one without VPR correction and another with the VPR correction previously developed by Zhang & Qi (2010). The new technique was found to provide significant improvements over the other two QPEs.

The next section describes the new VPR correction methodology. Case study results are then presented and the last section provides a summary.

METHODOLOGY

The main objective of the current study is to obtain high-resolution QPEs in northern California. The study domain (Fig. 1) extends from 38°N to 40°N in latitude and 123.5°W to 119.5°W in longitude. Radar QPE errors caused by vertical variations of the VPR include BB contamination (e.g. Zhang & Qi, 2010), radar beam overshooting (e.g. Joss & Waldvogel, 1990) and orographic enhancement (e.g. Kitchen *et al.*, 1994). The current study tested three QPE techniques: the first is the operational Weather Surveillance Radar – 1988 Doppler (WSR-88D) QPE (Fulton *et al.*, 1998) generated by the National Weather Service (NWS) in the USA; the second is a modified version of the so-called "tilt VPR" correction technique developed by Zhang & Qi (2010); and the third is the new VPR correction using the S-Band precipitation profiler (SPROF, White *et al.*, 2000; Matrosov *et al.*, 2006) data from the National Oceanic and Atmospheric Administration (NOAA) Hydrometeorological Test-bed (HMT, http://hmt.noaa.gov).

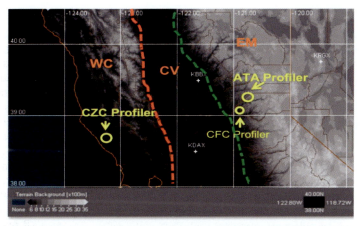

Fig. 1 Map of the study domain. The yellow circles indicate locations of three S-band precipitation profilers used in this study. Red and green dash lines divide this domain into three areas: west coast (WC), central valley (CV) and east mountainous area (EM).

Tilt VPR correction

The tilt VPR correction (Zhang & Qi, 2010) contains three steps:

(1) Computing azimuthal mean reflectivity from pre-defined precipitation areas with potential BB contamination and with radar sampling in the ice region. The areas are delineated according to the following criteria:
(i) VIL (vertically integrated liquid, Greene & Clark, 1972) < 6.5 kg/m^2;
(ii) Blockage < 50%;
(iii) Reflectivity > 15 dBZ and composite reflectivity > 30 dBZ (to capture BB peak) for BB areas;

(iv) Reflectivity > 15 dBZ or composite reflectivity >30 dBZ for ice regions.
(2) Deriving an idealized VPR by fitting the mean observed VPR with a linear model (Fig. 2). BB top, peak and bottom are identified and slopes (α, β, γ) of the linear VPR model are computed from the mean observed VPR.
(3) Applying a VPR correction to reflectivity pixels that meet the following criteria:
  (i) Height of the reflectivity bin is above the BB bottom.
  (ii) VIL < 6.5 kg/m^2;
  (iii) Reflectivity > 20 dBZ or composite reflectivity > 30 dBZ in the BB region;
  (iv) Reflectivity > 5 dBZ or composite reflectivity > 30 dBZ in the ice region.

An adjustment is made to the hybrid scan reflectivity according to the procedure, assuming a horizontally invariant VPR and a constant reflectivity below the BB bottom.

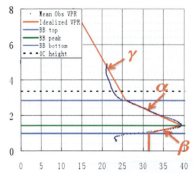

Fig. 2 The conceptual VPR model used in the current study. The blue dots represent the mean observed VPR, and the red line is the idealized VPR. The blue lines are BB top (upper) and bottom (below). The solid green and dashed black lines are BB peak and the 0°C height, respectively.

Parameterized VPR correction

Deriving the parameterized VPR Based on observations of thousands of SPROF VPRs and previous studies (e.g. Kitchen *et al.*, 1994), the tilt VPR model described above was extended to a 4-piece linear model (Fig. 3(a)). The fourth piece, namely the orographic precipitation enhancement process, was represented by a negative dZ/dh slope (η) below the BB.

The slope parameters in the current study are derived from three SPROF radars deployed during the NOAA/HMT 2005–2006 winter experiment. The radars were located in Cazadero (CZC) near the west coast of California, and in Alta (ATA) and Colfax (CFC) on the California Sierras (Fig. 1). These profiler radars measure time evolutions of the reflectivity structure along a vertical column with high temporal (1 min) and spatial (60 m) resolution (White *et al.*, 2000). The data used in the current study was 5-min averages of the 1-min observations.

Each 5-min averaged SPROF VPR was separated into four height layers corresponding to the linear model, which are above BB top (h_{top}), between BB top and peak (h_{peak}), between BB peak and bottom (h_{bttm}), and below BB bottom. The BB peak height in a given SPROF VPR is defined as the peak between pre-specified height limits of $h_{0C} - D$ to $h_{0C} + \delta$, where h_{0C} represents the freezing level height obtained from a sounding or a numerical weather prediction model, D (default = 700 m) represents the BB depth, and δ (default = 200 m) represents the uncertainty associated with the freezing level. The BB top (bottom) heights were found by searching the VPR for the nearest maximum curvature above (below) h_{peak}. Once the BB peak, top, and bottom heights were determined, the slopes of the VPR model were obtained through a linear least-squares fitting to the VPR data in each of the four corresponding height layers. Figure 3(b) shows the slopes derived from the SPROF data during the period 30 December 2005 to 1 January 2006. The slope of each piece derived from CFC is biggest among all three, and the other two are very close.

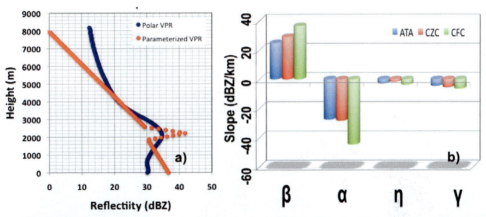

Fig. 3 (a) Parameterized VPR (red points) and polar VPR (blue points); (b) slopes from different VPR parts for different S-band precipitation profilers.

Applying reflectivity correction based on the parameterized VPR The analysis domain is divided into three regions: west coast (WC), central valley (CV) and east mountainous area (EM). For the WC area, the parameterized VPR derived from CZC data was applied and for the EM area, the parameterized VPRs from ATA and CFC were used. In the CV region, the volume scan-mean VPR from a WSR-88D radar, KDAX (Fig. 1), was used since no SPROF was deployed. The KDAX VPR was derived from volume scan reflectivity observations within an annular region (range: 20–80 km). The annular region was chosen as a balance between: (1) the need to be away from the radar origin to minimize the impact of potential ground clutter, and (2) the need to be close to the radar so that high vertical resolution can be represented. An example of KDAX volume scan mean VPR is shown in Fig. 3(a) (blue dots).

The VPR corrected reflectivity, $\hat{Z}(h_0)$, is obtained by:

$$\hat{Z}(h_0) = Z_{\text{obs}}(h) - 10\log_{10} \int_{x=h-\Delta h}^{x=h+\Delta h} \rho(x) f^4(x)\, dx \tag{1}$$

where h_0 is a low level reference height (the ground level), $Z_{\text{obs}}(h)$ is the observed reflectivity from a WSR-88D radar, h is the height of the beam centre, Δh is the half beam width, $f^4(x)$ is the two-way radar antenna gain function and ρ is the normalized VPR. It can be expressed as:

$$\rho(h) = Z(h)/Z(h_0) \tag{2}$$

where $Z(h)$ and $Z(h_0)$ are the reflectivity of parameterized or KDAX VPR (units: mm^6/m^3). The 4/3 Earth radius model is assumed for the beam propagation path. The BB height in the parameterized VPR is updated every hour and the KDAX VPR is updated every five minutes.

CASE STUDY RESULTS

A heavy rainfall event occurring between 12:00Z 30 December 2005 and 12:00Z 1 January 2006 in northern California was tested in the current study. Four WSR-88D radars (KBHX, KDAX, KRGX, and KMUX) were used for three radar QPE experiments. Figure 4(a) shows 24-h rainfall accumulations ending at 18Z 31 December 2005 without a VPR correction (Fig. 4a1), with the tilt VPR correction (Fig. 4a2), and with the parameterized VPR correction (Fig. 4a3). It is apparent that the VPR corrections alleviated some azimuthal discontinuities in the uncorrected rainfall field.

The tilt VPR correction (Fig.4b2) reduced overestimations in the bright band area (red dash line, Fig. 4b1) and improved the underestimation in the ice region (yellow dash line, Fig. 4b1).

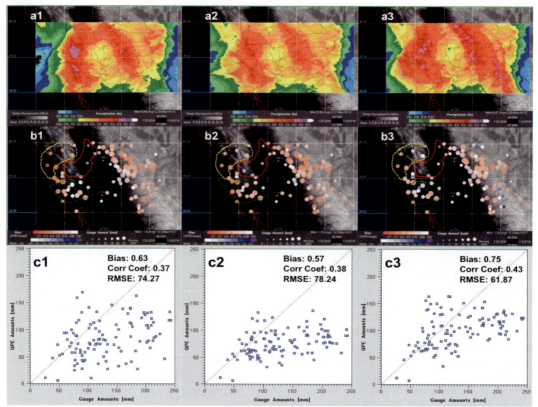

Fig. 4 24-h radar rainfall accumulations ending at 18Z on 31 December 2005 with no VPR correction (a1), tilt VPR correction (a2), and parameterized-polar VPR correction (a3). The bias ratio between the radar QPEs in (a1), (a2), and (a3) and gauge observations are shown in (b1), (b2) and (b3), respectively (a) (blue dots indicating radar overestimation and red indicating radar underestimation). (Row c) the corresponding radar QPE *vs* gauge scatter plots.

However, the QPE with the tilt VPR correction in the green dashed line area became worse than that with no VPR correction. In mountainous areas, the vertical structure of precipitation can vary in space due to the underlying topography. Using one VPR for radar QPE corrections across the whole radar domain may be problematic. The parameterized VPR correction improved the radar QPEs in all three regions (Fig. 4b3) over the other two. When compared with gauges (Fig. 4(c)), the 24-h rainfall estimate with the parameterized VPR correction had the best bias score (0.75 *vs* 0.63 and 0.57), highest correlation coefficient score (0.43 *vs* 0.38 and 0.37) and smallest root mean square error (RMSE) (61.87 *vs* 78.24 and 74.27 mm) among all three.

Figure 5 shows relative bias (a) and RMSE (b) scores for hourly radar QPEs from 18:00Z 30 December to 18:00Z 31 December 2005. For each hour, the parameterized VPR correction had a bias ratio closest to unity (Fig. 5(a)) among the three, and the tilt VPR correction had the worst underestimation. A detailed investigation indicated that, while the tilt VPR correction correctly accounted for the bright band effect, it failed to make adjustment for the orographic enhancement to the radar QPE. The tilt VPR was not able to capture the orographic enhancement because the four WSR-88D radars were relatively far away from the mid- to high-slopes of the mountains. Due to the underestimation, the tilt VPR correction resulted in slightly larger RMSEs in the hourly QPE than that with no VPR correction. The parameterized VPR correction, with a better representation of the orographic precipitation process than the tilt VPR, had the smallest RMSE on the hourly scale as well (Fig. 5(b)).

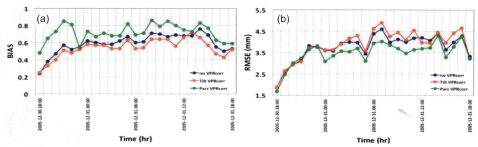

Fig. 5 The (a) relative bias and (b) root mean square error (RMSE) scores for radar precipitation.

SUMMARY

A new VPR correction algorithm has been developed and tested using one heavy rainfall event that occurred in northern California from 12:00Z 30 December 2005 to 12:00Z 1 January 2006. Region-specific VPRs are derived from S-band precipitation profiler data, and then applied to scanning radar QPE corrections across the complex terrain. The new VPR correction consistently improved the radar QPE compared to no or tilt VPR correction, while the tilt VPR correction brought adverse effects on the radar QPE due to a horizontal homogeneous assumption of VPR that is not suited for complex terrain. The new VPR correction algorithm showed good potential for improving radar-based QPEs both on hourly and 24-hourly time scales for complex terrain.

Acknowledgements Major funding for this research was provided under NOAA's Hydro-Meteorological Testbed (HMT) program and partial funding was provided under NOAA-University of Oklahoma Cooperative Agreement #NA17RJ1227.

REFERENCES

Andrieu, H. & Creutin, J. D. (1995) Identification of vertical profiles of radar reflectivity for hydrological applications using an inverse method. Part I: Formulation. *J. Appl. Meteor.* 34, 225–239.

Bellon, A., Lee, G.-W & Zawadzki, I. (2005) Error statistics of VPR corrections in stratiform precipitation. *J. Appl. Meteor.* 44, 998–1015.

Germann, U. & Joss, J. (2002) Mesobeta profiles to extrapolate radar precipitation measurements above the Alps to the ground level. *J. Appl. Meteor.* 41, 542–557.

Joss, J. & Lee, R. (1995) The application of radar-gauge comparisons to operational precipitation profile corrections. *J. Appl. Meteor,* 34, 2612–2630.

Joss, J. & Waldvogel, A. (1990) Precipitation measurement and hydrology. In: *Radar in Meteorology, Battan Memorial and 40th Anniversary Radar Meteorology Conference,* 577–597.

Kitchen, M., Brown, R. & Davies, A. G. (1994) Real-time correction of weather radar data for the effects of bright band, range and orographic growth in widespread precipitation. *Quart. J. Roy. Met. Soc.* 120, 1231–1254.

Koistinen, J. (1991) Operational correction of radar rainfall errors due to the vertical reflectivity profile. In: *Preprints, 25th Int. Conf. On Radar Meteor.,* Paris, France, Amer. Meteor. Soc. 91–94.

Matrosov, S. Y., Cifelli, R., Kennedy, P. C., Nesbit, S. W. & Rutledge, S. T. (2006) A comparative study of rainfall retrievals based on specific differential phase shifts at X- and S-band radar frequencies. *J. Atmos. Oceanic Technol.* 23, 952–963.

Vignal, B. & Krajewski, W. F. (2001) Large-sample evaluation of two methods to correct range-dependent error for WSR-88D rainfall estimates. *J. Hydrometeor.* 2, 490–504.

Vignal, B, Andrieu, H. & Creutin, J. D. (1999) Identification of vertical profiles of reflectivity from volume-scan radar data. *J. Appl. Meteor* 38, 1214–1228.

Vignal, B, Galli, G., Joss, J. & Germann, U. (2000) Three methods to determine profiles of reflectivity from volumetric radar data to correct precipitation estimates. *J. Appl. Meteor.* 39, 1715–1726.

White, A. B., Jordan, J. R., Marter, B. E. & Bartram, B. W. (2000) Extending the dynamic range of an S-band radar for cloud and precipitation studies. *J. Atmos. Oceanic Technol.* 17, 1226–1234.

Zhang, J. & Qi, Y. (2010) A real-time algorithm for the correction of bright band effects in radar-derived QPE. *J. Hydrometeor.* 11, 1157–1171.

Toward a physically-based identification of vertical profiles of reflectivity from volume scan radar data

PIERRE-EMMANUEL KIRSTETTER[1], HERVE ANDRIEU[2], BRICE BOUDEVILLAIN[3] & GUY DELRIEU[3]

1 *Laboratoire Atmosphères, Milieux, Observations Spatiales, 11, boulevard d'Alembert, 78280 Guyancourt, France*
  pierre-emmanuel.kirstetter@latmos.ipsl.fr
2 *Institut Français des Sciences et Technologies des Transports de l'Aménagement et des Réseaux, Department GER,
  Route de Bouaye BP 4129–44341, Bouguenais cedex, France*
3 *Laboratoire d'étude des Transferts en Hydrologie et Environnement, Domaine universitaire BP 53 38041,
  Grenoble cedex 09, France*

Abstract A method for identifying VPRs from volumetric radar data is presented that takes into account radar sampling. Physically-based constraints are introduced with a simple VPR model so as to provide a physical description of the vertical structure of rainfall over time-varying geographic domains in which the type of precipitation is homogeneous. The model parameters are identified in the framework of an extended Kalman filter, which ensures their temporal consistency. The method is assessed using the dataset from a volume-scanning strategy for radar quantitative precipitation estimation designed in 2002 for the Bollène radar (France). Positive results have been obtained; the physically-based identified VPRs: (i) present physically consistent shapes and characteristics considering beam effects, (ii) show improved robustness in the difficult radar measurement context of the Cévennes-Vivarais region, and (iii) provide consistent physical insight into the rainfield.

Key words rainfall estimation; vertical profile of reflectivity; Kalman filter; France

INTRODUCTION

Vertical variations of the radar reflectivity factor in the atmosphere are usually represented by the Vertical Profile of Reflectivity (VPR). The VPR influence is one of the major sources of error in the measurement of rainfall by weather radar. Nevertheless, the correction of radar data for the VPR influence is not yet fully satisfactory (Bellon *et al.*, 2005).This paper describes a VPR identification method based on the modelling of the vertical variations of the equivalent reflectivity factor using a limited number of physically-based parameters. A rain typing algorithm is used for an *a priori* separation of convective and stratiform regions within the rain field. The VPR inference is time-adjusted in the framework of an extended Kalman filter. This approach allows assimilation of new radar observations and to ensure the temporal consistency of the parameters which define the VPR. The manuscript is organized as follows. The next section describes the model of the vertical variations of the equivalent reflectivity factor, inspired by Boudevillain & Andrieu (2003), followed by a section which formulates the identification of the VPR in the framework of a nonlinear Kalman filter. The application of the VPR identification to the retained case study is then presented and the results obtained are discussed.

MODELLING THE VERTICAL VARIATIONS OF REFLECTVITY

The model proposed by Boudevillain & Andrieu (2003) serves to represent the vertical evolution of the equivalent radar reflectivity factor:

$$Z_e(h) = \frac{\lambda^4}{\pi^5 |K_w|^2} \int_0^\infty \sigma(D, \lambda, m(h))\, N(D,h)\, \mathrm{d}D \tag{1}$$

where h is altitude, σ is the backscattering cross-section of a hydrometeor which depends on the equivalent diameter D and the complex refractive index $m(h)$; the refractive index depends on the phase of the hydrometeors, their temperature and on the radar wavelength λ. $N(D,h)$ is the number of particles with diameters between D and $D + \mathrm{d}D$ per unit diameter range and per unit air volume at altitude h; $|K_w|^2$ is a constant depending on the refractive index for liquid water m_w. Equation (1) indicates that the equivalent radar reflectivity factor profile depends on (i) the phase of the hydrometeors, which drives their dielectric properties through a given diffusion model (T-matrix, Mie, Rayleigh), (ii) the particle size distribution (PSD), and (iii) implicitly on the vertical profile of temperature which constrains in particular the zero degree isotherm altitude and the melting layer. The atmospheric column is divided into three vertical layers. The upper layer contains particles of frozen water with air inclusions. In the lowest layer, the precipitation particles are raindrops. The intermediate layer is the melting layer in which particles are composed of a mixture of ice, air and liquid water. These three layers are defined by their altitude boundaries. The top of the precipitating cloud, provided by the radar echo top, is denoted h_T. The interface between solid and melting layers is denoted h_M; for stratiform precipitation h_M is denoted the freezing level. Δh_E is the melting layer's vertical extension. A reference level close to the ground denoted h_0 is considered at the bottom of the liquid layer. The temperature is assumed to decrease with altitude following the saturated adiabat. The scaling formalism initially proposed by Sempere Torres *et al.* (1994) is used to describe the relationship between the PSD (assumed gamma) and the equivalent radar reflectivity factor in the liquid phase, and to infer the PSD in the other layers.

The liquid layer

The liquid layer is defined between the reference level h_0 and the melting layer (level $h_M - \Delta h_E$). Hydrometeors are liquid drops and their constant is $|K_w|^2 \sim 0.93$. Variability of the equivalent radar reflectivity factor is driven by the DSD according to the radar reflectivity factor which varies linearly from Z_0 at h_0 to Z_m at the $h_M - \Delta h_E$ level, with a slope G_l.

The solid layer

In the solid layer, the hydrometeors are heterogeneous and described by a matrix of ice with inclusions of air. The "matrix inclusion" scheme (Klaassen, 1988) is used to retrieve the refractive index of hydrometeors and calculate their dielectric properties. The composition of a solid particle is parameterized using a density factor denoted D_g, varying between 0 (light snow) to 1 (hail) to cover the entire range of mass density of hydrometeors:

$$\rho(h) = \rho_{min}^{1-Dg} \rho_{max}^{Dg} \quad \text{with} \quad \rho_{min} = 5 \quad \text{and} \quad \rho_{max} = 900 \, \text{kg.m}^3 \tag{2}$$

The density factor drives the composition of the particles through the ice volume fraction of the total particle volume (Boudevillain & Andrieu, 2003):

$$f_{mat} = (\rho_{min} / \rho_{max})^{1-Dg} \quad \text{and} \quad f_{inc} = 1 - f_{mat} \tag{3}$$

where f_{mat} and f_{inc} are the matrix fraction and the inclusions fraction, respectively. The density factor D_g is part of the calculation of the complex refractive index m through the composition of particles and drives therein the dielectric properties of the particles. It is supposed to remain constant in the solid phase and the melting layer. The form of the VPR in the solid layer therefore depends on the PSD defined at the top of the liquid layer and on D_g.

The melting layer

The melting layer is a transitional zone in which the backscattering properties of precipitation particles change rapidly. The possible enhancing of the measured reflectivity by the radar, the bright band, occurs in this zone. The present study has adopted the simple and convenient scheme proposed by Hardaker *et al.* (1995) and used by Boudevillain & Andrieu (2003), which reproduces the high gradients of reflectivity by means of a reduced number of variables representative of the PSD, composition and dielectric properties in this zone. The continuity of the PSD at the solid/melting and melting/liquid transitions is ensured by assuming the PSD to be constant between solid particles and liquid raindrops (Boudevillain & Andrieu, 2003). Particles are composed of a mixture of liquid and solid water with inclusions of air. They are characterized by the melted mass fraction f_m increasing from 0 at the level h_M to 1 at the level $h_M - \Delta h_E$. A two-step processing of the Klaassen (1988) concept and the "matrix inclusion" scheme are applied. By driving the density and the dielectric properties of the particles, D_g controls the enhancement of the bright band. Values of D_g of about 0.8 simulate very light snowflakes with large air inclusions. These particles are more characteristic of stratiform precipitation, and the model simulates a large bright-band. Values of D_g of about 1.0 simulate denser particles more often met in convective precipitation. This simple melting layer model could be refined following the results of the series of papers devoted to the bright-band description (Szyrmer & Zawadzki, 1999).

SUMMARY

The vertical variations of the equivalent reflectivity factor according to the altitude can be represented using a model for the vertical variations of temperature, composition of hydrometeors and PSD. These vertical variations of the equivalent reflectivity factor can finally be written $Z_e(h;\phi)$, with $\phi = [G_l, h_T, h_M, \Delta h_E, D_g]$ is the vector grouping the five parameters of the VPR model. The vertical profile of reflectivity, defined as the equivalent reflectivity factor Z_e at altitude, normalized by its value at the reference level Z_0 is expressed: $z(h;\phi) = Z_e(h;\phi)/Z_0$.

THE VPR IDENTIFICATION METHOD

The data: rain-typed apparent VPRs

The volume radar data are divided into three sub-sets according to the rain separation typing which allows the detection of convective, stratiform and transition (or undetermined) rainfall (Delrieu *et al.*, 2009). It is supposed that all the vectors classified into a given type of rainfall are homogeneous and display the same VPR. The radar observations thus correspond to a series of vectors $Z_e(r,u)$ where r represents the distance and u rain-type. These observations are normalized by the equivalent reflectivity factor at the reference level which provides the so-called apparent VPR z_{ap}, calculated as follows:

$$z_{ap}(r,u) = \frac{\sum_{i=1}^{N(r,u,a_i,a_{ref})} Z_{em,i}(r,u)}{\sum_{i=1}^{N(r,u,a_i,a_{ref})} Z_{em,ref}(r,u)} \tag{4}$$

where $Z_{em,ref}$ and $Z_{em,i}$ are measured equivalent reflectivity factors for any pair of upper (a_i) and (a_{ref}) elevation angles. Details are provided in Kirstetter *et al.* (2010). The apparent VPRs obtained for various distances are finally regrouped in the vector $\underline{z_{ap}}$.

Application of the model for VPR identification

For rainfall type u, the VPR $z(h,u)$ for the domain D_u can be written as: $z(h,u) = \overline{Z(h,u)}/\overline{Z_0(u)}$ where $\overline{Z(h,u)}$ and $\overline{Z_0(u)}$ are the mean value of the equivalent reflectivity over D_u at altitude h and h_0, respectively. This VPR is modelled by the function $z(\overline{Z_0},\phi,h,u)$. For the elevation angle a_i, the mean reflectivity factor at distance r from the radar can be written:

$$\overline{Z_i(r,u)} = \int_{H^-(\theta_0,a_i)}^{H^+(\theta_0,a_i)} b^4(\theta_0,s)\, \overline{Z(s,u)}\, ds \tag{5}$$

where $b^4(\theta_0,r)$ is the two-way normalized power-gain function of the radar antenna at altitude h, corresponding to range r and elevation angle a_i; θ_0 is the 3-dB beamwidth while H^- and H^+ denote the lower and upper limits of the radar beam, respectively. The apparent VPR can be modelled as:

$$z_{ap}(r,u,a_{ref},a_i) = \overline{Z_i(r,u,\phi)} \big/ \overline{Z_{ref}(r,u,\phi)} \tag{6}$$

with

$$\overline{Z_i(r,u,\varphi)} = \int_{H^-(\theta_0,a_i)}^{H^+(\theta_0,a_i)} b^4(\theta_0,s)\, \overline{Z_0(u)}\, z(\overline{Z_0(u)},\varphi,s,u)\, ds \tag{7}$$

This model, denoted $z_{ap} = g(\phi)$, expresses the relationship between a series of apparent VPRs computed over the domain D_u and the physical parameters ϕ. It results from the coupling between the physically-based VPR model described previously and the radar sampling characteristics. The VPR identification consists of retrieving the parameters ϕ that best reconstitutes the observed values of apparent VPRs according to g.

The Extended Kalman filter: application to the VPR identification

The Kalman filter is a classical estimation method which has initially been developed for linear systems and further extended to nonlinear systems (Gelb, 1974). Here we use the Extended Kalman filter (EKF), adapted to weakly nonlinear systems. The EKF is applied for VPR identification as follows:

– the state vector is the vector which regroups the parameters defining the VPR characteristics
  $\phi = [G_l, h_T, h_M, \Delta h_E, D_g]$

– the observation vector is composed of the set of relative apparent VPRs; the observation error (Gaussian random with zero mean) is computed from Kirstetter *et al.* (2010);

– the observation equation is given by equation (12) which expresses the relative apparent VPRs as a function of the VPR characteristics;

– in the absence of dynamical modelling of VPR evolution, the VPR is supposed constant between two successive time steps. In these conditions, the state model is reduced to a steady state. This assumption means that the VPR representing a rain-type changes very slowly during a time increment. Its validity is tested by comparing the simulated relative apparent VPRs and observations. The test will be used to adjust the error of the steady-state system equation (Gaussian random with zero mean);

– the errors on parameters and the errors on data are independent.

The EKF enables system evolution, taking into account both the model prediction and the observations. The formulation of the VPR identification in the framework of a Kalman filter ensures the continuity of the VPR from one time-step to the next. The Kalman filter is appealing as a VPR identification method for two reasons: (i) its capability to identify the vector state consistent with observations at any time, (ii) its capability to take into account the temporal continuity of the process to be represented. Indeed, it is likely that the temporal evolution of VPR characteristics is not erratic, and displays slow variations between successive time-steps.

APPLICATION OF THE VPR IDENTIFICATION

Case study

A detailed description of the Bollène 2002 experiment can be found in Delrieu *et al.* (2009). This experiment was designed to evaluate the benefits of a radar volume-scanning strategy for radar quantitative precipitation estimation in mountainous regions. It allowed an enhanced sampling of the atmosphere at 10-min intervals. During the experiment, an exceptional mesoscale convective system (MCS) was sampled on 8–9 September 2002 (Delrieu *et al.*, 2005; Bonnifait *et al.*, 2009). Delrieu *et al.* (2005) distinguished three phases during the event: an initial period (8 September from 08:00 UTC to 22:00 UTC) where the MCS developed and became stationary in the northwest region of the city of Nîmes; a second phase (from 8 September 22:00 UTC to 9 September 04:00 UTC) where the mature MCS moved and stayed at the limit of the Cévennes mountain ridge; and a final phase (9 September from 04:00 UTC to 12:00 UTC) where a cold front swept the MCS out of the region. To illustrate the application conditions of the VPR identification, we will be using the radar data collected during this event characterized by a marked spatial heterogeneity.

The EKF was run from 11:00 UTC 8 September to 11:00 UTC 9 September at the same time-step as the volume scanning period. The initialization was the same for both convective and stratiform cases and the covariances of parameters were initialized with large values consistent with our lack of knowledge of the initial system state. A detailed sensitivity analysis of the method to the influential factors – *a priori* values of parameters, covariance errors on data and parameters – has been performed to determine the adapted application conditions.

RESULTS

The identified VPR distributions for the successive time steps of the 8–9 September 2002 event are displayed in Fig. 1. The rain-typed VPR populations (convective *versus* stratiform) are naturally very distinct. Note the stratiform VPR presents quite a large bright-band thickness compared to values mentioned in the literature (500–800 m) from vertically-pointing radar observations (e.g. Fabry & Zawadzki, 1995). This behaviour can probably be explained by the observation conditions (elevation angles) and the spatial variations of the bright-band within the domain of interest. Much less time variability may be noted from these VPR distributions than for

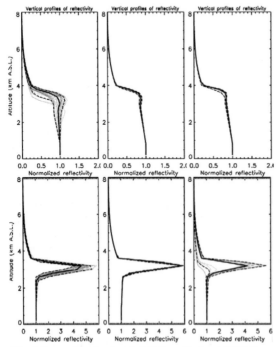

Fig. 1 Convective (top) and stratiform (bottom) VPR time series (grey curves) for 8–9 September 2002 rainfall event. From left to right, phases 1, 2 and 3 of the event (described in Delrieu *et al.*, 2005). The 10, 50 and 90% quantiles of the VPR distribution are displayed with dotted and continuous black lines.

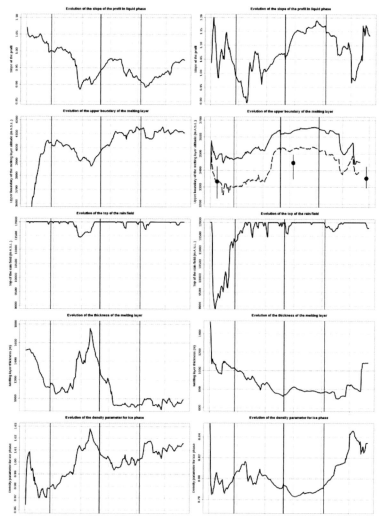

Fig. 2 Temporal evolution of the five identified parameters for the convective case (left) and the stratiform case (right) during 8–9 September rain event; from top to bottom the slope of the profile in the liquid layer G_1, the echo top h_T, the melting layer Δh_E, the upper boundary of the melting layer h_M and the density factor D_g. The vertical bars indicate phases 1, 2 and 3 of the event.

the distributions obtained with the previous method (Kirstetter *et al.*, 2010). Allowing assimilating new radar observations and ensuring the temporal consistency of the parameters which define the VPR in the method provides enhanced robustness in dealing with various conditions of sampling of the radar, typed VPRs, and in the case of strong or noisy fluctuations of the observed data. The time variations of VPR during the rain event are linked to the variations of the time-adjusted parameters $\phi = [G_l, h_T, h_M, \Delta h_E, D_g]$ shown in Fig. 2.

CONCLUSION

A VPR identification method based on a simple VPR model is presented. It takes into account the radar sampling and is applied on rain data of homogeneous type. Compared to a previous method

(Kirstetter *et al.*, 2010), new physically-based constraints are introduced and the number of parameters is considerably reduced, which makes the problem easier to solve and the solution more robust. The VPR identification in the framework of an extended Kalman filter ensures temporal consistency for the parameters. No first guess (like a range-weighted apparent VPR) is now necessary, nor any time aggregation of rain data to smooth the observations. The new method is therefore more robust and much less time consuming.

The method synthesizes radar data that provide heterogeneous information on the VPR, being differently affected by the range influence. It provides a physically-based description of the vertical structure of rainfall over time-varying geographic domains in which the type of precipitation is homogeneous. This enables to check the physical consistency of the retrieved VPR, and if necessary to add additional constraints to improve the consistency. Positive results have been obtained compared to the previous method, in so far as the physically-based identified VPR (i) presents physically consistent shapes and better characteristics than the previous VPR considering beam effects, (ii) shows improved robustness in the difficult radar measurement context of the Cévennes-Vivarais region, and (iii) provides consistent physical insight into the rainfield.

Acknowledgements This study was funded by the FP6 HYDRATE STREP (GOCE 037024) of the European Community.

REFERENCES

Andrieu, H. & Creutin, J. D. (1995) Identification of vertical profiles of radar reflectivity using an inverse method: 1 – Formulation. *J. Appl. Meteor.* 3, 225–239.

Andrieu, H., Delrieu, G. & Creutin, J. D. (1995) Identification of vertical profiles of radar reflectivity using an inverse method: 2 – Sensitivity analysis and case study. *J. Appl. Meteor.* 34, 240–259.

Bellon, A., Lee, G. W. & Zawadzki, I. (2005) Error statistics of VPR corrections in stratiform precipitation. *J. Appl. Meteor.* 44(7), 998–1015.

Bonnifait, B., Delrieu, G., Le Lay, M., Boudevillain, B., Masson, A., Belleudy, P., Gaume, E. & Saulnier, G.-M. (2009) Hydrologic and hydraulic distributed modelling with radar rainfall input - Reconstruction of the 8-9 September 2002 catastrophic flood event in the Gard region, France. *Adv. Water Resour.* 32, 1077–1089.

Boudevillain, B. & Andrieu, H., (2003) Assessment of vertically integrated liquid (VIL) water content radar measurement. *J. Atmos. Oceanic Technol.* 20(6), 807–819.

Delrieu, G., Ducrocq, V., Gaume, E., Nicol, J., Payrastre, O., Yates, E., Kirstetter, P.-E., Andrieu, H., Ayral, P. A., Bouvier, C., Creutin, J. D., Livet, M., Anquetin, S., Lang, M., Neppel, L., Obled, C., Parent-du-Chatelet, J., Saulnier, G. M., Walpersdorf, A. & Wobrock, W. (2005) The catastrophic flash-flood event of 8-9 September 2002 in the Gard region, France: a first case study for the Cévennes-Vivarais Mediterranean Hydro-meteorological Observatory. *J. Hydrometeor.* 6, 34–52.

Delrieu, G., Boudevillain, B., Nicol, J., Chapon, B., Kirstetter, P.-E., Andrieu, H. & Faure, D. (2009) Bollène 2002 experiment: radar rainfall estimation in the Cevennes-Vivarais region. *J. Appl. Meteor. Climatology* 48, 1422–1447.

Fabry, F. & Zawadzki, I. (1995) Long-term radar observations of the melting layer of precipitation and their interpretation. *J. Atmos Sci.* 52, 838–851.

Gelb, A. (1974) *Applied Optimal Estimation*. MIT Press, Cambridge, Massachusetts, USA.

Hardaker, P. J., Holt, A. R. & Collier, C. G. (1995) A melting-layer model and its use in correcting for the bright band in single-polarization radar echoes. *Quart. J. Roy. Met. Soc.* 121, 495–525.

Kirstetter, P.-E., Andrieu, H., Delrieu, G. & Boudevillain, B. (2010) Identification of Vertical Profiles of Reflectivity from Volumetric Radar Data using Rainfall Typing. *J. Appl. Met. & Climatol.* 49, 2167–2180.

Kirstetter, P.-E., Andrieu, H., Boudevillain B. & Delrieu, G. (2011) Toward a physically-based identification of Vertical Profiles of Reflectivity from volume scan radar data. *J. Appl. Met & Climatol.* (submitted).

Klaassen, W., (1988) Radar observations and simulation of the melting layer of precipitation. *J. Atmos. Sci.* 45, 3741–3752.

Sempere-Torres, D., Porr`a, J. & Creutin, J.-D. (1994) A general formulation for raindrop size distribution. *J. Appl. Met.* 33, 1494–1502.

Szyrmer, W. & Zawadzki, I. (1999) Modeling of the melting layer. Part One: Dynamics and microphysics. *J. Atmos. Sciences* 56, 3573–3592.

Vignal, B., Andrieu, H. & Creutin, J.-D. (1999) Identification of vertical profiles of reflectivity from voluminal radar data. *J. Appl. Meteor.* 38(8), 1214–1228.

Willis, T. W. & Heymsfield, A. J. (1989) Structure of the melting layer in mesoscale convective system stratiform precipitation. *J. Atm. Sci.* 46, 2008–2025.

Weather Radar and Hydrology
(Proceedings of a symposium held in Exeter, UK, April 2011) (IAHS Publ. 351, 2012).

201

Analysis of a scheme to dynamically model the orographic enhancement of precipitation in the UK

SELENA GEORGIOU, NICOLAS GAUSSIAT & HUW LEWIS

The Met Office, UK
selena.georgiou@metoffice.gov.uk

Abstract Gauge data in upland regions of the UK is sparse and often misrepresents intense precipitation events over small catchments. The production of flash flood warnings relies on high resolution input from the radar composite. It is therefore important that radar measurements of rainfall rate are as accurate as possible and account for the effects of orographic enhancements well. Within the Met Office, the Alpert & Shafir (1989) physically-based method of calculating the orographic enhancement of precipitation has recently replaced the previously operational climatology based one described by Hill (1983). The Alpert & Shafir model takes into account wind speed, wind direction, relative humidity, temperature and the topography of the region. The benefits of using a physical model are numerous. The corrections can be defined at much higher spatial resolution, with the possibility of introducing new fields, such as the vertical wind profile, and making further improvements to the physical model. The offline and operational trial results, as well as results from a post implementation analysis show that accuracy of the precipitation estimates is improved when using Alpert & Shafir's method.

Key words orographic enhancement; precipitation; vertical profile of reflectivity; seeder feeder

BACKGROUND AND MODEL DESCRIPTION

Hills and mountains have a significant influence on the amount and spatial distribution of rainfall. Orographic enhancement of precipitation is the increase in rainfall as a result of hills or mountains forcing the ascent of air. It has been estimated that average annual rainfall in England and Wales is up to 2.5 times as much over hills in the North and West of England as over nearby coastal areas, and may be as high as four times as much over the higher peaks (Hill, 1983).

There are several ways by which orographic enhancement can occur, including: rain drop displacement towards the lee slope (Robichaud & Austin, 1988; Bradley *et al.*, 1997); convection triggered by the forcing of air over topography (Gray & Seed, 2000); and formation of low level cloud (Bergeron, 1965; Kitchen & Blackhall, 1992). However, these processes alone rarely result in orographic enhancement. The Seeder-Feeder mechanism was proposed by Bergeron (1965) – forced accent of air forms the lower level "feeder" cloud. The air expands and cools as it rises. The temperature of the rising air adiabatically cools to its dew point, allowing for condensation. Given enough condensation, the cloud droplets become large enough to fall as precipitation. This is most likely to happen as a result of coalescence with droplets from a pre-existing "seeder" cloud above.

The previous scheme used in England and Wales used Hill's (1983) climatology-based method. This involved using an average rainfall intensity map calculated over England and Wales, and applying correction fields multiplied by a humidity dependant factor. 32 possible correction fields were based on four wind speed and eight wind direction categories at 5 km terrain resolution.

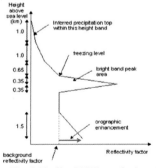

Fig. 1 Idealised VPR used when the cloud top height lies above freezing level (Kitchen *et al.*, 1994).

The Met Office VPR process

The vertical profile of reflectivity (VPR) is a function used to convert returned reflectivity into a sensible rainfall rate value. Variation in the profile arises from growth, evaporation and melting of precipitation at different altitudes. The Met Office uses an iterative VPR correction process described by Kitchen et al. (1994), which involves fitting an idealised vertical reflectivity factor profile to the radar data (Fig. 1). The orographic enhancement is converted from rainfall rate to reflectivity factor, and added to the background reflectivity factor in the 1.5 km above sea level.

Alpert and Shafir model description

The Alpert & Shafir (1989) model (AS) is a 3-D physical model, taking into account wind speed, wind direction, relative humidity, temperature and topography. They assume that the orographic enhancement equals moisture convergence resulting from uplift by hills and mountains.

The precipitation rate at height $Z_S(x,y)$ above the terrain is calculated as: $P = \rho q (V \cdot \nabla Z_S + W_l) + E$. This assumes that a constant boundary layer depth follows the topography. ρ is the mean air density; q is the specific humidity; V is the mean horizontal wind speed in the boundary layer; W_l is the synoptic scale vertical motion; E is the evaporation rate.

Within the Met Office implementation of the AS model, the orographic enhancements are calculated when relative humidity, $r > 85\%$, for which they found the evaporation contribution to be small, and therefore assume it can be ignored. The orographic enhancement due to local topography

is therefore calculated as: $P' = \varepsilon r e_s (Z_S) \dfrac{(u \frac{dz}{dx} + v \frac{dz}{dy})}{R_d T(Z_S)}$ where Z_S is the boundary layer depth; T is the

temperature at the top of the boundary layer. Here it is taken from level 6 of the Unified Model, 975 m above the land surface. It is assumed that the temperature decreases with elevation by the observed lapse rate (Alpert & Shafir, 1989); e_s is the saturated surface vapour pressure (a function of T); r is the relative humidity at the top of the boundary layer; u and v are the horizontal and vertical components of wind velocity at the top of the boundary layer. R_v is the gas constant for water vapour,

R_d is the gas constant for dry air and $\varepsilon = \dfrac{R_d}{R_v} = 0.622$; $(u \dfrac{dz}{dx} + v \dfrac{dz}{dy})$ is equivalent to $V \cdot \nabla Z$ and

represents the wind velocity at the top of the boundary level. To account for clouds that are upwind, and for advection, Alpert & Shafir (1989) calculate an area averaged enhancement for each grid box.

This is the weighted sum of the enhancements that were computed upstream: $P_0 = \left(\dfrac{f}{W} \right) \sum_{i=0}^{N} W_i P_i$

where, P_0 is the enhancement in each grid box; P_i is the enhancement computed for the ith upstream grid box; f is a precipitation efficiency factor, one of the tuning parameters we investigate within this

study. $W = \sum_{i=0}^{N} W_i$ and $W_i = \exp \left[\dfrac{-(x_i - x_0)^2}{2\sigma^2} \right]$ W_i is the weight applied to the enhancement in the ith

upstream grid box and scales the magnitude of the orographic enhancements applied upstream. The upland fetch, another of the tuning parameters, represents the maximum range and determines the number, N, of upstream grid boxes to which a weighted enhancement is applied. The Gaussian weighting function represented by W has a standard deviation given by $\sigma(W) = |V| t$ where $|V|$ is the

magnitude of the horizontal velocity vector at the top of the boundary layer, and t is the estimated cloud lifetime of a precipitation cloud, the final parameter tuned within the trial.

Radar and gauge data used in study

The motivation for this study stems from recent flooding resulting from river banks being breached during heavy rainfall events in the upland regions of North Pennines / Cumbria in NW England, and the mountainous regions of North and South Wales. The AS model parameters were tuned using data from 4 radars (Fig. 2) in these regions and the output was analysed within an offline trial. All UK network radars were included in the follow up operational trial. A post implementation analysis was carried out using data from 7 radars. Gauge data were used as ground truth (Fig. 2). Routinely received Environment Agency (EA) and Met Office collaboration project gauge data were used. Extra EA data were used within the trial for some heavy rainfall events.

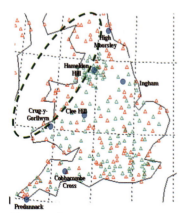

Fig. 2 Environment Agency (green) and Met Office (red) collaboration project raingauge locations in England and Wales. Highlighted area indicates the four radars involved in the initial offline trial.

METHODOLOGY

Initial sensitivity trial

An initial trial was carried out using a simplified model and synthetically generated terrain. This allowed for the model's sensitivity to terrain resolution and parameter values to be tested quickly and without a full model run through. 1 km topography resulted in the most realistic results, and suitable ranges for the key model parameters were derived for subsequent model testing.

Offline trial

Global statistics were produced, and comparisons were made with Hill's (1983) model. An offline trial was set up to enable comparison of the model output and efficiency with that of the existing method. An in depth statistical analysis was used to find the best tuned set of parameters. This involved looking at modelled surface rainfall at each gauge location as a function of accumulation threshold, range of gauge from the radar site and gauge elevation above sea level. The AS method requires input of wind, relative humidity and temperature fields. In the tests described here, each field was interpolated to 1 km resolution from the UK 4 km model. 1 km topography was used. Orographic enhancements were produced for each set of parameter values. The offline trial used was tested on heavy rainfall events that took place between June and November 2009. Values for the upwind fetch, efficiency factor and cloud lifetime model parameters were derived and subsequently used within an operational trial. The next stage was to produce the surface rainfall rate at 5-min time steps. This involved running the most recent orographic enhancement field, which is produced on an hourly basis, through the vertical profile of reflectivity (VPR) correction scheme (described by Kitchen, 1994). Using the trapezoidal rule, 13 5-min surface rainfall images were averaged over an hourly period to produce hourly accumulations at each gauge location. It was then possible to compare the results directly with the hourly gauge measurements, as well as with the accumulations modelled using the existing climatology based method (Hill, 1983).

Operational trial

It is important that any new operational model implementation performs at least as accurately as the previous implementation. In the offline analysis, the AS model was tested on archived data for heavy rainfall events using four radars. This provided a good indicator that the AS model is capable of providing improved rainfall detection in the upland catchment regions of Wales and northwest England. However, it is very important that it also works adequately across the higher topography encountered in Scotland. It was therefore necessary to initiate an operational trial to monitor the operational output of the Alpert & Shafir method across all UK radars. The operational trial period ran between 1 March 2010 and the implementation date of 1 May 2010.

OVERVIEW OF THE RESULTS

Rainfall pattern example

A snapshot example of the typical differences between the two schemes is shown in Fig. 3 for a heavy rainfall event on 29 March 2010. There was a prevailing southwesterly wind on this occasion. With reference to the upland regions marked A–C in Fig. 3(b), the following is observed when using the AS method: (i) The geographical extension of the enhancements applied in the Brecon Beacon and Dartmoor areas is increased. This indicates that enhancements downwind of the slope are at least as well and possibly better accounted for when using the Alpert and Shafir model. (ii) Enhancement intensity increases in the Snowdonia, Brecon Beacons and Dartmoor regions. (iii) The pattern of the enhancement corresponds well with the underlying topography.

Statistical results from the operational trial

The initial operation trial was run between 1 and 8 March 2010. Composite statistics were produced for this period, which largely consisted of low accumulation rainfall events (Table 1). The probability of detection (POD), false alarm rate (FAR) and critical success index (CSI) were also calculated. Many events involved drizzle conditions, with some higher accumulation events. For this period, a positive bias was found with both models, although relatively small at low accumulation thresholds. At all thresholds, improvements were found when using the AS method.

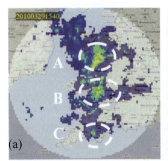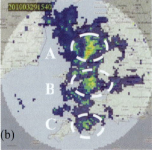

Fig. 3 Rainfall pattern in the region of the Crug-y-Gorllwyn radar, 29 March 2010. Shows comparison between (a) the Climatology method and (b) the Alpert and Shafir method for 5 km resolution data. Highlighted are the results in upland areas, A: Snowdonia, B: Brecon Beacons, C: Dartmoor.

Table 1 Composite threshold statistics for trial period 1 March 18:00 h to 8 March 09:00 h. An instance occurs where the radar measures an accumulation ≥ 0.005 mm h^{-1} and a gauge accumulation of ≥ 0.2 mm h^{-1}.

	Climatology based method				Alpert and Shafir method			
Threshold (mm h^{-1})	Instances	BIAS (mm h^{-1})	RMSE (mm h^{-1})	RMSF	Instances	BIAS (mm h^{-1})	RMSE (mm h^{-1})	RMSF
>0 mm	109	0.25	1	3.24	113	0.22	0.9	3.12
>0.2 mm	77	0.31	1.19	3.41	80	0.27	1.06	3.19
>1 mm	16	1.21	2.52	5.66	16	1.12	2.28	4.86
>4 mm	1	8.65	8.65	12.53	1	7.71	7.71	5.56

Statistical comparisons between the climatology and AS methods were carried out over all UK radars during the operational trial, and the results were similar to those shown in Table 1. After a successful trial, the AS model was implemented operationally in May 2010.

Statistical results from post implementation analysis

In the months following the model implementation, there were a number of heavy rainfall events, leading to flood warnings being issued by the Environment Agency in England and Wales. To verify the quality of the new implementation, further comparison between the models was performed using radar and gauge data from 20 such events within the period June to November 2010.

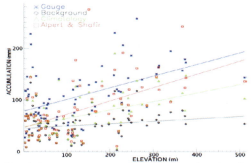

Fig. 4 Average gauge accumulations *vs* elevation for High Moorsley, over 15 heavy rainfall events. Comparisons are between the background (no orog. enh.), Hill (1983) and AS (1989) methods.

Table 2 Comparison between Hill (1983) and AS methods, November 2010. An instance occurs where the radar measures an accumulation ≥ 0.005 mm h^{-1} and a gauge accumulation of ≥ 0.2 mm h^{-1}.
(a) High Moorsley radar

| Threshold (mm h^{-1}) | Climatology-based method | | | | AS method | | | |
	Instances	Bias (mm h^{-1})	RMSE (mm h^{-1})	RMSF	Instances	Bias (mm h^{-1})	RMSE (mm h^{-1})	RMSF
>0.0	3734	0.41	1.19	3.27	3735	0.3	1.26	3.29
>0.2	3394	0.45	1.24	3.3	3411	0.32	1.32	3.31
>1.0	1674	0.83	1.73	3.71	1749	0.56	1.8	3.7
>4.0	161	3.05	3.92	3.84	199	1.69	3.85	3.76
>10.0	2	16.55	17.16	48.97	2	16.42	17.02	24.36

| (b) Crug-y-Gorllwyn radar | | | | | | | | |
Threshold (mm h^{-1})	Instances	Bias (mm h^{-1})	RMSE (mm h^{-1})	RMSF	Instances	Bias (mm h^{-1})	RMSE (mm h^{-1})	RMSF
>0.0	3269	0.69	1.46	4.48	3269	0.59	1.41	4.19
>0.2	2889	0.76	1.56	4.65	2918	0.65	1.5	4.31
>1.0	1295	1.48	2.28	6.12	1371	1.19	2.14	5.25
>4.0	127	4.55	5.51	9.27	129	4.12	5.27	7.01
>10.0	8	14.06	14.62	28.74	8	13.8	14.26	15.84

Table 2 shows comparisons between the two methods for the High Moorsley (NE England)and Crug-y-Gorllwyn (Wales) radars. Similar comparisons were made for the Clee Hill, Hameldon Hill, Cobbacombe Cross, Ingham and Predannack radars. Significant statistical improvements were calculated for each radar when using the AS model. The radar and gauge accumulations matched closer when using AS model (see Fig. 4). The AS method resulted in improvements in the bias, RMSE and RMSF statistics. The number of instances identified (Table 2) increased, in particular for higher accumulation thresholds. This indicates that implementation of the AS model has led to improved calculation of orographic enhancement in England and Wales, under heavy rainfall conditions. These improvements have been made with no detriment to the POD, FAR and CSI. With both schemes, there is sometimes significant radar underestimation when compared with gauge totals. Each model can only add an enhancement where the radar measures some precipitation in the first place. In many cases, the radar signal will overshoot both the seeder and feeder clouds, and miss any precipitation altogether, resulting in underestimation of precipitation.

FUTURE MODEL DEVELOPMENTS

Input model fields are currently derived from the 4 km model, soon to be replaced by the 1.5 km model. This higher resolution may make it possible to add in the vertical wind contribution.

A possible means of accounting for model errors in input fields would be to introduce a stochastic orographic enhancement by adding perturbations to the humidity, temperature and horizontal wind fields. An ensemble of the resulting orographic enhancement fields could be produced, allowing for uncertainty in the resulting enhancement fields to be better understood. A similar method to the Met Office Convection Diagnosis scheme could be used (Hand, 2002). This finds the probability that convection will take place within a grid square, and determines its structure, using NWP input model fields. To achieve this, the expected model errors in temperature and humidity are taken into account, and forcing from sub-scale mechanisms at 1 km is allowed. The probability of occurrence of each perturbation is used to determine forecast values within each grid square for the temperature. Variations in humidity are also accounted for.

CONCLUSIONS

The Hill (1983) and Alpert & Shafir (1989) methods of calculating orographic enhancements have been compared for a number of heavy rainfall events, using 1 km resolution topography and 1 km model fields. Significant statistical improvements are calculated when using the AS method.

With the climatology method, corrections were defined at 5 km resolution. This is unrealistic given local variations in terrain height. In contrast, the AS method enables corrections to be defined at much higher spatial resolution, and as a result there is the possibility of introducing new fields, such as the vertical wind profile, into future model developments. There is much scope for further changes to be implemented within the AS model as a result of future improvements in processing capabilities.

The AS model is quite simplified when compared with other models used to determine orographic enhancement. This is beneficial in terms of processing power and consequently its ability to be run in real time using present processing capabilities. However, as a result there are limitations associated with it. In the future, given increased processing and storage capacity, it should be possible to either extend the AS model to account for these limitations, or for replacement with a more comprehensive dynamic model. During the trial period, performance of the AS model using 1 km input fields was at least comparable with that of the climatology-based model, and in many cases significant statistical improvements were observed. The AS model was implemented operationally in May 2010 after successful trial periods. The results from a post-implementation analysis demonstrated statistical improvements as well as a greater number of instances of orographic enhancement being accounted for, with no detriment to the FAR.

REFERENCES

Alpert, P. & Shafir, H. (1989) A physical model to complement rainfall normals over complex terrain. *J. Hydrol.* 110, 51–62.

Bergeron, T. (1965) On the low-level redistribution of atmospheric water caused by orography. In: *Proc. Int. Conf. on Cloud Physics* (Tokyo), 96–100.

Bradley, S. G., Gray, W. R., Pigott, L. D., Seed, A. W., Stow, C. D. & Austin, G. L. (1997) Rainfall redistribution over low hills due to flow perturbation. *J. Hydrol.* 202, 33–47.

Gray, W. R. & Seed, A. W. (2000) The characterisation of orographic rainfall. *Meteorological Applications* 7, 105–119.

Hand, W. H. (2002) The Met Office convection diagnosis scheme. *Meteorological Applications* 9, 69–83.

Hill, F. F. (1983) The use of average annual rainfall to derive estimates of orographic enhancement of frontal rain over England and Wales for different wind directions. *J. Climate* 3, 113–129.

Kitchen, M. & Blackall, R. M. (1992) Orographic rainfall over low hills and associated corrections to radar measurements. *J. Hydrol.* 139, 115–134.

Kitchen, M., Brown, R. & Davies, A. G. (1994) Real-time correction of weather radar data for the effects of bright band, range and orographic growth in widespread precipitation. *Quart. J. Royal Met. Soc.* 120, 1231–1254.

Robichaud, A. J. & Austin, G. L. (1988) On the modelling of warm orographic rain by the seeder–feeder mechanism. *Quart. J. Royal Met. Soc.* 114, 967–988.

Weather Radar and Hydrology
(Proceedings of a symposium held in Exeter, UK, April 2011) (IAHS Publ. 351, 2012).

207

Impact of quality control of 3-D radar reflectivity data on surface precipitation estimation

KATARZYNA OŚRÓDKA, JAN SZTURC & ANNA JURCZYK

Institute of Meteorology and Water Management, 40-065 Katowice, ul. Bratków 10, Poland
katarzyna.osrodka@imgw.pl

Abstract In the paper the impact of quality control of 3-D radar reflectivity data on surface precipitation estimates is investigated. The developed processing chain for raw 3-D weather radar data aims at the data corrections due to non-meteorological echoes (e.g. from external interferences, specks) and disturbances in meteorological echoes (radar beam blockage, attenuation in rain). All the algorithms were worked out for single polarization radars. Precipitation rates were generated from uncorrected and corrected 3-D reflectivity data and compared in order to assess the algorithm efficiency. The investigation was performed on radars included in the Polish weather radar network POLRAD.
Key words radar; precipitation; quality; correction

INTRODUCTION

Weather radar measurements are burdened with numerous errors caused by both technical and meteorological reasons. Their nature is very different depending on their sources, so that quite different techniques must be applied to remove them. Many of them are very difficult to identify and remove, especially if non-meteorological and meteorological echoes are observed at the same time and place. Data from other sources would be useful for this purpose. However, in many cases they are not available in real time. Since the algorithms presented here were developed for operational purposes, the need for real-time availability was an essential criterion.

Weather radar data must be corrected before estimation of ground precipitation for hydrological applications. The corrections can be performed at two stages of radar information processing. Since the three-dimensional (3-D) data are commonly a starting point for all reflectivity-based products including ground precipitation rate, most corrections should be done at this stage. Further corrections must be performed at the 2-D product generation stage: adjustment to raingauge data, corrections due to distance to radar and related increase of the lowest beam height, etc. The presented works were focused on the 3-D reflectivity data correction.

TESTBED

Polish weather radar network POLRAD

The Polish network of weather radars operated by the Institute of Meteorology and Water Management (IMGW) consists of eight C-Band Doppler weather radars manufactured by Selex SI Gematronik. Two of them are dual polarimetric radars whilst the others are planned to be replaced. However, in the near future single polarimetric radars will still be a base for measurements, so the paper is limited to such types of radar.

Employed data

Source data of radar measurements are 3-D: so-called volumes obtained from several antenna elevations. The scan strategy used in the POLRAD network is described in Table 1.

Products related to surface precipitation are commonly used for hydrological purposes. Such 2-D products most often are delivered as CAPPI (Constant Altitude Plan Position Indicator) i.e. horizontal cut at preset altitude or SRI (Surface Rainfall Intensity) measured at a preset height above the ground based on a digital terrain map (Selex, 2010). The SRI products (at 1 km above ground) are tested here as they are more suitable for hydrology. Precipitation accumulations

Table 1 Scan parameters currently used in the weather radar network POLRAD.

Parameter	Value
Radar beam	1°
Number of azimuths	360
Maximum range	250 km
Distance between samplings along radar beam	1 km
Elevation angles	0.5; 1.4; 2.4; 3.4; 5.3; 7.7; 10.6; 14.1; 18.5; 23.8°

Table 2 List of correction algorithms used for weather radar data.

Code	Task	Algorithm	References
SPIKE	Removal of external interference	3-D reflectivity structure analysis	Zejdlik & Novák (2010)
SPECK	Removal of measurement noise	3-D reflectivity structure analysis	Michelson *et al.* (2000); Jurczyk *et al.* (2008)
PBB	Beam blockage correction	Using topography map	Bech *et al.* (2007); Selex (2010)
ATT	Rain attenuation correction	Reflectivity based: attenuation coefficient	Battan (1973); Selex (2010)

computed as the radar product PAC (Precipitation Accumulation) for different time periods constitute input to rainfall–runoff modelling.

SET OF CORRECTION ALGORITHMS

The challenges in corrections of radar observations investigated in the paper are summarised in Table 2. The correction chain consists of the set of algorithms related to the following overall tasks: removal of spurious, non-meteorological echoes (Spike, Speck) and correction of disturbed reflectivity measurements (PBB, ATT). The issues become more complicated if non-meteorological echoes interfere with meteorological ones.

REMOVAL OF NON-METEOROLOGICAL ECHOES

Removal of signals from external interference

Signals coming from external interference that overlap with the radar signal have become the source of non-meteorological echoes in radar images with increasing frequency. Their effect is similar to the spike generated by the Sun, but they are observed at all azimuths at any time and mainly at lower elevations. The echoes may be characterized by very high reflectivity.

The spurious echoes from the Sun and external interference, generally called spikes, are characterized by spatial structure that clearly differs from the precipitation field pattern. Recognition of such a kind of echo is not very difficult unless it interferes with the precipitation field. An algorithm developed for removal of such artefacts should deal with situations where the spike must be identified and separated from the precipitation field, then cut from it and replaced by more proper reflectivity values.

The algorithm developed in this scheme consists of three subsequent steps: (a) removal of "wide" spikes (with spread of several azimuths), (b) removal of "narrow" spikes (not wider than a few azimuths), (c) removal of all spurious echoes at high altitudes.

The first sub-algorithm is based on analysis of the horizontal and vertical structure of the given echo in terms of its statistical characteristics. The observation is that the variance of a spike along a radar beam is low, but high across it, and moreover, the feature is preserved within a large part of the given beam. The second sub-algorithm checks the variability of spikes within a narrower surrounding of the beam (e.g. no more than three pixels from the investigated one). The third one removes all echoes detected at high altitudes where no meteorological echoes can be found, and only spikes may appear. If the detected spike and meteorological echo interfere then reflectivities in the affected pixels are interpolated from the precipitation neighbourhood.

Examples of the algorithm running are presented in Figs 1 and 2, where a "wide" spike (Fig. 1) and "narrow" ones (Figs 1 and 2) are nearly completely removed. Efficiency of the algorithm depends on spike location relative to precipitation field: if the spike is located over non-precipitation areas then the efficiency is very high, especially in case of a "narrow" spike. On the other hand the efficiency can be lower for a not very smooth (especially along a beam) pattern of the "wide" spike. The task becomes more difficult if the spike lies on a precipitation echo; in this case the spike is significantly but not completely removed (Fig. 2). These examples are presented for the Legionowo radar, which is very strongly affected by spike-type echoes.

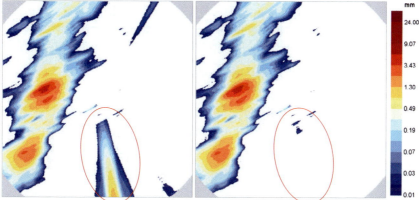

Fig. 1 Example of radar data before and after removal of a "wide" external signal. Legionowo radar, hourly PAC from SRI, 20-21 UTC 28 May 2010. "Narrow" spikes removal is switched on as well.

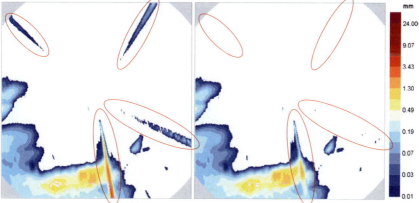

Fig. 2 Example of radar data before and after removal of "narrow" external signals. Legionowo radar, hourly PAC from SRI, 21-22 UTC 5 May 2010.

Speck and reverse speck removal

Speck or reverse speck is defined as an isolated pixel with non-echo surrounding (speck) or isolated non-echo pixel in echo surrounding (reverse speck). They both are considered as measurement noise: in the first case the echo is removed, whereas in the second one reflectivity is interpolated from neighbouring pixels. If precipitation data are generated only from one elevation, as for products such as SRI or CAPPI, then the specks are not significant; however, for e.g. MAX product (Maximum Display) they would be more crucial.

CORRECTIONS OF METEOROLOGICAL ECHOES

Beam blockage correction

The radar beam can be blocked by terrain, especially in a mountainous area, which results in a decrease of the signal power in the beam. This decrease is in proportion to the percentage beam blockage (*PBB*): therefore correction is made based on its value.

Correction due to partial beam blockage consists in applying a multiplicative correction factor (Bech *et al.*, 2007) for precipitation rate calculated from a specific *Z–R* formula:

$$R_{cor} = R \cdot (1 - PBB)^{-\frac{1}{b}} \tag{1}$$

where $b = 1.6$. The correction is employed if the *PBB* value is not higher than 0.7, but for higher *PBB* values it is proposed to take reflectivity from a neighbouring higher elevation.

In Fig. 3 an example for correction of blockage in a mountainous area is presented for the Pastewnik radar where very strong beam blockage is observed. It can be noticed that the partial and total blockage observed in the southwest was efficiently corrected.

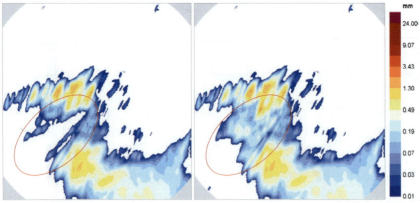

Fig. 3 Example of radar data before and after blocked signal reconstruction. Pastewnik radar, hourly PAC from SRI, 17-18 UTC 5 May 2010.

Correction of attenuation in rain

Attenuation is defined as a decrease in radar signal power after passing a meteorological target that results in underestimation of the measured precipitation rate *R*, such that:

$$A = 10 \cdot \log_{10} \frac{R_{cor}}{R} \tag{2}$$

where *A* is the specific attenuation (in dB km^{-1}) and R_{cor} is the non-attenuated precipitation rate.

The aim of the algorithm is to calculate the non-attenuated, i.e. corrected rain rate R_{cor}. Empirical formulas for specific attenuation can be found in the literature. For C-band radar the two-way attenuation *A* at 18°C can be estimated from the formula (Collier, 1989):

$$A = 0.0044 \cdot R^{0.17} \tag{3}$$

The computation should be performed iteratively (gate by gate) for each measurement gate, because the attenuation estimated at a given gate depends on corrected (i.e. non-attenuated) rain at all previous gates. A threshold for maximal correction is set up to avoid instability in the algorithm. The impact of the correction for a C-band radar is depicted in Fig. 4 where areas marked with red ovals show places where relatively high attenuation was detected behind intense precipitation.

EXAMPLE OF THE TOTAL SCHEME RUNNING

The total effect of the corrections on radar-based ground precipitation data is especially noticeable for longer accumulations (over 1 h). In Fig. 5, PAC data generated from SRI products for the whole of May 2010 are presented. The Legionowo radar covers lowland so that no effects related to beam blockage are observed here. However, intense external interference is detected mostly at the same azimuths, so it becomes more significant for the case of accumulations.

A similar analysis was performed for the Pastewnik radar located near the Sudeten Mountains where strong blockage is observed (Fig. 6). In the southwest part of the radar image the beam is totally blocked so precipitation is hardly detected here, and higher elevation data had to be employed. Significant improvement is noticeable, but is however not quite satisfactory.

Efficiency of the set of correction algorithms has been evaluated by analysis of statistical properties of the monthly accumulations. It is assumed that asymmetry and smoothness can be metrics of the data quality (Joe, 2011). They are quantified respectively through: (i) differences between values in pixels symmetrical with respect to the centre of the image, and (ii) ratio of squared mean and variance (in the certain grid around the pixel) averaged for the whole image.

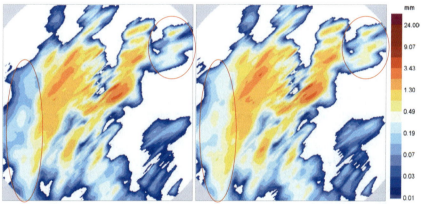

Fig. 4 Example of radar data before and after attenuation in rain correction. Legionowo radar, hourly PAC from SRI, 22–23 UTC 17 May 2010.

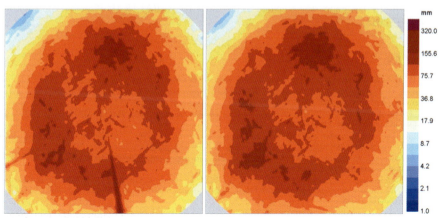

Fig. 5 Example of radar data before and after all corrections. Legionowo radar, monthly PAC from SRI, May 2010.

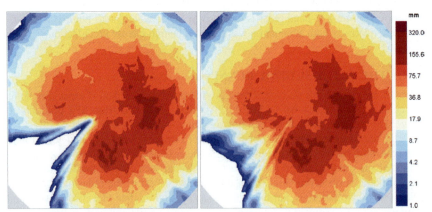

Fig. 6 Example of radar data before and after all corrections. Pastewnik radar, monthly PAC from SRI, May 2010.

The verification on May 2010 shows a decrease of the asymmetry metric from 0.12 to 0.10 for the Legionowo radar and from 0.38 to 0.34 for Pastewnik, and an increase of the smoothness metric for the radars from 135 to 160 and from 112 to 130, respectively. Therefore, both metrics show significantly higher quality of the data after corrections for the two radars are made.

The effect of Earth curvature causes lower accumulations far away from the radar site that are clearly noticeable in Figs 5 and 6 and should be corrected at the stage of 2-D data corrections by means of empirical formulae based on comparison with raingauge data.

FURTHER WORKS

The paper presents corrections performed on 3-D weather radar reflectivity data. Algorithms for the following corrections were developed: removal of external interference and measurement noise, as well as correction due to beam blockage and attenuation in rain. These algorithms significantly improve the data, however their efficiency depends on radar location, echo pattern and intensity, etc.

The scheme was optimized for data from the POLRAD weather radar network and works from 2011 in test mode. Further work will be devoted to developing corrections for 2-D reflectivity and precipitation rate data, especially ground estimates. Integration of the two correction stages and their verification on a rainfall–runoff model is planned.

Acknowledgement The paper was prepared in the frame of the BALTRAD Project (Baltic Sea Region Programme 2007-2013).

REFERENCES

Battan, L. J. (1973) *Radar Observations of the Atmosphere*. University of Chicago Press, Chicago, pp. 323.

Bech, J., Gjertsen, U. & Haase, G. (2007) Modelling weather radar beam propagation and topographical blockage at northern high latitudes. *Quart. J. Roy. Met. Soc.* 133, 1191–1204.

Collier, C. G. (1989) *Applications of Weather Radar System. A Guide to uses of Radar Data in Meteorology and Hydrology*. Ellis Horwood Limited, New York, USA.

Joe, P. (2011) The *WMO radar QC QPE inter-comparison Project RQQI*. WMO/CIMO/IOC-RQQI document (www.wmo.int/pages/prog/www/IMOP/meetings/RS/IOC-RQQI-1/Doc.2.pdf).

Jurczyk, A., Ośródka, K. & Szturc, J. (2008) Research studies on improvement in real-time estimation of radar-based precipitation in Poland. *Meteorol. Atmos. Phys*. 101, 159–173.

Michelson, D. B., Andersson, T., Koistinen, J., Collier, Ch. G., Riedl, J., Szturc, J., Gjersten, U., Nielsen, A. & Overgaard, S. (2000) *BALTEX Radar Data Centre products and their methodologies*. SMHI Reports Meteorology and Climatology, no. 90, Stockholm 2000, 76 pp.

Selex (2010) *Rainbow 5. Products and algorithms*. Release 5.31.0. Selex SI GmbH, Neuss, 442 pp.

Zejdlik, T. & Novák, P. (2010) Frequency protection of the Czech weather radar Network. In: *Proc. ERAD 2010. Advances in Radar Technology*, 319–321.

Weather Radar and Hydrology
(Proceedings of a symposium held in Exeter, UK, April 2011) (IAHS Publ. 351, 2012).

Real-time adjustment of radar data for water management systems using a PDF technique: The City RainNet Project

JOHN V. BLACK[1,2], CHRIS G. COLLIER[2], JOHN D. POWELL[1], RICHARD G. MASON[1] & ROD J. E. HAWNT[1]

1   *Hydro-Logic Ltd, Old Grammar School, Church Street, Bromyard HR7 4DP, UK*
    jblack@hydro-logic.co.uk
2   *UK National Centre for Atmospheric Science, School of Earth & Environment, University of Leeds, Leeds LS2 9JT, UK*

Abstract A key challenge of the project is to develop and implement a real-time, rainfall radar adjustment software system. This system will provide rainfall data of reliable accuracy, particularly in convective storm situations. The data produced by the system must have high enough accuracy and reliability to enable water companies and others to be confident in using it in the operation of water management systems. This project will deliver a technically-robust prototype of a commercially viable system, capable of delivering these objectives. The project involves three UK water companies (Yorkshire, Northumbrian and Scottish Water) who have, or will install raingauge networks on approximately 1 km × 1 km grids. The approach to the radar data adjustment reported in this paper is based upon using a Probability Matching Method (PMM). Each raingauge outstation comprises a weighing principle raingauge, the OttPluvio2, linked to an ISODAQ GPRS data logger manufactured by Hydro-Logic.

Key words radar; raingauge; probability matching; water companies

INTRODUCTION

In this paper we outline a Knowledge Transfer Partnership (KTP) R & D Project involving the University of Leeds and Hydro-Logic Ltd aimed at combining radar and raingauge data in real-time within urban areas. The project also involves three water companies, Yorkshire Water, Northumbrian Water, and Scottish Water. In Yorkshire the raingauge network has been installed in Scarborough, Filey, Flamborough and Bridlington. Northumbrian Water and Scottish Water will purchase dense raingauge networks for a city in each of their areas. For Scottish Water the city will be Dundee The motivation for the water companies is linked to their Asset Management Programmes (AMP) 5 concerned with developing storm intensity alarms before sewer flooding occurs, and post-event analysis through improved storm frequency statistics. For each incident of sewer flooding the UK Environment Agency may impose a fine of around £20k upon a water company.

At present, in the UK urban areas suffer from a lack of dense raingauge networks. Those tipping bucket raingauge networks that are available are costly to maintain, and often have uncertain calibrations. Also radar data in cities is sometimes of poor quality, particularly in short-period convective storms, Hence, the raingauge and radar data will be merged in real-time, with the raw data, the merged product being stored in a web-based archive. Prototype facilities will be provided for the water companies to access the quality controlled radar and raingauge data.

RAINGAUGE NETWORKS

The requirements for the raingauge networks in the cities were specified as:

–   low maintenance – annual visits only
–   lower cost of ownership than tipping bucket raingauges
–   wireless and battery powered
–   each connection to GSM/GPRS telemetry logger
–   proven measurement technology
–   a focus on medium to high intensity storms
–   good resolution and uncertainty

–   readily available sites having easy access
–   ease of installation

These requirements have been satisfied by the use of the OTT Pluvio2 weighing gauge, which was best in its class in the 2009 WMO raingauge trials. This raingauge is used with the Isodaq Frog telemetry unit. The water companies are funding the cost of the raingauge equipment, and installation / maintenance of a minimum of 36 gauges, which will be deployed in each city on a grid of minimum size area 25 km^2. Work in the project will ascertain the most appropriate grid configuration. The raingauge network installed in the Yorkshire Water area is shown in Fig. 1. The areas where the raingauge networks are, or will be, installed in relation to UK Met Office weather radar coverage is shown in Fig. 2.

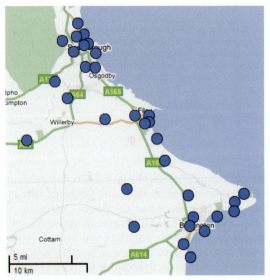

Fig. 1 Yorkshire Water raingauge network. The blue circles show the locations of the raingauges currently installed in Scarborough, Filey, Bridlington, and their surrounding areas. Note that an artefact of the placement of the symbols means that some raingauges appear to be located in the sea. At the time of writing 32 were installed. A total of 36 raingauges will be installed.

THE UK RADAR NETWORK

That part of the UK weather radar network relevant to the project comprises radar at the sites shown in Fig. 2. The radars operate at 5.4 GHz (C-band) with an antenna diameter of 3.7 m giving a beam width of nominally 1°. Radar does not measure rainfall (R) (or snowfall) directly but measures reflectivity (Z), which is then converted to rainfall (or snowfall) using a Z–R relationship. There are a number of "errors" associated with this process described by many authors, see e.g. Collier (1996). For the data to be used in this project many of these "errors" are removed or reduced by processing at the radar site and/or at the Met Office network centre at Exeter (Harrison *et al.*, 2009). The project will be working first with radar data over the Bridlington, Yorkshire, area from the Ingham, Lincolnshire radar. These data will be available over this town on a 2 km × 2 km grid.

RAINGAUGE ADJUSTMENT PROCEDURE

Many attempts have been made to remove the observed radar bias and reduce the Mean Absolute Error in radar measurements of rainfall using additional measurements provided by the raingauges

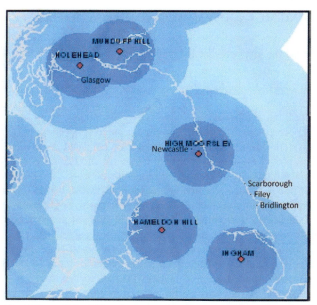

Fig. 2 Part of the UK radar network relevant to this project showing the three urban districts within which the dense raingauge networks are to be installed. Key to radar coverage: dark blue 1 km × 1 km grid, lighter blue 2 km × 2 km grid and light blue 5 km × 5 km grid (after Met Office).

(see e.g., Collier *et al.*, 1983). The motivation for much of this work arose from the demands of hydrologists for improved reliable, accurate rainfall estimates. Whilst considerable progress has been made in improving the quality of radar estimates of rainfall, including most recently, via the use of polarisation radar techniques, there remains a need to combine radar and raingauge data. However, particularly in highly variable convective rainfall, the use of raingauge adjustment can be detrimental depending upon the density of the raingauge network. Therefore many operational radars are only adjusted using an average over many raingauges close to the radar site with running daily (or longer) radar – raingauge comparisons. Procedures have been developed which derive individual radar (averaged over a number of grid squares) – raingauge ratios, and extrapolate the value of the grid providing correction factors (see for example the IRIS Product & Display Manual (February 2006) offered by Vaisala FM).

The success of this approach depends very much on the density of the raingauge network. Even if the network spacing is 1 km (say) in convective rainfall this can be insufficient. Figure 1 shows the raingauge network that has been installed in the Scarborough–Bridlington area of Yorkshire Water. Note that very large rainfall gradients in some locations change rapidly in this type of rainfall. Rosenfeld *et al.* (1994) proposed the combination of radar reflectivity and raingauge data using an approach called the Probability Matching Method (PMM). An adjustment function is derived from an analysis of historic radar and raingauge data. This can be updated and modified as more data are acquired in real-time and the rainfall type changes. One advantage of this approach is that a stable relationship between the raingauge measured rainfall and the radar reflectivity can be derived covering the full range of measured reflectivity values particularly if there are many raingauges available.

A further advantage is that the relationship can be used successfully when raingauges fail to operate in real-time, or are unavailable in specific locations because the raingauge network is not uniform due to siting difficulties. In such a circumstance Rosenfeld *et al.* (1994) noted that the adjustment function can be derived from relatively small reflectivity–raingauge samples. This system has not been fully tested in a real-time operational system. The current project has chosen to implement this system, the basis of which is described next.

THE PROBABILITY MATCHING METHOD (PMM) OF ADJUSTING RADAR ESTIMATES OF RAINFALL

The radar-raingauge samples for all the sites are listed together, and each value is converted to decibels (dBZ) as follows:

$$dBZ = 10 \log(200R^{1.6}) = 10(\log 200 + 1.6 \log R) \tag{1}$$

where R = radar or raingauge rainfall rate.

Here $Z = 200R^{1.6}$ has been used, but another relationship such as $Z = 400R^{1.4}$ could be used as there is only a small difference between them. The $400R^{1.4}$ is appropriate to convection, but $200R^{1.6}$ has been used in the Met Office system. The cumulative PDF values are tabulated being referred to as the "climatological" data sets. A plot of these values for both raingauge and radar separately will indicate the performance of the Ingham radar compared to the raingauge network. These radar data have the application of the Met Office raingauge adjustment procedure *excluded*, but do include all the other Met Office quality control procedures such as ground clutter removal, bright-band correction and attenuation correction. Atencia *et al.* (2010) pointed out in their study of this technique that problems with the tails of PDF distributions can be avoided by using a smoothing technique.

Figure 3 shows an example from a small sample of data of the PDF plots. Note the noise in the raingauge plot probably due to snow and ice impacting the measurements on the 24 December 2010. Removal of these obviously contaminated gauge data gives the data shown in Fig. 4. As the data set grows the need to do this should be removed. At the time of writing, the data set was too small to be reliable.

THE REAL-TIME PMM ALGORITHM

The real-time algorithm will start operating with the "climatological" datasets as a look-up table. Note that the reflectivity values will be entered in units of one half of one dBZ. As subsequent samples become available they will be added into this database, and the cumulative probability values for Z_R and Z_G will be re-calculated. This operation will be carried out every five minutes, although this frequency will be explored to investigate whether it can be done less often. However the incorporation of new data will use five minute values even if it proves possible to carry out the updating every 15 minutes for instance.

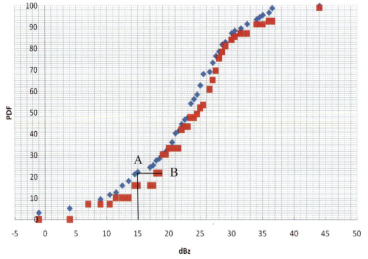

Fig. 3 Examples of PDF plots for three days of rainfall data.

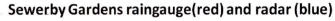

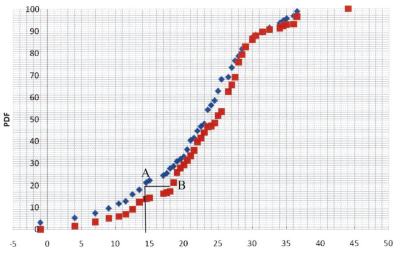

Fig. 4 Examples of PDF plots for the same data as in Fig. 3, but with the gauge data for 24 December removed.

REAL-TIME ADJUSTMENT OF THE RADAR DATA

To estimate the adjustment factor for each value of the radar rainfall rate (e.g. A), as shown in Fig. 4, the difference between gauge and radar is

$$AB \ dBZ = 10 \log_{10} (R/G)^{1.6} \qquad (2)$$

where AB is the distance between the two graphs and R/G is the gauge adjustment factor.

$$R/G = \text{Antilog}_{10} (AB / 16) \qquad (3)$$

Examples of adjustment factors are shown in Table 1.

Table 1 Examples of adjustment factors.

AB (dBZ)	R/G
1	1.2
2	1.3
3	1.5
4	1.8
5	2.1

AB is the distance between two graphs as shown in Figs 3 and 4. R/G is the raingauge adjustment factor.

In the plot shown in Fig. 4 the radar under-estimates the rainfall by about a factor of 1.8 for a radar rainfall of 15 dBZ. Applying this methodology beyond the location of the raingauge network should be done cautiously. It would not be expected to deliver substantial improvements in accuracy of the radar estimates of rainfall at significant distances from the raingauge network.

ASSESSING THE PERFORMANCE OF THE ADJUSTMENT PROCEDURE

The performance of the PMM technique will be assessed following the procedures outlined by Krajewski *et al.* (2010). Taking raingauge values as G_i, which have not been used to derive the PMM relationship, and the corresponding adjusted radar values as R_i then:

$$G_{av} = \Sigma\ G_i/N \text{ and } R_{av} = \Sigma\ R_i/N \tag{4}$$

$$\text{E}\ [G/R] = 100\ [\sigma\ [G/R]/\text{E}\ [G/R] \tag{5}$$

where $\sigma\ [G/R]$ = standard deviation of the ratios between raingauge measurements G and radar rainfall estimates R; the relative dispersion $\text{E}[G/R]$. The average differences = $100\Sigma[|G_i-R_i|/G_i]/N$; and average difference (storm bias removed), random component of the error = $100\Sigma\ |(G_i-R_i)|\text{E}[G/R]/G_i/N$. These parameters will be evaluated separately in summer, autumn, winter and spring. Independent Yorkshire Water and Environment Agency raingauges will be used for this assessment.

Comparison will also be made with the procedure for merging radar and raingauge data currently used in the Met Office Nimrod system similar to that described by Seo & Breidenbach (2002) and Seo *et al.* (2000). The performance of the PMM procedure will also be assessed against the analysis of the quality of the Nimrod radar data in upland areas reported by Lewis & Harrison (2007), and the procedure described by Vignal *et al.* (2000) in Switzerland.

REFERENCES

Atencia, A., Llasat, M. C., Garrote, L. & Mediero, L. (2010) Effect of radar rainfall time resolution on the predictive capability of a distributed hydrologic model. *Hydrol. Earth Syst. Sci. Discuss.* 7, 7995–8043.

Collier, C. G. (1996) *Applications of Weather Radar Systems. A Guide to Uses of Radar Data in Meteorology and Hydrology*, 2nd Edition. John Wiley & Sons, Chichester, UK, 390pp.

Collier, C. G., Larke, P. R. & May, B. R. (1983) A weather radar correction procedure for real-time estimation of surface rainfall. *Quart. J. R. Met. Soc.* 109, 589–608.

Krajewski, W. F., Villarini, G. & Smith, J. A. (2010) Radar-rainfall uncertainties. Where are we after thirty years of effort? *Bull. Am. Met. Soc.* 91, 87–94.

Harrison, D. L., Scovell, R. W. &. Kitchen, M. (2009) High-resolution precipitation estimates for hydrological users. In: *Proc. Inst. Civil Engng, Water Manage.* 162, Issue WM2, 125–135.

Lewis, M. W. & Harrison, D. L. (2007) Assessment of radar data quality in upland catchments. *Meteor. Apps.* 14, 441–454.

Rosenfeld, D., Atlas, D., Wolff, D. B. & Amitai, E. (1994) The window probability matching method for rainfall measurement with radar. *J. Appl. Met.* 33, 682–693.

Seo, D.-J. & Breidenbach, J. P. (2002) Real-time correction of spatially nonuniform bias in radar rainfall data using rain gauge measurements. *J. Hydromet.* 3, 93–111.

Seo, D.-J., Breidenbach, J., Fulton, R. & Miller, D. (2000) Real-time adjustment of range-dependent biases in WSR-88D rainfall estimates due to nonuniform vertical profile of reflectivity. *J. Hydromet.* 1, 222–240.

Vignal, B., Galli, G., Joss, J. & Germann, U. (2000) Three methods in determine profiles of reflectivity from volumetric radar data to correct precipitation estimates. *J. Appl. Met.* 39, 1715–1726.

Weather Radar and Hydrology
(Proceedings of a symposium held in Exeter, UK, April 2011) (IAHS Publ. 351, 2012).

219

Raingauge quality-control algorithms and the potential benefits for radar-based hydrological modelling

PHIL J. HOWARD, STEVEN J. COLE, ALICE J. ROBSON &
ROBERT J. MOORE

Centre for Ecology & Hydrology, Wallingford, UK
philhw@ceh.ac.uk

Abstract Raingauges and weather radar are essential sources of rainfall information for hydrological model-ling and forecasting. However, significant errors in raingauge time-series can drastically affect raingauge-only and combined radar-raingauge rainfall estimates. In turn, these errors can have a negative impact on hydrological model calibration, performance and failure diagnosis. This study considers the automated quality-control of 15-min rainfall totals obtained from 981 tipping-bucket raingauges across England and Wales. The Grid-to-Grid distributed hydrological model, now operated by the Flood Forecasting Centre in support of national flood warning, is used with gridded rainfall estimates to assess the utility of the raingauge quality-control procedures. Although a historical dataset is used here for demonstration and assessment purposes, the automated algorithms have been designed for implementation in real-time.

Key words quality control; raingauge errors; hydrological modelling; automated; flood forecasting; radar

INTRODUCTION

Good quality rainfall data are critical for accurate hydrological modelling. For operational application in real-time, for example in support of flood forecasting, there are competing demands on the timely availability of data and their quality. For raingauge data, operating agencies typically apply some form of manual quality-control, but this may be some time after data capture. To obtain improved real-time flood forecasts, much may be gained by applying automated quality-control algorithms to telemetry raingauge data.

Here we describe automated methods to improve the quality of raingauge data, such that a large number (981) of telemetry gauges may be included as live inputs to the Grid-to-Grid distributed hydrological model configured across England and Wales. Further to the requirement for real-time application, the size of the dataset alone makes the prospect of manual data quality-control daunting and impractically time consuming. It is not our intention to attempt to produce "perfect" rainfall time-series, but rather to robustly detect and effectively remove the most egregious faults in the raingauge network dataset.

Although time-of-tip records are available for many of the gauges in this study, they are not for all; hence the requirement for quality-control of a 15-minute interval time-series. This factor, combined with the spatial extent of the dataset and the long timescales considered here, leads to differences in our approach from those reported elsewhere (Jorgensen *et al.*, 1998; Steiner *et al.*, 1999; Stanzani *et al.*, 2000; González-Rouco *et al.*, 2001; Upton & Rahimi, 2003; Kondragunta & Shrestha, 2006).

The second part of this paper demonstrates the impact of the developed procedures on hydrolog-ical modelling, with examples from a historical study over the three years 2007–2009. Particular attention is given to the impact of the quality-control procedures when raingauge data are used in conjunction with radar rainfall data to obtain "best-estimates" of rainfall as input to the model.

QUALITY-CONTROL PROCEDURE

To enable real-time application and computational efficiency, all gauges are processed in parallel for each 15-min interval. A set of simple tests are performed on each gauge separately, before more involved comparisons to neighbours are made. Thus problems identifiable at the single gauge level should not adversely influence the quality-control of neighbours.

When a suspicious gauge value is detected, the value is flagged as missing and the details are logged. In the case of a historical study, past records are set to missing for the duration of the period over which the flag was set. For real-time application, it is thus necessary to run all checks at every time point, whereas for offline use, that requirement can be relaxed and long-term checks can be processed on a correspondingly less frequent basis (daily here).

In some operational situations, it may be desirable to substitute the suspicious records, for example by a Standard Average Annual Rainfall (SAAR)-weighted average of neighbouring values. However, this is neither necessary nor suitable for the present application where a gridded rainfall estimate calculated from functioning gauges already serves this purpose.

Single raingauge tests

For each raingauge, checks are made for exceedence of a threshold comfortably above historical maxima. First, each 15-minute value is checked against a 70 mm threshold, and then the accumulation over the last 24 h against a 350 mm threshold. Next, a set of checks which assess the likelihood of a partial blockage are performed. Complete blockages, where no tips at all are registered, can only be detected by comparison with neighbours.

A cumulative hyetograph for a blocked gauge is shown by the black line in Fig. 1, characterised by a gently curved profile when compared with neighbouring gauges. To distinguish such patterns without reference to neighbours, we identify long decreasing sequences of recorded rain (>2.5 h duration) noting that shorter sequences may well be the tail-end of a rainfall event, on the basis that a blocked gauge tends to drain more quickly the fuller it is (Upton & Rahimi, 2003). Equality is counted as a decrease within otherwise-decreasing runs, to account for lack of knowledge about individual tip-times within an interval. Because more rain may fall while a gauge drains, we also allow sequences to count as decreasing, so long as the number of decreases less the number of increases is greater than half the total sequence length.

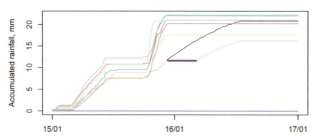

Fig. 1 Rainfall accumulations showing the effect of a partial blockage on the raingauge at Duston Mill (black line), from 15 to 16 January 2008. Also shown are its nearest seven neighbours (coloured lines), the quality-controlled accumulation (pink line) and the interval identified as suspicious by the tests (purple line). Note that records for the gauge at Litchborough (blue line) were missing for this period.

In the tail of such a blockage, there may be a sequence of 15-min time-steps with no recorded rain between two time-steps where single tips are recorded; knowing the tipping-bucket volume is important to identify a single tip. This is somewhat accounted for here by tracking the ongoing rate of tipping, with gaps between tips of up to 3 hours allowed. Further isolated tips are counted as independent events and may be detected later by multiple-gauge tests. Note that such small, sparse tips will have minimal impact on short-term hydrological modelling, though long-term water balances may be affected. Moreover, experience suggests that blockages can last for several weeks before being cleared: here, no attempt is made to maintain a gauge's blocked status and its responses to separate rainfall events are treated as independent.

We cannot expect to properly detect all problems. Figure 1 indicates the typical performance of the quality-control procedure: the blockage at Duston Mill is detected after the initial rainfall event has ended, but the blockage continues well beyond the detected range. Potential improvements to the current operation would be treating the gauge as missing over the entire event (or until what looks like a clearance occurs) and using a better classification of decreasing sequences.

Multiple-raingauge tests

Comparisons of each gauge's records against its neighbours' are performed once all single gauge tests have been completed for the current interval. Here the seven nearest gauges were used, seven being a compromise of proximity against inclusiveness suggested by Upton & Rahimi (2003). To avoid difficulties with particularly isolated gauges, a 50 km proximity threshold was applied. Also, a gauge is only compared against if it is deemed to have been functioning properly for at least two-thirds of the period under consideration.

Robust statistics (median and median absolute deviation) are used as the basis for detection of outliers, though other schemes could be considered (González-Rouco *et al.*, 2001). Use of these statistics somewhat mitigates the effect of multiple malfunctioning gauges in each other's proximity (Upton & Rahimi, 2003). Particular attention is given to detection of anomalously low values, since there is no simple and reliable method to distinguish intense localised rainfall from an error, other than the single-gauge tests already applied. This restriction could potentially be relaxed seasonally, as the likelihood of such events occurring reduces in winter.

Table 1 lists the quantities calculated for each gauge and the tests applied to distinguish outliers. For any quantity x, $[x]$ indicates the median of neighbours' values, x^+ the maximum, x^- the minimum and $\Delta(x)$ the median absolute deviation (MAD), that is $[|x - [x]|]$. τ_i is the tipping-bucket volume of gauge i. The thresholds are based on those found in (Upton & Rahimi, 2003).

Table 1 Multiple-raingauge quality-control tests.

Quantity calculated for gauge i	Criteria identifying fault
v, the last day's median value	$v < \min(v^- - \tau_i, \frac{1}{2}v^-)$ and $[v] - v > 10\Delta(v)$.
$\bar{v}$, the last day's maximum value	$\bar{v} < \min(\bar{v}^- - \tau_i, \frac{1}{2}\bar{v}^-)$ and $[\bar{v}] - \bar{v} > 10\Delta(\bar{v})$ or
	$\bar{v} > \max(\bar{v}^+ + 96\tau_i, 2\bar{v}^+)$ and $\bar{v} - [\bar{v}] > 10\Delta(\bar{v})$.
D, the total over the last 96 days	$D < \frac{1}{2}D^-$ and $[D] - D > 5\Delta(D)$, or
	$D > 2D^+$ and $D - [D] > 5\Delta(D)$.
χ, the mean of the cross-correlation of gauge i with its neighbours, for the last 96 days' totals	$\chi < \frac{1}{2}\chi^-$, $[\chi] > \frac{1}{2}$ and $[\chi] - \chi > 10\Delta(\chi)$.

A period of around 3 months was found to be suitable for studying long-term behaviour. Many detected faults persisted for durations significant on such a timescale (and longer in some cases). In practice, a lag of up to one day was allowed when determining the correlation between each pair of gauges and the largest value taken, to avoid problems with events/faults at the end of a day.

APPLICATION TO HYDROLOGICAL MODEL

The distributed hydrological model used to assess the impact of the quality-control procedure is the Grid-to-Grid (G2G) model (Bell *et al.*, 2007), which is in operational use across England and Wales by the Flood Forecasting Centre. This employs gridded rainfall, calculated from networks of raingauges and weather radars, to provide estimates of river flow on a 1 km scale across the model domain. Raingauge data, mostly in time-of-tip form, were pre-processed into consistent series of 15-min totals across the 3-year study period. Radar data were in the form of 5-min quality-controlled rain-rates.

Configuration and estimation of gridded rainfall inputs to the G2G model, using a multiquadric surface-fitting approach, is discussed in Cole & Moore (2008). That approach is applied here to the rainfall values from the network of 981 raingauges at each 15-minute time-step to obtain raingauge-only gridded rainfalls. Similar surface-fitting applied to a ratio of the gauge to the coincident radar value at each raingauge location is used to obtain raingauge-adjusted radar gridded rainfall.

Results of an offline case study

The spatial distribution of raingauge and radar locations is displayed in the left panel of Fig. 2. The centre and right panels give an overview of the long-term impact of the raingauge quality-control procedure across the whole of England and Wales. Several excessively high (white areas) and low (dark blue spots) rainfall accumulations were successfully removed during quality-control. Note the adverse effect that the sparseness of the raingauge network has over northwest Wales, and Anglesey in particular.

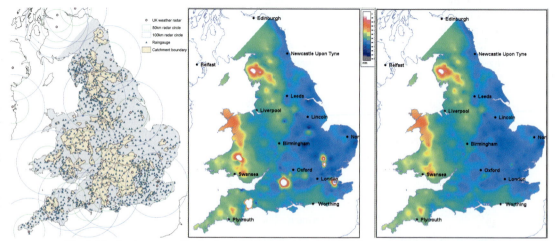

Fig. 2 Distribution across England and Wales of raingauges, weather radar and catchments used in case study (left). Rainfall accumulations for the 100 days up to 1 June 2009, for the multiquadric surface derived from 15 min totals of raw raingauge data (centre), and from quality-controlled data (right).

Summary statistics indicating the performance of G2G when using the different rainfalls as input are presented in Table 2. A three-month period from January to March 2008 was chosen to assess G2G performance over the set of 155 catchments whose boundaries are displayed in the left panel of Fig. 2. R^2 Efficiency and % bias (positive indicates overestimation) performance statistics, as a mean over all flow gauges, indicate the overall improvement obtained through raingauge quality-control. However, the signal is rather weakened by the averaging, since the majority of raingauges and time intervals experience no quality-control: 1.6% of values were identified as suspicious. For this period and set of catchments, the radar appears to significantly overestimate rainfall and use of either the raingauge-only or raingauge-adjusted radar rainfall estimates gives much improved G2G performance. Note that the G2G model has been calibrated using the quality-controlled raingauge-only rainfall estimates, as being the most consistent in time and space.

The benefit to hydrological modelling of raingauge quality-control is demonstrated much more clearly at the level of individual events. Figure 3 shows the accumulation over 15 days of rainfall in the vicinity of Blackburn, based on gauge-adjusted radar estimates with and without quality-control applied to the raingauges. There is clearly a severe fault with one nearby raingauge, which has recorded 30 times more rainfall than the average of its neighbours.

Table 2 Performance statistics for the Grid-to-Grid model over the three months from January to March 2008, using different rainfall estimates as input.

Gridded rainfall estimate	Mean R^2 Efficiency (Mean % Bias)	
	No raingauge quality-control	With raingauge quality-control
Raingauge-only	0.66 (−7.3)	0.68 (−5.9)
Raingauge-adjusted radar	0.63 (−3.2)	0.64 (−1.9)
Radar-only	0.47 (11)	N/A

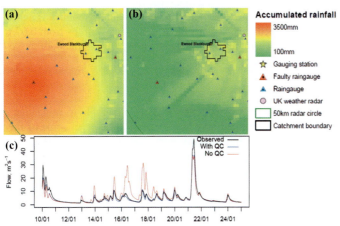

Fig. 3 Accumulated rainfall in the vicinity of Blackburn, over 15 days from 10 to 25 January 2008, using raingauge-adjusted radar: (a) without quality-control and (b) with quality-control of raingauge data. (c) Comparison of Grid-to-Grid modelled flow for the River Darwen at Ewood Blackburn using (a) as input (red line) and (b) as input (blue line).

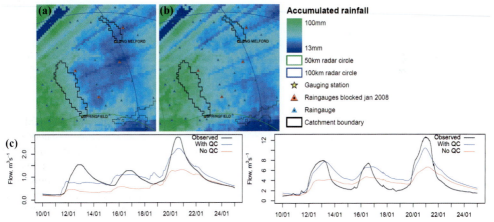

Fig. 4 Accumulated rainfall in the vicinity of Sudbury, over 15 days from 10 to 25 January 2008, using raingauge-adjusted radar: (a) without quality-control and (b) with quality-control of raingauge data. (c) Comparison of Grid-to-Grid modelled flow Chad Brook at Long Melford (left) and for the Chelmer at Springfield (right) using (a) as input (red line) and (b) as input (blue line).

Due to the large multiquadric offset parameter used in the gauge-adjustment process (Cole & Moore, 2008), this fault produces a large bias even over raingauges that are functioning well. Figure 3 shows observed and modelled flow for the River Darwen at Ewood Blackburn (drained area

39 km^2) for the same period. When gauge-adjusted radar rainfall is used as input to G2G, quality-control leads to an increase in the R^2 Efficiency of the resulting modelled flows, from 0.42 to 0.90.

A more subtle effect is produced by raingauges that are failing to record or underestimate the proper number of tips for some reason, such as full or partial blockage. Note that blockages may tend to cluster due to their environmental and seasonal nature. Figure 4 shows accumulations from gauge-adjusted radar rainfall, over the same period as Fig. 3, but for a much drier area near Sudbury. Several raingauges in the vicinity were detected as blocked and/or completely failing to register rainfall during this period. This leads to an overly low estimate of rainfall over the area when quality-control is not applied. The observed and modelled flow for two nearby river gauging stations is plotted in Fig. 4 and the improvement in modelling due to quality-control is again clear. For Chad Brook at Long Melford (drained area 47 km^2) the R^2 Efficiency climbs from 0.07 to 0.74 and for the River Chelmer at Springfield (drained area 190 km^2) it increases from 0.49 to 0.67.

CONCLUSIONS

It has been shown that raingauge-radar merging schemes can be vulnerable to raingauge errors. The automated quality-control routines presented here can successfully correct some of the worst of the faulty raingauge readings, leading to significantly improved performance for a distributed hydrological model that uses their derived rainfall estimates as input. There are clear benefits for historical archives of raingauge data and their use in forecast model calibration.

In real-time operation, the benefits of the quality-control algorithms presented here may be reduced, since it is no use identifying the previous month's readings as suspicious when only the most recent records may impact on flood forecasts. A refined procedure would have to focus more on detecting anomalous individual tips, perhaps via statistical comparison with historical records (González-Rouco *et al.*, 2001). There remains a problem with the spatial nature of the raingauge network since rainfall, or its absence, at one gauge may look suspicious, but 15 minutes later it becomes apparent that the storm responsible has moved to or from neighbouring gauge locations. Ideally, the algorithms would draw on radar data to obtain better criteria for detecting raingauge errors (Steiner *et al.*, 1999; Stanzani *et al.*, 2000).

Acknowledgements The Environment Agency and Met Office are thanked for making available the data used in this work. Science budget funding from the Centre for Ecology & Hydrology supported the research.

REFERENCES

Bell, V. A., Kay, A. L., Jones, R. G., Moore, R. J. & Reynard, N. S. (2009) Use of soil data in a grid-based hydrological model to estimate spatial variation in changing flood risk across the UK. *J. Hydrol.* 377, 335–350.

Cole, S. J. & Moore, R. J. (2008) Hydrological modelling using raingauge- and radar-based estimators of areal rainfall. *J. Hydrol.* 358(3–4), 159–181.

González-Rouco, J. F., Jiménez, J. L., Quesada, V. & Valero, F. (2001). Quality control and homogeneity of precipitation data in the southwest of Europe. *J. Climate* 14(5), 964–978.

Jorgensen, H., Rosenorn, S., Madsen, H. & Mikkelsen, P. (1998). Quality control of rain data used for urban runoff systems. *Water Sci. Tech.* 37(11), 113–120.

Kondragunta, C. R. & Shrestha, K. (2006). Automated real-time operational rain gauge quality-control tools in NWS hydrologic operations. In: *Proc. 20th AMS Conference on Hydrology*, P2.4, American Meteorological Society.

Moore, R. J., Cole, S. J., Bell, V. A. & Jones, D. A. (2006) Issues in flood forecasting: ungauged basins, extreme floods and uncertainty. In: *Frontiers in Flood Research* (ed. by I. Tchiguirinskaia, K. N. N. Thein & P. Hubert), 103–122. 8th Kovacs Colloquium, UNESCO, Paris, June/July 2006. IAHS Publ. 305. IAHS Press, Wallingford, UK.

Stanzani, R., Alberoni, P., Nanni, S., Mulazzani, C. & Pasquali, A. (2000). Raingauge and C-band radar monthly rainfall comparison in the Po plain area. *Phys. Chem. Earth (B)* 25(10–12), 981–984.

Steiner, M., Smith, J. A., Burges, S. J., Alonso, C. V. & Darden, R. W. (1999). Effect of bias adjustment and rain gauge data quality control on radar rainfall estimation. *Water Resour. Res.* 35(8), 2487.

Upton, G. J. G. & Rahimi, A. R. (2003). On-line detection of errors in tipping-bucket raingauges. *J. Hydrol.* 278(1–4), 197–212.

Weather Radar and Hydrology
(Proceedings of a symposium held in Exeter, UK, April 2011) (IAHS Publ. 351, 2012).

225

Blending of radar and gauge rainfall measurements: a preliminary analysis of the impact of radar errors

DANIEL SEMPERE-TORRES[1], MARC BERENGUER[1] & CARLOS A. VELASCO-FORERO[2]

1 *Centre de Recerca Aplicada en Hidrometeorologia, Universitat Politècnica de Catalunya. Gran Capità, 2–4 NEXUS-102, E-08034 Barcelona, Spain*
sempere@crahi.upc.edu
2 *Climate and Water Division, Bureau of Meteorology, GPO Box 727 Hobart, Tasmania 7001, Australia*

Abstract Several methodologies have been proposed to combine radar and raingauge measurements with the aim of generating improved quantitative precipitation estimates (QPEs). These methods are based on interpolating point raingauge measurements (implicitly assumed to be "the truth") and benefiting from the structure of the rainfall field as depicted by the radar. The use of a non-parametric approach based on radar measurements has been recently demonstrated, showing the benefits in the interpolation of raingauge measurements under the hypotheses of the Kriging approach. Several experiments have been carried out over a large number of cases and a variety of regions, Kriging with an external drift (i.e. the radar description of the rainfall field) being the approach showing more robust and (overall) better performance. Here, the impact of the discrepancies between two almost-collocated radars on the blended QPE fields was investigated.

Key words QPE; radar-raingauge blending; spatial variability of rainfall; radar errors; radar calibration

INTRODUCTION

Raingauge records are the most frequent direct measurement of rainfall, but remote sensing data are nowadays commonly available. Most notably, meteorological radars offer indirect information of the rainfall field with higher temporal and spatial resolutions than any operational network of raingauges. Despite the differences in the sampling characteristics of both systems, several authors have demonstrated the potential for generating improved rainfall estimates by combining their measurements. Historically, several approaches have attempted to combine radar and raingauge measurements. These range from the simplest methods based on multiplicative factors (e.g. Wilson & Brandes, 1979), to more sophisticated techniques based on multivariate analyses (Gabella *et al.*, 2000), on the adjustment of the marginal distributions of radar rainfall (Calheiros & Zawadzki, 1987), or on the interpolation of raingauge observations using geostatistical techniques (Krajewski, 1987; Creutin *et al.*, 1988; Seo, 1998; Sinclair & Pegram, 2005; Chumchean *et al.*, 2006; Goudenhoofdt & Delobbe, 2009).

A crucial aspect to face when using geostatistical techniques is the determination of the spatial variability model (variogram, see e.g. Goovaerts, 1997). Recently, Velasco-Forero *et al.* (2009) have proposed a Kriging approach in which a licit variogram is automatically (non-parametrically) estimated from radar measurements using the technique proposed by Yao & Journel (1998). This has been shown to be advantageous when interpolating the rainfall field (Velasco-Forero *et al.*, 2009; Schiemann *et al.*, 2011), as it allows the use of different variability models along a given event and depicting the spatial anisotropy of the rainfall field (i.e. avoiding the traditional subjective process of fitting a theoretical, frequently isotropic, variogram).

This work analyses the impact of errors in radar Quantitative Precipitation Estimates (QPEs) on the blended rainfall fields obtained with the method of Velasco-Forero *et al.* (2009). This analysis is done using an exceptional, real case study: we exploit the simultaneous data of two operational C-band radars, located within 4 km, and scanning the same domain during an intense rainfall event. Both radars show significant calibration errors that result in systematic biases in rainfall amounts. Although causes of these errors are well known and were identified years ago, their continuous evaluation and correction in real-time is an extremely complex and time-consuming task.

CASE STUDY

The case study analysed here occurred in the vicinity of Barcelona, Catalonia (Spain) between 8 October 2002 at 18:00 UTC and 10 October 2002 at 1200 UTC. This is a typical Mediterranean area where rain events frequently produce catastrophic floods over a wide range of river basins. The area is covered by two C-band radars within 4 km (see Fig. 1): one belonging to the Spanish Meteorological Agency (AEMET) and the other to the Catalan Weather Service (SMC – their technical details are summarized in Table 1). That is, these radars scan almost the same domain, providing two independent sets of reflectivity volume scans for the same event. This allows us to directly compare their estimates, avoiding the usual problems related to the different distances and location when comparing available simultaneous radar measurements.

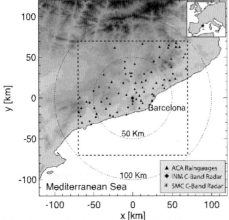

Fig. 1 Analysis domain. The dashed line square shows the area over which rainfall fields have been interpolated. Dotted-line circles are 50 km apart. The black diamond and star in the central part of the domain show the locations of the two C-band radars used in this study. The black triangles show the location of the raingauges used to interpolate the rainfall field.

Table 1 Main characteristics of the AEMET and SMC C-band radars.

	AEMET	SMC
Height (amsl)	664 m	629 m
Transmitted Power	250 kW (Magnetron)	8.9 kW (TWT)
PRF	250 Hz	Variable
Frequency	5.60 GHz	5.65 GHz
Beam width	0.9°	1.3°
Pulse duration	2 µs	2–20 µs
Number of azimuths	420	360
Antenna speed	6 rpm	4 rpm
Number of elevations	20 (between 0.5° and 25°)	17 (between 0.5° and 27°)
Time resolution	10 min	6 min

During this event, successive rain systems moved in from the sea with intense convective cells crossing the area of study. Data selected for our analysis consist of 36-hours of radar rainfall maps (rainfall accumulations from radar QPEs are presented in Fig. 2), and measurements from the 75 tipping-bucket raingauges of the Catalan Water Agency (ACA) in the area of study (with a maximum accumulation of 166 mm).

The merged radar-raingauge fields have been interpolated with a resolution of 1 km over the 140 × 140 km domain shown in Fig. 1.

Radar data processing

Radar-based QPE maps have been obtained using the basic radar processing usually applied in operational radars (as it was the case during the event). The non-meteorological echoes were suppressed using Doppler velocities (in our case combining the techniques of Sánchez-Diezma *et al.*, 2001 and Berenguer *et al.* 2006), and the sectors affected by beam blockage by the orography were identified using the algorithm of Delrieu *et al.* (1995). Finally, radar rainfall accumulation maps were generated considering rainfall motion and the evolution of rainfall intensities (Method 2 of Fabry *et al.*, 1994).

Advanced QPE as the one used by the SMC nowadays (the QPE processing of the Integrated Hydrometeorological Forecasting Tool, EHIMI; Corral *et al.*, 2009) that includes: (i) monitoring of signal stability and radome attenuation (Sempere-Torres *et al.*, 2003), (ii) reduction of the effects of beam blockage by the orography (using the algorithm of Delrieu *et al.*, 1995), (iii) clutter suppression by combining the techniques of Sánchez-Diezma *et al.* (2001) and Berenguer *et al.* (2006), (iv) extrapolation of elevated reflectivity measurements to the ground according to a double Vertical Profile of Reflectivity, VPR (as described by Franco *et al.*, 2006), and (v) conversion of reflectivity into rain-rate using a rainfall-type *Z-R* relationship (derived by Sempere-Torres *et al.*, 1997 for the area of study) was not used to ensure that the data used are as close as possible to the data operationally available during the event.

Calibration of radars

In this case study both radars showed significant biases in estimated rainfall amounts compared to raingauge measurements (as shown in the scatter-plots of Fig. 3): event accumulations show that the AEMET radar under-estimated rainfall amounts reported by raingauges, while the SMC radar clearly overestimated rainfall. These differences are mostly due to errors in the calibration of these radars, since at that time the calibration control of the radars was not monitored (although other non-systematic factors, such as attenuation by precipitation, the VPR or the Z-R conversion may have also had a significant impact).

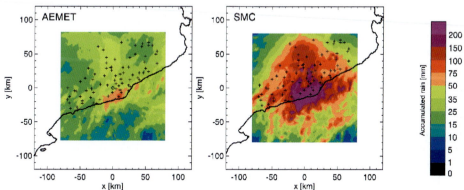

Fig. 2 Rainfall accumulations estimated from the measurements of the AEMET (left) and SMC (right) radars for the event of 8–10 October 2002 (36 h).

THE BLENDING TECHNIQUE

Kriging with an external drift

Some geostatistical methodologies to estimate rainfall fields by blending radar and gauge data using an automatic definition of the spatial variogram have been proposed and compared in previous studies (see e.g. Velasco-Forero *et al.*, 2009; Schiemann *et al.*, 2011). They showed that estimates obtained with Kriging with External Drift (hereafter referred to as KED) using the radar field to define the instantaneous, anisotropic and non-stationary spatial variability model are more accurate than those obtained with other estimators.

KED (a full description can be found in the literature, e.g. Goovaerts, 1997) can be used in an estimation process when exhaustive auxiliary information correlated with the target variable (in our case, rainfall) is available and models the estimates as a drift term plus a residual. Implementation of KED requires modelling the variogram of such residuals. In this application (KED_{OK} in the notation of Schiemann *et al.*, 2011) raingauge measurements are used as the primary variable and radar rainfall estimates are used as the secondary exhaustive variable. The necessary variogram for the residuals is obtained as follows:

(1)   The rainfall drift map is obtained by Ordinary Kriging interpolation of radar rainfall estimates at gauge locations (with the valid 2-D variogram obtained from the radar field).
(2)   A residual map is computed by subtracting the drift field from the radar rainfall field.
(3)   A valid 2-D variogram is computed from the residual map obtained in the previous step.

Automatic definition of the spatial variability model

In this study, the spatial variability models used in steps (1) and (3) of the KED recipe presented above are automatically estimated using a non-parametric approach based on a FFT-smoothing process proposed by Yao & Journel (1998). This automatic technique allows us to estimate two-dimensional (2-D) variograms without previous hypotheses and that can be updated at each time-step (given the low computational cost of the algorithm), showing considerable advantages compared with the conventional methods based on fitting parametric (frequently 1-D) variogram models (see e.g. Schiemann *et al.*, 2011).

RESULTS

Figure 3 shows that the two radars presented serious biases, which significantly contributed to the Root Mean Square Error (RMSE) when raingauge records are used as reference. When radar measurements are scaled with a constant multiplicative factor imposing radar accumulations at raingauges to be the same as measured by gauges (similarly as suggested by Wilson & Brandes, 1979), the biases are removed without worsening the correlation. This results in improved RMSEs (21.4 mm and 24.3 mm for the estimates obtained with the AEMET and SMC radars, respectively).

Figure 4 shows the accumulations resulting from implementation of the KED technique for the analysed event. In the figure, it is evident that the estimated fields show rainfall amounts that are now much more consistent between the two radars. This is especially so within the area well covered by the raingauge network. The figure shows that, besides the biases introduced by radar mis-calibration, some of the evident errors affecting radar QPE maps (underestimation with range – especially for the SMC radar – and less clear some underestimation due to beam blocking along the northeastern coast) have been clearly mitigated in the rainfall field resulting from the

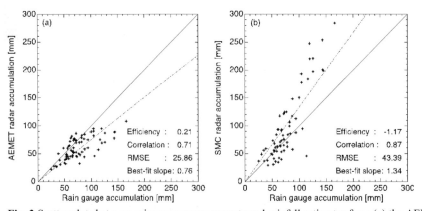

Fig. 3 Scatter plots between raingauge measurements and rainfall estimates from (a) the AEMET radar, and (b) the SMC radar for the event of 8–10 October 2002 (36 h).

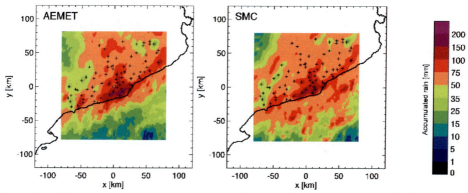

Fig. 4 Rainfall accumulations estimated with the technique of Velasco-Forero *et al.* (2009), based on Kriging with an external drift to combine raingauge records and the measurements of the AEMET (left) and SMC (right) radars for the event of 8–10 October 2002 (36 h).

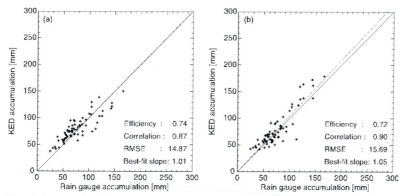

Fig. 5 Scatter plots between raingauge and KED accumulations for the cross-validation experiment described in the text for the event of 8–10 October 2002 (36 h). KED has been performed with radar measurements from (a) the AEMET radar, and (b) the SMC radar.

combination of both sources of information. Consequently, KED maps resemble each other more than the radar QPE maps of Fig. 2. This is not only thanks to the mitigation of biases, but the blending improves the representation of smaller scale patterns of the rainfall field as well. However, further analysis would be necessary to quantitatively assess the impact of the blending technique on the spatial scales of the rainfall field.

The performances of the KED technique implemented on the two radars have been evaluated by means of a leave-one-out cross-validation experiment (i.e. sequentially, one gauge record at a time is omitted from the calculations and estimated by KED using the rest of the gauges, which allows to compare the estimate with the non-used record). The results of this experiment are presented in Fig. 5, which shows the scatter plots between estimates based on both radars and actual records. The scatter plots show that the quality of the estimates seems to be almost independent of the radar used in the interpolation (the scores for both radars are almost identical): estimates show almost no bias (although no bias correction has been implemented), and the RMSE from the unbiased radar estimates mentioned above has been reduced by about 30%.

CONCLUSIONS

In this study, we have studied the performance of the technique proposed by Velasco-Forero *et al.* (2009) over a particular case for which two almost collocated radars produced significantly different rainfall amounts. These differences are attributed to such factors as radar mis-calibration, signal attenuation by precipitation, beam blockages or (less so) the Z-R relationships used.

The implemented technique is based on the interpolation of raingauge records by Kriging with an external drift, where the field is driven by the radar rainfall map. From the results found in this study, it seems valid to conclude that KED is able to compensate for severe errors affecting radar-based quantitative precipitation estimates. Although the impact of the blending is very clear on the mitigation of calibration errors, there is evidence that it also improves the depiction of smaller scale patterns of the rainfall field. It could be argued that this benefit should be expected, at least, up to scales comparable to the characteristic distance between raingauges.

Finally, it seems intuitive that the performance of the technique will depend on factors such as (i) the density of the raingauge network, (ii) the time window over which the blended radar and raingauge accumulations are computed, or (iii) on the quality of radar QPE (which decreases with range). The impact of such factors, as well as of the application of advanced radar processing, will be evaluated in a forthcoming paper.

Acknowledgements The authors thank the Spanish Meteorological Agency (AEMET), the Catalan Weather Service, and the Catalan Water Agency (ACA) for providing radar and raingauge data. This work has been carried out under the framework of the EC FP7 project IMPRINTS (FP7-ENV-2008-1-226555), and the Spanish Project ValGPM (ESP2007-62417).

REFERENCES

Berenguer, M., Sempere-Torres, D., Corral, C. & Sánchez-Diezma, R. (2006) A fuzzy logic technique for identifying nonprecipitating echoes in radar scans. *J. Atmos. Ocean. Tech.* 23, 1157–1180.

Calheiros, R. V. & Zawadzki, I. (1987) Reflectivity-rain rate relationships for radar hydrology in Brazil. *J. Climate Appl. Meteor.* 26, 118–132.

Chumchean, S., Sharma, A. & Seed, A. (2006) An integrated approach to error correction for real-time radar-rainfall estimation. *J. Atmos. Ocean. Tech.* 23, 67–79.

Corral, C., Velasco, D. Forcadell, D. & Sempere-Torres, D. (2009) Advances in radar-based flood warning systems. The EHIMI system and the experience in the Besòs flash-flood pilot basin. In: *Flood Risk Management: Research and Practice* (ed. by P. Samuels), London, UK.

Creutin, J. D., Delrieu, G. & Lebel, T. (1988) Rain measurement by raingauge-radar combination: a geostatistical approach. *J. Atmos. Ocean. Tech.* 5, 102–115.

Delrieu, G., Creutin, J. D. & Andrieu, H. (1995) Simulation of radar mountain returns using a Digitized Terrain Model. *J. Atmos. Ocean. Tech.* 12, 1038–1049.

Fabry, F., Bellon, A., Duncan, M. R. & Austin, G. L. (1994) High-resolution rainfall measurements by radar for very small vasins – the sampling problem reexamined. *J. Hydrol.* 161, 415–428.

Franco, M., Sánchez-Diezma, R. & Sempere-Torres, D. (2006) Improvements in weather radar rain rate estimates using a method for identifying the vertical profile of reflectivity from volume radar scans. *Meteor. Z.* 15, 521–536.

Gabella, M., Joss, J. & Perona, G. (2000) Optimizing quantitative precipitation estimates using a noncoherent and a coherent radar operating on the same area. *J. Geophys. Res.* 105, 2237–2245.

Goovaerts, P. (1997) *Geostatistics for Natural Resources Evaluation.* Oxford University Press.

Goudenhoofdt, E. & Delobbe, L. (2009) Evaluation of radar-gauge merging methods for quantitative precipitation estimates. *Hydrol. Earth Syst. Sci.* 13, 195–203.

Krajewski, W. F. (1987) Cokriging radar-rainfall and rain-gauge data. *J. Geophys. Res.* 92, 9571–9580.

Sánchez-Diezma, R., Sempere-Torres, D., Creutin, J. D., Zawadzki, I. & Delrieu, G. (2001) An improved methodology for ground clutter substitution based on a pre-classification of precipitaion types. In: *30th Int. Conf. on Radar Meteorology* (Amer. Meteor. Soc., Munich, Germany), 271–273.

Schiemann, R., Erdin, R., Willi, M., Frei, C., Berenguer, M. & Sempere-Torres, D. (2011) Geostatistical radar-raingauge combination with nonparametric correlograms: methodological considerations and application in Switzerland. *Hydrol. Earth Syst. Sci.* 15, 1515–1536.

Sempere-Torres, D., Porrà J. & Creutin, J. D. (1997) Characterization of rainfall properties using the drop size distribution: Application to autumn storms in Barcelona. In: *Int. Conf. on Cyclones and Hazardous Weather in the Mediterranean Area* (WMO-INM, Mallorca, Spain), 621–628.

Sempere-Torres, D., Sánchez-Diezma, R., Berenguer, M., Pascual, R. & Zawadzki, I. (2003) Improving radar rainfall measurement stability using mountain returns in real time. In: *31st Conf. on Radar Meteorology* (Seattle, Washington), 220–221.

Seo, D.-J. (1998) Real-time estimation of rainfall fields using rain gage data under fractional coverage conditions. *J. Hydrol.* 208, 25–36.

Sinclair, S. & Pegram, G. (2005) Combining radar and raingauge rainfall estimates using conditional merging. *Atmos. Sci. Lett.* 6, 19–22.

Velasco-Forero, C. A., Sempere-Torres, D., Cassiraga, E. F. & Gomez-Hernández, J. J. (2009) A non-parametric automatic blending methodology to estimate rainfall fields from raingauge and radar data. *Adv. Water Resour.* 32, 986–1002.

Wilson, J. W. & Brandes, E. A. (1979) Radar measurement of rainfall, a summary. *Bull. Amer. Meteor. Soc.* 60, 1048–1058.

Yao, T. & Journel, A. (1998) Automatic modeling of (cross) covariance tables using FFT. *Math. Geol.* 30, 589–615.

Weather Radar and Hydrology
(Proceedings of a symposium held in Exeter, UK, April 2011) (IAHS Publ. 351, 2012).

231

Application of radar-raingauge co-kriging to improve QPE and quality-control of real-time rainfall data

HON-YIN YEUNG[1], CHUN MAN[2], SAI-TICK CHAN[1] & ALAN SEED[3]

1 *Hong Kong Observatory, 134A Nathan Road, Kowloon, Hong Kong, China*
hyyeung@hko.gov.hk
2 *Department of Physics, the Chinese University of Hong Kong, Shatin, New Territories, Hong Kong, China*
3 *Centre for Australian Weather and Climate Research, Bureau of Meteorology, GPO Box 1289, Melbourne 3001, Australia*

Abstract Quantitative precipitation estimation (QPE) by weather radar often serves as an important input to hydrological and weather warning operations. Raingauge data are used by operational QPE systems for real-time bias adjustments and as ground truth in the verification of the rainfall estimates and forecasts. Raingauges are also subject to malfunction and quality-control is required before the data can be used quantitatively. A recently proposed procedure based on an analysis of differences between the radar rainfall estimate and the gauge observation and an interpolation of the local raingauges to the gauge site has been enhanced and is described in this paper.
Key words precipitation estimation; co-kriging; raingauge; quality control; Hong Kong

INTRODUCTION

Raingauge data and quantitative precipitation estimation (QPE) by weather radar serve as important inputs to hydrological and weather warning operations at the Hong Kong Observatory, e.g. the landslip and rainstorm warnings and flood alerts. Raingauge data are used in real-time to improve the quality of the radar rainfall estimates and act as ground truth for verification of the rainfall estimates and forecasts. Raingauges also experience malfunctions and therefore quality-control (QC) is required before the data are used in a real-time operational system. The raingauge is subject to both systematic and random errors that are caused by various factors including wind, wetting, evaporation, splashing, calibration, finite sampling, mechanical failure, funnel blockage, signal transmission interference, and power failure (Habib, 2001). Among the random errors, those arising from contamination during data transmission through radio telemetry systems and partial blockage of the raingauge itself by external obstructions, like insects nesting in the gauge or fouling by birds, are crucial in heavy rain monitoring and assessment. While telemetry errors usually contaminate a single observation with an extreme value, a blocked gauge will return zeroes or unreasonably small values that fall well within the climatological range and are therefore difficult to detect in real-time over short accumulation periods. A simple quality control procedure based on radar-raingauge co-kriging and an analysis of the difference between the gauge observation and the radar rainfall estimate was recently proposed by Yeung *et al.* (2010) and applied to short-duration rainfall accumulations (6-minutes). This paper presents improvements to the Yeung *et al.* (2010) method including: (i) use of a local climatological Z-R (reflectivity to rain-rate) relationship instead of the Marshall-Palmer Z-R relation $Z = 200R^{1.6}$, and (ii) introduction of QC criteria for 60-minute accumulations.

DATASETS

The datasets used in this study include raingauge and radar reflectivity data, covering the period March 2009 to February 2010. During the data acquisition period, there were a maximum of 157 tipping-bucket raingauges available in Hong Kong, covering a total land area of about 1100 km². The gauge distribution is uneven with an average separation of about 2.6 km. The raingauges (RG) have a resolution of 0.5 mm and record the data at 5 min intervals. The radar reflectivity data are provided as a mosaic of data from two S-band Doppler weather radars in Hong Kong, updated every 6 minutes. The reflectivity data at a constant altitude of 2 km above sea level were subjected to basic quality-control to remove clutter. They were then converted to rain-rate using the local climatological Z-R relation of $Z = 118R^{1.52}$. A data-thinning procedure was applied to the radar data so as to speed up the co-kriging analysis during the generation of raingauge-radar residuals, which resulted in an effective data resolution of around 5.3 km.

METHODOLOGY

Co-kriging techniques are very popular for spatial analysis in the field of geostatistics. For basic formulations of the problem, preparation of variograms and the general solutions to the co-kriging equations, readers are referred to the reference books by Wackernagel (1998) or Webster & Oliver (2001), as well as journal papers by Phillips *et al.* (1997) and Goovaerts (1998). For applications to QPE, reference could be made to Creutin *et al.* (1988), Schuumans *et al.* (2007) and Velasco-Forero *et al.* (2009). In Yeung *et al.* (2010), the use of ordinary co-kriging to combine radar rainfall estimates with gauge observations and to detect artefacts in the gauge data was explored. The key characteristics of the Yeung *et al.* (2010) scheme included: (a) speed – for computational efficiency use only gauges within a radius of 50 km of the point of interest, (b) stability – use average variograms instead of variograms that were estimated in real-time which could be ill-defined due to insufficient data samples, and (c) simplicity – based mainly on threshold checking. In this paper, results and discussions will be focused on the 60 min rainfall. Figure 1 shows the corresponding average variograms and cross-variogram obtained with the 2009–2010 datasets. The functional form and fitted parameters of the theoretical variograms/cross-variogram used for all subsequent co-kriging analysis are annotated on the same figure.

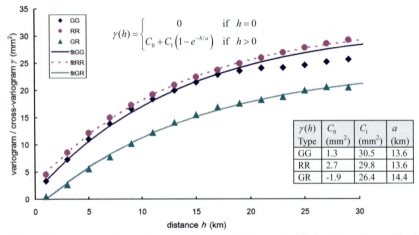

Fig. 1 The average variogram/cross-variogram of 60 min rainfall for Hong Kong. The legend labels "GG", "RR" and "GR" refer to gauge-gauge variogram, radar-radar variogram and gauge-radar cross-variogram respectively. The respective theoretical variograms/cross-variogram $\gamma(h)$ are shown as solid/dashed curves with fitted parameters for nugget (C_0), sill (C_1) and range (a) summarized in the inset table.

To see how the use of variograms/cross-variogram and co-kriging interpolation improves QPE in terms of spatial structure, we compare the rainfall map analysed by the co-kriging method with three other schemes available in the Observatory, namely Kriging (raingauge data only), Barnes (raingauge data only) and radar rainfall (reflectivity data only). Note that all rainfall maps are produced from gridded analysis of rainfall information. In gridded rainfall analysis, the analysis points refer to the grid box centres (with about 1.5 km spacing) and all raingauge data are used, even if co-located with the analysis points. Figure 2 shows an example of an intense rainband on 7 June 2008 approaching Hong Kong from the west. The rainfall isohyet maps (a)–(d) compare the results of the four types of QPE method mentioned above. At the time of analysis, the rainband started to affect the coastal areas posing a real challenge to the raingauge-only analysis schemes as there were no direct rainfall information available over the waters and raingauges are relatively sparse over Lantau Island (the largest island in the southwest corner of the map inside the dashed ellipse). With the subsidiary rainfall information derived from reflectivity, co-kriging QPE correctly analysed the rainfall distribution by putting the area with the heaviest rain (the red zone) to the west of Lantau. This is important for the accurate assessment of the likelihood of rainstorm flooding and landslides. Both types of hazards occurred in the subsequent few hours, paralysing the main traffic to the Hong Kong International Airport at Chek Lap Kok (the landmass adjacent to the north coast of Lantau) and causing a number of serious landslides and significant damage on Lantau Island.

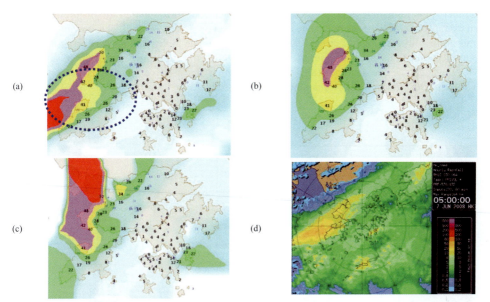

Fig. 2 Rainfall analyses prepared by four QPE schemes: (a) co-kriging; (b) Kriging; (c) Barnes and (d) radar rainfall estimation. In (a)–(c), the numbers are raingauge values and the green, yellow, pink and red colour filled contours refer to analysed rainfall in the ranges of 20–30, 30–40, 40–50 and 50–70 mm. In (d), the greenish and yellowish colours refer to rainfall in the 0.8–10 mm and 10–80 mm ranges respectively.

PROBABILITY DISTRIBUTIONS FOR RG-QPE RESIDUALS

The raingauge QC algorithm requires the probability distribution of the gauge-radar residuals. Three kinds of residuals – namely linear, logarithmic and standardized – are defined respectively as follows:

$$D = G - K \tag{1}$$

$$\xi = 10\log(G/K) = 10\log G - 10\log K \tag{2}$$

$$\delta = |G - K|/\sigma \tag{3}$$

where D is in units of mm, ξ is in decibels (dB) and δ is dimensionless. G is the raingauge observation, K is the co-kriging QPE at the location of the target gauge based on the neighbouring raingauges, and σ is the co-kriging estimation error. An arbitrarily small offset of 0.08 mm was added whenever either G or K was zero so as to make ξ well defined for all possible G and K values. Figure 3(a) shows the distribution of D. The small peaks and outliers shown up in the positive tail were taken to signify abnormally large raingauge values (such as those due to telemetry errors) as compared to their reference QPE. As the current dataset only covers one year, a conservative strategy was considered more appropriate and $D > 42$ was selected to identify spuriously large raingauge values (around 50 out of a total of 6 million samples).

False zeroes or unreasonably small raingauge data are rather subtle as they are typically buried well within the broad central peak of the distributions of D. Following Yeung $et\ al.$ (2010), the condition $\delta \leq 2$ was adopted to discard the uninteresting samples, i.e. those with a small $|D|$ as well as those bearing a relatively large estimation error. Figure 3(b) shows the resulting distribution of ξ with anomalous peaks in the tails. The peak around –19 dB signifies some unreasonably large negative residuals (G significantly smaller than K) which are attributable to false zeroes or unreasonably small gauge values. There is also an anomalous peak around +18 dB, hinting that some raingauge data are inconsistently larger than surrounding rainfall information. A known example of such curious raingauge reports occurred on 9 September 2010 following the strike by a rainstorm with very intense ground lightning activity. Some faulty stations kept reporting 5 mm of 5 min accumulations for a prolonged period even after the rainstorm had departed and other normal raingauges ceased to report any rainfall. Whether the positive anomalous peak could be another useful QC criterion will be examined and discussed later.

(a)   (b)

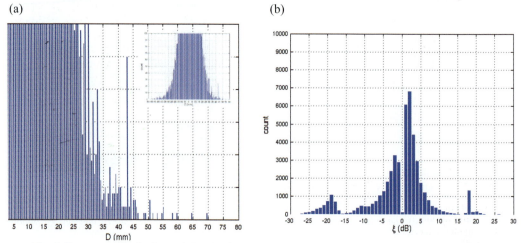

Fig. 3 Frequency distributions of the 60 min gauge-radar residuals during the period March 2009 to February 2010: (a) magnified view of the positive tail of the entire distribution (see inset) of linear residuals, and (b) logarithmic residual distribution reduced with $\delta \leq 2$ data samples discarded.

APPLICATIONS

Raingauge data QC

Following Yeung $et\ al.$ (2010) and based on the results in the last section, a simple QC procedure for 60 min rainfall accumulation reported by automatic raingauges is proposed as follows:

(i)   perform pre-QC screening to retain only those data lower than a threshold value G^c;
(ii) perform spatial consistency check on data retained in step (i) as follows:

- calculate a reference co-kriging interpolation QPE at the gauge location based on the neighbouring gauge observations,
- calculate the gauge-radar residuals (linear, logarithmic and standardized), and
- check against the corresponding threshold values (D^c, ξ^c and δ^c, respectively);

(iii) Assign QC flags "P" (for pass) or "R" (for reject) according to the following criteria:
- if $D > D^c = 42$ mm, flag "R",
- else if $\delta > \delta^c = 2$ and $\xi < \xi^c - = -19$ dB, flag "R",
- [Optional] else if $\delta > \delta^c$ and $\xi > \xi^c + = 18$ dB, flag "R", (superscripts "–" and "+" here refer to the peaks in the negative and positive tails, respectively);
- otherwise, flag "P".

The pre-QC screening in step (i) is necessary so as to avoid contaminating the co-kriging QPE with spuriously large rainfall observations. G^c was set to be the maximum instantaneous rain-rate of 513 mm/h which was recorded in Hong Kong in 1971. The other threshold values given in step (iii), namely $D^c = 42$ mm, $\delta^c = 2$, $\xi^c - = -19$ dB and $\xi^c + = 18$ dB, were deduced from the distribution of the gauge-radar residuals as explained in the preceding section. Note that the choices for $\xi^c -$ and $\xi^c +$ were relatively insensitive to the value of δ^c. Higher values may also be used at the expense of a lower error detection rate.

PERFORMANCE ASSESSMENT

The effectiveness of the proposed QC procedure was assessed through a rainfall event which occurred from 03:00 to 14:00 h local time on 10 June 2010. There were a total of 20 289 raingauge reports of 60 min rainfall available for testing. Besides the proposed objective QC procedure, subjective inspection by the authors was also performed for cross-validation. Three additional controlled tests using bogus data by setting $G \geq 10$ reports to zero, and zero reports to 10 mm and 47 mm were also conducted to simulate clogged raingauges, inconsistently large raingauge reports and data contamination during transmission through the telemetry systems, respectively.

Table 1 Test results for 60 min raingauge rainfall data QC.

R/G reports	QC method	Tests with different raingauge amounts, G (mm)					
		$G = 0.0$	$G \geq 0.5$	all G	bogus 0 ($G \geq 10$)	bogus 10 ($G = 0$)	bogus 47 ($G = 0$)
Subtotal	–	1967	18 322	20 289	2873	1653	1653
Rejected	co-kriging	57	0	57	2797	913	1652
	human	57	0	57	2873	1653	1653
Ratio of correct rejections by co-kriging method w.r.t. subtotal	–	–	–	–	97.4 %	55.2 %	99.9 %

Except for the test on inconsistently large raingauge reports, the proposed QC procedure was considered satisfactory as summarized in Table 1. The detection rate for a false observation of 10 mm is just 55%, and it is not possible to detect such errors reliably during convective rainfall events. Despite the high detection rate (99.9%) for abnormally large raingauge reports, it must be remarked that there exists a significant negative bias for the QPE in the case of extreme rainfall so that a raingauge report with residual exceeding the QC threshold $D^c = 42$ may well be a valid data if it occurs during an extreme rainfall event.

CONCLUDING REMARKS

Enhanced co-kriging QC schemes were developed and the performance was assessed through cross-validation and QC tests. The results showed that the QC scheme performed satisfactorily with a high error detection rate in general. However, it was found to be difficult to detect spurious gauge observations reliably in extreme rainfall events and when the errors were of a similar magnitude to the rainfall rates. For the size of Hong Kong (1100 km^2), the current co-kriging QPE and QC algorithms can be completed within a 5 min cycle on a commodity PC system and are therefore suitable for real-time operation.

Acknowledgements The authors wish to acknowledge and thank Mr C. M. Shun for his critical review and useful suggestions on the manuscript.

REFERENCES

Creutin, J. D., Delrieu, G. & Lebel, T. (1988) Rain measurement by raingauge-radar combination: a geostatistical approach. *J. Atmos. Oceanic Technol.* 5, 102–115.

Goovaerts, P. (1998) Ordinary co-kriging revisited. *Mathematical Geology* 30, 21–41.

Habib, E., Krajewski, W. F. & Kruger, A. (2001) Sampling errors of tipping-bucket rain gauge measurements. *J. Hydrol. Engng ASCE* 6, 159–166.

Phillips, D. L., Lee, E. H., Herstrom, A. A., Hogsett, W. E. & Tingey, D. T. (1997) Use of auxiliary data for spatial interpolation of ozone exposure in southeastern forests. *Environmetrics* 8, 43–61.

Schuurmans, J. M., Bierkens, M. F. P. & Pebesma, E. J. (2007) Automatic prediction of high-resolution daily rainfall fields for multiple extents: the potential of operational radar. *J. Hydromet.* 8, 1204–1224.

Velasco-Forero, C. A., Sempere-Torres, D., Cassiraga, E. F. & Gómez-Hernández, J. J. (2009) A non-parametric automatic blending methodology to estimate rainfall fields. *Adv. Water Resour.* 32, 986–1002.

Wackernagel, H. (1998) *Multivariate Geostatistics — An Introduction with Applications*, 2nd edition. Springer, Berlin.

Webster, R. & Oliver, M. (2001) *Geostatistics for Environmental Scientists*, Wiley, UK.

Yeung, H. Y., Man, C., Seed, A. & Chan, S. T. (2010) Development of a localized radar-rain gauge co-kriging QPE scheme for potential use in quality control of real-time rainfall data. In: *3rd WMO Int. Conf. on Quantitative Precipitation Estimation, Quantitative Precipitation Forecasting and Hydrology* (18–22 October 2010, Nanjing, China).

Combination of radar and raingauge observations using a co-kriging method

CHUNG-YI LIN[1] & TIM HAU LEE[2]

1 *Taiwan Typhoon and Flood Research Institute, 12F, No. 97, Sec. 1, Roosevelt Rd., Taipei 10093, Taiwan*
  evanlin@ttfri.narl.org.tw
2 *National Taiwan University, No. 1, Sec. 4, Roosevelt Rd., Taipei 10617, Taiwan*

Abstract A rainfall estimation algorithm using a co-kriging method which combines radar and raingauge measurements is presented in this study. In the first part, error-free raingauge data and four different known error-structured radar observations are used to examine the abilities of two ordinary co-kriging techniques and two universal co-kriging techniques to correctly estimate spatial distribution of rainfall. The four radar observation errors are: (1) additive white noise error (WN); (2) additive correlative error with bias (AE); (3) multiplicative correlative error (ME); and (4) trend error varying with radar range (TE). In the second part, one case study of true typhoon data is used to verify the capability of this method to utilize real data. The ordinary co-kriging (OCK) technique utilises the linear combination of all raingauge observations and the radar observation collocated with estimated grid. The modified ordinary co-kriging (MOCK) technique utilizes the radar observations on top of all raingauges in addition to the data used by OCK technique. The minimum error variance estimation of universal co-kriging (UCK) utilizes the gauge data only to form the covariance matrix. Based on the collocated true rainfalls and radar observations, and following a linear model assumption, the unbiased conditions are derived. UCKT is a UCK technique that includes satisfying the spatial trend unbiased condition. Case study results illustrate that OCK is the only technique that cannot avoid AE error from going into rainfall rate estimates. Both MOCK and UCKT can effectively prevent AE and TE error from entering the estimates, and reduce the influence of ME error. According to the statistics of the case studies, MOCK had the lowest root mean square error. The major advantage of UCK and UCKT is that it is not necessary to provide the semi-variograms involving radar data.

Key words co-kriging; ordinary Kriging; universal Kriging; rain-rate estimate; radar observation; gauge observation; data fusion; spatial interpolation; observing system experiment

INTRODUCTION

Radar has been used to observe weather phenomenon, especially rainfall, for some time due to its long and wide range of coverage. However, radar observation of rainfall is deeply affected by various non-meteorological conditions such as topography, distance of observation, and reflectivity to rainfall (Z-R) relationship. Therefore, unavoidably, there exist errors in radar observations. In this paper, four types of error are considered – additive white noise error, additive correlated error with bias, multiplicative correlated error and spatial trend error – along with combinations of these. Raingauges are also influenced by some problems, but they are competitively accurate compared to radar observations. Nevertheless, the sparse distribution of raingauges is their main defect for rainfall estimation. Consequently, much research concerning the combination of radar and raingauge observations has appeared in recent decades. The ratio-adjusted method and the co-kriging method are two of the most discussed methods on this topic.

Wilson & Brandes (1979) used the ratio of raingauge to radar observation of rainfall on the same grid to build a spatial function of point ratio through an objective analysis method. The co-kriging method uses raingauge data and coincident radar observation data to obtain their linear weighting on the estimated grid points, afterwards utilising minimum root mean square error and unbiased constraints to derive optimal weighting coefficients. The co-kriging method can be further divided into three sub-types: ordinary co-kriging (OCK), universal co-kriging (UCK) and disjunctive co-kriging (DCK). Krajewski (1987) and Creutin *et al.* (1988) used the OCK technique to combine radar and raingauge observations. The research indicates that radar has even larger observation errors than raingauges, but the OCK technique still has the ability to reduce radar observation error to obtain a rainfall distribution close to true rainfall. Schuurmans *et al.* (2007) use the universal co-kriging technique, and an unbiased condition which contains spatial trend, to build optimal interpolated constraints in estimating spatial rainfall patterns. Seo *et al.* (1990a,b) used two different raingauge densities and simulated rainfall data with non-normal distributions to evaluate OCK, UCK, and DCK techniques. They found that, when compared to the OCK method, the DCK method which focuses on a non-normal variable transformation (e.g. Steiger *et al.*, 1996) has no obvious advantage and incurs about 10 times more computing time than OCK.

In this study, the performance of OCK and UCK methods are investigated with regard to the four radar observation error types mentioned above. After that, a typhoon case study is tested to examine the estimation ability of these methods for a real case.

METHODOLOGY

Co-kriging method

Four different co-kriging techniques which deal with different weighting coefficients between radar and gauge and unbiased conditions are introduced in this section.

1. Ordinary Co-kriging technique (OCK) In the OCK method, data from N_G raingauges and radar observation data exactly above the estimated point are used to estimate rainfall for that grid point. The equation is:

$$\hat{P}_0 = \sum_{i=1}^{N_G} \lambda_{Gi} G_i + \lambda_{R0} R_0 \tag{1}$$

where $\hat{P}_0$ is the estimated rainfall, N_G is the number of raingauges near the estimated point, G_i is the observation for raingauge i, R_0 is the observation radar rainfall exactly above the estimated point, and λ_{Gi} and λ_{R0} are weighting coefficients for the gauge and radar observations. Therefore, the unbiased condition of the OCK technique can be derived as:

$$\lambda_{R0} + \sum_{i=1}^{N_G} \lambda_{Gi} = 1 \tag{2}$$

Following that, the weighting coefficients for gauges and radar can be achieved by the matrix of covariance or semi-variogram as:

$$
\begin{bmatrix}
\gamma_{1,1}^{GG} & \cdots & \gamma_{1,N_G}^{GG} & \gamma_{1,0}^{GR} & 1 \\
\vdots & \ddots & \vdots & \vdots & \vdots \\
\gamma_{N_G,1}^{GG} & \cdots & \gamma_{N_G,N_G}^{GG} & \gamma_{N_G,0}^{GR} & 1 \\
\gamma_{0,1}^{RG} & \cdots & \gamma_{0,N_G}^{RG} & \gamma_{0,0}^{RR} & 0 \\
1 & \cdots & 1 & 0 & 0
\end{bmatrix}
\begin{bmatrix}
\lambda_{G1} \\
\vdots \\
\lambda_{GN_G} \\
\lambda_{R0} \\
\nu
\end{bmatrix}
=
\begin{bmatrix}
\gamma_{1,0}^{GG} \\
\vdots \\
\gamma_{N_G,0}^{GG} \\
\gamma_{0,0}^{RG} \\
1
\end{bmatrix}
\tag{3}
$$

2. Modified Ordinary Co-kriging technique (MOCK) The major difference between the MOCK and OCK methods is that the MOCK method uses more radar data than the OCK method. Besides the radar data above the estimated point, radar data above every raingauge are used to calculate the rainfall for the estimated point. Its equation is:

$$\hat{P}_0 = \sum_{i=1}^{N_G} \lambda_{Gi} G_i + \lambda_{R0} R_0 + \sum_{i=1}^{N_G} \lambda_{Ri} R_i \tag{4}$$

where R_i is the radar observation above each raingauge and λ_{Ri} is the coefficient corresponding to each radar observation. Thus the unbiased conditions of MOCK are:

$$\sum_{i=1}^{N_G} \lambda_{Gi} = 1 \tag{5a}$$

$$\lambda_{R0} + \sum_{i=1}^{N_G} \lambda_{Ri} = 0 \tag{5b}$$

The covariance matrix is similar to equation (3) except there are additional radar to radar γ^{RR} in the matrix. Owing to space limitations, it is not shown here.

3. Universal co-kriging technique (UCK) The UCK method does not involve radar data directly in its estimation equation; on the contrary, it uses radar data as a constraint in the unbiased condition of weighting coefficients. Therefore, the estimation equation can be simply written as:

$$\hat{P}_0 = \sum_{i=1}^{N_G} \lambda_{Gi} G_i \tag{6}$$

Then the unbiased conditions are:

$$\sum_{i=1}^{N_G} \lambda_{Gi} = 1 \tag{7a}$$

$$\sum_{i=1}^{N_G} \lambda_{Gi} R_i = R_0 \tag{7b}$$

The most important advantage of the UCK method is that there is no need to calculate the semi-variogram of "gauge to radar" (γ^{GR}) and "radar to radar" (γ^{RR}) that often contain more error than γ^{GG}. Finally, the matrix to calculate weighting coefficients is:

$$\begin{bmatrix} \gamma_{1,1}^{GG} & \cdots & \gamma_{1,N_G}^{GG} & 1 & R_1 \\ \vdots & \ddots & \vdots & \vdots & \vdots \\ \gamma_{N_G,1}^{GG} & \cdots & \gamma_{N_G,N_G}^{GG} & 1 & R_{N_G} \\ 1 & \cdots & 1 & 0 & 0 \\ R_1 & \cdots & R_{N_G} & 0 & 0 \end{bmatrix} \begin{bmatrix} \lambda_{G1} \\ \vdots \\ \lambda_{GN_G} \\ v_1 \\ v_2 \end{bmatrix} = \begin{bmatrix} \gamma_{1,0}^{GG} \\ \vdots \\ \gamma_{N_G,0}^{GG} \\ 1 \\ R_0 \end{bmatrix} \tag{8}$$

4. Universal co-kriging technique with Trend (UCKT) If the radar observation contains errors dependant on the distance from radar site r, the advantage of the UCKT method is that it is able to add new constraints in the unbiased condition; for example, if the error varied with r linearly, the additional unbiased condition can be derived as:

$$\sum_{i=1}^{N_G} \lambda_{Gi} r_i = r_0 \tag{9}$$

where r_i is the distance from each gauge to the radar site, and r_0 is the distance from the estimation point to the radar site. Hence, if the radar observation error with distance could be investigated before estimation, the UCKT method can effectively remove those errors and acquire more accurate rainfall in the estimated field.

Ideal case data

In the ideal case study, it is first assumed the raingauge data are free of observation error, and then four ideal radar observation error types are introduced and one ideal rainfall spatial distribution is also presented. Due to the assumption, the radar rainfall observation can be formulated by the relationship with its true value as follows:

$$R_j = \phi_j P_j + \varepsilon_j \tag{10}$$

where P_j is the true rainfall at the same grid as the radar observation; ε_j and ϕ_j are additive and multiplicative error coefficients, respectively. By definition, four ideal radar observation error types are described as:

1. Additive white noise error, WN: $\phi_j = 1$, $E[\varepsilon_j] = 0$, $\text{var}[\varepsilon_j] = \sigma_\varepsilon^2$ and $\text{cov}[\varepsilon_i, \varepsilon_j] = 0$ when $i \neq j$.
2. Additive correlated error with bias, AE: $\phi_j = 1$, $E[\varepsilon_j] = \mu_\varepsilon$, $\text{var}[\varepsilon_j] = \sigma_\varepsilon^2$ and $\text{cov}[\varepsilon_i, \varepsilon_j] = \rho_{ij}^\varepsilon \sigma_\varepsilon^2$ when $i \neq j$.
3. Multiplicative correlated error, ME: $\varepsilon_j = 0$, $E[\phi_j] = \mu_\phi$, $\text{var}[\phi_j] = \sigma_\phi^2$ and $\text{cov}[\phi_i, \phi_j] = \rho_{ij}^\phi \sigma_\phi^2$ when $i \neq j$.
4. Spatial trend error, TE: $\phi_j = 1$, if the trend is a linear equation then $E[\varepsilon_j] = a_0 + a_1 r_j$, if the trend is a quadratic equation then $E[\varepsilon_j] = a_0 + a_1 r_j + a_2 r_j^2$, etc. TE is a kind of additive error, but ε_j is not considered in TE.

The "true" rainfall distribution of this ideal case consists of two parts: a deterministic term which is generated by a two-dimensional Gaussian function and a stochastic term which is created by the Kriging method. The different simulated radar observation data are derived by the definition specified in this section. The domain is 50 km × 50 km and there are 25 raingauges within it. The grid size is 1 km × 1 km. Both true and radar observation rainfall are on this gridded field. In the next section, each co-kriging technique is used to evaluate its ability to obtain the true rainfall from various kinds of radar observation error and error-free raingauge data.

Real case data

Data from Typhoon Morakot are used as a real case study. Morakot attacked Taiwan in August 2009, not only causing huge monetary damage but also taking the lives of hundreds of people. The estimated data time is 20:00 UTC 7 August 2009. There are four Doppler radars in Taiwan island and more than 600 raingauge stations. The resolution of radar data is 0.0125° of longitude and latitude from 118 to 123.5°E longitude and 20 to 27°N latitude. Only the UCK technique is used to examine the performance of the co-kriging method. The results are presented in the next section.

ESTIMATION RESULT

Ideal case discussion

Four co-kriging techniques are applied to evaluate their ability to combine radar and raingauge observation data in estimating the spatial rainfall distribution. Because this is an observation simulation system experiment, each grid has its "true" value; therefore, the root mean square error (RMSE) is used to calculate the estimation error for each experiment. Table 1 shows all RMSEs for each case and discussion and figures of each experiment are presented below.

Table 1 RMSE value (mm/h) for different radar error types and four co-kriging techniques.

Radar error type	Co-kriging technique Original error	OCK	MOCK	UCK	UCKT
WN	5.025	4.698	5.088	5.833	5.845
AE	7.327	7.008	4.345	4.198	4.169
ME	10.527	8.114	8.034	8.776	8.801
TE	40.212	17.540	0.814	6.439	1.357

1. Additive white noise error, WN For radar observation error type WN, the result show all four co-kriging techniques cannot recover the true rainfall; the RMSEs are slightly higher than the original radar error except for the OCK technique which is a little better. The error distribution of each method indicates that when the error type is WN, only those grid points near raingauges could be corrected; in contrast, estimated rainfall on those grids far away from raingauges are still close to the original radar observation which contains additive white noise error. It should be noticed that UCK and UCKT techniques have equal performance for WN, AE, and ME, because these three error types do not include a trend error, which is the major and only difference between UCK and UCKT; therefore the behaviour in UCKT is not discussed for these three error types.

2. Additive correlated error with bias, AE With the AE error type, the OCK technique has just slightly corrected the error from the original radar error (RMSE reducing from 7.327 to 7.008), while MOCK and UCK obviously lower the RMSE to around 4. Thus MOCK and UCK are able to reduce AE error type, but still cannot totally remove its effect. Meanwhile, the estimated error in MOCK and UCK techniques is close to zero for those grid points near raingauges, as shown previously for the WN error type.

3. Multiplicative correlated error, ME Table 1 shows the OCK and MOCK methods perform better than UCK for the ME error type, but the degree is not very significant. In the error distribution figures (not shown), the UCK technique can depress the overestimation error but deepen the underestimation error at the same time: this is reason why the RMSE of UCK is slightly higher than OCK and MOCK. Overall, the four techniques perform equally on error type ME and they cannot effectively remove the ME error across the whole domain.

4. Spatial trend error, TE Error distributions of the four techniques in error type TE are shown in Fig. 1. This demonstrates that MOCK and UCKT techniques can almost reduce the TE radar observation error and recover the true rainfall distribution. Because of the spatial trend error characteristic and considering the fact that UCKT is provided with the ability to remove trend error, the near zero estimated error of UCKT is to be expected. The UCK technique could decrease most errorS in the analytical domain, except towards the four corners where there are no raingauges.

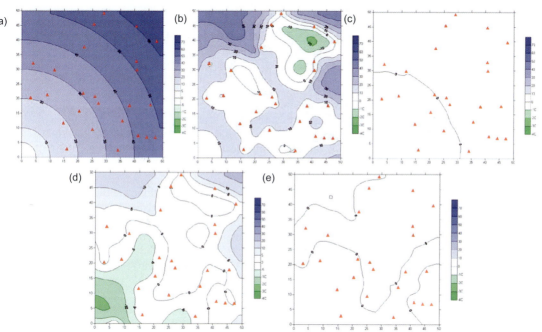

Fig. 1 Error distribution of TE: (a) original radar observation error, (b)–(e) estimated rainfall error from OCK, MOCK, UCK and UCKT techniques, respectively. Triangles indicate the locations of raingauges.

REAL CASE DISCUSSION

Figure 2 shows the rainfall distribution obtained from using raingauge data only, from radar observations only, and combined using the UCK technique. It illustrates that in this case, observation radar rainfall is obviously underestimated, especially in southern Taiwan. Because that area is covered with mountains, the radar elevation angle must be elevated and this results in underestimation of rainfall. In contrast, raingauges are able to capture the high rainfall information in this area, but lack a large range of observation domain such as the area over the sea. The results show that the UCK technique has the ability to combine both raingauge and radar observation information to produce a rainfall distribution that better describes the real rainfall over that area.

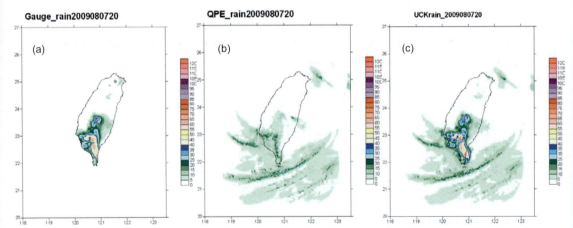

Fig. 2 Rainfall distribution from different estimated methods: (a) only raingauge data (b) only radar observations (c) using UCK technique.

CONCLUSION

This study first uses an observation simulation system experiment to evaluate four different combined radar and raingauge techniques to estimate the spatial rainfall distribution under the assumption of no error from raingauges, and then uses real data from typhoon Morakot in 2009 to examine the performance of the UCK technique for a real case. The result shows that the OCK technique can only adjust grids near to raingauges but the error from those grids far away from raingauges cannot be lowered.

The MOCK technique can effectively remove AE and TE error, and reduce the ME error. It often performs better than UCK and UCKT in this study. However, the major defect of MOCK is that it needs to calculate the radar-to-radar semi-variogram γ^{RR} which is always the most important error source for the co-kriging method. In addition, calculating weighting coefficients in MOCK is more time-consuming than for the other techniques.

For the ME error type, UCK and UCKT can avoid AE error and reduce TE error (UCKT), but underestimates rainfall in the area where there are no raingauges. UCK and UCKT's most advantageous point is it is not necessary to calculate two semi-variograms γ^{GR} and γ^{RR}: this can not only lessen the error contamination from radar observations, but also decrease the time of calculating.

For the real case, the results demonstrate that UCK has the ability to combine both information from radar and raingauges to generate a better estimated rainfall distribution. On this account, co-kriging could have the potential to improve quantitative precipitation estimation for the real case situation.

REFERENCES

Creutin, J. D., Delrieu, G. & Lebel, T. (1988) Rain measurements by rain gauge–radar combination: a geostatistical approach. *J. Atmos. Oceanic Technol.* 5, 102–115.

Krajewski, W. F. (1987) Co-kriging radar-rainfall and rain gage data. *J. Geophys. Res.* 92 (D8), 9571–9580.

Schuurmans, J. M., Bierkens, M. F. P., Pebesma, E. J. & Uijlenhoet, R. (2007) Automatic prediction of high-resolution daily rainfall fields for multiple extents: the potential of operational radar. *J. Hydromet.* 8, 1204–1224.

Seo, D. J., Krajewski, W. F., Azimi-zonooz, A. & Bowles, D. S. (1990a) Stochastic interpolation of rainfall data from rain gauges and radar using co-kriging. 2. Results. *Water Resour. Res.* 26, 915–924.

Seo, D. J., Krajewski, W. F. & Bowles, D. S. (1990b) Stochastic interpolation of rainfall data from rain gauges and radar using co-kriging. 1. Design of experiments. *Water Resour. Res.* 26, 469–477.

Steiger, B. von, Webster, R., Schulin, R. & Lehmann, R. (1996) Mapping heavy metals in polluted soil by disjunctive kriging. *Environ. Pollut.* 94(2), 205–215.

Wilson, J. W. & Brandes, E. A. (1979) Radar measurement of rainfall: a summary. *Bull. Am. Meteor. Soc.* 60, 1048–1058.

Weather Radar and Hydrology
(Proceedings of a symposium held in Exeter, UK, April 2011) (IAHS Publ. 351, 2012).

Comparison of different radar-gauge merging techniques in the NWS multi-sensor precipitation estimator algorithm

EMAD HABIB[1], LINGLING QIN[1] & DONG-JUN SEO[2]

1 *Department of Civil Engineering, University of Louisiana at Lafayette, PO Box 42991, Lafayette, Louisiana 70504, USA*
habib@louisiana.edu

2 *Department of Civil Engineering, The University of Texas at Arlington, Box 19308, Rm 438 Nedderman Hall, 416 Yates St, Arlington, Texas 76019-0308, USA*

Abstract This study performed an inter-comparison analysis of multi-level products of the radar-based multi-sensor precipitation estimation (MPE) algorithm. The main objective was to provide the user community and algorithm developers with insights on the potential value of increasing degrees of complexities in the algorithm in terms of bias removal and optimal merging with gauge observations. Different MPE products were considered: a gauge-only product, a radar-only product, a mean-field bias adjusted product, a local bias-adjusted product, and two products that are based on merging bias-adjusted products with gauge observations. The evaluation was conducted at the MPE native resolution (4×4 km² and hourly) using independent surface rainfall observations from a dense raingauge network in Louisiana, USA. The results demonstrate that some best-intended schemes for extensive radar and raingauge data processing do not lead to clear improvements and can even degrade the final products in some respects.

Key words rainfall; radar; multi-sensor; product; evaluation

INTRODUCTION

Weather radar systems provide rainfall observations with high spatial and temporal resolutions that are not available via traditional raingauges. Unfortunately, radar-based rainfall estimates are affected by a number of error sources. Thousands of different algorithms have been proposed over the last decades in order to reduce these uncertainties. For a meaningful assessment of these numerous efforts, it is crucial that their final effects be scrutinized in a comprehensive way. The most common evaluation techniques are limited to the bias and root-mean-square errors, computed by comparison of radar-rainfall estimates with single raingauges. This study attempts a much more complete analysis of different aspects of the uncertainties. It is applied to rainfall products that are delivered operationally by the National Weather Service (NWS) in the USA. The products can be based on radar only, gauges only, radar-adjusted for mean-field or local biases, or combinations of these techniques. The evaluation is conducted at the native resolution of the products (4 × 4 km² and hourly) using surface rainfall observations from an independent, dense raingauge network. The main focus of the analysis is: (1) to assess whether incremental algorithm complexities lead to actual performance improvement, and (2) to demonstrate how summary and conditional statistical metrics can distinguish varying levels of performance amongst different products. Findings from the current study will guide future enhancement of radar-based rainfall products and help the user on an informed usage of the products in a variety of research and operational applications.

STUDY SITE AND DATA SOURCES

The study site is the Isaac-Verot basin (~35 km²) located in the city of Lafayette in southern Louisiana (Fig. 1). The basin is frequently subject to frontal systems, air-mass thunderstorms, and tropical cyclones with mean-annual rainfall of about 140–155 cm and mean-monthly accumulation of as much as 17 cm. The area of the current study is fully within the boundaries of the NWS Lower Mississippi River Forecast Center (MRFC) service area.

Multisensor Precipitation Estimator (MPE) Products

The MPE produces hourly rainfall estimates on the approximately 4×4 km² HRAP grid. The MPE can produce as many as seven precipitation products based on different combinations of radar-gauge adjustment and merging (Seo *et al.*, 2010):

Radar-only mosaic (RMOSAIC) This product is based on a mosaic of the DPA products without any use of gauge observations.

Gauge-only analysis (GAGEONLY) This product is based on optimal interpolation of hourly raingauge observations within the service area of the RFC with adjustment, if necessary, for spatially non-homogeneous precipitation climatology.

Mean field bias-adjusted radar mosaic (BMOSAIC) This product is based on applying a radar-specific, time-varying but spatially uniform multiplicative adjustment factor to each pixel within the effective coverage of the radar in the DPA product.

Multi-sensor analysis based on BMOSAIC and raingauge data (MMOSAIC) This product is based on local merging of the mean field bias-adjusted radar mosaic (BMOSAIC) and the point raingauge observations using a co-kriging-like optimal estimation procedure.

Local bias-adjusted radar field (LMOSAIC) Unlike mean field bias correction, local bias adjustment corrects spatially non-uniform biases in the RMOSAIC field. The local bias adjustment is estimated and applied to all HRAP grid points in the RMOSAIC product using hourly raingauge observations within a fixed user-specified radius of influence.

Multi-sensor analysis based on LMOSAIC and raingauge data (MLMOSAIC) This product is similar to the MMOSAIC except that it is based on merging of the local bias-adjusted product (LMOSAIC) with the raingauge observations.

RFC-selected product for hydrologic operations (XMRG) This is not a new product in itself, but represents what the RFC forecaster decided to choose in real-time amongst the different MPE products described above and any manual corrections made to it.

Dense raingauge network

Independent observations from a dense raingauge network (Fig. 1) are used for MPE inter-product evaluation. The network is composed of 13 raingauge sites, with each site having two tipping-bucket gauges located side-by-side. Based on the network's spatial configuration, two MPE pixels are covered by multiple gauges (one pixel is covered by six gauges and another pixel is covered by four gauges). The high density of gauges within a single pixel provides a reliable and representative sample of surface rainfall, which is a pre-requisite for assessing the different MPE products (Habib *et al.*, 2009). Two years (2005 and 2006) of data were available for the study in which a total of 133 rainfall events were recorded.

METHODS

The estimation error of an MPE product, $e = R_{MPE} - R_S$, is defined as the difference between the MPE estimate, R_{MPE}, and the corresponding surface reference rainfall, R_S, estimated as the average of observations from multiple gauges within each of the two MPE pixels. The following metrics are used to inter-compare the different MPE products with respect to R_S:

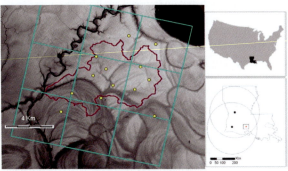

Fig. 1 Location of raingauges (represented by circles) and the 4×4 km^2 HRAP pixels grids of the MPE products in the Isaac-Verot basin, south Louisiana, USA. Large circles in the right-bottom graph show coverage of the nearest WSR-88D radars.

(a) Summary statistics. These include mean error or bias as a measure of systematic error, error standard deviation as a measure of random error, and two correlation coefficients (Pearson and Spearman rank) as measures of association.

(b) Conditional statistics. A parallel set of the summary statistics are developed in a conditional sense. The bias can be decomposed into three components: bias associated with successful detections or hits, bias due to rainfall misses, and bias due to false detections. The sum of these three components adds up to the total bias. The mean and the standard deviation of the estimation error can also be re-calculated by conditioning on varying values of R_S. If the estimation error, e, is defined as $e = R_{MPE} - R_S$, then the conditional error is formulated as $e|R_S = r_s$ and its conditional mean and standard deviation are defined as $\mu_e(r_s) = E[e \mid R_s = r_s]$ and $\sigma_e(r_s) = \sqrt{E[(e - \mu_e)^2 \mid R_s = r_s]}$, respectively. Following Ciach *et al.* (2007), we used a kernel regression approach to obtain non-parametric estimates of these two conditional statistics. Since the size of the conditional sample drops significantly with the increase of R_s, the conditional analysis was performed only up to $R_s = 25$ mm/h at which the sample size maintains 40 points or more. To assess the sample size impact on the estimated conditional statistics, we also report the bootstrap sampling distributions of these statistics.

RESULTS

Total and conditional bias

Figure 2 shows the total bias (expressed as a percentage relative to the total rainfall depth) and the three decomposed conditional components (hit, missed-rain, and false-rain biases). The most obvious feature in the total bias is the large overestimation of RMOSAIC (20.8%), which is mostly due to hit bias (18.2%). It is interesting to see that, while the overall bias of the GAGEONLY product is fairly small (–2.3%), its three components are relatively large, especially the hit and false rain (–10.2% and 14.1%, respectively). The detection problems (both false and failed) in the GAGEONLY product are probably due to the fairly low density and quality of raingauges that are available to the GAGEONLY analysis. The large overall bias in RMOSAIC was significantly reduced to 1.3% in the BMOSAIC product after removing the mean-field bias. The most significant bias reduction in BMOSAIC was in the hit-bias component which was almost completely eliminated (0.4%). This is not very surprising given that the mean field bias correction algorithm is designed specifically to address the hit bias. The next product (MMOSAIC) which merges BMOSAIC with individual gauges does not change the BMOSAIC total bias significantly, but it introduces some increases in the hit and false biases and eliminates some of the missed rain bias. This is consistent with the overall intent of MMOSAIC in that it focuses on reducing error variance rather than bias. Application of the local bias removal approach (LMOSAIC) instead of the mean-field bias adjustment has introduced a fairly significant negative overall bias (–8.4%), which is mostly attributed to the hit bias component. This result is not expected and suggests that there is significant room for improvement in the LMOSAIC algorithm. Merging LMOSAIC with individual gauge data further introduced more contribution of false rain bias, which cancelled out a comparable negative hit bias and resulted in a non-representative minimal value for the overall bias. The bias components of the final operational product XMRG reflect biases in the forecaster's choice in real time for what they consider to be the "best" product and various manual QC measures that they may take interactively during and/or after the product generation.

Next, we estimate the conditional bias for each of the MPE products (Fig. 3). It is clear that the GAGEONLY field consistently has a negative CB for all ranges of the surface reference rainfall, R_S, except for the low rainfall amounts (<3 mm/h) where a small positive bias is observed. The conditional underestimation in GAGEONLY deteriorates almost linearly with R_S where it exceeds – 10 mm/h for an R_S value of 25 mm/h. On the other hand, the radar-only product (RMOSAIC) has an opposite behaviour with a CB that overestimates for the entire range of R_S except for the very extreme values (R_S >22 mm/h). All the other MPE products fall between these two opposite behaviours; however, they all have a CB that consistently underestimates R_S except for the low

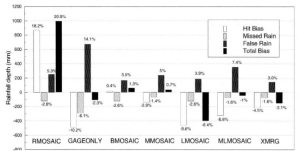

Fig. 2 Decomposition of total bias of different MPE products into its three components (hit bias, missed-rain, and false-rain) expressed as a percentage relative to the total rainfall depth.

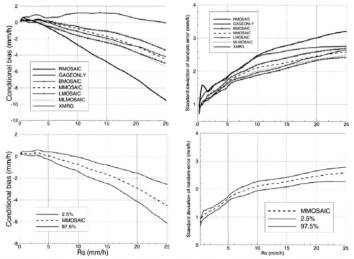

Fig. 3 Conditional bias (left) and standard deviation (right) as a function of the surface-rainfall rate (R_s) for all MPE products. The lower panels show the 2.5% and 97.5% confidence bounds for the estimated function based on bootstrap sampling (shown only for MMOSAIC as an example).

rainfall amounts (R_S <2.5–5 mm/h) where slight overestimation of CB is observed for most products. According to the CB bootstrap 95% bounds, it appears that the differences between GAGEONLY, RMOSAIC and the intermediate MPE products are statistically significant for most of the R_S values. However, the differences in CB amongst the intermediate MPE products are not as clear. Nevertheless, it is interesting to note that the smallest conditional bias was attained with the mean-field bias removal product (BMOSAIC). It is also interesting to see that the local-bias adjustment procedure resulted in an increased CB toward underestimation, especially for intermediate R_S values (LMOSAIC) and for high R_S values (both LMOSAIC and MLMOSAIC). This is an unexpected result and warrants further investigation. However, except for BMOSAIC, the differences among the other MPE products should be interpreted with caution since they do not appear to be statistically very significant based on analysis of the bootstrap distribution.

Random error

The standard deviation of the estimation error is calculated conditionally (i.e. by conditioning on R_S; Fig. 3). The unconditional analysis shows that, after GAGEONLY, the radar-only product has the highest error standard deviation. Applying any of the MPE bias and merging techniques (BMOSAIC, MMOSAIC, LMOSAIC and MLMOSAIC) resulted in statistically significant reduction in the error standard deviation compared to RMOSAIC (from 2 mm/h to about 1.5 mm/h). All four products have similar standard deviation except for LMOSAIC, which shows a slightly lower value though it does not appear to be statistically significant. The conditional analysis of the standard deviation of the error (Fig. 3) gives more insight into the product performance. The improvement in RMOSAIC compared to GAGEONLY appears to be mostly in the upper range of R_S ($R_S > 13$mm/h). Starting from RMOSAIC, the most reduction in the random error that is statistically significant appears to be achieved with the local bias-corrected product (LMOSAIC). Qualitatively, this result is not completely unexpected in that, compared to mean field bias correction, local bias correction tries to reduce errors from spatially faster-varying (i.e. higher-frequency) biases. Except for a slight reduction in the middle range of the distribution, merging of LMOSAIC with the gauge observation into MLMOSAIC does not bring about any further significant reduction for the most part of the distribution. Using a mean-field bias removal approach (BMOSAIC) appears to reduce the random error in RMOSAIC to a lesser extent than the LMOSAIC and further reductions are attained after merging with the gauge observations (MMOSAIC). The XMRG product has an average behaviour in comparison to the other four products.

Association statistics

The association between each MPE product and the reference rainfall R_S is assessed using two correlation measures: the Pearson's product-moment correlation coefficient (ρ) and the Kendall's rank correlation coefficient (τ) (Fig. 4). As expected, the gauge-only product shows the lowest levels of association ($\rho = 0.65$ and $\tau = 0.43$). According to the product-moment correlation coefficient, the RMOSAIC product has a slightly lower linear correlation with R_S than the other MPE products which all have similar correlation values ($\sim$0.9). Most of the inter-product differences in ρ are within the estimated sampling uncertainty bounds and therefore cannot be considered statistically significant. On the other hand, the rank-based correlation coefficient shows differences that are more significant. Starting with the radar-only product, RMOSAIC has a τ of 0.6, which stays unchanged with the application of mean-field or local bias removal techniques (BMOSAIC and LMOSAIC). Merging of the bias-adjusted products with individual gauge reports seems to have caused deterioration in the rank correlation where MMOSAIC and MLMOSAIC reported a τ of about 0.55 and 0.5, respectively. Seemingly an odd result, this is not unexpected in that the merging algorithms are conditional (given available observations) expectation operators, and hence smooth out, to a degree, the spatial details present in the radar rainfall data.

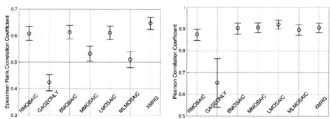

Fig. 4 Correlation measures for the different MPE products. The lower and upper bars represent the bootstrap confidence interval of the estimated statistics (2.5% and 97.5% bounds).

CONCLUSIONS AND CONCLUDING REMARKS

The results reported in this paper show that the most effective improvement of the rainfall products comes from applying the mean-field bias adjustment to the radar-only product. In comparison, the performance statistics of the other products showed marginal differences and, in a few cases, signs of deterioration are evident. The apparent lack of improvement by the rather involved procedure for local bias correction, as compared with the rather simple mean-field bias correction, is quite surprising. It warrants further investigation, including thorough scrutiny of the local bias correction methodology, as well as examination and further optimization of its algorithmic implementation procedure. The next level of complexity in the MPE algorithm comes from merging of BMOSAIC or LMOSAIC with individual gauge observations to result in MMOSAIC or MLMOSAIC. The intent of such merging is to reduce the random error in the final products (i.e. reduce the scatter). This outcome is partially fulfilled over the study site with the observed reduction in the conditional standard deviation of the random error in MMOSAIC and LMOSAIC. The potency of the multi-sensor merging technique depends on the availability of raingauge data. According to LMRFC records, the gauges available to the MPE algorithm over our study site include one gauge at a distance of ~7 km and four more gauges within 10–20 km (notice that observations from such gauges may not have been actually available and used during every operational hour). Such distances are larger than the correlation distances of hourly rainfall in this region. Future analysis involving testing of the MPE algorithm with different gauge densities may provide a quantitative recommendation on the gauge density requirements for the merging techniques to be more effective. We also note that the inter-product evaluation reported in this study was constrained to a single distance (~120 km) from the radar site in Lake Charles which the MPE estimates are mostly based on. Further analysis is necessary to investigate the range dependency of the performance of the different MPE products and their relative improvements across different ranges from the radar site. While the analysis was conducted using the native hourly temporal resolution of the MPE products, the statistical metrics provided evaluation information at both fine scale (e.g. results from the error standard deviation and the correlation coefficients) and longer aggregated scales (e.g. results from the bias).

There are two other factors that may have inhibited the potential benefits of the different MPE bias adjustment and merging techniques, i.e., the limited accuracy and quality of the operational gauge data available to the LMRFC forecasters. The RFCs rely on the satellite-based real-time data acquisition and distribution system known as the Hydrometeorological Automated Data System (HADS) as the main source of real-time gauge reports (Kim *et al.*, 2009). The accuracy of the GAGEONLY product over our site shows serious signs of underestimation (overall hit bias is –10.2%; event bias is within ±25%; conditional bias is around 40–50%). There are also clear symptoms of low-quality gauge data, as evident in the poor probability of detection and false detection by the GAGEONLY product. Since gauge observations represent a key data source in the multi-sensor rainfall estimation algorithm, further attention should be paid to develop effective and rigorous procedures for quality-control of raingauge measurements by the research and operational communities. Examples of such efforts are already under way such as the QC procedures recently implemented in the MPE algorithm (Kondragunta & Shrestha, 2006), and the recent QC algorithm developed by the National Severe Storm Laboratory (NSSL) as part of the NMQ-Q2 multi-sensor precipitation estimation (Zhang *et al.*, 2009). Finally, the analysis presented in this study indicates the value of the conditional statistics and their use in inter-product evaluation and verification. Conditional statistics provided deeper insight into the relative performance of the different products beyond what can be gained from typical summary statistics.

Acknowledgement This study was funded in part by a sub-award from the Louisiana NASA EPSCoR – DART 2 program. We would like to thank Jeff Graschel at LMRFC for providing the different MPE products and for numerous insightful discussions.

REFERENCES

Ciach, G. J. (2003) Local random errors in tipping-bucket rain gauge measurements. *J. Atmos. Oceanic Technol.* 20(5), 752–759.

Habib, E. Larson, B. & Graschel, J. (2009) Validation of NEXRAD multisensor precipitation estimates using an experimental dense rain gauge network in south Louisiana. *J. Hydrol.* 373, 463–478.

Kim, D., Nelson, B. & Seo, D.-J. (2009) Characteristics of reprocessed Hydrometeorological Automated Data System (HADS) hourly precipitation data. *Weather and Forecasting* 24, 1287–1296.

Kondragunta, C. & Shrestha, K. (2006) Automated real-time operational rain gauge quality controls in NWS hydrologic operations. Preprints, *20th AMS Conf. on Hydrology*, 29 Jan–2 Feb. 2006, Atlanta, GA, USA.

Seo, D.-J., See, A. & Delrieu, G. (2010) Radar and multisensor rainfall estimation for hydrologic applications. In: *Rainfall: State of the Science* (ed. by F. Y. Testik & M. Gebremichael), American Geophysical Union, Geophysical Monograph Series, Volume 191, 2010, 288 pages.

Zhang, J. *et al.* (2009) National mosaic and QPE (NMQ) system – description, results and future plan. In: *34th Conference on Radar Meteor* (Williamsburg, Virginia, USA).

Long-term evaluation of radar QPE using VPR correction and radar-gauge merging

EDOUARD GOUDENHOOFDT & LAURENT DELOBBE

Royal Meteorological Institute of Belgium, Avenue Circulaire 3 B-1180 Brussels, Belgium
edouard.goudenhoofdt@meteo.be

Abstract A new operational QPE algorithm based on C-band radar measurements has been developed. It is based on the computation of a mean apparent VPR. 24-h radar rainfall accumulations are combined with dense raingauge measurements using methods of various complexity. An independent raingauge network is used for verification. The relative performance of the methods is assessed using several statistics. A case analysis shows that the VPR QPE corrects for the high reflectivity circles seen on PCAPPI images. However, 2004–2010 statistics show that its benefit remains limited, especially after the application of merging methods. A seasonal analysis shows that the benefit of the radar is high in summer, while the VPR estimates have a slight positive or negative effect depending on the month and the method. The relative performance of the VPR estimates decreases with radar distance. These mitigated results suggest that a deeper analysis is needed to improve the method.

Key words C-band radar; VPR; QPE; merging; verification; Belgium

INTRODUCTION

Weather radar measurements are widely used to provide areal surface precipitation estimation over large areas. However, the accuracy of the measurements is limited due to various sources of uncertainty. One of the major constant sources of uncertainty is the height of the measurement for which the conditions generally differ from those at the ground. This effect leads to a non-uniform vertical profile of reflectivity (VPR). This is the case when growth, melting and evaporation of precipitation occurs (Joss & Waldvogel, 1990). In particular, the melting produces an enhancement of reflectivity known as the bright band which is more frequent in stratiform than in convective precipitation. Various methods have been proposed in the literature to correct the radar estimation from this effect. Most of the correction methods in the literature consist in estimating a representative VPR to extrapolate reflectivity measurements aloft towards ground level. It can be determined by using climatological profiles, local profiles at short distances from the radar (Kitchen *et al.*, 1994; Germann & Joss, 2002) or by means of an inverse method (Andrieu & Creutin, 1995). In the present work a simple VPR correction method, aimed for operational use, has been implemented (Vasquez Alvarez *et al.*, 2010). No other corrections such as clutter removal have been considered, but it is not supposed to have a significant impact on the long-term results.

Raingauges provide an accurate estimation of precipitation over a limited area (1 m^2) while weather radars provide precipitation estimation for larger areas (1 km^2) but with relatively less accuracy. Despite the significant difference between the two measurement scales, the idea of combining those two sources of rainfall estimation has been developed since the beginning of radar meteorology. Several methods of various complexity have been proposed in the literature, ranging from a simple factor correction to complex geostatistical interpolation. However, verification and comparison studies between different existing methods have been relatively limited and based mainly on cross-validation. In this study an independent network is used for a long-term verification.

RADAR AND GAUGE OBSERVATIONS

C-band radar measurements

Since 2001, RMIB has been operating a C-Band (5 GHz) weather radar located in Wideumont (southeast Belgium) at about 600 m above sea level. The radar, which has a range of 240 km, covers Belgium and Luxembourg and also parts of France, the Netherlands and Germany. It is a single polarisation radar with Doppler capabilities. The radar performs a scan at 5 elevations (0.3°, 0.9°,

1.8°, 3.3°, 6.0°) every 5 minutes. The radar sample volume has a resolution of 1° in azimuth and 250 m in range while the beam width is 1°. Every 15 minutes, the radar also performs a 10-elevation scan without Doppler filtering and a third scan, limited to 120 km range, which is used to retrieve radial velocities. More information regarding the radar characteristics and scanning strategy can be found in Delobbe & Holleman (2006). The volume data of the Wideumont radar have been archived at RMIB since 2002. Only the data from the first scan will be used in the present study.

Raingauge measurements

Two different networks of raingauges are used in this study (Fig. 1). The first one is operated by the hydrological service of the Walloon region (SPW) and is used to adjust the 24 h accumulations derived from radar observations. It consists of a dense and integrated network of 90 telemetric raingauges; most of them employ a tipping-bucket system providing hourly rainfall accumulations. For verification purposes, the RMIB climatological network is used. It includes 270 stations with daily measurements of precipitation accumulation at 08:00 h local time (LT). These stations are manual and the quality of the data is strictly controlled.

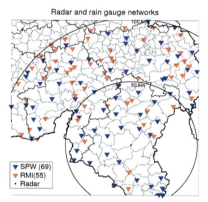

Fig. 1 The position of the Wideumont radar and the SPW and RMIB raingauge networks.

QPE ALGORITHMS

Vertical profile of reflectivity correction

The bright band is a layer of enhanced reflectivity caused by the scattering characteristics of the melting snow and their different velocities. Rings of higher reflectivity can be observed in the Pseudo Constant Altitude Plan Position Indicator (PCAPPI) product where it crosses the bright band. The different VPR correction methods aim not only to correct for this overestimation, but also to reduce the underestimation that occurs at long ranges, where the radar beams are quite high. This is done by using the shape of a representative VPR that has been estimated from volume data close to the radar. The shape of the VPR depends on the type of precipitation. VPRs corresponding to convective precipitation show more uniform values of reflectivity and have a significant vertical extension (up to 12 km). On the other hand, stratiform VPRs generally present the bright band effect. Therefore it is necessary to develop an algorithm to separate the convective and stratiform zones. The Steiner algorithm (Steiner *et al.*, 1995) has been used with a slight modification to be able to deal with the Belgian climatic conditions. The algorithm is applied at two different altitudes and a pixel must be convective for both to be labelled as convective. An objective validation of the algorithm was not possible but it tends to produce realistic output.

To extract the representative VPR for stratiform zones, the data between 10 and 70 km from the radar are considered so that the reflectivity along the vertical is well captured. The median average VPR (MAVPR) is obtained by using a moving average. A value at a certain height is the median value of all the measured reflectivity values located within a given vertical window (100 m). To filter out unrealistic profiles, the MAVPR will be used only if there is >70% of stratiform points within the specified range or >40% when a bright band has been identified. In any other case, a climatological profile with a slope of –2.0 dBZ/km is used as a representative VPR. The identification of the bright band is made by looking for a peak in the MAVPR and by checking the vertical gradients of reflectivity around the peak. The representative VPR (MAVPR or climatological profile) is then used to extrapolate the reflectivity measurements aloft down to a reference height as follows:

$$Z_{ext}(href) = Z(h) + Z_{VPR}(href) - Z_{VPR}(h) \tag{1}$$

where $Z_{ext}(href)$ is the extrapolated reflectivity value at reference height href, $Z(h)$ the measured reflectivity at height h, $Z_{VPR}(href)$ and $Z_{VPR}(h)$ the reflectivity values of the representative VPR at reference height and height h, respectively. For each stratiform pixel, the final reflectivity value at the reference height is the weighted average of all extrapolated values (maximum 5 values) obtained from the local reflectivity measurements at different heights. The weights are inversely proportional to the height. For the pixels labelled as convective, the final value at the reference height is the weighted average of the different reflectivity values of the local profile without any extrapolation based on the representative VPR. The reference height is set to 1 km. Finally, the final reflectivity at the reference height is converted into precipitation rate by using the Marshall-Palmer relation $Z = aR^b$ with a = 200 and b = 1.6.

Radar-gauge merging

In this study one raingauge interpolation method and four radar-gauge merging methods are used (Goudenhoofdt & Delobbe, 2009). For each method, the radar estimation (PCAPPI or VPR) associated with a raingauge location is the value of the single radar pixel covering this gauge.

(a) *Mean field bias (MFB)*: A correction factor, defined as the ratio between the total radar rainfall and the total raingauge rainfall, is applied to the radar data.
(b) *Range dependent adjustment (RDA)*: A distance dependent factor is applied based on a quadratic fit to the radar gauge ratios.
(c) *Brandes (BRA)*: A locally varying factor is applied based on the spatial interpolation of the radar gauge ratios.
(d) *Ordinary Kriging (KRI)*: For each location, a linear combination of the gauge value is constructed through the minimisation of the estimation error. No radar information is used.
(e) *External drift Kriging (EDK)*: same as Ordinary Kriging, but using the radar field as secondary information.

QPE VERIFICATION

Evaluation methodology

In Belgium there are several networks which cover the same region. It is then possible to use an independent network to evaluate and compare the different QPE methods. The RMIB climatological gauge network, which provides precipitation data every 24 h, will be used for this verification. Therefore the radar-gauge merging will be performed on 24 h accumulations. The verification gauges which are located too close (i.e. 2 km) from the merging network will be discarded. The range limit will be set between 10 and 100 km from the radar for both the merging and verification network. Only the days with at least 10 verification raingauge measurements above 1 mm will be considered. Two types of measure are used for each pair of estimate and gauge value: the multiplicative error (or ratio) and the additive error (or deviation). For both errors, a pair is valid if the gauge and all the estimate values exceed 1 mm. For all valid pairs, different statistics are computed (Goudenhoofdt & Delobbe, 2009) such as the mean bias (MB) and the mean absolute error (MAE).

Case analysis

Figure 2 shows that the rings of enhanced reflectivity which can be clearly seen on the PCAPPI image, have almost completely disappeared using the VPR correction. The mean absolute error (not shown) of the latter is also significantly lower than the one of the former. However, after the application of radar-gauge merging methods, the benefit of the VPR correction is relatively limited. Beside the good performance of the ordinary Kriging method, it is the external drift Kriging applied to the VPR-corrected radar estimates which performs best. This should be expected since a better correlation between radar and gauge data (after VPR correction) leads to larger weights for the radar values.

Overall statistics

Figure 3 shows the mean bias and the mean absolute error of the QPE algorithms based on six years of data. One can see that the VPR correction tends to slightly decrease the reflectivity since the mean bias decreases. All the merging methods succeed in reducing the mean bias. The mean absolute additive error (MAE) is the main statistic to compare the performance of the different algorithms. It can be seen that the MAE significantly decreases for all merging methods from 25% (mean field bias) to 40% (external drift Kriging). While the VPR-corrected radar estimate performs better than the classic radar PCAPPI estimate, its benefit is relatively small, especially

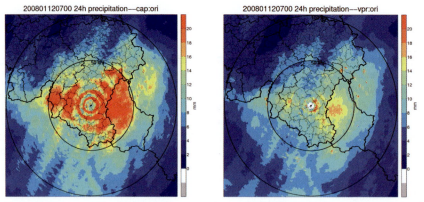

Fig. 2 24-h accumulation map based on two radar QPE algorithms: PCAPPI (left) and VPR (right).

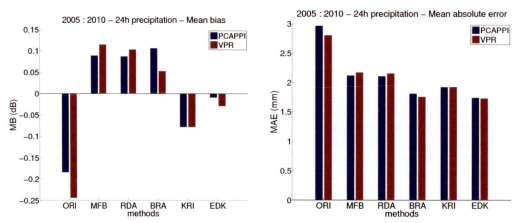

Fig. 3 6-years statistics of QPE: mean bias (left) and mean absolute error (right).

after the application of the radar-gauge merging methods. Indeed the MAE score tends to be slightly worse for the simple factor correction (mean field bias, range dependent adjustment) while slight improvements are obtained with the complex methods (Brandes and external drift Kriging).

Consequently the ranking of the radar-gauge merging method is not influenced by using the VPR method. The ordinary Kriging, based on the raingauge only, has a relatively good score. These results suggest that the VPR correction has, on average, a negligible effect on the final QPE estimation, independent of the merging method.

Temporal statistics

The spatial pattern of precipitation strongly depends on the period of the year, from widespread stratiform precipitation in winter to very local convective cells in summer. Looking at the mean bias (Fig. 4, left), we see that the radar tends to underestimate during the winter and overestimate during summer. All the merging methods succeed in limiting these biases, while the VPR method tends to reduce it. The MAE score (Fig. 4, right) exhibits a slight variation of the ranking during the year. The most significant change is the relatively bad performance of the raingauge interpolation method (KRI) in the summer, when the benefit of radar estimation is clear. The results obtained with the VPR correction tends to be limited for each month while there is a slight improvement in the winter for the Brandes method. It should be mentioned that the daily variability of the relative performance of the method is high, especially for the VPR-based methods. This further suggests that the VPR correction performs well only on a limited number of days.

Spatial statistics

The quality of the rainfall estimation depends on the distance between the radar and the measurements. The VPR correction aims at correcting for this effect. The radar estimation can also be affected in a specific region due to partial beam blocking and ground echoes (e.g. surrounding hills). It is then interesting to look at the statistics for each individual verification gauge, and as a function of the distance. Figure 5 shows that the performance (MAE score) of the VPR correction depends on the distance from the radar. The small benefit of the VPR correction at short distance tends to decrease for all merging methods when the distance increases. Its relative performance becomes negative for distances higher than 70 km, at which the climatological profile was used. At this range the variability of the VPR method performance (not shown) is also higher which suggests that the method is more unstable. Therefore the validity of the method at long distance must be further analysed.

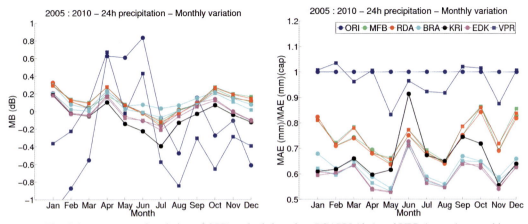

Fig. 4 6-years seasonal variation of QPE methods based on PCAPPI (dot) and VPR (square): mean bias (left) and mean absolute error (right).

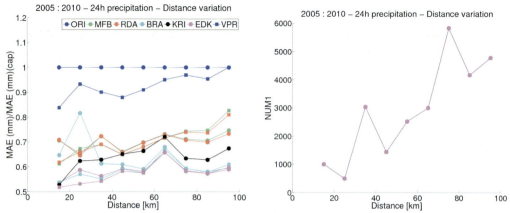

Fig. 5 Mean absolute error and number of valid pairs for QPE methods based on PCAPPI (dot) and VPR (square) as a function of distance from radar (10 km bins) based on 6 years of data.

CONCLUSION

Several QPE algorithms, including a VPR correction and different methods combining radar and raingauge data, have been tested. The recently implemented VPR correction scheme, suitable for operational use, aims to improve the radar estimation by extrapolating reflectivity measurements aloft toward the ground using a median apparent VPR. This correction scheme successfully removes the artificial circles due to the bright band compared to a basic PCAPPI estimate. A long-term verification from 2005 to 2010 of the 24 h accumulation field against an independent gauge network has been performed. The relative performance of the new VPR scheme and several radar-gauge merging methods has been assessed. The best method overall is the Kriging with external drift combining raingauge measurements and VPR-corrected radar data. However the relative positive impact of the VPR correction remains limited for all merging methods. A seasonal analysis shows that the radar added value is clear during summer when convective precipitation occurs. The spatial statistics shows that the performance of the VPR correction depends on the distance. While it has a better relative score for all merging methods close to the radar, the VPR method performance is worse at large distance (more than 70km). These results suggest that the VPR method adds some benefit to QPE estimation but that it should be further refined. Several improvements can be considered: the slope of the climatological profile can be adapted, the 10-elevation scan can be used instead of the 5-elevation scan and a more robust apparent VPR could be constructed using more than one single radar scan.

REFERENCES

Andrieu, H. & Creutin, J. D. (1995) Identification of vertical profiles of reflectivity using an inverse method. Part 1: Formulation. *J. Appl. Met.* 34, 225–239.

Delobbe, L & Holleman, I. (2006) Uncertainties in radar echo top heights used for hail detection. *Meteor. Appl.* 13, 361–374.

Germann, U & Joss J. (2002) Mesobeta profiles to extrapolate radar precipitation measurements above the Alps to the ground level. *J. Appl. Met.* 41, 542–557.

Goudenhoofdt, E. & Delobbe, L. (2009) Evaluation of radar-gauge merging methods for qualitative precipitation estimates. *Hydrol. Earth. Syst. Sc.* 13, 195–203.

Joss, J. & Waldvogel, A. (1990) Precipitation measurement and hydrology. In: *Radar in Meteorology* (ed. by D. Atlas), 577–606. American Meteorological Society.

Kitchen, M., Brown, R. & Davies, A. G. (1994) Real-time correction of weather radar data for the effects of bright band, range and orographic growth in widespread precipitation. *Quart. J. Roy. Met. Soc.* 120, 1231–1254.

Steiner, M., Houze, R. A. J. & Yuter, S. E. (1995) Climatological characterization of three dimensional storm structure from operational radar and rain gauges data. *J. Appl. Met.* 34, 1978–2007.

Vazquez Alvarez, M., Goudenhoofdt, E. & Delobbe, L. (2010) Implementation and evaluation of VPR correction methods based on multiple volume scans. In: *Proc. ERAD 2010.*

A 10-year (1997–2006) reanalysis of Quantitative Precipitation Estimation over France: methodology and first results

PIERRE TABARY[1], PASCALE DUPUY[1], GUY L'HENAFF[1],
CLAUDINE GUEGUEN[1], LAETITIA MOULIN[1], OLIVIER LAURANTIN[2],
CHRISTOPHE MERLIER[2] & JEAN-MICHEL SOUBEYROUX[3]

1 *Centre de Météorologie Radar, DSO, Météo France, Toulouse, France*
pierre.tabary@meteo.fr
2 *Division Coordination Etudes et Prospective, DSO, Météo France, Toulouse*
3 *Direction de la Climatologie, Météo France, Toulouse, France*

Abstract In order to provide a common reference for hydrologists (e.g. for calibrating model parameters, assessing the added value of inputting high space-time resolution data in hydrological models), Météo France is currently running a national collaborative project aimed at producing a high-resolution (1 km^2), 10-year reference database (1997–2006) of hourly Quantitative Precipitation Estimations (QPE) covering the entire French metropolitan territory with no spatial nor temporal gaps. The input data that are used are the individual 5 min 512 × 512 km^2 pseudo-CAPPI radar reflectivity images of the French radar network and quality-controlled hourly and daily (from 6 UTC to 6 UTC) raingauges. Several validation exercises have been performed to validate the various steps of the processing chain. In particular, the final product – 1 km^2 composite hourly accumulation maps – has been evaluated with independent raingauge data over one year in two different geographical / meteorological contexts.

Key words radar Quantitative Precipitation Estimation; kriging; radar–raingauge merging

INTRODUCTION

In order to provide a common reference for hydrologists (e.g. for calibrating model parameters, assessing the added value of inputting high space-time resolution data in hydrological models), the French national weather service is currently running a national collaborative project aimed at producing a 10-year reference database of Quantitative Precipitation Estimations (QPE). The initiation of this work stems back to the previous Weather Radar and Hydrology Conference (WRAH2008, Grenoble, 2008), where the need for reanalysis of QPE was clearly identified during a workshop (Delrieu *et al.*, 2009). Similar projects have been conducted or are currently underway in the radar hydrometeorology community (e.g. Overeem *et al.*, 2009; Nelson *et al.*, 2010). The objective is to make optimum use of all available information in the operational archives in order to obtain the best surface precipitation accumulation estimation over France with no gaps and to provide associated uncertainties at the hourly time-step and 1 km^2 spatial resolution. The various modules of the processing chain are described hereafter. The final product – 1 km^2 composite hourly accumulation maps – has been evaluated with independent raingauge data over one year in two different geographical / meteorological contexts.

DATA USED AND PERIOD OF ANALYSIS

Taking into account the evolution of the radar network, the availability of radar products and the need to cover a period of at least 10 years, a decision was made to focus on the 1997–2006 time period. This time period will be extended to current time in the future. In 1997, the French operational network consisted of 13 radars. A further 11 radars have been deployed over the period 1997–2006, raising the total number of operational radars to 24 in 2006. The very large variation over time of the radar coverage is one of the numerous reasons why dynamic quality codes are so important. The scan strategy of the radars over the considered time period typically consisted of 1 (flat areas) to 4 (mountainous areas) elevation angles revisited every 5 minutes.

Radar data that are used for the reanalysis are single-radar 5 min, 1 km^2, 512 × 512 km^2, pseudo-CAPPI reflectivity images. These data are the only ones that have been continuously archived since 1997. They are not corrected for (1) partial beam blocking (referred to as PBB

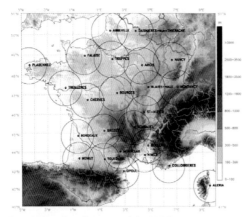

Fig. 1 French radar network in 2006.

hereafter), (2) vertical profile of reflectivity (VPR) effects, (3) advection effects, (4) attenuation by gases, precipitation or wet radome, (5) clear-air (insects / birds / chaff). Ground-clutter (hereafter referred to as GC) is theoretically corrected for, even though the state-of-the-art GC identification methods used at the beginning of the 1997–2006 time period was not perfect. Reflectivity data are coded as follows: <8 dBZ, 8–16 dBZ, 16–20 dBZ, 20–21 dBZ, 21–22 dBZ, … The coarse resolution of the coding at low levels is a limiting factor for the precise estimation of precipitation at low rain rates. On the raingauge side, hourly and daily (from 6 UTC on one day to 6 UTC on the following day) data are available in the operational databases. These data are routinely checked by experts and – if needed – corrected for. The typical number of hourly raingauges over France (550 000 km^2) is 1000, compared to 4000 daily.

RADAR DATA PROCESSING

Radar data pre-processing turned out to be absolutely necessary before considering merging them with raingauge data. A number of modules have been developed – based upon the operational experience of radar data processing at Météo France (to address the various error sources that have been identified with the data). The principles that governed the choice of the various algorithms are the following: simplicity, robustness, efficiency, interoperability. Because the project is working on a tight schedule (the aim being to deliver a V1 version of the re-analysis database by the first quarter of 2012) limited time was available to specify and test each module. The assumptions and limitations of each algorithm are acknowledged and perspectives regarding their improvement are mentioned.

Establishment of GC maps for all [radar;year] couples

Occurrence frequency maps are computed for each [radar;year] couple. The thresholds of 25 dBZ (S-band radars) and 15 dBZ (C-band radars) have been used to compute the occurrence frequency. Pixels having an occurrence frequency exceeding some threshold (determined subjectively by an expert, typically 3–12%) are classified as GC and *never* used for the considered year. This may appear as a drastic approach, but emphasis was put on minimizing the rate of unfiltered clutter that may corrupt the radar–raingauge analysis ("better have no data than risk introducing bad data"). Notice that anomalous propagation GC is not filtered by the proposed approach, which is a problem for some radars (e.g. Bordeaux) of the network that are very frequently subject to anomalous propagation. The reason for re-establishing the GC map for each year stems from the fact that the scan strategy of the radar may have changed (faster antenna rotation rates, more elevation angles in the volume coverage pattern, etc.). GC maps could be updated more frequently, but would require more time and effort.

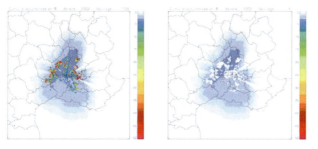

Fig. 2 Bollène (S-band) radar 2002: occurrence frequency map (512×512 km^2) without (left) and with (right) application of a 4% threshold (512×512 km^2).

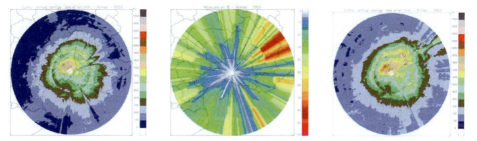

Fig. 3 Nîmes (S-band) radar 2002: raw yearly accumulation (left), PBB map (centre) and corrected yearly accumulation map (right).

Establishment of PBB maps for all [radar;year] couples

For each [radar;year] couple, a yearly rainfall accumulation map is computed using the GC-identified Cartesian pseudo-CAPPI reflectivity images converted into rainfall rates using the Marshall-Palmer *Z-R* relationship ($Z = 200R^{1.6}$). This accumulation map is then converted into polar coordinates. Accumulation curves (functions of the azimuth) are then computed for various classes of distances (0–10 km, 10–20 km, etc.). These curves are then filtered with a running 10° filter that replaces each value with the upper 95% percentile value. Once this is done, the original curve is compared to the filtered curve and the PBB rate is obtained for each [distance;azimuth] couple. The aim of this procedure is to identify narrow masks, with the assumption that such masks have an extension that is less than 10°. Wider masks (e.g. arising from mountains) will not be captured by this approach. However, wide masks are assumed to be identified and corrected for through the daily comparison with raingauges and the daily calibration factor maps (see further down). The retrieved PBB rates are converted into a 512×512 km^2 Cartesian map for further application to the raw 5 min reflectivity pseudo-CAPPIs. This empirical approach to PBB was preferred over using a simulation tool (e.g. Delrieu *et al.*, 1995) because it takes into account simultaneously orogenic and non-orogenic masks, potential biases in the antenna's pointing angles and coupling between PBB and Vertical Profiles of Reflectivity (VPR) effects (see quantification of that effect in Tabary (2007)).

Clear-air / weak signals processing and computation of hourly radar rainfall accumulations

The approach that was taken to eliminate clear-air echoes (most likely birds and insects), whose frequency and intensity are known to be quite high on the S-band radars located in southern France during the autumn and spring seasons, simply consists in keeping only radar pixels with a reflectivity above a certain threshold Z_{MIN}. Based upon operational experience, Z_{MIN} was taken equal to 20 dBZ at S-band and 16 dBZ at C-band. Notice that technologies such as polarimetry, volumetric scans, high-resolution and frequent (5 min) satellite imagery were not yet operationally available over the considered time period of re-analysis (1997–2006); hence the proposed (rather brutal) approach. Pixels with a reflectivity value less than Z_{MIN} are considered as "weak" and their reflectivity is temporarily set to Z_{MIN} (i.e. the maximum value a "weak" pixel can take). At each pixel, the hourly radar rainfall accumulation of the "weak" values within the hour (ACC_{WEAK}) is then compared to the hourly accumulation of the "non-weak" values within the same hour (ACC_{NOWEAK}). If ACC_{WEAK} is found to be much smaller than ACC_{NOWEAK}, then ACC_{WEAK} is considered to be negligible and the hourly accumulation is taken equal to ACC_{NOWEAK}. Otherwise, the hourly accumulation is considered to be unavailable and set to WEAK_VALUE. In that case, the sum of $ACC_{WEAK}+ACC_{NOWEAK}$ is kept in memory for further exploitation (see next sub-section). In other words, the proposed approach is such that radar data are not used to provide the "no-rain" information. The 5-min, 512×512 km^2 Cartesian reflectivity pseudo-CAPPI are converted into rainfall rate maps using Marshall-Palmer. Two-dimensional advection fields are then computed using a standard cross-correlation approach (as in Tuttle & Foote (1990)) between two successive images, spaced apart by 5 minutes. The advection fields are subsequently used to over-sample the rainfall rates maps (every minute) and produce smooth hourly accumulation maps (see Tabary (2007) for a detailed description of the approach).

Production of daily accumulations and computation of radar/raingauge calibration factor map

The 512×512 km^2 radar hourly accumulations are subsequently accumulated over 24 h (from 6 UTC to 6 UTC the following day). The exact same approach is taken as at the hourly time-step to process "weak" and "no-weak" hourly accumulations. The radar-based 24 h rainfall accumulation map, wherever it is available (i.e. outside GC classified areas, high PBB areas and "weak" areas), is then confronted with 24 h raingauges. A radar/raingauge calibration factor field is computed as follows:

–   a circular neighbourhood (with a radius of 30 km) is moved successively over each 1 km^2 pixel of the 512×512 km^2 radar domain;
–   for each new position of the neighbourhood, the raingauges inside the neighbourhood having reported more than 0.6 mm in 24 h are paired with the corresponding radar pixels (in cases where radar rainfall accumulation is not classified as GC, high PBB or weak);
–   a number N of (radar, raingauge) 24 h accumulations couples are established; wherever N is higher than 3, the median value of the N radar/raingauge ratios is computed and attributed to the central pixel of the neighbourhood;

The calibration factors are then applied to the daily radar accumulation, wherever possible. Where the calibration factor cannot be computed, the resulting daily accumulation is given by ordinary Kriging of daily raingauges.

GENERATION OF THE BEST DAILY ACCUMULATION FROM RADAR AND RAINGAUGES OVER EACH RADAR DOMAIN

In order to obtain the best daily estimation of precipitation, an extra step consists in merging the calibrated daily radar accumulation map with daily raingauges using Kriging with external drift (KED). The calibrated radar accumulation (the external drift) in itself is already a good estimation of daily precipitation. The main goal of this step is to ensure that the raingauge accumulations are retrieved (at the location of the gauges) in the final result.

The description of KED equations can be found in Hengl *et al.* (2003), as well as the description of the regression-Kriging method that is the one actually used in the project, which is shown to lead to the same results.

GENERATION OF THE BEST HOURLY ACCUMULATIONS FROM RADAR AND RAINGAUGES OVER EACH RADAR DOMAIN

This step (temporal disaggregation) consists in deriving hourly precipitation from the best daily precipitation accumulation estimation. This is achieved by distributing the 24 h accumulation over the 24 h composing the day as follows:

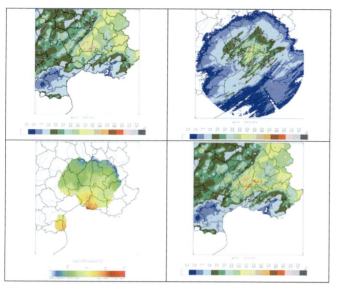

Fig. 4 Nîmes (S-band) – 21 October 2002 –24 h accumulation map from Kriged raingauges (top left), raw radar rainfall accumulation (top right), radar/raingauge calibration factor map (expressed in log10, bottom left) and calibrated radar rainfall accumulation map (bottom right).

– Hourly radar rainfall accumulations are first corrected using the calibration factors established in part 3. Because of all the criteria that are imposed (on the number of reporting gauges, the quality of the radar data, etc.), the calibration factors are not available everywhere. An extrapolation algorithm is therefore applied in order to propagate the values that could be computed all over the radar domain.

– Hourly precipitation accumulation fields are then computed from available hourly (calibrated) radar and raingauge data. The method used to compute these temporary fields is here again KED. As KED requires the drift (the hourly radar accumulation) to be available all over the domain, missing radar data are replaced by hourly ordinary Kriging values.

For a given point of the 512×512 km^2 radar domain, letting h_i ($i \in [1;24]$) be the hourly estimation derived from merging hourly radar and raingauge data, σ_i the Kriging estimation error, H the sum of the 24 h_i and D the best daily estimation of precipitation (see above); then we define the weight $w_i = h_i/H$ and the final hourly estimation $w_i.D$, with an uncertainty approximated to $\sigma_i.D/H$. Special attention is paid to some particular cases (where $H = 0$).

RADAR COMPOSITING AND GENERATION OF THE BEST HOURLY COMPOSITE ACCUMULATION MAP OVER FRANCE

The final step is to generate a map all over France by compositing the different local estimates of precipitation, which are available on Cartesian 512×512 km^2 domains centred on the available radars. Notice that because of the size of each individual radar domain (512×512 km^2) and the number and location of radars in operation at any time between 1997 and 2006, this approach allows a complete coverage of the French territory. The estimation of hourly precipitation and its uncertainty for one point can be provided by different local estimates. It has been decided to use the uncertainties as weights in the combination of estimations in overlapping areas.

For a given point, letting h_i and e_i be the hourly rainfall estimation (in mm) and its uncertainty (also in mm) given by an individual estimation i, then the result of the weighted linear combination is (in terms of hourly QPE H (in mm) and uncertainty E (in mm)):

$$H = (h_1/e_1 + h_2/e_2 + ... + h_n/e_n) / (1/e_1 + 1/e_2 + ... + 1/e_n) \tag{1}$$

$$E = (\sqrt{n}) / (1/e_1 + 1/e_2 + ... + 1/e_n). \tag{2}$$

Table 1 Left: Reanalysis *vs* ordinary Kriging over the (Abbeville,Arcis,Trappes) domain in 2001. Nearly 57 000 observations were used to compute the scores. Right: Reanalysis *vs* ordinary Kriging over the (Bollène,Nîmes) domain in 2002. Nearly 34 200 observations were used to compute the scores. HRG: hourly raingauge accumulation (mm).

(Abbeville,Arcis,Trappes) domain in 2001	CORR HRG > 0 mm / HRG > 2 / HRG > 5 mm	NB HRG > 0 mm / HRG > 2 / HRG > 5 mm	(Bollène,Nîmes) domain in 2002	CORR HRG > 0 mm / HRG > 2 / HRG > 5 mm	NB HRG > 0 mm / HRG > 2 / HRG > 5 mm
Ordinary Kriging	0.67 / 0.40 / 0.27	0.78 / 0.58 / 0.43	Ordinary kriging	0.69 / 0.53 / 0.38	0.82 / 0.67 / 0.57
Re-analysis	0.73 / 0.56 / 0.45	0.84 / 0.69 / 0.59	Reanalysis	0.75 / 0.63 / 0.54	0.87 / 0.75 / 0.68

RESULTS

In order to evaluate the final composite 1 km^2 hourly QPE, some raingauge data have been removed from the whole process and left aside for independent validation purposes. At the current state of the project, only two different sets of data have been produced. The first one corresponds to the [Abbeville,Arcis,Trappes] radar triplet (north of France, see Fig. 1) for the year 2001, the second one to the [Bollène,Nîmes] couple (southeast of France, see Fig. 1) for the year 2002.

Table 1 presents some first results (correlation CORR and normalised bias NB = ΣQPE_i / ΣHRG_i) for various hourly raingauge accumulation thresholds (0, 2 and 5 mm in 1 hour), in comparison with ordinary Kriging of raingauges on the same domains. The normalized bias is defined here as the ratio of the total QPE accumulation over the total observed accumulation.

CONCLUSIONS AND OUTLOOK

A processing chain has been developed in order to produce a high-resolution (1 km^2), 10-year reference database (1997–2006) of hourly Quantitative Precipitation Estimations (QPE) covering the entire French metropolitan territory with no spatial or temporal gaps. The chain uses the individual 5-min 512 × 512 km^2 pseudo-CAPPI radar reflectivity images of the French radar network and quality-controlled hourly and daily (from 6 UTC to 6 UTC) raingauges as inputs. Simplicity, robustness, efficiency and interoperability are the principles that have governed the decisions regarding the various modules. Several validation exercises have been performed to validate the various steps of the processing chain. In particular, the final product – 1 km^2 composite hourly accumulation maps – has been evaluated with independent raingauge data over one year in two different geographical / meteorological contexts. The V1 of the database (1997–2006) is expected to be delivered by the end of the first quarter of 2012. Later on, the database will probably be extended from 2006 onwards and improvements will be made to several modules.

REFERENCES

Delrieu, G., Creutin, J. D. & Andrieu, H. (1995) Simulation of radar mountain returns using a digitized terrain model. *J. Atmos. Oceanic Technol.* 12, 1038–1049.

Delrieu, G. Braud, I., Berne, A., Borga, M., Boudevillain, B., Fabry, F., Freer, J., Gaume, E., Nakakita, E., Seed, A., Tabary, P. & Uijlenhoet, R. (2009) Weather Radar and Hydrology (preface). *Adv. Water Resour.* 32, 969–974.

Hengl, T., Geuvelink, G. B. M. & Stein, A. (2003) Comparison of kriging with external drift and regression-kriging. Technical note, ITC, Available on-line at http://www.itc.nl/library/Papers 2003/misca/hengl comparison.pdf.

Nelson, B. R., Seo, D.-J. & Kim, D. (2010) Multisensor precipitation reanalysis. *J. Hydrometeor.* 11(3), 666–682.

Overeem, A., Holleman, I. & Buishand, A. (2009) Derivation of a 10-year radar-based climatology of rainfall. *J. Appl. Meteor. Climatol.* 48, 1448–1463.

Tabary, P. (2007) The new French radar rainfall product. Part I: methodology. *Weath. Forecasting* 22(3), 393–408.

Tuttle, J. D. & Foote, G. B. (1990) Determination of boundary layer airflow from a single Doppler radar. *J. Atmos. Oceanic Tech.* 7, 218–232.

Weather Radar and Hydrology
(Proceedings of a symposium held in Exeter, UK, April 2011) (IAHS Publ. 351, 2012).

261

Temporal and spatial variability of rainfall at urban hydrological scales

I. EMMANUEL[1], E. LEBLOIS[2], H. ANDRIEU[3] & B. FLAHAUT[1]

1 *PRES L'UNAM, Ifsttar, Département GER, CS4, 44341 Bouguenais, France*
isabelle.emmanuel@ifsttar.fr
2 *CEMAGREF, 3 B Quai Chauveau, 69009 Lyon, France*
3 *PRES L'UNAM, Ifsttar, Département GER and IRSTV FR CNRS 2488 Bouguenais, France*

Abstract The main objective of this paper is to characterize the spatial and temporal variability of rainfall at scales that are consistent with urban hydrological applications. In this way, a total of 24 rain periods have been analysed according to a geostatistical approach. This analysis has focused on the non-zero rainfall variogram. The studied rain periods were recorded by the weather radar of Treillières (10 km north of Nantes, France) in 2009. This radar device provides rainfall radar images with a high level of spatial resolution (250×250 m^2) and instantaneous temporal resolution. Results indicated four different types of rainfall fields, which display very different variability scales, including double structures within the same field. This study highlights the benefit of radar images featuring high temporal and spatial resolution, which in turn allow studying small-scale variability.

Key words rainfall structure; geostatistics; hydrological scales

INTRODUCTION

Rainfall is the key term involved in many hydrological processes; it is particularly important in the field of urban hydrology, where rainfall response is dominated by runoff occurring on the impervious surfaces of small and medium-sized catchments, which thus display shorter response times (typically less than an hour or two). Niemczynowicz (1999) noticed that rainfall input is often a weakness in urban hydrology. This issue has mainly been addressed in terms of the density of raingauge networks required to capture the spatial and temporal variability of rainfall at spatial and temporal scales consistent with the response time of urban catchments (Schilling, 1991). The interest in weather radar to analyse the spatial variability of rainfall really started when the accuracy and the reliability of weather radars began to be established. Different studies have confirmed that weather radar images constitute valuable data for assessing rainfall field variability over a range of spatial and temporal scales, including those in urban hydrology (Miniscloux *et al.*, 2001; Berne *et al.*, 2004). Nevertheless, it appears that the literature devoted to this topic is not yet extensive and moreover does not provide a clear assessment of rain period variability covering a wide array of meteorological situations. The present paper will contribute to this research topic by characterizing the spatial and temporal features of different types of rain periods recorded during 12 days in 2009. The study herein is based on weather radar images with radar pixels of 250×250 m^2. Such high spatial resolution allows analysis of the small-scale variability of rainfall, which is consistent with the smaller surface areas of many urban catchments.

PRESENTATION OF THE CASE STUDY

Radar data

Radar data have been provided by the C-band weather radar of Treillières, with a 3-dB beam width of 1.25°. This radar is located 10 km north of Nantes, in the southern part of Brittany (France). The scanning strategy of the Treillières radar consists of three successive volume scans, with each scan lasting 5 min and composed of four elevation angles. The PPI operated at the three lowest elevation angles (i.e. 0.4°, 0.8° and 1.5°) are contained in all three scans and therefore are repeated every 5 min. Polar measurements are projected every 5 min into a 128×128 km^2 Cartesian grid with a spatial resolution of 250×250 m^2. The data from each radar pixel are instantaneous.

The radar data used for our purposes were not operational and have solely been processed by the French Meteorological Office (Météo France) as follows:
(a) Dynamic identification of ground clutter based on the pulse-to-pulse fluctuation of radar reflectivity. This identification scheme is dynamic and incorporates anomalous propagations as well as rainfall over ground clutter (Tabary, 2007).
(b) Static performance, for each scan, of measurements of each elevation angle. Measurement of the lowest angle (which is not ground clutter) is the one used.
In addition, we implemented the three following processing steps:
(c) Filtering of isolated pixels. A pixel is assumed to be isolated once its reflectivity value exceeds the average reflectivity of its eight neighbours by a reflectivity difference of 20 dBZ. This reflectivity value is then replaced by the average of its eight neighbours.
(d) Reflectivity-to-rainfall rate conversion, according to the Marshall-Palmer Z-R relationship (i.e. $Z = 200\ R^{1.6}$, with Z being the radar reflectivity in mm^6/m^3 and R the rainfall rate in mm/h).
(e) Identification of the partial beam blocking areas. Three areas of rainfall underestimation have been invalidated as a result of visualizing the total rainfall amount image over the 12 recorded days.
In polar coordinates, the width of the radar bin increases with range. In order to maintain the spatial resolution of $250 \times 250\ m^2$ as being representative of actual radar sampling, the study zone has been limited to a circle of 20-km radius centred on the Treillières radar.

Recorded rain periods

A total of 12 days, recorded between May and November 2009, have been analysed. The total amount of rain for each day and the standard deviation computed between the various radar pixels inside our study zone show that the sampled days are of different types. Daily rainfall amounts range between 2.2 and 27.5 mm, with standard deviations varying from 1 to 13.6 mm.

Partitioning of the recorded days into rain periods and groups

A meteorological analysis of the 12 recorded days has been conducted over our study zone with support provided by the French Meteorological Office. This analysis, associated with the visualization of radar images, has allowed selecting 24 rain periods over the 12-day recording, such that each selected rain period is homogeneous in terms of type of rain. These periods last between 35 min and 3 h 5 min.

These 24 rain periods were then sorted into four groups: Group 1 contains 9 periods of light rain; Group 2 comprises 7 shower periods; Group 3 combines 4 periods of relatively unorganized storms; and Group 4 has 4 storm periods organized into rain bands.

Validation of the composition of homogeneous groups

Two kinds of analyses have served to verify the relevance of the protocol for separating the recorded days into 24 rain periods and then sorting these periods into four groups.

A Principal Component Analysis (PCA) (Pearson, 1901) was performed first by considering the rainfall radar images of rainfall intensities for all 24 rain periods (i.e. a total of 405 radar images) as the subjects. Four variables were chosen to characterize each radar image (within the study zone): (1) percentage of zeros in the radar image, and, for positive rain data, (2) the average, (3) the standard deviation, and (4) the median. A mean centring PCA was introduced since the studied variables do not have the same units. This PCA reveals that the studied rain periods are composed of similar radar images as regards the considered variables. Moreover, the PCA highlights that radar images of Group 1 are characterized by an average and standard deviation less than those of the images from other groups. Group 2 images display, compared to the other groups, a high average and standard deviation, in association with a high percentage of zeros. On the whole, images from Groups 3 and 4 have a higher median (associated with a higher average);

their separation however into two distinct groups is not justifiable according to the considered PCA.

A second analysis, based on descriptors other than the percentage of zeros, average, standard deviation and median, thus turns out to be necessary. For this, the minimum distance separating each radar pixel of positive intensity from a pixel of zero intensity is computed. This second analysis enables distinguishing Group 3 and Group 4 rain periods. The Group 3 rainfall areas reveal maximum intensities located both at the centre and on the edge of these areas. In contrast, Group 4 rainfall areas are better organized with maximum intensities located preferentially at the centre.

Both previous studies have shown the homogeneity of the considered rain periods. They have also confirmed that the four groups differ from one another and that, moreover, each group is composed of rain periods with similar characteristics.

STRUCTURAL ANALYSIS

Presentation

Rainfall field variability can be described using different methods. In the following a geostatistical approach has been adopted. In the classical approach, the rainfall process $w(x,t)$ is analysed as a random process, w being a rainfall intensity value, falling at the location x and at time t (Matheron, 1965). If we consider the fields $w(x,t)$ as the realization of the random function W, the structure function of this random function can be expressed by the variogram γ as follows:

$$\gamma(du) = 1/2 \, E[(W(u) - W(u + du))^2] \tag{1}$$

where du represents a spatial shift (or a temporal increment), u at the location x or the time t. E denotes the mathematical expectation. For further details on the variogram, the reader may refer to Journel & Huijbregts (1978).

The variogram is used to characterize the variability of precipitation and has mainly been computed from raingauge measurements, though radar data have started to be introduced for the purpose of characterizing the precipitation structure in an urban context or during flash flooding.

A rain period can be associated with a multi-realization context, whereby each radar image is considered as a single realization. When the studied phenomenon has a defined variance, the variogram reaches a sill at a given range. In this case, the sill is equal to the variance. It then becomes possible to normalize the variogram, by the respective variance of each realization, and ultimately to average this normalized variogram over all realizations. This variogram is known as the climatological variogram (Berne *et al.*, 2004) and is characterized by a sill equal to one. It yields a structure representative of a set of realizations displaying a similar structure by assigning an equivalent weight to each realization. It also offers the opportunity to remove the bias that often affects radar data. The range corresponds to the decorrelation distance, i.e. the distance from which two measurement points exhibit independent statistical behaviour.

A rainfall field is composed of the alternation of zones where it rains (i.e. zone of non-zero rainfall) and zones of zeros where it does not rain. In this study, we have focused on the non-zero variogram: each point used to compute this variogram has an intensity strictly higher than 0 mm/h.

The multi-directional spatial and temporal non-zero variograms have been computed for all four groups and studied through an analysis of their shapes and ranges.

Spatial and temporal structure of non-zero rainfall

Spatial structure The instantaneous climatological and multi-directional spatial variograms of non-zero rainfall for each group are presented in Fig. 1.

The variograms of Groups 1 and 2 present a unique structure with a range of 17 km and 5 km, respectively, which means that for the Group 2 rain periods, two points 5 km apart exhibit independent behaviour. The Group 3 variogram is unusual for featuring two main structures and decreasing after the second sill. The first structure corresponds to a 2.5-km range, while the second corresponds to a 15-km range. The Group 4 variogram is also unusual in that it reaches a

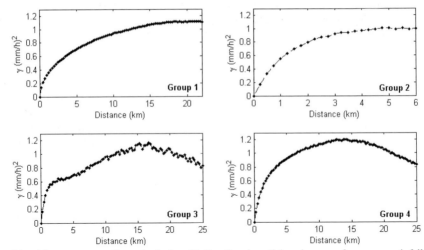

Fig. 1 Instantaneous climatological multi-directional spatial variogram of non-zero rainfall of the four groups zoomed in order to best visualize the rise and the shape of each variogram.

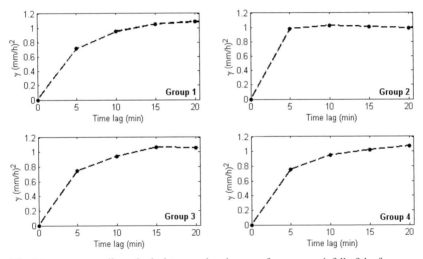

Fig. 2 Instantaneous climatological temporal variogram of non-zero rainfall of the four groups.

maximum value at 12 km, after which variability decreases. No explanation has been found to rationalize the decrease in the variograms of Groups 3 and 4 for high shifts. It must be emphasized that all four variograms do not display a "nugget effect" and differ significantly from variograms calculated with raingauge data.

Temporal structure The instantaneous climatological temporal variograms of each group are presented in Fig. 2.

The temporal variograms of Groups 1 and 2 have a unique range of 15 min and 5 min, respectively. Even though the spatial variograms of Groups 3 and 4 are unusual, the temporal variograms have a unique temporal range of around 15 min and 20 min, respectively.

This variogram analysis is consistent with a visualization of the radar images of rain periods belonging to each group. The Group 1 rain periods are not highly variable over the study zone. Group 2 is composed of rain periods containing small variable clusters. Moreover, the special structures of Group 3 and Group 4 spatial variograms can be explained by the presence of small variable clusters inside less variable patterns of larger spatial extent. In the following section, we will attempt to isolate these two structures and analyse them separately.

Particular case of double structures The idea here is to separate, for Groups 3 and 4, the pixels of each radar image into two classes; the first class corresponds to small variable clusters, while the second corresponds to the larger and less variable patterns. To complete this task, the Self-Organizing Map (SOM) algorithm (also called Kohonen's algorithm) has been adopted. All classification algorithms enable sorting pixels into the most homogeneous and/or distinct classes possible. The unique feature of the SOM algorithm is its ability to build the various classes within a predefined representative space (a sphere in our case), where it places priority on the similarity between adjacent classes. For a more complete description, the reader is referred to Kohonen (2001).

A visualization of radar images reveals that the small variable clusters are characterized by higher intensities than the less variable patterns. Each radar pixel therefore is described by two variables: the mean intensity of its eight neighbours, and the associated standard deviation (indicative of variability). An initial classification of radar pixels into 25 classes was completed by focusing on both variables. It is worthwhile to note that these two variables are positively correlated. Subsequently, the 25 classes were categorized into two "superclasses" (Bennani, 2006): the first class assembles the more intense and variable pixels (Class 1), and the second is devoted to the less intense and less variable pixels (Class 2). This grouping into two classes has been simplified by the continuity existing between neighbouring classes, and it is this continuity that characterizes the SOM algorithm.

This classification was performed for each rain period contained in Groups 3 and 4, and all radar pixels of non-zero rainfall were associated with one of the two classes. The variograms were then computed according to the same approach as before, i.e. a first time by considering just pixels from Class 1 and a second time by considering just the Class 2 pixels (Figs 3 and 4).

The structural analysis of Class 1 and 2 pixels indicates that Group 3 rain periods are composed of clusters, with a 1.5-km spatial range and 5-min time decorrelation (Class 1), in association with less variable patterns characterized by 11-km and 15-min ranges (Class 2). Two distinct structures can also be identified during the Group 4 rain periods: the first is characterized by clusters with ranges of 5 km and 5 min (Class 1), and the second by patterns of 15-km and 15-min ranges (Class 2). Recall that Class 1 contains more intense pixels than Class 2.

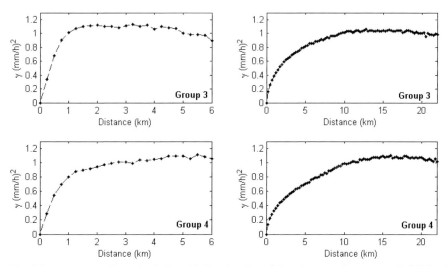

Fig. 3 Instantaneous climatological multi-directional spatial variogram of non-zero rainfall for Class 1 (left) and for Class 2 (right).

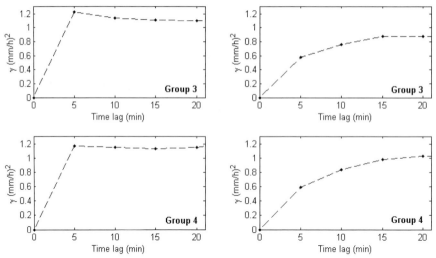

Fig. 4 Instantaneous climatological temporal variogram of non-zero rainfall for Class 1 (left) and for Class 2 (right).

CONCLUSION

The analysis has focused on the non-zero rainfall variogram. Results exposed four distinct rainfall structures with very different variability scales. The first structure had a decorrelation distance of 17 km and a decorrelation time of 15 min; it corresponded to homogenous rainfall areas. The second structure was differentiated by ranges of 5 km and 5 min and corresponded to highly variable and intense rainfall areas. The third and fourth structures were double structures, composed by associating small and intense clusters (ranges of 5 min and less than 5 km), which were located inside less variables areas (ranges of 11–15 km and 15 min). The third structure appeared to be relatively disorganized, whereas the fourth structure has been identified as rainfall bands with maximum intensities located at their centres. This study has highlighted the benefit of radar images with fine spatial resolution. A later study seems to be useful to analyse how those different rainfall structures affect hydrological modelling of small and fast catchments.

Acknowledgements The authors would like to thank Météo France, which provided all the radar data, and particularly Abdel-Amin Boumahmoud and Pierre Tabary who offered valuable assistance during the data reading phase.

REFERENCES

Bennani, Y. (2006) *Apprentissage Connexionniste.* Lavoisier, Paris, France, 143–155.

Berne, A., Delrieu, G., Creutin, J.-D. & Obled, C. (2004) Temporal and spatial resolution of rainfall measurements required for urban hydrology. *J. Hydrol.* 299, 166–179.

Journel, A. & Huijbregts, C. J. (1978). *Mining Geostatistics.* Academic Press, London, UK.

Kohonen, T. (2001) *Self-Organizing Maps.* Springer Series in Information Sciences, Vol. 30, Springer, Berlin, Germany.

Matheron, G. (1965) *Les Variables Régionalisées et leur estimation.* Masson et Cie, Paris, France.

Miniscloux, F., Creutin, J.-D. & Anquetin, S. (2001) Geostatistical analysis of orographic rainbands. *J. Appl. Met.* 40, 1835–1854.

Niemczynowicz, J. (1999) Urban hydrology and water management – present and future challenges. *Urban Water* 1(1), 1–14.

Pearson, K. (1901) On lines and planes of closest fit to systems of points in space. *Philosophical Magazine,* Series 6, 2, 559–572.

Schilling, W. (1991) Rainfall data for urban hydrology: what do we need? *Atm. Res.* 27 (1–3), 5–21.

Tabary, P. (2007) The new French operational radar rainfall product. Part 1: methodology. *Weather Forecast.* 22, 393–408.

3 Rainfall forecasting (nowcasting & numerical weather prediction)

Weather Radar and Hydrology
(Proceedings of a symposium held in Exeter, UK, April 2011) (IAHS Publ. 351, 2012).

269

Comparison of optical flow algorithms for precipitation field advection estimation

THOMAS PFAFF & ANDRÁS BÁRDOSSY

Hydrology and Geohydrology, Institute of Hydraulic Engineering, University of Stuttgart, Germany
thomas.pfaff@iws.uni-stuttgart.de

Abstract Estimates of the advection field as derived from successive weather radar images are not only an essential piece of information for precipitation nowcasting, they can also be of value in order to improve the quality of radar-based precipitation estimates themselves. In order to develop a correction scheme for radar accumulations using advection information, three different methods to determine the optical flow between two radar images were tested. The main criteria for algorithm selection were: (a) execution speed to allow application in an operational setting, (b) the quality of the estimated advection field, assessed by visual inspection and common error measures like RMSE and MAE, and (c) the robustness of the algorithm, i.e. the dependence of the estimation quality on the choice of their governing parameters. A simple block matching algorithm, an optical flow algorithm based on image intensity gradients and an approach that uses information on multiple image scales to optimize the search pattern of an extended block matching method were considered. All three methods were reasonably fast for calculating the advection fields and showed a similar distribution of their error measures. The last algorithm showed the most robust behaviour, the estimated advection field being virtually independent of the parameter choice. Applying the accumulation correction scheme using advection fields calculated by this last algorithm, improved the agreement between radar estimates and station measurements for the majority of the stations.

Key words weather radar; advection; optical flow

INTRODUCTION

Estimation of advection fields from weather radar images has long been of interest for meteorological analyses of storm motions and the internal structure of storms. It complements the radial velocity fields gathered by Doppler information, but is not dependent on it and can be done, even if Doppler information is not available. The first approach to use flow fields estimated from two successive weather radar images to analyse in-storm motions resulted in the TREC-algorithm (Rinehart & Garvey, 1978). This algorithm and its successor COTREC (Li *et al.*, 1995) have been used successfully in several hydrometeorological studies (Tuttle & Gall, 1999; Trier *et al.*, 2000; Berenguer *et al.*, 2005). More recently, advection fields estimated from radar reflectivity images were used in nowcasting of precipitation (Bowler *et al.*, 2006). These fields are also being used to improve the precipitation estimates retrieved from weather radar data (Sinclair, 2007). Approaches to do so are based on the fact that one radar image is an instantaneous snapshot of the precipitation field. The development and movement of this field is not captured until the next snapshot. If these images are accumulated to a common time scale, in order to compare them with gauge data, this can lead to errors, if precipitating cells move very fast. An example of this problem can be seen later in Fig. 2.

One way to account for the movement of the precipitation field during accumulation is to produce intermediate representations of the field by determining the advection vectors and translating the intensity along these vectors for time steps between two images. In order to provide a smooth transition between the two bounding images, both are used for each intermediate time step with different weights according to their temporal distance from the point in time at which they were taken, as given in:

$$\text{Acc} = \frac{1}{\sum_{j,i} w_{j,i}} \sum_{i=0}^{n} \left[w_{0,i} \cdot \text{adv}\left(I_{t_0}, \vec{v}, i \cdot \tfrac{t_1 - t_0}{n}\right) + w_{1,i} \cdot \text{adv}\left(I_{t_1}, -\vec{v}, i \cdot \tfrac{t_1 - t_0}{n}\right) \right] \tag{1}$$

Here, Acc is the accumulation over the time period $[t_0; t_1]$ between two successive radar images. The number of discretisation steps between the two images is denoted by n. Between the time steps the image intensities I at time t_0 are advected forward in time by the advection function adv(I, v, dt), while the image at time t_1 is propagated backward in time (represented here by the negative advection vector field $-v$). The contribution of each of these new intermediate images to the final accumulation is weighted by weights $w_{i,j}$, which are calculated to decrease linearly giving the weight 1 to the image at the time that it was taken and giving the weight 0 once the image has been advected by n sub-steps.

Figure 1 shows a schematic representation of both the advection process and the weight that is given to the respective intermediate images. The actual images taken at times t_0 and t_1 are depicted with a grey background.

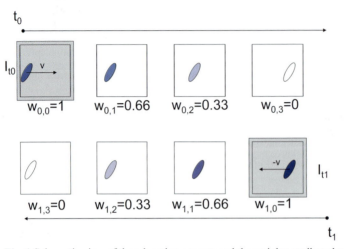

Fig. 1 Schematic view of the advection process and the weights attributed to each advected image.

ADVECTION ESTIMATION ALGORITHMS

Block Matching

The Block Matching (BM) technique is one of the simplest ways to determine the movement of objects between two images. First the initial image is divided into blocks of a certain size. These blocks may or may not overlap. Then, for each block in the initial image, the second image is searched for the block that would give the highest correlation. The vector from the initial block centre to the centre of the found block becomes this initial block's advection vector estimate. This is the method used in the TREC procedure (Rinehart & Garvey, 1978). It is very easy to implement, but may become time-consuming if large blocks or search radii have to be used. There are three parameters that can affect estimation quality. These are the block size, the shift size and the maximum search radius. An increasing block size decreases the probability of finding a wrong block due to similarities at small scales; it decreases, however, the speed of the estimation as more image pixels have to be compared. Additionally, if the block size is too large it becomes more difficult to estimate the correct advection vector if the objects in the image show a rotating motion (as is often the case for large-scale meteorological patterns) or become sheared. The shift size defines the amount of overlap between blocks in the initial image. The smaller the shift size, the more detailed the estimated advection field will be at the cost of more searches. The number of blocks n_{blocks} in one image dimension of size *is* depends on the block size (*bs*) and shift size (*ss*) according to $n_{blocks} = ((is - bs) / ss) + 1$.

If the maximum search radius selected is too small, it may result in the correct matching block not being found. If the algorithm is implemented to calculate all block correlations up to the maximum search radius, computation time increases quadratically with search radius.

Optical flow using intensity gradient information

This method was developed by Horn & Schunck (1981). It uses an analogue of the continuity equation and applies it to the pixel intensities, in order to calculate a flow vector for each pixel of an image. An additional smoothness constraint is imposed on the flow field to make flow vectors of neighbouring pixels become similar in magnitude and direction. For this study, the implementation, which is part of the Open Computer Vision (OpenCV) library (Bradski & Kaehler, 2008) was used. The three major parameters for this algorithm that were examined in this study were: the parameter λ, which controls the importance given to the smoothness constraint, the maximum number of iterations allowed to the algorithm to satisfy the smoothness constraint and the window size of an averaging pre-processing, which is recommended in the original algorithm description to treat large displacements in an image correctly. This algorithm, which is the basis of the optical flow calculations for the STEPS nowcasting suite (Bowler *et al.*, 2004, 2006), will be denoted HS in the following.

Optical flow using an efficient block search on multiple image scales

This algorithm was presented by Bouguet (2000) and is based on a method developed by Lucas & Kanade (1981), but is applied to several levels of image resolution, using the results from the coarser resolutions as first guesses for the calculations at the finer resolution, down to the original image resolution. The Lucas Kanade algorithm mainly tries to improve the searching efficiency of the BM method by calculating a guess for the displacement vector from image intensity gradient information.

The Bouguet algorithm only calculates the optical flow for a limited number of image pixels, which satisfy some conditions necessary for the Lucas Kanade algorithm to work. The advection vectors for all other pixels were interpolated using inverse distance weighting. As the different levels of image resolution go under the name of "pyramids" in the image processing literature, this algorithm has been abbreviated PyrLK. It can be found under that abbreviation in the OpenCV library, from where its implementation was taken for this study.

For this algorithm also, three parameters were varied: the size of the search window (analogous to the block size in the BM method), the maximum number of iterations used in the Lucas and Kanade algorithm to determine the optimal displacement vector, and the minimum distance between two tracked pixels.

METHODOLOGY

Advection accuracy

The accuracy of the estimated advection vector fields was assessed by comparing the image taken at time t_1 with the image at time t_0 after it was translated according to the advection field. The comparison was summarized by calculating the root mean square error (RMSE) and mean absolute error (MAE) from all image pixels. This was done for each of the three algorithms and for each parameter combination.

Measures of location like the mean or the median of RMSE and MAE calculated from several successive image pairs would show the average accuracy of each algorithm.

Robustness

As each of the considered algorithms has to be adjusted by setting certain parameters, it was of interest, how sensitive the advection field estimation would be to the choice of these parameters. This would be especially important in an operational context, where a robust algorithm, which would never fail completely, might be preferred, even if it is not always producing the best results.

Measures of spread like minimum, maximum, standard deviation and skew of RMSE and MAE should show whether an algorithm would constantly produce good results or if it might fail completely from time to time.

Improvement of precipitation estimates

While it appears plausible that taking advection into account should improve the precipitation estimates when accumulating precipitation from radar images, the magnitude of this effect should nevertheless be quantified. This was done by applying the advection correction to hourly accumulations of radar data for one day and comparing them to the hourly measurements at gauges that were available for this area. The comparison was done using RMSE, MAE and correlation coefficient between radar and gauge values. As a reference, the same comparison was done using the "naïve" accumulations, which just add the values of all radar images for one hour.

The difference in RMSE, MAE and correlation between advection-corrected and uncorrected radar accumulations were compared.

RESULTS

Accuracy and robustness

The accuracy and robustness analyses were conducted on a data set of 13 successive radar images of the German Weather Service (DWD) radar Dresden for the time period 23 June 2008 01:50 UTC+00 to 23 June 2008 02:50 UTC+00. During this period a set of showers moved over the area northwest of the radar at high speeds (cf. Fig. 2).

The three advection algorithms were applied to this series of images with all combinations of the following parameters:

-   BM: block size: 5, 10, 20 pixels; shift size: 5, 10, 20 pixels; max. search radius: 5, 10, 20 pixels.
-   HS: λ: 0.001, 0.01, 0.1, 1.0, 10; maximum iterations: 32, 64, 128, 256, 512; pre-smoothing window: 1, 7, 17, 33 pixels.
-   PyrLK: window size: 3, 5, 10 pixels; maximum iterations: 100, 500, 1000; minimum distance: 5, 10, 20 pixels.

Tables 1 and 2 summarise the results of these calculations. It can be seen that the PyrLK algorithm outperforms the others with respect to almost all statistics. Except for the minimum RMSE value, which is achieved by the HS algorithm, the PyrLK method provides the best estimates on average, as well as showing the smallest standard deviation. When looking at MAE, it performs best with respect to all statistics.

Table 1 Statistics of root mean square error (RMSE).

Algorithm	Min	Max	Median	Mean	Std	Skew
BM	3.787	6.420	4.835	4.858	0.533	0.602
HS	3.228	5.161	4.191	4.234	0.454	0.045
PyrLK	3.261	4.091	3.850	3.743	0.289	–0.345

Table 2 Statistics of mean absolute error (MAE).

Algorithm	Min	Max	Median	Mean	Std	Skew
BM	20.586	92.874	29.841	33.079	16.345	2.828
HS	19.025	36.938	26.529	27.113	4.584	0.027
PyrLK	16.243	26.136	20.699	21.544	3.076	0.010

The skew of the distributions was given to show whether the majority of the values tend more towards the maximum or the minimum. Positive skew could be interpreted as a tendency of the algorithm to fail completely in the case of a wrong parameter choice. This is the case for the BM method. The HS algorithm shows the most symmetric error behaviour, while the PyrLK shows a slight negative skew when looking at RMSE. It has to be noted that the PyrLK algorithm produced the same advection fields for a certain pair of images, regardless of the parameter choices stated above. Given the good performance of these fields, the method appears to be very robust.

The statistics of execution times, for the estimation of the advection field for one radar image of the robustness study data set, are presented in Table 3. It shows that the HS algorithm is the fastest of the three methods. The BM method shows the longest execution times, also with the largest spread. The PyrLK method performs in between the other two, being, on average, a factor of five slower than HS and a factor of four faster than the BM method. The statistics were aggregated from the timings for each of the images and all parameter combinations.

Improvement of precipitation estimates

Apart from the questions of accuracy and robustness of an advection algorithm, it is of interest, if this information could be used to improve the radar precipitation estimates in terms of agreement between radar and gauge accumulations. Figure 2 gives an example of how the inclusion of advection information can at least lead to a visually more plausible accumulation image. While just adding the individual images, as done in the left part of Fig. 2, leads to a ripple pattern, the advection-aware accumulation more reasonably distributes the precipitation intensities along the paths of the respective shower cells.

The numerical analysis as described earlier was conducted on a set of 1463 radar images from the DWD radar Dresden, with a time resolution of 5 minutes for the two months of June and July 2008. The images were accumulated to hourly resolution to be matched with gauge observations. The advection fields were calculated using the PyrLK algorithm.

Figure 3 shows the distributions of the differences in RMSE, MAE and correlation coefficient between simple accumulations (subscript *acc*) and advection-corrected accumulations (subscript *vacc*). For the majority of stations the inclusion of advection information leads to a better agreement

Table 3 Execution time statistics.

Algorithm	Min (s)	Max (s)	Median (s)	Mean (s)	Std (s)
BM	374.9	704.1	699.4	592.8	153.5
HS	27.26	27.17	27.23	27.22	0.022
PyrLK	146.6	150.4	148.5	148.5	1.121

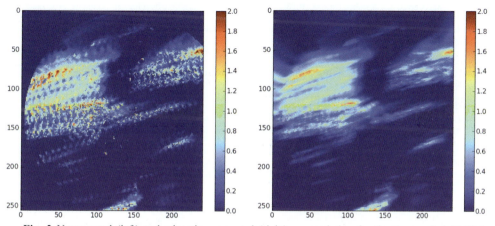

Fig. 2 Uncorrected (left) and advection-corrected (right) accumulation for the time period 01:50 to 02:50 UTC 23 June 2008. Radar Dresden. *X* and *Y* axis in km. Colour scale in mm.

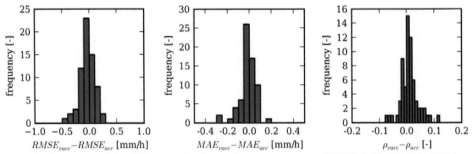

Fig. 3 Histograms of the differences in the performance measures RMSE (left), MAE (middle) and correlation coefficient (right) between advection-corrected and uncorrected radar accumulations when compared to gauge data.

between radar and gauge accumulation (smaller RMSE and MAE, larger correlation). Yet, there are still a significant number of stations where the correction appears to make things worse.

SUMMARY

Three algorithms for the estimation of advection fields were compared in this study, two of which are similar to those used by well-established methods for storm movement analysis and radar-based precipitation nowcasting. The third, which shows a remarkable degree of robustness, also provided very good estimates of the advection field, performing best in almost all statistics considered.

While correcting radar images for the effect of advection during accumulation seems to be a reasonable approach, the correction applied in this study did not always produce better results when compared to accumulations of the snapshots themselves. The reason for this may lie in errors in the estimation of the advection field. Longer-term analyses may be used to check whether this effect is diminishing in the long run or whether it remains as a systematic error.

Acknowledgements This study was conducted as part of the OPAQUE project, which was funded by the German Federal Ministry of Education and Research (BMBF) under grant number 0330713A.

REFERENCES

Berenguer, M., Corral, C., Sánchez-Diezma, R. & Sempere-Torres, D. (2005) Hydrological validation of a radar-based nowcasting technique. *J.. Hydromet.* 6(4), 532–549.
Bouguet, J. Y., *et al.* (2000) Pyramidal implementation of the Lucas Kanade feature tracker – description of the algorithm. *Intel Corporation, Microprocessor Research Labs* 1(2), 1–9.
Bowler, N. E., Pierce, C. E. & Seed, A. W. (2004) Development of a precipitation nowcasting algorithm based upon optical flow techniques. *J. Hydrol.* 288(1–2), 74–91.
Bowler, N. E., Pierce, C.E. & Seed, A. W. (2006) STEPS: A probabilistic precipitation forecasting scheme which merges an extrapolation nowcast with downscaled NWP. *Quart. J. Roy. Met. Soc.* 132(620), 2127–2155.
Bradski, G. & Kaehler, A. (2008) *Learning OpenCV: Computer Vision with the OpenCV Library*, O'Reilly Media, Sebastopol, USA.
Horn, B. K. P. & Schunck, B. G. (1981) Determining optical flow. *Artificial Intelligence* 17(1–3), 185–203.
Li, L., Schmid, W. & Joss, J. (1995) Nowcasting of motion and growth of precipitation with radar over a complex orography. *J. Appl. Met.* 34(6), 1286–1300.
Lucas, B. D. & Kanade, T. (1981) An iterative image registration technique with an application to stereo vision. In: *7th International Joint Conference on Artificial Intelligence* (ed. by P. J. Hayes), 674–679. William Kaufmann.
Rinehart, R. E. & Garvey, E. T. (1978) Three-dimensional storm motion detection by conventional weather radar. *Nature* 273(5660), 287–289.
Sinclair, S. (2007) Spatio-temporal rainfall estimation and nowcasting for flash flood forecasting. PhD thesis, University of KwaZulu-Natal, Durban, South Africa.
Trier, S. B., Davis, C. & Tuttle, J. (2000) Long-lived mesoscale convective vortices and their environment. Part I: Observations from the Central United States during the 1998 Warm Season. *Monthly Weather Rev.* 128(10), 3376–3395.
Tuttle, J. & Gall, R. (1999) A single-radar technique for estimating the winds in tropical cyclones. *Bull. Am. Met. Soc.* 80(4), 653–668.

Weather Radar and Hydrology
(Proceedings of a symposium held in Exeter, UK, April 2011) (IAHS Publ. 351, 2012).

275

Extending a Lagrangian extrapolation forecast technique to account for the evolution of rainfall patterns over complex terrain

PRADEEP V. MANDAPAKA, URS GERMANN, LUCA PANZIERA & ALESSANDRO HERING

146 via ai Monti, Locarno Monti, Switzerland
pradeep.mandapaka@meteoswiss.ch

Abstract In this study, we employed a Lagrangian extrapolation scheme (MAPLE) to obtain short-term (lead times <5 h) rainfall forecasts over a large region broadly centred on Switzerland. The high-resolution forecasts from MAPLE were then evaluated against the radar observations for 20 summer rainfall events using categorical and continuous verification techniques. The verification results were then compared with Eulerian extrapolation forecasts. In general, Lagrangian persistence forecasts outperformed Eulerian persistence forecasts. Although MAPLE performed well for short lead times, the performance deteriorated rapidly with increase in lead time. Results also showed that the predictability of the MAPLE model depends on the spatial correlation structure and temporal evolution of the rainfall events.
Key words radar-rainfall; predictability; lifetime; MAPLE

INTRODUCTION

Hydrological applications such as flash-flood and debris-flow forecasting require reliable quantitative precipitation forecasts (QPF) at high spatial and temporal resolutions and with short lead times (<5–6 h). Since weather radars provide good areal coverage with high resolution in space and time, many radar-based short-term QPF techniques have been developed over the years (e.g. Austin & Bellon, 1974; Mecklenburg *et al.*, 2000; Germann & Zawadzki, 2002; Seed, 2003; Bowler *et al.*, 2007). At MeteoSwiss, our aim is to develop a short-term forecasting tool that is capable of forecasting storm advection, growth, and dissipation at high resolutions, and short lead times with reasonable accuracy. Radar-echo extrapolation is an attractive avenue towards achieving the above goal as short-term high-resolution QPF fields can be generated fast enough to provide emergency management authorities sufficient time to issue an alert. Although radar-echo extrapolation techniques have a long history of development and application, no rigorous evaluation of an extrapolation technique was carried out in the past over a complex orographic region such as Switzerland for the time scales in the order of minutes. The objective of the study is to provide the reader *quantitative* information in terms of skill scores and lifetimes for a highly complex orographic region using a state-of-the-art extrapolation system.

The extrapolation scheme selected for the analysis is MAPLE (McGill Algorithm for Precipitation nowcasting using Lagrangian Extrapolation). The MAPLE algorithm was originally developed by the radar group at the McGill University, Canada. The algorithm and its performance in forecasting rainfall events over the continental United States has been documented by Germann & Zawadzki (2002) and Germann *et al.* (2006b). In this study we also compare the performance of MAPLE with the corresponding Eulerian forecasts. Figure 1 shows the study region (620×620 km^2) along with the complex orography and the location of three MeteoSwiss radars.

Radar-rainfall estimation errors

It is well known that radar-rainfall estimates are affected by uncertainties from various sources (e.g. Germann *et al.*, 2006a; Villarini & Krajewski, 2010). Germann *et al.* (2006a) discussed in detail various challenges associated with radar-rainfall estimation in mountainous regions. Villarini & Krajewski (2010) provided an exhaustive review of various sources of uncertainty in

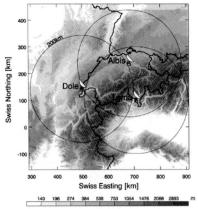

Fig. 1 Map showing the complex orography of the study region and location of the three radars within the 620×620 km^2 domain.

radar-rainfall estimates. Over the years, several studies have evaluated radar-rainfall products, proposed models for residual errors, and employed the error models to represent radar-rainfall observational uncertainties in the form of ensembles: see Mandapaka & Germann (2010) for a review. Likewise, the ensemble framework was widely used to characterize the uncertainties in the QPF model (for example, errors in the initial conditions, model structure). However, much work needs to be carried out regarding the propagation of the radar-rainfall residual errors through the QPF models. In this study we focus on deterministic forecasts from MAPLE and assume radar-rainfall observations to be the reference rainfall fields. Propagation of rainfall measurement uncertainties through the forecast chain and quantifying the overall errors in the QPF model output is beyond the scope of this study.

Lagrangian persistence (MAPLE)

In this section, we provide only a brief description of the MAPLE algorithm. The forecast set-up primarily consists of two steps: (1) velocity field estimation using the variational echo-tracking (VET) algorithm, and (2) extrapolation of the current radar image honouring the motion field estimated in the above step. The VET algorithm was proposed by Laroche & Zawadzki (1994) to estimate the three-dimensional wind field from the single Doppler clear-air echoes. Germann & Zawadzki (2002) later modified the technique to obtain the velocity field from the radar-rainfall composites. We adopted the same version of the VET algorithm as presented in Germann & Zawadzki (2002) and estimated the velocity field at a resolution of 25×25 km^2. The procedure was repeated every 5 min using 30 min of past radar-reflectivity composites. All the above computations were carried out using radar-reflectivity fields in dBZ scale to avoid any effects of the skewness of the precipitation field (in rain-rate scale) on the velocity field estimation.

The velocity vectors obtained from the VET technique can be used for extrapolation in four different ways: (1) constant vector forward scheme, (2) constant vector backward scheme, (3) semi-Lagrangian forward scheme, and (4) semi-semi-Lagrangian backward scheme. The technical details, merits and limitations of all the four options were discussed in detail in Germann & Zawadzki (2002). We employed the semi-Lagrangian backward scheme in the current study. In the semi-Lagrangian backward scheme, the origin of a parcel that would end up at a particular grid point in the forecast field is determined by following the streamlines upstream in space and backward in time. In this manner, we allow for the rotation in the displacement vector for the extrapolation up to a certain lead time τ. Given the velocity field $\boldsymbol{u}$ at the initial time step t_0 from the VET algorithm, the lead time τ is divided into N time-steps of length Δt, and the displacement α is iteratively obtained using equation (13) of Germann & Zawadzki (2002).

Eulerian persistence

In the Eulerian persistence scheme, the current radar-rainfall image is taken as a forecast for all the lead times. The approach is simple and is usually taken as a baseline system in the forecast evaluation studies.

RADAR-RAINFALL DATA

Radar-rainfall data for 20 summer (June, July and August) rainfall events between the years 2005 and 2010 are used to evaluate the forecasts from MAPLE. The instantaneous radar reflectivity measurements from a network of three radars undergo several steps of data processing and quality control resulting in 5-min rain-rate composites on a rectangular grid of 1-km spatial resolution. Some of the major steps in the data processing and quality control are the polar-Cartesian grid transformation, identification and mitigation of ground clutter, correction for the vertical variability of tradar-reflectivity, and minimizing the bias between the radar- and gauge-measured rainfall values (e.g. Joss & Lee, 1995; Germann *et al.*, 2006a).

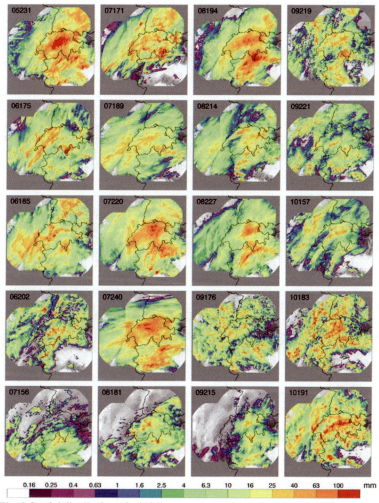

Fig. 2 Spatial distribution of rainfall accumulation for each of the selected events over a 620×620 km^2 domain.

The data selected for this study consists of a total of 854 hours of rainfall. Figure 2 shows the spatial distribution of event-scale accumulation fields. Hereafter the events will be referred to by the names indicated on each panel of Fig. 2. From Fig. 2, it can be noticed that there is quite a bit of variability among the events. For example, Event 10183 is composed of several small and intense convective showers lasting of short duration, while Event 05231 displays quasi-stationary behaviour with widespread rainfall for several hours. Two of the selected events (05231 and 07220) caused catastrophic flooding.

FORECAST VERIFICATION

Several verification measures were proposed in the literature to characterize the forecast performance. However, no single verification measure gives a complete picture of the forecast performance. We selected a combination of categorical and continuous skill scores to evaluate MAPLE and Eulerian forecasts. The evaluation was performed at the finest space and time resolutions.

Categorical verification

We limit the categorical evaluation to binary rain/no-rain patterns. All the pixels with rainfall (R) below certain threshold (R_t) were considered as non-rainy and those with $R \geq R_t$ were considered as rainy. Three different thresholds were considered: 0.1 mm/h (reflectivity $Z = 10$ dBZ), 1.0 mm/h ($Z = 25$ dBZ), and 5.0 mm/h ($Z = 35$ dBZ). The four possible outcomes were then arranged in the form of a 2×2 contingency table containing hits, false alarms, misses, and correct negatives. For a fixed lead time, a contingency table was created by applying the threshold and pooling all the forecast-observation pairs at finest resolution (1 km in space every 5 min). The categorical scores used in this study are probability of detection (POD), false alarm ratio (FAR), and the logarithm of odds ratio (OR). The probability of detection and false alarm ratio are widely used in the literature and their definition can be found in most books on forecast verification (e.g. Jolliffe & Stephenson, 2003). Used widely in fields related to medicine, the odds ratio is relatively new to the field of weather forecasting. It is described in detail in Stephenson (2000). If the odds ratio ≤ 1, the forecasts have no skill and for the perfect forecast, it is equal to ∞. One of advantages of the odds ratio is that it is independent of the base rate of the precipitation. Also, an odds ratio greater than 1 indicates that the forecasts have added-value compared to the climatological information.

Continuous verification

Continuous verification implies use of the entire spectrum of the intensity scale to characterize the forecast performance. Similar to the categorical skill scores, there are several continuous verification measures which can be used to assess the quality of forecasts. Some of the most widely used measures are correlation, mean absolute error, and root mean square error. We selected correlation coefficient for this study. For an exponential decay of correlation coefficient with lead time, Germann & Zawadzki (2002) defined predictability or the lifetime of the forecast as the lead time at which the correlation drops to 1/e. We followed the same definition and compared the lifetime of MAPLE forecasts and Eulerian forecasts.

RESULTS AND DISCUSSION

To get insight into the storm-to-storm variability in forecast performance, we estimated the average, minimum, and maximum skill score (using 20 skill scores for 20 events) for each lead time. Figure 3 shows the variation of the POD, FAR, and LOR with lead time for three thresholds. The solid dark line indicates the average skill score from the MAPLE forecasts, while the grey lines indicate the minimum and maximum scores. As expected, POD decreases and FAR increases

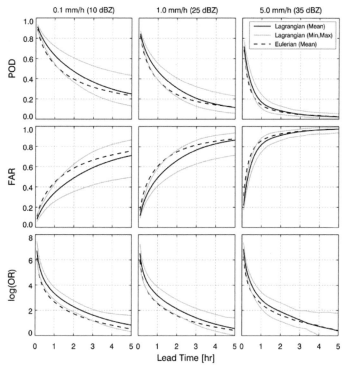

Fig. 3 Probability of detection (POD), false alarm ratio (FAR), and logarithm of odds ratio (OR) for Lagrangian and Eulerian forecasts as a function of lead time for three rain/no-rain thresholds.

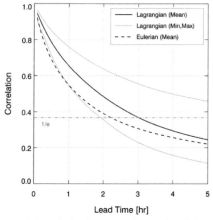

Fig. 4 Correlation between observed and (Lagrangian and Eulerian) forecasted fields as a function of lead time.

with lead time for all thresholds. Likewise, the performance of MAPLE decreased with the increase in the R_t value. Figure 3 also shows the comparison with Eulerian forecast skill (dashed lines). The skill in the Eulerian forecasts as characterized by POD, FAR and LOR was found to be lower than for the MAPLE forecasts for $R_t = 0.1$ mm/h and 1.0 mm/h. The behaviour was similar for the higher threshold of 5 mm/h, but only up to a lead time of 3 h. Beyond a lead time of 3 h, the skill scores of MAPLE and Eulerian forecasts were almost identical (Fig. 3). The logarithm of odds ratio for both the Eulerian and MAPLE forecasts is mostly above zero, suggesting that the forecasts indeed have added-value compared to the climatological forecasts (Fig. 3).

The correlation coefficient was estimated for forecast fields for all the lead times and for each event. The variation of correlation coefficient with lead time is shown in Fig. 4. The lifetime for the MAPLE forecasts varied from about 2 h to >5 h (Fig. 4).

CLOSING REMARKS

Short-term quantitative precipitation forecasts at high space–time resolutions were obtained by extrapolating the current radar-reflectivity patterns in Lagrangian space. The MAPLE model was implemented over a large region characterized by complex orography. Forecasts were then evaluated for 20 summer-time rainfall events using skill scores such as probability of detection, odds ratio, and correlation coefficient. Initial evaluation results suggest good skill in the MAPLE forecasts up to a lead time of around 3 h. Forecasts based on Lagrangian extrapolation outperformed the Eulerian forecasts. Over the USA, the life-time of forecasts from the same model was reported to be around 5 h. The decrease in the life-time over Switzerland compared to that for the USA could be due to several factors, such as meteorological situations, geographical set-up and also the characteristics of the radar-reflectivity input. However, we believe the main reason is the presence of strong orographic forcing in Switzerland. The MAPLE model by construction does not take the storm growth and dissipation into account, leading to large QPF errors. The quantitative information from this study such as forecast life-time, will help forecasters and emergency management authorities to better manage decision-support systems. At MeteoSwiss, we are working towards incorporating the growth and dissipation of storms into the extrapolation model MAPLE to increase the forecast life-time.

REFERENCES

Austin, G. & Bellon, A. (1974) The use of digital weather radar records for short-term precipitation forecasting. *Quart. J. Roy. Met. Soc.* 100(426), 658–664.

Bowler, N., Pierce, C. & Seed, A. (2007) STEPS: A probabilistic precipitation forecasting scheme which merges an extrapolation nowcast with downscaled NWP. *Quart. J. Roy. Met. Soc.* 132(620), 2127–2155.

Germann, U. & Zawadzki, I. (2002) Scale-dependence of the predictability of precipitation from continental radar images. Part I: Description of the methodology. *Monthly Weather Rev.* 130(12), 2859–2873.

Germann, U., Galli, G., Boscacci, M. & Bolliger, M. (2006a) Radar precipitation measurement in a mountainous region. *Quart. J. Roy. Met. Soc.* 132(618), 1669–1692.

Germann, U., Zawadzki, I. & Turner, B. (2006b) Predictability of precipitation from continental radar images. Part IV: Limits to prediction. *J. Atmos. Sci.* 63(8), 2092–2108.

Jolliffe, I. & Stephenson, D. (2003) *Forecast Verification: A Practitioner's Guide in Atmospheric Science.* John Wiley & Sons, Chichester, UK.

Joss, J. & Lee, R. (1995) The application of radar-gauge comparisons to operational precipitation profile corrections. *J. .Appl. Met.* 34 (12), 2612–2630.

Laroche, S. & Zawadzki, I. (1994) A variational analysis method for retrieval of three-dimensional wind field from single-doppler radar data. *J. Atmos. Sci.* 51, 2664–2682.

Mandapaka, P. V. & Germann, U. (2010) Radar-rainfall error models and ensemble generators. *Rainfall: State of Science, Geophysical Monograph Series* 191, 247–264.

Mecklenburg, S., Joss, J. & Schmid, W. (2000) Improving the nowcasting of precipitation in an alpine region with an enhanced radar echo tracking algorithm. *J. Hydrol.* 239, 46–68.

Seed, A. (2003) A dynamic and spatial scaling approach to advection forecasting. *J. Appl. Met.* 42(3), 381–388.

Stephenson, D. (2000) Use of the "odds ratio" for diagnosing forecast skill. *Weather & Forecasting* 15, 221–232.

Weather Radar and Hydrology
(Proceedings of a symposium held in Exeter, UK, April 2011) (IAHS Publ. 351, 2012).

281

Nowcasting of orographic rainfall by using Doppler weather radar

L. PANZIERA, U. GERMANN, A. HERING & P. MANDAPAKA

MeteoSvizzera, via ai Monti 146, CH-6605 Locarno Monti, Switzerland
luca.panziera@meteoswiss.ch

Abstract A novel radar-based heuristic tool for nowcasting orographic precipitation is presented. The system benefits from the strong relation, due to the orographic forcing, between mesoscale flows, air-mass stability and rainfall patterns. The system is based on an analogue approach: past situations with mesoscale flows, air mass stability and rainfall patterns most similar to those observed at the current instant are identified by searching in a large historical data set. Deterministic and probabilistic forecasts are then generated every five minutes as new observations are available, based on the rainfall observed by radar after the analogous situations. This approach constitutes a natural way to incorporate evolution of precipitation into the nowcasting system and to express forecast uncertainty by means of ensembles. A total of 127 days of long-lasting orographic precipitation constitutes the historical archive in which the analogous situations are searched. The system is originally developed for the Lago Maggiore region in the southern part of the European Alps, but it can be extended to other mountainous regions given the availability of radar data and the presence of a strong orographic forcing. An evaluation of the skill of the system shows that the heuristic tool performs better than both Eulerian persistence and the COSMO2 numerical model.

Key words Alpine radar; nowcasting; analogues; orographic precipitation; mesoscale flows; air mass stability; IMPRINTS

INTRODUCTION

Orographic rainfall is responsible for most of the floods that affect the Alpine region. During intense precipitation events, meteorological and hydrological services need frequently updated very short-term predictions of the intensity, location and evolution of orographic rainfall. As recently demonstrated by the project MAP D-PHASE (Rotach *et al.*, 2009) and IMPRINTS, tools for nowcasting precipitation assume high importance in the last stage of the weather forecasting process.

The complexity of the meteorological phenomena occurring over the orography and the difficulty to obtain detailed and precise observations of rainfall makes challenging the task of nowcasting precipitation in mountainous regions: errors of a few kilometres in the location of the predicted rainfall, if negligible for large catchments, can have a large impact on river runoff of small basins. The large spatial and temporal variability of orographic precipitation requires monitoring systems capable of measuring rainfall with high spatial and temporal resolution. Weather radars are designed for this scope, but for use in a routine manner weather radars in mountainous regions require proper solutions (Germann *et al.*, 2006).

In the smallest steep Alpine catchments, heavy orographic precipitation combined with rapid runoff can lead to flash floods. For example, in Verzasca Valley, a small and steep catchment (186 km^2) located in the southern Alps (see Fig. 1), the time between a precipitation impulse and the main runoff response at the outlet is of the order of 1–2 h (Germann *et al.*, 2009). This means that coupling radar observations with a runoff model results in a 1–2 h lead time for predicting a flash flood peak, even if no precipitation forecast is available. Nowcasting of rainfall in small alpine catchments is aimed at extending such short lead times. In particular, nowcasting methods should provide forecasts of precipitation up to 6–8 h ahead, with an updating frequency not larger than 1 h and a spatial resolution of a few kilometres.

In this paper we present NORA (Nowcasting of Orographic Rainfall by means of Analogues), a novel analogue-based heuristic nowcasting tool for short-term forecasting of precipitation. Since the mountains act as an external forcing to the movement of the air masses that does not change over time, we propose to exploit the orographic forcing for nowcasting of precipitation; in fact, it dominates the patterns of rainfall in the mountains (e.g. Houze *et al.*, 2001, Panziera & Germann

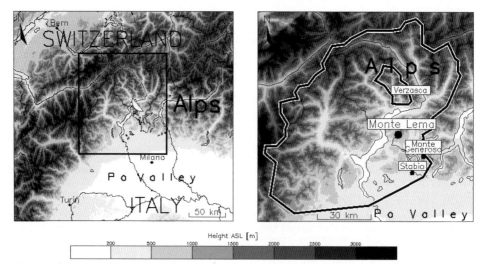

Fig. 1 Left: Orographic map of the Central Alps between Italy and Switzerland. The location of MeteoSwiss Monte Lema radar (1625 m above sea level) is indicated by the radar symbol. Right: Zoom into the Lago Maggiore region, with the location of the radar and of the ground stations considered in this study: Stabio (353 m) and Monte Generoso (1608 m). The black lines denote the Maggiore Lake catchment (7830 km²) and Verzasca Valley (186 km²).

2010, hereafter PG10). The basic idea of NORA is to search in the past for analogues, i.e. the situations with mesoscale flows, air mass stability and rainfall patterns most similar to those observed at the current instant. Quantitative precipitation forecasts are then based on the rainfall accumulated in the hours following the most similar past situations. The tool, which produces precipitation forecasts every 5 minutes up to several hours ahead, can be targeted for specific user requirements. Since the forecast is built on past fields of accumulated precipitation, growth and decay of rainfall, which are not accounted for by persistence methods, are automatically taken into account by NORA. A limitation of the analogue model is that the system is not able to predict rainfall fields not observed in the past. For this reason the current radar image is also taken as one of the final analogues, so that NORA forecasts also include the information given by the Eulerian persistence approach (which means considering the last radar image as the forecast). The system is originally developed for the Lago Maggiore region; however, it can also be extended to other mountainous regions, given the availability of local radar data, nearby ground stations at different altitudes and the presence of a strong orographic forcing.

DATA AND METHODS

Radar and ground stations data

The MeteoSwiss weather radar located on the top of Monte Lema (1625 m above sea level, see Fig. 1), one of the southern-most mountains of the Alpine ridge in the Lago Maggiore region, is used to estimate rainfall at the ground and mesoscale winds. Information on rainfall is obtained by the operational MeteoSwiss radar product for quantitative precipitation estimation. Such a product represents the best estimate of precipitation at the ground, and it is retrieved through a weighted mean of all the radar observations aloft. The horizontal spatial resolution of the precipitation map is 1 km x 1 km, the temporal resolution is 5 minutes. For an evaluation of the radar performances in the Lago Maggiore region, see Section 2.1 of PG10. The mesoscale wind is estimated by Doppler velocity measurements, which have a spatial resolution of 1 degree in azimuth and 1 km in range and are also updated every 5 minutes.

A pair of adjacent automated meteorological stations is used by NORA to estimate the stability of the lower atmosphere: Stabio (353 m) and Monte Generoso (1608 m). These stations are located on the first slopes of the Alps very close to the Po Valley (see Fig. 1) at different altitudes, thus permitting an estimate of the mean stability of the atmospheric layer comprised between the two stations.

Selection of historical cases

The historical archive in which NORA searches for the analogues results from a trade-off between the following three requirements. First, the data set in which the analogues are searched should theoretically be as large as possible. Second, the archive has also to be homogeneous to avoid artefacts of instrumental changes and different data-processing techniques. Third, long-lasting and widespread events typically caused by a large-scale supply of moisture towards the Alps should be selected. Isolated convection and air-mass thunderstorms were excluded from the historical data set, as these are less extended in space and time and thus rarely result in critical amounts of rainfall. Seventy-one precipitation events observed in the Lago Maggiore region from January 2004 to December 2009, corresponding in total to 127 days of precipitation (3050 hours), are thus considered. Details about the choice of the rainfall events are given in Section 2.2 of PG10.

Terminology

To specify the temporal period for which the forecast is produced, we make use of the following terminology. A *t*-hour forecast means a prediction of the rainfall accumulated in the temporal period of *t* hours. Following the American Meteorological Society glossary, lead time is defined as the length of time between the issuance of the forecast and the occurrence of the phenomena that were predicted. Thus, 1-h forecast with zero lead time means the prediction of rainfall from 0 to 1 h ahead, 1-h forecast with 3 h lead time means the prediction from 3 to 4 h ahead, etc.

Predictors

The quantities estimated in real-time and used by NORA to find the analogues in the historical archive (predictors) are four mesoscale flows, air mass stability, wet area ratio and image mean flux. A rationale for the choice of the regions in which the mesoscale flows are estimated, illustrated in Fig. 2, is given in Section 5 of PG10, while in Section 2.4 of the same paper the methodology by which these winds are estimated from radar Doppler measurements is described. The four mesoscale flows are:

(1) Low Level Flow (LLF): it is estimated upstream of the Alps southeast of Maggiore Lake between 1.5 and 2 km above sea level. It monitors the low level wind which is forced to ascend over the mountains. During orographic rainfall, LLF blows typically from south to southeast.

(2) Mid Level Flow (MLF): a layer extending from 2.5 to 3.5 km was chosen to estimate this wind, which represents the wind ascending over the terrain above the Lago Maggiore region. Also MLF is typically southerly when precipitation is observed in the Lago Maggiore area.

(3) Cross Barrier Flow (CBF): this flow is retrieved south of the Alpine crest between 3 and 4 km above sea level in the northwesterly Lago Maggiore area, in a region where the wind is typically observed to change direction (from southerly to northerly) in correspondence with frontal passages.

(4) Upper Level Flow (ULF): this flow is estimated between 4 and 5 km in a wide ring around the radar site and it represents the upper level synoptic flow. During orographic precipitation the wind in this region blows usually from south to southwest.

The moist Brunt-Väisälä frequency derived from Stabio and Monte Generoso ground stations is also taken as a predictor of orographic rainfall by NORA. Section 3.2 of PG10 gives the motivation for the choice of this quantity.

WAR, defined as the fraction of wet area in the radar image, and IMF, the average rainfall of the image, are also used as predictors by NORA. Both these quantities are estimated only within the Lago Maggiore catchment, with a threshold for WAR of 0.16 mm/h.

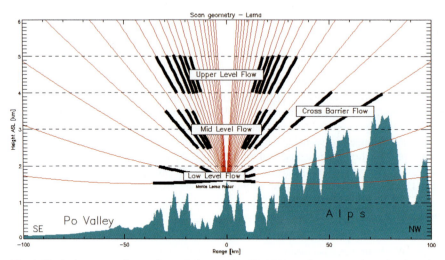

Fig. 2 Vertical cross-section performed along a line SE–NW passing across the radar site, showing the 20 elevation angles of the Monte Lema radar scan strategy (thin lines) and the regions selected for the flow estimates (thick lines).

The analogue method

The historical situations most similar to the current one, i.e. the analogues, are identified through a two-step process. First, 120 analogues from the predictors which describe the orographic forcing are obtained (forcing analogues); second, 12 analogues are selected from the first ones by considering radar rainfall features (final analogues).

In the first step, mesoscale flows and air mass stability are used to determine the analogues in the historical archive. The analogue model employed by NORA makes use of the Euclidean distance computed in a multidimensional space whose coordinates are given by the zonal and meridional components of the mesoscale flows and air mass stability. The past conditions with the smallest distances are assumed to be the most similar to the current situation. Since the performance of the wind estimation algorithm depends on the features of the wind field itself (see Section 2.4 of PG10), it is not always possible to estimate all four mesoscale winds every 5 min. Therefore, the number of dimensions of the space in which the analogues are searched may change, depending on the number of the available mesoscale flows. If the estimate of all the four mesoscale winds fails, the NORA forecast is not produced.

The second step of the process by which NORA searches for the analogues aims to identify the forcing analogues with the rainfall patterns most similar to the current one. Therefore, in this step WAR, IMF and their variation in the previous 2 h are used to calculate the euclidean distance between the current radar image and each one of the 120 meteorological analogues. The 12 samples which have the smallest Euclidean distances are assumed to be the situations most similar in rainfall pattern, and are the final analogues taken into account by NORA. The past rainfall fields observed every 5 min after the final analogues, according to the required lead time and temporal period, are used to produce the forecast. By averaging these fields, 12 ensemble members, one for each analogue, are obtained. If the rainfall fields observed after the analogue and necessary to produce the forecasts are not available because the analogue is too close to the current instant, the average of the rainfall measured from the analogue to the current instant is taken as the corresponding ensemble member. Since the current radar image is one of the final analogues by construction, it is always one of the 12 ensemble members, independently of the lead time of the forecast. NORA is capable of producing both deterministic and probabilistic forecasts. Deterministic predictions are obtained simply by averaging the ensemble members. Probabilities of exceeding particular rain rate thresholds are based on the probability density function derived from the 12 ensemble members.

NORA FORECAST VERIFICATION

The comparison between hourly NORA and Eulerian persistence predictions is presented in Fig. 3. The temporal resolution of the verification data set is 5 min, for a total of 32 126 samples (2677 h, 111.5 days). NORA forecasts were verified within the Lago Maggiore catchment, which is 7830 km^2 (see Fig. 1). For computational reasons the original spatial resolution of the precipitation field was reduced to 9 km^2 by aggregating 9 pixels of 1 km^2. In order to evaluate the skill of NORA in

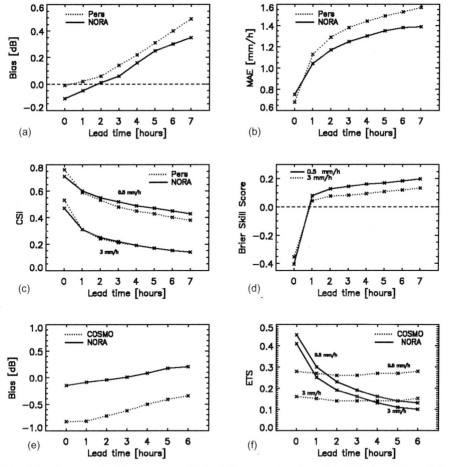

Fig. 3 Skill scores for 1-h forecasts with lead times ranging from 0 to 7 h for NORA, Eulerian persistence and COSMO2 forecasts. Bias and MAE describe the evaluation of deterministic forecast, CSI and ETS of binary forecasts for the two thresholds indicated. Brier Skill Score is relative to the verification of probabilistic forecasts, with Eulerian persistence as the reference forecasting method. See text for details about the verification data set.

real-time mode, the 24 h following the sample for which the forecast was made were excluded from the historical archive in which the analogues are searched.

For a definition of the verification scores presented in Fig. 3, see Wilks (1995). The figure shows that Eulerian persistence forecast skill is similar to that of NORA for forecasts with zero lead time. For larger lead times, however, NORA generally outperforms Eulerian persistence, even if the accuracy of the forecasts decreases with increasing lead time. However, binary NORA forecasts exceeding 3 mm/h, do not constitute an improvement with respect to persistence predictions (panel (c)). Panel (d) of Fig. 3 shows that probabilistic NORA forecasts constitute an improvement with respect to Eulerian persistence forecasts for lead times larger than zero (positive values of the Brier Skill score).

The performance of NORA was also compared with that of the numerical model COSMO2, operationally used at MeteoSwiss. The comparison between NORA and COSMO2 was performed for a subset of our archive, given by the precipitation events that occurred after 25 May 2007, for a total of 14 535 samples. The comparison between NORA and COSMO2 was performed for a subset of our archive, given by the precipitation events that occurred after 25 May 2007. It could be argued that the better performance of NORA with respect to COSMO2 is due to the fact that NORA forecasts are derived from radar archive, whereas the numerical model output gives only a little weight to the radar image through the latent heat nudging. However, the results shown in panels (e) and (f) of Fig. 3 denote a discrepancy between COSMO2 and NORA forecasts, which is unlikely to be caused by the dependence of NORA forecasts on the radar archive. This is further confirmed by the visual comparison between COSMO2, NORA forecasts and radar observation fields: the low performance of the numerical model is clear and it is due to problems in forecasting the location and the amplitude of the rainfall. Figure 3(e) shows that COSMO2 suffers from a large negative bias, which influences the binary verification presented in panel (f).

CONCLUSIONS

A novel heuristic analogue-based nowcasting tool for very short-term forecasting of orographic precipitation (NORA) was presented. The system can be developed and improved by adding new historical cases and predictors. The presented skill scores show that for forecasts with lead time equal or larger than 1 h, both probabilistic and deterministic NORA forecasts are better than Eulerian persistence. The skill of NORA is also larger than that of the numerical model COSMO2 in the first 4–5 h of forecast. The analogue model used by NORA to produce rainfall forecasts leads to a natural methodology to create an ensemble of precipitation fields. In fact, by taking the past situations most similar to the observed one, an ensemble of forecast precipitation fields is naturally found. Since they result from precipitation observed in the past, these fields have by construction realistic statistical space–time properties. The development of NORA in mountainous regions other than the Lago Maggiore area is possible, given the availability of good quality radar and ground station data, and the presence of a strong orographic forcing. However, the skill of NORA in other regions vary depending on the efficiency of the orographic forcing itself, which ultimately depends on the shape of the orography and on the repeatability of the mesoscale flows leading to precipitation.

REFERENCES

Germann, U., Galli, G., Boscacci, M. & Bolliger, M. (2006) Radar precipitation measurement in a mountainous region. *Q. J. R. Met. Soc.* 132, 1669–1692.

Germann, U., Berenguer, M., Sempere-Torres, D. & Zappa, M. (2009) REAL – ensemble radar precipitation estimation for hydrology in a mountainous region. *Quart. J. Roy. Met. Soc.* 135, 445–456.

Houze, R., Houze, R. A., Jr, James, C. & Medina, S. (2001) Radar observations of precipitation and airflow on the Mediterranean side of the Alps: Autumn 1998 and 1999. *Quart. J. Roy. Met. Soc.* 127, 2537–2558.

Panziera, L. & Germann, U. (2010) The relation between airflow and orographic precipitation on the southern side of the Alps as revealed by weather radar. *Quart. J. Roy. Met. Soc.* 136, 222–238.

Rotach, M. W. *et al.* (2009) MAP D-PHASE: Real-time demonstration of weather forecast quality in the Alpine region. *Bull. Am. Met. Soc.* 90, 1321-1336.

Wilks, D. S. (1995) *Statistical Methods in the Atmospheric Sciences.* Academic Press, 467 pp.

The relationships between the upstream wind and orographic heavy rainfall in southwestern Taiwan for typhoon cases

LEI FENG[1], PAO-LIANG CHANG[2] & BEN JONG-DAO JOU[3]

1 *Taiwan Typhoon and Flood Research Institute, 11F, No. 97, Sec. 1, Roosevelt Road, Taipei 10093, Taiwan*
  fenglei@ttfri.narl.org.tw
2 *Central Weather Bureau, 64 Gongyuan Road, Taipei 10048, Taiwan*
3 *Department of Atmospheric Science, National Taiwan University, No. 1, Sec. 4, Roosevelt Road, Taipei 10617, Taiwan,*
  APEC Research Center for Typhoon and Society (ACTS)

Abstract Typhoon Morakot (2009) landed on northern Taiwan and then moved toward the northwest. Extreme heavy rainfall occurred in the mountainous region of southwest Taiwan. It was noticed that there were very strong horizontal westerly flows upstream of the mountain in southwest Taiwan. The relation between this upstream horizontal westerly wind and the heavy rain over the mountain is the major focus of this study. The 24-h maximum rainfall produced by Morakot was >1500 mm, and >20 stations in the area measured rainfall >1000 mm in 24 h. An algorithm was proposed to predict the extreme orographic heavy rain over southwestern Taiwan using radar-derived low-level horizontal winds. The Chigu radar is located 80 km upstream (westerly wind) of the mountainous regions. The EVAD technique was applied to retrieve the horizontal winds. The averaged horizontal winds between 0.5 and 3.0 km height are treated as the upstream low-level flow impinging on the mountain. A very good relationship between the low-level averaged speed and the hourly rainfall amount was achieved and the linear correlation coefficient is near 0.88. A similar algorithm was applied to two other typhoons: Haitang and Talim both in 2005; linear correlation coefficients of 0.80 and 0.84 were obtained, respectively. It is suggested that the upstream velocity of the flow determined the amount of heavy rainfall over the mountainous region in the strong wind regimes.

Key words typhoon; orographic heavy rain; Doppler radar; horizontal wind speed upstream of the mountain

INTRODUCTION

Typhoon Morakot (2009) landed on the northeast part of Taiwan and then moved slowly in a northwesterly direction. There was a strong westerly wind in southwestern Taiwan associated with the typhoon circulation, and this produced extremely high rainfall in the mountainous regions. This heavy rainfall caused a severe disaster in southern Taiwan. The floods and mudslides induced by typhoon Morakot caused the death of nearly 700 people in southern Taiwan. The 24-h maximum rainfall produced by Morakot was >1500 mm, and >20 stations recorded >1000 mm. The event total rainfall is about 3000 mm in the attacking period (95 h), which is about two-thirds of the annual average rainfall for mountainous regions of southern Taiwan.

For tropical cyclones affecting Taiwan, the importance of orography on modulating mesoscale rainfall distributions has been well known, e.g. Chang *et al.* (1993) used surface observations from 82 typhoons occurring during a 20-year period to document the effects of Taiwan's terrain on the surface features of typhoons. Their analyses showed that the location of the typhoon circulation centre relative to topography was crucial in controlling precipitation patterns over Taiwan. Lee *et al.* (2006) analysed rainfall data from conventional surface stations and automatic raingauges for 58 typhoons affecting Taiwan during 1989–2001. Their results showed that a large average typhoon rainfall occurred generally over the mountainous regions and that the averaged rainfall amount considerably increased with surface station elevation. Nevertheless, owing to the relatively coarse observations of conventional raingauges, these previous studies provide only a gross view of the relationship between the typhoon's precipitation and topography, and in particular, their associated physical processes are poorly documented.

The formation mechanisms of orographic rain have been proposed in previous studies (e.g. Smith, 1979; Houze, 1993; Lin, 1993). The majority of heavy orographic rainfall events over mesoscale topography are caused by either (1) upslope rain in conjunction with conditional or potential instability, or (2) leeside convective rain advected in from the upwind slope region (which is produced by sensible heating in the vicinity of the mountain peaks or triggered by convergence *in situ*). In order to trigger the instability, the orographic lifting should be strong enough to force air parcels to ascend to their level of free convection. Thus, the occurrence of orographic rainfall may be determined by the wind velocity perpendicular to the mountain range (U), the moist static stability (N_w) of the incoming airstream, and the mountain height (h_m). In fact, these three factors can be consolidated into a single non-dimensional parameter, i.e. the moist Froude number ($F_w = U/N_wh_m$), which may be used to predict the location and propagation of the orographically-induced convective systems.

There are only a few radar observational studies on the heavy rainfall over mountainous regions associated with the typhoon circulation in Taiwan. Yu & Cheng (2008) use the measurement from two ground-based Doppler radars located in northern Taiwan to document the detailed aspects of the precipitation distribution as Xangsane moved northward immediately off the eastern coast of Taiwan. They focus on the observed spatial and temporal variations of precipitation related to the topographic features (one is approximately 3-D mountain barrier and height near 1000 m, another is relatively lower, narrower 2-D mountain range), upstream airflow, and typhoon precipitation. For this study, there are many differing aspects to the Xangsane typhoon case.

The mountainous regions of this study are different to the Xangsane typhoon case, one in southwestern Taiwan and another in northern Taiwan. The southwestern Taiwan case has a higher mountain range (>3000 m) than the northern Taiwan case (about 1000 m), and the prevailing wind was a westerly wind and northerly wind, respectively. In this paper, we try to build a simple method to predict the magnitude of orographic rainfall for southwestern Taiwan induced by typhoon.

METHODOLOGY

Collier (1975) developed a numerical "parameterization" model to predict precipitation amount over hilly terrain in North Wales. He used the larger-scale model output as an input to calculate the vertical velocity profile resulting from orographic and large-scale baroclinic disturbances. The results show that the simple formulation may predict the catchment rainfall with a high degree of accuracy on occasions for particular weather conditions.

Doswell *et al.* (1996) proposed the essential ingredients for flash floods over a flat surface. Lin *et al.* (2001) extended the "ingredient-based methodology" for heavy orographic rainfall forecasting over mesoscale mountains and summarized that heavy orographic rainfall requires significant contributions from any combination of the following common ingredients: (1) high precipitation efficiency of the incoming airstream; (2) the presence of a moist, moderate to intense LLJ; (3) steep orography to help release the instability; (4) favourable (e.g. concave) mountain geometry and a confluent flow field; (5) strong environmentally forced upward vertical motion; (6) the presence of a high moisture flow upstream; (7) a pre-existing large-scale convective system; (8) slow (impeded or retarded) movement of the convective system; and (9) a condition-ally or potentially unstable upstream airflow.

Lin *et al.* (1996) proposed using $U(\partial h/\partial x)q$ as a rainfall intensity index, where U is the low-level flow velocity perpendicular to the mountain range, $\partial h/\partial x$ is the mountain slope parallel to the basic flow, and q the low-level mixing ratio of water vapour. This index can help to predict the occurrence of heavy orographic rainfall over mountains. For the Taiwan and Japan cases (belonging to the tropical cyclone category), the proposed index is roughly proportional to the observed maximum rainfall rate.

Following this concept, we simplify the index to the U item only, but add a low-level average. The benefit of the new index is that the formula is reduced to use only one parameter. However, this new index can only be suitable for a typhoon situation and particular area. The low-level flow velocity was derived from the Chigu Doppler radar using the EVAD (extended velocity azimuth display) technique. Then, we extract the U component (perpendicular to the mountain range) from the wind velocity and average between 0.5 km and 3 km. Figure 1 shows the topography of southern Taiwan, the position of the Chigu Doppler radar site and the distribution of raingauges. Since the

Chigu radar site is located at the upstream side and about 70–80 km from mountain ranges, the derived VAD wind from this radar would be considered as an undisturbed upstream wind. The rectangle depicted by a dashed line is the rainfall prediction target area with area about 3300 km^2. The averaged rainfall of the target area is calculated as an arithmetic average of the 29 raingauges inside the rectangle.

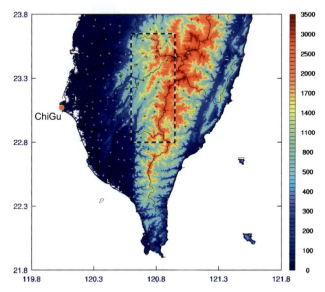

Fig. 1 The topography of southern Taiwan. The red circle represents the position of Chigu Doppler radar site. The rectangle depicted by the dashed line is the rainfall prediction target area; there are 29 raingauges within this area of about 3300 km^2.

ANALYSIS

The warning period of typhoon Morakot was from 00:00 h 6 August to 06:00 h 10 August 2009 LST (total 95 h). Figure 2 shows the distribution of surface rainfall analysed from raingauge data. Figure 2(a) is the maximum hourly rainfall that occurred in the 95-h period (the maximum hourly rainfall for different stations might occur at different times). Figure 2(b) is the maximum value of continuing 12-h rainfall accumulations occurring in the 95-h time-interval. Figure 2(c) is total rainfall amount for the 95-h period. The rainfall analysis shows clearly that heavy rainfall occurred over southwest Taiwan, and the hourly maximum rainfalls >100 mm/h are more spread in space than for the other figures. The maximum 12 h rainfall accumulation reaches 965 mm, and it means that the hourly rain-rates for this station are >80 mm/h in the continuing 12-h period, which is an extremely high amount of rain. In the typhoon warning period, the maximum total rainfall is 2767.5 mm. When the typhoon left, it induced the southwest flow and >350 mm rainfall occurred in the same place on the next day. If this rainfall is added to the storm total, the highest maximum rainfall is 3059 mm at A-lisan station.

Chigu Doppler radar is located upstream of the target area. In order to retrieve the low-level wind component (U) which is perpendicular to the mountain range, we use the EVAD (Extended Velocity Azimuth Display) analysis technique to generate the horizontal wind for different heights at 500-m intervals starting from 500 m AGL. For comparison with U and average rainfall, we define a rectangular area shown in Fig. 1: this target area is about 3300 km^2 and 29 raingauge stations lie within the box. The correlation between the low-level flow velocity and hourly average rain-rate is quite good. It is shown that the simple index $U(\partial h/\partial x)q$ proposed by Lin *et al.* (2001) should work for this typhoon and the U component might be of the most importance.

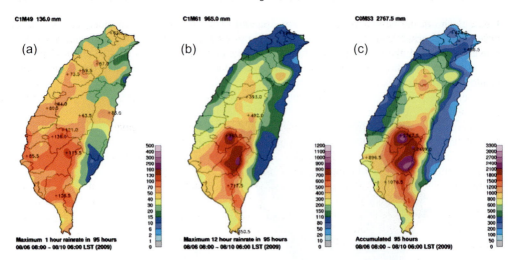

Fig. 2 The rainfall distributions of typhoon Morakot in the typhoon warning period (95 h); the rainfall data are from the surface raingauge network: (a) maximum hourly rainfall; (b) maximum 12-h rainfall accumulation; (c) event total rainfall.

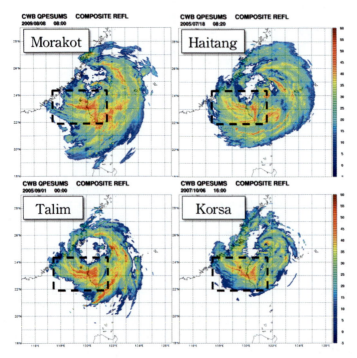

Fig. 3 The similar radar reflectivity pattern of the upstream long-lived rain bands from Morakot (2009), Haitang (2005), Tailam (2005) and Korsa (2007) typhoons.

For a particular area, the slope of the terrain ($\partial h/\partial x$) is the same, so can be treated as a constant. The low-level mixing ratio of water vapour (q) is difficult to measure above the ground. For typhoon cases, especially when the target area has heavy rainfall, there appears to be a very similar radar reflectivity pattern. The most obvious signature is the upstream stationary long-lived rain bands.

Figure 3 shows the similar radar reflectivity pattern of these upstream long-lived rain bands for different typhoons. We could assume the value of q was the same when the special radar reflectivity signature appeared: then the index could reduce to only the U item for a particular weather situation.

The upstream long-lived rain bands can be observed by the CWB radar network, with the rain bands persisting for more than 31 h in southwestern Taiwan for the Morakot typhoon case. The rain bands can be simulated very well by a mesoscale numerical model according to results obtained under the MEFSEA project.

A conceptual model is proposed for the prediction of the orographic heavy rainfall for the target area of southwestern Taiwan. When the typhoon is in the vicinity of northern Taiwan and with a slow moving speed, it could generate east–west direction rain bands in southwestern Taiwan and in the upstream side of mountain ranges. Rain cells embedded in the upstream rain bands will move to the mountainous regions by the strong westerly wind. In this situation, the mountainous regions will have very heavy rainfall. We find a very good linear relationship between the low-level average wind and the average hourly rainfall for Morakot typhoon: the correlation coefficient is near 0.88. The other two typhoons, Haitang (2005) and Talim (2005), also have a similar linear relationship, with correlation coefficients of 0.80 and 0.84, respectively (Fig. 4).

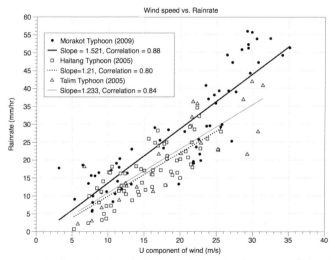

Fig. 4 The linear relationship between the low-level average wind and the average hourly rainfall for Morakot (2009), Tailam (2005) and Haitang (2005) typhoons.

In the vertical cross-section of the rain band, shown in Fig. 5, there are many rain cells embedded in upstream rain bands. As seen by the time series of this cross-section picture, the rain cells moved to mountainous regions.

It was expected that a rain cell carried a large amount of medium-size raindrops to mountainous regions and should mix with extreme large-density small drops produced by the upslope flow. We believe this is a very efficient enhancement process of orographic rainfall for a typhoon circulation and fixed terrain as in southwestern Taiwan. The enhanced rainfall was at low level and very difficult to be observed by horizontal scanning radar. More observations are needed to verify this hypothesis, especially using vertical pointing radar to observe the low-level enhancement.

SUMMARY

We proposed an orographic heavy rainfall conceptual model for a typhoon situation over southwestern Taiwan. When the typhoon moved into the vicinity of northern Taiwan and with a

Lei Feng et al.

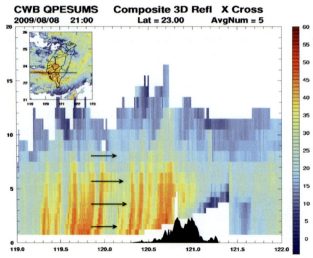

Fig. 5 The radar reflectivity cross-section of the upstream rain band of the Morakot typhoon. The arrows express the direction of movement of rain cells embedded in the upstream rain band.

slow speed, it would form an obvious convergence zone in the southwest part of Taiwan. In the convergence zone it will generate an east–west direction rain band in the upstream side of the mountain range. Rain cells embedded in this upstream rain band moved to mountainous regions. The cells carried a large amount of medium-size raindrops to a mountain area and these should mix with extreme large density small drops produced by the upslope flow. It is suggested that this is a very efficient enhancement process of orographic rainfall for a typhoon circulation and fixed terrain, as in southwestern Taiwan. A very good linear relationship was found between the low-level average wind and the average hourly rainfall for Morakot, Haitang and Talim typhoons; the correlation coefficients are 0.88, 0.80 and 0.84, respectively. It is proposed that the low-level wind component (U) perpendicular to the mountain range should be used to predict the hourly rain-rate on mountainous regions of southwest Taiwan for typhoons which generate a stationary upstream rain band situation.

Acknowledgements The Chigu Doppler radar and raingauge data used in this study were provided by the Central Weather Bureau. The study is supported by the National Science Council of Taiwan under Grants NSC 99-2625-M-492-002-MY3.

REFERENCES

Chang, C.-P., Yeh, T.-C. & Chen, J. M. (1993) Effects of terrain on the surface structure of typhoons over Taiwan. *Monthly Weather Rev.* 121, 734–752.

Collier, C. G. (1975) A representation of the effects of topography on surface rainfall within moving baroclinic disturbances. *Quart. J. Roy. Met. Soc.* 101, 407–422.

Doswell, C. A., III, Brooks, H. & Maddox, R. (1996) Flash flood forecasting: an ingredient-based methodology. *Weather Forecast.* 11, 560–581.

Houze, R. A. (1993) *Cloud Dynamics.* Academic Press, San Diego, California, USA.

Lee, C.-S., Huang, L.-R., Shen, H.-S. & Wang, S.-T. (2006) A climatology model for forecasting typhoon rainfall in Taiwan. *Nat. Hazards* 37, 87–105.

Lin, Y.-L. (1993) Orographic effects on airflow and mesoscale weather systems over Taiwan. *Terr. Ocean. Atmos.* 4, 381–420.

Lin, Y.-L., Chiao, S., Wang, T.-A., Kaplan, M. L. & Weglarz, R. P. (2001) Some common ingredients for heavy orographic rainfall. *Weather Forecast.* 16, 633–660.

Nishiwaki, N., Misumi, R. & Maki, M. (2010) Relationship between orographic enhancement of rainfall rate and movement speed of radar echoes: case study of Typhoon 0709. *JMSJ* 88, 931–936.

Rotunno, R. & Houze, R. A., Jr. (2007) Lessons on orographic precipitation from the Mesoscale Alpine Programme. *Quart. J. Roy. Met. Soc.* 133, 811–830.

Smith, R. B. (1979) The influence of mountains on the atmosphere. *Advances in Geophysics* 21, Academic Press, 87–230.

Yu, C.-K. & Cheng, L.- W. (2008) Radar observations of intense orographic precipitation associated with Typhoon Xangsane (2000). *Monthly Weather Rev.* 136, 497–521.

Weather Radar and Hydrology
(Proceedings of a symposium held in Exeter, UK, April 2011) (IAHS Publ. 351, 2012).

293

Use of ensemble radar estimates of precipitation rate within a stochastic, quantitative precipitation nowcasting algorithm

CLIVE PIERCE[1], KATIE NORMAN[1] & ALAN SEED[2]

1 *Met Office, FitzRoy Road, Exeter EX1 3PB, UK*
  clive.pierce@metoffice.gov.uk

2 *Australian Bureau of Meteorology, The Centre for Australian Weather and Climate Research, GPO Box 1289, Melbourne, Victoria 3001, Australia*

Abstract Several techniques for the generation of ensembles of radar observations are described and evaluated. These have been combined to generate ensemble estimates of surface precipitation rate for use in conjunction with the Short Term Ensemble Prediction System. STEPS is an operational, quantitative precipitation nowcasting algorithm developed jointly by the Met Office and the Australian Bureau of Meteorology. It generates ensemble nowcasts of precipitation rate and accumulation by scale-selectively blending a weather radar-based, extrapolated analysis of surface precipitation rate with a recent precipitation forecast from a high-resolution configuration of the Unified Model, and a time series of synthetically generated precipitation fields (noise) with space–time statistical properties inferred from radar. Currently, STEPS incorporates an observation uncertainty algorithm based upon on analysis of Z-R errors. In this paper, the performance of STEPS precipitation nowcast ensembles, generated using radar ensembles, is compared with that of operational STEPS precipitation nowcasts, produced using unperturbed observations.

Key words radar; observation error; nowcast; ensembles

BACKGROUND

In recent years, the Met Office has invested significant effort in quantifying the uncertainties inherent in hydro-meteorological nowcasts and forecasts and communicating these to users. This strategy recognizes the impact of nonlinear error growth on the accuracy of high-resolution forecasts and the needs of customers in relation to decision making and the management of risks associated with severe weather. In support of this strategy, the Short Term Ensemble Prediction System (STEPS; Bowler *et al.*, 2006) was implemented in the Met Office in 2008.

Radar-inferred estimates of surface precipitation rate are subject to errors from a variety of sources. Failure to properly account for these observation errors within STEPS will limit the predictive skill of the resulting ensemble nowcasts in the first hour or so. To date, STEPS treatment of observation error has been confined to the modelling of errors arising from the use of a fixed Z-R relationship. The effects of other significant errors, including, for example, vertical profile corrections, have been ignored until now.

Recent collaborative work undertaken by the Met Office and Australian Bureau of Meteorology has explored several models of radar observation error, based upon techniques described in the literature. These have been used to generate experimental ensembles of radar observations for use within STEPS (Norman *et al.*, 2010). This paper reviews the formulation of the radar ensemble generators developed by Norman *et al.* (2010) and presents preliminary results of experiments designed to explore the impact of these radar ensembles on the performance of STEPS.

FORMULATION AND EVALUATION OF TWO RADAR-BASED ENSEMBLE GENERATORS

Context

Radar errors broadly fall into three categories: physical biases, measurement biases and random sampling errors. For the purposes of optimizing deterministic estimates of precipitation rate at the surface, much effort has historically been invested in correcting physical (e.g. ground clutter and beam blockage) and measurement (e.g. Z-R conversion) biases. More recently, a growing number of researchers have focused their attention on the treatment of random sampling errors and how these can be employed to improve hydro-meteorological nowcasting.

There are two main approaches to the modelling of random sampling errors in weather radar observations. One entails a statistical description of the difference between the radar estimates and a reference (e.g. Ciach *et al.*, 2007; Llort *et al.*, 2008; Germann *et al.*, 2009). A second involves modelling the characteristics of individual sources of error (e.g. Jordan *et al.*, 2003; Lee & Zawadzki, 2005a,b, 2006; Lee *et al.*, 2007). The challenge with the first approach is the requirement for a reference field: this is usually derived from a dense network of raingauges. The difficulty with the second approach is that the true error structure of radar observations can vary significantly, depending on the meteorological conditions, and is therefore largely unknowable.

A description of the radar ensemble generators

Noman *et al.* (2010) implemented two radar-based ensemble generators following the methodologies proposed by Germann *et al.* (2009; hereafter GM09) and by both Lee *et al.* (2007) and Jordan *et al.* (2003; hereafter Seed10). These are described below.

Implementation of GM09

A statistical model of the errors in radar-inferred, sub-hourly estimates of surface precipitation accumulations was built using historical raingauge data. This model was used as a proxy for the error in surface estimates of instantaneous precipitation rate. Four attributes of the radar errors were derived: a mean error vector (the systematic error), the variance of the errors (magnitude of the random error), and the spatial and temporal correlations in the error fields. A covariance matrix was used to store information about the standard deviation and spatial correlation of the errors.

These four attributes were used to produce a time series of perturbation fields with which to generate an ensemble of fields of instantaneous precipitation rate from a single, best estimate of the distribution of surface precipitation rate inferred from radar. Since the covariance and mean error can only be measured at gauge locations, these point values were interpolated in space. A second-order auto-regressive (AR-2) model was used to impose temporal correlations on the perturbations.

Implementation of Seed10

Seed10 modelled radar errors from two specific sources: those arising from the use of a fixed Z-R relationship, and those originating from assumptions made about the vertical profile of reflectivity (VPR). Perturbation fields representing the random component of these two error sources were combined with a best estimate of the radar-inferred surface precipitation field to produce ensembles. Since the two error models were developed for use within the STEPS (Bowler *et al.*, 2006) their implementation exploits the STEPS scale (cascade) decomposition algorithm outlined later in this paper. This entails combining the scale-decomposed perturbation fields with scale-decomposed radar observations.

The statistical distribution of the Z-R perturbations was largely modelled on the findings of Lee & Zawadzki (2005a,b). Using collocated radar and disdrometer measurements, these authors found significant coherence in Z-R errors over time periods ≤1 h. However, except in those areas in close proximity to a radar, errors arising from assumptions made about the VPR are likely to dominate over those due to departures from an assumed Z-R relationship. This is particularly so when the bright-band is close to ground. Details of the VPR error model can be found in Norman *et al.* (2010).

Comparison of GM09 and Seed10

The GM09 and Seed10 radar ensemble generators differ in several key respects. Firstly, the Seed10 models for Z-R and VPR errors exclude any consideration of systematic errors (bias), whereas the GM09 model includes a bias correction. Secondly, since the GM09 model uses raingauge reports as a reference, it is prone to the impact of gauge measurement errors. Furthermore, its reliance on raingauges dictates that its useful application is restricted to regions with a dense raingauge network. By contrast, the Seed10 models are limited by the coverage of the weather radar network.

Performance comparison of the radar ensemble generators

The verification statistics presented below are similar to those described in Norman *et al.* (2010). They are based upon five case studies representing a range of winter and summer convective and widespread precipitation events. For each case, the GM09 and Seed10 algorithms were used to generate 30 member ensembles of one hour precipitation accumulation for a portion of the UK weather radar network (the Midlands) supported by a dense raingauge network. Tipping bucket raingauge measurements were integrated over a 60-min period and used as the verification reference. A number of ensemble verification measures were employed to compare the performances of the two algorithms. These included the reliability (attributes) diagram, the Relative Operating Characteristic (ROC) and the spread–skill relationship.

A perfect ensemble generator should generate event probabilities that match the observed frequencies of those events. A skilful ensemble will produce a reliability curve lying between the dashed *no skill* line (Hit Rate = False Alarm Rate) and the diagonal perfect reliability line. The attributes diagrams in Fig. 1(a)–(b) shows Seed10 (a) to be under-confident for events with low observed frequencies and over-confident for events with higher observed frequencies. GM09 (b) is consistently over-confident; this is indicative of insufficient ensemble spread, though the reliability curve has a slope closer to the perfect reliability line. Both ensemble generators exhibit significant skill.

ROC curves for a range of thresholds (not shown) demonstrate that GM09 was slightly more skilful than Seed10. However, both algorithms exhibit an ability to discriminate between events and non-events (i.e. an exceedence *versus* a non-exceedence of a threshold) and show skill relative to the unperturbed radar estimates of surface precipitation accumulation at thresholds ≤2 mm/h.

For a perfectly calibrated ensemble, the spread about the ensemble mean should equate to the error in the ensemble mean. Figure 2(a),(b) compares spread–skill plots for the GM09 (a) and Seed10 (b) against spread–skill plots for two reference ensembles, one with no skill (near horizontal line), the other with perfect spread (the diagonal line with a slope of 1). These plots are generated by computing the average spread–skill relationship in bins containing equal numbers of spread-error data samples. Both ensembles were under-spread. This was more pronounced in the case of Seed10. Overall, GM09 was marginally more skilful than Seed10.

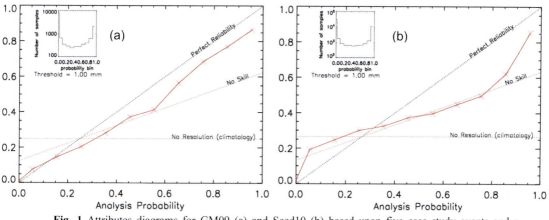

Fig. 1 Attributes diagrams for GM09 (a) and Seed10 (b) based upon five case study events and a threshold hourly accumulation of 0.5 mm.

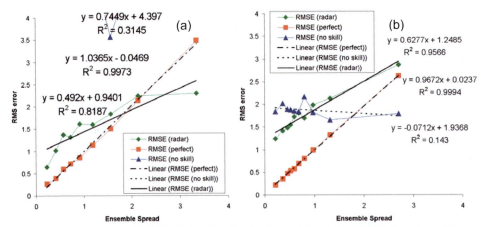

Fig. 2 Ensemble spread–skill relationships for the GM09 (a) and Seed10 (b) algorithms. The spread–skill relationships derived from 30 member ensembles (labelled RMSE (radar)) are compared with two reference ensembles, one with no skill (labelled RMSE (no skill)), the other with perfect spread (labelled RMSE (perfect)). Straight line fits to each of these relationships are plotted.

THE IMPACT OF RADAR ENSEMBLES ON THE PERFORMANCE OF STEPS

Overview of STEPS

STEPS (Bowler *et al.*, 2006) generates control and ensemble nowcasts of precipitation rate and accumulation by blending a radar-based extrapolation nowcast with a high-resolution precipitation forecast from the most recent run of a high-resolution configuration of the Met Office's NWP model (the Unified Model). This blending is performed on a hierarchy of scales using a cascade modelling framework. The weights assigned to the two forecast components vary with estimates of their current skill on each cascade level at each forecast step. Synthetically generated precipitation (noise) with space–time statistical properties inferred from weather radar are used to downscale the NWP forecast component and perturb the blended cascade to generate ensembles. The scheme incorporates a Z-R observation error model similar to that implemented within Seed10.

Implementation of GM09 and Seed10

Norman *et al.* (2010) implemented the GM09 and Seed10 radar ensemble generators as standalone algorithms. These have been integrated to produce a single ensemble generator combining the statistical, gauge-based error model with the Z-R and VPR error models. The integrated scheme relies solely upon GM09 perturbations in regions of the UK with adequate gauge density. Elsewhere reliance is placed on the Seed10 perturbations. Since STEPS runs on a domain larger than the coverage of the UK weather radar network, perturbed estimates of observed surface precipitation were derived from a recent, 4-km resolution UM precipitation forecast in areas with no radar coverage.

The evaluation methodology and verification statistics

Twenty member ensembles of instantaneous precipitation rate were produced every 15 minutes for the duration of three case study events. These included episodes of showers and widespread precipitation. STEPS was run in two configurations using these data. In the first configuration, a control nowcast and a 20 member ensemble nowcast of precipitation rate and accumulation were generated from the unperturbed radar and UM-based best estimate of surface precipitation rate. In the second configuration, similar products were produced, but in this case, each STEPS ensemble member was derived from a separate, perturbed analysis of surface precipitation rate.

A selection of ensemble verification statistics similar to those compiled by Norman *et al.* (2010) are presented below. These focus on the performance of T+1 h STEPS nowcasts of one hour precipitation accumulation, since the impact of observation errors on performance should be confined to the first hour of a nowcast. In Figs 3 and 4, the left-hand graphic (a) relates to the performance of STEPS ensembles generated from the unperturbed observations, and the right-hand graphic (b) represents the performance of equivalent nowcast ensembles generated using the ensemble of perturbed observations. The verification reference used throughout was tipping bucket raingauge accumulations.

The attributes diagrams for the two configurations of STEPS ensemble nowcasts shown in Fig. 3(a),(b) relate to a threshold of 0.5 mm/h. The two configurations exhibit similar levels of skill and these are on par with those reported for the radar observation ensembles in the previous section. Both ensembles tend to be under-confident for events with low observed frequencies and over-confident for events with higher observed frequencies.

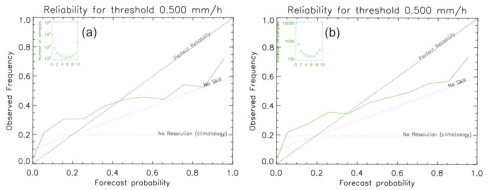

Fig. 3 Attributes diagrams for STEPS ensemble runs from unperturbed observations (a) and equivalent ensemble runs from perturbed observation (b) based upon three case study events and a threshold hourly accumulation of 0.5 mm.

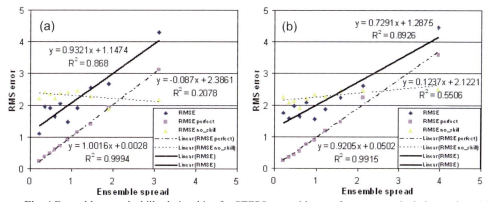

Fig. 4 Ensemble spread–skill relationships for STEPS ensemble runs from unperturbed observations (a) and equivalent ensemble runs from perturbed observations (b). The spread–skill relationships derived from 20 member ensembles (labelled RMSE) are compared with two reference ensembles, one with no skill (labelled RMSE no_skill), the other with perfect spread (labelled RMSE perfect).

The ROC curves (not shown) suggest that the abilities of the two configurations of nowcast ensembles to discriminate between events and non-events (i.e. an exceedence *versus* a non-exceedence of a threshold) were similar, though the ensemble generated from perturbed observations was slightly superior in terms of ROC score. Neither algorithm was as skilful as the control (unperturbed) member nowcast at thresholds of 2 mm/h and 4 mm/h. This finding suggests an issue with the calibration of the observation error models, since it was not replicated in equivalent verification statistics for T+2h and beyond (not shown).

Figure 4(a),(b) compares spread–skill plots for STEPS ensemble runs from unperturbed observations (a) and equivalent ensembles from perturbed observations (b) with spread–skill plots for two reference ensembles, one with no skill (near horizontal line), the other with perfect spread (the diagonal line). Both ensembles are under-spread although the ensembles generated from perturbed observations exhibit a closer relationship between spread and error for larger errors.

CONCLUSIONS

The impact of an improved treatment of radar observation errors on the performance of STEPS ensemble precipitation nowcasts has been explored using a small number of case study events. Preliminary verification results suggest some beneficial impact on the ensemble spread–skill relationship in the first hour. However, ROC verification statistics indicate that the observation error model may not be well calibrated for threshold hourly accumulations ≥ 2 mm/h. An extended trial of the radar ensemble generators is needed to confirm these conclusions.

REFERENCES

Bowler, N. E., Pierce, C. E. & Seed, A. W. (2006) STEPS: A probabilistic precipitation forecasting scheme which merges an extrapolation nowcast with downscaled NWP. *Quart. J. Roy. Met. Soc.* 132, 2127–2155.

Ciach, G. J., Krajewski, W. F. & Villarini, G. (2007) Product-error-driven uncertainty model of probabilistic quantitative precipitation estimation with NEXRAD data. *J. Hydrometeorology* 8, 1325–1347.

Germann, U., Berenguer, M., Sempere-Torres, D. & Zappa, M. (2009) REAL – ensemble radar precipitation estimation for hydrology in a mountainous region. *Quart. J. Roy. Met. Soc.* 135, 445–456.

Jordan, P. W., Seed, A. W. & Weinmann, P. E. (2003) A stochastic model of radar measurement errors in rainfall accumulations at catchment scale. *J. Hydrometeorology* 4, 841–855.

Lee, G. W., Seed, A. W. & Zawadzki, I. (2007) Modelling the variability of drop size distributions in space and time. *J. Appl. Met. & Climatol.* 46, 742–756.

Lee, G. W. & Zawadzki, I. (2005a) Variability of drop size distributions: time-scale dependence of the variability and its effects on rain estimation. *J. Appl. Met.* 44, 241–255.

Lee, G. W. & Zawadzki, I. (2005b) Variability of drop size distributions: noise and noise filtering in disdrometric data. *J. Appl. Met.* 44, 634–652.

Lee, G. W. & Zawadzki, I. (2006) Radar calibration by gage, disdrometer, and polarimetry: theoretical limit caused by the variability of drop size distribution and application to fast scanning operational radar data. *J. Hydrol.* 328, 83–97.

Llort, X., Velasco-Forero, C., Roca-Sancho, J. & Sempere-Torres, D. (2008) Characterization of uncertainty in radar-based precipitation estimates and ensemble generation. In: *Proc. Fifth European Conf. on Radar in Meteorology and Hydrology (ERAD 2008)*.

Norman, K., Seed, A. & Pierce, C. (2010) A comparison of two radar rainfall ensemble generators. In: *Proc. Sixth European Conf. on Radar in Meteorology and Hydrology (ERAD 2010)*.

Weather Radar and Hydrology
(Proceedings of a symposium held in Exeter, UK, April 2011) (IAHS Publ. 351, 2012).

299

Probabilistic forecasting of rainfall from radar nowcasting and hybrid systems

SARA LIGUORI & MIGUEL RICO-RAMIREZ

University of Bristol, Department of Civil Engineering, Bristol BS8 1TR, UK
s.liguori@bristol.ac.uk

Abstract The use of Quantitative Precipitation Forecasts (QPFs) from either Numerical Weather Prediction (NWP) or radar nowcasting models in flood forecasting systems extends the time available to issue warnings and take actions. However, uncertainty in the rainfall input affects the accuracy of flow predictions. Radar nowcasts have a higher skill at short lead times, whereas NWP models produce more accurate forecasts at longer lead times. Hybrid systems, merging NWP and radar-based forecasts, have been developed to produce more skilful forecasts than either independent component (i.e. NWP/radar nowcasting). This study aims at assessing radar nowcasts and hybrid forecasts provided by the state-of-the-art model STEPS. The forecasts were run on a 1000 km × 1000 km domain covering the UK, at 2-km spatial and 15-min temporal resolutions. Results show that the forecasting system benefits from the blending with the NWP forecasts.

Key words QPFs; ensemble forecasting; STEPS; nowcasting; hybrid forecasts

INTRODUCTION

Flooding, often originated by exceptional rainfall events, has caused lots of damage in the UK in recent years. The use of precipitation forecasts as input to hydrological models is crucial in the efficiency of flood forecasting and warning systems as it extends the forecast lead time. However, the predictability of flood and flash flood events is limited by the uncertainty in the rainfall input.

Quantitative Precipitation Forecasts (QPFs) can be achieved by radar nowcasting or Numerical Weather Prediction (NWP) models. Nowcasting is the process of extrapolating a sequence of radar images in order to produce short-range rainfall forecasts. NWP models forecast the future state of the atmosphere at longer lead times by solving mathematical equations. Radar nowcasts are affected by a rapid loss of information as the forecasting time increases. NWP models, on the other hand, suffer by the imperfect assimilation of the initial state as well as by resolution issues. However, NWP forecasts tend to have a higher skill than radar nowcasts on longer lead times. Potentially, more skilful forecasts could be achieved by hybrid systems, merging radar nowcasts and NWP forecasts (Golding, 1998). Lin *et al.* (2005) compared the average skill of radar nowcasts with 1 h precipitation accumulation forecasts from four different NWP models using the nowcasting method developed by Germann & Zawadzki (2002). They identified at 6 h the threshold time after which NWP models perform better than radar nowcasts.

This work aims at contributing to the assessment of the performance of rainfall hybrid forecasting systems against pure nowcasting systems. It provides a discussion on the results of the implementation of the state-of-the-art forecasting system STEPS (Short-Term Ensemble Prediction System), developed by the UK Met Office (UKMO) and the Australian Bureau of Meteorology (Bowler *et al.*, 2006). The STEPS model was implemented for both nowcasts and hybrid forecasts. On this occasion, rainfall forecasts with 2-km spatial resolution were produced at 15-min intervals. The STEPS model employs a stochastic ensemble generation approach to provide probabilistic nowcasts and hybrid forecasts. This approach has also been implemented for the purpose of probabilistic quantitative precipitation estimation (PQPE) from radars by Germann *et al.* (2009).

MODELS AND DATA

The STEPS model

The probabilistic hybrid model STEPS, described by Bowler *et al.* (2006), produces hybrid forecasts by merging radar nowcasts and NWP forecasts. The nowcasts are performed with a modified version of the S-PROG model (Seed, 2003), combining spectral decomposition of the precipitation field into a multiplicative cascade, estimation of the advection field and temporal evolution of precipitation. The advection velocity is determined through an improved version of the optical flow method described by Bowler *et al.* (2004). An extrapolation forecast cascade is merged to a NWP forecast to obtain deterministic hybrid forecasts. Uncertainties related to the forecasting of the temporal evolution of precipitation are modelled through stochastic noise, in accordance with the temporal persistence of scale-filtered features and in order to compensate for the loss of predictability of extrapolated small-scale precipitation features. Probabilistic forecasts are generated from different realizations of the noise for a number of ensemble forecasts. Uncertainties related to the motion of the precipitation field are modelled through a perturbation field added to the extrapolation advection velocity at each ensemble realization.

Radar data

Radar data were provided by the UKMO through the British Atmospheric Data Centre at 1-km and 5-min spatial and temporal resolution, respectively. The original radar data have been averaged on a 2 km × 2 km grid over a 1000 km × 1000 km domain covering the UK, at 15-min temporal resolution. Correction methodologies have been applied by the UKMO radar processing system to account for the different sources of error in the precipitation estimation from weather radar (Harrison *et al.*, 2000).

The MM5 model

The NWP forecasts to be merged with the radar nowcasts in order to produce hybrid forecasts with the STEPS model have been obtained through the Penn State/NCAR (National Centre for Atmospheric Research) mesoscale model MM5 (Dudhia, 1993). The model requires initial and lateral boundary conditions to be provided by a coarse-resolution global model. On this occasion, the meteorological input data to initialize the MM5 model were provided by the ECMWF (European Centre for Medium Range Weather Forecasts) global model. Meteorological fields were retrieved from the deterministic forecasts of the ECMWF model, with a 0.5 × 0.5 lat/lon resolution and 3-h temporal resolution. The selected meteorological fields comprised ground temperature and mean sea level pressure plus horizontal wind components, temperature, relative humidity and geopotential height fields retrieved at different pressure levels between 1 and 1000 hPa. In the MM5 model, these fields are horizontally interpolated from a latitude–longitude grid to mesoscale, rectangular gridded domains and vertically interpolated from pressure levels to terrain following coordinates required by the numerical integration. Sub-grid non-resolvable processes, such as radiation, moisture fluxes, turbulence, convection, condensation, evaporation and surface heat are parameterized.

The NWP model was set up with a four-nested domain configuration. As this work is preparatory to the assessment of the implementation of short-term hybrid rainfall forecasts in the context of rainfall–runoff modelling, the domains were centred on the Upper Medway catchment located in the southeast of England (UK), which has been selected as case study for an ongoing hydrological assessment. The domains have increasing resolution from 54 km (largest domain) to 2 km (smallest domain) with respect to the nesting ratio required by the MM5 model, which is 3:1. In order to run the STEPS model, the MM5 four original domains were blended into a single domain at 2-km resolution over a 1000 km × 1000 km domain matching the radar domain. Figure 1 shows the radar domain (a) and the NWP blended domain (b) used for forecast and verification.

SIMULATIONS SET-UP

Radar nowcasts and hybrid forecasts were generated for a number of events that occurred during 2007. For a single event, a single 24-h deterministic forecast was performed with the MM5 model, and the results were taken at 15-min intervals. The STEPS model was used to run deterministic and probabilistic forecasts. However, the analysis of the results discussed for the purpose of this work focuses on the performance of the probabilistic forecasts. A quantitative comparison between deterministic and probabilistic hybrid forecasts is provided by Liguori *et al.* (2011), where an assessment of the STEPS model for rainfall and flow prediction over a small urban area is discussed.

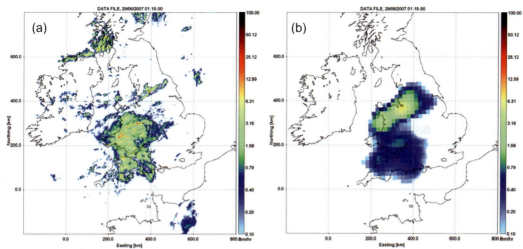

Fig. 1 Radar domain (a) and NWP blended domain (b) with 2-km spatial resolution. The size is 1000 km × 1000 km.

For a single event, the STEPS model was set up to produce a probabilistic forecast comprising 20 equally likely ensemble members. Given the starting time of the NWP forecast for a given event, t_0, the STEPS model was configured to start a forecast every hour on a 24-h window, between $t_0 + 1$ h and $t_0 + 18$ h. A 15-min forecasting time step was adopted to run 6-h lead time nowcasts and hybrid forecasts on the forecast domain shown in Fig. 1 at 2-km resolution. Nowcasts and hybrid forecasts will be hereafter referred to as NC and HF, respectively.

RESULTS

Both ensemble HF and ensemble NC have been analysed through a probabilistic verification measure. Relative Operating Characteristic (ROC) curves have been computed for all ensemble HF and NC generated with the STEPS model by taking into account the model output at all points in the forecast domain. Therefore, the forecasts output has been taken at 2-km resolution, with the size of the verification domain being 500 × 500 grid squares. The ROC is a graphical forecast verification measure (Wilks, 2006), that can be used to evaluate detection and forecasting systems, both deterministic and probabilistic, and characterizes the ability of the system to correctly discriminate the occurrence and the non-occurrence of pre-defined events (Mason & Graham, 2002). The curves display the relation between hit rate (HR) and false alarm rate (FAR) for a pre-defined event as decision criterion varies. For a given event (e.g. rainfall exceeding a given threshold), the HR is defined as the ratio of the number of correct forecasts to the number of occurrences of the event in the verification dataset, while the FAR is the ratio of the false alarms to the number of non-occurrences of the event within the verification dataset.

On this occasion, the curves have been computed for rainfall thresholds between 0 mm/h (i.e. $R > 0$ mm/h, occurrence of rainfall) and 5 mm/h (i.e. $R \geq 5$ mm/h, occurrence of rainfall equal to or above 5 mm/h). The ROC curves have been used to describe the performance of the forecasting system for both nowcasts and hybrid forecasts through the associated index of forecast accuracy (i.e. the area A under the curve), at lead time between 15 min and 6 h. The analysis has been performed on a single event basis, in order to highlight differences and consistencies in the performance of the forecasting system among various events. A single ROC curve, valid for a single event and rainfall threshold at a given lead time, is computed by taking into account all forecasts initialized at various starting times for that event.

Based on the Met Office (National Climate Information Centre) series from 1914, the summer 2007 (period June–August) was the wettest for England and Wales since 1912, with extreme rainfall events being more frequent and widespread. Figure 2 shows the ROC curves computed for the ensemble HF (upper row) and the ensemble NC (lower row) initialized between 01:00 UTC and 18:00 UTC on 20 July 2007. The 20 July 2007 was a heavy frontal event with embedded convective elements, caused by a slow-moving depression centred over southeast England. ROC curves in Fig. 2 are for lead times of 60, 180 and 360 min and rainfall thresholds between 0 and 5 mm/h. The accuracy of both HF and NC decreases with lead time, as the value of the area A under the ROC curve decreases as the lead time increases from 60 min to 6 h for all rainfall thresholds. The area A also generally decreases as the rainfall intensity threshold increases. The ensemble HF has a slightly higher performance than the ensemble NC at all lead times and rainfall thresholds. However, the STEPS model performs relatively well in predicting this event with both ensemble HF and NC, as the values of the area A are around 0.8 or higher for rainfall thresholds up to 3 mm/h, and above 0.7 for the 5 mm/h threshold at all lead times.

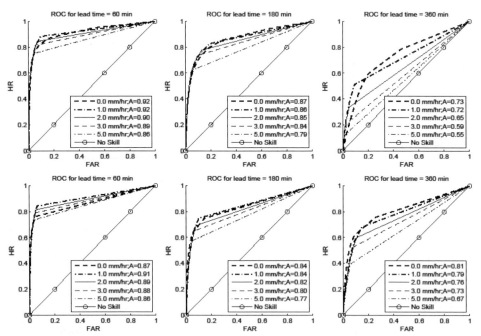

Fig. 2 ROC curves for the 20 July 2007 event. Ensemble HF (upper row) and ensemble NC (lower row) compared at 1 h, 3 h and 6 h forecasting lead time.

Figure 3 shows the ROC curves computed for the ensemble HF and the ensemble NC initialized between 01:00 UTC and 18:00 UTC on 29 June 2007, similarly to Fig. 2. The 29 June 2007 event was a large frontal event with predominant convective elements. The ROC curves show that the performance of the ensemble HF decreases with lead time and rainfall intensity threshold. The area A computed for $R > 0$ mm/h and $R \geq 1$ mm/h is respectively equal to 0.88 and 0.86 at 60 min, and decreases to 0.79 and 0.77 at 180 min. Therefore, at these lower thresholds, the accuracy of the hybrid forecasting system decreases with lead time, but remains fairly good even at 3 and 6 h. At higher-intensity thresholds, the performance of the hybrid forecasting system is good at 60 min, but decreases very rapidly, with values of the area A below 0.7 at 3 h indicating very poor performance. The values of the area A computed for the ensemble HF are very close to those computed for the ensemble NC. However, small differences show that the ensemble HF also performed slightly better for this event than the ensemble NC.

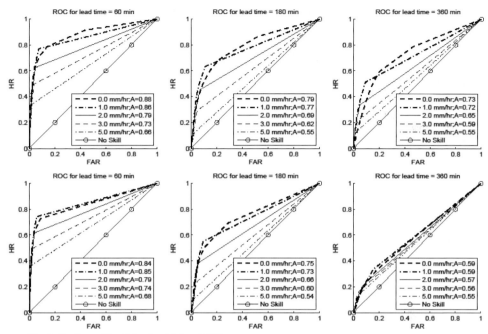

Fig. 3 Similar to Fig. 2 for the 29 June 2007 event.

Figure 4 is similar to Figs 2 and 3 for HF and NC initialized between 12:00 UTC on 17 January 2007 and 05:00 UTC on 18 January 2007 at rainfall thresholds between 0 and 3 mm/h. This was a frontal event with higher intensity rainfall over some localized regions in the UK. The ensemble HF performs better than the ensemble NC also for this event, and the difference between the ensemble HF and the ensemble NC in terms of values of the ROC area A computed at a given threshold increases with increasing lead time. The accuracy (area A) of the ensemble HF ranges in between 0.82 and 0.86 at 1 h and decreases to values below 0.8 at 3 h. At all lead times the area A computed for $R > 0$ mm/h is lower than the corresponding value computed for higher thresholds. Also, from 60 min ahead, the area A computed for $R \geq 0$ mm/h is equal or lower than the corresponding value computed for higher thresholds. This result can be explained by looking at the values of the FAR computed for corresponding probability thresholds at various rainfall intensity thresholds. Since the ROC combines HR and FAR, if, for a series of probability thresholds, the HR remains the same and the FAR decreases with increasing intensity threshold, then the ROC curve computed for a lower intensity threshold can happen to be located below the curve computed for a higher intensity threshold.

CONCLUSIONS

The results of the analysis of the probabilistic HF and NC for the 20 July, 29 June and 17 January 2007 events summarized by the ROC curves show that the blending of NWP forecasts and radar nowcasts generally improves the performance of the STEPS model at all lead times and for all rainfall thresholds, regardless of the different nature of the simulated events. However, differences between the performance of the hybrid forecasting system and the performance of the nowcasting system vary depending on the event and the forecasting lead time. The discrimination accuracy of the STEPS model generally decreases with increasing rainfall intensity. Therefore, the performance of the model in predicting rainfall decreases with increasing rainfall intensity threshold, and this is consistent for both HF and NC. However, the prediction of the 17 January 2007 event was affected

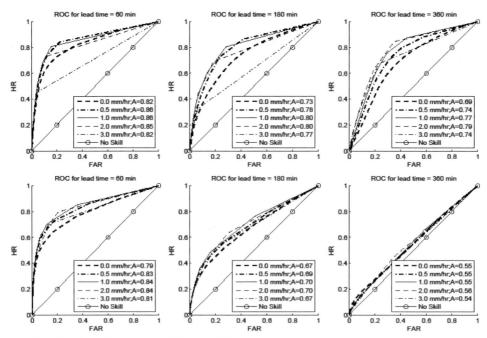

Fig. 4 Similar to Fig. 2 for the 17 January 2007 event.

by a large number of false alarms at low rainfall thresholds. In a hydro-meteorological forecasting approach, any improvement in the rainfall forecasts can potentially enhance the performance of the system in providing flood forecasts and warnings. However, the comparative assessment of HF and NC for flow and flood prediction is out of the scope of this paper.

Acknowledgements The authors would like to thank ECMWF, BADC, and the UK Met Office for providing some of the data sets used in this study. The authors would also like to thank the UK Met Office for providing the STEPS and the nowcasting models. The authors acknowledge the support from the Engineering and Physical Sciences Research Council (EPSRC) via grant EP/I012222/1.

REFERENCES

Bowler, N., Pierce, C. E. & Seed, A. (2006) STEPS: A probabilistic precipitation forecasting scheme which merges and extrapolation nowcast with downscaled NWP. *Quart. J. Roy. Met. Soc.* 132, 2127–2155.

Dudhia, J. (1993) A nonhydrostatic version of the Penn State-NCAR Mesoscale Model: Validation tests and simulation of an Atlantic cyclone and cold front. *Monthly Weather Rev.* 121, 1493–1513.

Germann, U. & Zawadzki, I. (2002) Scale dependence of the predictability of precipitation from continental radar images. Part I: Description of the methodology. *Monthly Weather Rev.* 130, 2859–2873.

Germann, U., Berenguer, M., Sempere-Torres, D. & Zappa, M. (2009) REAL-Ensemble radar precipitation estimation for hydrology in a mountainous region. *Quart. J. Roy. Met. Soc.* 135, 445–456.

Golding, B. (1998) Nimrod: a system for generating automated very short range forecasts. *Meteorological Applications* 5, 1–16.

Harrison, D. L., Driscoll, S. J. & Kitchen, M. (2000) Improving precipitation estimates from weather radar using quality control and correction techniques. *Meteorological Applications* 7, 135–144.

Liguori, S., Rico-Ramirez, M. A., Schellart, A. N. A., Saul, A. J. (2012) Using probabilistic radar rainfall nowcasts and NWP forecasts for flow prediction in urban catchments. *Atmospheric Research* 103, 80–95.

Lin, C., Vasic, S., Kilambi, A., Turner, B. & Zawadzki, I. (2005) Precipitation forecast skill of numerical weather prediction models and radar nowcasts. *Geophys. Res. Lett.* 32. doi:10.1029/2005GL023451.

Mason, S. & Graham, N. (2002) Areas beneath the relative operating characteristics (ROC) and relative operating levels (ROL) curves: Statistical significance and interpretation. *Quart. J. Roy. Met. Soc.* 128, 2145–2166.

Seed, A. (2003) A dynamic and spatial scaling approach to advection forecasting. *J. Appl. Met.* 42, 381–388.

Wilks, D. (2006) *Statistical Methods in the Atmospheric Sciences*. Academic Press.

Weather Radar and Hydrology
(Proceedings of a symposium held in Exeter, UK, April 2011) (IAHS Publ. 351, 2012).

305

PhaSt: stochastic phase-diffusion model for ensemble rainfall nowcasting

N. REBORA & F. SILVESTRO

CIMA Research Foundation, Via Magliotto 2, 17100 Savona, Italy
nicola.rebora@cimafoundation.org

Abstract Hydrometeorological hazard management often requires the development of reliable statistical rainfall nowcasting systems. Ideally, such procedures should be capable of generating stochastic ensemble forecasts of precipitation intensities on scales of the order of a few kilometres, up to a few hours in advance. Ensemble rainfall nowcasting allows for characterizing the uncertainty associated with nowcasting procedures by providing a probabilistic forecast of the future evolution of an event. Here we discuss an ensemble rainfall nowcasting technique, named PhaSt (Phase Stochastic), based on the extrapolation of radar observations by a diffusive process in Fourier space. The procedure generates stochastic ensembles of precipitation intensity forecast fields where individual ensemble members can be considered as different possible realizations of the same precipitation event. The model is tested on a data set of rainfall events measured by the C-POL radar of Mt Settepani (Liguria, Italy) and its performance verified in terms of standard probabilistic scores.

Key words nowcasting; ensemble; probabilistic forecast; rainfall

INTRODUCTION

Nowcasting of intense precipitation fields has several important applications for hydrometeorological risk management, such as flash flood warning. Research has led to the development of a large number of deterministic nowcasting techniques which use radar fields as initial conditions for local precipitation estimation. However, a probabilistic approach is crucial both for evaluating uncertainties that arise in the forecasting process and for generating useful products for hydrological application (e.g. Krzysztofowicz, 2001). While some of the existing probabilistic methods are based on Bayesian methods (Xu, 2003; Fox, 2005), stochastic processes are commonly used to evolve individual rain cells or the entire field (e.g. Anderson & Ivarsson, 1991; Mellor, 2000; Grecu & Krajewski, 2000; Seed, 2003; Xu, 2005). The probabilities of occurrence of intense precipitation events can be derived in a probabilistic way (e.g. Germann & Zawadzki, 2004) or from ensembles of stochastic realizations.

The predictability of precipitation at a short temporal range is strongly dependent on the spatial scale of the considered structures (see e.g. Wilson *et al.*, 1998; Germann & Zawadzki, 2002), an observation which has motivated the development of stochastic nowcasting methods operating in the spectral Fourier domain, such as the method proposed in Xu (2005). In these techniques an autoregressive stochastic process is used to evolve directly in time the Fourier coefficients of the precipitation or reflectivity fields; the spectral representation allows one to take naturally into account scale dependencies of the parameters. While these techniques have been shown to be skilful up to 1 h, high-intensity features can be smoothed in the forecast. In addition to scale dependency of predictability, there are other important properties of spatio-temporal precipitation fields which can be used to improve on current models: (a) both the amplitude distribution and the correlation structure of the fields, i.e. their spectral power-spectra, are frequently only weakly varying with time (see e.g. Zawadzki, 1973); (b) precipitation structures are often persistent in a Lagrangian framework, leading to a good nowcasting skill of simple Lagrangian persistence nowcasting methods applied either to the entire field or to individual precipitation structures. These two observations motivated the development of a new spectral-based nowcasting procedure (Metta *et al.*, 2009). The procedure preserves in time both the initial one-point distribution and the initial power-spectrum of the field to nowcast, using a stochastic process for the evolution of the spectral phases. In order for the model to account for phase velocities correlated in time, we use a simple Langevin-type stochastic model for the phases, evolving phase velocities using an Ornstein-Uhlenbeck process.

THE PHAST ALGORITHM STEP BY STEP

The extrapolation of precipitation fields in time is done by preserving the initial spectral Fourier amplitudes, and evolving in time only the Fourier phases. This approach, proposed by Metta et al. (2009), distinguishes this model from other stochastic spectral models (e.g. Xu, 2005) in which the complex Fourier coefficients themselves are evolved using a stochastic process. The simplest possible time-evolution model for Fourier phases which allows for time-correlated phase velocities is represented by a Langevin-type model. This type of stochastic model has been used widely in turbulence literature as a closure model for isotropic turbulence (Herring & Kraichnan, 1972), to model turbulent Lagrangian velocities in dispersion problems (see e.g. van Dop et al., 1985) and to create synthetic turbulent velocity fields (Kraichnan, 1970; Fung & Vassilicos, 1997). This model represents a generalization which also includes the simple case of uncorrelated phase velocities, corresponding to a Markovian evolution of spectral phases. Specifically, the independent evolution in time of Fourier phases, $\phi(k_x, k_y, t)$ at each wave number $k_s^2 = k_x^2 + k_y^2$, can be written, using an Ornstein-Uhlenbeck stochastic process for spectral phase velocities as:

$$\begin{cases} d\phi_{k_s} = \omega_{k_s}\, dt \\ d\omega_{k_s} = -\dfrac{(\omega_{k_s} - \omega_{k_s}')}{T}\, dt + \sqrt{\dfrac{2\sigma_s^2}{T}}\, k_s\, dW \end{cases}$$

where $\omega_{ks} = \omega(k_x, k_y, t)$ is phase velocity, $\omega'_{ks} = \omega'(k_x, k_y)$ a reference average phase velocity, T a decorrelation time, $\sigma^2(k_x, k_y)$ a scale-dependent variance of the phase velocities and W a random variable derived from a Wiener process. While the first term represents Lagrangian persistence, with a relaxation over the time scale T to reference phase velocities, the second term represents a diffusion in phase velocity which accounts for uncertainty growing with time and is scale dependant. A simple model of the scale dependence of uncertainty growth is obtained by choosing a simple isotropic linear of $\sigma(k_x, k_y)$ from wave numbers (k_x, k_y).

Using this approach, there is no distinction in modelling Eulerian advection and Lagrangian development of the precipitation fields. Both the two components are considered together in the Fourier-phase evolution in the spectral domain.

In the following the equations are reported, in their discrete version, to be used for operational implementation of the PhaSt algorithm. The operational nowcasting process is thus composed of the following steps:

(1) The two most recent radar images (r_{t0}, r_{t0-1}) are Gaussianized. This is achieved by applying a technique similar to that adopted by Thelier et al. (1992) for the AAFT (Amplitude Adjusted Fourier Transform), with the empirical relation Ψ used to Gaussianize the precipitation field. Two different Ψ functions for t_0 and t_{0-1} are derived:

$$g_{t_0}(x, y) = \Psi_{t_0}\left[r_{t_0}(x, y)\right]$$
$$g_{t_0-1}(x, y) = \Psi_{t_0-1}\left[r_{t_0-1}(x, y)\right]$$

(2) The Fourier transform of the two Gaussianized fields is performed. In this way, a 2-D Fourier spectrum is obtained i.e. a 2-D power spectrum and corresponding Fourier phases:

$$F_{t_0-1}(k_x, k_y) = \int_{-\infty}^{+\infty} g_{t_0-1}(x, y) e^{ik_x x + ik_y y} dxdy = [E_{t_0-1}(k_x, k_y)]^{1/2} e^{i\phi_{t_0-1}(k_x, k_y)}$$

$$F_{t_0}(k_x, k_y) = \int_{-\infty}^{+\infty} g_{t_0}(x, y) e^{ik_x x + ik_y y} dxdy = [E_{t_0}(k_x, k_y)]^{1/2} e^{i\phi_{t_0}(k_x, k_y)}$$

(3) The phases of the Gaussianized observed field at $t=t_0$ field are evolved in time according to a Langevin equation with parameters T and σ:

$$\omega(k_x, k_y, t_0) = \frac{d\phi(k_x, k_y)}{dt} \approx \frac{\phi_{t_0}(k_x, k_y) - \phi_{t_0-1}(k_x, k_y)}{\Delta t}$$

$$\omega^*(k_x, k_y) = \omega(k_x, k_y, t_0) + W(t_0)\sqrt{\frac{2\sigma^2}{T}\Delta t\left(k_x^2 + k_y^2\right)}$$

$$W(t_0) \in N(0,1)$$

(4) A predicted Gaussian field is obtained by Fourier anti-transforming, at each time step $t > t_0$, the power spectrum estimated at $t = t_0$ associated with the phases obtained from the Langevin process. The Fourier phases derived, for observed radar fields, at time t_0 and t_{0-1} are evolved in time according to the following expressions:

$$\phi(k_x, k_y, t_{0+1}) = \phi(k_x, k_y, t_0) + \omega(k_x, k_y, t_0)\Delta t$$

$$\vdots$$

$$\omega(k_x, k_y, t_{0+1}) - \omega(k_x, k_y, t_0) = -\frac{\omega(k_x, k_y, t_0) - \omega^*(k_x, k_y)}{T}\Delta t + W(t_{0+1})\sqrt{\frac{2\sigma^2}{T}\Delta t(k_x^2 + k_y^2)}$$

$$W(t_{0+1}) \in N(0,1)$$

$$\downarrow$$

$$\phi(k_x, k_y, t_{0+2}) = \phi(k_x, k_y, t_{0+1}) + \omega(k_x, k_y, t_{0+1})\Delta t$$

$$\vdots$$

$$\omega(k_x, k_y, t_{0+2}) - \omega(k_x, k_y, t_{0+1}) = -\frac{\omega(k_x, k_y, t_{0+1}) - \omega^*(k_x, k_y)}{T}\Delta t + W(t_{0+2})\sqrt{\frac{2\sigma^2}{T}\Delta t(k_x^2 + k_y^2)}$$

$$W(t_{0+2}) \in N(0,1)$$

$$\downarrow$$

$$\phi(k_x, k_y, t_{0+3}) = \phi(k_x, k_y, t_{0+2}) + \omega(k_x, k_y, t_{0+2})\Delta t$$

$$\vdots$$

$$\omega(k_x, k_y, t_{0+n-1}) - \omega(k_x, k_y, t_{0+n-2}) = -\frac{\omega(k_x, k_y, t_{0+n-2}) - \omega^*(k_x, k_y)}{T}\Delta t + W(t_{0+n-1})\sqrt{\frac{2\sigma^2}{T}\Delta t(k_x^2 + k_y^2)}$$

$$W(t_{0+n-1}) \in N(0,1)$$

$$\downarrow$$

$$\phi(k_x, k_y, t_{0+n}) = \phi(k_x, k_y, t_{0+n-1}) + \omega(k_x, k_y, t_{0+n-1})\Delta t$$

(5) A Gaussian field is then obtained by applying the Fourier anti-transform using the evolved phases array $\phi(k_y, k_y, t)$ and the original spectral amplitude E_{t0}, so:

$$g_n(x,y,t) = \int_{-\infty}^{+\infty} \left\{ [E_{t_0}(k_x,k_y)]^{1/2} e^{i\phi(k_x,k_y,t)} \right\} e^{-ik_x x - ik_y y} dk_x dk_y$$

where:

$$t > t_0$$

(6) The nth member of the predicted rainfall ensemble, $p_n(x,y,t)$, is then obtained by applying the nonlinear transformation Ψ^{-1} (inverse Gaussian transformation) to the predicted Gaussian field:

$$p_n(x,y,t) = \Psi_{t_0}^{-1}[g_n(x,y,t)]$$

Steps (3)–(6) are repeated to obtain an ensemble of possible nowcasted fields. Each field, indicated by the subscript n, will start from the same initial condition but will develop differently due to the stochastic nature of the Langevin process. We refer to Ferraris *et al.* (2003a,b) and Rebora *et al.* (2006) for further details on how a nonlinear transformation of a linearly correlated field, such as the one applied in step 6 of the procedure, generates scaling properties in resulting precipitation fields.

EXAMPLE OF APPLICATION

After fixing the T and σ parameters, either by direct estimation from a recent sequence of observed fields or by choosing them from a library, possibly conditioned on large-scale synoptic conditions, a certain number of realizations of the precipitation process can be built. The result is an ensemble of forecast stochastic fields from which a probability of exceedence of a fixed precipitation threshold can be estimated. In Figs 1 and 2 we show, for two values of σ, namely $\sigma = 0.1$ and $\sigma = 0.01$, and by fixing $T = 2$, the behaviour of PhaSt statistics for the precipitation event observed in September 2007 by the Mt Settepani C-Pol radar. For the same event an example of PhaSt output is reported in Fig. 3. A simple way of estimating parameter T is by calculating the correlation time of the main precipitation structures, while the estimate of σ can be related to the level of predictability of the observed rainfall event. Low levels of σ (e.g. $\sigma = 0.01$) are related to more coherent and widespread events, while larger values (e.g. $\sigma = 0.1/0.5$) can be used when convective activity is the dominant component.

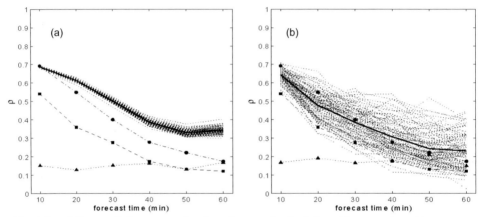

Fig. 1 Correlation coefficient, ρ, as a function of nowcasting lead-time: (a) nowcasting with parameters $T = 2$ and $\sigma = 0.1$; (b) nowcasting with parameters $T = 2$ and $\sigma = 0.01$. Dotted lines represent PhaSt ensemble, the black continuous line is the ensemble mean, dashed line with black points represents a nowcast made using Lagrangian persistence, dashed line with squares represents the Eulerian persistence case. Dotted line with triangles represents a random field for evaluating the no-skill values.

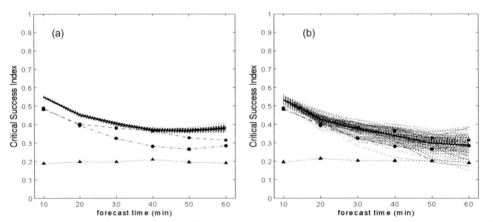

Fig. 2 Critical Success Index (CSI) as a function of nowcasting lead time. (a) nowcasting with parameters $T = 2$ and $\sigma = 0.1$; (b) nowcasting with parameters $T = 2$ and $\sigma = 0.01$. Dotted lines represent PhaSt ensemble, the black continuous line is the ensemble mean, dashed line with black points represents a nowcast made using Lagrangian persistence, dashed line with squares represents the Eulerian persistence case. Dotted line with triangles represents a random field for evaluating the no-skill values.

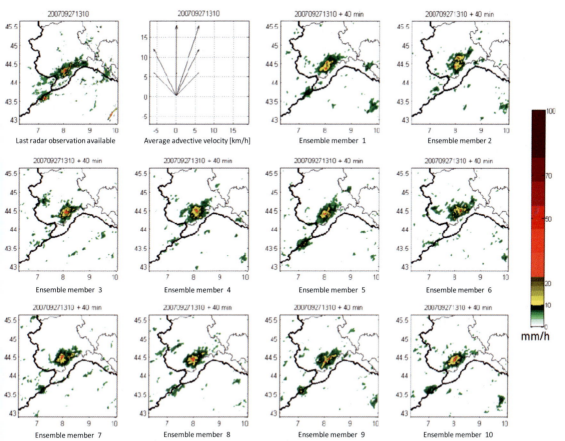

Fig. 3 Example of PhaSt operational output (with $T = 2$ and $\sigma = 0.1$) for the precipitation event observed on 9 September 2007 by the Mt. Settepani C-Pol weather radar (Piemonte/Liguria Regions, Italy). In this figure are reported (from top-left): (i) the last radar observation, (ii) the average advective velocity for each ensemble member, (iii) 10 possible realizations of precipitation field with 40 min lead time issued by the operational version of PhaSt.

CONCLUSIONS

Model skill assessed using the correlation coefficient and CSI index shows good skill for this model up to forecast lead times around 60 min, in line with the skill reported in the literature for alternative nowcasting procedures. The technique presented here has the advantage of allowing the generation of stochastic nowcasting ensembles from which quantitative estimates for the probability of occurrence of precipitation can be derived.

The PhaSt algorithm is operationally used for flash flood monitoring and nowcasting by the Civil Protection Service of Liguria Region (Italy) (Silvestro & Rebora, 2011). Further work will aim to verify the long-term operational skill of these probabilistic forecasts and to determine the best procedures for estimating the model free parameters (T and σ).

REFERENCES

Andersson, T. & Ivarsson, K.-I. (1991) A model for probability nowcasts of accumulated precipitation using radar. *J. Appl. Met*. 30, 135–141.

Ferraris, L., Gabellani, S., Parodi, U., Rebora, N., von Hardenberg, J. & Provenzale, A. (2003a) Revisiting multifractality in rainfall fields. *J. Hydromet*. 4, 544–551.

Ferraris, L., Gabellani, S., Rebora, N. & Provenzale, A. (2003b) A comparison of stochastic models for spatial rainfall downscaling. *Water Resour. Res.* 39, 1368–1384.

Fox, N. I. & Wikle, C. K. (2005) A Bayesian quantitative precipitation nowcast scheme. *Weather Forecasting* 20, 264–275.

Fung, J. C. H. & Vassilicos, J. C. (1998) Two-particle dispersion in turbulent-like flows, *Phys. Rev (E)*. 57, 1677–1690.

Germann, U. & Zawadzki, I. (2002) Scale-dependence of the predictability of precipitation from continental radar images. Part I: Description of the methodology. *Monthly Weather Rev.* 130, 2859–2873.

Germann, U. & Zawadzki, I. (2004) Scale-dependence of the predictability of precipitation from continental radar images. Part II: Probability forecasts. *J. Appl. Met.* 43, 74–89.

Grecu, M. & Krajewski, W. (2000) A large-sample investigation of statistical procedures for radar-based short-term quantitative precipitation forecasting. *J. Hydrol.* 239, 69–84.

Herring, J. R. & Kraichnan, R. H. (1972) Comparison of some approximations for isotropic turbulence. In: *Statistical Models and Turbulence*, Lecture Notes in Physics 12, 148–194. Springer.

Kraichnan, R. H. (1970) Diffusion by a random velocity field. *Phys. Fluids* 13, 22–31.

Krzysztofowicz, R. (2001) The case for probabilistic forecasting in hydrology. *J. Hydrol.* 249, 2–9.

Mellor, D., Sheffield, J., O'Connell, P. E. & Metcalfe, A. V. (2000) A stochastic space–time rainfall forecasting system for real time flow forecasting I: Development of MTB conditional rainfall scenario generator. *Hydrol. Earth System Sci* 4, 603–615.

Metta, S., von Hardenberg, J., Ferraris, L., Rebora, N. & Provenzale, A. (2009) Precipitation nowcasting by a spectral-based nonlinear stochastic model. *J. Hydromet.* 10, 1285–1297.

Rebora N., Ferraris, L., von Hardenberg, J. & Provenzale, A. (2006) RainFARM: rainfall downscaling by a filtered autoregressive model. *J. Hydromet.* 7, 724–738.

Seed, A. W. (2003) A dynamical and spatial scaling approach to advection forecasting. *J. Appl. Met.* 42, 381–388.

Silvestro, F. & Rebora, N. (2011) Ensemble nowcasting of river discharge by using radar data: operational issues on small and medium size basins. In: *WRaH 2011 Proceedings*.

Thelier, J., Eubank, S., Longyin, A., Galdrikian, B. & Farmer, J. (1992) Testing nonlinearity in time series: the method of surrogate data. *Physica D* 58, 77–94.

van Dop, H., Nieuwstadt, F. T. M. & Hunt, J. C. R. (1985) Random walk models for particle displacements in inhomogeneous unsteady turbulent flows. *Phys. Fluids* 28, 1639–1653.

Wilson, J. W., Crook, N. A., Mueller, C. K., Sun, J. & Dixon, M. (1998) Nowcasting thunderstorms: a status report. *Bull. Am. Met. Soc.* 79, 2079–2099.

Xu, G. & Chandrasekar, V. (2005) Operational feasibility of neural-network-based radar rainfall estimation. *IEEE Geosci. Remote Sens. Lett.* 2, 13–17.

Zawadzki, I. I. (1973) Statistical properties of precipitation patterns. *J. Appl. Met.* 12, 459–472.

Weather Radar and Hydrology
(Proceedings of a symposium held in Exeter, UK, April 2011) (IAHS Publ. 351, 2012).

311

Ensemble radar nowcasts – a multi-method approach

ALRUN TESSENDORF & THOMAS EINFALT

Hydro & Meteo GmbH & Co. KG, Breite Straße 6-8, D-23552 Lübeck, Germany
a.tessendorf@hydrometeo.de

Abstract Radar nowcasting has for a long time been a competition between individual approaches with their strengths and weaknesses. The introduction of ensembles makes it possible to benefit from several techniques and can help in forecast applications by providing statistical information. This study focuses on how to prepare results of ensemble forecasts for risk assessment in real-time warning applications. A set of ensembles, combining runs from four forecast methods with perturbed initial conditions, is constructed and the results are evaluated using six different criteria. For predicting the current forecast quality from the ensemble spreading, quality parameters based on the contingency table were derived from the ensemble forecasts.

Key words rainfall forecast; radar; ensembles; nowcasting; risk assessment; forecast quality

INTRODUCTION

With its high variability at small temporal and spatial scales, precipitation is a difficult forecast parameter. Precipitation nowcasts based on radar images have been produced since the first weather radars were used. Former studies (e.g. Bowler *et al.*, 2006; Germann *et al.*, 2009) have underlined the potential to exploit ensembles in order to provide realistic bandwidths of the uncertainties in measurement and forecast. The approach presented here combines several forecast methods and the uncertainty of observed initial parameters to obtain an ensemble spreading wide enough to reflect nowcast uncertainty.

For real-time forecast and warning purposes, a full analysis of a big number of ensemble runs is not always appropriate. Therefore we propose a way to use the ensemble spreading for a prediction of the current forecast quality, making use of the common quality parameters POD (probability of detection) and FAR (false alarm rate).

METHODS AND DATA

For this study, data from radar Hamburg (DX product of the German Weather Service) are used. Data processing and correction is done with the software SCOUT (Hydro & Meteo, 2009) which has implemented numerous data quality items from COST 717 (Michelson *et al.*, 2005). The forecast is done for the full area covered by Hamburg radar with a maximum lead time of 1–3 h. For further analysis and quality tests, a 70 km × 70 km section in the centre of the radar-covered area was chosen as test area. The time period chosen for evaluating the forecast results is from 1 to 5 August 2010. Several precipitation events occurred during this time in the test area, with high intensities and high daily sums.

Forecast evaluation

To objectively evaluate precipitation forecasts is not a trivial task. Verification can be done with dichotomous tests or one can consider the height of the difference between forecast and observation. In order to evaluate the quality of the forecast methods used in this study we use six different parameters. The parameters probability of detection (POD), false alarm rate (FAR), probability of false detection (POFD) and Peirce skill score (PSS) can be derived from the contingency table (Donaldson *et al.*, 1975). The contingency table classifies an occurred and predicted event as "hit".

A not predicted and not occurred event is classified as "correct negative" and not predicted or not occurred events as "misses" and "false alarms" (see Table 1).

Table 1 The contingency table for forecast verification: events are classified as rain or no rain events for both observation and forecast.

Forecast:	Observation:	
	Rain	No rain
Rain	Hit	False alarm
No rain	Miss	Correct negative

Table 2 The class table for forecast verification: rain sums in mm denote the upper bound of the corresponding class interval.

Rain sum												
<(mm)	0.05	0.5	1	3	6	12	20	40	60	100	180	320	500
(class)	0	1	2	3	4	5	6	7	8	9	10	11	12

In this study a radar-measured rain intensity higher than 0.5 mm/h averaged over one grid point of 5 km × 5 km is specified as a positive observed event. The event at the grid point is compared with the predicted event and classified into the appropriate category in the contingency table (see Table 1). This is done for each of the 14 × 14 grid points in the test area and, with the total number of hits, misses, false alarms and correct negatives, the quality parameters are calculated (Donaldson *et al.*, 1975):

$$POD = \frac{hits}{hits + misses}, \qquad FAR = \frac{false\ alarms}{hits + false\ alarms} \qquad (1)$$

and according to Flueck (1987) and Doswell (1990):

$$POFD = \frac{false\ alarms}{correct\ negatives + false\ alarms} \qquad (2)$$

$$PSS = POD - POFD \qquad (3)$$

For an additional measure of the difference between radar observation and forecast, 1 h rain sums were divided into the classes listed in Table 2. Based on this table the mean error (ME) and root mean square error (RMSE) were calculated in class units.

Applied forecast methods

The four selected forecast methods combine physical approaches and image-processing methods in different ways. The basis of all methods is an image-processing method, tracking cells from one image to the next. The cell centres are located using a correlation approach. The features of the recognised cells (velocity vector, growth rate) are saved and used for extrapolation into the future, with a forecast produced every 5 min.

–   Method 1: Cell-tracking, linear extrapolation. Every cell is displaced independently based on its velocity vector, determined by following the way of the cell centre of gravity, cell shapes remain stable.
–   Method 2: Semi-Lagrange method. A spatial interpolation is done with all identified cell displacement vectors to get the two-dimensional velocity field used for the forecast. The velocity field is assumed to be stable during the lead time.
–   Method 3: Same as Method 2, but with a velocity field that at every forecast time step is displaced by the mean velocity vector.
–   Method 4: Same as Method 2, but uses the growth rates of the recognized cells to extrapolate for each cell the growth or decay (size and intensity) into the future.

Table 3 Verification results of the forecast methods used for the test period; lead time is 1 h.

Forecast method	POD	FAR	POFD	PSS	ME (class)	RMSE (class)
1	0.79	0.28	0.06	0.68	0.45	0.86
2	0.80	0.41	0.10	0.61	0.59	1.05
3	0.80	0.41	0.10	0.61	0.59	1.06
4	0.80	0.39	0.10	0.62	0.58	1.04

Construction of ensembles

Mean values and standard deviations of the cell displacement vectors and growth rates over the previous five time steps are calculated. Assuming standard normal distributions, these values are used to generate new random displacement vectors and growth rates for each ensemble run. Perturbation is done separately for each of the recognised cell vectors, before interpolating the 2-D velocity field, taking into account model uncertainty in recognising the cells as well as real variations in cell speed. For the following evaluation the four methods, each with the same number of ensemble runs, were used to construct one forecast ensemble.

RESULTS

Comparison of forecast methods

From 1 to 5 August 2010, forecasts were calculated every 5 min and compared to the radar measurements producing the quality parameters described above. The 5-day mean results are shown in Table 3. The POD is similar for all methods, in the 5-day-average none of the methods seems to be superior in detecting rain cells. The FAR shows a clear difference between Method 1 and the other methods. This difference can be explained by the fact that Methods 2–4 produce numerical diffusion on the forecast grid and tend to overpredict the area coverage of rainfall. The integrated cell growth and decay in Method 4 can improve the FAR compared to Method 2. Method 1 obtains better results with POFD and PSS and this effect can also be observed with the mean error (ME) and RMSE. Overall, the method differences of the quality parameters are relatively small compared to temporal variations of the forecast quality. As an example, the POD, averaged over the all methods, decreases from 0.87 to 0.42 within 1 h on 1 August. At the same time the FAR increases from 0.19 to 0.60 (compare Fig. 5, 16:30–17:30 UTC). The use of forecast ensembles for estimating the temporal variation of the forecast uncertainty is analysed next.

Results of ensemble runs

A set of ensemble forecasts (20 runs from four methods) was produced every 5 min for 1 August 2010. On each of the 14 × 14 grid points, a 1 h precipitation time series was calculated. Examples of the time series starting at 17:00 UTC and at 20:30 UTC are shown in Fig. 2. The different colours mark runs of the different forecast methods. The corresponding radar measured time series is shown in black.

At 17:00 UTC, the spread of the different ensemble time series is large: rain sums between 0 and 9 mm are predicted. Figure 3 shows the corresponding radar images (a) at the beginning of the forecast time interval, (b) after 30 min, (c) after 60 min and also (d–f) the forecast images of one forecast run using Method 4. From 17:00 to 18:00 UTC the rain cell that lies slightly below the middle of the image section in Fig. 3(a) changes its shape visibly. The cell that can be seen in the radar image at 18:00 UTC in the lower right shrunk strongly during the previous 60 min. These changes were not picked up in the forecast run; therefore, radar observation and forecast differ considerably. Averaged over all ensemble runs the POD is 0.40 and the FAR is 0.73. From 20:30 to 21:30 UTC the rain sums from the different runs are much closer to each other; only two runs are distinct outliers (see Fig. 2). Also the precipitation forecast image for 21:30 UTC (Fig. 4(f))

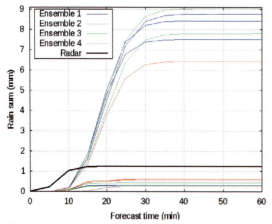

Fig. 1 Precipitation time series of ensemble runs and radar observation (black line) for one grid point over the 60-min forecast interval, starting at 1 August 2010, 17:00 UTC.

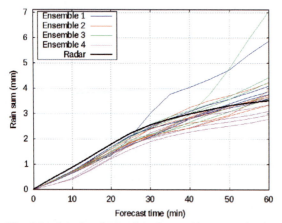

Fig. 2 Precipitation time series of ensemble runs and radar observation (black line) for one grid point over a 60 minutes forecast interval, starting at 1 August 2010, 20:30 UTC.

corresponds much better to the radar image (Fig. 4(c)). The average POD at this time is 0.70 and the FAR is 0.32.

Ensemble based prediction of forecast quality

A parameter was searched to predict the quality of a single forecast, derived from the set of ensemble forecasts. In analogy to the verification, forecast quality cannot be objectively assessed with only one parameter. Thus, the same approach as for the evaluation of the forecast quality was chosen, using the contingency table (see Table 1). We concentrated on the parameters FAR and POD as they allow for a quick estimation of the forecast quality. The quality parameters predicted from the ensembles are referred to here as equivalent POD and equivalent FAR. For deriving the equivalent parameters, the rain sum after the 60 min forecast was calculated at each grid point and the median and the mean deviation from the ensemble median were determined.

For the classification in the contingency table, forecast and observation are replaced by the ensemble median and a single forecast, differing from the median by the mean deviation. From the total number of hits, misses, false alarms and correct negatives in the test area, the equivalent POD and the equivalent FAR are calculated as formulated by equations (1) and (2).

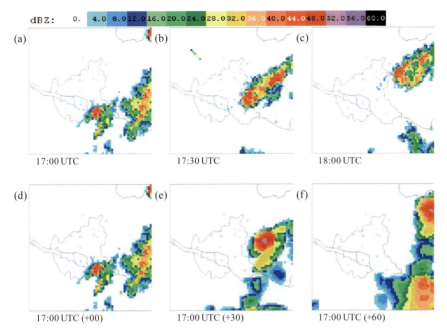

Fig. 3 (a)–(c) Images from radar Hamburg on 1 August 2010 from 17:00 to 18:00 UTC. (d)–(f): The corresponding forecast images, using forecast Method 4.

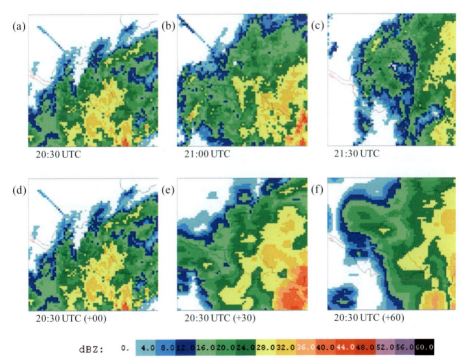

Fig. 4 (a)–(c): Images from radar Hamburg on 1 August 2010 from 17:00 to 18:00 UTC. (d)–(f): The corresponding forecast images, using forecast Method 4.

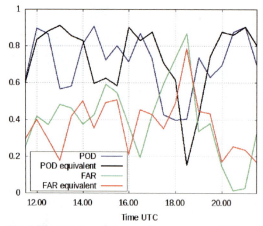

Fig. 5 Time series of the quality parameters POD and FAR and the equivalent quality parameters derived from the ensemble forecasts.

The comparison of the quality parameter results is shown in Fig. 5 as 10-h time series (1 August 2010) with values displayed every 30 min. During the whole time interval, the parameters are defined because rain was measured and forecast somewhere in the test area. The POD and FAR show considerable variations and the equivalent POD and FAR show a similar behaviour. The correlation between POD and equivalent POD is 0.42 and the correlation between FAR and equivalent FAR is 0.68; both correlations are significant within the 90% confidence interval. The equivalent POD and FAR for the forecast times described above at 18:00 UTC and 21:30 UTC show again a lower than average and a higher than average forecast quality, respectively.

SUMMARY AND CONCLUSION

With four presented forecast methods and the uncertainties in the initial parameters (displacement vectors and growth rates) sets of ensemble forecasts were constructed. These can give a better estimate of the current weather situation than a single forecast. From the ensemble runs, quality parameters were derived (equivalent POD and equivalent FAR) that allow a quick estimation of the forecast quality and that can be helpful for the release of warnings. The method employed can be extended to other quality parameters, depending on which are useful for the particular warning purpose. Before using the parameters for a warning system, they have to be tested on longer time intervals and additionally for extreme events. It can be expected that the quality of the predicted uncertainty parameters in particular depends on how good the forecast ensemble reflects the spectrum of the possible precipitation distributions.

REFERENCES

Bowler, N. & Seed, A. W. (2006) STEPS: A probabilistic precipitation forecasting scheme which merges an extrapolation nowcast with downscaled NWP. *Quart. J. Roy. Met. Soc.* 132, 2127–2155.

Donaldson, R. J., Dyer, R. M. & Krauss, M. J. (1975) An objective evaluator of techniques for predicting severe weather events. In: *Preprints: 9th Conf. Severe Local Storms*, Norman, Oklahoma, Amer. Met. Soc., 321–326.

Doswell, C. A., Davies-Jones, R., Keller, D. L. (1990) On summary measures of skill in rare event forecasting based on contingency tables. *Weather & Forecasting* 4, 97–109.

Flueck, J. A. (1987) A study of some measures of forecast verification. In: *Preprints, 10th Conf. Probability and Statistics in Atmospheric Sciences*. Edmonton, Alberta, Amer. Met. Soc., 69–73.

Germann, U., Berenguer, M., Sempere-Torres, D. & Zappa, M. (2009) REAL – ensemble precipitation estimation for hydrology in a mountainous region. *Quart. J. Roy. Met. Soc.* 135, 445–456.

Hydro & Meteo (2009) The SCOUT Documentation, version 3.30. Lübeck, 69 pp.

Michelson, D., Einfalt, T., Holleman, I., Gjertsen, U., Friedrich, K., Haase, G., Lindskog, M. & Jurczyk A. (2005) Weather radar data quality in Europe – quality control and characterization. Review. COST Action 717, Working document, Luxembourg 2005.

Weather Radar and Hydrology
(Proceedings of a symposium held in Exeter, UK, April 2011) (IAHS Publ. 351, 2012).

317

Application of Error-Ensemble prediction method to a short-term rainfall prediction model considering orographic rainfall

EIICHI NAKAKITA[1], TOMOHIRO YOSHIKAI[2] & SUNMIN KIM[2]

1 *Disaster Prevention Research Institute, Kyoto University, Gokasho, Uji 611-0011, Kyoto, Japan*
nakakita@hmd.dpri.kyoto-u.ac.jp
2 *Graduate School of Engineering, Kyoto University, Kyoto-Daigaku-Katsura 615-8510, Kyoto, Japan*

Abstract In order to improve the accuracy of short-term rainfall predictions, especially for orographic rainfall in mountainous regions, a conceptual approach and a stochastic approach were introduced into a radar image extrapolation using a Translation Model. In the conceptual approach, radar rainfall measurements are separated into orographic and non-orographic rain fields by solving physically-based equations, including additional atmospheric variables, such as vertical wind velocity. In the stochastic approach, mean bias of current prediction errors was estimated and used to adjust mean prediction bias. Furthermore, the vertical wind velocity was updated with the mean bias for convective rainfall. As a result, 1-h prediction accuracy in mountainous regions was much improved for the case study. In the future, improved updating procedures can be expected to allow more accurate predictions.
Key words short-term rainfall prediction; orographic rainfall; ensemble forecasting prediction; prediction error

INTRODUCTION

Many mountainous regions exist in Japan, making orographic rainfall events common. Orographic rainfall persists and may cause localized heavy rainfall to induce flash floods and sediment disasters. However, the translation model, a short-term rainfall prediction model proposed by Shiiba *et al.* (1984), does not distinguish orographic rainfall in observed radar rainfall measurements. Therefore, the model does not anchor predicted rainfall in the mountainous regions. In order to prevent disasters and reduce damage due to orographic rainfall, orographic rainfall must be identified and its mechanism introduced into a short-term rainfall prediction procedure.

Of the many studies related to short-term rainfall prediction and orographic rainfall, Nakakita & Terazono (2008) proposed a physically-based method in order to consider orographic rainfall; they developed the Tatehira Model (Tatehira, 1976) and applied this new scheme to the Translation Model. This model improved prediction accuracy in mountainous regions by separating radar rainfall measurements into orographic and non-orographic rain fields. In another study, Kim *et al.* (2009) introduced a stochastic prediction method, the Error-Ensemble Prediction Method, using a prediction error structure found in the Translation Model. They estimated future prediction error using characteristics of current prediction error. This procedure improved prediction accuracy by modifying the spatial prediction bias derived from the Translation Model.

In this study, a new method is proposed, combining Nakakita & Terazono's (2008) physical scheme with the stochastic scheme introduced by Kim *et al.* (2009). The error ensemble scheme of Kim *et al.* (2009) may be very useful in reducing the uncertainty of the previous method by Nakakita & Terazono (2008). This utility results from the fact that the error sources are specified, and the error is related to orographic rainfall extracted by analysing the prediction error. Therefore, the physical model is updated using the value of orographic error, and the orographic rainfall is predicted more accurately. The concern here is not with cases in which there is rapid growth and decay of rainfall fields because the method proposed in this study has limitations. Therefore, the model was applied to a typhoon rainfall event in June 2004 as a case study. The target area is Japan's Kinki region. The performance of the new method is discussed, focusing on the Odaigahara mountainous region and the southwestern part of the Kii peninsula.

METHODOLOGY

Translation Model

In this study, the Translation Model by Shiiba *et al.* (1984) is used for deterministic prediction of short-term radar rainfall. In this model, the horizontal rainfall intensity distribution, $R(x, y)$ with the spatial coordinate (x, y) at time t is defined as follows:

$$\frac{\partial R(x,y)}{\partial t} + u_r(x,y)\frac{\partial R(x,y)}{\partial x} + v_r(x,y)\frac{\partial R(x,y)}{\partial y} = \delta(x,y) \tag{1}$$

where $(u_r(x, y), v_r(x, y))$ are advection velocities along (x, y), respectively, and $\delta(x, y)$ is the rainfall growth-decay rate with time. As with other similar equations for the rainfall intensity distribution, characteristics of the Translation Model are defined by the vectors $u_r(x, y)$, $v_r(x, y)$ and $\delta(x, y)$, which are specified on each grid as follows:

$$u_r(x,y) = c_1 x + c_2 y + c_3, \quad v_r(x,y) = c_4 x + c_5 y + c_6, \quad \delta(x,y) = c_7 x + c_8 y + c_9 \tag{2}$$

The parameters c_1–c_9 sequentially are optimized using the square root information filter and rainfall observations. Equation (1) is approximated by the centred difference scheme on a regular horizontal grid, with $\Delta x \times \Delta y$ (e.g. 3×3 km) being the grid size and Δt (e.g. 5 min) the time resolution. The Translation Model provides expected rainfall movement under an assumption that the vectors $u_r(x, y)$ and $v_r(x, y)$ are time invariant for the next several hours. It further assumes that there is no growth–decay of rainfall during that time (i.e. $\delta(x, y) = 0$ for all x and y).

Physically-based methodology for orographic rainfall

Tatehira (1976) proposed a physically-based method for calculating orographic and non-orographic rain fields from observed radar rainfall measurements. In this method, the orographic effect is calculated based on the seeder-feeder mechanism. In other words, non-orographic hydrometeors are produced in upper layers, and the hydrometeors capture cloud droplets produced by the orographic effect in low layers. The strong rain-bands stagnated near the mountain top (orographic rainfall) in a radar-observed rainfall distribution are estimated using additional atmospheric variables. The flux of cloud water content L (g/m^3) during air parcel ascent along with an observed wind is expressed by the following equation:

$$\frac{dL}{dt} = -cL - a(L - L_c) + WG - WL(\partial \ln \rho_v / \partial z) \tag{3}$$

where ρ_v is the density of water vapour (g/m^3), c is the ratio of cloud drops captured by seeder hydrometeors (non-orographic rainfall), a is the ratio of precipitation particles to cloud drops, L_c is the threshold amount of water content before conversion into precipitation begins (g/m^3) and G is the amount of saturated water vapour ρ_s increased by the ascent of a saturated air parcel (g/m^4) (i.e. $-d\rho_s / dz$). Finally, W is the vertical wind velocity (m/s), which is estimated as an inner product of horizontal wind and gradient of height (i.e. $\nabla h \cdot \mathbf{u}$).

In this study, representative values of these atmospheric variables are estimated in the seven layers at heights of 200, 400, 1000, 2000, 3000, 4000, and 5000 m in the σ-vertical coordinate system. The initial conditions of the atmospheric variables are estimated in each layer according to the method of Nakakita *et al.* (1997). Air temperature, horizontal wind, and relative humidity are estimated from GPV (Grid Point Value) data, and surface horizontal wind is estimated from AMeDAS (Automated Meteorological Data Acquisition System) data from the Japan Meteorological Agency (JMA). The data sets are interpolated into every 10 min and 3 km horizontal resolution and the vertical layers. Nakakita and Terazono provide further details regarding the estimation method (2008).

In equation (3), the first and second terms on the right-hand side show that the amount of water content decreased. The third term expresses the condensation of the water vapour as the air parcel ascends a unit distance. The last term changes vertically due to the influence of atmospheric compressibility. The last term can be ignored and the equation integrated over time along the GPV wind flow. In this way, cloud water content is analytically calculated from atmospheric variables in each layer and grid.

Nakakita & Terazono (2008) calculated the orographic rainfall intensity R_o (mm/h) from the radar observation rainfall R_{radar} (mm/h). Here, it is calculated as follows:

$$R_o = \frac{L_{in} + WG\Delta t - L_{out}}{\Delta t} \times 3.6 \times h_k, \qquad (k = 1, 2, \dots 7) \tag{4}$$

where h_k is the thickness of each layer (m), Δt is a timescale (s) during which an air parcel passes one mesh, and L_{in} and L_{out} are amounts of cloud water content (g/m^3) in an inflow and outflow side mesh, respectively. They assumed that the ratio c of cloud drops captured by raindrops is as follows:

$$c = 0.6778 R_n^{0.731} \times 10^{-3} \tag{5}$$

The radar observation rainfall R_{radar} is interpreted to be the summation of orographic rainfall R_o and non-orographic rainfall R_n, so that the following equation holds true:

$$R_{radar} = R_o + R_n \tag{6}$$

The orographic rainfall is supposed to be a function of non-orographic rainfall and is calculated by solving the simultaneous equations (4)–(6) in multi-atmospheric layers.

Translation Model with consideration of orographic rainfall

Nakakita & Terazono (2008) introduced the physically-based method to the Translation Model. Figure 1 shows the procedure for short-term rainfall prediction by using the Translation Model with consideration of orographic rainfall. First, cloud water content analytically is calculated from atmospheric variables in each layer and grid. Supposing that the radar observation rainfall (R_{radar}) is that of the lowest layer (200 m height), it is separated into orographic (R_{o1}) and non-orographic (R_{n1}) rainfall by solving equations (4)–(6). It is assumed that the non-orographic rainfall (R_{n1}) can be expressed as the sum of the orographic rainfall (R_{o2}) and non-orographic rainfall (R_{n2}) in the upper layer. In this way, orographic rainfall of each layer can be calculated from the lowest to the highest layer repeatedly. Second, under the assumption that non-orographic rainfalls are not affected by orographic effects, only the non-orographic rain fields are advected to lead time (60 min ahead). At the lead time, new atmospheric variables are calculated from GPV and AMeDAS data. This method is utilized because atmospheric variables, such as water vapour and wind, are expected to change, even if orographic rain fields are stagnant for several hours. Supposing that the advected non-orographic rain fields are of the top layer, new orographic rain fields are calculated by solving equations (4)–(6) from the top layer to the lower layer. Lastly, prediction rain fields are calculated by combining the orographic and non-orographic rain fields.

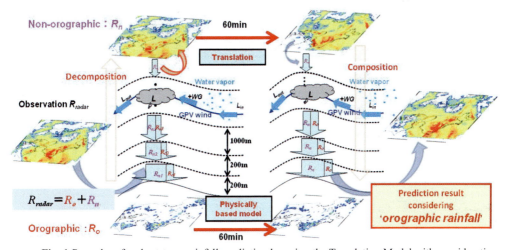

Fig. 1 Procedure for short-term rainfall prediction by using the Translation Model with consideration of orographic rainfall.

Ensemble forecasting method by using prediction error field

Kim *et al.* (2009) introduced a stochastic prediction method considering the spatial distributions of error in prior prediction rain fields in the Translation Model. The spatial distribution of an absolute prediction error E_k was considered with the forms:

$$E_k = R_{o,k} - R_{p,k} \qquad (k = 1, 2, ..., n) \tag{7}$$

They are calculated and statistically analysed, where n is the number of error fields, $R_{o,k}$ is an observed radar rainfall field, and $R_{p,k}$ is a predicted rainfall field using the Translation Model. The calculation in equation (7) is conducted for each corresponding grid cell of $R_{o,k}$ and $R_{p,k}$. In this study, 11 consecutive error fields (5-min resolution) are calculated, and the prediction lead time is 60 min, as shown in Fig. 2. The current characteristics of the 60-min prediction error can be presented by basic statistics fields, such as the ensemble mean field $\overline{E}$. If the spatio-temporal characteristics of the prediction error last for 60 min, the statistical characteristics of the error on the prediction lead time $t_0 + \Delta T$ are similar to the characteristics of the current error. In this study, the bias-modified prediction field $R_{p,0}$ is calculated by adding the ensemble mean field $\overline{E}$ to the prediction field R_p at time $t_0 + \Delta T$:

$$R_{p,0} = R_p + \overline{E} \tag{8}$$

RESULTS AND ANALYSIS OF THE ERROR FIELDS

Data sets and target area

The radar data come from the C-band composite radar, provided by Japan's MLIT (Ministry of Land, Infrastructure and Transportation) at a resolution of 1 km and 5 min. Composite data primarily are produced from measurements at the lowest elevation angle. We assumed this radar data as the lowest layer's data (at 200 m height). Topology data come from the digital elevation model (DEM) produced by Japan's Geospatial Information Authority at a resolution of 1 km. The target area is the Kinki region of Japan. Using the Gauss-Krueger projection method, the datasets are projected onto 123 × 123 grid cells (369 × 369 km with 3-km resolution) (Masaharu, 2002). The water vapour and wind data are estimates from GPV and AMeDAS data by JMA.

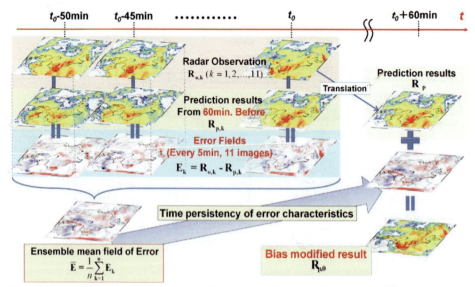

Fig. 2 Schematic drawing of the error-ensemble prediction algorithm (Kim *et al.*, 2009).

This study utilizes the 2004 rainfall event for Typhoon 0406. The initial time of the prediction is 11:00 h 21 June 2004 (JST). The prediction lead time is 60 min, and the prediction error fields at 10:10 h to 11:00 h (every 5 min) were used for generating the ensemble mean. Attention was focused on the southwestern part of the Kii peninsula.

RESULTS

Figure 3 shows observed rainfall and prediction results using each methodology at 12:00 h in the Typhoon 0406 event. It was found that the bias adjustment method (e) improves prediction accuracy, especially for the mountainous area where orographic rainfall could not be predicted using only a physically-based model (c). Table 1 shows the validation of prediction accuracy in the southwestern part of Kii peninsula with Root Mean Square Error and Correlation Coefficient. The accuracy of 60-min prediction in the mountainous region was improved by combining the physical and stochastic methodologies.

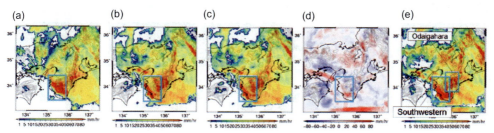

Fig. 3 Examples of prediction results by each method in the June 2004 event. (a) the observed rainfall at prediction lead time; (b) prediction result by simple Translation Model (w/o. orographic); (c) prediction result by the physically-based model (w/orographic); (d) error mean field by the ensemble method (w/orographic); (e) bias adjustment prediction result by using the error mean field (w/orographic).

Table 1 Root mean square error (RMSE) and correlation coefficient (CC) in southwestern part of the Kii peninsula.

Method	Translation	Physically-based	Bias adjustment (translation)	Bias adjustment (physically-based)
RMSE	27.13	26.26	26.08	25.83
CC	0.04	0.23	0.44	0.49

FURTHER MODIFICATION (*W*-UPDATING METHOD)

This study has proposed a new method combining Nakakita and Terazono's (2008) conceptual scheme with the stochastic scheme by Kim *et al.* (2009). In the ensemble forecasting method, we simply added the error mean bias to the prediction result. However, Typhoon rainfall events are mainly convective rainfall, which was not considered in the Tatehira model. Therefore, in order to modify the prediction bias using a physically-based procedure, we attempted to update the vertical wind velocity W with the mean bias so that convective rainfall could be considered.

We assumed that the vertical wind velocity consists of two kinds of wind: geographical wind velocity Wg and convective wind velocity Wc. We also assumed that positive error mean bias in the mountainous region is caused by convective orographic rainfall. Therefore, we tried to get information from the error mean bias to improve the accuracy of convective wind velocity Wc. The error mean value was distributed according to the thickness of layers h_k as follows:

$$R_{oi,j,k} + \overline{E}_{i,j} \times \frac{h_k}{H} = \left(\frac{L_{in\ i,j,k} + Wg_{i,j,k}G\Delta t - L_{out\ i,j,k}}{\Delta t} + Wc_{i,j,k}G \right) \times 3.6 \times h_k \tag{9}$$

where H is the total layer thickness (m). Figure 4 shows the results of applying the W-updating method. Using the W-updating method, vertical wind velocity greatly is increased along with the orographic rainfall in the mountainous region. In order to compare the two methods, we calculated the error mean bias using both methodologies. It was found that simple moving averages of ensemble mean bias in the Odaigahara region was 28%, reduced by the W-updating method. Therefore, we can reduce prediction bias by using the W-updating methodology. In the future, more accurate physically-based predictions can be expected via further improvement of the W-updating method.

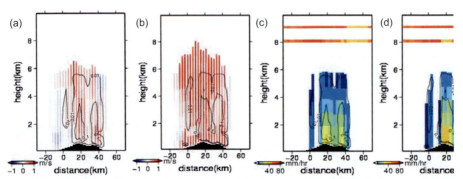

Fig. 4 Vertical profiles of vertical wind velocity and orographic rainfall calculated by physically-based method ((a), (c)) and W-updating method ((b), (d)).

CONCLUSIONS

In order to improve the short-term prediction accuracy of rainfall, especially for orographic rainfall in mountainous regions, a conceptual approach and a stochastic approach were introduced into a radar image extrapolation method with a Translation Model. We found the following:

–   Combining the physical and stochastic methodologies improved the accuracy of 60-min prediction in mountainous regions.
–   The amount of orographic rainfall in mountainous region increased, and the W-updating method improved prediction accuracy.
–   More accurate physically-based predictions can be expected through improvement of the W-updating method.

Acknowledgements The authors are grateful to the Kinki Regional Development Bureau, Ministry of Land, Infrastructure, Transport and Tourism, Japan for providing radar rainfall data. We further extend our gratitude to the anonymous reviewers for their suggestions and comments, which led to the improvement of this paper.

REFERENCES

Kim, S., Tachikawa, Y., Sayama, T. & Takara, K. (2009) Ensemble flood forecasting with stochastic radar image extrapolation and a distributed hydrologic model. *Hydrol. Processes* 23, 597–611.

Masaharu, H. (2002) Deviation of the formula of the Gauss-Kruger projection – an attempt to clarify required preliminary mathematical knowledge. *Map* 39(4), 31–37 (in Japanese).

Nakakita, E., Ikebuchi, S., Nakamura, T., Kanmuri, M., Okuda, M., Yamaji, A. & Takasao, T. (1996) Short-term rainfall prediction method using a volume scanning radar and GPV data from numerical weather prediction. *J. Geophys. Res.* 101(D21), 26181–26197.

Nakakita, E. & Terazono, M. (2008) Short-term rainfall prediction taking into consideration nonlinear effect of non-orographic rainfall on orographic rainfall. *Ann. J. Hydraul. Engng*, JSCE 52, 331–336 (in Japanese).

Shiiba, M., Takasao T. & Nakakita, E. (1984) Investigation of short-term rainfall prediction method by a translation model. *Ann. J. Hydraul. Engng*, JSCE 28, 423–428.

Tatehira, R. (1976) Orographic rainfall computation including cloud-precipitation interaction. *Tenki* 23, 95–100 (in Japanese).

Weather Radar and Hydrology
(Proceedings of a symposium held in Exeter, UK, April 2011) (IAHS Publ. 351, 2012).

323

On the DWD quantitative precipitation analysis and nowcasting system for real-time application in German flood risk management

TANJA WINTERRATH[1], WOLFGANG ROSENOW[2] & ELMAR WEIGL[1]

1 *Deutscher Wetterdienst, Department of Hydrometeorology, Frankfurter Straße 135, 63067 Offenbach, Germany*
tanja.winterrath@dwd.de
2 *Deutscher Wetterdienst, Department of Research and Development, Michendorfer Chaussee 23, 14473 Potsdam, Germany*

Abstract Quantitative precipitation analyses and forecasts with high temporal and spatial resolution are essential for hydrological applications in the context of flood risk management. Therefore, the Deutscher Wetterdienst, together with representatives of the water management authorities of the German federal states have developed high-resolution quantitative precipitation analysis and nowcast products based on the combination of surface precipitation observations and weather radar-based precipitation estimates. Gauge adjustment is performed hourly, making use of 16 operational radar systems and approximately 1300 conventional precipitation measurement devices. The nowcast algorithm is based on the advection of precipitation elements based on the mapping of precipitation patterns in successive image data. The subsequent quantification makes use of the latest adjustment process. Additional information about the precipitation phase, required for the determination of the discharge efficiency of precipitation, is retrieved by combining various observational and model data with the radar-based forecasts. The nowcasting system is supplemented by a qualitative hail forecast.

Key words radar; precipitation; gauge adjustment; nowcasting; quantification; precipitation phase; real time; risk management; DWD; Germany

INTRODUCTION

With the installation of a weather radar network and an automatic precipitation station network, completed in 2000, the Deutscher Wetterdienst (DWD) laid the basis for temporally and spatially highly-resolved quantitative precipitation analyses and forecasts in real time. Radar online adjustment and nowcasting systems combine these two data sets and supply quantitative precipitation analyses and quantitative precipitation forecasts, respectively.

PRECIPITATION ANALYSIS

The operational weather radar network of the DWD currently comprises of 16 sites where precipitation scans are performed every 5 min. The reflectivity fields are combined into a high-resolution composite covering Germany and transformed into precipitation amounts by applying a categorized Z-R relationship. Gauge adjustment is performed hourly making use of approximately 1300 conventional precipitation measurement devices. Currently, three different adjustment approaches are applied: the adjustment of radar data applying factors and differences to the gauges, respectively, and a merging algorithm (the latter not described within the scope of this paper). Based on a real-time verification using 20% of the available raingauges, a pixel-wise weighted combination of the differently adjusted products is performed in order to provide the "best-adjusted" QPE product.

The method

The derivation of precipitation from radar is not based on a direct measurement of the near-ground precipitation, but on the reflected signals of hydrometeors at higher altitudes. The radar reflectivity Z can be transformed into precipitation rates R by applying empirical Z-R relationships. However, the quality of precipitation data derived purely from radar measurements is insufficient for quantitative applications, e.g. in flood risk management. The RADOLAN method combines the advantages of two measurement principles: gauges record the ground precipitation at the measurement sites, while weather radar provides information about the areal distribution of the precipitation. The numerical method producing a synthesis of both data sources is called gauge-adjustment. The data base for the online gauge-adjustment is formed by the operational weather radar network with 16 C-band systems (see Fig. 1) and the common gauge network of the DWD and the German federal states with currently approximately 1100 automatic raingauges (see Fig. 2). The retrieval of quantitative precipitation information is performed on an hourly basis using the precipitation sum of both measurement devices. In addition, online data of hourly precipitation from about 200 sites in neighbouring countries are incorporated.

Prior to the actual adjustment several pre-processing steps are applied. In detail, these are the correction of orographic shading, the application of a categorized Z-R relationship, the quantitative composition of the single scans with radii of 150 km to a mosaic of 900 km × 900 km and a grid length of 1 km covering Germany, the statistical suppression of clutter, the smoothing of gradients, and the pre-adjustment by a factor mapping the ratio of the maxima of the gauge and the radar data. The gauge precipitation data are compared to the corresponding precipitation values of the pre-adjusted radar composite. Two different adjustment procedures are applied in the following:

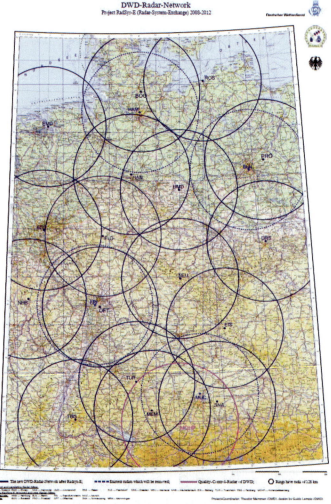

Fig. 1 The radar network of the Deutscher Wetterdienst. The precipitation nowcasting system is based on the German high-resolution radar composite.

one using factors and the other differences between the gauge and the radar data. For this, the radar value with the smallest deviation from the gauge value is taken from the corresponding radar pixel and the adjacent eight pixels. The resulting factors and differences for each measurement site, respectively, are interpolated onto the composite grid and applied to adjust the radar precipitation data. To determine which of the two adjustment procedures is the better one, the absolute difference to the particular adjusted field is calculated for so-called control stations that have been separated from the data collective beforehand. The accordingly calculated weighted sum of the two adjusted precipitation fields provides the best gauge-adjusted product, which is named RW of RADOLAN (Bartels *et al.*, 2008). This is provided to the clients in real time with a delay of less than 30 minutes.

Case study On 26 July 2008, in a warm humid air mass over western Germany, local thunderstorms developed that led to catastrophic precipitation. The city centre and western areas of Dortmund were especially badly affected. At the station Dortmund University, operated by the private company Meteomedia, thus not used in the operational adjustment procedure, the 3-hourly total measured rainfall between 12:51 and 15:50 UTC exceeded 200 mm (Schenk & Wehry, 2008). The spatial distribution of this precipitation event can only be represented by quantitative gauge-adjusted radar precipitation data (see Fig. 3). The original radar data overestimated the gauge-measured precipitation by nearly 100 mm/3 h, while the operational RADOLAN product shows an underestimation of 70 mm/3 h, 60 mm of which can be attributed to the second hour alone. Table 1 lists the precipitation sums (RR) derived from the different methods. The gauge-adjusted precipitation data agrees well with the station data for the first as well as the last hour of the considered time interval. The application of the interpolated adjustment values derived on the

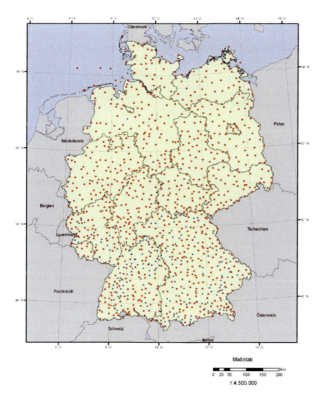

Fig. 2 Automatic raingauges of the Deutscher Wetterdienst (red dots) and the German federal states (blue triangles) that are operationally used in the online gauge-adjustment procedure RADOLAN.

Interpolated gauges   **Unadjusted radar**   **Gauge-adjusted radar**

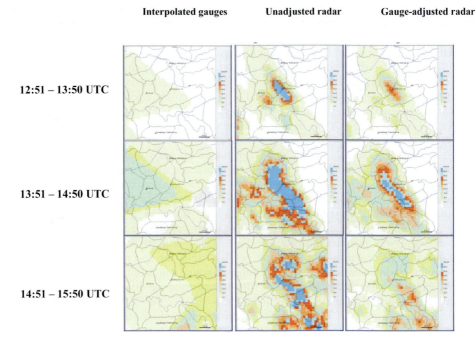

12:51 – 13:50 UTC

13:51 – 14:50 UTC

14:51 – 15:50 UTC

Fig. 3 1-hour precipitation sums for the intense severe weather period on 26 July 2008 retrieved from interpolated gauge data (left), radar data with application of a Z-R relationship (centre) and gauge-adjusted radar data (right) for an area of approx. 40 km × 40 km with Dortmund in the centre.

Table 1 Summary of accumulated rainfall (RR) for the severe weather event in Dortmund on 26 July 2008.

Data source	RR (mm/h) (12:50–13:50)	RR (mm/h) (13:51–14:50)	RR (mm/h) (14:51–15:50)	RR (mm/3 h) (12:51–15:50)
Radar	100.1	126.9	71.9	298.9
Gauge-adjusted	43.1	66.5	19.0	128.6
Gauge	50.3	125.2	24.7	200.2

basis of several online gauges, however, leads to an underestimation of the precipitation amount for the second hour that is most likely due to the attenuation of the radar beam in the centre of this extraordinarily intense event.

PRECIPITATION NOWCAST

The high-resolution radar composites serve as input for a tracking algorithm providing precipitation nowcasts for a lead time of up to 2 h. The tracking algorithm is based on the advection of precipitation elements using the displacement vector field derived from the mapping of precipitation patterns in successive image data. The forecast amounts of stratiform precipitation are quantified by making use of the most recent gauge-adjustment procedure, assuming persistence of the precipitation frequency distribution, resulting in quantitative precipitation forecasts for the next two hours with an update frequency of 15 min. Additional information about the precipitation phase, required for the determination of the discharge efficiency of precipitation, is retrieved by combining various observational and model data with the radar-based forecasts. The nowcasting system is supplemented by a qualitative hail forecast. The quantitative temporal and spatial high-resolution forecasts are provided to clients within a few minutes for use in flood risk management.

The development of the nowcast system is performed within the project RADVOR-OP (radar-based realtime precipitation forecast for operational use: precipitation nowcast system).

Tracking

The core of the nowcasting procedure is a tracking algorithm, which estimates the motion field from the latest sequential radar images and subsequently extrapolates the current analysed precipitation fields along these vectors into the future. For this purpose, structures of different spatial scales are identified and mapped in sequential radar images. The interpolation of the resulting vectors as well as the usage of a weighted mean of current and recent vector fields additionally guarantee a continuous displacement. While the focus is on the meso-β scale (spatial extent: 25–250 km) covering mainly stratiform precipitation, the application of the algorithm on the meso-γ scale (spatial extension: 2.5–25 km) also allows the detection of shorter-lived convective structures. Every 5 min, the tracking algorithm provides qualitative-quantitative forecasts of 5-min precipitation totals, as well as hourly totals derived from summing up the 12 individual forecasts, with subsequent smoothing to avoid herringbone patterns for fast moving cells. To minimize the impact of clutter pixels, an additional progressive clutter filter is applied (Winterrath & Rosenow, 2007).

Quantification

As described in the previous chapter, the quality of radar-derived precipitation data without gauge-adjustment is insufficient for use in flood risk management. To optimize the quality of forecast precipitation sums, the forecast products of the tracking algorithm are quantified, making use of the latest gauge-adjustment procedure. The basic assumption is the persistence of the frequency distribution of the precipitation since the most recent gauge-adjustment. The frequency distributions of the current gauge-adjusted as well as the forecast precipitation fields are approximated by functions of the Weibull type. Due to the asymptotic behaviour of the Weibull function a linear approximation is applied for large intensities. The resulting transformation function between the two frequency distributions is applied to the forecast 1-h precipitation distribution. This quantification method is justified if the dynamical change of the precipitation with time is smaller than the change due to the gauge-adjustment, as is typical for stratiform events. However, the quantification is suspended in the case of convective events determined by a cell detection algorithm.

Precipitation phase

At the heart of forecasting the aggregate phase of precipitation is the so-called satellite and radar weather system. These methods basically combine two approaches applying empirical decision criteria: ground-based point observations of the weather (the synoptic code) are spatially interpolated according to areas of similar reflectance derived from the MSG infrared channel. In addition, NWP forecasts of the height of the freezing level in relation to the underlying topography determine an estimate of the snow line. The qualitative determination of the expected aggregate state is applied to the radar-based precipitation nowcasts. In conclusion, a supplementary product giving the percentage of solid precipitation is provided in addition to the quantitative precipitation nowcasts in real time every 15 min for use in flood risk management, as well as for warning issues.

Case study On 6 January 2011, a precipitation band with regions of snow as well as freezing rain passed through Germany followed by a rise in temperature and subsequent rain. Figure 4 shows the predicted percentage of solid precipitation (colour coding) with reference to the predicted precipitation sum (isolines) given in mm/h for the 1-h forecast performed at 08:30 UTC. In addition, the observed weather at 09:00 UTC is given as symbols. The displacement of the precipitation band is very well represented by the nowcast system. Beside the regions of solid precipitation in the northern and southern parts marked by the purple-coloured areas, the area of freezing rain is also clearly captured by the forecast as liquid precipitation.

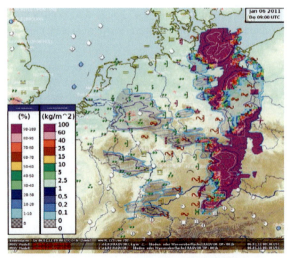

Fig. 4 Quantitative 1-h precipitation nowcast for 6 January 2011, 09:30 UTC (isolines); forecast percentage of solid precipitation (colour coding); synoptic observation at 09:00 UTC (pictograms).

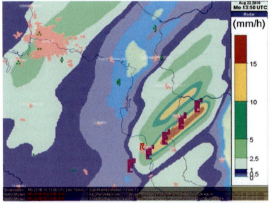

Fig. 5 Quantitative 1-h precipitation nowcast for 23 August 2010, 13:50 UTC (colour coding); forecast possibility for hail (purple areas); synoptic observation at 13:00 UTC (pictogram).

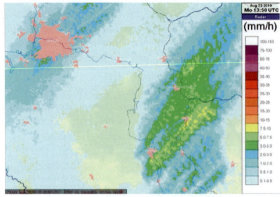

Fig. 6 Gauge-adjusted radar-based 1-h precipitation analysis for 23 August 2010, 13:50 UTC. Note the different colour tables in Figs 5 and 6.

Hail warning

During wintertime, the forecast of hail is determined based on the satellite weather described above. For convective hail storms in summer, a different approach has been developed based on radar-based precipitation intensity and lightning activity. In this case, if a certain probability for the occurrence of hail is exceeded, the potentially affected area is flagged.

Case study On 23 August 2010, a hailstorm passed the city of Cottbus, approx. 100 km southeast of Berlin, in a northeasterly direction. The quantitative 1-h nowcast of precipitation performed at 12:45 UTC with a valid time of 13:50 UTC (Fig. 5) predicts the location, timing and intensity of the precipitation event well, in good agreement with the gauge-adjusted precipitation analysis (Fig. 6). The purple areas mark the forecast of the thunderstorm track with possible hail, with the five separated areas being due to the limited temporal resolution of 15 min. The predicted thunderstorm was verified by a synoptic observation at 13:00 UTC at Cottbus, although no hail was reported in this case.

OUTLOOK

Within the scope of the current project RadSys-E, the Doppler C-band radar systems are successively being replaced by dual polarisation radar systems accompanied by an extension of the network to 17 operational radar sites within the scope of the next years (see Fig. 1). In addition, the extension of the considered area to the whole hydrological catchment of Germany by integrating foreign radar systems as well as gauge data is continuously being promoted in cooperation with the neighbouring national weather services. The nowcast system is continuously being improved, e.g. by the introduction of sounding information for the determination of the aggregate state, while new methods using polarisation information will open new possibilities in the future.

Acknowledgements The projects RADOLAN and RADVOR-OP have been financed by the water management authorities of the German federal states (Länderarbeitsgemeinschaft Wasser).

REFERENCES

Bartels, H., Weigl, E., Reich, T., Lang, P., Wagner, A. Kohler, O. & Gerlach, N. (2004) Projekt RADOLAN: Routineverfahren zur Online-Aneichung der Radarniederschlagsdaten mit Hilfe von automatischen Bodenniederschlagsstationen (Ombrometer). http://www.dwd.de/RADOLAN (in German).

Schenk, F. & Wehry, W. (2008) Der Unwetter-Regen im Ruhrgebiet vom 26. Juli 2008. Beiträge zur Berliner Wetterkarte. http://wkserv.met.fu-berlin.de/edu/wetterkarte/beilagen/2008/unwetterruhr.pdf (in German).

Winterrath, T. & Rosenow, W. (2007) A new module for the tracking of radar-derived precipitation with model-derived winds. *Adv. Geosciences* 10, 77–83.

Aspects of applying weather radar-based nowcasts of rainfall for highways in Denmark

M. R. RASMUSSEN[1], S. THORNDAHL[1] & M. QUIST[2]

1 *Aalborg University, Department of Civil Engineering, Sohngaardsholmsvej 57, DK-9000 Aalborg, Denmark*
mr@civil.aau.dk
2 *Danish Road Directorate, Thomas Helstedsvej 22, DK-8660 Skanderborg, Denmark*

Abstract This work investigates three different approaches to nowcasting rainfall for highways. The simplest method is based on using the observed precipitation field at the beginning of the trip. The most developed nowcast is based on a COTREC nowcaster, which is dynamically adjusted to online raingauges. The nowcasts are performed with a lead time of up to 2 h. The average speed on Danish highways varies between 110 and 130 km/h. As a result, the performance of the nowcast is dependent on the direction of the precipitation and the direction and speed of the road users, as well as the type of precipitation.

Key words nowcast; highway; traffic conditions; weather radar

INTRODUCTION

The Danish road network consists of 73 331 km of roads. Of these, 3790 km are state roads and are considered major lines of transportation. Although these only represent 5% of the total network, 45% of all traffic moves along them. Poor visibility, or even the possibility for aquaplaning due to heavy rain, can therefore jeopardise the safety of many road users. Poor visibility is not just a question of rain intensity, but also of drop size.

The winter road weather has been of great interest for many years, due to the risk of icing and general slippery roads (Symons & Perry, 1997; Pedersen *et al.*, 2010). Weather radar has been used to predict wetness and visibility (Seliga & Wilson, 1995). The consequences of subsequent flooding of the road is also of great relevance (Versini *et al.*, 2010).

The purpose of the project reported here is to investigate the potential of a warning system that can warn motorists of severe rainfall ahead affecting the traffic. The warning should help motorists lower their speed, so they can better handle low visibility and the risk of aquaplaning. In order to achieve this, a system for nowcasting dedicated to road applications is under development in Denmark. The system utilises weather radars in combination with the COTREC method for nowcasting (Li *et al.*, 1995). The model used here has previously been used for forecasting runoff in storm drainage systems (Thorndahl *et al.*, 2009, 2010). The system is modified to forecast along lines (roads) and take the movement of the motorist relative to the rain into account. Two very different events (stratiform and convective) are used to illustrate the potential of this approach.

The strategy for warning motorists about severe weather depends on the type of information system used. It can generally be split into two different approaches:

- Static warnings from electronic boards placed along the road, where motorists get general information on the traffic and weather conditions ahead.
- Continuous information streamed to individual cars, e.g. through the car's GPS and Traffic Management Channel (TMC) or through smartphone applications.

The first approach requires a nowcast which shall cover the time before the next electronic board is visible, while the latter approach can be continuously updated.

METHOD

Radar-based rainfall extrapolation methods uses cross-correlation between two or more adjacent radar images to predict the future rain. Even with advanced methods for predicting the dynamics of the storm, the nowcast will always be less accurate than the actual radar observation. For stratiform storms the algorithm performs better than for convective storms. This is primarily due to the difference in precipitation variability.

The challenge of radar-based nowcast algorithms, such as COTREC used here, is that the available lead time is restricted by the operational range of the radar and the type of storm. The C-band radars operated by the Danish Meteorological Institute (DMI) have a maximum range of 240 km (Gill *et al.*, 2006). However, only within a 75-km range of the radar can data be used for quantitative precipitation estimation. The acceptable lead time for COTREC-based nowcast is, by experience, around 0.5–2 h. Local storm cells can appear and disappear within 30 min. The predictive capability is therefore less than 30 min in these cases. Precipitation intensity is used here as the primary variable, as it affects both aquaplaning and visibility. When predicting the rain intensity at one fixed point, the accuracy is to some extent decided by the type of storm. Due to the dynamics of rain, radar data beyond a range of 15–20 km can suffer attenuation in convective conditions, and may only prove to be useful for more stratiform events. However, some types of convective cells can tend to succeed each other in a way that even nowcasts for convective events can be helpful.

For a moving target, e.g. a car on the highway, the situation is slightly different. Besides the storm moving with a velocity (e.g. 30 km/h), the car will move up to 130 km/h. If the storm and the car are moving in opposite directions, this yields a relative velocity of 160 km/h. Within the 30-min time limit of the nowcast model for convective storms, the car would have travelled 80 km relative to the storm. For this reason, the travelling speed and direction can influence the quality of the nowcast for individual cars. As the GPS knows the exact position of the car, it is possible to automatically relate the position of rain to the position of the car. This approach is already used to warn about traffic accidents ahead or severe traffic congestion.

It should also be noted that a 5-min warning on extreme precipitation in many cases will be sufficient for motorists to lower their speed.

The nowcast strategies used here are divided into three classes depending on the complexity of the nowcast performed:

- The *"One Shot"* method assumes that the precipitation pattern is fixed at the start of the trip and will not change during the duration of the trip. The situation is equal to the motorist looking at the latest weather radar images before the trip and observing the location and intensity of the precipitation.
- The *COTREC* method uses a nowcast of the precipitation field at the start of the trip based on the motion for the precipitation pattern. Two or more consecutive radar images are used to calculate the vector field for precipitation. It is assumed that this motion will be the same during the trip. It is also assumed that the original distribution of high and low intensity areas will not change, but just be advected according to the vector field.
- The *updated* method assumes that the motorist has access to a GPS with TMC (Traffic Management Channel) technology, which is continuously updated. This could also be a Smartphone (Android, iPhone, Windows mobile or other) with built-in GPS and data access. Here, the data are streamed out to the motorist's mobile application every 10 min, depending on the update rate of the radar. The data can either be used as static images which will represent the next 10 min, or could be forecast with COTREC the next 10 min before a new radar image is available.

In all three cases it is assumed that the motorist travels with a specified velocity from one position to another, so the simple Taylor hypothesis can be assumed:

$$x = \overline{U}t \tag{1}$$

where x is the distance travelled, $\overline{U}$ is the average speed on the distance travelled and t is the time since the start of the trip. So the distances are directly related to the time the motorist has been under way. Therefore 10 min is equivalent to 20 km, if the car travels at 120 km/h. This is the reason why distance is chosen over time as the variable when comparing methods (in figures presented later).

It should be mentioned that neither of the methods compensates for the time lag between the measurement of the precipitation and the delivery of data. This can typically be of the order of 5–10 min, depending on the scanning strategy of the radar and calculation load for the nowcast algorithms. For this reason, the updated method will serve as a reference for the "One Shot" and COTREC methods.

The test examples used here are for 28 May 2010 when a number of showers passed over the northern part of Denmark, and a more stratiform event on 18 November 2009 (see Fig. 1).

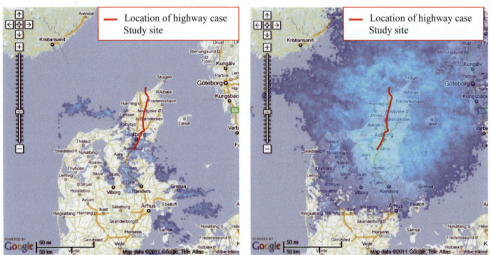

Fig. 1 Precipitation moving north at 02:00 h 28 May 2010 and 10:30 h 18 November 2009 (Google Maps, 2011).

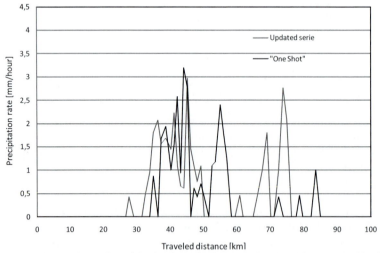

Fig. 2 Comparison of precipitation intensity between the "One Shot" method and the updated method along the E45 from Hirtshals towards Århus on 28 May 2010.

The precipitation on 28 May 2010 moves along the E45 from Århus to Hirtshals. On 18 November the rain is more evenly distributed and the extent of the rain in Fig. 1 in part reflects the range of the Sindal radar. It is assumed in both cases that a car travels along the E45 at an average speed of 120 km/h from Hirtshals towards Århus along a 100-km stretch of highway. Data are from the C-band radar at Sindal operated by the Danish Meteorological Institute (Gill *et al.*, 2006). The spatial resolution is 2 km × 2 km and the time between images is 10 min.

RESULTS AND DISCUSSION

As can be seen from Fig. 2, the "One Shot" method has only limited capability in this case. The onset of rain along the trip is predicted 10 min later than observed. Also, in the later part (after 50 km) of the trip, both timing and intensity is not well predicted.

If the same trip is predicted using COTREC, a more positive result is observed (Fig. 3). This result indicates that both timing and intensity levels are better predicted than with the "One Shot" method. It is worth noticing that the last part of the prediction the COTREC method is located approximately 80 km from the starting point, which is equivalent to a 40 min nowcast. Neither method is able to predict the rise in intensity which is observed in the last part of the event.

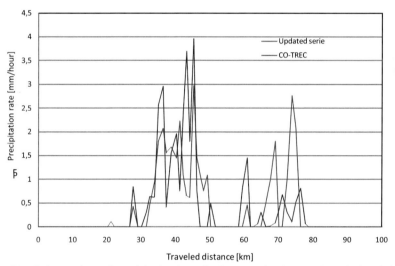

Fig. 3 Comparison of precipitation intensity between the COTREC method and the updated method along the E45 from Hirtshals towards Århus on 28 May 2010.

The primary reason for the mis-timing of the "One Shot" method is that the precipitation is moving north, when the traffic is moving south. The precipitation is moving at approximately 30 km/h along the highway. This accounts for the approximately 10–15 min time shift in the first part of the trip. Later, the precipitation field turns towards the east and the time shift is thereafter constant.

In the case of the more stratiform event (Figs 4 and 5), the COTREC method, surprisingly, only performs slightly better than the "One Shot" method. Over the first 20 km of the trip, the three methods all predict the same, as it is based on the observation for the first 10 min. In the middle part, some higher-intensity spikes occur. The "One Shot" method predicts two spikes, but the precipitation level rather poorly. The COTREC method gives a better average value, but fails to predict the high intensity spike. For the last half of the trip, all three methods agree much better. This is more the expected situation for more stratiform rainfall. However, the COTREC method fades in the very last part of the trip. This is due to only one weather radar being used for this test case, with data for the area outside the radar measuring range (where there are no rain data) being extrapolated over the last part of the highway stretch. This can be avoided using a mosaic from several radars.

The test also indicates that applying the COTREC method within the 10-min interval between radar images could be a benefit. Although a 10-min nowcast is not very long, it represents a travelled distance of 20 km. By using the nowcast method to forecast every minute, the

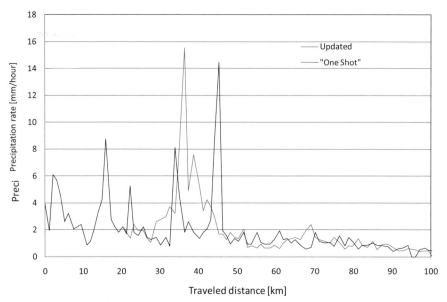

Fig. 4 Comparison of precipitation intensity between the "One Shot" method and the updated method along the E45 from Hirtshals towards Århus on 18 October 2009.

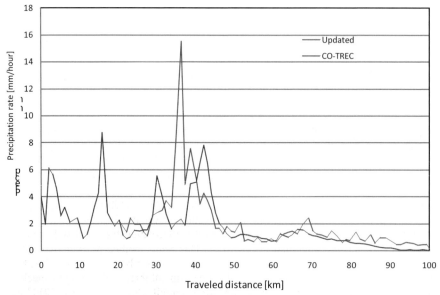

Fig. 5 Comparison of precipitation intensity between the COTREC method and the updated method along the E45 from Hirtshals towards Århus on 18 October 2009.

corresponding distance would be 2 km, which would be more in balance with the resolution of the radar images.

The results are not verified against ground observations, due to the lack of suitable raingauges along the highway. An alternative could be to measure the rain intensity from a moving car. However, most raingauges are not suited to measure rain intensities at 130 km/h wind speeds. The only verification which might be used could be a mobile vertical pointing radar (Rasmussen *et al.*, 2008) or a mobile disdrometer. This approach will be pursued further in the project. Finally, this test case has only been illustrated for two distinctly different rainfall situations. The data amount used is presently very small. As a consequence, an analysis of several years of data is ongoing.

One weakness of the methods is that rainfall intensity is not directly understandable for the average motorist. It is proposed that the rain intensity is translated into a resulting visibility. This should be improved by an estimate of the water depth on the road surface. A high water depth results in both potential aquaplaning and a decrease in visibility due to water spray.

CONCLUSION

The two examples illustrate that it is possible to give fairly accurate nowcasts of precipitation to a motorist moving along a highway. Both the updated method and the COTREC perform similarly. However, it has to be considered whether the methods could be improved by a finer nowcast time resolution.

Acknowledgement The authors acknowledge the Danish Meteorological Institute for the radar data used in this study.

REFERENCES

Gill, R. S., Overgaard, S. & Bøvith, T. (2006) The Danish weather radar network. In: *Proc. Fourth European Conf. on Radar in Meteorology and Hydrology.*

Li, L., Schmid, W. & Joss, J. (1995) Nowcasting of motion and growth of precipitation with radar over a complex orography. *J. Appl. Met.* 34(6), 1286–1300.

Pedersen, T. S., Petersen, C., Sattler, K., Mahura, A. & Sass, B. (2010) Physiographic data for road stretch forecasting. In: *15th SIRWEC Conference* (Québec, Canada).

Rasmussen, M. R., Vejen, F., Gill, R. & Overgaard, S. (2008) Adjustment of C-band radar with mobile vertical pointing radar. In: *WRaH 2008, Int. Symp. Weather Radar and Hydrology* (Grenoble, France).

Seliga, T. A. & Wilson, L. L (1995) Applications of the NEXRAD radar system for highway traffic management. In: *Proc. 27th Conf. on Radar Meteorology* (Vail, Colorado, USA), 176–178.

Symons, L. & Perry, A. (1997) Predicting road hazards caused by rain, freezing rain and wet surfaces and the role of weather radar. *Met. Appl.* 4(1), 17–21.

Thorndahl, S., Rasmussen, M. R., Grum, M. & Neve, S. L. (2009) Radar based flow and water level forecasting in sewer systems – a Danish case study. In: *Proc. 8th Int. Workshop on Precipitation in Urban Areas* (St. Moritz, Switzerland).

Thorndahl, S. L., Rasmussen, M. R., Nielsen, J. E. & Larsen, J. B. (2010) Uncertainty in nowcasting of radar rainfall: a case study of the GLUE methodology. In: *Proc. Advances in Radar Technology, Sixth European Conf. on Radar in Meteorology and Hydrology* (Sibiu, Romania).

Versini, P.-A., Gaume, E. & Andrieu, H. (2010) Application of a distributed hydrological model to the design of a road inundation warning system for flash flood prone areas. *Nat. Hazards Earth Syst. Sci.* 10, 805–817.

336

Weather Radar and Hydrology
(Proceedings of a symposium held in Exeter, UK, April 2011) (IAHS Publ. 351, 2012).

Use of radar data in NWP-based nowcasting in the Met Office

SUSAN BALLARD[1], ZHIHONG LI[1], DAVID SIMONIN[1], HELEN BUTTERY[1], CRISTINA CHARLTON-PEREZ[1], NICOLAS GAUSSIAT[2] & LEE HAWKNESS-SMITH[1]

1 *Met Office, Dept of Meteorology, University of Reading, Reading RG6 6BB, UK*
sue.ballard@metoffice.gov.uk
2 *Met Office, FitzRoy Road, Exeter EX31 3PB, UK*

Abstract The Met Office is developing an hourly cycling 1.5 km resolution NWP-based nowcast system (0–6 h), principally for prediction of convective storms for flood forecasting. Test suites were run on a domain covering southern England and Wales nested in a UK 4 km domain. These have used 3D-Var or 4D-Var in combination with latent heat nudging of radar-derived precipitation rates and humidity nudging based on 3D cloud cover analyses. An example shows the precipitation forecast compared to the current extrapolation nowcast system. The results of a trial, showing positive impact of Doppler radar winds out to about 5 h on forecasts of precipitation from the 3D-Var system, are presented. The paper also discusses work underway to allow assimilation of rain-rates and radar reflectivity within the variational schemes and the potential to measure the low-level humidity impact on radar refractivity as an additional source of data to improve flood forecasting.

Key words NWP; variational data assimilation; radar; flood forecasting; UK; nowcasting; Doppler winds; reflectivity

INTRODUCTION

Increasing availability of computer power and nonhydrostatic models has made Numerical Weather Prediction (NWP) at convective scales of 1–4 km resolution a reality for national meteorological services in the past few years.

The Met Office is developing an hourly cycling high-resolution (1.5 km) NWP forecasting system for nowcasting over southern England (for domain see Fig. 2) and already has an operational 4 km model for the UK and a routine 1.5 km UK model, both with 3-hourly cycling 3D-Var and 6-hourly forecasts to 36 h. These are based on the Met Office Unified Model (Davies *et al.*, 2005) and variational data assimilation system (Lorenc *et al.*, 2000; Rawlins *et al.*, 2007) including latent heat and moisture nudging (Macpherson *et al.*, 1996; Jones & Macpherson, 1997; Dixon *et al.*, 2009).

Nowcasts need to be produced rapidly (ideally to be available to customers within 15 min of data time) and to match the observations as closely as possible in the early hours of the forecast, so are more challenging for data assimilation. Currently the nowcasting forecasts are produced from the UKPP (UK Post-Processing) system using STEPS (Bowler *et al.*, 2006) at 2 km resolution using extrapolation forecasts in the first few hours merged with downscaled 4 km resolution NWP forecasts at later times.

The NWP-based nowcast system is principally for prediction of convective storms for flood forecasting but also to take over from existing nowcasting and site-specific forecasting techniques. Boundary conditions will be provided by the 6 hourly 1.5 km resolution forecast system. NWP has the potential to improve nowcasts of precipitation through its ability to predict the full life-cycle of convection, and other forms of precipitation, from initiation, growth and decay using the equations of motion and representation of physical processes.

An hourly analysis and forecast system has been run over a limited number of cases of summer rain and convection using conventional data and 3D-Var or 4D-Var. The current aim is to use 4D-Var, if affordable and beneficial, in a real-time continuously running NWP-based nowcast system to exploit the high temporal and spatial coverage of radar Doppler radial winds and reflectivity or derived surface rain-rates directly within the variational analysis scheme. Direct use in 4D-Var should allow optimum extraction of information through interaction with other data sources and the potential to modify the dynamical and physical forcing of precipitation and convective storms. Research is also proceeding to investigate the background errors, balances and control variables required for use in convective-scale data assimilation. This paper presents initial results from this work planned to deliver a prototype nowcasting system by 2012.

COMPARISON OF NWP-BASED NOWCASTING AND UKPP FORECASTS

Figure 1 shows UKPP forecasts of surface rain-rate valid at 21:00 UTC 3 June 2007 at range 3, 2 and 1 h compared with the radar-derived surface rain-rates. At T+3 the UKPP forecast mainly comes from the UK 4 km NWP forecast which tends to produce convective cells at too large a scale and not enough light banded convection, as can be seen in the forecast of the band of convection in the east. At T+2 the STEPS scheme has produced the line of convection in the east, but it is too narrow, presumably due to convergence in the diagnosed advection. By T+1 a reasonable forecast has been produced.

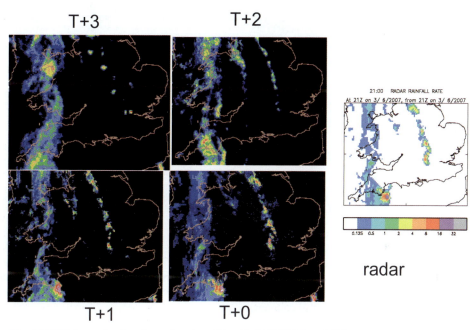

Fig. 1 UKPP forecasts of surface rain-rate valid at 21:00 UTC 3 June 2007.

Figure 2 shows the 1.5 km NWP-based nowcast forecasts of surface rain-rate for this case valid at 21:00 UTC 3 June 2007. This has used latent heat nudging based on 15-min radar-derived rain-rates and nudging of hourly humidity derived from 3D cloud cover analyses used in conjunction with hourly cycles of 4D-Var of conventional observations over 1-h time windows. The NWP forecasts improve at shorter lead times due to the benefit of data assimilation, in particular latent heat nudging of surface precipitation rates derived from radar data available every 15-min in the 1 h data assimilation window. In comparison with the UKPP forecasts, the 1.5 km NWP forecast has a better representation of the rain band in the east at both T+3 and T+2. The representation of the rain in the southwest of England is poorer than from UKPP. However, for the first case compared with UKPP, and without optimization of the data assimilation scheme and exploitation of more frequent conventional and radar Doppler radial wind observations, this is a very promising result. The forecast in the southwest can be improved by assimilation of 15-min time frequency GPS water vapour data, but that at present is only available 90 min after data time, so cannot be used in a nowcast system. In future it is hoped that GPS data can be available closer to data time, but another potential source of water vapour information is from radar refractivity exploiting the interaction of the radar beam with ground clutter: work is underway with Reading University to investigate the potential for obtaining this information from the UK weather radar network.

T+3   **T+2**

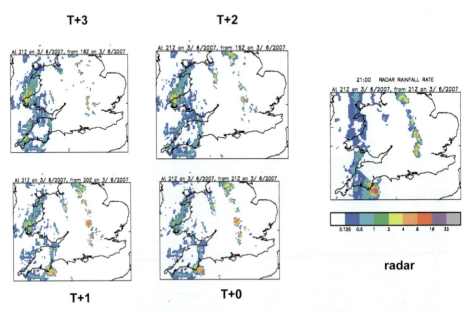

radar

T+1   **T+0**

Fig. 2 4D-Var assimilation with latent heat nudging of radar-derived surface rain rates. Fields valid at 21:00 UTC 3 June 2007. The forecasts show the full domain of the prototype NWP-based nowcast system.

Direct assimilation of the radar-derived surface precipitation rates within 4D-Var is being investigated, as well as direct assimilation of the radar reflectivity or indirect assimilation of radar reflectivity through derived temperature and humidity increments derived from external 1D-Var of multiple-beam elevations in vertical columns.

RADAR RADIAL DOPPLER WINDS

Weather radar potentially provides a high-resolution source of wind data from the Doppler returns from hydrometeors and insects. Four radars currently produce Doppler radial winds operationally in the south of England every 5 min when there is precipitation (see Fig. 3). Code has been developed to allow their processing, quality control, monitoring and super-obbing, and data assimilation. Trials have been run to investigate their impact on forecasts using 3D-Var. It was hoped to make use of Doppler radial wind scans valid at analysis time operational in the UK models in 2011 (i.e. 3-hourly data in the 3-hourly 3D-Var cycles replacing use of hourly VAD winds from the same radars). Work with Reading University was undertaken to look at the potential for use of winds derived from insect returns in fine weather (Rennie *et al.*, 2010) and this work will continue in collaboration with CAWCR in Australia. Radar returns only give radial winds (i.e. in the direction of the radar beam) rather than 3D wind components, so the additional information in areas of overlapping radars (dual Doppler) may increase the impact of winds in those locations.

Radar-derived surface rain-rates 21.00 UTC 8/1/08   Radar radial Doppler winds m/s towards radar

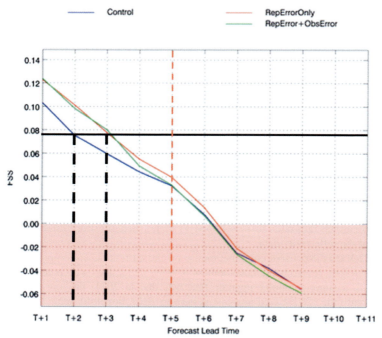

Fig. 3 Example of radar radial Doppler wind coverage over the UK on 8 January 2008.

Fig. 4 ΔFSS for 0.2 mm hourly precipitation accumulation threshold at a scale of 55 km. Positive values of ΔFSS represent skill from the forecast. Blue is the control forecast, red includes Doppler radial winds with a specified observation error derived from O-B statistics and referred to as representativeness error, green includes Doppler radial winds with the representativeness error plus the superobservation standard deviation as the observation error.

The Met Office has for a while had radial winds available from four radars in southern England. The radars each have five elevation scans – the majority being 1, 2, 4, 6 and 9 degrees and one near London being 1, 2, 4, 5 and 5.5 degrees. The Doppler winds are available out to about 100 km range and this provides a small amount of dual or triple Doppler overlap in southern England. Much work has been done on specification of observation errors and investigating the impact of superob variances and/or representativeness errors.

The impact of the data have been assessed in the 1.5 km southern England hourly cycling 3D-Var prototype nowcasting system. For the initial tests only the radar scans closest to the analysis hour were selected from each radar for assimilation. Initial subjective and objective verification looks promising. The location and coverage of rain is affected and improved in some situations. Figure 4 shows the increase in fractional skill score (Roberts & Lean, 2008) of hourly precipitation accumulations, due to assimilation of radial wind from the four Doppler radars in the nowcasting domain for southern England, calculated over four case studies of about 10–19 cycles, each using hourly cycling 3D-Var and 11 hour forecasts. Figure 4 shows an hours gain in skill in the earliest hours of the forecasts and a positive impact out to 6 h, the period of impact being limited by the small size of the domain and the spread of information from the boundary conditions into the domain.

In the future more work will be done on the specification of observation error and to test the impact of hourly and higher time frequency data in the 4D-Var prototype nowcasting system.

The impact and areal influence of observations in an analysis depends on the background error correlation and covariances (i.e. the short-range forecast error) at the analysis time in addition to the observation error itself. The forecast quality, as well as its sensitivity to inclusion of different or new observation types, greatly depends on the assumed background errors; so work is underway to try to define improved background errors for the 1.5 km resolution forecasts in terms of both correlations between variables, length scales and error variances. These need to extract longer time-scale, synoptic-scale information as well as short-scale information observed in high spatial and time resolution observations such as radar. This is very challenging work.

CONCLUSIONS AND FUTURE WORK

The 1.5 km resolution NWP is showing promise for very short-range prediction of convection over the UK. This paper has shown the potential of use of radar-derived rain-rate through latent heat nudging on top of 4D-Var and the benefit of radar Doppler radial wind in 3D-Var. 4D-Var has the potential to exploit higher time-frequency observations and to extract more information from them than 3D-Var. Therefore, research is continuing on use of high time-frequency Doppler radial winds and direct use of radar-derived surface rain-rate and direct and indirect use of multi-elevation volume scan reflectivity in 4D-Var. Although latent heat nudging is still showing benefit in forecasts, it cannot correctly represent resolved convection where latent heat release occurs in different locations to surface rain-rate; so it is hoped to obtain benefits from direct 4D-Var or indirect 1D-Var use of the reflectivity data.

Unfortunately 4D-Var is expensive computationally on the current super-computer available to the Met Office. Therefore, research and development is being undertaken with both 3D-Var and 4D-Var systems. With a super-computer upgrade that was due in 2011, the aim was to start running a prototype real-time NWP-based nowcast system in 2012, hopefully with 4D-Var if the upgrade provides sufficient computer resources.

Due to the tight time constraints on forecasts, it may be necessary to move away from use of a time window centred on the analysis time to one finishing at the analysis time; especially as data sources such as GPS providing information on low-level humidity are currently only available 90 min after data time from the most accurate processing system, although less accurate but more timely data may become available. There are many sources of information on different variables (e.g. GPS, radar refractivity, satellite imagery and surface observations for low-level humidity) and the usefulness of the different data sources will be investigated to provide an optimum system. Currently experiments have been undertaken nested in the UK 4 km NWP forecast system, but ultimately the nowcasting system will be embedded in the UK 1.5 km NWP forecast system.

The whole UK network of weather radars will gradually be updated to produce Doppler radial winds and also dual-polarization data and radar refractivity measurements. The use of radar data in NWP high-resolution variational data assimilation has the potential to improve on current extrapolation-based nowcasts. To achieve this we need high-quality radar data and good quality control (see companion papers by Georgiou *et al.*, 2012 and Harrison & Curtis, 2012), fast processing (techniques and computer power), careful specification of observation and forecast background error covariances and correlations through the scientific design of the data assimilation system and a good representation of the dynamical and microphysical processes in the NWP forecast model. In future it is hoped to exploit ensemble techniques in both the data assimilation and production of forecasts. If there is sufficient computer power available for hourly NWP forecasts to 12 h, that will provide potential for 6 h of 1-hourly lagged ensemble forecasts and a measure of the predictability of the nowcast forecasts.

REFERENCES

Bowler, N. E., Pierce, C. E. & Seed, A. W. (2006) STEPS: A probabilistic precipitation forecasting scheme which merges an extrapolation nowcast with downscaled NWP. *Quart. J. Roy. Met. Soc.* 132, 2127–2155.

Davies, T., Cullen, M. J. P., Malcolm, A. J., Mawson, M. H., Staniforth, A., White, A. A. & Wood, N. (2005) A new dynamical core for the Met Office's global and regional modelling of the atmosphere. *Quart. J. Roy. Met. Soc.* 131, 1759–1782.

Dixon, M, Li, Z, Lean, H., Roberts, N. & Ballard, S. P. (2009) Impact of data assimilation on forecasting convection over the United Kingdom using a high-resolution version of the Met Office Unified Model. *Monthly Weather Rev.* 137, 1562–1584.

Georgiou, S., Ballard, S., Gaussiat, N. & Harrison, D. L. (2012) Quality monitoring of the UK network radars using synthesised observations from the Met Office Unified Model. In: *Weather Radar and Hydrology* (Proc. 8th Int. Symp., Exeter). IAHS Publ. 351. IAHS Press, Wallingford, UK (this volume).

Harrison, D. & Curtis, A. (2012) Use of long-term diagnostics to monitor and asses the effectiveness of radar data processing techniques and the quality of precipitation estimates. In: *Weather Radar and Hydrology* (Proc. 8th Int. Symp., Exeter). IAHS Publ. 351. IAHS Press, Wallingford, UK (this volume).

Jones, C. D. & Macpherson, B. (1997) A latent heat nudging scheme for the assimilation of precipitation data into an operational mesoscale model. *Meteorol. Appl.* 4, 269–277.

Lorenc, A. C., Ballard, S. P, Bell, R. S., Ingleby, N. B., Andrews, P. L. F., Barker, D. M., Bray, J. R., Clayton, A. M., Dalby, T., Li, D., Payne, T. J. & Saunders, F. W. (2000) The Met Office global 3-dimensional variational data assimilation scheme. *Quart. J. Roy. Met. Soc.* 126, 2991–3012.

Macpherson, B., Wright, B. J., Hand, W. H. & Maycock, A. J. (1996) The impact of MOPS moisture data in the U.K. Meteorological Office mesoscale data assimilation scheme. *Monthly Weather Rev.* 124(8), 1746–1766.

Rawlins, F. R., Ballard, S. P., Bovis, K. R., Clayton, A. M., Li, D., Inverarity, G. W., Lorenc, A. C. & Payne, T. J. (2007) The Met Office global 4-dimensional data assimilation system. *Quart. J. Roy. Met. Soc.* 133, 347–362.

Rennie, S. J., Dance, S. L., Illingworth, A. J., Ballard, S. P. & Simonin, D. (2011) 3D-Var assimilation of insect-derived doppler radar radial winds in convective cases using a high-resolution model. *Monthly Weather Rev.* 139, 1148–1163.

Roberts, N. M. & Lean, H. W. (2008) Scale-selective verification of rainfall accumulations from high-resolution forecasts of convective events. *Monthly Weather Rev.* 136, 78–97.

Quality monitoring of UK network radars using synthesised observations from the Met Office Unified Model

SELENA GEORGIOU[1], NICOLAS GAUSSIAT[1], DAWN HARRISON[1] & SUE BALLARD[2]

1 *The Met Office, Exeter, UK*
selena.georgiou@metoffice.gov.uk

2 *Advanced Nowcasting Research Group, Met Office, Department of Meteorology, Univ. Reading, Reading RG6 6BB, UK*

Abstract The Met Office radar processing system delivers quality-controlled radar reflectivities to NWP. Quality information and radar reflectivity data are then passed to the Observation Processing System (OPS) where synthetic observations are calculated using model fields interpolated at the exact observation locations. Long-term statistical comparison between synthetic and real observations has the advantage of identifying individual radar calibration problems through relative comparisons with other radars. The effectiveness of the forward modelling of the reflectivity can also be evaluated through absolute statistical comparisons. Presented here is an analysis of statistical information derived from the quality monitoring system. Included is a description of the contribution made to the radar signal bias with range as a result of the combined effects of the bright band, attenuation by rain and clouds and beam broadening. The results are used to demonstrate that the atmospheric gaseous attenuation makes a significant contribution to the overall range bias, and it is therefore beneficial to account for this within the radar site processing.

Keywords quality control; unified model; data assimilation; model verification; gaseous attenuation

BACKGROUND

Description of the radar data processing in the Observation Processing System (OPS)

For the purpose of comparing radar and model data, it is necessary to average the rays and gates of the radar data to a similar resolution as that of the model. Within the OPS, this is achieved by superobbing the observations, by spatially averaging the difference between the real observations and their synthetic equivalents produced using the model background fields. It is important to note that in the current forward model, the attenuation due to atmospheric gases, rain and cloud is not accounted for. The bright band and beam broadening effects are also not currently simulated. At present only mixing ratios for rain and ice are used, however, the forward model could be extended to incorporate graupel mixing ratios. Snow is treated in the same way as ice within the model. Figure 1 shows the processing that occurs within the OPS setup to produce the output presented in Fig. 2 and the netCDF files used to produce the statistical analysis of the O-B values.

Quality flags, produced by the radar pre-processing, are used in the superobbing process such that only observations that pass the minimum quality criteria are taken into account. Gaussiat (2008a) illustrates the benefits of using clutter flags and describes the importance of superobbing

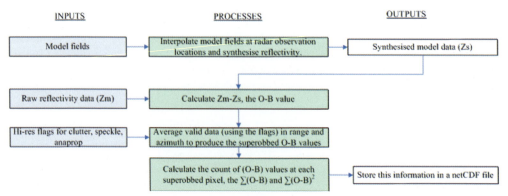

Fig. 1 Ops reflectivity monitoring process flow diagram.

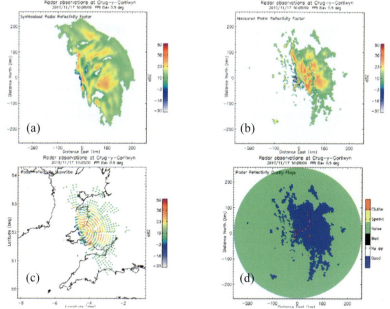

Fig. 2 Examples of the modelled and observed reflectivity values at the Crug-y-Gorllwyn radar on 17 November 2010. (a) Modelled reflectivity factor. (b) Measured radar reflectivity factor. (c) Superobs of the measured radar reflectivity. (d) Quality flags associated with the measured radar reflectivity.

to reduce the amount of available data and minimise representative errors by best matching the sampling volume of the model and superobbed data. The key steps in producing the pre-processed radar reflectivities are detailed in Gaussiat (2008b). This involves flagging of radar bins affected by ground clutter, spikes and partial beam blockage. Figure 2 shows the observed and modelled observations, as well as the quality flags used and the superobbed reflectivities.

Development of the OPS monitoring statistics for different purposes

The superobbed differences between real (radar) reflectivity measurements and the synthesised (model) reflectivity (the O-B value) are useful for three key areas of analysis: data assimilation (initialising the model with high resolution radar data), model verification and radar quality control (QC). Each key area of analysis uses the same information over different time windows and observation domains. Currently, the O-B statistics (bias and standard deviation) are calculated for each superobbed cell, and netCDF files including this information are produced eight times a day. For the purpose of data assimilation, the statistics are used to characterise observation representativeness errors. Therefore, short time windows are used to match the data assimilation window. Although scans are produced every 5 min, only the 3 hourly scans valid for the specific monitoring times are used. Model verification is carried out by calculating the long term averaged O-B value over each model grid point. For example, by looking at statistical differences it might be possible to quantify differences in the way the model acts over the sea compared with over land. With a sufficiently long archive of O-B values, misrepresentation of the precipitation under different conditions could also be determined.

Recently the use of the OPS monitoring system has been extended to radar QC. Harrison & Curtis (2011) describe the development of the Met Office Radar Data Quality Management System (RDQMS) in more detail. The OPS reflectivity monitoring system is under development and forms part of this project. Currently, problems that may exist with individual radars are inferred by looking at their relative calibration to other radars in the UK network. This is achieved by averaging the O-B values over each range gate to determine how the bias varies with range.

METHODOLOGY

Finding the O-B value for one data time step alone is not useful as there are too few observations, and large bias variations between the data used at different times. As a result, any relative comparisons between radars would be messy and unrepresentative of anomalies. It is therefore important for the bias and other statistics such as the RMSE to be calculated using longer periods. Long term averaged results identify underlying issues regarding radar functionality, such as calibration problems. Biases arising from the bright band effect, beam broadening and attenuation can also be observed using long term averaged data. Calculation of the optimal averaging periods is detailed within this section.

Determining the optimum time period for averaging the observations over

The average O-B value has been calculated over time periods, T, ranging from 5 to 140 days. T varies inversely with the variance in the averaged data (Fig. 3). However, as T increases, there is a trade off between the time required to produce the statistics and the practical usefulness of the information produced. It is therefore important to derive a minimum value for T, which produces data with acceptably low variance. Figure 3 shows that the variance between the radars is <1dB when time periods greater than 80 days are used. The variance is ~3dB when T = 5 to 10 days. When the variance is minimised by using a sufficiently long time period, the relative variance between the radars is largely representative of their differences in calibration.

By extrapolation of the data in Fig. 3(a), the variance becomes constant at around 0.5dB (T ~200 days), which shall be taken as the optimal average variance. Over such a long period, the relative calibration between the radars is less useful as any radar calibration issues may already have been identified using statistics calculated over a shorter time period. Instead these statistics could be used as a long term average measure of the relative difference between radar and model reflectivity profiles. The shorter term averaged results for each radar could be assessed against this benchmark. If there are any significant anomalies, then it may be evident of a recent calibration error arising at that particular radar. It is therefore important that various time periods are used to calculate the statistics on a daily rolling basis. The shortest sensible time period has been determined as ~30 days, where the average variance is around 1.5 dB (Fig. 3(a)). In order to further reduce the response time, the use of shorter time periods, e.g. 5 days, could be useful for identifying large changes in relative calibration that result from radar hardware changes.

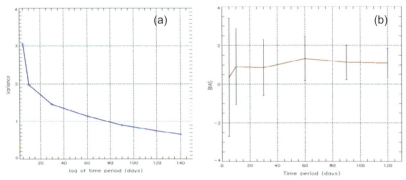

Fig. 3 (a) Variance between observed and modelled reflectivities *vs* the period of archived data used at 0.5 degree scan elev. (b) Corresponding bias between modelled and observed reflectivities and its associated error.

ANALYSIS

Within this section, we describe the usefulness of the OPS monitoring system in identifying radar calibration issues and potential limitations to the current system of forward modelling.

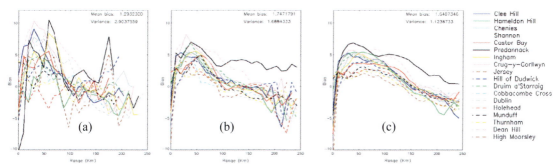

Fig. 4 Corrected bias between model and radar observations at 0.5 degree scan elevation calculated for a single window starting in May 2010 over time periods: (a) 5 days, (b) 30 days, (c) 90 days.

Identification of radar calibration issues

Figure 4(c) shows an example of the averaged bias over 90 days. Here, data are smoothed over a long enough time period such that the relative calibration of the radars can be assessed within 1dB.

The radar that shows the greatest relative difference in calibration is Predannack, situated in Cornwall, for which the majority of observations are over the sea. In the UK, the prevailing wind direction is from the southwest. It is therefore important that this radar is well calibrated such that the fronts heading towards the UK can be observed as accurately as possible. The relative difference in bias is significantly greater for this radar when compared with the others (greater than the standard deviation). It is possible that part of the relative bias might be due to possible differences between how the model performs over the sea and how it performs over land. The model used to use a different precipitation efficiency over land and sea. However this has not been investigated and may only account for a small part of the observed difference in calibration. Recently the precipitation efficiency has been made a function of a prognostic aerosol referred to as "murk" so it will be interesting to see if this has resulted in any change in bias at Predannack. Ingham, High Moorsley and Dublin are also showing evidence of slight mis-calibration when compared with the other radars. Polar plots of the accumulated O-B values will be developed to identify where exactly the greatest anomalies exist and to better assess why they are arising.

The effect of atmospheric gaseous attenuation on bias with range

The radar signal is affected by varying attenuation due to clouds and hydrometeors and also by constant attenuation due to gases present in the atmosphere. Gaseous attenuation is an absorption process and results in the radar signal amplitude being reduced by the gases present in its transmission path. This attenuation varies directly with frequency. It is also temperature, pressure and humidity dependent (Bean & Dutton, 1968; Liebe, 1985).

The Met Office C-band weather radars typically operate at a frequency of 5.625 GHz (~5.3 cm wavelength), at which the gaseous attenuation is mainly due to oxygen. Battan (1973) derived a one way gaseous attenuation value of approximately 0.008 dB km^{-1} due to oxygen and 0.0007 dB km^{-1} due to water vapour at 5.3 cm wavelength (20°C temperature and a pressure of one atmosphere). This amounts to a two way gaseous attenuation of approximately 0.0174 dB km^{-1}. At 100 km range from the radar, this will result in 1.74 dB attenuation due to atmospheric gases, increasing to 4.35 dB at 250 km. This value is significant and demonstrates that consideration should be given to adding a gaseous attenuation factor within the radar site processing. Here, we determine the range biases that affect the signal received by UK radars, but which are not currently accounted for within the OPS reflectivity processing. These biases are due to the combined effect of the forward modelling being too simplistic (the two-way attenuation from atmospheric gases, clouds and precipitation, beam broadening and the bright-band effect are not simulated) and the precipitation vertical profile being misrepresented in the model.

The effect of the attenuation, in particular, is very noticeable when the difference between modelled and observed reflectivity values are averaged over a relatively long time period of 90 days

Table 1 The variance in the bias and the range bias factor due to attenuation are calculated for results averaged over different time periods.

Time period (days)	Variance in the bias (dB)	Range bias factor (dB km^{-1})
5	3.05	0.0344
10	1.97	0.0322
30	1.45	0.0299
60	1.14	0.0296
90	0.9	0.0287
120	0.75	0.0280

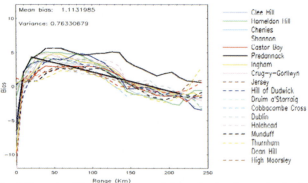

Fig. 5 Example of the bias between model and radar observed reflectivities, using data averaged over 120 days at 0.5° scan elevation. The black linear fit line shows approximate signal decrease due to attenuation.

or longer (Table 1). The OPS reflectivity monitoring statistics have been used to derive a range bias value by finding the average bias over all radars as a function of range at 0.5° scan elevation. A linear fit was added to these data (Fig. 5), between 20 and 210 km, and its equation calculated. The average gradient of the linear fit was calculated over 5 windows of 120 days time period (Table 1). This is approximately equivalent to the range bias factor. Using this method a range bias value of 0.028 dB km^{-1} is calculated. The gaseous attenuation value presented by Battan (1973) of 0.0174 dB km^{-1} is approximately two thirds the value we have calculated (Table 1).

The bias due to gaseous attenuation is linear, and can therefore be subtracted from the value calculated in Table 1, resulting in an bias value which has been averaged over range of 0.0106 dB km^{-1}. This is due to the combined effect of beam broadening, the bright band effect and precipitation attenuation, which cannot be treated in a linear way. As a first approximation, the attenuation due to atmospheric gases is often calculated as the product of a hardware specific constant and the range from the radar, and is commonly corrected for within the signal processor (Hannesen, 2001). However for the purpose of monitoring radar reflectivity against model fields, the attenuation due to the atmospheric gases would be more accurately accounted for by incorporating it within the forward modelling. By eliminating this bias source, the bias due to the bright band effect, beam broadening and attenuation by cloud/rain could be quantified in the same way with greater precision, making the bias due to other anomalies easier to distinguish.

FUTURE WORK

Visualisation of OPS reflectivity monitoring

This work forms part of the larger RDQMS project (Harrison & Curtis, 2011), the aim of which is to produce a comprehensive system of monitoring and displaying various aspects of the radar data quality. When the OPS monitoring system is running on a fully operational basis, the results will be displayed in a fully integrated way with other elements of the RDQMS system.

Mapping anomalies with greater accuracy

One of the next stages of this work will be the development of polar plots displaying the long-term O-B bias (using at least 80 days worth of data to achieve an O-B variance below 1 dB). This would be a useful tool for mapping exactly where anomalies in the data exist. Using this data along with the underlying topography, it should be possible to establish where and why deficiencies in both the NWP model and the radar calibration may arise. Anomalies resulting from: residual clutter; residual beam blockage; radar processing errors; model orographic enhancement; and sea/land model parameterisations could be investigated and quantified to some extent using such an approach.

Inclusion of other statistics

Some further modifications to the OPS reflectivity code are required to incorporate additional statistics such as the POD, FAR and CSI. As opposed to the UK-wide verification statistics, aimed at identifying model deficiencies, these measures will be available for each radar and will be useful for identifying further differences in radar performance. Significant temporal changes in the average CSI at a particular radar could be indicative of radar sensitivity changes. Anomalies in the FAR or POD statistics could be due to noise or residual clutter.

CONCLUSIONS

The OPS reflectivity monitoring has useful applications within data assimilation, model verification and radar QC. The optimum time periods that the O-B data should be calculated over have been determined, and the optimal variance in the data is approximately 0.5dB if using 200 days of data and 1dB using 80 days. Any significant deviations from these values could be indicative of radar calibration errors. The variance is significantly larger when using much shorter time periods making it more difficult to directly infer conclusions about calibration issues. Averaging the data over long time periods is useful for identifying long-term calibration errors, but will smooth more recent or short term calibration issues. The short term bias (~30 days) should be used for identifying the shorter term issues.

It has also been found that the attenuation of the radar signal as a result of atmospheric gases is significant. A bias factor, averaged over range, due to a combination of: the bright band effect; beam broadening and the attenuation due to rain, clouds and gas, has been calculated for UK radars as 0.028 dB km^{-1} using data from a 120-day time period. Gaseous attenuation is responsible for approximately two thirds of this value. It is therefore recommended that the effect of gaseous attenuation is accounted for within the radar site processing. It could also be accounted for within the forward modelling process.

Acknowledgement The authors would like to thank Tim Darlington for his advice regarding atmospheric gaseous attenuation of the radar signal.

REFERENCES

Battan, L. J. (1973) *Radar Observation of the Atmosphere.* The University of Chicago Press, USA.

Bean, B. R. & Dutton, E. J. (1968) *Radio Meteorology.* Dover Publications.

Harrison, D. & Curtis, A. (2011) Use of long-term diagnostics to monitor and asses the effectiveness of radar data processing techniques and the quality of precipitation estimates. In: *Proc. 5th Symp. on Weather Radar and Hydrology.* Exeter, UK.

Gaussiat, N. (2008a) Comparisons of radar reflectivities with synthesised observations from NWP model output. *Met Office Internal Technical Report.*

Gaussiat, N. (2008b) Pre-processing of radar reflectivities for NWP. *Met Office Internal Technical Report.*

Gun, K. L. S. & East, T. W. R. (1954) The microwave properties of precipitation particles. *Quart. J. Roy. Met. Soc.* 80, 522–545.

Hannesen, R. (2001) Quantitative precipitation estimation from radar data –a review of current methodologies. Gematronik GmbH.

Hitschfeld, W. & Marshall, J. S. (1954) Effect of attenuation on the choice of wavelength for weather detection by radar. In: *Proc. IRE* 42(7), 1165–1168.

Holleman, I. & Beekhuis, H. (2004) *Weather Radar Monitoring Using the Sun.* KNMI.

Ippolito, L. J. (1998) *Propagations Effects Handbook for Satellite Systems Design.* 5th edn, Stanford Telecom ACS, Jet Propulsion Laboratory.

Liebe, H. J. (1985) An updated model for millimetre wave propagation in moist air. *Radio Sci.* 20, 1069–1089.

Sauvageot, H. (1991) *Radar Meteorology.* Atrech House Publishing.

348

Weather Radar and Hydrology
(Proceedings of a symposium held in Exeter, UK, April 2011) (IAHS Publ. 351, 2012).

Operational radar refractivity retrieval for numerical weather prediction

J. C. NICOL[1], K. BARTHOLEMEW[1], T. DARLINGTON[2], A. J. ILLINGWORTH[1] & M. KITCHEN[2]

1 *University of Reading, Reading, UK*
j.c.nicol@reading.ac.uk
2 *UK Met Office, Exeter, UK*

Abstract This work describes the application of radar refractivity retrieval to the C-band radars of the UK operational weather radar network. Radar refractivity retrieval allows humidity changes near the surface to be inferred from the phase of stationary ground clutter targets. Previously, this technique had only been demonstrated for radars with klystron transmitters, for which the frequency of the transmitted signal is essentially constant. Radars of the UK operational network use magnetron transmitters which are prone to drift in frequency. The original technique has been modified to take these frequency changes into account and reliable retrievals of hourly refractivity changes have been achieved. Good correspondence has been found with surface observations of refractivity. Comparison with output of the Met Office Unified Model (UM) at 4-km resolution indicate closer agreement between the surface observations and radar-derived refractivity changes than those represented in the UM. These findings suggest that the assimilation of radar-derived refractivity changes in Numerical Weather Prediction models could help improve the representation of near-surface humidity.

Key words radar refractivity; humidity; NWP

INTRODUCTION

In this paper, we describe the implementation and evaluation of radar refractivity retrieval on one of the radars of the UK operational weather radar network. Particular considerations regarding the implementation of refractivity retrieval on these radars are discussed. The retrieval of hourly changes compare well with surface observations of refractivity as measured at two sites within the domain of ground clutter coverage. The representation of refractivity, as a proxy for humidity, in the Met Office Unified Model is also investigated.

BACKGROUND

Radar refractivity retrieval is a relatively new application of weather radar measurements requiring the measurement of the phase of ground clutter returns, originally presented in Fabry *et al.* (1997). Refractivity (N) is a convenient measure of the refractive index (n) of air, where $N = (n–1) \times 10^6$ in parts per million (ppm). This technique utilises the phase change between two times of returns from stationary ground clutter targets. The refractivity change between these two times will produce a particular phase change as a function of range. By measuring the gradient of the phase change with respect to range over short distances, spatial maps of near-surface refractivity changes may be derived in regions with sufficiently stationary ground clutter. At C-band wavelengths, a refractivity change of 1 ppm results in a phase change gradient of 13°/km with respect to range. As radar refractivity is closely related to humidity (1 ppm $\approx$ 1% RH @ 20°C), it is anticipated that such measurements will provide valuable insights into the dynamic variability of water vapour and may be a valuable new data source for assimilation into Numerical Weather Prediction models, particularly with respect to the initiation of convection.

The refractivity technique has previously been demonstrated for radars with klystron transmitters. Klystron transmitters are very stable in terms of frequency. Weather radars in the UK use magnetron transmitters, for which the transmitted frequency is prone to drift. These frequency drifts are primarily caused by changes in the ambient temperature (Skolnik, 1990) and changes in the average input power (e.g. change in pulse duration or PRF). Changes in the transmitted frequency (experienced by radars with magnetron transmitters) during the time taken for Doppler radar measurements are negligibly small, however they become significant when considering phase measurements made at considerably different times and therefore must be treated for radar refractivity retrieval using magnetron transmitters. The role of the transmitted frequency on absolute phase measurements has not been well-understood. It was originally maintained that in order to apply radar refractivity retrieval to magnetron radars, the transmitted frequency would be needed to be measured in real-time with an accuracy of at least 1 ppm (Fabry *et al.*, 1997). It has since been proposed (Parent du Chatelet and Boudjabi, 2008) that phase changes primarily occur due to STALO frequency changes, rather than transmitted frequency changes. Indeed, phase changes must be corrected for any changes in the frequency of local oscillators (Nicol *et al.*, 2011), with an accuracy of at least 1 ppm (i.e. 5.6 kHz at C-band). However, it was also shown that transmitted frequency changes can be a limiting factor in refractivity retrievals when a long pulse length is used.

UK OPERATIONAL RADAR NETWORK

The UK operational weather radar network currently comprises 16 magnetron-based C-band (5-cm wavelength) radars. The coverage of ground clutter throughout the UK is indicated in Fig. 1(a). This represents the possible coverage of refractivity retrievals from the entire network. The testing and development of refractivity retrievals on the operational radars has focused on an operational radar at Cobbacombe in southwest England. The topography surrounding Cobbacombe from a digital terrain model is shown in Fig. 1(b).

Phase and phase variability data are collected at each gate along with the LO frequency for each PPI at the lowest operational elevation angle (0°), which are repeated every 5 minutes. A relatively long pulse is employed for low-elevation scans (2 μs, 300 m). The radar transmits with a pulse repetition frequency (PRF) of 300 Hz and scans at 1.2 rpm or 7.2°/s.

CONSIDERATIONS FOR REFRACTIVITY RETRIEVAL

Frequency-dependence of phase measurements

Transmitted and local oscillator frequency changes must be considered independently regarding phase change measurements at two significantly different times (Nicol *et al.*, 2011). The two effects described below combine additively. The local oscillators (LO) frequency is considered to be the

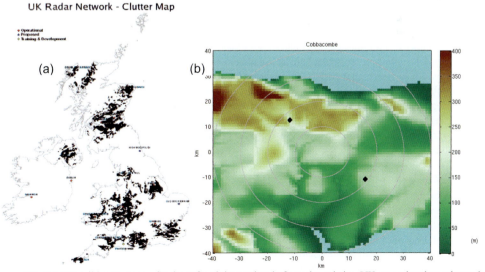

Fig. 1 (a) Possible coverage of radar refractivity retrievals from the existing UK operational weather radar network, (b) Topography surrounding the radar at Cobbacombe, indicating the surface-observation stations at Liscombe (NW of radar) and Dunkeswell (SE of radar).

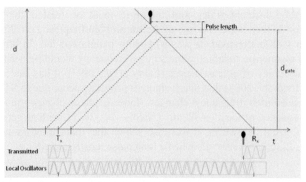

Fig. 2 Illustration of the dependence of phase measurements from stationary targets on the transmitted and local oscillator frequencies at two times (red and black waves).

sum of the local oscillator frequencies (e.g. STALO + COHO or STALO + Numerically-Controlled Oscillator for analogue and digital radar receivers, respectively). LO frequency changes cause a phase change error which is proportional to the time between transmission (Tx) and sampling of the received signal (Rx). This is equivalent to the distance from the radar to the centre of a particular range-gate. This steady phase change with range results in an additive refractivity error in retrievals, if uncorrected. Represented graphically in Fig. 2, the LO frequency at two times (red and black waves) are depicted relative to the transmission and reception of a finite pulse. The contribution to the phase change from changes in the LO frequency depends only on the time between Tx to Rx and the change in LO frequency between the two times.

In contrast, transmitted frequency changes cause a phase change which is proportional to the target distance from the centre of the range-gate. In Fig. 2, upon transmission the radar pulse propagates away from the radar, is reflected back from a target and the returned signal is sampled at Rx. One may infer that the phase of the received signal depends on the transmitted frequency and path difference relative to the centre of the pulse (2× distance of the target from the centre of the range-gate). Thus, the phase change between two times (red and black waves) depends on the transmitted frequency change and the distance of the target from the centre of the range-gate. This results in an additive phase change error depending on the exact target locations relative to the range-gate centre and not a refractivity bias. If we assume that targets are uniformly-distributed across each range-gate, a transmitted frequency change of 100 kHz would result in phase change errors of about 20° with a 300 m pulse length. Similar errors would occur for refractivity changes of about 20 N due to the uncertainty of the exact ground clutter target location (Nicol *et al.*, 2011).

Particularly with long pulses at shorter weather radar wavelengths, these effects combined with other sources of phase change error, such as target motion, can prevent reliable refractivity retrievals. The use of a relatively long pulse (300 m in range) for refractivity retrieval implies that performance will be degraded when either large transmitted frequency or refractivity changes occur. For these reasons, refractivity changes can only be reliably extracted over limited periods of time. For the current radar configuration, we consider hourly refractivity changes as a candidate for data assimilation in NWP.

Spreading targets

Refractivity retrieval requires returns from many independent targets, however, some very strong backscattering ground clutter targets may dominate over many successive range-gates. After correction for LO frequency changes, the phase change from these targets is proportional to the transmitted frequency change and not the refractivity change. Unless excluded from refractivity retrievals, such targets will bias refractivity retrievals for both magnetron and klystron radars towards the fractional change in transmitted frequency and towards zero, respectively. For the operational weather radars in the UK, the LO frequency is set to match the transmitted frequency (measured in real-time from the transmit pulse) immediately prior to each PPI. It has been shown that returns from spreading targets may be used to check the accuracy with which transmitted and LO frequency changes are measured and recorded (Nicol *et al.*, 2011). This has confirmed that LO frequency changes are known to better than 1 kHz, or equivalently, resulting refractivity errors will be less than 0.2 N and may be neglected.

IMPLEMENTATION AND VALIDATION

It has been shown that both refractivity and transmitted frequency changes may result in large phase change errors when a long pulse length is used (Nicol *et al.*, 2011). In addition, large refractivity changes can lead to phase change aliasing and problems arising from smoothing the phase change field. These problems are most pronounced using long pulses at short wavelengths. Therefore, the use of a reference phase map to estimate refractivity (Fabry *et al.*, 1997), rather than refractivity changes, is not achievable for the radar specifications considered. To maintain reliable retrievals, the time between PPIs needs to be limited (e.g. hourly changes).

A measurement of pulse-to-pulse phase variability (PQI; Nicol *et al.*, 2009) allows stationary targets to be identified in real-time. An example of a PQI field and the corresponding dBZ image are shown in Fig. 3(a) and (b), respectively. A PQI threshold of –5 dB is used to eliminate poor quality targets such as non-stationary clutter and precipitation. Spreading targets may be identified by examining the phase change correlation across adjacent range-gates between times when significant refractivity and frequency changes have occurred (Nicol *et al.*, 2011). They may then also be excluded from retrievals. For the remaining targets, a phase change correction for LO frequency changes (Δf_{LO}) must be added to the raw phase change measurements using equation (1). This correction is proportional to the range-gate distance (d_{gate}).

$$\Phi(d_{gate}) = -\frac{4\pi d_{gate}\Delta f_{LO}}{c} \tag{1}$$

Apart from this correction, the formulation of radar refractivity measurements is essentially the same as the original formulation for which both the transmitted and LO frequencies are constant in time (i.e. equation (2) from Fabry, 1997). Strictly speaking, one must correct for LO rather than transmitted frequency changes, contrary to the implication in Fabry (1997). Although the LO frequency is typically adjusted to track the transmitted frequency in magnetron-based radar systems, this is a subtle though important distinction to make when considering radar refractivity retrievals (Nicol *et al.*, 2011). A 2D-Gaussian function (truncated at 3 × std. dev.) is used to spatially-average the corrected phase changes on a gate-by-gate basis (std. dev. (range) = 375 m; std. dev. (azimuth) = 750 m). To estimate refractivity changes, phase change gradients with respect to range(over 3 range-gates = 900 m) are also averaged using a 2D-Gaussian function (std. dev. = 1.5 km). Thus, the resulting maps of hourly refractivity changes have a resolution of about 3 km. Refractivity errors are estimated from the standard deviation of these phase change gradients within regions covered by the truncated 2D-Gaussian function. Examples of the refractivity change (between 12:50 and

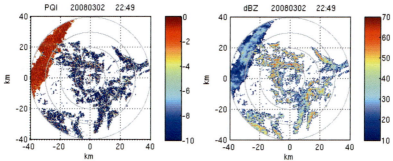

Fig. 3 Phase Quality Indicator (a) and reflectivity (b) 22:50 UTC 2 March 2011 clearly depicting the ground clutter field within 40 km of the radar and a narrow band of precipitation to the NW.

13:50 UTC 7 March 2008) and corresponding error estimate are shown in Fig. 4(a) and (b), respectively.

Radar refractivity retrievals have been validated using surface observations of temperature, pressure and RH. Data from two stations shown in Fig. 1(b) (Liscombe and Dunkeswell) were available for comparisons from March to August 2008. Comparisons suggest that eliminating measurements with error estimates greater than 1.5 N largely excludes poor quality retrievals. Although refractivity changes are not necessarily available at all times at a given location due to the elimination of poor quality targets, the accumulated hourly refractivity retrievals at times show excellent agreement with surface observations. Figure 5(a) and (b) show the refractivity change relative to the beginning of the period (9 July 2008–16 July 2008) from surface observations (black lines) at Liscombe and Dunkeswell, respectively. Also shown is the corresponding radar-derived refractivity change (red lines), obtained by accumulating the individual hourly changes throughout the 7-day period (made up of 168 hourly changes). Hourly radar refractivity changes have a correlation of about 0.6 with respect to surface observations during the study period.

REFRACTIVITY IN NUMERICAL WEATHER PREDICTION

The Unified Model (UM) of the UK Met Office is moving to higher spatial resolution. The horizontal resolution is currently at 4-km and soon to move to 1.5-km. Output from the UM (4-km) for a 10-day period (25 July 2008–3 August 2008) has been selected to analyse the representation of refractivity (humidity) in the UM under a variety of synoptic conditions. An example of a refractivity field calculated from model variables (T, RH, p) is shown in Fig. 6(a). Hourly changes have been calculated throughout this period, an example of which is shown in Fig. 6(b). Both UM and radar-derived hourly refractivity changes have been compared with surface observations made at Liscombe and Dunkeswell. The daily correlations of hourly refractivity changes with surface observations indicate that the radar refractivity retrievals consistently outperform the Unified Model throughout

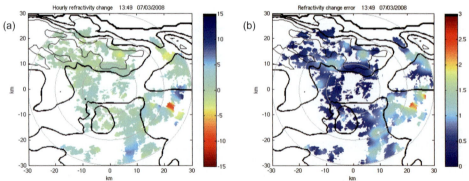

Fig. 4 An example of the refractivity change (a) between 12:50 UTC and 13:50 UTC 7 July 2008 with corresponding error estimate (b). Height contours at 0, 50, 150, 250 and 350 m.

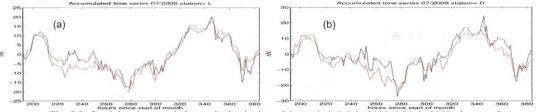

Fig. 5 Refractivity change relative to the beginning of the period (9 July 2008–16 July 2008) from surface observations (black lines) at Liscombe (a) and Dunkeswell (b). The corresponding radar-derived refractivity change (red lines), obtained by accumulating the individual hourly changes (168 at each site) throughout the period.

this period, as shown in Fig. 7. The correlation of hourly changes between the UM and synoptic stations is weaker for humidity (0.13) than for temperature (0.55) and pressure (0.61) suggesting that humidity is relatively poorly represented in the UM.

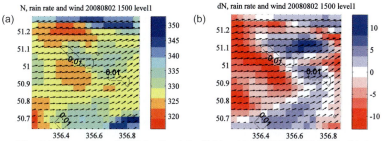

Fig. 6 (a) Examples of a UM refractivity field at 15:00 UTC 02/08/2008, (b) the refractivity change over the previous hour. Contours depict modelled rain rate.

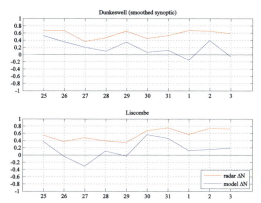

Fig. 7 Correlation of the daily time series, each based on 24 successive hourly changes, of UM (red) and radar-derived (blue) refractivity changes with respect to surface observations during the 10-day UM study period (25 July 2008–3 August 2008) at Dunkeswell (top) and Liscombe (bottom).

CONCLUSIONS

Radar refractivity retrievals have been developed for radars of the UK operational weather radar network. Various considerations which have been discussed require that the time between PPIs used for retrievals is limited to less than a few hours for the current configuration of these radars.

Radar retrievals of hourly refractivity changes show consistently better agreement than the Unified Model, in comparison with synoptic station measurements. Radar refractivity retrievals should benefit data assimilation as the representation of near-surface humidity in the Unified Model is relatively poor. A quasi-operational refractivity retrieval processing system is currently under testing and development within the Met Office as refractivity data are being collected by an increasing number of radars in the operational network throughout 2011.

REFERENCES

Fabry, F., Frush, C., Zawadzki, I. & Kilambi, A. (1997) On the extraction of near-surface index of refraction using radar phase measurements from ground targets. *J. Atmos. Oceanic Technol.* 14, 978–987.

Nicol, J., Bartholomew, K. & Illingworth, A. (2009) A technique for deriving the humidity of air close to the surface using operational rain radar. *Proc. 8th Symposium on Tropospheric Profiling*, 18–23 October 2009, Delft, Netherlands..

Nicol, J., Illingworth, A., Darlington, T. & Kitchen, M. (2011) The consequences of frequency changes for radar refractivity retrieval. *J. Atmos. Oceanic Technol.* (submitted).

Parent-du-Chatelet, J. & Boudjabi, C. (2008) A new formulation for a signal reflected from a target using a magnetron radar: Consequences for Doppler and refractivity measurements. In: *Proc. 5th European Conf. on Radar Meteorology and Hydrology (ERAD)* (Helsinki, 30 June–4 July).

Skolnik, M. (1990) *Radar Handbook*. (2nd edition), McGraw-Hill, 1200 pp.

354

Weather Radar and Hydrology
(Proceedings of a symposium held in Exeter, UK, April 2011) (IAHS Publ. 351, 2012).

Assessment of radar data assimilation in numerical rainfall forecasting on a catchment scale

JIA LIU[1], MICHAELA BRAY[1,2] & DAWEI HAN[1]

1 *Water and Environmental Management Research Centre, Department of Civil Engineering, University of Bristol, Bristol BS8 1TR, UK*
jia.liu@bristol.ac.uk

2 *Institute of Environment and Sustainability, School of Engineering, Cardiff University, Cardiff CF24 0DE, UK*

Abstract Numerical Weather Prediction (NWP) model is gaining popularity among the hydrometeorological community for rainfall forecasting. However, data assimilation of the NWP model with real-time observations, especially the weather radar data, is still a challenging problem. The NWP model has its advantage in modelling the physical processes of storm events, while its accuracy is negatively influenced by the "spin-up" effect and the errors in the model driving. To fully utilise the available information and to improve the performance of the NWP model, observations need to be assimilated in real-time. This study focuses on a small catchment located in southwest England with a drainage area of 135.2 km^2. The Weather Research and Forecasting (WRF) model and the three-dimensional variational (3DVar) data assimilation system are applied for the assimilation of radar reflectivity together with surface and upper-air observations. Four 24-h storm events are selected, with variations of rainfall distribution in time and space. The improvement in rainfall forecasts caused by data assimilation is examined for four types of events. For a better assimilation, a radar correction ratio is further developed and applied to the radar data.

Key words numerical rainfall forecasting; WRF; 3DVar data assimilation; radar reflectivity; radar bias correction

INTRODUCTION

Flood forecasting in small catchments with short concentration times depends on accurate rainfall forecasts to extend the forecast lead time. Weather radars and numerical weather prediction (NWP) models complement each other with their own unique strengths and weaknesses in rainfall forecasting. At present, most nowcasting models are based on the extrapolation of radar echoes, which utilise statistical or empirical approaches that rapidly lose their accuracy with increasing lead time (Ebert *et al.*, 2004). Modelling using high-resolution NWP models has the advantage of developing detailed precipitation fields and allowing for the development of new storms. However, the accuracy of the NWP forecasts is negatively influenced by the "spin-up" effect (Daley, 1991) and errors which exist in the initial conditions required to drive the model. Data assimilation can help improve the performance of the NWP model with respect to the spin-up effect and the errors in the model driving. Recent investigations have shown that the performance of the NWP model can be significantly improved for the next several hours by assimilating real-time observations, especially the radar reflectivity and radar-derived Doppler velocity into the model (e.g. Milan *et al.*, 2008; Dixon *et al.*, 2009; Salonen *et al.*, 2010; Sokol, 2010). In this study, the assimilation of the radar reflectivity together with surface and upper-air observations is investigated on a catchment scale for the improvement of the forecasted rainfall quantity in order to assist in more efficient real-time flood forecasting.

The Weather Research and Forecasting (WRF) model and its 3-D variational (3DVar) data assimilation system are adopted and the study area is chosen as the Brue catchment, located in southwest England (51.08°N and 2.58°W) with a drainage area of 135.2 km^2. Four 24-h storm events are selected with different characteristics of the rainfall distribution in time and space. The assimilation results of the four events are quantitatively verified against the rainfall accumulations observed by a raingauge network. Considering the uncertain quality of the radar data, a correction ratio is developed to improve the radar measurements based on the gauge observations. The assimilation results of the corrected radar data are compared with the original assimilation results and the limitations of the correction ratio are further discussed.

MODEL AND DATA SOURCES

WRF and the 3DVar data assimilation system

Data assimilation experiments in this study are carried out with the Advanced Research WRF model (ARW) Version 3.1 (Skamarock *et al.*, 2008). Triple nested domains are designed with 28 vertical pressure levels (Table 1). In order to reduce the computing time and to make the results directly compatible with the lumped hydrological model for flood forecasting, the grid spacing of the innermost domain is set to be 10 km so that the study catchment (135.2 km^2) can be mostly covered by a single grid. Rainfall forecasts on this single grid are treated as the catchment average rainfall. Among the physical parameterisations provided by the WRF model, the most widely applied ones are used in this study. The ECMWF operational forecast data with spatial resolution of 2.5°×2.5° is used to drive the WRF model for the deviation of the initial and lateral boundaries.

Table 1 Settings of the triple nested domains in the WRF model.

	Time step (h)	Grid spacing (km)	Grid no.	Domain size (km)	Downscaling ratio*
Dom1	3	250	15 × 15	3750 × 3750	–
Dom2	3	50	15 × 15	750 × 750	1:5
Dom3	1	10	5 × 5	50 × 50	1:5

* Downscaling ratio = the grid size of the children domain divided by that of the mother domain.

Data assimilation is the technique by which observations are combined with a NWP product (the lateral or boundary conditions) and their respective error statistics to provide an improved estimate of the atmospheric state. In this study, the 3-D variational data assimilation system (Barker *et al.*, 2004) provided by the WRF package (WRF-3DVar) is adopted in each of the triple nested domains for assimilating the radar reflectivity and other observations. The NCEP background error covariance estimated in grid space by the NMC method (Parrish & Derber, 1992) is adopted in this study. The control variables are estimated with the differences of 24 h and 48 h GFS forecasts with T170 resolution valid at the same time for 357 cases distributed over a period of one year. The pixel-based radar reflectivity data are assimilated directly in the WRF-3DVar system, by stating the latitude and longitude of the pixel centre and the height of the radar beam above that pixel. The following equation is used as the observation operator to calculate the model-derived reflectivity to compare with the assimilated observations (Sun & Crook, 1997):

$$Z = 43.1 + 17.5 \log(\rho q_r) \tag{1}$$

where Z is the reflectivity in dBZ, ρ is the air density in kg/m^3 and q_r is the rainwater mixing ratio. The total water mixing ratio is chosen as the moisture control variable and a warm rain parameterisation (Dudhia, 1989) is included to assist the partitioning of moisture and hydrometeor increments. Since the error covariance of the radar observations is difficult to measure in real-time, no measurement error is defined in this study, in order to see how good the assimilation results could be by using the raw radar data.

Weather radar data and surface/upper-air observations

The radar data used in WRF-3DVar are from a C-band radar, the Wardon Hill radar (50.49°N, 2.33°W) which gives a complete coverage of the Brue catchment. The Wardon Hill radar is located at an elevation of 255 m above the sea level and 30 km to the south of the Brue catchment. The radar cycles through four different scan elevations (0.5°, 1.0°, 1.5°, 2.5°) every 5 minutes up to a range of 210 km. The 3dB radar beam width is 1.0°. For the lowest scan elevation of 0.5°, the radar beam height above the Brue catchment is approx. 0.4 km. The Brue catchment is located in a radar sector free of beam blocking and ground clutter for all the four scans. Software at the radar site converts measurements of reflectivity from radial grids to Cartesian grids with resolutions of 2 km (lowest scan) and 5 km (all four scans). The rainfall reflectivity from the lowest scan on the

2 km Cartesian grid (a 76 × 76 grid of 2 km square pixels covering a radius of 76 km) is assimilated by WRF-3DVar in this study.

The Brue catchment contains a network of 49 Casella 0.2 mm tipping bucket raingauges. The catchment average rainfall obtained from the raingauge network by the Thiessen polygon method is then used for a preliminary examination of the radar data quality and later treated as the "ground truth" for the WRF forecasted rainfall. For the comparison with the gauge observations, the radar reflectivity has to be transformed into the rainfall rate by using the following Z-R relation:

$$Z = 200 \times R^{1.6} \tag{2}$$

where Z and R are the reflectivity ($mm^6 m^{-3}$) and the rainfall rate (mm/h), respectively. Table 2 shows the durations of the selected four 24 h storm events and the rainfall accumulations measured by the raingauge network and the Wardon Hill radar. Figure 1 illustrates the comparison on the time series bars of the catchment average rainfall measured by the gauge network and the weather radar. Obvious underestimation by the Wardon Hill radar can be noted for all the four storm events in Table 2. As pointed out by Borga *et al.* (2002), the major factors affecting the rainfall estimation of the Wardon Hill radar in the Brue catchment are the non-uniform vertical profile of reflectivity, the orographic enhancement of precipitation, the anomalous propagation of the radar beam, the radar calibration stability effects, and the uncertainty in Z-R conversion. It has been found that around 20% underestimation of the Wardon Hill radar in the Brue catchment can be explained by the systematic and drift errors in radar calibration and the biased Z-R relationship. The remaining bias might be due to the beam overshooting and the orographic enhancement effect. However, Fig. 1 shows that the rainfall occurrence and the variance of the rainfall intensity observed by the Wardon Hill radar are quite consistent with the raingauge observations. For this reason, the assimilation of the radar reflectivity is also expected to have some positive effect on improving the rainfall forecasts of the WRF model.

Table 2 The four storm events and their 24-h rainfall accumulations.

ID	Start time	End time	Gauge (mm)	Radar (mm)	Radar/Gauge
a	1999/10/24/00:00	1999/10/25/00:00	29.38	10.36	0.3527
b	1995/09/06/18:00	1995/09/07/18:00	31.97	10.20	0.3190
c	2000/04/02/18:00	2000/04/03/18:00	31.12	12.22*	0.3926
d	1994/08/03/12:00	1994/08/04/12:00	22.30	10.30	0.4622

* Due to the lack of radar data, the radar accumulation for Event c is for only 10 hours.

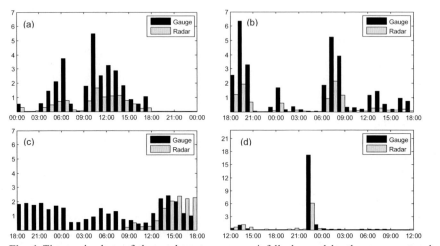

Fig. 1 Time series bars of the catchment average rainfall observed by the gauge network and the Wardon Hill radar for the 24 h durations of (a) Event a, (b) Event b, (c) Event c and (d) Event d.

Besides the radar reflectivity, two types of NCAR archived data (SYNOP and SOUND, see http://dss.ucar.edu/) are also assimilated by the WRF-3DVar, which provide surface and upper-level observations of pressure, temperature, humidity and wind from fixed and mobile stations. The NCAR observations are pre-processed by OBSPROC for the conversion into the "LITTLE-R" format before they can be used in WRF-3DVar. The US Air Force (AFWA) OBS error file which is provided by the WRF-3DVar system and contains various instrumental and sensor errors is used to define the measurement error of the NCAR observations.

RESULTS

Rainfall forecasting results after data assimilation by WRF-3DVar

For the four 24-h storm events, Event a is found to have evenly-distributed rainfall in space, but uneven and discontinuous rainfall in time. The rainfall distribution of Event b is neither even in space nor continuous in time. Event d is an extreme case of Event b, whose rainfall is highly concentrated in a small area of the catchment and happens in a quite short time period. Event c is the most even case among the events, which has a 2-D evenness of the rainfall distribution in both time and space. It has been found in previously studies that when using the ERA-40 re-analysis data, WRF has the best performance in reproducing Event c, while it totally fails in capturing the highly concentrated rainfall of Event d in either time or space. When using the ECMWF operational data for real-time forecasting, the result might be different and it is interesting to see whether data assimilation can help improve the rainfall forecasts downscaled by WRF for the four types of storm events.

The cumulative curves of the catchment average rainfall after the assimilation of radar reflectivity and NCAR observations are shown in Fig. 2 for the four events. The original runs (run1, run6 of Event a; run1, run4, run7 of Event b; run1, run6 of Event c; run1, run4 of Event d) are represented by solid curves which are the original results downscaled from the ECMWF forecast data. Data-assimilation runs (initialised with circles) are started with solid curves and then become dashed after 6 h when new runs are commenced with new observations assimilated. From the cumulative curves of the four events, obvious improvements can be seen for Event a, Event b and Event c. The updated curves follow similar trends as the gauge observed curves (the solid black ones) for Event a and Event c, which are the events with the spatial evenness of the rainfall distribution. For Event b with 2-D rainfall unevenness, although the total rainfall amount has been increased, it is mainly caused by a sharp increase of rainfall happening during the second 6-h period. However, this is not consistent with the gauge observations. The improvement of the trickiest case, Event d, is almost negligible compared to the other three events, and is dislocated. On the one hand, this is due to the poor boundary conditions provided by the ECMWF forecasts: no rain is generated in the original runs of Event d. On the other hand, the heavy convective storm, as Event d, may develop very quickly without preceding precipitation being previously detected in the surrounding regions. Sometimes the observations are too weak to trigger a storm process in the model because the state of the model variables does not support the development of such convective storms (Sokol, 2009). Consequently, even if radar reflectivity is assimilated into the model, rainfall forecasts in Event d are not improved. In that case, data containing information on the cloud development that precedes the formation of precipitation may be helpful, e.g. data provided by the satellites. Furthermore, from the gauge observed curves of Event d, it is found that the storm actually happens in the middle of the second 6-h period. A shortened interval of the assimilation time thus might help in capturing the process of the storm.

The 24-h accumulations of the WRF forecasted results before and after data assimilation are summarised in Table 3 for the four storm events. Values in brackets represent the accumulative errors in the percentage of the respective amounts measured by the raingauges. Events a, b and c have obvious improvement after data assimilation with the accumulative error decreased by 87%, 34% and 19%, while for Event d the accumulative amounts is only improved by 1%.

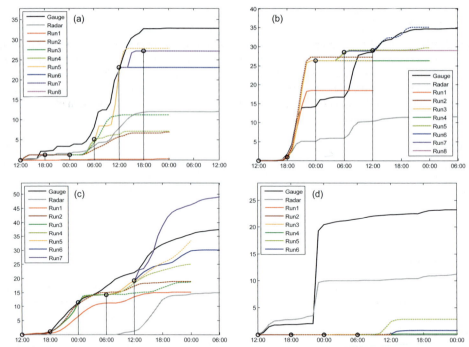

Fig. 2 Rainfall cumulative curves after data assimilation for the four storm events.

Table 3 Cumulative amounts of the WRF forecasted rainfall for the 24-h durations of the four events (mm).

	Before assimilation	After assimilation
Event a	0.15 (–99%)	25.95 (–12%)
Event b	17.21 (–46%)	28.18 (–12%)
Event c	18.68 (–40%)	37.60 (21%)
Event d	0.06 (–100%)	0.12 (–99%)

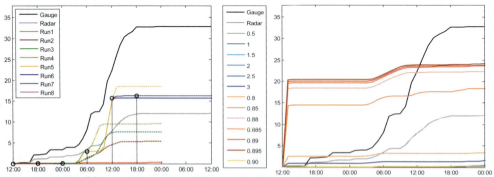

Fig. 3 Rainfall cumulative curves of Event a for (a) data assimilation results using the corrected radar reflectivity and (b) a trial of applying different radar correction ratios to Run2.

A radar correction ratio and its effect on data assimilation

Considering the underestimation of the radar data, a correction ratio is designed to correct the radar bias at each data-assimilation time based on the gauge observations, as shown below:

$$correcting\_ratio = \frac{R6h_{\text{gauge}}}{R6h_{\text{radar}}} \qquad (3)$$

where $R6h_{\text{gauge}}$ and $R6h_{\text{radar}}$ represent the accumulations of the catchment average rainfall (in mm) during the antecedent 6 h of the assimilation time, based on the observations from the raingauge and the Wardon Hill radar, respectively. The correction ratio is directly multiplied to the 2 km-pixel based rainfall rate estimated by the Wardon Hill radar at the data assimilation time and then converted into the reflectivity before being assimilated by WRF-3DVar. It had been expected that the assimilation results using the corrected radar data would offer further improvement. The assimilation results of Event a are shown in Fig. 3(a) however, little improvement is found in comparison to Fig. 2(a). Figure 3(b) shows the variance of the cumulative curve of Run2 in Event a with the correction ratio changing from 0.5 to 3. A sudden increase of the curve happens when the ratio ranges between 0.8 and 0.9, which shows the instability of the correction ratio. This might be caused by a conflict happening to the moisture control variable in the vertical direction by assimilating the radar reflectivity from the one single scan elevation. Moreover, the correction ratio is limited by extrapolating the catchment scale radar error to the whole scan area. Using more observations available beyond the Brue catchment, a more reasonable correction ratio is expected to be found.

CONCLUSIONS

The assimilation of radar reflectivity, together with surface and upper-air observations is realised using WRF-3DVar for rainfall forecasting of four storm events. Significant improvements are seen with events of evenly-distributed rainfall across the catchment; for highly convective storms where rainfall is concentrated in a small area and happens in a short time period, the improvement is not obvious. It should be noted that the improvements are only obviously seen after 6 h since the data assimilation (see the cumulative curves in Fig. 2). This means to solve the spin-up problem and to make the NWP model as competitive as nowcasting methods, more efficient assimilation is needed. We hypothesise that a shortened assimilation time interval and additional observations containing the cloud development information might be able to assist in capturing the highly convective storm. For more efficient utilisation of the radar reflectivity, data from various scan elevations should be considered for assimilation together. Better quality control of the radar data is needed and a more reliable correction method is to be found in the future. After these, the next step is to compare the results with nowcasting methods to see whether there is any advantage of the NWP model in short lead-time rainfall forecasts after data assimilation.

REFERENCES

Barker, D. M., Huang, W., Guo, Y. R., Bourgeois, A. & Xiao, Q. N. (2004) A three-dimensional variational data assimilation system for MM5: implementation and initial results. *Monthly Weath. Rev.* 132, 897–914.

Borga, M., Tonelli, F., Moore, R. J. & Andrieu, H. (2002) Long-term assessment of bias adjustment in radar rainfall estimation. *Water Resour. Res.* 38, 1226, doi:10.1029/2001WR000555.

Daley, R. (1991) *Atmospheric Data Analysis*. Cambridge University Press.

Dixon, M., Li, Z., Lean, H., Roberts, N. & Ballard, S. (2009) Impact of data assimilation on forecasting convection over the United Kingdom using a high-resolution version of the Met Office Unified Model. *Mon. Wea. Rev.* 137, 1562–1584.

Dudhia, J. (1989) Numerical study of convection observed during the winter monsoon experiment using a mesoscale two dimensional model. *J. Atmos. Sci.* 46, 3077–3107.

Ebert, E. E., Wilson, L. J., Brown, B. G., Nurmi, P., Brooks, H. E., Bally, J. & Jaeneke, M. (2004) Verification of nowcasts from the WWRP Sydney 2000 Forecast Demonstration Project. *Weather Forecast* 19, 73–96.

Milan, M., Venema, V., Schuettemeyer, D. & Simmer, C. (2008) Assimilation of radar and satellite data in mesoscale models: a physical initialization scheme. *Meteorol. Z.* 17, 887–902.

Parrish, D. F. & Derber, J. C. (1992) The national meteorological center's spectral statistical-interpolation analysis system. *Monthly Weather Rev.* 120, 1747–1763.

Salonen, K., Haase, G., Eresmaa, R., Hohti, H. & Järvinen, H. (2010) Towards the operational use of Doppler radar radial winds in HIRLAM. *Atmos. Res.* doi:10.1016/j.atmosres.2010.06.004.

Skamarock, W. C., Klemp, J. B., Dudhia, J., Gill, D. O., Barker, D. M., Duda, M. G., Huang, X. Y., Wang, W. & Powers, J. G. (2008) A description of the advanced research WRF Version 3. *NCAR Technical Note, NCAR/TN-475+STR*.

Sokol, Z. (2010) Assimilation of extrapolated radar reflectivity into a NWP model and its impact on a precipitation forecast at high resolution. *Atmos. Res.* doi:10.1016/j.atmosres.2010.09.008.

Sun, J. & Crook, N. A. (1997) Dynamical and microphysical retrieval from Doppler radar observations using a cloud model and its adjoint, Part I: Model development and simulated data experiments. *J. Atmos. Sci.* 54, 1642–1661.

Convective cell identification using multi-source data

ANNA JURCZYK, JAN SZTURC & KATARZYNA OŚRÓDKA

Institute of Meteorology and Water Management, 40-065 Katowice, ul. Bratków 10, Poland
anna.jurczyk@imgw.pl

Abstract Identification of convective cells is an important issue for detecting severe meteorological phenomena and precipitation nowcasting. The proposed model that classifies each individual radar pixel as convective or stratiform was developed based on multi-source data and applying a fuzzy logic approach. For both classes (stratiform or convective), membership functions for all investigated parameters were defined and aggregated as weighted sums. Comparison of the weighted sums decides which category a considered radar pixel belongs to. Each membership function was determined for selected parameters from: weather radar network, satellite Meteosat 8, lightning detection system, and numerical weather prediction (NWP) model. Then convective pixels were clustered to obtain individual cells, assuming that cells with a small distance between their maxima are joined.

Key words precipitation; convection

INTRODUCTION

The paper presents a scheme for identification of convective cells based on data derived from remote sensing measurements and numerical weather prediction (NWP). The scheme has been developed as a starting point for precipitation nowcasting with particular emphasis placed on convective phenomena.

The first step in convective precipitation nowcasting is to detect the area of convection. The simplest method devised to identify convective area is the background-exceedence technique in which a preset threshold of reflectivity is considered as a criterion for convection. Steiner *et al.* (1995) improved the approach by introducing the additional parameter: difference from averaged background reflectivity. This algorithm called SHY95 was modified by involving an analysis of the vertical structure of radar reflectivity (Biggerstaff & Listemaa, 2000; Rigo & Llasat, 2004; and others). Gagne *et al.* (2009) combined k-means clustering to partition the reflectivity field into clusters with a decision tree to classify them using parameters related to storm morphology. Other approaches like using bright band fraction (Rosenfeld *et al.*, 1995) or artificial neural network (Anagnostou, 2004) were also tested.

Generally, the presented scheme for detection of convective cells is divided into three modules: (a) determination of convection areas, (b) division of the area into particular cells, and (c) classification of precipitation structure.

The proposed model, that segments the radar field into convective or stratiform areas, was developed using multi-source data delivered by a weather radar network, a satellite Meteosat 9, lightning detection system, and a NWP model, operated by Institute of Meteorology and Water Management (IMGW). The detected convective areas were split into convective cells using a geometrical approach applied to reflectivity data. Verification was performed by a meteorological forecaster as a human expert. The investigation was carried out for the territory of Poland with 1-km resolution data from 2007.

INPUT METEOROLOGICAL DATA

Weather radar information

Weather radar data are generated by the Polish radar network POLRAD that consists of eight C-band Doppler radars of Selex SI Gematronik. Two-dimensional (2-D) radar products are produced by Rainbow 5 software with 1-km spatial resolution every 10 minutes. Radar products (according to denotation introduced in Rainbow 5 software; Selex, 2010) employed for the analyses are listed in Table 1.

The horizontal structure of the radar reflectivity field turned out to be a useful factor to distinguish between convective and stratiform precipitation. Therefore two additional parameters listed in Table 2 were introduced. Both parameters were calculated as a ratio of the value in a considered pixel to the linear average of the rain pixels from the surrounding background of 11-km radius.

Table 1 Weather radar products employed in the algorithm.

Parameter	Radar product	Units	Configuration
Maximum of reflectivity (Z_{max})	MAX(Z)	dBZ	Range of height: 1–15 km
Height of radar echo top (*EchoTop*)	EHT(Z)	km	Echo threshold: 4 dBZ
Water content in atmosphere (*VIL*)	VIL	mm	Range of height: 1–10 km

Table 2 Weather radar-based parameters employed in the algorithm.

Parameter	Units	Source radar product
Difference of reflectivity ($\Delta Z = Z_{max} / Z_{mean}$)	dBZ	MAX(Z)
Difference of VIL ($\Delta VIL = VIL / VIL_{mean}$)	mm	VIL

Table 3 Meteorological satellite products employed in the algorithm.

Parameter	Units	Description
Cloud Type (*CT*)	(classes)	–

Table 4 Lightning detection system products employed in the algorithm.

Parameter	Units	Description
Cloud-to-ground lightning (*CG*)	Number	Reports about each lightning generated every 1 min

Meteorological satellite data

Meteorological data are generated by the EUMETSAT Meteosat 9 satellite using the SEVIRI visible (VIS) and infrared (IR) imager. The Meteosat pixel size depends on geographical location and is about 3 × 3 km for the territory of Poland. Cloud Type (*CT*) is a product that informs about type of clouds according to their height and transparency (Table 3). Classification considering division into stratiform or cumuliform type is to be implemented in the future. One class out of 15 presently available classes (*CT* = 14) is treated with high probability as convective, 8 others give ambiguous information, and the others are classified as non-convective.

Lightning detection system

The lightning locations in Poland are detected by the PERUN system (based on Vaisala SAFIR 3000) which comprises nine detectors. Reports of the system are generated with 1-min frequency and include records about each detected lightning. Both intercloud (IC) and cloud-to-ground (CG) lightning are detected; however, the quality of intercloud information is not satisfactory, especially outside of the network central area. Therefore, only CG lightning data were employed to calculate lightning density for the analyses (Table 4). Accuracy of location of CG lightning is about 1 km.

NWP information

Numerical weather prediction (NWP) data in Poland are delivered by the COSMO-PL model initialized every 12 hours. Forecasts are generated with 1-h time step and 7-km spatial resolution. The two convection parameters *CAPE* and *TTI* listed in Table 5 were employed in the algorithm for separation of convective and stratiform precipitation.

Table 5 Numerical weather prediction products employed in the algorithm.

Parameter	Units	Description
Convective Available Potential Energy (*CAPE*)	J·kg⁻¹	Based on thermodynamic diagram
Total Totals Index (*TTI*)	°C	Based on thermodynamic diagram

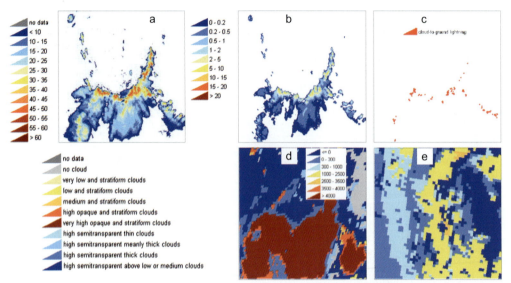

Fig. 1 Example of input data for Legionowo (12 Aug. 2007, 14:00 UTC): radar products (a) MAX (dBZ); (b) VIL (mm); (c) cloud-to-ground lightning; (d) Meteosat Cloud Type; (e) COSMO-PL CAPE (J·kg⁻¹).

Example of meteorological data

An example of selected parameter variables, out of those meteorological input data listed above, for the range of the Legionowo radar (12 August 2007, 14 UTC) is shown in Fig. 1. High values of particular parameters suggest possible convection, especially high radar reflectivity, VIL, presence of lightning and very high opaque clouds indicate strong convection.

CONVECTION AREA DETECTION

Algorithm description

The algorithm consists of the determination of convection areas using meteorological data from different sources and a fuzzy logic approach. A fuzzy logic approach was applied to categorize radar pixels into convective or stratiform precipitation. For both classes (S – stratiform or C – convective), membership functions for all selected parameters were defined and aggregated as weighted sums:

$$P_x = \sum_{i=1}^{n} P_{xi} \cdot W_{xi} \tag{1}$$

where x is the precipitation class (S or C), i is the parameter number, n is the number of parameters, P_{xi} is the membership function for ith parameter, and W_{xi} is the weight of ith membership function. Comparison of the weighted sums decided which category a considered radar pixel belongs to.

Each membership function was determined using multi-source meteorological data from 2007. Firstly the whole dataset was subjectively labelled by hand as convective or stratiform areas. In this way two subsets were created with all gathered parameter values for each class.

Membership functions (one- or two-dimensional) were empirically established for each parameter. For 1-D membership functions, analyses of scatter diagrams for the particular parameters allowed threshold values to be obtained, above or below which values of corresponding membership functions were attributed to 0 or 1, and linear (or nonlinear) approximation with values of membership functions within range (0, 1). In the case of 2-D membership function the uncertainty area (membership function range between 0 and 1) was assumed between 95 and 75 percentiles of all parameters' pairs gathered in appropriate data subset. The radar pixel was classified as a convective or stratiform one depending on for which class the weighted sum of membership functions was higher.

Example of detection

The performance of the fuzzy logic classifier for separating convective regions from stratiform ones is demonstrated in Fig. 2. As a case study example, the event from Fig. 1 was chosen. The area marked in red was identified as convective.

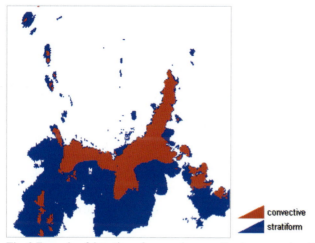

convective

stratiform

Fig. 2 Example of detection of convective area (Legionowo radar, 12 Aug. 2007, 14:00 UTC).

DIVISION INTO CONVECTIVE CELLS

Algorithm description

Convective area was divided into cells clustered around the cell cores that were identified as local maxima in the radar reflectivity field. The question is, if all the maxima are actually related to the cell cores, what is the range of each individual cell. A geometrically-based algorithm was applied to solve the problem. The following steps were performed to divide the whole convection area into particular cells:

Determination of cell centres The centres were determined by searching for local maxima in the Z_{max} field (radar reflectivity obtained from MAX product). The maxima set was verified by analysis of the maxima positions relative to each other, and then they were considered as centres of convection cells.

Convective pixels clustering The clustering was performed for each pixel x by finding a local maximum i for which the value of function u is minimal:

$$u(i,x) = \lambda \cdot grad(i,x) + (1-\lambda) \cdot dist(i,x) \tag{2}$$

where x is the analysed pixel, i is local maximum, λ is the constant $\lambda \in (0, 1)$, *grad* is the sum of negative gradients of reflectivity values along the path from the given pixel x to the considered local maximum i, and *dist* is the distance between the pixel and the maximum. The given pixel x was finally assigned to the convective cell (cluster) with the maximum i.

Correction of the clustering The convective cells were joined if two criteria were fulfilled: (a) there were no pixels with significantly lower reflectivity values between cells' maxima considering their values, (b) a distance between them was below the preset threshold.

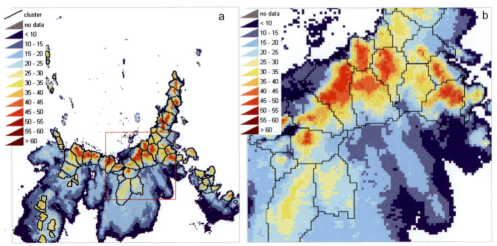

Fig. 3 (a) Example of division into particular convective cells; (b) excerpt from the map with convective cells over radar reflectivity (MAX product in dBZ) (Legionowo radar, 12 Aug. 2007, 14:00 UTC).

Example of division

The performance of the division of convection area into particular cells is presented in Fig. 3(a) for the event depicted in Fig. 1. Borders of all clustered convective cells are drawn here. In Fig. 3(b) an excerpt from the whole radar range image is shown with maximum radar reflectivity as the background.

CLASSIFICATION OF PRECIPITATION STRUCTURE

Algorithm description

A precipitation structure was identified based on investigation of the convection area detected in the first step of the scheme. As a starting point the classification proposed by Rigo & Llasat (2004) has been adapted. The two features of the precipitation structures have been taken into account: (a) percentage of convective pixels in the whole precipitation area, (b) size of the area. The following precipitation structures have been distinguished:

– Mesoscale convective system (MCS) is detected if over 10% of precipitation area is identified as convective and moreover covers a large area.
– Multicell systems (MUL) is a structure similar to MCS in terms of convective pixels' percentage, but not observed over such a large area.
– Isolated convection (IND) is detected if only small, separated, and independent of each other convective cells are observed.
– Convection embedded in stratiform rainfall (EST-EMB) is defined as an area of stratiform precipitation with convective nucleons where the area of convection does not exceed 10% of the whole precipitation structure.
– Stratiform (EST) is recognized if the convective pixels' percentage is lower than 1%.
– The whole investigated precipitation field is divided into the above classes by examining each precipitation object according to these criteria.

Example of classification

Classification into particular classes of precipitation structure for event from Figs 1–3 is presented in Fig. 4. The largest precipitation object (in yellow) with numerous convective cells (see Fig. 3(a)) was recognized as a mesoscale convective system, the smaller one (in orange) as a multicell system and moreover some isolated cells (in red) were found here.

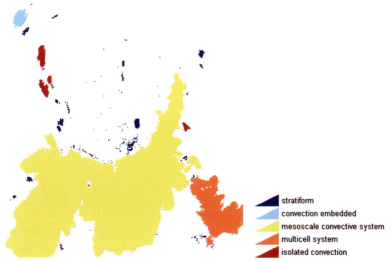

stratiform
convection embedded
mesoscale convective system
multicell system
isolated convection

Fig. 4 Example of classification of precipitation structure (Legionowo radar, 12 Aug. 2007, 14:00 UTC).

FURTHER WORKS

The presented three modules of the convective cells' detection algorithm are parts of a precipitation nowcasting model taking into account convective phenomena. At present, two further modules are being developed. The first module will apply a conceptual model for convection cell lifecycle based on a discrimination of its stages: introduced by Hand (1996) and employed in GANDOLF nowcasting model (Pierce *et al.*, 2000). The second one is to nowcast convective precipitation structures based on both extrapolation and evolution of convective cells in time.

The algorithms are a part of convective precipitation nowcasting model which will be included into INCA-PL nowcasting model.

Acknowledgement The paper was prepared in the frame of INCA-CE Project (Central Europe) and Polish Ministry of Science and Higher Education grant NR14-0003-06. The authors wish to thank Monika Kaseja from Meteorological Forecast Bureau of IMGW in Kraków for help in evaluation of the proposed algorithms.

REFERENCES

Anagnostou, E. N. (2004) A convective/stratiform precipitation classification algorithm for volume scanning weather radar observations. *Meteorol. Appl.* 11, 291–300.

Biggerstaff, M. I. & Listemaa, S. A. (2000) An improved scheme for convective/stratiform echo classification using radar reflectivity. *J. Appl. Meteor.* 39, 2129–2150.

Gagne, D. J., McGovern, A. & Brotzge, J. (2009) Classification of convective areas using decision trees. *J. Atmos. Oceanic Technol.* 26, 1341–1353

Hand, W. H. (1996) An object-oriented technique for nowcasting heavy showers and thunderstorms, *Met. Apps.* 3, 31–41.

Pierce, C. E., Collier, C. G., Hardaker, P. J. & Haggett, C. M. (2000) GANDOLF: a system for generating automated nowcasts of convective precipitation. *Met. Apps.* 7, 341–360.

Rigo, T. & Llasat, M. C., 2004. A methodology for the classification of convective structures using meteorological radar: Application to heavy rainfall events on the Mediterranean coast of the Iberian Peninsula. *Nat. Hazards Earth Syst. Sci.* 4, 59–68.

Rosenfeld, D., Emitai, E. & Wolf, D. B. (1995) Classification of rain regimes by the three-dimensional properties of reflectivity fields. *J. Appl. Meteor.* 34, 198–211.

Selex (2010) *Rainbow 5. Products and algorithms.* Release 5.31.0. Selex SI GmbH, Neuss, pp. 442.

Steiner, M., Houze, R. A. & Yuter, S. E. (1995) Climatological characterization of three-dimensional storm structure from operational radar and rain gauge data. *J. Appl. Meteor.* 34, 1978–2007.

4 Uncertainty Estimation

Weather Radar and Hydrology
(Proceedings of a symposium held in Exeter, UK, April 2011) (IAHS Publ. 351, 2012).

369

Error model for radar quantitative precipitation estimates in a Mediterranean mountainous context

GUY DELRIEU[1], LAURENT BONNIFAIT[1], PIERRE-EMMANUEL KIRSTETTER[1,2] & BRICE BOUDEVILLAIN[1]

1 *Laboratoire d'étude des Transferts en Hydrologie et Environnement, Grenoble, France*
guy.delrieu@ujf-grenoble.fr
2 *National Severe Storms Laboratory, Norman, Oklahoma, USA*

Abstract Characterizing the error structure of radar quantitative precipitation estimation (QPE) is recognized as a major issue for applications of radar technology in hydrological modelling. This topic is further investigated in the context of the Cevennes-Vivarais Mediterranean Hydrometeorological Observatory dedicated to improving observation and modelling of extreme hydrometeorological events in the Mediterranean. The reference rainfall problem is firstly addressed: after quality-control of the raingauge measurements, various interpolation techniques (isotropic and anisotropic Ordinary Kriging, Universal Kriging with external drift) are implemented and compared through a cross-validation procedure. Then, the block Kriging technique allows the estimation and selection of reference values for a series of time-steps (1–12 h) and hydrological mesh sizes (5–50 km^2). The conditional distributions of the residuals between radar and reference values are modelled using generalized additive models for location scale and shape. The distributions are analysed for the operational real-time radar products and the Observatory re-analysed products, the latter being by construction less affected by conditional bias. As expected, the error model is dependent on the space and time scales considered. The hourly raingauge network is found to be not dense enough for providing reliable spatial estimations for sub-daily time-steps.

Key words Mediterranean heavy precipitation; weather radar; quantitative precipitation estimation; error model; space and time scales

CONTEXT

Within the PreDiFlood project of the French National Research Agency (ANR), several past Mediterranean heavy precipitation events that produced major traffic disruptions in the Gard Department, France, have been selected to develop a nowcasting tool for road flooding. This tool is based on radar quantitative precipitation estimates (QPEs), a distributed hydrological model and an *a priori* vulnerability assessment of the road network in the region. Early developments of the approach are described by Versiani *et al.* (2010a,b). The present paper focuses on the radar QPE and its assessment over a range of temporal and spatial scales; it follows a previous contribution dedicated to the QPE assessment over 1 km^2 radar pixels (Kirstetter *et al.*, 2010). Figure 1 displays the region of interest and the rainfall observing systems available, which include: (i) three weather radar systems of the Météo France ARAMIS network at Nîmes (S-band), Bollène (S-band) and Sembadel (C-band); (ii) an hourly raingauge network with 252 raingauges complemented with 162 daily raingauges; and (iii) two disdrometers for drop-size distribution measurements. The three events considered in the PreDiFlood project occurred on 29–30 September 2007, 19–22 October 2008, 1–2 November 2008. Due to its extraordinary magnitude, the catastrophic event of 8–9 September 2002 (Delrieu *et al.*, 2005; Bonnifait *et al.*, 2009) is included as well in the following analyses. Regarding radar QPE, two products are considered with (i) the Météo France operational product called PANTHERE hereafter (Tabary, 2007), and (ii) the OHM-CV off-line product (Delrieu *et al.*, 2009). The two radar data processing systems essentially cope with the same error sources (permanent ground clutter, beam blockages, vertical profile of reflectivity and the associated range effects; see Delrieu *et al.* (2009) and Villarini & Krajewski (2010) for a description of the main radar error sources) with slightly different approaches. Regarding bias correction, the PANTHERE product makes use of a standard Z-R relationship and the QPEs are raingauge-adjusted using a uniform calibration factor updated every hour. For the OHM-CV products, an "effective" Z-R relationship is optimized so as to minimize the bias and the conditional bias between radar and raingauge estimates at the event time scale following the approach proposed by Bouilloud *et al.* (2010).

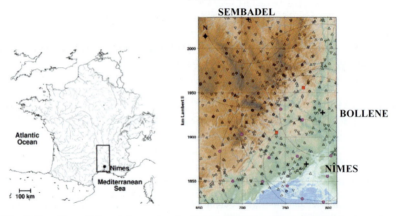

Fig. 1 Location of the study region in France (left) together with the topography and the rainfall observation system (right) available in the Cévennes region. In the present work, the radar datasets come from the Bollène and Nîmes Météo France weather radar systems (cross and 50-km range markers) and the reference rainfall is derived from the hourly raingauges (triangles pointing up).

REFERENCE RAINFALL

In line with studies performed in other countries (e.g. Seo & Krajewski, 2011), the present work aims at extending the error model proposed by Kirstetter *et al.* (2010) by assessing the radar QPE over a range of temporal and spatial scales. We denote the true unknown rainfall amount over a given area A, centred at a given point $\underline{x}$, for a given time interval T and centred at time t, as:

$$R_{AT}(\underline{x},t) = \frac{1}{A}\frac{1}{T}\iint R(\underline{u},v)\,d\underline{u}\,dv \tag{1}$$

where R denotes the true rainfall amount at a given location and time. The radar QPE products are gridded, typically with a good spatial resolution of 1 km², so that the radar QPE over domain A may be expressed as:

$$R_{AT}^*(\underline{x},t) = \frac{1}{N_A}\sum_{i=1}^{N_A} R_T^*(a_i,t) \tag{2}$$

where a_i represents a radar pixel, N_A is the number of pixels covering domain A and R_T^* the radar-estimated rain amount at time t during a time interval T.

The block Kriging technique (Goovaerts, 1997) was used to establish the reference rainfall, denoted $R_{AT}^{ref}(\underline{x},t)$, from the raingauge network with:

$$R_{AT}^{ref}(\underline{x},t) = \sum_{i=1}^{N_g}\lambda_i G_T(\underline{x}_i, t) \tag{3}$$

where $G_T(\underline{x}_i, t)$ is the raingauge amount at point $\underline{x}_i$ and time t during T and N_g is the number of raingauges accounted for in the estimation. The Kriging technique utilizes the variogram function $\gamma_T(h)$ to model the spatial correlation of the rain fields:

$$\gamma_T(h) = \frac{1}{2}E(R_T(\underline{x},t) - R_T(\underline{x+h},t))^2 \tag{4}$$

where h denotes the raingauge interdistance. From raingauge data collected in the Cévennes region, Lebel *et al.* (1987) have established the following empirical model for the decorrelation distance (range of the variogram) as a function of the rainfall duration:

$$d = 25\,T^{0.3} \tag{5}$$

with d in km and T in hours, leading to decorrelation distances of 25, 43 and 64 km for durations of 1, 6 and 24 h, respectively; this result indicates the spatial representativity of raingauge measurements increases with the rainfall duration.

The OHM-CV raingauge data were critically analysed using the empirical variogram function of the raingauge measurements in order to detect defective raingauges at the event time-step (Kirstetter *et al.*, 2010). This being done, the following Kriging techniques were implemented in cross-validation mode in order to assess their relative predictive skills: (i) Ordinary Kriging (OK) with the climatological variogram (equation (5)) supposed to be isotropic (**clim**); (ii) OK with isotropic variogram inferred for each time-step from radar data (**iso**); (iii) OK with anisotropic variogram inferred for each time-step from radar data (**ani**); and (iv) Universal Kriging with external drift (**ked**). The latter method uses the TRADHy radar product to define the so-called drift, i.e. the spatial evolution of the rain field mean supposed to be non-stationary. Figure 2 presents the results obtained by the various interpolation techniques in terms of the Nash-Sutcliffe efficiency between estimated and observed values at gauged points for the four rain events and various integration time-steps ranging from 1 to 24 hours. The performance of the TRADHy product (**radar**) is displayed as well by comparing the raingauge observed values and the corresponding radar pixel values through the same assessment criterion.

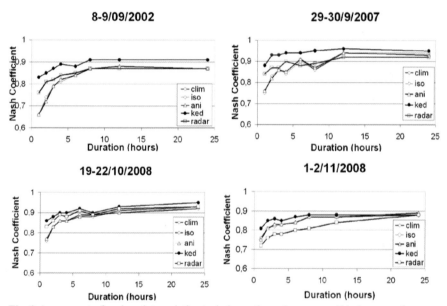

Fig. 2 Assessment of various interpolation techniques through a cross-validation procedure; see text for details.

Figure 2 calls for the following comments: (i) the three OK methods (clim, iso, ani) have very similar performance; this indicates there is little advantage in making the variogram inference for each time-step and/or in accounting for the anisotropy of the rain fields in the present context; (ii) for 3 events out of 4, the radar products obtain better assessment criteria than the OK techniques; for the remaining event, the radar products are known to be affected by residual ground clutter in the region mostly affected by the rain event; (iii) the most striking is probably that the KED estimator systematically outperforms the other techniques, whatever the respective performance of the radar products and the raingauge OK products. This latter result confirms the findings of Velasco-Forero *et al.* (2009) about the relevance of the KED technique for merging quality-checked radar and raingauge data.

It also casts some doubt on the idea of establishing a reference rainfall with the raingauge network available in this study. However, in the following, the error model is built on a validation exercise different from the cross-validation exercise: for a given landscape discretised into hydrological meshes of size A, we select a number of reference meshes, based on the OK estimation standard deviation, for which the estimation is good on account of a good local raingauge coverage. Figure 3 displays the maps of the normalized estimation standard deviation for two spatial discretisations (5 and 50 km^2) obtained with the 1-h climatological normalized variogram (variogram range: 25 km; sill: 1). Unsurprisingly, the meshes with the lowest normalized estimation standard deviations are those containing – or located in the vicinity of – raingauges.

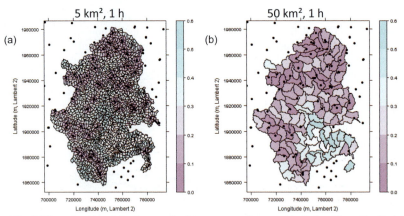

Fig. 3 Discretisation of the three main Cévennes catchments (Ardèche, Cèze, Gardon) into hydrological meshes of 5 km^2 (a) and 50 km^2 (b). The hourly raingauge network is displayed as well. The coloured scale refers to the spatial evolution of the normalized block OK estimation standard deviation for the 1 h time-step (variogram range: 25 km, variogram sill: 1).

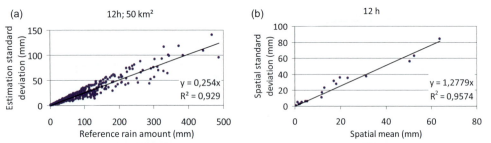

Fig. 4 Example of linear relationship between the field standard deviation and mean at the 12 h time-step (a); example of scaled estimation standard deviations as a function of the reference rain amount for the 12 h and 50 km^2 space–time scales (b).

The reference estimation standard deviations are denormalised with the following procedure: (i) hydrological meshes with normalised estimation variances σ^2_{refN} less than 0.1 are selected as reference meshes; (ii) since the variogram sill is theoretically equal to the field variance, the reference estimation standard deviations are obtained through the following expression: $\sigma^2_{ref}(\underline{x},t)=\sigma^2_f(t)\,\sigma^2_{refN}(\underline{x},t)$, where σ_f is the field standard deviation; (iii) for practical estimation, σ_f is derived from a linear regression between the field standard deviation and the field mean (Fig. 4(a)): $\sigma_f(t)=a_T\,m_f(t)$ and (iv) a local mean $m_A(t)$ is used in the denormalisation to account for the non-stationarity of the rain field: $\sigma^2_{ref}(\underline{x},t)=a^2_T\,m_A(t)^2\,\sigma^2_{refN}(\underline{x},t)$. An example is displayed in Fig. 4(b); a linear regression is found to satisfactorily model the evolution of reference standard deviation as a function of the reference rain amount.

Radar QPE error model

The discussion here is focused on the respective merits of the PANTHERE and OHM-CV QPE products with respect to the OK reference rainfall for the two 2008 rain events for which both QPE products are available. The error model is based on a detailed analysis of the residuals between the radar and reference rain amounts $\Delta_{AT}^{ref}(\underline{x},t) = R_{AT}^{*}(\underline{x},t) - R_{AT}^{ref}(\underline{x},t)$, conditioned by the reference rain amounts $R_{AT}^{ref}(\underline{x},t)$, as initially suggested by Ciach *et al.* (2007). Kirstetter *et al.* (2010) showed the radar residuals obtained for 1 km^2 integration areas to be symmetrically distributed and rather well described by the Gaussian or preferably the Laplace (double exponential) probability density functions (pdfs). In this work, the conditional distributions of the residuals were modelled in the framework of Generalized Additive Models for Location, Scale, and Shape (Rigby & Stasinopoulos, 2004) whose implementation is available in the gamlss package (http://gamlss.org/) in R. Such semi-parametric models consist of two components: a parametric pdf given each value of R^{ref} and a non-parametric relationship between the pdf parameters over the definition domain of the reference rainfall. The conditional densities are assumed to have the same parametric form for all value of the reference rainfall. The gamlss package offers a wide range of two-parameter (Gaussian, reverse Gumbel, gamma, log-normal, etc.) and three-parameter (exponential Gaussian, power exponential, t-family, etc.) continuous pdfs, as well as a number of non-parametric fitting techniques (cubic splines, penalized splines, loess function, etc.) for the second component of the model. The goodness-of-fit of a given model is assessed by investigating the generalized Akaike information criterion (GAIC), a penalized function of the log-likelihood function to be minimized in the fitting procedure. The reverse Gumbel pdf, a moderate positive skewed pdf, was found to produce the best fitting scores in the present study. The resulting mean and standard deviation of the radar residuals are displayed in Figs 5 and 6, respectively, for the (1 h, 5 km^2), (1 h, 50 km^2), (12 h, 5 km^2) and (12 h, 50 km^2) space–time scales. In Fig. 6, the OK estimation standard deviation is displayed as an indicator of the reference quality. The following comments can be listed:

(i) By construction, the conditional bias of the OHMCV products is negligible for the 12 h duration; however, it is noticed to be significant for the 1 h duration, indicating the time dependency of the Z-R relationship. Although gauge-adjusted, the PANTHERE QPE products are more significantly biased, notably for high rain amounts.

(ii) In terms of standard deviation of the residuals, the OHM-CV and PANTHERE products have very similar performance; the most striking feature is probably associated with the respective

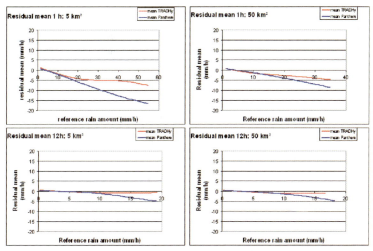

Fig. 5 TRADHy and PANTHERE QPE error models for the two 2008 rain events: residual mean (red and blue curves) as a function of the reference rain amount.

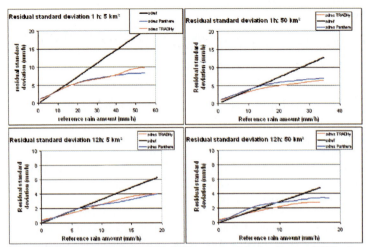

Fig. 6 TRADHy and PANTHERE QPE error models for the two 2008 rain events: residual standard deviation (red and blue curves) and reference estimation standard deviation (black line) as a function of the reference rain amount.

values of the radar residual standard deviation and the estimation standard deviation of the reference rainfall. Since the latter is equal to or even higher compared to the former (especially for the shortest space-time scales), the reference rainfall, based on the sub-daily raingauge network, cannot be considered as very reliable for the spatial estimations considered here. An opposite result has been obtained for 1 km^2 meshes containing a raingauge in Kirstetter *et al.* (2010).

REFERENCES

Bonnifait, L., Delrieu, G., Le Lay, M., Boudevillain, B., Masson, A., Belleudy, P., Gaume, E. & Saulnier, G.-M. (2009) Hydrologic and hydraulic distributed modelling with radar rainfall input: Reconstruction of the 8–9 September 2002 catastrophic flood event in the Gard region, France. *Adv. Water Resour.* 32, 1077–1089.

Bouilloud, L., Delrieu, G., Boudevillain, B. & Kirstetter, P. E. (2010) Radar rainfall estimation in the context of post-event analysis of flash floods. *J. Hydrol.* 394, 17–27.

Ciach, G. J., Krajewski, W. F. & Villarini, G. (2007) Product-error-driven uncertainty model for probabilistic quantitative precipitation estimation with NEXRAD data. *J. Hydromet.* 8, 1325–1347.

Delrieu, G., Ducrocq, V., Gaume, E., Nicol, J., Payrastre, O., Yates, E., Kirstetter, P. E., Andrieu, H., Ayral, P. A., Bouvier, C., Creutin, J. D., Livet, M., Anquetin, S., Lang, M., Neppel, L., Obled, C., Parent-du-Chatelet, J., Saulnier, G. M., Walpersdorf, A. & Wobrock, W. (2005) The catastrophic flash-flood event of 8-9 September 2002 in the Gard region, France: A first case study for the Cevennes-Vivarais Mediterranean Hydrometeorological Observatory. *J. Hydromet.* 6, 34–52.

Delrieu, G., Boudevillain, B., Nicol, J., Chapon, B., Kirstetter, P.-E., Andrieu, H. & Faure, D. (2009) Bollène 2002 experiment: radar rainfall estimation in the Cevennes-Vivarais region. *J. Appl. Met. and Climatology* 48, 1422–1447.

Goovaerts, P. (1997) *Geostatistics for Natural Resources Evaluation.* Oxford University Press, 483 pp.

Kirstetter, P. E., Delrieu, G., Boudevillain, B. & Obled, C. (2010) Toward an Error Model for Radar Quantitative Precipitation Estimation in the Cévennes-Vivarais Region, France. *J. Hydrol.* 394, 28–41.

Lebel, T., Bastin, G., Obled, C. & Creutin, J. D. (1987) On the accuracy of areal rainfall estimation: a case study. *Water Resour. Res.* 23 (11), 2123–2134.

Rigby, R. A. & Stasinopoulos, D. M. (2004) Generalized additive models for location, scale and shape. *Appl. Statist.* 54, 1–38.

Seo, B.-C. & Krajewski, W.F. (2011) Investigation of the scale-dependent variability of radar-rainfall and rain gauge error correlation, *Adv. Water Resour.* 34(2), 152–163.

Tabary, P. (2007) The new French operational radar rainfall product. Part 1: Methodology. *Weather and Forecasting* 22(3), 393–408.

Velasco-Forero, C. A., Sempere-Torres, D. Cassiraga, E. F. & Gomez-Hernandez, J. J. (2009) A non-parametric automatic blending methodology to estimate rainfall fields from rain gauge and radar data. *Adv. Water Resour.* 32 (7), 986–1002.

Versini, P.-A., Gaume, E. & Andrieu, H. (2010) Application of a distributed hydrological model to the design of a road inundation warning system for flash flood prone areas. *Nat. Hazards Earth Syst. Sci.* 10, 805–817.

Versini, P.-A., Gaume, E. & Andrieu, H. (2010) Assessment of the susceptibility of roads to flooding based on geographical information – test in a flash flood prone area (the Gard region, France). *Nat. Hazards Earth Syst. Sci.* 10, 793–803.

Villarini, G. & Krajewski, W. F. (2010) Review of the different sources of uncertainty in radar-based estimates of rainfall. *Surveys in Geophys.* 31, 107–129.

Investigating radar relative calibration biases based on four-dimensional reflectivity comparison

BONG-CHUL SEO[1], WITOLD F. KRAJEWSKI[1] & JAMES A. SMITH[2]

1 *IIHR-Hydroscience & Engineering, The University of Iowa, Iowa City, Iowa 52242, USA*
bongchul-seo@uiowa.edu
2 *Department of Civil and Environmental Engineering, Princeton University, Princeton, New Jersey 08544, USA*

Abstract A methodology to compare radar reflectivity data observed from two different ground-based radars is proposed. This methodology is motivated primarily by the need to explain relative differences in radar-rainfall products and to establish sound merging procedures of multi-radar observing networks. The authors compare radar reflectivity for well-matched radar sampling volumes viewing common meteorological targets. While spatial and temporal interpolation is not performed in order to prevent any distortion arising from the averaging scheme, the authors considered temporal separation and three-dimensional matching of two different sampling volumes based on the original polar coordinates of radar observation. Since the proposed method assumes radar beam propagation under the standard atmospheric condition, we do not consider anomalous propagation cases. The reflectivity comparison results show some systematic differences year to year, but the variability of those differences is fairly large due to the sensitive nature of radar reflectivity measurement. The authors performed statistical tests to check reflectivity difference consistency for consecutive periods.
Key words radar reflectivity; radar-rainfall; radar calibration bias

INTRODUCTION

A well-calibrated and reliable observation system is an essential prerequisite for quantitative rainfall estimation and its application to hydrological fields. Using radar measurements instead of sparsely distributed raingauges to describe the main aspects of rainfall fields yields higher resolution information in space and time. Radar calibration uncertainty (i.e. calibration offset; see e.g. Atlas & Mossop, 1960) might affect the accuracy of rainfall estimates due to lack of information on the absolute calibration procedures and schedule. The systematic bias may present the most significant practical challenge when multiple radar data are considered for a hydrologic unit represented by a river basin covered by multiple radars (e.g. Smith *et al.*, 1996; Brandes *et al.*, 1999).

The primary objective of this study is to investigate the differences in radar reflectivity observed by two ground-based radars for overlapping areas and to detect systematic temporal patterns that reveal differences in radar calibration. This requires development of a methodology that matches coincident sampling volumes from two different ground-based radars. We collected eight years of radar volume scan data which contain the recent upgrade of data resolution (called super-resolution) from two WSR-88D (NEXRAD) radars in the state of Iowa, USA. Our analysis uses volume data match-ups that consider temporal and spatial coincidence in the hope that this can show relative biases for common target locations. In this paper, we present large sample statistics of the matched reflectivity pairs observed. The results show high variability of reflectivity differences matched without revealing a clear systematic behaviour of the differences.

METHODOLOGY

To investigate the calibration-related measurement differences in the absence of operational information regarding dates of radar calibration procedures, we use volume data match-ups that consider temporal and spatial coincidence for common target locations. Our approach avoids spatial interpolation and grid projections to eliminate the effects of data smoothing or distorting.

First, we consider temporal coincidence to match two radar beams from radars separated by a significant distance. The NEXRAD radars are not synchronized, although they operate using several scanning strategies (e.g. Crum & Alberty, 1993). We obtain an observation time for every radar ray in a specific elevation angle from the volume scan information included operationally in the data file headers. As determining the exact matching time is almost impossible, we introduce a parameter of 30-s, which represents tolerance in time separation between two radar ray observations from different radars. The effect of such temporal differences depends on the storm velocity, but in general allows comparison of volumes that are within spatial scales that are commensurate with the data resolution (about 1 km).

For spatial match-up, one needs to account for the horizontal locations represented in spherical coordinates and vertical heights of radar sampling volumes. Since the geographic coordinates of all radar sites and the spherical coordinates of radar sampling volumes are known, the spherical coordinates that represent the centre (C_2 in the top panel of Fig. 1) of a sampling volume from one radar can be easily transformed with respect to the other radar. The differences (dθ, dr, and dh in Fig. 1) in azimuth, range, and height, respectively, between two sampling volumes characterize how close and well matched the two radar sampling volumes are. These three spatial parameters can be described by the proportion over radar beam width, sampling bin size, and vertical beam width to consider the variability of sampling volume size along the distance from the radar. We used a 3% tolerance limit for all three spatial parameters in this study. Although two centre points of radar sampling volumes are sufficiently close and satisfy the tolerance limits defined above, the size of the sampling volumes might differ depending on the distance from both radars. Therefore, a matching zone of the volumes should be defined within near range (i.e. 3 km) from the equidistance line between radars. This conditional requirement for the equidistance zone may reduce the concern of range dependent biases for this match-up methodology.

A main assumption of our methodology is that a radar beam propagates under the standard atmospheric condition. To address the situations where this is not the case, we employed a separate procedure to remove echoes due to anomalous propagation (AP) conditions in the vicinity of the radar. To this effect, two-dimensional AP clutter maps generated using an adaptation of the algorithm by Steiner & Smith (2002) are used to remove the cases that violate the standard atmosphere assumption.

The smaller sampling volume of the super-resolution data may (or may not) reduce concerns about volume mismatch based on the spatial matching procedure discussed above and decrease the systematic uncertainty of radar measurement differences. Different sampling volume size leads to different matching tolerances for the super-resolution data. For example, the time separation tolerance for super-resolution should be smaller than that for legacy-resolution because the required time for a storm to pass over a super-resolution pixel is definitely shorter. Therefore, we performed sensitivity analysis with respect to the variation of both temporal and spatial tolerances.

DATA SOURCES

For a reflectivity data comparison between two ground-based radars, we collected eight years (2003 through May 2010) of Level II radar volume data from the Des Moines and Davenport WSR-88D radars (KDMX and KDVN, respectively) in Iowa. This period includes the recent

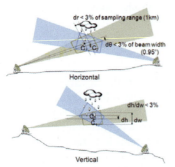

Fig. 1 Spatial match-up of two radar sampling volumes in the horizontal plane (top panel) and in the vertical height (bottom panel).

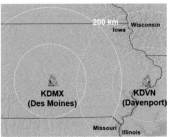

Fig. 2 Locations of the KDMX and KDVN radars. Circles represent every 100 km range from the individual radars.

resolution change of Level II data and can be divided into two periods of the legacy-resolution (2003–May 2008) and super-resolution (June 2008–2010) data. The new super-resolution data (Torres & Curtis, 2007) provide enhanced resolution of 0.5° in azimuth and 250 m in range compared to the legacy-resolution data of 1.0° by 1 km. The distance between the two radars is about 260 km, and the elevation difference between two radar sites is about 70 m. Figure 2 shows the locations of the KDVN and KDMX radars, and each range ring in Fig. 2 represents every 100 km distance centred on both radars. Since the radar beam propagation under non-standard atmospheric conditions might considerably affect reflectivity measurement values, information on the anomalous propagation (AP) for every radar volume data was constructed using an adaptation of the algorithm by Steiner & Smith (2002). Radar volume data consist of about nine elevation angle data, but only the lowest three elevation angle data are compared to obtain matched pairs near the ground.

REFLECTIVITY COMPARISON

The comparison results are organized by the period of radar volume data resolution (legacy- and super-resolution), as applying different tolerance values might affect the sample size and statistical properties of the obtained results.

Legacy-resolution

The legacy-resolution period for the reflectivity comparison is defined as 2003 through May 2008, since the radar Level II data format changed in early May and June of 2008 for the KDMX and KDVN radars, respectively. The legacy-resolution data for the KDVN and the super-resolution data for the KDMX coexist from early May through early June of 2008. This transition period was excluded from the comparison analysis due to the complexity and difficulty of matching the different sizes of radar sampling volumes.

To match sampling volumes from the two radars, we applied four parameters (30-s for time separation and 97% agreement for azimuth, range, and height tolerances) and initially obtained 5278 matched pairs over the legacy-resolution data period. To reduce the possibility of including mismatch cases arising from anomalous beam propagation, we used two-dimensional AP clutter maps generated using every volume scan data over the entire period. Since certain atmospheric conditions (e.g., temperature inversion) in the vicinity of radar can be a source of anomalous beam propagation, AP clutter maps were used to compute the AP fraction for the range of 20–30 km from the radar. The range within 20 km was not considered in order to exclude ground clutter caused by radar side-lobes. We computed the AP fraction for all matched cases obtained and removed some of the cases where the fraction value is greater than 0.2 based on the preliminary analysis for the variation of the standard deviation of reflectivity differences with respect to the computed AP fraction. Figure 3 shows the scatter plots after eliminating anomalous radar beam propagation cases from matched samples. The annual mean values of reflectivity difference (defined by KDMX-KDVN) in Table 1 indicate that the relative calibration bias changed somewhat from year to year. For example, there was approximately a 1.5 dBZ change in mean

Table 1 Annual statistics values of reflectivity (dBZ) differences (defined by KDMX-KDVN).

Year	Legacy–resolution						Super–resolution		
	2003	2004	2005	2006	2007	2008	2008	2009	2010
Correlation	0.92	0.93	0.91	0.94	0.93	0.95	0.78	0.84	0.87
Mean	0.01	0.96	0.23	0.46	−1.17	−0.95	−1.94	−2.73	−1.56
Std.	3.23	3.29	3.48	3.55	2.92	2.48	6.59	6.03	5.04

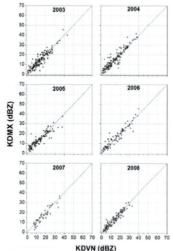

Fig. 3 Scatter plots of reflectivity values of the matched pairs after eliminating anomalous radar beam propagation cases (legacy-resolution period).

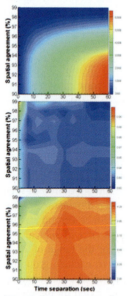

Fig. 4 Contour maps for statistical properties of reflectivity differences with spatial and temporal tolerance variation.

difference values between 2006 and 2007. While Table 1 demonstrates that the KDMX radar was hotter (on average) than the KDVN until 2006, the situation was changed in 2007. A 1 dBZ difference (e.g. in 2007 and 2008) may result in a significant difference in rainfall estimation for strong convective storm cases. For example, a 1 dBZ difference causes at least a 14 mm/h difference in rainfall intensity at ranges higher than 50 mm/h (corresponding to 49 dBZ in radar reflectivity) when using the NEXRAD Z-R relation (Fulton *et al.*, 1998).

Super-resolution

Sensitivity analysis Due to the smaller sampling volume than that of legacy-resolution, one might consider using enhanced tolerance for time separation to obtain the coincidence of common meteorological targets within the smaller horizontal extent (250 m). Since the smaller sampling volume may offer an opportunity to reduce the uncertainty arising from volume mismatch using the same spatial tolerances used for legacy-resolution, we investigate the sensitivity of temporal and spatial tolerances with regard to the statistical properties of matched samples. Based on the parameter (tolerance) sensitivity analysis, the reasonable range of temporal and spatial tolerances should be applied for the super-resolution reflectivity data comparison. For the spatial agreement, we considered all three-dimensional components of radar sampling volume, which are tolerances in azimuth, range, and vertical directions. We used the maximum range of 60-s for time separation and 90% agreement for spatial tolerance (all three components have at least 90% agreement) and obtained a large (about 670 000) matched sample dataset. From this obtained sample data, we eliminated AP cases defined as more than 20% AP contamination at the range of 20–30 km from both radars, reducing the sample size to about 68 000 over a two-year super-resolution data period.

The statistical features we consider include sample size (the number of matched pairs obtained), the correlation coefficient, and the standard deviation of reflectivity differences. Figure 4 shows the contour map for the number of matched pairs, correlation coefficient, and standard deviation of reflectivity differences with the variation of tolerances. We estimated the representative values of spatial agreement (vertical axis in Fig. 4) using the mean of three spatial tolerance (azimuth, range, and vertical height) values. The matched sample size (top panel) increases with a longer time separation and lower (poor) spatial agreement, as expected. The sample size becomes sensitive to the variation in time separation as the spatial agreement value decreases, which implies that the sample size does not change much with time separation at higher values of spatial agreement (for example, 99% agreement). For the distribution of the correlation coefficient (middle panel) between paired reflectivity values for given temporal and spatial consideration, time separation within 5 s looks significant. It is likely that mismatch cases cause the irregular distribution of the lower correlation patterns. For the standard deviation of reflectivity differences (bottom panel), a similar aspect to correlation distribution (middle panel) is observed in the upper-left corner. Since the standard deviation value is >3 dBZ even in a very strict tolerance range, the contour map shows the large variability of reflectivity differences.

Relative bias Although the sensitivity analysis results show that a temporal separation smaller than 5 s can offer better statistical agreement between reflectivity data observed from different radars, selecting such a small value significantly reduces the number of matched pairs. Therefore, we applied a 10-s separation and 97% agreement for the super-resolution data as a compromise between statistical agreement of reflectivity differences and obtained sample size. Figure 5 eliminates AP cases and shows the scatter plots of the obtained super-resolution samples. The quantitative information on this comparison is presented in Table 1. The mean difference values in Table 1 reveal that reflectivity values of the KDVN radar were consistently greater than those of the KDMX radar for the super-resolution period. The observed difference from the legacy-resolution data in 2007 and 2008 also showed that the KDVN radar was hotter. However, super-resolution shows much higher variability in reflectivity differences than legacy-resolution due to a rough tolerance for time separation and perhaps other reasons.

We also performed a statistical consistency test (two-sample t-test; Moore, 2003) for the change of reflectivity differences over time. The statistics (mean, standard deviation, and sample size) used for the two-sample t-test were estimated month by month to provide statistical significance based on a reasonable sample size. The null-hypothesis for the test (two-sided) is that the mean reflectivity differences between the current month and the previous period are consistent. Due to the small sample size (there is no matched sample for several months), the legacy-resolution period was not included in this analysis.

Figure 6 illustrates the sample size, the average difference and the standard deviation for individual and accumulated months (accumulation is defined as sample aggregation of consecutive months whose statistical properties are consistent), and break points of the consistency in reflectivity difference over the 24 months of the super-resolution period (June 2008–May 2010). Since monthly mean difference and standard deviation (the second panel from the top in Fig. 6) seem largely variable, we accumulated samples from the test which turned out to be consistent. Consistent samples of consecutive months, thus, are accumulated, and the new statistics values (mean, standard deviation, and sample size) for the accumulated months are computed to perform the two-sample t-test. In the next step of this analysis, the test compares the statistics values of the accumulated months with those of the subsequent month. As shown in Fig. 6, two break moments over 24 months were observed. At those moments (between October and November 2008; February and March 2010), the statistical consistency (null hypothesis) was violated because the monthly mean of reflectivity differences was abruptly changed for various reasons (for example, radar calibration). However, we have not verified that the observed change of reflectivity differences was caused by radar calibration due to lack of information on calibration procedures and the schedule for both radars.

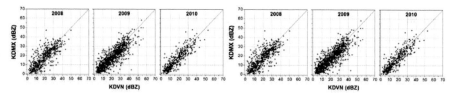

Fig. 5 Scatter plots of reflectivity values of the matched pairs after eliminating anomalous radar beam propagation cases (super-resolution period).

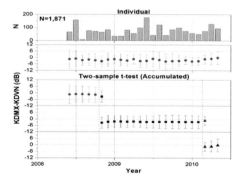

Fig. 6 Sample size, reflectivity difference mean and standard deviation for individual months, and the result of the statistical consistency test for accumulated months.

SUMMARY AND DISCUSSION

This study yielded development of a methodology that matches and pairs the collocated and coincident sampling volumes from two different ground-based radars. We considered temporal separation and 3-D spatial agreement (in azimuth, range, and vertical height) of sampling volumes from two different radars. Eight years of radar volume scan data which contain the recent upgrade of data resolution (called super-resolution) was collected from the KDVN and KDMX radars in

Iowa. Since the proposed method assumes radar beam propagation under the standard atmospheric condition, anomalous propagation cases were eliminated from the matched data sample using an adaptation of the algorithm by Steiner & Smith (2002).

Since the radar sampling volume of the super-resolution data is eight times smaller than that of legacy-resolution (see e.g. Torres & Curtis 2007; Seo & Krajewski 2010), we used different tolerances for the volume matching. For example, when considering storm velocity and terminal velocity of raindrops within smaller sampling volumes of super-resolution, the same time separation parameter of legacy-resolution (i.e. 30-s) might lead to a temporal mismatch of sampling volumes. We performed sensitivity analysis of statistical properties of matched reflectivity data with regard to the variation of temporal and spatial tolerances. The analysis showed that a stricter tolerance range in time and space provides better statistical agreement of matched reflectivity data, but it results in an extremely small number of samples. As a compromise, 10-s for time separation and 97% for spatial agreement parameters were used for the super-resolution data matching.

The scatter plots (Figs 3 and 5) of matched samples for both legacy- and super-resolution show somewhat clear systematic differences. Due to the small sample size obtained from the legacy-resolution data, we only used the super-resolution data to statistically test the temporal consistency of reflectivity differences. From the two-sample t-test, we found two inconsistent moments of reflectivity differences characterized by monthly statistics (sample size, mean, and standard deviation). These two break points could be related to calibration procedures for one of the radars (KDVN and KDMX). However, we could not confirm what caused the observed change of reflectivity difference due to the lack of operational information on calibration procedures and the schedule for both radars.

Since this is a preliminary study motivated by multiple radar data merging, strategies on how to apply the estimated relative biases and how to consider systematic features (e.g. range-dependent bias) of radar observations remain to be developed.

Acknowledgements This work was supported by the US National Science Foundation as part of the development of the Hydro-NEXRAD 2 software system.

REFERENCES

Atlas, D. & Mossop, S. C. (1960) Calibration of a weather radar by using a standard target. *Bull. Amer. Meteor. Soc.* 41(7), 377–382.

Brandes, E. A., Vivekanandan, J. & Wilson, J. W. (1999) A comparison of radar reflectivity estimates of rainfall from collocated radars. *J. Atmos. Oceanic Technol.* 16(9), 1264–1272.

Crum, T. D. & Alberty, R. L. (1993) The WSR-88D and the WSR-88D operational support facility. *Bull. Amer. Meteor. Soc.* 74(9), 1669–1687.

Fulton, R. A., Breidenbach, J. P., Seo, D. -J., Miller, D. A. & O'Bannon, T. (1998) The WSR-88D rainfall algorithm. *Wea. Forecasting* 13(2), 377–395.

Moore, S. D. (2003) *The Basic Practice of Statistics.* W. H. Freeman and Company, New York, USA.

Seo, B. -C. & Krajewski, W. F. (2010) Scale dependence of radar-rainfall uncertainty: initial evaluation of NEXRAD's new super-resolution data. *J. Hydrometeor.* 11(5), 1191–1198.

Smith, J. A., Seo, D. J., Baeck, M. L. & Hudlow, M. D. (1996) An intercomparison study of NEXRAD precipitation estimates. *Water Resour. Res.* 32, 2035–2045.

Steiner, M. & Smith, J. A. (2002) Use of three-dimensional reflectivity structure for automated detection and removal of nonprecipitating echoes in radar data. *J. Atmos. Oceanic Technol.* 19(5), 673–686.

Torres, S. M. & Curtis, C. D. (2007) Initial implementation of super-resolution data on the NEXRAD network. Preprints. In: *23rd Conf. on International Interactive Information and Processing Systems (IIPS) for Meteorology, Oceanography, and Hydrology* (San Antonio, Texas, Amer. Meteor. Soc.).

A quality evaluation criterion for radar rain-rate data

CHULSANG YOO, JUNGSOO YOON, JUNGHO KIM, CHEOLSOON PARK & CHANGHYUN JUN

School of Civil, Environmental and Architectural Engineering, College of Engineering, Korea University, Seoul 136-713, Korea
envchul@korea.ac.kr

Abstract This study proposed a radar rain-rate quality criterion (RRQC), a measure of goodness for the radar rain-rate. The RRQC proposed is based on the similar concept of total variance in the statistical analysis of variance, which considers both the bias and variability of radar rain-rate with respect to the raingauge rain-rate. The RRQC was estimated for three storm events with the raw radar data, along with improved versions based on G/R correction and merging by co-Kriging. Additionally, these radar data were applied to the runoff analysis of the Choongju Dam Basin, Korea. By investigating the relation between the RRQC in the rain-rate input and the errors in the runoff output, a minimum quality of radar rain-rate applicable to the rainfall–runoff analysis was explored.

Key words radar rain-rate; RRQC; G/R ratio; co-Kriging; rainfall–runoff analysis

INTRODUCTION

Due to the temporal and spatial variability and intermittency of rain-rate, it is not easy to measure the rain-rate field with sufficiently good quality. The ground raingauge measurement is generally assumed to be the "truth", which is nothing but point values measured discretely in space. However, the radar can provide rain-rate information continuously in space, and also the temporal gap between consecutive measurements is small enough to be assumed continuous in its application. Ideally, the radar can overcome all the problems that the raingauge network has (Wilson & Brandes, 1979; Yoo *et al.*, 2010). However, the high potential of radar for rain-rate measurement is deteriorated in actual practice, especially for flood forecasting (Krajewski & Smith, 2002). There are many problems to be overcome regarding rain-rate observed using radar (Seo, 1998; Chumchean *et al.*, 2006). It is obvious that by simply introducing the radar, not only the accuracy of rain-rate measurements, but also their usefulness in a rainfall–runoff analysis may not be guaranteed.

Another practical problem for the use of radar measurements lies in the runoff analysis. Selection of a rainfall–runoff model and estimation of its model parameters, as well as the intrinsic nonlinearity of the rainfall–runoff processes and error involved in the runoff measurements, make the use of radar measurements more complicated. Lack of a proper measure for the evaluation of radar rain-rate is also a serious problem, which, especially, hinders the basic evaluation of the usefulness of radar rain-rate in its application to the runoff analysis. The quality of radar rain-rate should be evaluated by considering the runoff characteristics like the total volume, peak flow, and peak time of runoff.

This study proposes a measure of goodness for the evaluation of radar rain-rate. This measure considers both the bias and variability of radar rain-rate with respect to the raingauge rain-rate. Additionally, radar measurements of three different storm events are quantified by the measure proposed in this study, which is then analysed in relation to the runoff result in three different sub-basins of the Choongju Dam Basin, Korea.

Quality evaluation methodology of radar rain-rate

This study sets the raingauge rain-rate as the basis for the evaluation of the radar rain-rate, i.e. that all the raingauge rain-rates are assumed to be true. If the radar rain-rate is perfect, pairs of radar and raingauge rain-rate will make a line with unity slope. If the radar rain-rate is not perfect but still has a strong linear relation with the raingauge rain-rate, the slope between them will obviously be different from one. Most pairs of radar and raingauge rain-rate data may be located very near the trend line in this case, and thus the mean deviation of these data pairs from the trend line would be very small. However, it could be larger if the relation between the radar and raingauge rain-rate is weak, even in the case that the slope of the trend line is near unity.

The above explanation concerning the radar rain-rate is the basic idea for its evaluation. Especially, the quality of radar rain-rate is targeted for the application of rainfall–runoff analysis. Both the bias (or the difference in their means) and the variability (or the uncertainty) of radar rain-rate much affect the runoff analysis. Thus, the quality criteria should consider both the bias and variability. In this study, the bias is quantified by the concept of "Sum of Squares Bias (SSB)" and the variability "Sum of Squares Error" (SSE). This concept of data evaluation is very similar to that in the analysis of variance (ANOVA). The SSB and SSE in this study are defined as follows:

$$SSB = \sum_{i=1}^{n} (\hat{x}_i - \bar{x}_{45})^2 \tag{1}$$

$$SSE = \sum_{i=1}^{n} (x_i - \hat{x}_i)^2 \tag{2}$$

where n represents the number of pairs of radar and raingauge rain-rate data, x_i is the radar rain-rate data, $\hat{x}_i$ is the value on the trend line (between radar and raingauge rain-rate) corresponding to the radar rain-rate, and $\bar{x}_{45}$ is the value on the 1:1 line (45 degree line) corresponding to the radar rain-rate.

It is also possible to introduce the "Mean Squares Bias" (MSB) and "Mean Squares Error" (MSE), which are the SSB and SSE divided by their degrees of freedom. Also, the MSB and MSE constitute the "Mean Square Total" (MST), which could be a measure of quality for the radar rain-rate considering both the bias and variability. These concepts are summarized in Table 1, similar to the ANOVA table.

Table 1 Evaluation table of radar rain-rate.

	Degree of freedom	Sum of squares	Mean of squares
Bias	1	SSB	MSB
Error	n – 2	SSE	MSE
Sum	n – 1	SST	MST

The above MST measure, however, cannot be applied to compare different storm events with different numbers of data pairs. Even though MSB and MSE are estimated by considering their degree of freedom, the MSB is basically proportional to the number of data pairs. Also, the value of MST itself is too big to be used as a measure of data quality. Thus, in this study, we proposed a standardized measure of goodness for the radar rain-rate, which is named the radar rain-rate quality criterion (RRQC):

$$RRQC = \left(1 - \frac{MST_{radar}}{MST_{x=0}}\right) \times 100\% \tag{3}$$

where MST_{radar} is the same as the MST in Table 1, which is the sum of MSB and MSE. Also, $MST_{x=0}$ means the MST when all the radar rain-rates are assumed zero. In this case, the MSE becomes zero and only the MSB remains. This MSB can be calculated by equation (1) for the line x = 0 instead of the trend line derived from the radar and raingauge rain-rates observed, i.e.

$$MST_{x=0} = MSB_{x=0} = \sum_{i=1}^{n} \bar{x}_{45°}^2 = \sum_{i=1}^{n} y_i^2 \tag{4}$$

where n is the number of data pairs and y_i represents the raingauge rain-rate data observed.

STORM EVENTS, STUDY BASIN AND QUALITY OF RADAR RAIN-RATE

Storm events and study basin

This study considers three storm events for the evaluation of radar rain-rate, and also for the runoff analysis (Table 2). Storm event #1 is the typhoon Rusa in 2002, #2 a convective storm in 2003, and #3 a frontal storm in 2003. The radar rain-rate data analysed were limited to those within the Choongju Dam Basin in Korea to ensure consistency in both the rain-rate and runoff data analysis. The Choongju Dam Basin is located in the middle of the Korean Peninsula, its basin area is 6656.6 km^2 (Fig. 1(c)). A total of 53 raingauges within and near the Choongju Dam Basin were considered in this study, of which 26 raingauges were used for the mean-field bias correction and the data merging by co-Kriging. The remaining 27 raingauges were used for the evaluation of the radar rain-rate data. The radar data used were those observed by the Mt Kwanak Radar, about 100 km distant from the Choongju Dam Basin. CAPPI radar data at 1.5 km height were used and reflectivity values were converted to rain-rate by the Z-R relationship (a = 200, b = 1.6).

This study divided the Choongju Dam Basin into three sub-basins for the runoff analysis (Fig. 1). The smallest one has an area of 2302.2 km^2 and the largest 6656.6 km^2. Sub-basin 1 in the upstream part of the Choongju Dam Basin is located in the mountain area, and sub-basin 3, the largest one, is the Choongju Dam Basin itself. The purpose of this basin subdivision is to consider the effect of basin area on the runoff analysis, which seems more significant especially when the size of a basin is rather small.

Table 2 Storm events considered for the evaluation of radar rain-rate.

Storm event	Period	Duration (h)
#1	2002.08.31 01:00 h ~ 2002.09.01 07:00 h	31
#2	2003.06.27 01:00 h ~ 2003.06.27 23:00 h	23
#3	2003.08.24 08:00 h ~ 2003.08.27 10:00 h	74

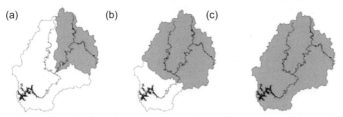

Fig. 1 Chungju Dam Basin and its basin subdivision. (a) Sub-basin 1 (2302.2 km^2), (b) Sub-basin 2 (4791.2 km^2), (c) Sub-basin 3 (6656.6 km^2).

Quality of radar rain-rate

The raw radar rain-rate data were found to be very poor with their RRQC under 10% (Table 3). However, the quality of radar rain-rate data corrected by applying the G/R ratio and those merged by applying co-Kriging was found to be significantly improved. Especially, the radar rain-rate data merged by applying co-Kriging gave more than 70% for RRQC in all three storm events considered. On the other hand, the RRQC of the data corrected by applying the G/R ratio was varied. It is obvious that the co-Kriging considering both the mean and spatial correlation structure of rain-rate is the more effective method for quality improvement of radar rain-rate data than the G/R ratio considering only the mean. Also, storm event #2 showed the best data quality, but storm event #3 the worst. Especially storm event #3 for sub-basin 1 showed very low quality with its RRQC less than 10%. The relatively low quality of radar rain-rate for sub-basin 1, located in the upstream and mountain area of the basin, may partially be due to the orographic effect.

Table 3 The RRQCs estimated for three sub-basins and three storm events.

Sub-basin	Storm event	Raw	G/R	co-Kriging
1	#1	7.1	57.4	91.1
	#2	8.0	92.5	97.8
	#3	0.4	9.0	76.9
2	#1	7.3	57.4	88.5
	#2	9.0	95.9	95.9
	#3	2.4	27.2	69.2
3	#1	8.2	65.1	93.1
	#2	9.7	95.8	97.6
	#3	2.0	25.7	69.7

QUALITY OF RADAR RAIN-RATE IN RUNOFF ANALYSIS

Runoff analysis

ModClark, a hydrological distributed rainfall–runoff analysis model, was used for the runoff analysis in this study. The model parameters used were those estimated by Yoo & Shin (2010) for the same basin. Figure 2 shows the runoff simulation results using the radar rain-rate evaluated in the previous section. Obviously the radar rain-rate merged by co-Kriging produced the best results. The raw radar rain-rate produced the worst results, as was expected.

These runoff analysis results show that there is a strong correlation between the rain-rate input and runoff output. Many researches have reported similar findings (Sharif *et al.*, 2004; Vischel & Lebel, 2007); however, it is rare to relate the quality of radar rain-rate with that of runoff.

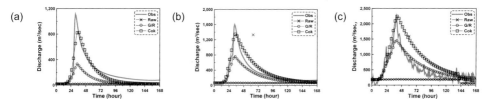

Fig. 2 Runoff simulation results for storm event #2: (a) Sub-basin 1, (b) Sub-basin 2, (c) Sub-basin 3.

Relation between rain-rate error and runoff error

This study considered the following three runoff errors of total volume, peak flow, and peak time. These are defined as follows:

$$Error_{volume}(\%) = \left| \frac{\sum_{i}(Q_{R_i} - Q_{Obs_i})}{\sum_{i} Q_{Obs_i}} \times 100 \right| \qquad (5)$$

$$Error_{peak}(\%) = \left| \frac{P_R - P_{Obs}}{P_{Obs}} \times 100 \right| \qquad (6)$$

$$Error_{peak\,time}(\%) = \left| \frac{T_R - T_{Obs}}{T_{Obs}} \times 100 \right| \qquad (7)$$

where Q_{R_i} is the runoff simulated at time i and Q_{Obs_i} is the runoff observed at time i. P_R is the peak flow simulated and P_{Obs} is the peak flow observed. Finally, T_R is the peak time simulated

Table 4 Runoff errors estimated for the total volume, peak flow and peak time.

Sub-basin	Storm event	Total volume (%)			Peak flow (%)			Peak time (%)		
		Raw	G/R	co-Kriging	Raw	G/R	co-Kriging	Raw	G/R	co-Kriging
1	#1	85.8	45.3	47.4	95.2	25.2	119.3	11.1	11.1	0.0
	#2	89.7	63.8	17.6	98.0	70.1	24.4	12.9	9.7	12.9
	#3	78.8	62.3	45.5	93.6	76.0	49.6	9.1	9.1	0.0
2	#1	91.6	62.0	22.3	98.2	60.3	11.1	9.4	9.4	0.0
	#2	80.1	30.3	15.8	95.7	52.1	16.0	6.1	0.0	3.0
	#3	74.0	44.4	31.0	92.0	61.9	38.4	9.4	3.1	0.0
3	#1	81.7	42.5	9.8	95.9	34.0	16.0	9.4	12.5	3.1
	#2	72.6	1.4	43.9	93.1	36.7	2.1	2.6	0.0	2.6
	#3	72.5	28.4	17.1	87.3	38.7	5.1	5.7	0.0	5.7

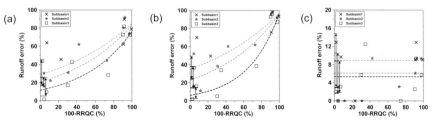

Fig. 3 The relation between the errors in the radar rain-rate and the runoff: (a) Total volume, (b) Peak flow, (c) Peak time.

and T_{Obs} is the peak time observed. Table 4 summarizes the runoff error estimated for the three sub-basins and three storm events considered.

Figure 3 shows the relation between the rain-rate error and the runoff error. The rain-rate error was quantified by 100-RRQC, which was introduced to make the relation between rain-rate error and the runoff error be proportional.

As can be seen in Fig. 3(a), the relation between the rain-rate error 100-RRQC and the error in the total volume of runoff (hereafter, the total volume error) is not linear. Even though the general relation is satisfied, such that the bigger rainfall error leads to the bigger total volume error, a lower limit of the total volume error obviously exists. This lower limit of the total volume error seems to be from the runoff model selection, model parameters, and/or runoff data observed. Even though only three sub-basins were compared in this study, it was obvious that the smaller the sub-basin the larger the lower limit of the total volume error. This trend was also the same as in that between rainfall error and the error in the peak flow (hereafter, called the peak flow error). However, the relation between rain-rate error and error in the peak time (hereafter, called the peak time error) was rather odd, which seems insensitive to RRQC.

When using the radar rain-rate for the runoff analysis, it is very important to get an idea about the quality of radar rain-rate required. However, the acceptable maximum runoff error should firstly be decided. As can be seen in Fig. 3, the minimum total volume error and the minimum peak flow error of sub-basin 1 are all about 30% with their uncertainty range about +15 to –10% for the significance level of 5%. Assuming the acceptable maximum error is 45% considering the minimum error 30% with its uncertainty 15%, sub-basin 1 should have RRQC higher than 58%. This also becomes 66% if considering the peak flow error.

The total volume error and peak flow error for sub-basins 2 and 3 are much lower than that of sub-basin 1. These smaller errors seem to be due to the larger sub-basin area, i.e. the possible variability of rain-rate in space and the uncertainty in the runoff processes seem to be compromised to make the overall uncertainty in runoff calculation become much less. For sub-basin 2, the lower limit of total volume error was estimated to be about 26%. To secure this amount of error along with its uncertainty, sub-basin 2 should have RRQC higher than 83%. This also becomes 65% if considering the peak flow error. For sub-basin 3, RRQC should be higher than 58%, which also becomes 68% if considering the peak flow error.

The above results are only for the three sub-basins and three storm events considered in this study. As the number of sub-basins and storm events are very small, it may be hard to draw any general guidelines on the quality criteria of radar rain-rate for its application to the runoff analysis. However, it is obvious that an upper quality limit on runoff estimation exists, so simply perfect radar rain-rate may not lead to perfect runoff estimation. In that sense, the radar rain-rate information with moderate quality could lead to a satisfactory runoff estimation result. In this application example, the radar rain-rate for sub-basins 1, 2, and 3 should have RRQC higher than 66%, 83%, and 68%, respectively.

CONCLUSIONS

This study proposed a radar rain-rate quality criterion (RRQC), a measure of goodness for the evaluation of the radar rain-rate. The proposed RRQC is based on the similar concept of total variance used in the statistical analysis of variance. RRQC was estimated for three storm events with raw radar rain-rate along with improved versions based on G/R correction and merging with raingauge data by co-Kriging. Additionally, these radar data were applied to the runoff analysis of the Choongju Dam Basin, Korea. By analysing the relation between the RRQC in the rain-rate input and the errors in the runoff output, a minimum quality of radar rain-rate applicable to the rainfall–runoff analysis was explored.

The RRQC of the raw radar rain-rate data was estimated at worst to be <10%, and when merged with raingauge data by co-Kriging was estimated at best to be >70% for all three storm events considered. However, the RRQC of the data corrected by applying the G/R ratio was varied. It is obvious that co-Kriging, which considers both the mean and spatial correlation structure of radar rain-rate, is a more effective method for the quality improvement of radar rain-rate data. In the runoff simulation, the radar rain-rate merged with the raingauge rain-rate by the co-Kriging method also produced the closest correspondence with observed runoff.

It was found that larger rainfall error leads to greater runoff error; however, a lower limit of the runoff error obviously exists. This lower limit seems to derive from the runoff model selection, model parameters, and/or runoff data observed. As an upper quality limit of runoff estimation exists, the radar rain-rate information with moderate quality could lead to a satisfactory runoff estimation result. In this example application to the Choongju Dam Basin in Korea, the radar rain-rate with its RRQC of >70% would lead to a satisfactory runoff simulation result.

Acknowledgements This work was supported by the National Research Foundation of Korea (NRF) grant funded by the Korea government (MEST) (no. 2010-0014566).

REFERENCES

Chumchean, S., Seed, A. & Sharma, A. (2006) Correcting of real-time radar rainfall bias using a Kalman filtering approach. *J. Hydrol.* 317, 123–137.
Krajewski, W. F. & Smith, J. A. (2002) Radar hydrology: rainfall estimation. *Adv. Water Resour.* 25, 1387–1394.
Seo, D. J. (1998) Real-time estimation of rainfall fields using radar rainfall and rain gage data. *J. Hydrol.* 208, 37–52.
Sharif, H. O., Ogden, F. L., Krajewski, W. F. & Xue, M. (2004) Statistical analysis of radar rainfall error propagation. *J. Hydrometeorol.* 5, 199–212.
Vischel, T. & Lebel, T. (2007) Assessing the water balance in the Sahel: impact of small scale rainfall variability on runoff. Part 2: Idealized modeling of runoff sensitivity. *J. Hydrol.* 333, 340–355.
Wilson, J. W. & Brandes, E. A. (1979) Radar measurement of rainfall: a summary. *Bull. Am. Met. Soc.* 60, 1048–1058.
Yoo, C. & Shin, J. W. (2010) Decision of storage coefficient and concentration time of observed basin using Nash model's structure. *J. Korea Water Resour. Assoc.* 43(6), 559–569.
Yoo, C., Yoon, J. & Ha, E. (2010) Sampling error of area average rainfall due to radar partial coverage. Stoch. *Environ. Res. Risk Assess.* 24(8), 1097–1111.

388

Weather Radar and Hydrology
(Proceedings of a symposium held in Exeter, UK, April 2011) (IAHS Publ. 351, 2012).

Radar Quality Index (RQI) – a combined measure for beam blockage and VPR effects in a national network

JIAN ZHANG[1], YOUCUN QI[2,3], CARRIE LANGSTON[2] & BRIAN KANEY[2]

1 *National Severe Storms Lab, 120 David L Boren Blvd., Norman, Oklahoma 73072, USA*
jian.zhang@noaa.gov

2 *Cooperative Institute for Mesoscale Meteorological Studies, University of Oklahoma, 120 David L Boren Blvd, Norman, Oklahoma 73072, USA*

3 *Nanjing University of Information Science and Technology, Nanjing, China*

Abstract The next-generation multi-sensor quantitative precipitation estimation (QPE), or "Q2", is an experimental hydrometeorological system that integrates data from radar, raingauge, and atmospheric models and generates high-resolution precipitation products on a national scale in real-time. The quality of the Q2 radar QPE varies in space and in time due to a number of factors, which include: (1) errors in measuring radar reflectivity; (2) segregation of precipitation and non-precipitation echoes; (3) uncertainties in Z–R relationships; and (4) variability in the vertical profile of reflectivity (VPR). In the current study, a Radar QPE Quality Index (RQI) field is developed to present the radar QPE uncertainty associated with VPRs. The RQI field accounts for radar beam sampling characteristics (blockage, beam height and width) and their relationships with respect to the freezing level. A national RQI map is generated by mosaicking single radar RQI fields. The radar quality information is useful to hydrological users and can add value in radar rainfall applications.

Key words radar QPE quality; beam blockage; VPR; national radar network

INTRODUCTION

The next-generation multi-sensor QPE ("Q2", Vasiloff *et al.*, 2007; Zhang *et al.*, 2011) is an experimental hydrometeorological system that integrates data from radar, raingauge, and atmospheric models and generates high-resolution precipitation products on the national scale in real-time. One of the Q2 products is the radar-based precipitation estimation (called "Q2rad" hereafter), which is generated by mosaicking precipitation fields derived from individual radars in the United States Weather Surveillance Radar – 1988 Doppler (WSR-88D) network. The Q2rad process includes reflectivity quality controls to remove non-precipitation echoes and a precipitation classification to segregate convective, stratiform, hail, tropical rain, and snow in radar reflectivity fields. Different Z-R relationships are applied for different precipitation types. The Q2rad products are generated for the Conterminous United States (CONUS) every 5-min and has a spatial resolution of ~1 km. The product suite contains precipitation rate, and 1- to 72-h accumulations.

The quality of the Q2rad product varies in space and in time due to a number of factors, which include: (1) errors in measuring radar reflectivity, e.g. the calibration bias; (2) contaminations from non-precipitation echoes, e.g. ground clutter due to anomalous propagations; (3) uncertainties in Z–R relationships; and (4) variability in the vertical profile of reflectivity (VPR). The radar QPE error due to factor (1) can be minimized through a close monitoring and vigorous maintenance of the radar network. Radar QPE errors due to factor (2) may be minimized with dual-polarization capabilities (e.g. Ryzhkov *et al.*, 2005; Park *et al.*, 2009). The uncertainty associated with Z-R relationships can be significant, and the understanding of this uncertainty requires observations of drop size distributions in different precipitation regimes. The error associated with Z-R uncertainties is expected to decrease with dual-polarization capabilities as well, since additional radar variables can be used for hydrometeor classifications (e.g., Ryzhkov *et al.*, 2005; Giangrande & Ryzhkov, 2008).

The radar QPE uncertainty associated with VPRs is closely related to sampling strategies of scanning radars and may not be reduced by the dual-polarization capabilities. Specifically, this uncertainty is related to the height and width of radar beams as well as their relationships with the vertical structure of precipitation. The current study attempts to develop a real-time national Radar QPE Quality Index, or RQI, that can represent relative qualities of the radar QPE in space and in time. One objective is to use the RQI field as a weighting factor when merging the radar QPE with other QPEs on the national scale. Previous studies such as Pellarin *et al.* (2002) had attempted to quantify radar QPE errors by modelling VPR effects. However, uncertainties still exist with the models, given assumptions about the spatial uniformity of the VPR. For practical reasons of a real-time and national implementation, a simple mathematical formulation similar to Friedrich *et al.* (2006) is adapted in the current study.

The mathematical formulation of the RQI is described in the next section. Then example RQI fields and their relationships with the radar QPE accuracy in a national network are presented. A summary and discussions about the future work are provided in the last section.

METHODOLOGY

The dependency of radar QPE accuracies on VPRs is illustrated in Fig. 1. Assuming an idealized stratiform precipitation is horizontally uniform and produces the same amount of rainfall everywhere at the surface. Then the VPR (blue lines in Fig. 1) of the precipitation is also horizontally uniform. It has a negative dZ/dh slope in the ice region, a peak (bright band) near the freezing level and a zero dZ/dh slope below the bright band.

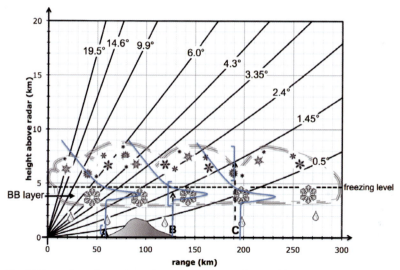

Fig. 1 Illustration of scanning radar sampling issues for stratiform precipitation in complex terrain. The blue lines represent the VPR of the precipitation. The radar is located at the origin, and the bold black lines represent axis of various radar tilts. The numbers denote the corresponding elevation angles. The blue square brackets represent the 3-dB beam widths of the 1st tilt at point A, and the 2nd tilt at points B and C.

Assuming a radar is located at the origin and scans in a mode with nine elevation angles (Fig. 1): at point A, radar data from the 1st tilt is used to estimate rainfall. At points B and C, radar QPEs are derived from the 2nd tilt because of a significant blockage in the 1st tilt at ~80–100 km range. Assuming an accurate Z-R relationship is applied to the unblocked radar reflectivity closest to the ground, the radar rainfall estimate at point A would be accurate. However, the radar QPEs at point B would be overestimating because the data was obtained within the bright band. At point C, the radar QPE would be underestimating due to (1) decreasing of reflectivity in ice region and (2) the radar beam is partially overshooting the cloud top.

Partially blocked data are commonly used for radar QPEs if the blockage is not significant (e.g. data with up to 60% blockages are used in the operational WSR-88D radar QPE after a blockage correction is applied; see Fulton *et al.*, 1998). The amount of blockages is usually pre-determined using static terrain data assuming a standard atmospheric refraction condition. When anomalous propagation occurs, the actual blockage can be much different than the pre-calculated and the radar QPE quality could be degraded.

Based on the discussions above, the RQI field is developed using two simple mathematical formulations:

$$RQI = RQI_{blk} \cdot RQI_{hgt}: \tag{1}$$

$$RQI_{blk} = \begin{cases} 1; & blk \leq 10\% \\ \left(1 - \dfrac{blk - 0.1}{0.4}\right); & 10\% < blk \leq 50\% \\ 0; & blk > 50\% \end{cases} \tag{2}$$

$$RQI_{hgt} = \begin{cases} 1; & h_{0C} > D_{bb} \cap h_a < h_{0C} - D_{bb} \\ \exp\left[-\dfrac{(h_a - h_{0C} + D_{bb})^2}{H^2}\right]; & h_{0C} > D_{bb} \cap h_a \geq h_{0C} - D_{bb} \\ \exp\left[-\dfrac{h_a^2}{H^2}\right]; & h_{0C} \leq D_{bb} \end{cases} \tag{3}$$

The variables in equations (1)–(3) are defined as: RQI_{blk}: RQI based on blockages (dimensionless); RQI_{hgt}: RQI based on radar beam height (dimensionless); h_a: height of the beam axis (in metres above radar level [m ARL]); h_{0C}: height of 0°C (in m ARL); D_{bb}: depth of the bright band layer (in meters; default = 700 m); blk: beam blockages (dimensionless; a value of 0.5 is equivalent to a 50% power blockage); H: a height scale factor (in metres; default = 1500 m).

The simple linear and exponential functions are chosen because of their computationally efficiencies for real-time implementation in a national system. The formulations represent the general characteristics of the radar QPE errors (~0–1st order) associated with the blockages and VPR effects. At any given time, the errors are generally larger in areas with low freezing levels than those with high freezing levels, and the errors are generally larger in complex terrain (large blockages) than in flatlands (no blockages). However, the current RQI does not provide a direct quantitative measure of radar QPE errors. To obtain a quantitative relationship between the two, radar QPE error fields will be derived through comparisons with gauge observations. The error fields will be compared with the RQI fields and quantitative relationships are developed. The advantage of the RQI product is that it is generated every 5-min in real-time, and each radar precipitation rate field has an associated RQI field. Range-dependent bias maps derived from long-term VPRs may not be useful for such small time scales.

RQI PRODUCTS FROM A NATIONAL NETWORK

Radar QPEs are commonly calculated from the lowest radar bins that are not significantly blocked. Those bins constitute a 2-D polar grid called "hybrid scan" (O'Bannon 1997; Fulton *et al.*, 1998). Figure 2 shows an example of such hybrid scan from KDAX radar located in northern California, USA. Due to the complex terrain (Fig. 2(a)) around the radar, there exist significant blockages on the lowest two elevation angles (Fig. 2(b),(c)) of the WSR-88D radar Volume Scan Pattern (VCP) #12 (http://www.ofcm.gov/homepage/text/pubs.htm: *Federal Meteorological Handbook no. 11*). Thus the hybrid scan for VCP 12 (Fig. 2(d)) of this radar consists of data from the 1st to 3rd tilts.

Example RQI fields from a single radar (KDAX) are shown in Fig. 3. The RQI_{blk} (Fig. 3(a)) does not change with time as long as the radar scanning strategy remains the same. On the other hand, RQI_{hgt} can change from volume scan to volume scan because the freezing level height varies with time. A KDAX RQI_{hgt} field on the hybrid scan of VCP 12 and for a freezing level height of 3 km (ARL) is shown in Fig. 3(b). The corresponding RQI field (Fig. 3(c)) indicates that the KDAX radar QPE from this time has relatively good quality within 150 km of range along the northwest – southeast direction, which corresponds to the central valley region (Fig. 2(a)). To the east, the high quality range decreases to ~100 km due to some blockages in the region. The radar QPE quality to the west is much poorer than other areas because of severe blockages at ~40 km west of the radar.

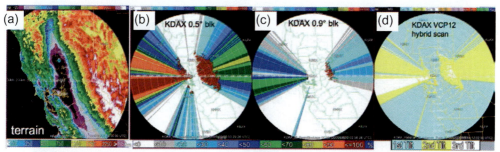

Fig. 2 Terrain height (a) around KDAX radar, and the radar's beam blockages at 0.5° (b) and 0.9° (c) elevation angles. The hybrid scan for VCP 12 is shown in panel (d). The radar site is located at the centre of each image.

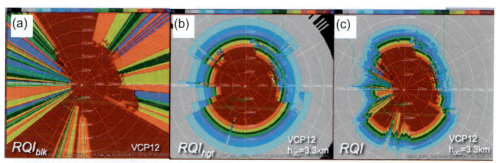

Fig. 3 KDAX RQIblk field for hybrid scans of VCP 12 and RQIhgt (b) for hybrid scan of VCP 12 with a 0°C height of 3.3 km ARL. The corresponding RQI field is shown in panel (c), which is valid at 16:21 UTC on 13 January 2006.

Single radar RQI fields from the WSR-88D network are mosaicked in real-time to generate a national RQI product in the Q2 system (nmq.ou.edu). Figure 4 shows two real-time national RQI maps, one valid at 17:00 UTC on 28 August 2010 (Fig. 4(a)) and another at 17:00 UTC on 10 February 2011 (Fig. 4(b)). The corresponding surface temperature fields are shown in Fig. 4(c) and (d), respectively. The two maps clearly show that the radar QPE quality is better in the warm season than in the cool season. Further, apparent radar coverage gaps exist in the western USA even in the warm season. The national RQI field is updated every 5-min and reflects the real-time radar QPE quality distributions across CONUS under different synoptic regimes, radar scanning strategies, and even radar outage situations.

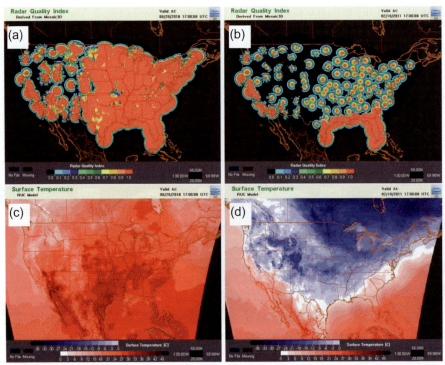

Fig. 4 National RQI fields valid at 17:00 UTC on 28 August 2010 (a) and 17:00 UTC 10 February 2011 (b). The corresponding surface temperature fields are shown in panels (c) and (d), respectively.

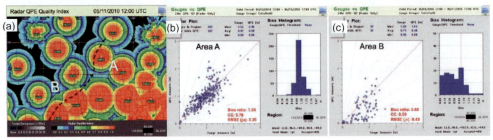

Fig. 5 National RQI field valid at 12:00 UTC on 11 May 2010 (a) and scatter plots of the 24-h radar QPEs ending at 12:00 UTC on 11 May 2010 in areas A (b) and B (c) *vs* gauge observations.

A preliminary evaluation of the Q2 radar QPE accuracy under different RQI distributions indicated a good correlation between the accuracy and RQI. Figure 5(a) shows the RQI field in north central USA region during a storm event on 10–11 May 2010. RQI values in the southeast half of the domain (area "A", Fig. 5(a)) are apparently higher than those in the northwest half (area "B", Fig. 5(a)). The difference in the RQI field is reflected in the radar QPE accuracy when the 24-h radar rainfall estimates from the two areas are compared with gauge observations. For area A, the radar QPE average was within 6% of the gauge observed mean, and had a correlation coefficient of 0.78 and a root-mean-square-error (RMSE) of 0.25 in (Fig. 5(b)). For area B, the radar QPE had a 35% underestimation (Fig. 5(c)), and a much lower correlation coefficient (0.59 *vs* 0.78) and larger RMSE (0.43 *vs* 0.25 in) than in area A. Further studies are underway to refine the RQI product and to develop quantitative relationships between the RQI and the radar QPE errors over the CONUS domain.

SUMMARY AND FUTURE WORK

A Radar QPE Quality Index, or RQI, field was developed in an attempt to show relative qualities of the Q2 national radar QPE in different seasons and different regions. The RQI field accounts for radar beam sampling characteristics and their relationships with respect to the atmospheric freezing level. A national RQI map is generated in real-time with 1-km resolution and 5-min frequency. The RQI field reflects the real-time radar QPE accuracies associated with synoptic environments, radar scanning strategies, and even radar outage situations. It also shows radar coverage voids and provides guidance for the deployment of gap-filling radars.

Future work will include the development of a quantitative relationship between the RQI and the Q2 radar QPE errors. Rainfall data from a national gauge network will be quality controlled and radar QPE errors with respect to the gauge observations derived. The error distributions will be compared with the RQI fields and quantitative relationships between the two will be developed. The error-related RQI field will provide added values for radar QPE product users. Additionally, the real-time RQI can be used as a weighting factor for the radar QPE when it is merged with QPEs from different sensors (e.g. satellite).

REFERENCES

Friedrich, K., Hagen, M. & Einfalt, T. (2006) A quality control concept for radar reflectivity, polarimetric parameters, and Doppler velocity. *J. Atmos. Ocean. Tech.* 23, 865–887.

Fulton, R., Breidenbach, J., Seo, D.-J., Miller, D. & O'Bannon, T. (1998) The WSR-88D Rainfall Algorithm. *Weath. Forecasting* 13, 377–395.

Giangrande, S. E. & Ryzhkov, A. V. (2008) Estimation of rainfall based on the results of polarimetric echo classification. *J. Appl. Meteorol. Clim.* 47, 2445–2462.

O'Bannon, T. (1997) Using a 'terrain-based' hybrid scan to improve WSR-88D precipitation estimates. In: *Preprints, The 28th International Conference on Radar Meteorology*, 7–12 September 1997, 506–507.

Park, H.-S., Ryzhkov, A. V., Zrnić, D. S. & Kim, K.-E. (2009) The hydrometeor classification algorithm for the polarimetric WSR-88D: description and application to an MCS. *Wea. Forecasting* 24, 730–748.

Pellarin, T., Delrieu, G., Saulnier, G. M., Andrieu, H., Vignal, B. & Creutin, J. D. (2002) Hydrologic visibility of weather radar systems operating in mountainous regions: case study for the Ardeche Catchment (France). *J. Hydrometeorology* 3, 539–555.

Ryzhkov, A. V., Schuur, T. J., Burgess, D. W., Heinselman, P. L., Giangrande, S. E. & Zrnic, D. S. (2005) The joint polarization experiment: polarimetric rainfall measurements and hydrometeor classification. *Bull. Amer. Meteorol. Soc.* 86, 809–824.

Vasiloff, S., Seo, D.-J., Howard, K., Zhang, J., *et al.* (2007) Q2: Next generation QPE and very short-term QPF. *Bull. Amer. Met. Soc.* 88, 1899–1911.

Zhang, J., Howard, K. *et al.* (2011) National Mosaic and multi-sensor QPE (NMQ) system: description, results and future plans. *Bull. Amer. Met. Soc.* 92, 1321–1338.

394

Weather Radar and Hydrology
(Proceedings of a symposium held in Exeter, UK, April 2011) (IAHS Publ. 351, 2012).

Probabilistic rainfall warning system with an interactive user interface

JARMO KOISTINEN[1], HARRI HOHTI[1], JANNE KAUHANEN[1], JUHA KILPINEN[1], VESA KURKI[1], TUOMO LAURI[1], ANTTI MÄKELÄ[1], PERTTI NURMI[1], PIRKKO PYLKKÖ[1], PEKKA ROSSI[1] & DMITRI MOISSEEV[2]

1 *Finnish Meteorological Institute, PO Box 503, FI-00101 Helsinki, Finland*
jarmo.koistinen@fmi.fi
2 *University of Helsinki, Department of Physics, PO Box 64, FI-00014 Helsinki, Finland*

Abstract A real-time 24/7 alert system is under development. It consists of gridded forecasts of the best estimate rainfall and exceedence probabilities of rainfall class thresholds over a continuous time range of 30 minutes to 5 days. Nowcasting up to 6 h employs a 51 ensemble member extrapolation of weather radar measurements together with lightning location and satellite data. From approximately 2 h to 2 days a Poor man's Ensemble Prediction System (PEPS) will be used, employing the NWP models HIRLAM and AROME. The longest forecasts use ECMWF EPS data. The mixing of the ensemble sets is performed through mixing of accumulations with equal exceedence probabilities. Alert dissemination employs SMS messages via mobile phones. The interactive user interface facilitates free selection of alert sites and warning thresholds at any location in Finland.

Key words rainfall; probabilistic forecasting; radar; NWP; mobile services

INTRODUCTION

The damages due to heavy rainfall and flooding are often amplified by the lack of targeted and probabilistic warnings for risk management across all forecast lead-times from minutes to days. In many applications, socio-economic losses due to rainfall could be lowered by means of automatic and fast dissemination of alerts with individually-tuned threshold criteria for each customer and application. Quantitative risk forecasts require that a tool for estimating exceedence probabilities of accumulated precipitation is available. The usual solution is ensemble prediction. Convective precipitation patterns, which are frequently responsible for cases of urban flooding, evolve rapidly in time and space implicating relatively large errors in any forecasts of them. Nevertheless, radar-based extrapolative nowcasts provide considerable benefits in flood forecasting for small catchments (e.g. Werner & Cranston, 2009).

Depending on the spatial scale of target areas and structure of the precipitating weather system, the skill of extrapolation techniques becomes low within 1–6 h from the time of the remotely-sensed measurement (radar, satellite, lightning locator). With longer forecasts, ensembles from numerical weather prediction (NWP) models must be used. A major challenge in ensemble prediction from short to long lead-times is to guarantee smooth continuity of forecasts in the overlapping time zones where predictions from one data source (e.g. radar extrapolations) are blended with those from a second source (e.g. NWP). A further challenge is the calibration of probabilities obtained from an ensemble prediction system. The forecasting systems provide relative probabilities of various precipitation thresholds which do not necessarily agree with the statistics or climatology of the real precipitation. The final difficulty is probabilistic estimation of flooding risks, i.e. the ways of coupling meteorological ensemble predictions to produce hydrological ensembles attached with error estimates.

At the Finnish Meteorological Institute (FMI) a three year national project is underway where we provide localized warnings of rainfall risk. The main output products from the warning system consist of nationwide gridded high-resolution forecasts of the best estimate rainfall and exceedence probabilities of rainfall accumulations across a continuous lead-time range of 30 minutes to 5 days. Nowcasting up to 6 h employs a 51 member ensemble of extrapolated weather radar measurements together with parallel ensembles from lightning location and satellite data. For lead-times up to 2 days a Poor man's Ensemble Prediction System (PEPS) will be used from the high-resolution NWP model HIRLAM. The longest forecasts employ rainfall ensembles available in the form of ECMWF EPS data.

During summer 2010 we tested warning dissemination of radar-based nowcasts of rainfall accumulation over the next 2 h. The dissemination used SMS messages via mobile phones. In the interface, users could freely select the locations of the warnings, but the rainfall classes and their risk thresholds were fixed by the service providers. The number of customers was some thousands. We aim to extend the service to cover all lead-times and multiple accumulation periods. It will also include a dedicated user-interface whereby an advanced customer can set individual warning criteria for rainfall at a point or in an area. Presently, the ensemble forecasting system is devoted to precipitation (and fair weather) estimation and warnings. Operational coupling to hydrological models and services will hopefully follow in the future. So far, such coupling has been tested successfully in a parallel R&D project on a case study basis. This paper gives a preliminary overview of the methodologies applied in the Finnish forecasting and warning process.

NOWCAST SOURCE DATA

Weather radars

The national composite of 500 m pseudo-CAPPI data of radar reflectivity factor (dBZ) from 8 C-band radars every 5 minutes is used as the input for the nowcast ensemble system. As the main information source of the nowcasting engine is weather radar, good quality-control (QC) of radar data is necessary. The data used in the system are in most cases of rather good quality and their availability is high (>99%). Doppler filtering removes ground clutter and many clear air echoes are weaker than the lowest threshold of 0 dBZ used in nowcasting of rain. In addition to Doppler filtering we apply fuzzy-logic pattern recognition (Peura, 2002) which removes a major part of the non-meteorological patterns, such as external emitters. However, hail as well as part of sea and ship clutter and attenuation remain in the filtered data and introduce quality problems, especially in cases of large reflectivities. Currently FMI is in the process of updating its conventional radars to dual-polarization systems. During the summer 2010 nowcast campaign the most significant source of error was intense ship and sea clutter in anomalous propagation situations. A number of false alarms were generated, as the nowcasting system extrapolated the clutter echoes to the coastal areas of Finland. Therefore the use of dual-polarization information will have an important role in the near future quality-control of our radar-derived rainfall data.

Satellite data

In addition to the weather radar network, the nowcasting system uses satellite-based very short-range forecasts of convective rainfall. Convection is classified using Spinning Enhanced Visible and Infrared Imager (SEVIRI) data. SEVIRI on board of Meteosat Second Generation satellites makes observations every 15 minutes and it covers the entire Earth disk visible at 0° longitude. SEVIRI has 12 channels, eight of which are in the thermal infrared, three in visible, one in the ozone absorption band (all with 3 km resolution) and one high-resolution visible (1 km resolution). Convective rain-rate (CRR) is calculated from the various SEVIRI channels using statistical calibration tables from the weather radar network together with checking the patterns of rain to distinguish between stratiform and convective rainfall. For processing the SEVIRI data to convective rain-rate, the Nowcasting SAF MSG software package has been implemented and adapted to the high latitudes. For further details see Algorithm Theoretical Basis Document for "Convective Rainfall Rate" (CRR-PGE05 v3.1), NWC SAF at http://www.eumetsat.int.

Lightning data

The observation that thunderstorms produce both lightning and heavy precipitation can be used to estimate rainfall (Pessi & Businger, 2009). The rainfall-lightning-relationship (RLR) has been studied in various places globally. The relationship varies climatologically (latitudinal dependence) and regionally (water/land dependence). Precipitation estimates based on the lightning location data of the Nordic Lightning Information System (NORDLIS) has been obtained in the following way: (1) weather radar-based rainfall rates per 15 minutes are compared against cloud-to-ground lightning data in 10 km × 10 km fixed squares over Finland, (2) a relationship of the form $R = aN^b$ is fitted between rainfall and lightning flash rate (R is the rainfall [mm/15 min/100 km^2], N is the flash rate [ground flashes/15 min/100 km^2]), (3) the relationship is used to produce estimates of the rainfall according to lightning location data for every 15 min. In short, lightning data are converted into pseudo-precipitation fields, which mimic the "actual" precipitation fields obtained from weather radar. This way the lightning-based precipitation fields can be used similarly to radar data in the ensemble forecasts.

Creation of ensemble forecasts

The method of nowcasting precipitation for 0 to 6 h in the future, developed at FMI (Hohti *et al.*, 2000), employs a modified correlation-based atmospheric motion vector (AMV) system developed by EUMETSAT (Holmlund, 1998). It provides a sophisticated automatic quality-indicator (QI) of the vectors which is useful in application to radar images. The five latest 500 m pseudo-CAPPI reflectivity fields, combined from eight radars in Finland at time intervals of 5 min, are used as the input to the AMV system. Therein each 16 × 16 km^2 grid box is compared to the neighbouring grid-boxes from the previous time-step, and the best autocorrelation is chosen to show the area of origin of the precipitation patterns in the grid-box. Individual vectors of movement will pass a quality-control and the vector field is smoothed. Each motion vector field is assumed to be static as a function of time. Reversed trajectories are then calculated from each point of the future field back to the last observed reflectivity field. Speed and direction inaccuracies along the trajectories are cumulatively summed as a function of time. The inaccuracies establish orthogonal error bars along the trajectory which define an elliptic source area of the trajectory. The source ellipse grows around each starting point of a trajectory as a function of time. Reflectivities within the source area are then applied to calculate the most probable precipitation intensity, intensity class probabilities and predicted accumulation at the end point of the trajectory. Trajectories are presently calculated for 5 h onwards. Notice that these trajectories propagate backwards, because the future precipitation at each grid point is presently observed in the source area located upstream in the latest measured precipitation field.

In satellite data the CRR product is conceptually comparable to the radar information: thus the nowcast of precipitation probabilities is calculated with the same method as that for radar. The motion vectors for each convective rain object are calculated using a 15 minute time-step. The CineSat software, which is used operationally at FMI, is able to provide the motion vectors. The method to calculate the vectors is basically the same as in the NWC SAF MSG software High Resolution Winds. Once the motion vectors have been calculated they, together with the CRR observation images, are fed into the radar-based ensemble derivation system. From that step onwards the procedure of satellite- and lightning-based ensemble nowcasting of rainfall probabilities is identical to the weather radar nowcasting. However, we aim to perform the extrapolation predictions of satellite- and lightning-based precipitation up to 12 h. The main reason for such a long lead-time is the fact that organized large convective patterns, such as mesoscale convective systems (MCS), tend to have life-cycles of 8–12 h. During the life-cycle the convective systems often move long distances so that at the initial time of the forecast they are still located outside the coverage of the radar network. On the other hand, NWP forecasts are not always applicable, as cases of MCS have been detected in Finland where the operational high-resolution models completely missed the inception of a severe MCS which developed only 6 h after the initial time of running the model.

For forecast lead-times from 0 to 120 h, the intention is to also produce probabilistic precipitation forecasts from different sources of NWP output. At the moment, ECMWF EPS output is the only probabilistic source of information that is operationally available. ECMWF data contain the undisturbed control run, which uses the same boundaries as the deterministic IFS-model, and 50 disturbed forecasts. The disturbed forecasts are generated using the singular vector technique to find the most rapidly growing disturbances. Together, these 51 individual forecasts are used to calculate the probabilities of rainfall accumulations. While ensemble forecasts from a limited area high-resolution NWP are currently missing, deterministic NWP data are available and can be used to produce poor-man ensembles (PEPS) with lagged analysis cycles or one can apply a neighbourhood method (Theis, 2005) for single model output. At FMI there are two operational versions of HIRLAM (RCR and MBE) and the output of the AROME model available. Both PEPS and true EPS-based short-range forecasts are in the development phase at FMI. Thus ECMWF EPS data are the only source of NWP ensembles applied in the summers of 2010 and 2011.

Calculating the probabilities of 6-h accumulated rainfall from ECMWF ensembles is straightforward, although the climatological probability distribution function (PDF) of the ensemble forecasts does not necessarily match with the true PDF of precipitation. NWP figures for shorter periods than 6 h are obtained assuming constant rainfall intensities during the period. For each accumulation period and for each lead-time the ensemble forecasts from radars, satellite and lightning data, and from the NWP models are stored in the form of complementary cumulative distribution functions (CCDFs), i.e. in the form of exceedence diagrams. As ensembles from each data source consist of 51 members, the resolution of the CCDF is approximately 2% (1/51). For example, the fifth largest precipitation in the array of ranked ensemble predictions is the precipitation amount which is exceeded with a probability of 10%. Employing the CCDF, it is easy to interpolate the exceedence probability for any precipitation threshold given by the customer, except for extreme values with probabilities much less than 2%.

Blending of various sets of ensemble forecasts

A major challenge in ensemble forecasting is blending of precipitation patterns from several sources of data. This is a significant issue as we aim to provide smoothly continuous probabilities in the wide forecast time-range of 30 min to 5 days. For example, a 6 h nowcast of extrapolated radar-based precipitation exhibits often completely different patterns and intensities than those obtained from a NWP-based very short-range forecast. Although techniques of mixing of patterns exist, such as optical flow, the mixed result will be arbitrary from the point of meteorologically reasonable structures if the correlation between the two original patterns is low. Although sophisticated hydrologic modelling would need spatially correct rainfall distributions we omit, for the time being, such a requirement. The reasons are simplicity in derivation and the fact that a large majority of customers do not care how realistic the precipitation patterns are that occur in the forecasts, as their application requires data from a single grid-point. Therefore we follow the simple scheme in simultaneous predictions from radars, satellite and lightning data, and NWP. Blending of the ensemble forecast sets from these sources is performed using mixing of precipitation amounts of equal exceedence probabilities without any adjustments to the resulting precipitation patterns. In this way the different forecasts are blended into a single integrated ensemble from which the user-products are subsequently derived. At each forecast grid-point, this blending is performed so that a weighted-average of the precipitation amount representing an equal exceedence probability at each set of ensembles is derived. In order to achieve a reasonable integrated ensemble, the weights must be given in a way that takes the spatial and temporal characteristics of different forecast sources into account. These characteristics are quantified by associating a metric we call "quality" with each forecast source, and modelling the metric so that it reflects the spatial and temporal uncertainty of the forecast source. Quality serves as a measure of relative reliability of each forecast source for a given location and forecast time.

The quality metric for each forecast source has two components, called climatological and adaptive quality. Climatological quality is the heuristically obtained background quality which is used as the first assumption of the final quality. The climatological quality metric is defined in three dimensions, namely grid location and forecast time. In spatial dimensions, both the coverage and spatial resolution of the forecast source are taken into account. In the temporal dimension, the relative weight of the forecast, i.e. validity in terms of predictive skill, is modelled as a function of forecast time. Adaptive quality is a modification applied to the climatological quality when there is a reason to assume the latter is not valid. For example, when the model forecasts the onset of convection in the next few hours, an adaptive quality trigger would limit the temporal weight of the radar forecast and rely more on model forecasts at a time earlier than indicated by the climatological quality metric. Currently, adaptive quality triggers are applied only in the temporal dimension. The blended forecast is obtained as a weighted average of the individual forecasts, with quality metrics serving as weights. Blending is performed at each grid-point by ranking the ensemble members of each forecast source according to the amount of rainfall (exceedence probability) and calculating a quality-weighted average of the members with equal rank number. This way N ensemble members from M forecast sources are each combined into a single set of N blended ensemble members. The blended ensemble serves as the input data for product algorithms.

Output products, user interface and warning services

The radar forecast engine uses one megapixel radar reflectivity composites of 1 km resolution for motion vector analysis. Motion vectors are applied in the initial composite to extrapolate 51 ensemble members of radar reflectivity factor (intensity of precipitation) having 1 min temporal resolution. During the first summer, radar-based warnings were disseminated covering the next 2 h, although the derivation of ensembles extends up to 5–6 h. A new ensemble forecast is computed every 5 mins. The forecast engine code is designed for threading. The input composite can be divided into an arbitrary amount of slices for parallel processing. With 2.8 GHz processors using eight threads it takes about 20 seconds to generate 51 ensemble members in a grid of 760 × 1226 points. The uncompressed initial output datasets are written to file for each member, having a total size of 10 GB. Ensemble data from other sources (satellite, lightning, NWP) are converted to the same geometry as the radar data and blended as explained above. The time-step in the ensemble time series increases as a function of forecast length, being 1 min in the shortest nowcast and 6 h in the longest ones. Probabilities of precipitation can then be generated for arbitrary time periods by using the difference of accumulation between the end and the beginning of the selected forecast period. Finally, these files are converted to one HDF5 file in which the amount of data can be compressed to about 10% of the raw output data.

During summer 2010, FMI Commercial Services (KAP) successfully developed and piloted services using ensembles of radar nowcast data. The task was divided into three parts: (1) defining service, data sources and interfaces; (2) implementing and developing the service on production and distributing system; and (3) launching and piloting the service. The service was a short message service (SMS) based alert for rain (0–2 h). The service was launched together with the teleoperator DNA Ltd in June 2010. Subscribing was done by SMS and end-users paid a small fee for every alert. The business model was also tested and people were requested whether they were willing to pay for this kind of service. The pilot service was promising, although the number of rain periods was small compared to the average summer. The pilot service was tested by some 3000 users. A user was able to select the location from a list of 50 000 site names in Finland. However, the exceedence probabilities could not be requested by the user, but instead were classified by the service provider to only two categories ("occurs", "will not occur") and to four accumulation categories (no rain, weak, moderate and heavy rain). The feedback was in general positive. Some problems occurred with radar data related to non-meteorological data, to the accuracy of the user's location and to wishes for additional information and animations. Among ordinary citizens the most popular interest was not heavy rainfall and potential flooding, but actually the "risk" of fair weather, i.e. the probability that no precipitation will occur.

For the summer season of 2011 it is planned that animations will be included in the alerts. Also, it is planned to define a "pro"-version of the service for dedicated users, e.g. in hydrology. In the service, the end-user will be able to order more customized alerts for accumulation periods 1, 6 and 24 h, as well as estimations of the onset and ending times of precipitation. Areal accumulation probabilities will be provided for some hydrological test users, e.g. for the wastewater treatment works of greater Helsinki. A verification system utilizing telemetry raingauges will be established.

The Continuous Ranked Probability Score (CRPS) and its skill derivative (CRPSS) will be the principal quality measures for exceedence probabilities of rainfall class thresholds. Regarding the eventual deterministic end-user products, some of the novelty verification measures like EDI (Extremal Dependence Index) and SEDS (Symmetric Extreme Dependency Score) (see Ferro & Stephenson, 2011) are applied along with the more common verification metrics for categorical predictands. Until proper verification results are available we are not able to quantify the effectiveness and accuracy of the system.

In the national project there is no actual hydrologic component. However, in a collaborative project, urban hydrological and hydraulic models have been tested, coupling them to our ensemble nowcasts of rain. The results were quite encouraging but will be reported elsewhere.

REFERENCES

Ferro, C. A. T. & Stephenson, D. B. (2011) Extremal Dependence Indices: improved verification measures for deterministic forecasts of rare binary events. *Weather & Forecasting* 26, 699–713.

Hohti, H., Koistinen, J., Nurmi, P. & Holmlund, K. (2000) Precipitation nowcasting using radar-derived atmospheric motion vectors. *Phys. Chem. Earth (B)* 25, 1323–1329.

Holmlund, K. (1998) The utilization of statistical properties of satellite-derived atmospheric motion vectors to derive quality indicators. *Weather and Forecasting* 13, 1093–1104.

Pessi, A. T. & Businger, S. (2009) Relationships among lightning, precipitation, and hydrometeor characteristics over the North Pacific Ocean. *J. Appl. Meteorol. Clim.* 48, 833–848.

Peura, M. (2002) Computer vision methods for anomaly removal. In: *2nd European Conf. Radar in Meteorology and Hydrology, Delft, The Netherlands*, 312–317.

Theis, S. E., Hense, A. & Damrath, U. (2005) Probabilistic precipitation forecasts from a deterministic model: a pragmatic approach. *Meteorol. Appl.* 12, 257–268.

Werner, M. & Cranston, M. (2009) Understanding the value of radar rainfall nowcasts in flood forecasting and warning in flashy catchments. *Meteorol. Appl.* 16, 41–55.

400

Weather Radar and Hydrology
(Proceedings of a symposium held in Exeter, UK, April 2011) (IAHS Publ. 351, 2012).

Impact of small-scale rainfall uncertainty on urban discharge forecasts

A. GIRES[1], D. SCHERTZER[1], I. TCHIGUIRINSKAIA[1], S. LOVEJOY[2], C. ONOF[3], C. MAKSIMOVIC[3] & N. SIMOES[3,4]

1 *Université Paris-Est, Ecole des Ponts ParisTech, LEESU, 6-8 Av Blaise Pascal Cité Descartes, Marne-la-Vallée, 77455 Cx2, France*
auguste.gires@leesu.enpc.fr
2 *McGill University, Physics Department, Montreal, PQ, Canada*
3 *Imperial College London, Department of Civil and Environmental Engineering, UK*
4 *Department of Civil Engineering, University of Coimbra, Coimbra, Portugal*

Abstract We used a multifractal characterization of two heavy rainfall events in the London area to quantify the uncertainty associated with the rainfall variability at scales smaller than the usual C-band radar resolution (1 km^2 × 5 min) and how it transfers to sewer discharge forecasts. The radar data are downscaled to a higher resolution with the help of a multifractal cascade whose exponent values correspond to the estimates obtained from the radar data. A hundred downscaled realizations are thus obtained and input into a semi-distributed urban hydrological model. Both probability distributions of the extremes are shown to follow a power-law, which corresponds to a rather high dispersion of the results, and therefore to a large uncertainty. We also discuss the relationship between the respective exponents. In conclusion, we emphasize the corresponding gain obtained by higher resolution radar data.

Key words multifractals; rainfall downscaling; urban hydrology; power law

INTRODUCTION

This paper implements multifractal techniques (see Lovejoy & Schertzer, 2007 for a recent review), which are standard tools to analyse and simulate geophysical fields, e.g. rainfall, that are extremely variable over a wide range of scales, and in urban hydrology to quantify the impact of small-scale unmeasured rainfall variability. Indeed numerous hydrological studies (see Singh, 1997 for a review) show that rainfall variability has an impact on the modelled flows, which is more or less significant according to the rainfall event and the catchment size and features. In urban areas the effects are enhanced because of greater impervious coefficients and shorter response times (Aronica & Cannarozzo, 2000; Segond *et al.*, 2007).

In this paper we study the mainly urban 910-ha Cranbrook catchment, situated in the London Borough of Redbridge, and known for regular local flooding (Gires *et al.*, 2012). The rainfall data are obtained from the Nimrod composites, a radar product of the Met Office in the UK (Harrison *et al.*, 2000), whose resolution is 1 km in space and 5 min in time. Here a winter frontal rainfall event (9 February 2009) and a summer convective one (7 July 2009) are investigated. Square areas of size 64 km^2 during 21 h, centred on the heaviest rainfalls of these events, are analysed. Figure 1 displays the total rainfall depth for both events. The very localized rainfall cells of the convective July event are clearly visible.

In the next sections, first the multifractal properties of the rainfall fields are analysed for both events. Then an ensemble of realistic spatially downscaled (to the scale of 125 m) rainfall fields is generated with the help of universal multifractal cascades, and the corresponding ensemble of hydrographs is simulated. The variability among these ensembles is used to characterize the uncertainty due to small-scale unmeasured rainfall variability, mainly on the peak flow.

MULTIFRACTALS AND RAINFALL DOWNSCALING

In this section, the multifractal analysis of the rainfall events and the implemented downscaling technique are briefly presented. More details can be found in Gires *et al.* (2012).

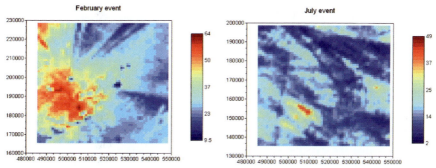

Fig. 1 Map of the total rainfall depth (in mm) over the studied area for the February (left) and July (right) events. The coordinate system is the British National Grid (unit: m)

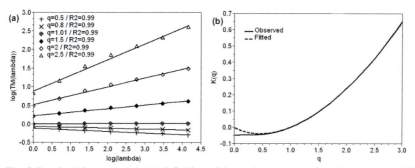

Fig. 2 For the February event: (a) definition of the scaling moment function (equation (1)) and (b) plot of $K(q)$.

Multifractals

In this paper the rainfall field is investigated with the help of universal multifractals, which have been extensively used to analyse geophysical fields that are extremely variable over a wide range of scales (Lovejoy & Schertzer, 2007; Royer *et al.*, 2008; Nykanen, 2008; Gires *et al.*, 2010). They basically rely on the concept of multiplicative cascades. In that framework the statistical moment of arbitrary qth power of the rainfall field R_λ at the resolution λ (=L/l, the ratio between the outer scale of the phenomenon and the observation scale) exhibits a scaling behaviour:

$$< R_\lambda^{\ q} > \approx \lambda^{K(q)} \tag{1}$$

where $< >$ denotes ensemble average (over the different time steps analysed), and $\approx$ asymptotic equivalence. $K(q)$ is the scaling moment function and quantifies the scaling variability of the rainfall field. Figure 2(a) displays equation (1) in a log–log plot for the February event. The straight lines (the coefficient of determination are all >0.99), whose slopes are $K(q)$, indicate a good scaling behaviour and show the relevance of this analysis. Similar curves are found for the July event. $K(q)$ for the February event is plotted in Fig. 2(b).

In the specific framework of universal multifractals (Schertzer *et al.*, 1997), $K(q)$ is described by three scale independent parameters (UM parameters): C_1 the mean intermittency ($C_1 = 0$ for a uniform field), α the multifractality index ($\alpha = 0$ for a monofractal field, and $\alpha = 2$ for the extreme log-normal case), and H the non-conservation ($H = 0$ for a conservative field). $K(q)$ is given by:

$$K(q) = \frac{C_1}{\alpha - 1}\left(q^\alpha - q\right) + Hq \tag{2}$$

Greater values of C_1 and α correspond to strong extremes. The parameters are estimated with the help of the DTM technique and spectral slope (Lavallée *et al.*, 1993). Here the numerical values of the UM parameters are quite different for both events. The estimates of α, C_1 and H are 1.62, 0.14 and 0.56, respectively, for the February event and 0.92, 0.49 and 0.57, respectively, for the July event. $K(q)$ plotted with the estimated UM parameters for the February event is displayed in Fig. 2(b). The agreement with the empirical curve is very good, and the discrepancies for small moments are explained by a multifractal phase transition associated to the influence of the numerous zeros (i.e. a pixel of a time step with no rain) (Gires *et al.*, 2010). The rainfall events exhibit two different statistical behaviours (Hubert *et al.*, 1993): indeed for the February event $\alpha > 1$ which indicates that the extreme values are not bounded whereas for the July event $\alpha < 1$ indicates bounded extreme values. In the multifractal framework, the probability distribution of the extreme values observed on a given dataset are expected to follow a power-law (or equivalently, the statistical moments cannot be estimated for moments greater than the power law exponent) whose exponent strongly depends on C_1. Here C_1 for the July event is more than three times greater than C_1 for the February event, indicating a lower theoretical exponent. The effect is partially compensated by the greater value of α for the February event. This means that extreme values will be observed more frequently for the July event.

Multifractal spatial downscaling of the rainfall field

A stochastic downscaling technique is used to generate realistic high-resolution rainfall fields (Venugopal *et al.*, 1999; Deidda, 2000; Olsson *et al.*, 2001; Ferraris *et al.*, 2002; Rebora *et al.*, 2006). The framework of the cascade process is well suited to this problem (Biaou *et al.*, 2003), since continuing the cascade beyond the observation resolution enables generating a realistic high-resolution rainfall field. More precisely, for each pixel of 1 km × 1 km three steps of discrete UM cascades (Pecknold *et al.*, 1993) are simulated with the UM parameters estimated on the available data (i.e. on a range of scale from 1 to 64 km). The final resolution is 125 m × 125 m. This process is illustrated in Fig. 3(a). An example of downscaling (which generates rainfall variability inside a radar pixel) over the modelled area for a time step of the February event is displayed in Fig. 3(b). The downscaling of two consecutive time steps is independent, but the obtained variables remain dependant because they are generated from larger structures that are dependent.

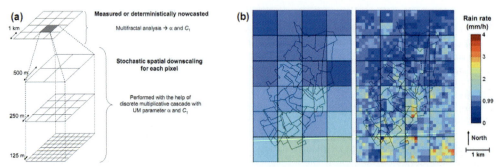

Fig. 3 Illustration (a) and example over the Cranbrook area for a time step of the February event (b) of the downscaling technique implemented.

RAINFALL–RUNOFF MODEL

For the Cranbrook catchment, Infoworks CS (Wallingford Software, 2009) was used by Thames Water Utilities (2002) to calibrate a semi-distributed (i.e. the area is divided into 51 sub-catchments of size ranging from 1 to 62 ha) model that includes the major surface water sewers (Fig. 4). Sub-catchments are defined by sewer nodes and are considered as being homogeneous. Their discharge is computed from the rainfall with the help of a double linear reservoir model. The main parameters are the slope and the length (represented with dashed lines in Fig. 4). The simulation parameters were maintained unchanged for all simulations. The total rainfall depth over the 51 sub-catchments for the July event is also displayed in Fig. 4. It ranges from 4 to 14 mm, with the heaviest rainfall situated in the south near the outlet. A similar distribution with values ranging from 16 to 23 mm is observed for the February event. Conduits drain the water from one (for upstream conduits) or several sub-catchment(s) (for downstream conduits), and their characteristic length L_{da} is defined as the square root of the area drained by the conduit. In this paper the hydrographs of 10 conduits with L_{da} ranging from 370 to 2910 m are analysed (Fig. 4). This enables us to study the impact of the size of the studied area on the uncertainty associated with small-scale rainfall variability.

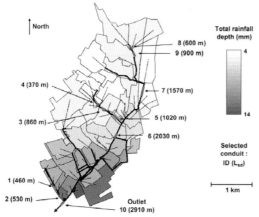

Fig. 4 The Cranbrook Catchment and its underground sewer system. The total rainfall depth over the 51 sub-catchments for the July event is also displayed.

QUANTIFYING THE IMPACT OF SMALL-SCALE UNMEASURED RAINFALL VARIABILITY

An ensemble of 100 realistic downscaled rainfall fields and the corresponding ensemble of hydrographs were simulated. The uncertainty associated with small-scale unmeasured rainfall variability is quantified by investigating the variability among the ensembles.

Variability among the ensemble of rainfall fields

To give an insight of the uncertainty on the rainfall fields, for each conduit we evaluated the maximum average rainfall intensity (R_{max}) over the drained area for each sample of downscaled fields. The histogram of these values (one per sample) for the conduit 2 and the February event is shown in Fig. 5(a). Similar curves are found for the other conduits and for the July event. It appears that the extreme cases (i.e. the right part of the histogram) exhibit a power-law behaviour of the form:

$$\Pr(R_{max} > x) \approx x^{-k} \tag{3}$$

Indeed, this relation in a log–log plot is displayed in Fig 5(b), and the determination coefficient of the straight line describing the fall-off of the probability distribution is very good (0.98 ± 0.01 and 0.97 ± 0.01 for the February and the July event, respectively, according to the conduit). The values k_{rain} found are plotted against L_{da} in Fig. 6. First it appears that k_{rain} increases with L_{da}, which implies a thinner probability fall-off for larger drained area. It means that the effect of small-scale rainfall variability is damped for larger area, which was expected. Second, in general k_{rain} is smaller for the July event, which reflects the fact that the variability among the ensemble of downscaled rainfall fields is greater for the convective event than for the frontal event, which was expected due to greater values of C_1.

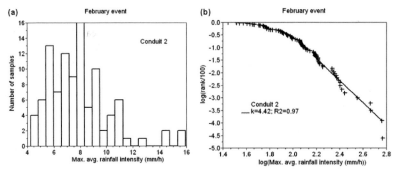

Fig. 5 Histograms (a) and determination curve of k_{rain} (b) for the maximum average rainfall rate of the conduit 2 for the February event.

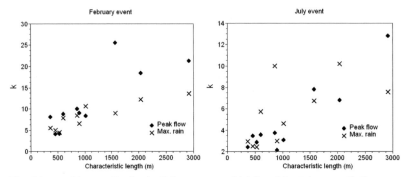

Fig. 6 k_{rain} and k_{flow} vs. L_{da} for the February event (right) and the July event (left).

Variability among the ensemble of simulated peak flow

A similar analysis was performed on the ensemble of simulated hydrographs. For each selected conduit, the peak flow and its time of occurrence was retrieved. The first point is that no significant influence was found on the time of occurrence. Concerning the peak flow, as for the maximum average rainfall, the probability distribution exhibits power-law fall-off (curves similar to Fig. 5(b) but for the peak flow are found, with R^2 equal to 0.97 ± 0.01 for both events according to the conduit). The power-law behaviour means that the uncertainty on the simulated peak flow associated to unmeasured small-scale rainfall variability cannot be neglected. It is striking to see that the uncertainties are significant despite the coarse resolution of the hydrological model (the average square root of sub-catchment area is 380 m) with regard to the resolution of the downscaled rainfall fields (125 m). Nevertheless further investigations with hydrological models with greater spatial resolution and taking into account the interaction between surface flow and the sewer system (El Tabach *et al.*, 2009; Maksimovic *et al.*, 2009) would be needed to fully take advantage of the spatial downscaling. Figure 6 displays the exponent k_{flow} vs the characteristic length of the conduit L_{da} for both events. As for k_{rain}, k_{flow} tends to increase with L_{da}. Independently of the event and L_{da}, the variability among the ensemble of rainfall fields is transferred to the ensemble of peak flow with the same qualitative features (i.e. a power-law fall-off of the probability distribution). Quantitatively it appears that k_{flow} is often (i.e. most of the conduits for the February event and some for the July event) greater than k_{rain}, which would mean that the rainfall–runoff process slightly dampens the variability of the ensemble of rainfall fields. Finally it appears that the variability observed for the July event is similar to the one observed up to 1 km for the February event.

CONCLUSION

Multifractal cascades are used to generate an ensemble of realistic downscaled rainfall fields of a convective and a frontal rainfall event (after validating this framework for both events). The probability distribution of the generated rainfall extremes exhibits a power-law fall-off, which is also retrieved on the peak flow of the corresponding simulated ensemble of hydrographs. It means that the uncertainty associated with small-scale rainfall variability cannot be neglected. As a consequence it is recommended to either take into account this uncertainty in the real-time management of sewer networks or to improve the resolution of rainfall data in urban areas by implementing X-band radars whose spatial resolution is roughly 100 m. Concerning the numerical values of the characteristic exponents of the power-law fall-off, it seems that they are greater for peak flow than for rainfall, and for larger areas. They are greater for the frontal event than for the convective one. Nevertheless further investigations with other case studies and rainfall events are required to clarify this point.

Acknowledgements A. Gires greatly acknowledges the Université Paris-Est and Imperial College London for financial and partial financial support. A. Gires, D. Schertzer and I. Tchiguirinskaia greatly acknowledge partial support from the Chair "Hydrology for Resilient Cities" (sponsored by Veolia) of Ecole des Ponts ParisTech (http://www.enpc.fr/hydrologie-pour-une-ville-resiliente). The authors thank the Met Office for the Nimrod data, MWH Soft and Thames Water Utilities Ltd. N. Simões acknowledges the financial support from the Fundação para a Ciência e Tecnologia – Ministério para a Ciência, Tecnologia e Ensino Superior, Portugal [SFRH/BD/37797/2007].

REFERENCES

Aronica, G. & Cannarozzo M. (2000) Studying the hydrological response of urban catchments using a semi-distributed linear non-linear model. *J. Hydrol.* 238, 35–43.

Biaou, A., Hubert, P., Schertzer, D., Tchiguirinskaia, I. & Bendjoudi, H. (2003) Fractals, multifractals et prévision des précipitations. *Sud Sciences et Technologies* 10, 10–15.

Deidda, R. (2000) Rainfall downscaling in a space-time multifractal framework, *Water Resour. Res.* 36, 1779–1794.

El Tabach, E., Tchiguirinskaia, I., Mahmood, O. & Schertzer, D. (2009) Multi-Hydro: a spatially distributed numerical model to assess and manage runoff processes in peri-urban watersheds, Proceedings Final conference of the COST Action C22 Urban Flood Management, Paris 26/27.11.2009, France.

Ferraris, L., Gabellani, S., Rebora, N. & Provenzale, A. (2003) A comparison of stochastic models for spatial rainfall downscaling. *Water Resour. Res.* 39, 1368–1384.

Gires, A., Tchiguirinskaia, I., Schertzer, D. & Lovejoy, S. (2010) Analyses multifractales et spatio-temporelles des précipitations du modèle Méso-NH et des données radar. *Hydrol. Sci. J.* 56(3), 380–396.

Gires, A., Onof, C., Maksimovic, C., Schertzer, D., Tchiguirinskaia, I. & Simoes N. (2012) Quantifying the impact of small scale unmeasured rainfall variability on urban hydrology through multifractal downscaling: a case study. *J. Hydrol.* (submitted).

Harrison, D. L., Driscoll, S.J. & Kitchen, M. (2000) Improving precipitation estimates from weather radar using quality control and correction techniques, *Met Appl* 7(2),135–144.

Hubert, P., Tessier, Y., Lovejoy, S., Schertzer, D., Schmitt, F., Ladoy, P., Carbonnel, J. P. & Violette, S. (1993) Multifractals and extrem rainfall events. *Geophys. Lett.* 20, 931–934.

Lavallée, D., Lovejoy, S. & Ladoy, P. (1993) Nonlinear variability and landscape topography: analysis and simulation. In: *Fractals in geography* (ed. By L. De Cola & N. Lam), 158–192. Prentice-Hall, New York, USA.

Lovejoy, S. & Schertzer, D. (2007) Scale, scaling and multifractals in geophysics: twenty years on. In: *Nonlinear Dynamics in Geosciences* (ed. By A. A. Tsonis & J. Elsner), 311–337. Springer.

Maksimović, Č., Prodanović, D., Boonya-aroonnet, S., Leitão, J. P., Djordjević, S. & Allitt, R. (2009) Overland flow and pathway analysis for modelling of urban pluvial flooding. *J. Hydraul. Res.* 47(4), 512–523.

Nykanen, D. K. (2008) Linkages between orographic forcing and the scaling properties of convective rainfall in mountainous regions, *J. Hydromet.* 9, 327–347.

Pecknold, S., Lovejoy, S., Schertzer, D., Hooge, C. & Malouin, J. F. (1993) The simulation of universal multifractals. In: *Cellular Automata: Prospects in astrophysical applications* (ed. by J. M. Perdang & A. Lejeune), 228–267. World Scientific.

Olsson, J., Uvo, C. B. & Jinno, K. (2001) Statistical atmospheric downscaling of short-term extreme rainfall by neural networks, *Phys. Chem. Earth* (B) 26(9), 695–700.

Pandey, G., Lovejoy, S. & Schertzer, D. (1998) Multifractal analysis including extremes of daily river flow series for basins one to a million square kilometres. *J. Hydrol.* 208(1–2), 62–81.

Rebora, N., Ferraris, L., von Hardenberg, J. & Provenzale, A. (2006) The RainFARM: Downscaling LAM predictions by a Filtered AutoRegressive Model, *J. Hydromet.* 7, 724–737.

Royer, J.-F., Biaou, A., Chauvin, F., Schertzer, D. & Lovejoy, S. (2008) Multifractal analysis of the evolution of simulated precipitation over France in a climate scenario, *C. R. Geoscience* 92(D8), 9693–9714.

Schertzer, D., Lovejoy, S., Schmitt, F., Tchiguirinskaia, I. & Marsan, D. (1997) Multifractal cascade dynamics and turbulent intermittency. *Fractals* 5(3), 427–471.

Segond, M-L., Wheater, H.S. & Onof, C. (2007) The significance of small-scale spatial rainfall variability on runoff modelling, *J. Hydrol.* 173, 309–326.

Singh V. P. (1997) Effect of spatial and temporal variability in rainfall and watershed characteristics on stream flow hydrograph. *Hydrol. Processes* 11, 1649–1669.

Thames Water Utilities Ltd Engineering (2002) Surface water model of Cranbrook and Seven Kings Water for London Borough of Redbridge Appendix B, Model Development Report.

Venugopal, V., Foufoula Georgiou, E. & Sapozhnikov, V. (1999) A space-time downscaling model for rainfall. *J. Geophys. Res.* 104(D16), 19705–19721.

Wallingford Software (2009) Infoworks CS Help documentation.

5 Hydrological impact and design studies

Joint analysis of radar observation and surface hydrological effects during summer thunderstorm events

P. P. ALBERONI, M. CELANO, R. FORACI, A. FORNASIERO, A. MORGILLO & S. NANNI

ARPA Servizio Idro-Meteo-Clima, Viale Silvani, 6 Bologna, Italy
palberoni@arpa.emr.it

Abstract During summer 2010 a special project was carried out in Emilia-Romagna (north Italy) focused on the issuing of warning associated with severe weather effects on local territory (e.g. sewer systems, road, small catchments and urban hydrological problems). This project has been set up to understand limits and, hopefully to improve, actual capabilities in the operational issuing of warning procedure. The first result, as expected, was the separation into two main classes of situation which caused such types of problem. On the one hand we had weather events where the synoptic forcing was well defined and where numerical modelling was able to correctly forecast the occurrence of such events. On the other extreme we had weather events where numerical models fail to forecast due to a number of causes like: weak or wrong synoptic forcing, very localised intense precipitation events, thunderstorms. This work focuses on an analysis of some of the events included in the second category, to highlight the link between observations and local problems and to understand how radar observations can play an important role in warning emission for this type of event. To tackle this ambitious goal, radar QPE is analysed along with the geo-localisation of surface problems occurring during the event; further particular attention is paid to the time evolution of surface precipitation pattern. An analysis of Limited Area Model performance will be carried out to highlight if, and in which circumstances, a very high resolution run could improve forecasting capability in such an event.

Key words warning; severe events

INTRODUCTION

It is a well known that the forecast of severe convective events is limited by the ability of the models to reproduce small-scale phenomena, with the correct space-temporal phase, and consequently to provide valuable information for the issuing of warning concerning these types of event. The ARPA-EMILIA-ROMAGNA procedures for forecasting and monitoring the hydrological risk connected with meteorological events include the emission of a warning with a lead-time of 24–36 h for the flood formation and propagation monitoring. These procedures do not include severe convective events, since their predictability has a much smaller time-scale and since their effects are more rapid than the mentioned floods. Within this framework, the Civil Protection Agency of Emilia Romagna region has proposed dedicated procedures to directly alert the local administrations, in cases of not forecasted severe events, based on monitoring activities.

Starting from these considerations an experiment was carried out in summer 2010 with the aim to improve the identification of severe convective events, which are prone to produce surface effects in urban area infrastructure and small catchments, and to verify the ability to deliver warnings for such events from the joint system Weather Service-Civil Protection.

As part of the experiment the inventory of damage – with the affected places, the forecast activity, the monitoring actions, the registered floods, the recorded landslides, the synoptic and local characteristic of the phenomena, and the amount of precipitation with its severity – were collected for each event.

A number of relevant case studies has been analysed to define the adequate instruments for the forecasting and nowcasting of severe events and to identify a useful time warning. Based on the analyses and the information collected, an evaluation of the efficiency of the forecast and civil protection activity was given.

To highlight the critical nature of the warning procedure the outcomes from two case studies are presented here, taken as reference cases for the typical problems to be addressed in the warning emission procedure. The events occurred on 25 May, in the late morning, and on 9 September, which had major effects in the first hours of the afternoon.

The two events were not correctly forecasted by the models and consequently the warning emission procedure did not foresee any hydrogeological risk and consequently local responsible structures were not sufficiently prepared to protect human activities and populations. In the first case the major surface effect was flooding of the hospital in the city of Copparo and in the second case it was the closure of the highway due to flash floods and hail, which affected several municipalities causing damage to crops and properties (lands, farms, buildings, etc.). In the first part of the work the two reference cases were analysed. The meteorological evolution is discussed in order to highlight key aspects of each event and their interaction with surface effects. Then a sensitivity test of the impact of using a very high-resolution meteorological model is briefly discussed in the context of its possible role in an updated warning procedure. Finally, in the last part of the paper the conclusions are drawn.

THE COPPARO FLOOD, 29 MAY 2010

The thunderstorm event of 29 May 2010 could be classified as an air-mass thunderstorm. The synoptic situation was characterized by a high positioned over the Atlantic, which extends its effect up to the western Mediterranean Sea. A direct effect of this is that a weak synoptic wave travelling on its northeastern boundary propagates in the central Mediterranean area, slightly affecting Italy, mainly over its central–south part. North Italy was under a weak divergence zone with a cold pool linked to the synoptic disturbance above it (Fig. 1). During the rest of the day the geopotential field over Italy was slowly rising according to the passage of the synoptic disturbance.

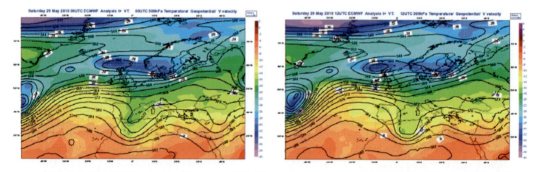

Fig. 1 29 May 2010: analysis of geopotential height and temperature at 500 hPa for 00:00 and 12:00 UTC.

At surface level the mesoscale analysis shows a relatively flat msl pressure field which, starting from 1017 hPa in the morning, deepens up to 17:00 UTC (Fig. 2). The wind analysis highlights that for the whole of the morning a sea-breeze pattern dominates the circulation over eastern North Italy. Thermodynamic analysis of atmospheric profiles, as well as the pattern of surface pseudo-adiabatic equivalent potential temperature, do not suggest the possibility of occurrence of severe convection in the vicinity of Copparo. A more prone area is located north of the Po River where a clear signal is developing during the morning of 29 May.

Accordingly to the rule for warning emission and the quite weak situation, no warning was issued from ARPA and so no monitoring procedure was activated for Emilia Romagna.

Starting from 9:30 UTC some convective systems were developing in Emilia Romagna over the Apennines ridge, as could be expected since the circulation is dominated by breeze, as well as on the southern flank of the Alps. This is the area prone to develop convection and a more scattered system quickly appears above it. At 11:30 UTC scattered convective systems were mainly over the Veneto region, with the more southern placed just south of the Po River, close to Copparo. As explained before, the situation is dominated by very weak synoptic forcing and the thunderstorm moves very slowly, causing a consistent precipitation over the affected area (as seen in the radar images displayed in Fig. 3). Due to this convective system, local flooding occurred in the town of Copparo with problems to traffic circulation and flooding at the hospital, where the first aid unit was affected.

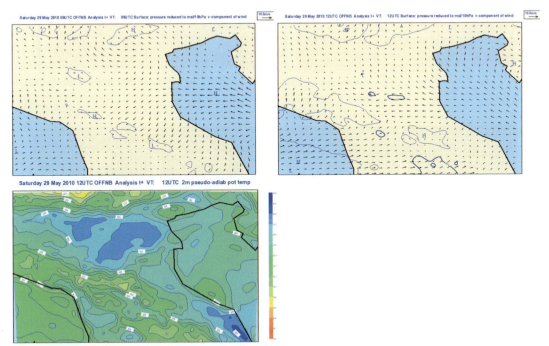

Fig. 2 29 May 2010: msl pressure and 10 m wind at 09:00 (top-left) and 12:00 UTC (top-right), pseudoadiabatic potential temperature at 12:00 UTC (bottom).

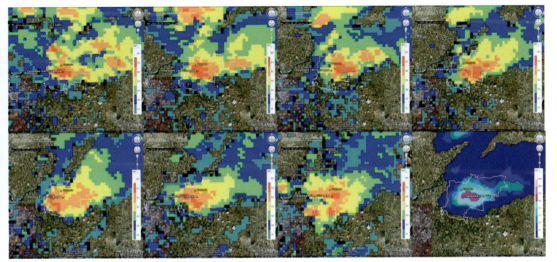

Fig. 3 29 May 2010: Radar reflectivity maps every 15 min (from 11:45 to 13:45 UTC) and 1h-accumulated precipitation at 12:45 UTC (right-bottom panel). The Copparo municipality, the hospital and the raingauge location (in yellow) are also shown.

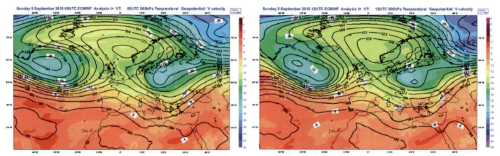

Fig. 4 5 September 2010: analysis of geopotential height and temperature at 500 hPa for 00:00 and 12.00 UTC.

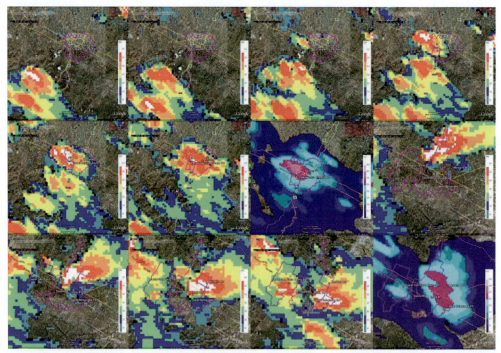

Fig. 5 5 September 2010: radar reflectivity maps every 15 min (from 13:00 to 14:15 UTC) and 1h-accumulated precipitation at 14:45 UTC for the cell affecting the Casalecchio municipality. Radar reflectivity maps every 15 min (from 14:00 to 14:45 UTC) and 1h-accumulated precipitation at 15:00 UTC for the cell affecting Imola and neighbouring municipalities.

A raingauge close to Copparo registered a rainfall of 92 mm in a 6 h period, with 56.8 mm in 1 h, corresponding to a return period >100 years. Radar QPE agrees with this amount. The last image in Fig. 3 shows the radar hourly QPE together with flooded areas within Copparo city.

HIGHWAY PROBLEMS DUE TO SEVERE EVENTS, 5 SEPTEMBER 2010

The synoptic situation is dominated by an omega configuration, in the 500 hPa geopotential height, with an African ridge and two minima, one over the ocean south of Greenland, and the other over northeastern Europe. The weather over Italy is determined by this pattern, which also caused a cold advection over north Italy in the morning of 5 September (Fig. 4).

At mesoscale, the surface flow is mainly from the east, entering from the Adriatic Sea over the Po Valley, with a convergence over the central-western mountains of the Emilia-Romagna region. The situation is therefore favourable to the development of thunderstorms.

In fact, thunderstorms developed in this area around 11:00 UTC (Fig. 5) and propagated mainly in a southeasterly direction. The outflow, blowing downstream on the northern side of the mountains, triggers the development of new systems. They are particularly severe near the boundary between Modena and Bologna provinces where hail-falls hit the towns of Montese, Zocca and Savigno. Following its descending path, the flow is channelled in the Reno Valley causing the generation of a severe convective system near Bologna. The heavy rainfall caused flooding of the highway and its temporary closure.

On the same day in the plain near Ferrara, thunderstorms began around 9:30 UTC. Systems slowly moved southward. Starting from 13:00 to 13:30 UTC the system clearly assumes supercell characteristics. It propagates quite fast in a south-southeast direction causing heavy rain and hail along its path. The traffic on the highway was temporarily blocked. The event was not correctly forecasted by the Limited Area Model (LAM) at either 7 or 2.8 km grid-mesh size, while the precipitation over the Apennines was captured. Propagation of the systems downstream of the mountains was not forecasted as well as the development of the supercell storm.

OUTCOMES FROM VERY HIGH-RESOLUTION MODEL

Convection is a key aspect of the operational forecast especially when, as in the summer season, it represents a major source of variability. Unfortunately there are still some difficulties in representing the convection in the high-resolution LAM because the actual spatial resolution of most of the LAM domain is about 2–3 km and it is well known that at this spatial resolution the convection is partly resolved and partly a subgrid process. Increasing the spatial resolution enables the model to represent the convection explicitly. An attempt to run the COSMO model at 0.01° grid length (about 1 km) of spatial resolution has been made for these convective case studies (29 May 2010 and 9 September 2010) to explore how and if increasing model resolution could improve the quality of the forecast. Both case studies were run nesting the 1 km model (hereafter COSMOI1) over the COSMOI7 forecasts. The COSMOI7 model is the operational model and it runs with a spatial resolution of about 7 km and with the convection scheme parameterised. The initial conditions (IC) are provided by a previous 12 h assimilation cycle, while the boundary conditions are provided by the IFS forecast. The COSMOI1 model runs with the convective scheme switched off and without any kind of previous assimilation cycle. The COSMOI2 model has the same configuration of COSMOI1 with the exception of its spatial resolution which has a grid length of about 2.8 km.

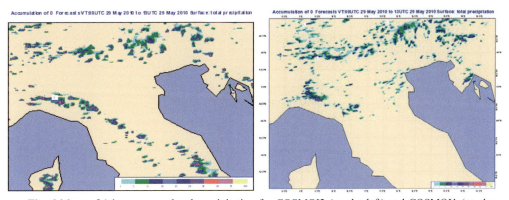

Fig. 6 Maps of 1 hour accumulated precipitation for COSMOI2 (on the left) and COSMOI1 (on the right) at 13:00 UTC 29 May 2010. The map on the right shows the COSMOI1 convective activity near the target area of Copparo.

Both of them are nested over the COSMOI7 forecasts, so the comparison between the two models is clear. The first trial shows benefits in terms of "initiation of convection" (the time in which the first precipitation begins to appear) and in representation of the convective cells. For example, for the case of 29 May 2010, in which the atmospheric forcing was too weak, neither COSMOI7 nor COSMOI2 are able to create the convective cell that hits the city of Copparo, while COSMOI1 shows at least a convective activity (see Fig. 6) near the target area (this information could be useful in giving an early warning).

CONCLUSIONS

The analyses carried out on severe convective events occurring during summer 2010 highlighted that, in the case of weak synoptic forcing where there is not a correct prediction from models, only the monitoring tools could play a crucial role in the warning emission procedure.

If we consider, for example, a plain flood situation then evolution of the synoptic signal is "enough" to inform us of the potential risk that will take place. Model skill (global as well as LAM) for these events is quite good so that a forecaster can rely on this system and does not really need additional information.

The two cases presented here belong to a completely different category of event. In this case the operational LAM does not provide any useful insight, in terms of time and space localization as well as the severity of such events. Synoptic analyses just showed a weak cold advection (for one of them) and no other significant signature. This suggests that the weak and scarce LAM capability to forecast such an event is not only a LAM fault in itself, but a problem mainly related to the very low predictability of convective events.

From these two events we can emphasize that the main step to increase for monitoring and nowcasting concerns the early identification of generating convection in the area involved. The main and effective instruments for this goal are radar, satellite and lightning detector. The most important products are: radar reflectivity images and their sequences, cell travel-speed, radar QPE, raingauge precipitation, satellite images and cell characterisation.

Further, in order to give helpful support to the nowcaster, a detailed geo-location tool as well as an overlap capability for critical points in the infrastructure are key resources to be implemented for a warning emission procedure.

Nevertheless, for events like the one affecting Copparo (a quasi-stationary severe thunderstorm), the critical aspect is the persistency of precipitation because only the whole event accumulation defines the danger severity of the event.

Weather Radar and Hydrology
(Proceedings of a symposium held in Exeter, UK, April 2011) (IAHS Publ. 351, 2012).

415

Observations of hailstorms by X-band dual polarization radar

SHIN-ICHI SUZUKI, KOYURU IWANAMI, TAKESHI MAESAKA, SHINGO SHIMIZU, NAMIKO SAKURAI & MASAYUKI MAKI

National Research Institute for Earth Science and Disaster Prevention, 3-1 Tennoudai, Tsukuba, Ibaraki 305-0005, Japan
ssuzuki@bosai.go.jp

Abstract Weather spotters reported small hail associated with convective storms during 2008–2010 in the Kanto area, Japan, and several of the storms were observed by an X-band dual polarization radar located in Ebina city, near Yokohama, Japan. Observed reflectivity Z_H and polarimetric parameters (e.g. differential reflectivity Z_{DR}, the specific differential phase K_{DP}, and the correlation coefficient ρ_{HV}) were analysed in terms of hail detection by radar. Attenuation correction was applied to Z_H and Z_{DR} using the self-consistent method, because this correction is essential for X-band radar data. In several reflectivity cores, the rainfall rate estimated from K_{DP} was smaller than that estimated from Z_H, especially in regions where $Z_H > 50$ dBZ. This finding indicates the presence of hail, because K_{DP} is insensitive to hailstones. In many cases, the occurrence of small (but non-zero) values of K_{DP} and small values of ρ_{HV} (<0.9) indicate the presence of wet hail or a mixture of hail and rain. Z_{DR} was smaller than that expected from Z_H in the case of rain. These results are consistent with those from S-band radars, suggesting the potential for detecting hail by X-band dual polarization radar.

Key words hail; X-band; dual polarization

INTRODUCTION

Hailstorms in Japan cause damage to crops, cars, houses, etc. Some severe weather phenomena, such as downbursts and tornadoes, are accompanied by hail. The forecasting of such storms is important for human safety; however, it is difficult to provide sufficient warning of such events because of their short lifetime and small horizontal scale compared with synoptic-scale phenomena, which are well forecasted. The detection of hailstones in the air may enable adequate warnings of hailstorms and severe weather as a nowcasting technique. Hail detection may help to avoid false flood warning from measurements of radar reflectivity.

Weather radars are primary tools in detecting hydrometeors. Many researchers have attempted to detect hail using weather radars. Generally, a high value of reflectivity Z_H indicates hail; however, there is no clear threshold value in discriminating hail from heavy rain. Multi-parameter (MP) radars have the potential to detect hailstones. For example, Feral *et al.* (2003) used two single-wavelength radars, one at S-band (near 10 cm) and the other C-band (near 5 cm) to detect hailstorms.

Polarimetric radar measurements have been used to discriminate hailstones from rain (e.g. Aydin *et al.*, 1986; Vivekanandan *et al.*, 1990; Heinselman & Ryzhkov, 2006). The value of differential reflectivity Z_{DR} for dry hail is expected to be near 0 dB because such hailstones are close to spherical in shape and tumble through the air, whereas oblate raindrops yield positive Z_{DR} values. The specific differential phase K_{DP} is also expected to be close to zero for hailstones, because they are tumbling with no alignment. The relatively small value of dielectric constant, about 3, for ice particles, whereas that of liquid water is about 80, also make KDP very small (Battan, 1979; Aydin *et al.*, 1995; Straka *et al.*, 2000).

S-band polarimetric weather radars have been used to identify and analyse falling hail (Balakrishman & Zrnic, 1990; Aydin *et al.*, 1995; Depue *et al.*, 2007). Straka *et al.* (2000) summarized the typical values of polarimetric radar data for hail at S-band. An advantage of S-band radars is the low attenuation in heavy precipitation. In recent years, X-band (near 3 cm) radars and observation networks of such radars have been developed for rainfall estimations, because they have the advantages of fine resolution, small antennas compared with S-band, high mobility, and low cost. Polarimetric radars at X-band have another important advantage: K_{DP} is larger than that at C- or S-band for the same rainfall rate. This result allows X-band K_{DP} to be applied to much weaker rainfall events for a given error in K_{DP} estimation from the differential propagation phase ϕ_{DP} (Maki *et al.*, 2005). The rainfall rate has been successfully estimated using K_{DP} for heavy rainfall (Kato & Maki, 2009).

Reflectivity Z_H and Z_{DR} have also been used as parameters in discriminating hail from heavy rainfall; however, attenuation resulting from rain is significant at X-band, especially for high-reflectivity echoes such as strong rain. The detection of hailstones is expected to be improved using X-band polarization radars if attenuation is well corrected.

Weather spotters reported small hailstorms associated with convective storms from 2008 to 2010 in the Kanto area, Japan. They uploaded photographs of hailstones onto their blogs. Several of these storms were observed by X-band dual polarization radar. With the aim of precisely detecting hailstones, we analysed the observed polarimetric variables of hailstorms.

OBSERVATIONS AND ANALYSIS METHOD

Radar observations

The observation data used in this study were collected by a dual-polarization radar at X-band, operated in Ebina city (35.4°N, 139.4°E), Japan, by the National Research Institute of Earth Science and Disaster Prevention (NIED). The main specifications of this radar (herein, EBN radar) are listed in Table 1. The radar simultaneously transmits and receives the horizontally and vertically polarized scatter signals. Three-dimensional data were collected by volume scans at 12 elevation angles, ranging from 0.7° to 10.4°, every 5 minutes. All PPI scan data were mapped to a Cartesian grid with a horizontal grid spacing of 0.0045° × 0.0055° and a vertical grid spacing of 500 m, using Cressman interpolation, after attenuation correction for every ray (see the following subsection).

Table 1 System characteristics of the EBN radar.

Frequency	9.375 GHz
Antenna	2.1 m diameter parabolic antenna
Antenna gain	41.6 dB
Beam-width	1.3°
Transmission tube	Magnetron
Peak power	50 kW
Polarization	Horizontal and vertical
Observation range	80 km

Attenuation correction

The correction method employed for Z_H and Z_{DR} is a key factor in the present analysis. We used the self-consistent method (Bringi *et al.*, 2001) for X-band radar (Park *et al.*, 2005b). Z_H and Z_{DR} are corrected using K_{DP}. Note that the correction is valid for rain and not for ice particles. It is possible that Z_H and Z_{DR} can be estimated to be smaller than the actual value if the ice particle is present and if the hail is missed. A zero lag cross-correlation coefficient, ρ_{HV}, is also corrected according to the signal-to-noise ratio.

RESULTS

9 August 2008

Falling hail was reported on 9 August 2008 at Mt. Fuji (35.36°N, 138.73°E and 3776 m height) in Yamanashi prefecture. The mountaintop was covered with ice particles, despite it being the middle of summer. Figure 1(a) shows the CAPPIs (Constant Altitude PPIs) at 3 km above sea level (ASL), and Fig. 1(b) shows vertical cross-sections at 35.37°N for Z_H, Z_{DR}, K_{DP}, and ρ_{HV} at 0440 UTC. Cell A was located near Mt. Fuji at this time, and Z_H exceeded 65 dBZ at the cell centre (138.8°E). Z_H before correction was less than 40 dBZ, which is too small to recognize the echo as hail. There is still uncertainty in correcting Z_H and Z_{DR}, and they may actually be larger if hail is present; however,

Z_H is large enough to indicate the presence of hailstones after the attenuation correction. Z_{DR} at 3 km ASL was relatively low, considering the value of Z_H. In contrast, K_{DP} is not small at this height. ρ_{HV} was small compared with that expected for pure rain. In the vertical section, the region of strong Z_H (i.e. $Z_H > 50$ dBZ) extended up to 6 km ASL; however, Z_{DR} was large only below about 3 km ASL and was negative at elevations >4 km ASL. The distribution of K_{DP} is similar to that of Z_{DR}. ρ_{HV} is <0.85 at around 4 km ASL. The 0°C height at 00:00 UTC was 4.6 km at Hamamatsu city, about 100 km away to the southwest to the Mt. Fuji. These observations suggest the presence of hail, at least above 4 km ASL, with the hail melting below this level.

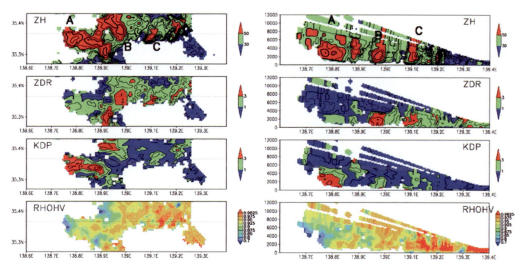

Fig. 1 (a) CAPPIs of Z_H, Z_{DR}, K_{DP}, and ρ_{HV} at 3 km ASL, and (b) vertical cross-sections at 35.37°N. Dashed lines in (a) indicate the position of vertical sections.

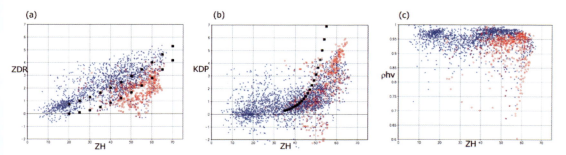

Fig. 2 Scatter diagrams of (a) Z_H–Z_{DR}, (b) Z_H–K_{DP}, and (c) Z_H–ρ_{HV} at 2 km ASL. Blue symbols indicate data in the region 35.2°N–35.4°N and 138.9°E–139.3°E, around Cells B and C; red symbols indicate data in the region 35.2°N–35.4°N and 138.7°E–138.8°E, around Cell A. The black squares in (a) and (b) are explained in the text.

Other cells were located east of Cell A. Cell B was located at 138.97°E, for which Z_H exceeded 60 dBZ. As with Cell A, Z_{DR} for Cell B was relatively large below 3 km ASL and was negative above 4 km ASL. K_{DP} was almost zero at altitudes above 4 km ASL, which is a relatively small value for rain (compared with the large value of Z_H). Therefore, the cell is expected to have produced hail, even below 4 km ASL. Cell C, which was located east of Cell B, showed similar distributions of Z_H, Z_{DR}, and K_{DP} to those observed for Cell B.

Figure 2 shows scatter diagrams for the relations Z_H–Z_{DR}, Z_H–K_{DP}, and Z_H–ρ_{HV} observed at 2 km height, at the same time as the observations shown in Fig. 1. The black squares in Fig. 2(a) indicate the upper and lower boundaries of the range in Z_{DR}, which contains 70% of the observed radar data for rain in July 2009, for a given Z_H. The values of Z_{DR} observed around Cells B and C (blue symbols in the figure) are similar to, or larger than those for rain. Z_{DR} around Cell A (red symbols) is lower than that for rain. In the Z_H–K_{DP} plot, black squares indicate the mean value of K_{DP} for rain, as obtained according to the R–Z and R–K_{DP} relations at X-band (Park *et al.*, 2005b). The values are about three times larger than those at S-band (Straka *et al.*, 2000). K_{DP} is lower than the value for rain around Cell A (red symbols in Fig. 2(b)) and around Cells B and C (Fig. 2(b)). Hydrometeor around cell B and C might be wet or horizontally oriented hail. Observed ρ_{HV} (Fig. 2(c)) is also low, especially around Cell A (red symbols) in the range of high reflectivity (i.e. $Z_H > 40$ dBZ). This finding suggests the falling of wet hail around Mt Fuji. These features from polarimetric data are similar to those of hail observed at S-band, as summarized by Straka *et al.* (2000), and suggest the presence of wet hail below 4 km ASL and dry hail aloft.

Note that the correction method applied here is valid for rain, but not for ice particles. The attenuation correction of Z_H and Z_{DR} for Cells A and B may be insufficient because the value of K_{DP} is small for ice particles to correct reflectivity. It is very difficult to correct them precisely, however, the current correction shows us the possibility to identify hailstones in this case. If the correction algorithm is improved, reflectivity will be getting larger and more hydrometeors may be recognized as hail.

18 October 2008

Weather spotters and TV stations reported small hail at Shinjuku, Tokyo on 18 October 2008. Convective clouds were observed by the EBN radar. Figure 3 shows CAPPIs at 1 km ASL and vertical cross-sections at 35.699°N, near Shinjuku, of Z_H, Z_{DR}, K_{DP}, and ρ_{HV} at 0820 UTC. Z_H was large around 35.74°N and 139.76°E in the CAPPI (Fig. 3(a)), as were Z_{DR} and K_{DP}. This precipitation core is expected to be rain with sleet. Another precipitation core was observed southwest of the precipitation core of the rain, at around 35.69°N, 139.69°E. The maximum value of Z_H was larger than 50 dBZ, and Z_{DR} was larger than 3 dB, which corresponds to the values for rain; however, K_{DP} was close to zero. In the vertical sections of the precipitation core (Fig. 3(b)), Z_H and Z_{DR} are also large, but K_{DP} is low. The strong precipitation should be recognized as hail. In addition to the above features, ρ_{HV} at the core (in the vertical sections) is low, indicating pure rain. Scatter diagrams showing the relations Z_H–Z_{DR}, Z_H–K_{DP}, and Z_H–ρ_{HV} (Fig. 4) reveal that K_{DP} and ρ_{HV} are low in the

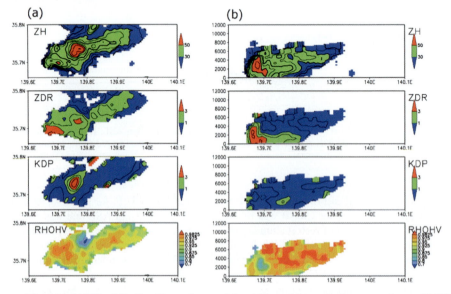

Fig. 3 (a) CAPPIs of Z_H, Z_{DR}, K_{DP}, and ρ_{HV} at 1 km ASL, and (b) vertical cross-sections at 35.699°N.

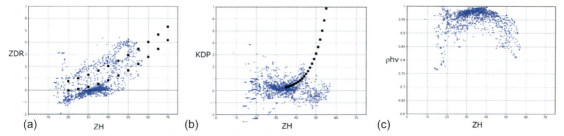

Fig. 4 Scatter diagrams of (a) Z_H–Z_{DR}, (b) Z_H–K_{DP}, and (c) Z_H–ρ_{HV} in a vertical section at 35.699°N. Black squares in (a) and (b) are the same as those in Fig. 2.

range for which $Z_H > 40$ dBZ, and Z_{DR} has intermediate values. These features suggest the presence of wet hail that was melting whilst falling from aloft. The hailstones may have comprised ice cores surrounded by liquid water. This case provides a good example of the excellent performance of X-band MP radar in capturing hailstones.

26 July 2010

Strong convective storms occurred in the Kanto area of Japan on 26 July 2010. Large hailstones (~3 cm in diameter) were reported by a weather spotter in Saitama prefecture. The strong echo (maximum Z_H above 60 dBZ) was observed by EBN radar, even though the storm was located near the edge of the observation range. The characteristics of the polarimetric data were almost identical to those described in the previous cases: K_{DP}, Z_{DR}, and ρ_{HV} had small values in the precipitation cores of large Z_H.

Figure 5(a) shows CAPPIs at 2 km ASL at 0720 UTC. The echo (i.e. the high-Z_H region) extended from WNW to ESE. Z_{DR} is large and positive on the south side of the echo and negative on the north side. K_{DP} showed a peak at slightly north of the maximum Z_H. ρ_{HV} is small on the north side of the echo. The structure of the echo is seen more clearly in the vertical section (Fig. 5(b)). At 7 km ASL on the south side of the echo, Z_H exceeds 50 dBZ and both Z_{DR} and K_{DP} are small, indicating hailstones. Below 4 km ASL, on the south side of the Z_H maximum, Z_{DR} is large; on the north side,

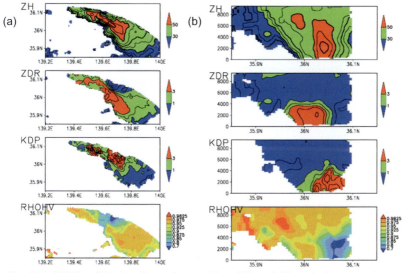

Fig. 5 (a) CAPPIs of Z_H, Z_{DR}, K_{DP}, and ρ_{HV} at 2 km ASL, and (b) vertical cross-sections at 139.65°E.

Z_{DR} is negative, K_{DP} is intermediate, and ρ_{HV} is low. These features suggest the presence of large raindrops below 4 km ASL and hailstones aloft, produced by strong updraft on the south side of the echo. Falling wet hail and sleet are suggested for the north side of the echo.

DISCUSSION

The X-band MP radar of NIED observed several hailstorms in the Kanto area of Japan, demonstrating its ability to detect hail. The characteristics of the observed polarization data from hailstorms are similar to those at S-band, except that K_{DP} is about three times larger for rain than that at S-band. The use of K_{DP} in combination with Z_H is expected to detect hailstones at X-band. If Z_H is large (i.e. >50 dBZ) and K_{DP} is much lower than that for rain, hydrometeors are expected to be hail; however, even if Z_H is large enough to indicate hail, hydrometeors are expected to be rain if K_{DP} is large enough to indicate rain. Considering the insensitivity of K_{DP} to ice particles and the high performance of K_{DP} for rainfall estimates and attenuation corrections for X-band MP radar, K_{DP} should be used rather than Z_{DR}. Z_{DR} has potential to be a useful parameter; however, it needs to be suitably corrected using K_{DP}, which does not require attenuation correction.

The present results suggest that the correction of Z_H works well enough for use in the detection of hailstones from rain. The problem with X-band radars in terms of rainfall attenuation has been solved by using the self-consistent method for attenuation correction; however, the correction for Z_H may be insufficient for the hailstones themselves, because K_{DP} is insensitive to ice particles. It is possible to reach a conclusion of rain in the case of hail if Z_H is low due to attenuation by hail and if K_{DP} is also close to zero, even if the intrinsic value of Z_H is large. The attenuation correction for ice particles requires further work; however, the use of an observation network of X-band radars would enable the reliable detection of hail. Even if the attenuation correction is insufficient in the region of hail, X-band radars can detect the edge of the hail region, where Z_H is expected to be well corrected.

The threshold value for K_{DP} (in distinguishing hail from rain) is expected to be a function of Z_H, and needs to be examined precisely based on comparisons between ground observations and radar observations of hail. K_{DP} is expected to be close to zero for dry hail, but to have a larger value for wet hail (Straka *et al.*, 2000). The obtained threshold may be utilized as a parameter of a membership function used in a fuzzy logic method of a hydrometeor classification, in addition to being a parameter that directly distinguishes hail from rain.

REFERENCES

Aydin, K., Seliga, T. A. & Balaji, V. (1986) Remote sensing of hail with a dual linear polarization radar. *J. Climate Appl. Met.* 25, 1475–1484.
Aydin, K., Bringi, V. N. & Liu, N. (1995) Rain-rate estimation in the presence of hail using S-band specific differential phase and other radar parameters. *J. Appl. Met.* 34, 404–410.
Balakrishnan, N. & Zrnic, D. S. (1990) Estimation of rain and hail rates in mixed-phase precipitation. *J. Atmos. Sci.* 47, 565–583.
Battan, L. J. (1973) *Radar Observation of the Atmosphere.* University of Chicago Press, 323 pp.
Bringi, V. N., Keenan, T. D. & Chandrasekar, V. (2001) Correcting C-band radar reflectivity and differential reflectivity data for rain attenuation: self-consistent method with constraints. *IEEE Trans. Geosci. Remote Sens.* 39, 1906–1915.
Depue, T. K., Kennedy, P. C. & Rutledge, S. A. (2007) Performance of the hail differential reflectivity (HDR) polarimetric radar hail indicator. *J. Appl. Met. Climate* 46, 1290–1301.
Feral, L., Sauvageot, H & Soula, S. (2003) Hail detection using S- and C-band radar reflectivity difference. *J. Atmos. Oceanic Tech.* 20, 233–248.
Heinselman, P. L. & Ryzhkov, A. V. (2006) Validation of polarimetric hail detection. *Weather & Forecasting* 21, 839–850.
Kato, A. & Maki, M. (2009) Localized heavy rainfall near Zoshigaya, Tokyo, Japan on 5 August 2008 Observed by X-band polarimetric radar – preliminary analysis. *SOLA* 5, 89–92.
Maki, M., Park, S.-G. & Bringi, V. N. (2005) Statistical error of rain rate estimations due to natural variations of raindrop size distributions for 3-cm wavelength polarimetric radar. *J. Meteor. Soc. Japan* 83, 871–893.
Park, S.-G., Bringi, V. N. , Chandrasekar, V., Maki, M. & Iwanami, K. (2005a) Correction of radar reflectivity and differential reflectivity for rain attenuation at X-band. Part I: theoretical and empirical basis. *J. Atmos. Oceanic Tech.* 22, 1621–1632.
Park, S.-G., Maki, M., Iwanami, K., Bringi, V. N. & Chandrasekar, V. (2005b) Correction of radar reflectivity and differential reflectivity for rain attenuation at X-band. Part II: evaluation and application. *J. Atmos. Oceanic Tech.* 22, 1633–1655.
Straka, J. M., Zrnic, D. S. & Ryzhkov, A. V. (2000) Bulk hydrometeor classification and quantification using polarimetric radar data: synthesis of relations. *J. Appl. Met.* 39, 1341–1372.
Vivekanandan, J., Bringi, V. N. & Raghavan, R. (1990) Multiparameter radar modeling and observations of melting ice. *J. Atmos. Sci.* 47, 549–564.

Weather Radar and Hydrology
(Proceedings of a symposium held in Exeter, UK, April 2011) (IAHS Publ. 351, 2012).

421

Multifractal study of three storms with different dynamics over the Paris region

I. TCHIGUIRINSKAIA[1], D. SCHERTZER[1], C. T. HOANG[1] & S. LOVEJOY[2]

1 *Université Paris-Est Ecole des Ponts ParisTech LEESU, 6-8 Av Blaise Pascal Cité Descartes, Marne-la-Vallee, 77455 Cx2, France*
ioulia@leesu.enpc.fr
2 *McGill University, Physics Department, Montreal, PQ, Canada*

Abstract Research is now triggered by the permanent need to better relate the measured radar reflectivity to surface rainfall. Knowledge of flow structure within cloud formation systems and the associated convective–stratiform separation may provide useful information in this respect. We will first discuss how stochastic multifractals can handle the differences of scales and measurement densities of the raingauge and radar data; and help to merge information from these data. Mosaics from the Météo-France ARAMIS radar network are used that correspond to horizontal projections of radar rainfall estimates for a 1 km × 1 km × 5 min grid over France. In particular, three storm events with different dynamics over the Paris region were selected to illustrate the efficiency of the multifractal framework. In spite of the difficulty that usually the same precipitation field comprises both stratiform and convective formations, their respective scaling properties allow the deciphering and classification of the radar data.

Key words multifractals; convective-stratiform formations; rainfall extremes; power law

INTRODUCTION

Precipitation constitutes a natural phenomenon which has very strong socio-economic impacts, especially as heavy rainfall is usually associated with storms. Climate change and sprawling urbanization are two drivers that would increase the extremes. Both already put into question the usual implicit approximation of statistical stationarity, and therefore classical operational notions. To take this aspect into account, the hydrological systems of the rainfall alert and forecast need to provide the communities with more detailed space–time information. Ideally, this should cover time scales from 1 s and spatial scales ranging from millimetres to about 50 km, i.e. from drop sizes to Paris region scale. This range of scales is crucial for urban hydrology and it also corresponds to the lesser known scale range for precipitation fields. There are many ongoing discussions on possible rainfall scaling breaks below 30 km (e.g. Gires *et al.*, 2011). A deeper understanding of the rainfall process across scales seems to be necessary for bridging the scale gap between radar and *in situ* measurements, which has been a long-lasting problem for radar calibration (Austin, 1987). In this paper we explore the scaling behaviour of horizontal projections of the radar rainfall estimates by properly dealing with the strong intermittency of rainfall that is more obvious at small scales. We first discuss interlinks between what could be termed convective/stratiform rainfall and a possible rainfall scaling break. Then we illustrate how the multifractal methods handle the differences of scales and measurement densities of the raingauge and radar data that are available from a 1 km × 1 km × 5 min grid. Finally we underline how merging the information from these data would increase our understanding of rainfall extremes. Empirical parameters allow an implementation of a procedure for simulation and nowcasting of rainfall fields and for cross-fertilization with operational applications.

DEFINITIONS ACROSS SCALES

Initially triggered by climatological studies, the discussion of convective *vs* stratiform rainfall was widely enlarged by the aspects of radar rainfall measurement. Most precipitation systems can be sorted as convective or stratiform, while there is no consensus on a precise definition of the terms convective/stratiform (see e.g. Steiner & Smith, 1998); notably their distinction is not always sharp. The most common cinematically-based classification into convective/stratiform formations appeals to the characteristic scale and magnitude of the vertical velocity field. Indeed, Houze (1993) underlined that stratiform rainfall is more variable in the vertical than the horizontal, since hydrometeors (snow or rain) in stratiform clouds grow primarily by descent through a widespread updraft whose velocity magnitude is <1 m s^{-1}. Thus, purely stratiform rainfall would result from mid-latitude frontal systems. Convective rainfall is associated with regions where rainfall remains equally variable in the horizontal and in the vertical directions, so the magnitudes of upward air motion would be sufficiently large in respect to the fall velocity of the hydrometeors. Steiner & Smith (1998) advocated that the definition of what might be termed convective/stratiform rainfall should hold across all scales. As a step in this direction, Seed (2004) investigated the extent of the vertical-horizontal anisotropy, depending on the physics dominating the rainfall production process. He proposed use of variograms of radar reflectivity in the vertical and horizontal dimensions to distinguish between convective and stratiform rainfall, although he admitted that this method requires very large samples of data. In particular, it requires vertical profiles of radar reflectivity that are not actually available among the ARAMIS products of Météo-France (Gilet et al., 1983).

The same precipitation field often comprises formations of both types (Houze, 1997). In spite of this difficulty, meteorologists often select samples of predominantly stratiform or convective rainfall fields aiming to investigate their respective scaling properties. For example, Chumchean (2004) pre-classified over 800 fields of hourly rainfall accumulation over Sydney (Australia). He used the spectral analysis on 256 × 256 km^2 rainfall fields of 1 km resolution. The spatial spectrum $E(k)$ is the density of "energy" of the field (i.e. the squared field) per unit wave number k (km^{-1}). A random field is *scaling* when its spectrum follows a power law, $E(k) \propto k^{-\beta}$. Chumchean (2004) found differences in the estimates of the power law exponent, β, often called the "spectral slope", for stratiform and convective rainfall, with respective average values of 2.3 and 2.1. However, the significance of this 10% difference could be put into question due to uncertainties. In fact, rainfall is a complex space–time process. Hence, contrary to the usual approaches that take into account only spatial fluctuation estimators, we propose to analyse the time evolution of this estimator to classify convective and stratiform rainfalls. We therefore discuss the capacity of a more general space and time scaling analysis, often resulting in non-stationary estimates of scaling parameters. Then the time series of scaling exponents could be directly used, for example, to automatically sort out the convective rainfall episodes among numerous radar data archives.

We use the ARAMIS data of Météo-France (Parent du Châtelet, 2003) for three storms over the Paris region, each lasting 4 h. These storms were *a priori* classified by meteorologists as two mixed events (on 11 January 2008 and on 9 February 2009 from 00:12 h to 00:16 h) and a stratiform rainfall on 13 February 2009 (from 00:01 h to 00:05 h). However, this classification remains only an indicative one for the selected periods and location. We started from the (spatial)

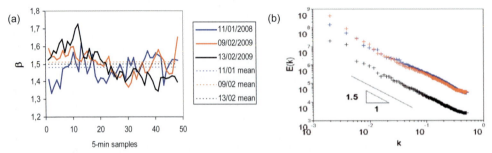

Fig. 1 (a) Temporal evolution of the spectral slopes obtained on power spectra in space (1 km resolution) of consecutive samples of 5 minute rain accumulation; (b) power spectra in space, averaged over the duration of an event: 11 January 2008 (blue), 9 February 2009 (red) and 13 February 2009 (black). All events yield the average spectral slope $\beta = 1.5$ (in space and time), independently of the physics that dominated the rainfall production process.

spectral analysis of 5-min time consecutive rainfall accumulation maps of 512×512 km^2 with a 1 km resolution. For all three storm events, the time series of the spatial power law exponents fluctuate around the mean value of $\beta = 1.5$ (Fig. 1(a)), independently from the physics that dominated the rainfall production process. The same value of the β exponent characterises the spatial power spectra averaged over the duration of the event (Fig. 1(b)). Although the stratiform rainfall is notably less intense (i.e. the spectrum is lower than the two convective ones), there are no significant differences in scaling behaviour at larger scales of these spectra. The power law behaviour confirms the scaling nature of the rainfall field, although Fig. 1(a) points out the problem of fluctuations or non-stationarity of the power law exponent estimates (and more generally, of scaling parameters) during each event. These fluctuations put into question the possibility to distinguish in a reliable manner convective *vs* stratiform situations with only the help of the spectral slope, although it was quite appealing.

FRACTALS AND CONVECTIVE-STRATIFORM CLASSIFICATION

The simplest scaling feature of radar images, i.e. rainfall clustering over a fractal support, can be easily established with the help of the straightforward box-counting algorithm. The boxes are defined as the pixels of a regular 2-D grid covering the radar image, where a mesh size l is consecutively increased by a given factor (usually 2). Then the fractal dimension Df of the rain support is the exponent of the scaling law of the number of pixels $n(l)$ containing at least one positive rainfall measurement: $n(l) \propto l^{-Df}$. Obviously a homogeneous 2-D field will correspond to $Df = 2$, and lower and lower fractal dimensions correspond to more and more inhomogeneous fields. Figure 2(a)–(c) displays the temporal evolution of the two fractal dimensions estimated on national radar images (512×512 km^2) of 5 min rainfall accumulation, over two distinct sub-ranges of scales. Indeed, we detected a change in the scaling behaviour at about 20 km that corresponds to low values of the coefficient of determination estimated over the full range of scales, and was particularly visible on 11 January 2008. This implies that two distinct fractal dimensions are necessary over the small scale range (1–16 km) and over the large scale range (16–512 km), respectively. The small scale analysis (i.e. performed over the small scale range) confirms a general dependence of the support fractal dimension on the percentage of rainfall under the threshold of radar detection (i.e., the "zero rainfall"): the fractal dimension tends to decrease with an increase of zero rainfall, e.g. it decreases from 1.93 to 1.88 during the episode of 9 February 2009 when the zeros increase from 30% to 41%. Although the same $Df \approx 1.8$ is observed for two types of rainfall events: the convective type on 11 January 2008 with about 55% of zero rainfall during the episode and the stratiform type on 13 February 2009 with 70% of zeros. Therefore, a small scale fractal dimension would not be sufficient to sort out radar data. In contrast, a comparison of small scale and large scale fractal dimensions gives a striking dissimilarity of the first two rainfall events (11 January 2008 and 9 February 2009) with respect to the third one (13 February 2009). While the large scale fractal dimensions remain close to 2 for the first two events and hence, remain distinct from the corresponding small scale fractal dimensions. In contrast, on 13 February 2009 the estimates of the large scale fractal dimension fluctuate around the small scale value of $Df \approx 1.8$. This result is in agreement with the fact that the horizontal extent of vertical air motions during a stratiform situation may grow up to hundreds of kilometres. Hence, the stratiform and decaying convective rainfall systems would naturally result in a more widespread precipitation, which in turn would result in similar estimates of fractal dimension over small and large scales. In contrast, a convective situation is characterized by significant vertical air motions over horizontal scales ranging from several metres to several kilometres. So, over a larger area, a convective rainfall system can be understood as a homogeneous distribution ($Df \approx 2$) of smaller scale precipitation cells, all having a lower fractal dimension. Therefore, the fractal concept provides a simple basis to automatically sort out convective rainfalls from radar data. It is worth mentioning that for both rainfall types, the estimates of fractal dimension at small scales Df

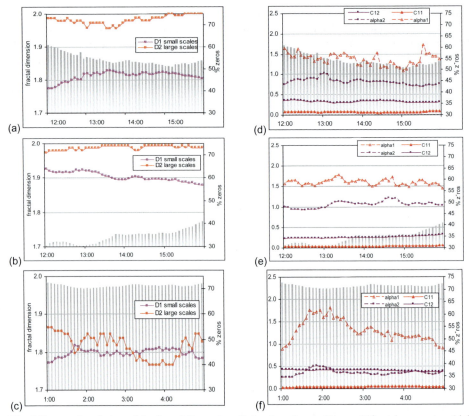

Fig. 2 Temporal evolution of the fractal dimensions (left column) and of the multifractal parameters (right column), all estimated on 48 images (512×512 km^2) of 5 min rainfall accumulations: 11 January 2008 (a), (d), 9 February 2009 (b), (e) and 13 February 2009 (c), (f). Grey histogram indicates the percentage of pixels with "zero-rainfall", pink and red lines correspond to the estimates over the small scales (1–16 km), orange and purple lines correspond to the estimates over the large scales (16–512 km).

remain smaller than 2. This highlights that homogeneous distributions of rainfall would be particularly inappropriate for urban modelling. Indeed, at smaller scales, the proportion of pixels having non-zero rainfall should decrease as $l^{(2-Df)}$ to cope with reality.

PHYSICAL SIGNIFICANCE OF MULTIFRACTAL PARAMETERS

Different estimates of the fractal dimension exponent Df could be obtained over the same range of scales by simply changing the threshold that defines the non-zero rainfall: the rainfall intensities, therefore, correspond to a multifractal field (Schertzer & Lovejoy, 1987; Biaou *et al.*, 2003). Universal multifractals (Schertzer & Lovejoy, 1997) can be defined with the help of a very limited number of parameters that have a strong physical significance and can be evaluated either theoretically or empirically. The parameter α measures the multifractality of the field. When $\alpha = 0$, the field is fractal and could be defined by a unique fractal dimension. C_1 is the co-dimension of the mean field and measures its mean fractality. A homogeneous field that fills the embedding space has $C_1 = 0$. When this parameter increases, the field becomes more and more intermittent, i.e. is concentrated on smaller and smaller fractions of the total area. The parameter estimation methods are based on the upscaling of the field raised to various powers (Lavallée *et al.*, 1993). Figure 2(d)–(f) displays the temporal evolution of the multifractal parameters α and C_1

that are estimated on the same radar data and also separately over the same small and large ranges of scales, as for the fractal dimension estimates of Fig. 2(a)–(c). The multifractality of rainfall fields is stronger ($\alpha > 1$) over small scales (i.e., over the range of 1–16 km). Over larger scales (i.e. over the range of 16–512 km), the high percentage of zero-rainfall reduces the multifractality while increasing the mean co-dimension C_1. The rainfall field appears to be more intermittent over large scales. This phenomenon is sometimes considered as the "unsolved issues regarding the treatment of the zero values" (Seed, 2004).

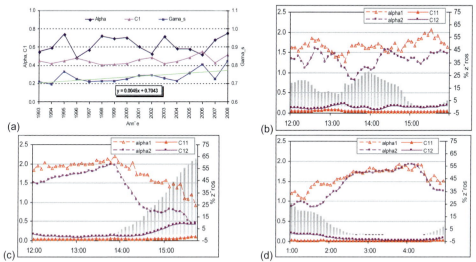

Fig. 3 Temporal evolution of multifractal parameters: (a) raingauge time series at Orly with α (dark blue), $C1$ (pink), γs (blue) and the linear fit of the estimates of γs (green line); (b)–(d) centred radar images of 5 min rainfall accumulation, 11 January 2008, 12:00–16:00 h (b), 9 February 2009, 12:00–16:00 h (c) and 13 February 2009, 01:00–05:00 h (d). Grey histograms indicate the percentage of zero-rainfall pixels, dashed lines for α, dotted lines for C1 (red and purple for the small (1–16 km) and large (16–128 km) scale ranges, respectively).

TENDENCY IN THE EVOLUTION OF RAINFALL EXTREMES

Over large scales, $\alpha < 1$ and C1 ≈ 0.25–0.5 (Fig. 2(d)–(f)) are qualitatively close to those estimated on continuous raingauge time series. This opens ways of combining the analysis of raingauge and radar data, in spite of their differences in scales and measurement densities. Figure 3(a) displays an example of the temporal evolution of multifractal parameters estimated on time series at Orly raingauge in the region of Paris during the period 1993–2008. The stochastic multifractal parameter estimates reflect the long-range dependencies and clustering of rainfall extremes, often having power law probability distributions. A simultaneous increase of both parameters directly implies an increase of extremes. Otherwise, the evolution of extremes can be investigated with the help of the maximum observable singularity: $\gamma_s = C_1 \alpha /(\alpha - 1)(C_1^{(1-\alpha)/\alpha} - 1/\alpha)$. As for Orly's time series, a slight positive slope of the linear regression of its temporal evolution (Fig. 3(a)) indicates a possibly more important increase of the maximum intensity of rainfall, $R_{max}(\tau) \propto \tau^{-\gamma_s}$, which should not be neglected for small time scales τ. If such behaviour would be only observed on radar data, it could be attributed to a change in the QPE system during 2004. This illustrates a potential usefulness of completing the scaling analysis of radar data with the help of an analysis of available raingauge time series.

It is worthwhile noting that the difference in estimates between small and large scales becomes less obvious when reducing the initial domain to 128 × 128 km², compatible with the surface area of the Paris region (12011 km²). The storm events, selected for this paper, were indeed essentially located over this region. Reducing the initial domain, we therefore limited our analysis to the area of intense rainfall. Hence, the multifractal parameters obtained (Fig. 3(b)–(d)) are in good agreement with those estimated over the small scale range of the initial domain (Fig. 2(d)–(f)), and furthermore are compatible with those obtained for heavy rainfall episodes extracted from local raingauge time series (Hoang, 2008).

CONCLUSIONS

The aim of this study is to explore the scaling behaviour of horizontal projections of the radar rainfall estimates by properly dealing with the strong intermittency of rainfall. We start from the investigation of possible relations between radar rainfall scaling breaks and associated convective–stratiform separation. The differences of scales and measurement densities of the raingauge and radar data are easily handled in the framework of stochastic multifractals. This framework embodies all possible probability tail behaviours and clustering of the extremes, including possible long-range dependencies: two most important statistical features to fit empirical quantile estimates. Multifractal methods allow rainfall quantiles to be evaluated dynamically. For universal multifractals, the parameters α and C_1 fully define their statistics. This explains why the observed compatibility between the multifractal parameters for raingauge measurements and radar data is essential to merge the two types of data. Increased reliability of the parameters reduces the uncertainty in the predicted regional hydrological extremes. The results obtained illustrate a slight tendency for the hydrological extremes to become particularly local and intense in the Paris area. These preliminary results encourage more extensive use of radar archives for better detection of local climate trends.

Acknowledgements The authors thank Météo-France for providing radar data, and J. Parent du Châtelet and C. Augros for stimulating discussions. The authors thank the county council of Val-de-Marne for providing rainfall time series and P. Bompard for highlighting discussions. Financial support from the interdisciplinary research project R2DS "Water management in the Paris region in the context of climate change" (GARP-3C) and the Chair "Hydrology for Resilient Cities" of Ecole des Ponts ParisTech, sponsored by VEOLIA, is highly acknowledged.

REFERENCES

Austin, P. M. (1987) Relation between measured radar reflectivity and surface rainfall, *Mon. Wea. Rev.* 115, 1053–1070.
Biaou, A., Hubert, P., Schertzer, D., Tchiguirinskaia, I. & Bendjoudi, H. (2003) Fractals, multifractals et prévision des précipitations. *Sud Sciences et Technologies* 10, 10–15.
Chumchean, S. (2004) Improved use of radar rainfall estimation for use in hydrological modelling. PhD Thesis, Dept. Civil Engineering, University of New South Wales, Australia.
Gilet, M., Ciccione, M., Gaillard, C. & Tardieu, J. (1983) Projet ARAMIS: le réseau Français de radars météorologiques. In: *Hydrological Applications of Remote Sensing and Remote Data Transmission*, 295–305. IAHS Publ. 145. IAHS Press, Wallingford, UK.
Hoang, C. T. (2008) Analyse fréquentielle classique et multifractale des séries pluviométriques a haute résolution. MSc Thesis, University P&M Curie, Paris, France.
Gires, A., Tchiguirinskaia, I., Schertzer, D. & Lovejoy, S. (2011) Analyses multifractales et spatio-temporelles des précipitations du modèle Méso-NH et des données radar, *Hydrol. Sci. J.* 56(3), 380–396.
Houze, R. A., Jr. (1993) *Cloud Dynamics.* Academic Press, 573 pp.
Houze, R. A., Jr. (1997) Stratiform precipitation in regions of convection: a meteorological paradox? *Bull. Am. Met. Soc.* 2179–2196.
Lavallée, D., Lovejoy, S., Schertzer D. & Ladoy, P. (1993) Nonlinear variability and landscape topography: analysis and simulation. *Fractals in Geography* (ed. by L. De Cola & N. Lam), 171–205. Prentice-Hall.
Marshall, J. S. & Palmer, W. M. (1948) The distribution of raindrops with size. *J. Met.* 5, 165–166.
Parent du Châtelet, J. (2003) ARAMIS, le réseau Français de radars pour la surveillance des précipitations, *La Météorologie* 40, 44–52.
Schertzer, D. & Lovejoy, S. (1987) Physical modeling and analysis of rain and clouds by anisotropic scaling of multiplicative processes. *J. Geophys. Res.* D8(8), 9693–9714.
Schertzer, D. & Lovejoy, S. (1997) Universal multifractals do exist! *J. Appl. Met.* 36, 1296–1303.
Seed, A. (2004). Modelling and forecasting rainfall in space and time. In: *Scales in Hydrology and Water Management*, 137–152. IAHS Publ. 287. IAHS Press, Wallingford, UK.
Steiner, M. & Smith, J. A. (1998) Convective versus stratiform rainfall: an ice-microphysical and kinematic conceptual model. *Atmos. Res.* 47–48, 317–326.

6 Hydrological modelling and flood forecasting

Weather Radar and Hydrology
(Proceedings of a symposium held in Exeter, UK, April 2011) (IAHS Publ. 351, 2012).

429

Weather radar and hydrology: a UK operational perspective

ROBERT J. MOORE, STEVEN J. COLE & ALICE J. ROBSON
Centre for Ecology & Hydrology, Wallingford OX10 8BB, UK
rm@ceh.ac.uk

Abstract Weather radar forms an essential and integral tool for water management in the UK, especially for monitoring and warning of flooding: the main focus of this perspective paper. An overview is first given of the radar network and its associated rainfall data products used by the environment agencies responsible for flood defence. The Hyrad (HYdrological RADar) system is deployed to receive, visualise and analyse these products, and to further process them for use within flood forecasting systems. Regional systems employ networks of models configured to make forecasts at specific locations. Very recently, countrywide systems employing an area-wide G2G (Grid-to-Grid) hydrological model have been implemented. Both types of system, used operationally in a complementary way, are reviewed in relation to their use of, and demands for, weather radar-related data. Activity on implementing probabilistic approaches to flood forecasting which benefit from using radar in ensemble rainfall prediction is outlined, and future prospects discussed.

Key words weather radar; hydrology; rainfall; flood; forecasting; distributed hydrological model

USE OF RADAR IN REGIONAL FLOOD FORECASTING SYSTEMS

Weather radar forms a key tool for water management across the UK, especially in support of flood warning, on account of its timely data availability and detailed spatial coverage. Quality-controlled rain-rates are available as "observation data" every 5 min on a 1 km grid formed as a composite from a network of 18 C-band radars over the British Isles (Harrison *et al.*, 2012). The rain-rate fields are used in forming "forecast data" as deterministic nowcasts of rainfall out to 6 h: fields are advected using an optical flow algorithm and merged with numerical weather prediction (NWP) rainfalls according to their relative accuracy at different lead-times (Bowler *et al.*, 2006). Rainfall accumulations for 15-min intervals that account for storm movement are available, both as observation and forecast data, for use with rainfall–runoff models for flood forecasting in real-time.

These gridded rainfall products are disseminated by the Met Office to environment agencies in England and Wales, and Scotland, who employ the Centre for Ecology & Hydrology's Hyrad system for real-time data receipt, processing, archiving and display (Moore *et al.*, 2004). The CatAvg component of Hyrad is configured to calculate catchment average time-series from the observed and forecast radar rainfall products (and NWP rainfalls for longer lead-times) for onward transmission to the National Flood Forecasting System (NFFS) for England and Wales and to "FEWS Scotland", both based on the Delft-FEWS open system architecture (Werner *et al.*, 2009). The NFFS is implemented as a set of regional flood forecasting systems which make flood forecasts at specific locations in support of flood warning for each region, and using configurations of models representing the river network (FEWS Scotland is configured to warning scheme areas). A variety of model types are used, including rainfall–runoff, transfer function, snowmelt, hydrological routing, and hydrodynamic routing (Moore *et al.*, 2005). Forecast updating methods employ observed river flows available at the time of forecast construction to improve accuracy. The radar rainfall forecasts out to 6 h are critical for making extended lead-time forecasts, beyond the lag time of the catchment for a given forecast location. For longer forecast lead-times of a day or more, use is made of the NWP rainfalls. In the future, these predictions will increasingly benefit from radar data assimilation when initialising the weather model (Ballard *et al.*, 2012).

The utility of radar is judged to be less when observation-based rainfall estimates are required to maintain the water balance of rainfall–runoff models of catchments. This applies both for offline hydrological model calibration and in real-time up to the time the forecast is made. Experience has shown that raingauge data, although providing only point estimates of rainfall at gauge locations, can provide a more robust estimator than radar, which can suffer from transient errors (Cole & Moore, 2008). Even when radar rainfall is used in combination with raingauge network data through a merging procedure, these errors can prove pervasive and suppressed only to a limited degree. As a consequence, it is common that the preferred rainfall estimator for use with rainfall–runoff models is based on raingauge data at times when observational data are available. However, availability of raingauge data in real-time is constrained by the polling regime of the telemetry schemes, which in the UK predominantly employ PSTN technology: charging tariffs (and sometimes battery life) can inhibit routine frequent polling.

Radar rainfall data do not suffer from such availability issues, and therefore are configured to be used as the default for times when limited or no polled raingauge data are available. The greater utility of raingauge network data over use of radar rainfall data of course depends on circumstance: the relative sparsity of a raingauge network to the small scale of convective storms is a common argument forwarded in support of weather radar for areal rainfall estimation. There are over 900 raingauges providing 15-min rain accumulations available to NFFS over England and Wales, and more than 250 for the system used over Scotland, with areas of about 151 000 and 79 000 km^2, respectively but gauge density is very varied.

USE OF RADAR FOR COUNTRYWIDE FLOOD FORECASTING

The regional flood forecasting systems are primarily configured to make flood forecasts at river gauging station locations, and their station flow records are used for model calibration. Forecasts of river flow for ungauged locations are required in these regional systems, for example where lateral inflows are input to hydrological channel flow routing models and hydrodynamic river models. The methods used vary from simple scaling of flows from nearby gauged locations (based on area and possibly rainfall climatology for the catchment) through to simple methods of rainfall–runoff model transfer from gauged sites. The latter approach can benefit from radar through using forecast rainfalls over the ungauged lateral inflow area.

A comprehensive study of methods of flood forecasting for ungauged locations (Moore *et al.*, 2006, 2007) recognised shortcomings in conventional methods: it argued the case for pursuing a distributed grid-based hydrological modelling approach. This allows a storm pattern to be shaped into a flood over space and time using information on properties of the terrain, soil/geology and land cover to configure the distributed model, allowing flows to be forecast everywhere in a physical-conceptual way. In particular, there is no need to work with aggregated "catchment characteristics" in this approach but to deal directly with landscape properties at the grid-scale of the model. A 1 km grid coincident with the radar grid is judged appropriate for current flood forecasting requirements, with soil and land-cover gridded datasets also available at this resolution.

A special kind of distributed hydrological model was conceived for flood forecasting application that differed from conventional physics-based distributed models that employ detailed representation of soil water movement in the vertical, and are typically not well supported by available soil information. This physical-conceptual distributed model, called the Grid-to-Grid or G2G Model, is of depth-integrated form and suitable for use with national datasets of terrain/soil/geology/land-cover properties along with dynamic gridded rainfalls derived from radar and raingauge observations and weather models.

The G2G was prepared and trialled for operational use within NFFS under the R&D project "Hydrological Modelling using Convective Scale Rainfall Modelling" (Environment Agency, 2009), with particular emphasis on its use for probabilistic flood forecasting using high resolution ensemble rainfall forecasts from STEPS and forms of NWP product. During the course of the project "The Pitt Review" of the Summer 2007 floods (Cabinet Office, 2008) recognised the need for a consistent countrywide approach to flood forecasting capable of providing forecasts "everywhere" out to 5 days, and that the new distributed grid-based model could fulfil this need. Regional case studies of G2G over the southwest and Midlands were, as a consequence, extended in the final phase of the project to include a G2G configuration with England and Wales coverage for fluvial rivers. This was followed by an "Operational Implementation of G2G on NFFS" project which culminated in the G2G being used within the newly-formed joint Environment Agency and Met Office Flood Forecasting Centre when preparing the Flood Guidance Statement for England and Wales (Price *et al.*, 2012). The role of this national G2G model was seen as complementary to the more detailed NFFS regional models that are typically calibrated to make forecasts at specific gauged locations. Of particular importance is the coherent spatial picture of flooding and its evolution over time that the G2G can provide, not possible with the regional network models.

Much experience was gained in the use of radar rainfall data, in both observation and forecast form, through these G2G developments. National calibration of the G2G model exposed difficulties with the routine use of gridded rainfall estimators based on radar, either used alone or in combination with raingauges. Long-term average gridded radar rainfall maps exposed the usual problems of beam blockage and discontinuities when compositing data from different radars. Transient errors in radar rainfall were detected in the G2G river flow simulations, when compared with river gauging station records, and proved hard to diagnose and remove. Whilst tipping-bucket raingauge records from some 981 stations could be affected by anomalous single values, over-recording due to tip doubling or by missing values reported as zero, these proved easier to diagnose and account for in an automated way (Howard *et al.*, 2012). The outcome of these exploratory investigations was that national calibration of G2G was best achieved using a gridded raingauge-only rainfall estimator: it proved more robust and free of the transient radar rainfall errors that could seriously corrupt efforts at model calibration. The operational model configuration was similarly configured to employ, as first priority, the raingauge-only rainfall estimator in its hierarchy of sources (Price *et al.*, 2012), but would commonly default to a merged gauge-radar or radar-only estimator due to constrained real-time access to polled raingauge telemetry data.

The G2G was subsequently configured to Scotland to meet requirements of the newly formed Scottish Flood Forecasting Service, operated jointly by the Scottish Environment Protection Agency and Met Office. Whilst the poorer and more uneven coverage of raingauges over Scotland initially suggested radar rainfall would prove of greater value for G2G model calibration, the greater robustness of the raingauge records won through, as they did for England and Wales; as a result, a similar rainfall source hierarchy has been adopted for operational use. Cranston *et al.* (2012) provides further information on the use of weather radar and countrywide flood forecasting in Scotland, noting the significant step-change in warning capability the G2G facilitates: moving from a few flood warning schemes to complete coverage for fluvial rivers.

USE OF RADAR IN PROBABILISTIC FLOOD FORECASTING

At present, flood forecasting and warning practice in the UK is primarily deterministic in nature. Active steps are being taken to explore the benefits of probabilistic methods that take forecast uncertainty into account, especially those associated with the rainfall predictions used to extend the lead-time of flood forecasts. Progress is being made possible through the availability of STEPS, which in addition to a deterministic rainfall forecast out to 6 h (based on radar extrapolation merged with NWP rainfalls) can provide an ensemble of equally-likely forecasts. An initial demonstration of its use with the PDM catchment rainfall–runoff model was provided by Pierce *et al.* (2005).

Approaches for accounting for uncertainty in the NFFS regional model networks was addressed in the "Risk-based Probabilistic Fluvial Flood Forecasting for Integrated Catchment Models" R&D Project; an overview is provided by Laeger *et al.* (2010) whilst the pathway to operational implementation remains an open question and the subject of further investigation. Of particular relevance to weather radar is an approach that combines "model" uncertainty (embracing errors in the model state, structure and its observed inputs) with rainfall forecast uncertainty, and is able to utilise the STEPS forecast rainfall ensemble for the latter. Standard ARMA (AutoRegressive Moving Average) model theory applied to the flood model errors, using a log transform to approximate a normality assumption, allows model uncertainty limits to be calculated for different lead times using estimates of the ARMA parameters and the residual error variance. The further uncertainty associated with using the STEPS forecasts can be captured by producing spaghetti plots of the forecast hydrographs using each ensemble member as the rainfall forecast, along with the ARMA model error uncertainty limits. This can be further simplified to obtain forecasts of flow at different lead-times corresponding to a given quantile (percentage exceedance) value and the associated model uncertainty limits at this value. Figure 1 plots the flow values for a specified quantile and different forecast lead times bracketed by the model uncertainty, along with the observed flows. Comparing the graphs obtained for high (90%), medium (50%) and low (10%) quantiles allows the uncertainty introduced by the rainfall forecasts to be appreciated separately from the model uncertainty delineated by the grey-shaded bands. Here, the PDM rainfall–runoff model is employed as the flood model for the Calder catchment to Todmorden and a 9-h (padded out to 20 h with zero rainfall) STEPS ensemble rainfall forecast is used with a time-origin at 06:00 h 21 January 2008.

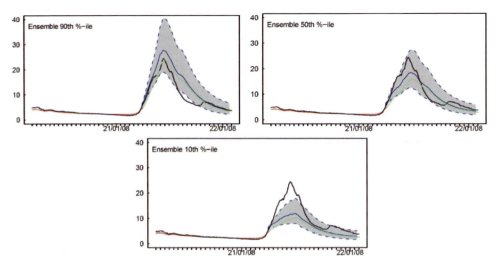

Fig. 1 Flood forecast uncertainty due to (i) model error: grey bands indicate 95% probability envelope, and to (ii) rainfall forecast error: indicated across the graphs by the high, medium and low (90, 50 and 100) percentile flows, calculated from the PDM rainfall–runoff model ensemble flow forecasts with STEPS forecast rainfalls (time origin 06:00 h 21 January 2008) as input. The percentile flows are green for simulated and blue for ARMA-updated forecasts. Observed and simulated flows are black and red lines, respectively.

An important purpose of the "Hydrological Modelling using Convective Scale Rainfall Modelling" R&D Project was to explore the use of probabilistic flood forecasting using ensemble rainfall forecasts, from STEPS and from future high-resolution NWP ensemble rainfall forecasts under development within the Met Office. It was recognised that the distributed nature of the G2G Model was particularly appropriate for use with rainfall forecast ensembles on account of its sensitivity to storm position, a major source of uncertainty in convective storms. Figure 2 shows the evolution of forecast maps of flood risk, here portraying the probability of exceeding the 10-year flood flow over the course of a flood event affecting the Avon and Tame catchments in the English Midlands. G2G is employing a STEPS ensemble rainfall forecast made at 09:00 h 20 July 2007, with forecasts beyond 6 h padded out with zero rainfall so as to facilitate tracking the movement of water down the river network in the G2G model forecasts. It can be seen how the "hotspots" of flood risk move from headwater streams down to confluences and larger rivers over the duration of the flood. Whilst the probabilities await a thorough assessment, it is clear that this approach has real value as an indicator of relative risk in space and time that can guide flood preparedness. A future R&D Project will explore the use of G2G to provide a new capability for flood forecasting of rapidly responding catchments, employing a new Blended Ensemble rainfall forecast out to 24 h that blends a high-resolution (2.2km) NWP with STEPS radar rainfall and noise extrapolation.

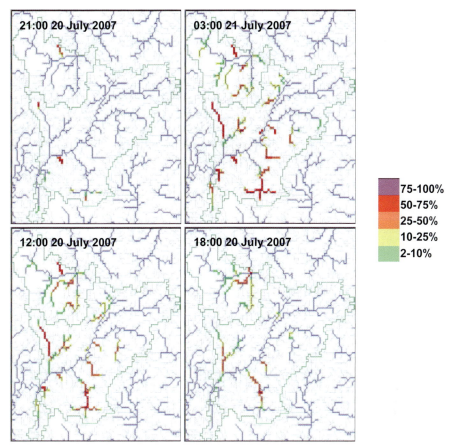

Fig. 2 Evolution of forecast flood risk for a summer 2007 flood event over the Avon and Tame catchments (English Midlands) obtained using the G2G Model and a STEPS ensemble rainfall forecast, showing progressive movement of "hotspots" from small headwater rivers to confluences and larger rivers. Bright (red and pink) colours indicate high probabilities (>50%) of exceeding the 10-year flood. Grey: 1 km river network; Blue: river network with drainage area >20 km²; Green: boundary of modelled area.

FUTURE PROSPECTS

Whilst this perspective paper has focused on flood forecasting applications, there are other important developments in progress concerning the use of radar in operational water management. One area is the need to assess storm rarity in relation to monitoring urban drainage system compliance to decide whether storm water overflows are indicative of design exceedance or a result of system failure requiring mitigating action. A development of the Hyrad system will integrate the Flood Estimation Handbook assessment of rainfall rarity, using an intensity-duration-frequency method, with radar rainfall to produce map displays of rainfall rarity and integrated estimates for areas contributing runoff to urban drainage systems (Cole *et al.*, 2012).

The Pitt Review highlighted the importance of pluvial (surface water) flooding and the lack of a suitably tailored warning service. In response, an Early Rainfall Alert service has been developed based on rainfall rarity exceedance which makes use of radar rainfall data. A future opportunity is to explore the use of modelled surface runoff from the G2G distributed hydrological model: this not only accounts for rainfall intensity, but also the condition of the receiving ground in relation to its land-cover, soil/geology properties and changing wetness. Dynamic maps of surface water flood risk with reference to impacts might be developed as operational products guiding management decisions. A further opportunity concerns bathing water regulations and the need to give warning of potentially unsafe conditions. The area-wide coverage provided by both radar rainfall products and the G2G model offers the prospect of modelling the heterogeneous pollution response of differing source areas leading to improved bathing water quality forecasts in real-time. Modelling the combined morphological and flood responses to intense convective rainfall is an additional challenge that will benefit from the use of radar rainfall estimates.

An on-going renewal programme providing dual-polarisation capability to the UK radar network, and advanced processing procedures, aims to improve the quality of radar rainfall for hydrological use. An assessment is proposed aimed at ensuring the hydrological benefits are fully appreciated, exploited and pulled through to operational use.

REFERENCES

Ballard, S., Zhihong, L., Simonin, D., Buttery, H., Charlton-Perez, C., Gaussiat, N. & Hawkness-Smith, L. (2012) Use of radar data in NWP-based nowcasting in the Met Office. In: *Weather Radar and Hydrology* (ed. by R. J Moore, S. J. Cole & A. J. Illingworth) (Proc. Exeter Symp., April 2011), IAHS Publ. 351. IAHS Press, Wallingford, UK (this issue).

Bowler ,N.E., Pierce, C.E. & Seed, A.W. (2006) STEPS: A probabilistic precipitation forecasting scheme which merges an extrapolation nowcast with downscaled NWP. *Quart. J. Roy. Met. Soc.* 132(620), 2127–2155.

Cabinet Office (2008) The Pitt Review: Lessons learned from the 2007 floods. http://webarchive.nationalarchives.gov.uk/20100807034701/http://archive.cabinetoffice.gov.uk/pittreview/thepittreview/final_report.html.

Cole, S. J., McBennett, D., Black, K. B. & Moore, R. J. (2012) Use of weather radar by the water industry in Scotland. In: *Weather Radar and Hydrology* (ed. by R. J Moore, S. J. Cole & A. J. Illingworth) (Proc. Exeter Symp., April 2011), IAHS Publ. 351. IAHS Press, Wallingford, UK (this issue).

Cole, S. J. & Moore, R. J. (2008) Hydrological modelling using raingauge-and radar-based estimators of areal rainfall. *J. Hydrol.* 358(3–4), 159–181.

Cranston, M., Maxey, R., Tavendale, A., Buchanan, P., Motion, A., Cole, S., Robson, A., Moore, R. J. & Minett, A. (2012) Countrywide flood forecasting in Scotland: challenges for hydrometeorological model uncertainty and prediction. In: *Weather Radar and Hydrology* (ed. by R. J Moore, S. J. Cole & A. J. Illingworth) (Proc. Exeter Symp., April 2011), IAHS Publ. 351. IAHS Press, Wallingford, UK (this issue).

Environment Agency (2010) Hydrological modelling using convective scale rainfall modelling – phase 3. Project: SC060087/R3, Authors: J. Schellekens, A. R. J. Minett, P. Reggiani, A. H. Weerts (Deltares); R. J. Moore, S. J. Cole, A. J. Robson & V. A. Bell (CEH Wallingford). Research Contractor: Deltares and CEH Wallingford, Environment Agency, Bristol, UK, 231pp. http://nora.nerc.ac.uk/13829/.

Harrison, D. L., Norman, K., Pierce, C. & Gaussiat, N. (2012) Radar products for hydrological application in the UK. *Water Management* 165(2), 89–103.

Howard, P. J., Cole, S. J., Robson, A. J. & Moore. R. J. (2012) Raingauge quality-control algorithms and the potential benefits for radar-based hydrological modelling. In: *Weather Radar and Hydrology* (ed. by R. J Moore, S. J. Cole & A. J. Illingworth) (Proc. Exeter Symp., April 2011), IAHS Publ. 351. IAHS Press, Wallingford, UK (this issue).

Laeger, S., Cross, R., Sene, K., Weerts, A., Beven, K., Leedal, D., Moore, R. J., Vaughan, M., Harrison, T. & Whitlow, C. (2010) Probabilistic flood forecasting in England and Wales – can it give us what we crave? *BHS Third International Symposium*, Role of Hydrology in Managing Consequences of a Changing Global Environment, Newcastle University, UK, 19–23 July 2010, British Hydrological Society, 8pp.

Moore, R. J., Bell, V. A., Cole, S. J. & Jones, D. A. (2007) Rainfall–runoff and other modelling for ungauged/low-benefit locations. Science Report – SC030227/SR1, Research Contractor: CEH Wallingford, Environment Agency, Bristol, UK, 249pp. http://nora.nerc.ac.uk/1134/.

Moore, R. J., Bell, V. A. & Jones, D. A. (2005) Forecasting for flood warning. *C. R. Geosciences* 337(1–2), 203–217.

Moore, R. J., Cole, S. J., Bell, V. A. & Jones, D. A. (2006) Issues in flood forecasting: ungauged basins, extreme floods and uncertainty. In: *Frontiers in Flood Research* (ed. by I. Tchiguirinskaia, K. N. N. Thein & P. Hubert), 103–122, 8th Kovacs Colloquium, UNESCO, Paris, June/July 2006. IAHS Publ. 305. IAHS Press, Wallingford, UK.

Moore, R. J., Jones, A. E., Jones, D. A., Black, K. B. & Bell, V. A. (2004) Weather radar for flood forecasting: some UK experiences. *6th Int. Symp. on Hydrological Applications of Weather Radar*, 2-4 February 2004, Melbourne, Australia, 11pp.

Pierce, C., Bowler, N., Seed, A., Jones, A., Jones, D. & Moore, R. (2005) Use of a stochastic precipitation nowcast scheme for fluvial flood forecasting and warning. *Atmos. Sci. Letters* 6(1), 78–83.

Price, D., Hudson, K., Boyce, G., Schellekens, J., Moore, R.J., Clark, P., Harrison, T., Connolly, E. & Pilling, C. (2012) Operational use of a grid-based model for flood forecasting. *Water Management* 165(2), 65–77.

Werner, M., Cranston, M., Harrison, T., Whitfield, D. & Schellekens, J. (2009) Recent developments in operational flood forecasting in England, Wales and Scotland. *Met. Appl.* 16(1), 13–22.

Weather Radar and Hydrology
(Proceedings of a symposium held in Exeter, UK, April 2011) (IAHS Publ. 351, 2012).

435

On the accuracy of the past, present, and future tools for flash flood prediction in the USA

JONATHAN J. GOURLEY[1], ZACHARY L. FLAMIG[2], YANG HONG[2] & KENNETH W. HOWARD[1]

1 *National Severe Storms Laboratory, 120 David L. Boren Blvd., 73072, Norman, Oklahoma, USA*
jj.gourley@noaa.gov

2 *Atmospheric Radar Research Center, University of Oklahoma, 120 David L. Boren Blvd., 73072, Norman, Oklahoma, USA*

Abstract The skill of the USA National Weather Service's flash flood guidance tool has been quantified from 2006 to 2008 using a combination of flash flood observations from spotter reports, automated stream discharge measurements, and witness reports from the public. A 15-year radar-based rainfall archive was used to run a distributed hydrologic model, thus enabling the estimation of flood frequencies at every 4-km grid cell. Exceedances of these return period flows were considered as predictors of flash flooding, and were validated using the same aforementioned datasets to establish the skill of present flash flood guidance. Significant improvements were realised using the forward modelling approach. Given the advent of 1-km^2, 5-min radar rainfall observations, distributed hydrologic models, and increased computing power, all of which are commensurate with the scales of flash flooding, it is now possible to directly forecast the probability of flash flooding over the conterminous USA in real time.

Key words flash flood; radar; distributed hydrologic model

INTRODUCTION

Flash flooding is the world's costliest weather-related natural hazard, causing loss of life and damage to infrastructure. Despite the significance of flash-flooding impacts on society, the tools used to detect and forecast them have not experienced the rapid development seen with radar-based algorithms designed to forecast tornadoes, hail, and wind events. The advent of weather radar networks in many countries worldwide revealed newly observed reflectivity signatures with hail storms, descending cores associated with microbursts and high surface wind events, and rotation in mesocyclones and tornadoes from Doppler velocity couplets. The contribution of weather radar to flash flood forecasting has been its capability to estimate rainfall at resolutions of approximately 1 km^2/5 min. The primary characteristic that separates flash floods from the other weather hazards is their dependence on terrestrial properties such as terrain slope, land cover type/roughness, soil type, depth, and degree of saturation; flash flooding is both a meteorological and hydrological phenomenon. Therefore, improvements in rainfall estimation on flash flood forecasting are conditioned on the ability of a secondary tool or model to accurately portray impending impacts from ponding surface water and small streams overreaching their banks.

While novel hydrometeorological models utilising flow observations have been developed for site-specific flash flood prediction (e.g. Georgakakos, 1987) decades ago, in this study we evaluate three different tools that have been used operationally in the past, present, or proposed to the USA National Weather Service (NWS) to provide guidance to forecasters on the likelihood of impending flash floods. First, establishing the benchmark performance of the legacy tool is needed in order to develop and refine contemporary approaches. We utilise observations of streamflow in basins with catchments <260 km^2 to define flash flooding events within the Arkansas-Red River basin in the south-central USA (see Fig. 1). The analysis provided herein complements the study of Gourley *et al.* (2011) in that we now include results from a modified threshold-frequency method proposed by Reed *et al.* (2007). In the final section of this paper we discuss limitations of existing methods and provide future directions in the development of flash flood forecasting tools.

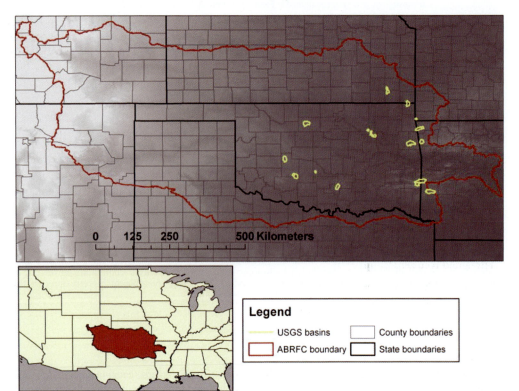

Fig. 1 Study domain showing Arkansas-Red River basins outlined in red and the gauged flash flood basins in yellow.

RADAR-BASED FLASH FLOOD FORECASTING METHODS

Flash Flood Guidance (FFG)

Since the 1970s, the primary tool used in the USA NWS for flash flood prediction has been FFG. FFG is the threshold rainfall required to initiate flooding on small streams that generally respond to rainfall in less than 6 h (Georgakakos, 1986). FFG is computed by first defining the excess rainfall (i.e. the effective volume of rain that is completely transformed into surface runoff) that results in flooding at each grid cell. This excess rainfall, or Thresh-R, is computed once offline by dividing an estimate of bankfull discharge (assumed to be the 2-year return period flow) by the unit hydrograph peak flow (Carpenter et al., 1999). Then, forecasters run the Snow-17 model to compute snowmelt contribution and the Sacramento Soil Moisture Accounting model (SAC-SMA) under different rainfall scenarios in order to produce rainfall–runoff curves (Burnash, 1995). SAC-SMA is a lumped-parameter model that is run 1–3 times per day on basins with catchment areas on the order of 1000 km^2. The rainfall–runoff curves are used to determine the 1-, 3-, and 6-h basin-average rainfall rates required to exceed Thresh-R given the current soil moisture conditions; this is FFG. Finally, forecasters compare FFG values to real-time radar-based rainfall estimates as the primary decision-support tool to warn the public of impending flash floods.

Gridded Flash Flood Guidance (GFFG)

The derivation of high-resolution GFFG was largely motivated by the FFG scale mismatch between the rainfall–runoff calculations on approximately 1000-km^2 basins and application downscale to basins as small as 5 km^2 (Schmidt et al., 2007). GFFG is computed similarly to FFG in that it relies on exceedence of static, raster-based Thresh-R values. The Natural Resources Conservation Service (NRCS) Curve Number (CN) method is used to estimate a basin's maximum potential storage based on spatially distributed maps of land use and four hydrologic soil groups. The CN is adjusted in real-time according to modelled soil saturation based on simulations from the NWS Hydrology Laboratory Research Distributed Hydrologic Model (HL-RDHM) (Koren *et al.*, 2004). Then, the basin's maximum potential retention (S) is computed as follows:

$$S = \frac{1000}{CN} - 10 \tag{1}$$

The value for S is then substituted into the following NRCS equation:

$$R = \frac{(P - 0.2S)^2}{(P + 0.8S)} \tag{2}$$

where R is accumulated direct runoff in mm (i.e. Thresh-R), and P is the accumulated rainfall in mm (i.e. GFFG). The equations are solved for P nominally at 1-, 3- and 6-h accumulation periods to yield GFFG.

Threshold frequency approach

Reed *et al.* (2007) demonstrated a flash flood forecasting approach called Distributed Hydrologic Model – Threshold Frequency (DHM-TF). First, a long-term record of gridded precipitation is used to force the HL-RDHM distributed hydrologic model to yield a model-simulated climatology of discharge at each 4-km grid point. A flood frequency analysis is then performed on the time series of simulated flows at each grid point to compute the simulated 1-, 2-, 5-year, etc. return period flows. These flows are then stored and used as thresholds for warning levels during the forecast period. In forecast mode, the HL-RDHM model is forced with real-time radar-based rainfall on an hourly basis. Exceedence of forecast flows over the corresponding return period flows provides warning of impending flash floods.

The method assumes that the uncalibrated simulations yield the same ranked histograms of discharge compared to observed flows during the hindcast period. In this sense, the method is unaffected by model bias. However, there can be erroneous performance due to consistent timing errors. We have thus created a modified version of DHM-TF for application to gauged basins, which shifts the simulated flows during the hindcast period so that they are matched with the observed flows. A Pearson linear correlation coefficient is used to guide the time shift that is applied to the simulated flows. It is then assumed the same time shift will be valid and applicable during the forecast period.

INTERCOMPARISON OF METHODS

In order to evaluate the three methods, we first computed annual exceedence probabilities using the full record of observed discharge (minimum of 10 years) in the 15 gauged basins shown in Fig. 1 assuming the annual maximum flows follow a log-Pearson Type III distribution. These probabilities were converted to return period flows, and we defined observed flash floods as events in which discharge exceeded its 2-year return period flow. Flash flood guidance and GFFG were evaluated in Gourley *et al.* (2011) from 1 September 2006 to 31 August 2008 by considering instances in which rainfall exceeded the guidance value for exceedence ratios ranging from 0.1 to 3.0. The results indicate the best skill, defined by the critical success index (CSI), was with 3-h FFG when considering an exceedence ratio of 1.5 for individual grid points and 1.0 when considering basin-average rainfall. The CSI values were 0.29 and 0.34; these values serve as the benchmark performance describing the flash flood forecasting tools used in present-day NWS operations. In Europe, Norbiato *et al.* (2008) evaluated FFG and found CSI values of 0.43 on calibrated basins, which deteriorated to 0.22 when parent basin parameters and soil moisture states were transposed to interior basins.

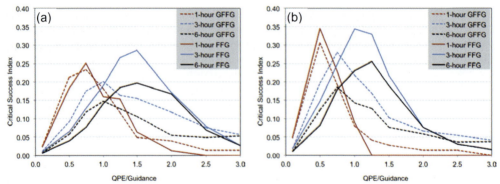

Fig. 2 Skill of flash flood forecasting tools used in the US National Weather Service based on exceedence of 2-year return period flows in basins shown in Fig. 1. Critical success index is computed for different ratios of (a) basin-maximum exceedence and (b) basin-mean exceedence (from Gourley *et al.*, 2011).

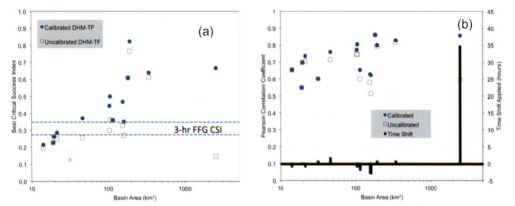

Fig. 3 (a) Skill of flash flood forecasts using calibrated and uncalibrated distributed hydrologic model simulations and (b) correlation coefficient that was used to shift simulated discharge time series to match observations. The time shift (in h) is plotted against the secondary ordinate as vertical bars.

To evaluate DHM-TF values, first we obtained the complete archive beginning in 1996 of gridded (4 km), hourly rainfall estimated from radar and raingauges in the study region. We then ran HL-RDHM from January 1996 to December 2002 at the same resolution as the rainfall forcing (4 km/1 h) and computed simulated return period flows at each grid point. Observed flash floods were defined as above (i.e. discharge exceeded 2-year return period flow) and forecast flash floods were classified when simulated flows exceeded their 2.2-year return period flow. The 2.2-year return period flow yielded an optimised CSI of 0.39 during the same 2-year time period used to evaluate FFG and GFFG; it outperformed both methods when using the exact same evaluation datasets.

In order to examine the DHM-TF approach more completely, we extended the validation period to include streamflow observations from January 2003 to September 2009 and added the 342 km^2 Ft. Cobb and 2484 km^2 Illinois basins to the study basin dataset. Then, we developed a modification to the original DHM-TF method where we time-lagged (i.e. calibrated) the simulations to match observations by maximising the linear correlation coefficient. No adjustments were made to the magnitude of simulated flows. The CSI shown in Fig. 3(a) indicates the calibrated DHM-TF method outperformed 3-h FFG for basins >20 km^2. The uncalibrated method beat the operational tools for basins >160 km^2, with the exception of the largest basin. In Fig. 3(b), we can see that a rather significant time shift of 35 h was applied to the simulated time series to match the observed flows. This timing error was found to be a result of poorly estimated routing parameters. More recently, the two routing parameters have been more accurately estimated using digital elevation derivatives including hillslope and drainage area. Because the timing of uncalibrated simulated flows was greatly offset from observations, the CSI (0.15) was heavily penalised during validation. Smaller basins were less susceptible to this timing error because they have shorter response times. The major advantage of the uncalibrated DHM-TF method is that it can be applied at any grid point where there is a climatological archive of rainfall; i.e., there is no need for a stream gauge at the basin outlet. Both methods indicate worsening skill as a linear function with decreasing basin area. Similar results were found in Carpenter & Georgakakos (2004) and Reed *et al.* (2007), and were attributed primarily to model parametric and radar rainfall uncertainty. Another possible reason for this skill reduction with smaller basins is the relatively coarse resolution of the rainfall forcing and model-simulated flows. We can see from Fig. 3(a) that forecasts from DHM-TF (calibrated and uncalibrated) were worse than 3-h FFG for basins <20 km^2, which is approximately one model grid cell.

CONCLUSIONS AND FUTURE DIRECTIONS

In interpreting the results from the different flash flood forecasting tools, it should be noted that the skill associated with each product is subject to uncertainties from the radar-based rainfall inputs and the streamflow observations, and thus lower CSI scores result. All tools are subject to the same uncertainties so that their skill scores can be compared in a relative sense. In a retrospective analysis, we have shown that flash floods can be predicted more accurately than operational tools using the modified DHM-TF method, which is based on distributed hydrologic model simulations exceeding simulated return period flows. This method can be applied anywhere globally, even in ungauged catchments, as long as there is a representative archive of gridded rainfall. We also developed an efficient alternative to model calibration where we merely time-lagged the simulations so that they matched observed flows. This method of course requires observed streamflow and thus applies to gauged catchments. However, recent model developments have yielded *a priori* estimates of the two routing parameters that will likely mitigate the timing errors shown in this study, and will broaden its application to ungauged catchments.

These preliminary results have prompted us to further develop the method into a conterminous USA-wide flash flood prediction system using real-time rainfall forcing from the National Severe Storms Laboratory's National Mosaic and Multisensor QPE system (NMQ) (http://nmq.ou.edu) at 5-min/1-km^2 resolution. An ensemble of simulations from two distributed hydrologic models will be run in hindcast for the period of available record of radar-based rainfall to derive the simulated flow warning levels. Then, the hydrologic models will be run at the same resolution as the rainfall forcing to produce predictions of flash flooding. Additional studies presently underway are examining the potential improvement of flash flood predictions by assimilating soil moisture and streamflow observations, the utility of storm-scale precipitation forecasts as forcing to increase lead time, incorporation of satellite data in multi-sensor rainfall products where radar coverage is inadequate, and detailed validation using observations from trained spotters, the public, and remote-sensing data from space. All efforts are aimed at improving the tools made available to operational forecasters so that their forecasts of flash flooding will improve and subsequent impacts on lives and property will be reduced.

REFERENCES

Burnash, R. J. C. (1995) *Computer Models of Watershed Hydrology.* Water Resources Publication, 1144 pp.

Carpenter, T. M. & Georgakakos, K. P. (2004) Impacts of parametric and radar rainfall uncertainty on the ensemble streamflow simulations of a distributed hydrologic model. *J. Hydrol.* 298, 202–221.

Carpenter, T. M., Sperfslage, J. A., Georgakakos, K. P., Sweeney, T. & Fread, D. (1999) National threshold runoff estimation utilizing GIS in support of operational flash flood warning systems. *J. Hydrol.* 224, 21–44.

Georgakakos, K. P. (1986) On the design of national, real time warning systems with capability for site-specific flash flood forecasts. *Bull. Am. Met. Soc.* 67, 1233–1239.

Georgakakos, K. P. (1987) Real-time flash flood prediction. *J. Geophys. Res.* 92, 9615–9629.

Gourley, J. J., Erlingis, J. M., Hong, Y. & Wells, E. (2011) Evaluation of tools used for monitoring and forecasting flash floods in the US. *Weath. Forecasting* (accepted).

Koren, V., Reed, S., Smith, M., Zhang, Z. & Seo, D.-J. (2004) Hydrology Laboratory Research Modeling System (HL-RMS) of the US National Weather Service. *J. Hydrol.* 291, 297–318.

Norbiato, D., Borga, M., Degli Esposti, S., Gaume, E. & Anquetin, S. (2008) Flash flood warning based on rainfall depth-duration thresholds and soil moisture conditions: an assessment for gauged and ungauged basin. *J. Hydrol.* 362, 274–290.

Reed, S., Schaake, J. & Zhang, Z. (2007) A distributed hydrologic model and threshold frequency-based method for flash flood forecasting at ungauged locations. *J. Hydrol.* 337, 402–420.

Schmidt, J. A., Anderson, A. J. & Paul, J. H. (2007) Spatially-variable, physically-derived flash flood guidance. In: *AMS 21st Conference on Hydrology* (San Antonio, Texas, pp. 6B.2).

Real-time radar-rainfall estimation for hydrologic forecasting: a prototype system in Iowa, USA

WITOLD F. KRAJEWSKI, RICARDO MANTILLA, BONG-CHUL SEO, LUCIANA CUNHA, PIOTR DOMASZCZYNSKI, RADOSLAW GOSKA & SATPREET SINGH

IIHR-Hydroscience & Engineering, The University of Iowa, Iowa City, Iowa 52242, USA
witold-krajewski@uiowa.edu

Abstract The estimation procedure to generate rainfall maps in real-time consists of Level II radar volume data collection, quality checks of the acquired data, and rainfall estimation algorithms such as non-meteorological target detection, advection correction, Z-R conversion, and grid transformation. The rainfall intensity map that is generated using data from seven radars around the State of Iowa is updated at nominal 5-min intervals, and the accumulation map is produced based on 15-min, 1-h, and daily intervals. These rainfall products are fed into a physically-based flood forecasting model called CUENCAS that uses landscape decomposition into hillslopes and channel links. The authors present preliminary results of analysis done on real-time radar-rainfall products using raingauge data and hydrological simulations from flood events in 2008 and 2009. They also show how differences in rainfall forcing affect peak flow discharge.

Key words precipitation; radar-rainfall; flood forecasting

INTRODUCTION

In 2008, eastern Iowa in the United States of America faced record flooding events described as a 500-year flood for some locations. In order to mitigate the effects of similar disasters in the future, a real-time regional flood forecasting system is being developed by the Iowa Flood Center (IFC), an academic research unit at The University of Iowa. The system includes a rainfall monitoring system that provides storm/rainfall information in near real-time for the state. The rainfall monitoring system offers high-resolution radar-derived rainfall information in time and space that captures key features of rainfall fields and enables post-data analysis of the storms. The rainfall product covers all of Iowa using data from WSR-88D radars, which were recently upgraded to provide a resolution of $0.5°$ in azimuth and a range of 250 m.

The IFC disseminates the rainfall maps in near real-time to the public. The "state-wide rainfall intensity and accumulation map" provided by the IFC is generated by combining data from seven radars (two located in Iowa and five surrounding it). These rainfall products are fed into a physically-based flood forecasting model called CUENCAS (Mantilla & Gupta, 2005; Cunha *et al.*, 2011) that uses landscape decomposition into hillslopes and channel links. The hillslopes have the spatial scale of about 0.1 km^2, closely matching the radar-rainfall resolution of 0.25 km^2. The model predicts the hydrologic response of the landscape to rainfall inputs and provides predictions for the entire state without the need for extensive calibration. Web-based tools allow visualization of the real-time rainfall monitoring system and communication of the results to the general public. In this paper, we describe the structure and main components of the system and present a preliminary analysis of the results for the rainfall product evaluation and model prediction.

REAL-TIME RADAR-RAINFALL ESTIMATION

Real-time radar-rainfall estimation involves three steps: (1) Level II radar volume scan data collection and quality check; (2) rainfall estimation using associated algorithms/modules; and (3) rainfall product publishing and delivery for subsequent hydrological forecasting. Figure 1 illustrates the structure and sequential procedures related to the module components of the system. The first step involved in collecting data and performing a quality check is referred to as "Harvesting" in Fig. 1. The harvester acquires radar volume scan data in real-time and provides qualified data into the quantitative rainfall estimation algorithms/modules.

Level II data collection

The data collection step consists of obtaining Level II streaming data from the Unidata Local Data Manager (LDM; http://www.unidata.ucar.edu/software/ldm/), the conversion of the LDM-received data into a readable format with data-processing and rainfall estimation modules, and a partial file recovery and data quality check when the LDM format conversion fails. The data quality check procedure is necessary due to incomplete or corrupted (damaged) files related to network stability or random losses. We will discuss this issue in more detail later in this section.

Local data manager The LDM data feed has been configured to receive Level II data in real-time from seven WSR-88D radars (KDVN in Davenport, Iowa, KDMX in Des Moines, Iowa, KARX in La Crosse, Wisconsin, KMPX in Minneapolis, Minnesota, KOAX in Omaha, Nebraska, KFSD in Sioux Falls, South Dakota, and KEAX in Kansas City, Missouri) that cover the entire State of Iowa, as shown in Fig. 2. Later, we will document the problems of combining data from seven radars.

Data format conversion Since the file format of LDM-received data is not compatible with the rainfall estimation modules that use the National Aeronautics and Space Administration (NASA) Radar Software Library (RSL; http://trmm-fc.gsfc.nasa.gov/trmm_gv/software/rsl/) to read and manipulate radar volume scan data, a file format conversion step is required. Some files acquired through the LDM are incomplete or damaged due to network losses or other random errors. In such cases, this conversion step fails, and a subsequent conversion step is needed to recover the partially-packed data. Based on our monitoring of transferred data completion with the help of Unidata, we found a significant proportion of the error rate (approximately 15%) could be recovered by an additional procedure (Singh, 2010).

Data quality check To check the availability of the recovered data, we considered the structure of radar volume data. Since our quantitative radar-rainfall estimation does not need the full information of radar volume data, the quality check procedure inspects the completeness of only the reflectivity volume data for several of the lowest elevation angles. Incomplete/damaged data in reflectivity fields are discarded. As a result of the data quality check, the recovery step of partially-packed data considerably improved data availability, and the error rate was reduced to approximately 1.5% (Singh, 2010).

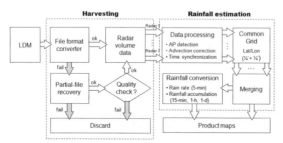

Fig. 1 Structure of the real-time radar-rainfall estimation procedure.

Fig. 2 Location of seven radars and the NWS COOP raingauges.

Rainfall estimation algorithms/modules

Generating real-time rainfall inputs in the monitoring system is a fundamental step for hydrological forecasting. Radar-rainfall map production in the system involves several independent and optional modules that perform radar reflectivity data processing, merge multiple radar data onto pre-defined common spatial domains (to cover the entire Iowa), and transform radar-measured reflectivity to rainfall amounts (rain rate/accumulation), as shown in Fig. 1. In this section, we briefly describe all of the components of the rainfall monitoring system. For more detailed information on algorithms implemented in the system, refer to Seo *et al.* (2011).

Detection and removal of non-meteorological target The 3-D structure of radar volume data allows the detection of non-precipitation echoes (e.g. ground clutters) by estimating the likelihood of Anomalous Propagation (AP) conditions. The procedure proposed by Steiner & Smith (2002) is used to construct a 2-D AP map. The binary mask map is created using the horizontal and vertical radar echo structure that is assessed for the computation of AP occurrence. After applying this AP procedure, a 2-D radar reflectivity map is generated using the kernel weighting proposed in Seo *et al.* (2011).

Advection correction The intermittent sampling span of radar observation, which is dependent on the Volume Coverage Pattern (VCP) that typically ranges from 4- to 10-min, could cause a significant radar-rainfall accumulation error. The approach proposed by Fabry *et al.* (1994) is applied to generate radar reflectivity fields at a nominal time stamp (i.e. every 1-min) using computed velocity vectors from two (prior) consecutive reflectivity maps. Another benefit of using this module is that advection enables time-synchronization of radar data among the seven radars producing volume scan data at different time stamps. If the advection module is inactive due to its considerable processing time, another time-synchronization module that produces reflectivity maps every 5-min by linearly-interpolating two consecutive maps can be applied (see Seo *et al.*, 2011).

Grid transformation The radar-centred 2-D reflectivity fields represented by radial azimuth and range are remapped onto a common Lat/Lon geographic grid (approx.. 0.5×0.5 km^2) before individual fields from seven radars are merged into a single field. Such high resolution of rainfall products is compatible with the spatial scale of the hydrological model used by the IFC for flood forecasting. The geographic coordinates do not cause distortion arising from map projection and are easy to transform into any projected coordinate system.

Multiple radar data merging With a common spatial basis, all available individual reflectivity maps (sometimes, data from some radars are missed due to radar maintenance or network delay/loss) are merged into a single reflectivity field. For Level II data reception and the generation of time-synchronized products, the waiting interval specified in the system due to unexpected network delay is 25-min (Singh, 2010). If data from certain radars do not arrive within this time frame, only available time-synchronized reflectivity maps are delivered to the merging module. Each individual field is spatially combined using range-dependent weights estimated by the double exponentially decaying function (see e.g., Seo *et al.*, 2011).

Z-R conversion and rainfall accumulation To convert a merged radar reflectivity field (mm^6/mm^3) to rainfall rate (mm/h), we apply a typical power law relationship ($Z = 300R^{1.4}$) used for the WSR-88D radars (e.g. Fulton *et al.*, 1998). Since two parameters of the above equation are adaptable, we can specify other values for those parameters according to precipitation type (see e.g. Marshall & Palmer, 1948; Rosenfeld *et al.*, 1993). Rain-rate maps generated every 5-min are integrated to publish rainfall accumulation maps over specific time durations (15-min, 1-h, and 1-d). Daily accumulation is computed from midnight to midnight based on local time in Iowa.

Rainfall product publishing and delivery

We have developed web-based tools, as illustrated in Fig. 3, to visualize the outcome of the real-time rainfall monitoring system and to communicate results to the general public. Rain-rate maps (current and animation for several past hours) are updated every 5-min, and daily rain totals are provided through the website, as shown in Fig. 3. The "ASCII" product files are delivered and stored at the specified URLs to be fed into the hydrological model. Products older than 6-h are compressed to reduce the capacity of data storage.

RAINFALL PRODUCT EVALUATION

Currently, rainfall maps generated and published through the aforementioned processes are all radar-only products, implying that there is no raingauge adjustment. Since numerous studies have reported a variety of error sources in radar-rainfall estimates and the accuracy of rainfall inputs to a hydrological model significantly affects the model's performance, an evaluation procedure for the rainfall products is required to understand the model outcome.

The authors collected daily raingauge data for the period August–October 2010 from the US National Weather Service Cooperative Observer Program (NWS COOP) network that covers the entire State of Iowa. Figure 2 shows the 104 raingauges that were used to evaluate rainfall products generated in real-time. No advection correction was applied for the product used in evaluation as the advection module was only implemented after the collected period. For another comparison between radar-based rainfall products, the US National Stage IV QPE product (Lin & Mitchell, 2005), which are multi-sensor (radar + gauge) precipitation estimates, was used for the same period as the collected raingauge data. Figure 4 shows the results of radar-gauge (left panel) and radar-radar (right panel) comparisons. As Fig. 4 shows, our real-time radar–rainfall products look systematically-biased (the bias value represented by G/R is 0.77) and overestimated on average when compared to raingauge data which are assumed as true rainfall. For the radar-radar product comparison based on hourly rainfall estimation, a similar biased pattern was observed. The mean difference (mm/h) conditioned on positive rainfall values for both products is 0.9, and the correlation coefficient is 0.78.

Fig. 3 Real-time rainfall product map.

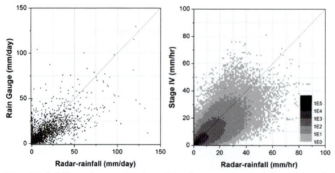

Fig. 4 Rainfall product evaluation with raingauge data and Stage IV radar-based rainfall estimates.

HYDROLOGIC APPLICATION

The application of real-time radar rainfall products for flood forecasting requires a better understanding of the expected uncertainties and how these uncertainties propagate through hydrological models. Since weather radar provide rainfall estimates over large regions essentially continuously in space and time, and there are multiple locations for which flood forecasting is desired, we need hydrological models that perform well across a range of scale, i.e. "everywhere". Models that rely on calibration using historical data ensure a certain level of performance only at the scale for which they were calibrated. This is clearly inadequate. For example, in Iowa, the National Weather Service provides streamflow forecasts at some 100 points, while the flood forecasting system we are developing requires predictions at some 500 communities and 400 other points of interests (bridges, gauges, etc.). Clearly, a different approach is needed.

The main component of our framework is a multi-scale physics-based hydrological model. The spatial discretization scheme adopted in our model provides a realistic description of the landscape that is decomposed into runoff generation areas (hillslopes) connected by the stream and river networks. Runoff generation is parameterized at the hillslope scale using the Soil Conservation Curve Number method. Mass and momentum balance equations are then used to simulate transport processes at these specific scales, and the across-scale link is provided by the river network extracted in detail from the Digital Elevation Model. Calibration is avoided through the use of parameters that can be directly (e.g. from independent field scale studies) or indirectly (e.g. via remote sensing) observed for the scale for which they are considered in the model formulation.

Our framework has been applied to simulate the 2008 Iowa flood using high-resolution bias-corrected radar rainfall products. Without calibration, the model simulated hydrographs across multiple scales. Since the model provides discharge information for each stream (link) throughout the river network, we also obtain the event peak flow scaling (see Mandapaka *et al.*, 2009 and Gupta *et al.*, 2010 for relevant background and illustration). Figure 5(a) presents the simulated scaling plot with observed values and simulated and observed hydrographs for three locations with a drainage area equal to 152 (b), 1033 (c), and 13 322 (d) km^2.

Future analyses will take advantage of this framework to investigate the benefits of real-time radar data for flood prediction. We will also investigate how different data processing techniques can improve forecast accuracy, even in the absence of ground-based information. Advection

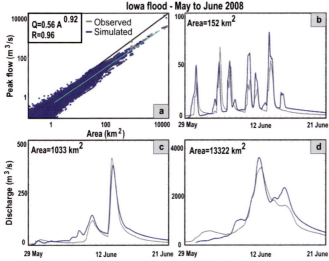

Fig. 5 Simulation of 2008 flood event. (a) Simulated (blue) scaling of peak flow. Observed values are plotted in grey observed values; (b), (c) and (d) observed (grey) and simulated (blue) hydrographs for 152, 1033 and 13322 km^2 drainage area basins.

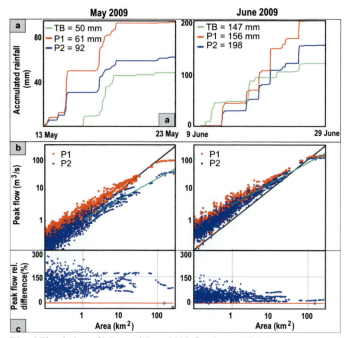

Fig. 6 Simulation of May and June 2009 flood event. (a) Average accumulated rainfall in the basin using tipping bucket data (green), P1 (red) and P2 (blue); (b) simulated scaling of peak flow – model forced by P1 (red) and P2 (blue). (c) Peak flow relative difference obtained by P1 and P2 rainfall products.

correction is one example of a data technique that has the potential to improve information for hydrological applications, especially for small basins. This method takes into consideration the movement and evolution of storms, potentially minimizing the rainfall accumulation errors caused by sampling intervals and improving accuracy of the storms' spatial distribution. Small errors in the location of storms might result in large discrepancies in flood prediction, since precipitation is being used as a forcing to the wrong hillslope/watershed.

Initial experiments were performed for Clear Creek, a 251-km^2 basin located northwest of Iowa City, Iowa. While we are processing a long-term (8 year) radar-rainfall dataset for the basin to perform continuous simulation, we have initially selected flood events that occurred in May and June of 2009. In Fig. 6(a), we compare the basin average accumulated rainfall obtained by different radar–rainfall processing algorithms (P1: no advection correction and P2: advection correction) and tipping bucket (TB). Since the advection correction algorithm was only recently implemented in our system, rainfall products were generated off-line for the flood events of 2009. For the two events, P2 presented accumulated values closer to the TB values, while P1 seems to present higher bias. However, no systematic behaviour could be observed when comparing the radar products with gauge information, and no general conclusion can be made at this point. Future studies will include a large number of events and larger study areas.

The two products were used to force the hydrological model with the goal of evaluating the sensitivity of peak flow discharge to differences in rainfall forcing. Figure 6(b) presents the scaling of peak flow for P1 (blue) and P2 (red), which demonstrates how different rainfall forcing introduced at the hillslopes scale (~0.01 km^2) propagates to middle-size watersheds (~250 km^2). Figure 6 (c) presents the relative difference in peak discharge obtained by forcing the hydrological model with P1 and P2. For the May event, a difference in rainfall accumulation on the order of 33% resulted in differences in peak flow as high as 250% for the hillslope and 100% for the largest basin area. Similar behaviour is observed for the June event. These preliminary results indicate the importance of highly accurate radar rainfall products for flood prediction.

CONCLUSIONS AND DISCUSSION

We discuss development of a real-time flood forecasting system to serve the public in a large region of the USA. The system is an ambitious undertaking as it combines federal, state, and academic resources and is a hybrid of time-proven technologies and untested concepts. We hope that the undertaking will shorten the path from research ideas to practical implementation for the benefit of the public.

The proven technology is radar-rainfall estimation. While it is certainly far from being error-free, it provides an unprecedented view of rainfall's variability at the space and time scale commensurate with the scale of physical processes taking place in the landscape. We have yet to comprehensively study radar-rainfall uncertainty propagation through our hydrological model. Within the structure of the model, the input uncertainty is filtered out by the water transport over the river network. Therefore, as long as there are no systematic biases introduced in the inputs or runoff generation, there exists a scale where the random errors average out (see Mandapaka *et al.*, 2009). The goal of the uncertainty propagation is to find out that scale.

The untested concept is a hydrological model not needing extensive calibration. Early results are promising, but much remains to be explored. The model requires considerable computational resources to simulate sizable (~50 000 km^2) river basins. We are developing efficient, scalable parallel algorithms to accomplish the task and organize long-term datasets to fully evaluate the model and its performance.

Acknowledgements This work was supported by the US National Science Foundation and the Iowa Flood Center of The University of Iowa.

REFERENCES

Cunha, L. K, Krajewski, W. F. & Mantilla, R. (2011) A framework for flood risk assessment under nonstationary conditions or in the absence of historical data. *J. Flood Risk Manage.* 4(1), 3–22.

Fabry, F., Bellon, A., Duncan, M. R. & Austin, G. L. (1994) High resolution rainfall measurements by radar for very small basins: the sampling problem reexamined. *J. Hydrol.* 161(1–4), 415–428.

Fulton, R. A., Breidenbach, J. P., Seo, D. -J., Miller, D. A. & O'Bannon, T. (1998) The WSR-88D rainfall algorithm. *Wea. Forecasting* 13(2), 377–395.

Gupta V. K., Mantilla R., Troutman B. M., Dawdy D. & Krajewski W .F. (2010) Generalizing a nonlinear geophysical flood theory to medium-sized river networks. *Geophys. Res. Lett.* 37, L11402.

Lin, Y. & Mitchell, K. E. (2005) The NCEP Stage II/IV hourly precipitation analyses: development and applications. Preprints, 19th Conf. on Hydrology, San Diego, USA, Amer. Meteor. Soc.

Mandapaka P. V., Krajewski W. F., Mantilla R. & Gupta V. K. (2009) Dissecting the effect of rainfall variability on the statistical structure of peak flows. *Adv. Water Resour.* 32 (10), 1508–1525.

Mantilla, R. & Gupta, V. K. (2005) A GIS numerical framework to study the process basis of scaling statistics in river networks. *IEEE Geosci. Remote Sens. Lett.* 2(4), 404–408.

Marshall, J. S. & Palmer, W. McK. (1948) The distribution of raindrops with size. *J. Atmos. Sci.* 5(4), 165–166.

Rosenfeld, D., Wolff, D. B. & Atlas, D. (1993) General probability-matched relations between radar reflectivity and rain rate. *J. Appl. Meteor.* 32(1), 50–72.

Seo, B. -C., Krajewski, W. F., Kruger, A., Domaszczynski, P., Smith, J. A. & Steiner, M. (2011) Radar-rainfall estimation algorithms of Hydro-NEXRAD. *J. Hydroinform.* 13(2), 277–291.

Singh, S. H. (2010) A system for generation of near real-time feeds of user-customized hydrometeorology data-products from NEXRAD radar-data. MS Thesis, The University of Iowa, Iowa City, Iowa, USA.

Steiner, M. & Smith, J. A. (2002) Use of three-dimensional reflectivity structure for automated detection and removal of nonprecipitating echoes in radar data. *J. Atmos. Oceanic Technol.* 19(5), 673–686.

Which QPE suits my catchment best?

M. HEISTERMANN, D. KNEIS & A. BRONSTERT

University of Potsdam, Institute for Earth and Environmental Sciences, 14476 Potsdam, Germany
maik.heistermann@uni-potsdam.de

Abstract We often seek to identify from a set of available QPE products the one with the least error for a particular catchment. However, point-based verification approaches such as cross-validation do not inform the user about the spatial representativeness of the error. Instead, we can force a hydrological model with different QPE products and select the QPE which best reproduces the observed discharge. In order to reduce effects of model calibration on the outcome of such a "hydrological verification", we propose a Monte Carlo-based approach. We applied this approach in a case study for two catchments in southeast Germany and found that hydrological verification and cross-validation can, in fact, usefully complement one another.

Key words weather radar; quantitative precipitation estimation; verification; rainfall–runoff modelling

INTRODUCTION

Typically, hydrologic modellers seek to force their model with a QPE product which has the least error. Such products can be derived, e.g. from raingauge or radar observations or by merging multiple precipitation sensors. However, choosing a suitable product from a set of available products is a challenging task. Any QPE approach has its specific drawbacks and the error of a QPE might vary in space and time as well as it might depend on the spatial and temporal scale of application. Traditionally, a QPE is verified against observations at independent raingauges. However, point-based verification does not inform the user about the spatial representativeness of the error which has been identified on a particular reference point. Consequently, we cannot assign a particular error to sub-domains of our study region (such as catchments).

In a different approach, the quality of a QPE can be assessed by using it as input to a rainfall–runoff (RR) model. We would typically expect that a better agreement between simulated and observed runoff indicates a higher quality of the respective QPE used to force the model. The major advantage of this approach is that a catchment's runoff represents the area-integrated system response. Thus, the verification is representative for the scale of model application and not just for selected points. However, using conceptual RR-models to benchmark different QPE approaches, we face an additional problem: systematic errors in a QPE could be offset by model calibration. If we were to compare different QPE methods in a benchmark test, the actual differences in their quality might be obscured if the model was calibrated separately for each particular QPE. On the other hand, when using a single common parameter set, we might favour a particular QPE over others.

Gourley & Vieux (2005) have addressed the above problem by generating a large collection of parameter sets for a given RR model. For each QPE of interest, they carried out runoff simulations using all parameter sets of the collection. Subsequently, they computed goodness-of-fit measures for every single model run by comparing the simulated runoff to observations. That way, they obtained a distribution of the goodness-of-fit for each considered QPE method. This distribution provided the basis for a comparison of the QPEs under consideration. In the following, we will show how we adopted the approach of Gourley & Vieux (2005) and then applied the ranking approach to a real-world case study under quasi-operational conditions in two small mountain catchments in Germany. We benchmark six different QPE methods over a continuous time period of over four years. We will compare the results of the "hydrological verification" procedure to a conventional point-based cross-validation approach in order to reveal potential consistencies or inconsistencies.

METHODOLOGY

Hydrological verification framework

When we apply a rainfall runoff (RR) model to benchmark the quality of different QPE, model calibration becomes a major concern: It might be possible to produce runoff simulations of equivalent goodness-of-fit from QPE products of very different quality by cleverly tuning the RR model to each of the products. As a solution, Gourley & Vieux (2005) suggested a Monte Carlo approach that abandons the idea of calibration. A ranking of the QPE methods established in this way is assumed to be "fair" because the error associated with a specific method cannot be systematically compensated by the choice of the RR model's parameters. The procedure as adopted in this study includes three steps:

(1) Create a large number of random model parameter sets (e.g. 1000) by Latin Hypercube sampling. Parameters should include the RR model's sensitive parameters and the sampling should cover the full plausible range of each selected parameter.
(2) For each of the QPE data sets to be benchmarked, carry out runoff simulations using the same model parameter sets from step (1). Save the goodness-of-fit (*here*: the root mean squared error, RMSE) for each of the Monte Carlo ensemble members.
(3) For each QPE, compute the mean RMSE over those 10 out of 1000 ensemble members which show the best performance. Rank the QPE methods based on their mean RMSE taking into account whether the differences are significant. Repeat the procedure with the best 10, 25, 50 and 100% of the best performing members.

According to this approach, a particular QPE is superior to another one if its mean RMSE over the selected percentage of ensemble members is significantly lower than the mean RMSE of the competing QPE. We consider the difference in the mean RMSE for two competing QPEs to be significant if the 95% confidence intervals of the two mean RMSEs do not overlap. For each QPE, we estimate these confidence intervals by combining the technique of block sub-sampling (Politis, 2003) with traditional bootstrapping: First, each of the N selected ensemble members is partitioned into M contiguous blocks. Second, we compute the RMSE for each block. Third, bootstrapping is applied to infer the confidence interval of the mean RMSE from the set of $N \times M$ RMSE values.

Once the confidence limits of the mean RMSEs are determined for all QPE methods, a ranking can be established. The rank of a specific method equals 1 plus the number of competing methods, for which a significantly lower RMSE was determined. According to this rule, the best QPE method is always assigned a rank of 1, and the rank with the highest number is associated with the worst method(s) of QPE.

The same ranking approach, including block-subsampling and bootstrapping, is applied to the series of residuals obtained from point-based verification (leave-one-out cross-validation based on the available hourly raingauge observations). The results from the cross-validation will be compared to the results of the hydrological verification.

Case study set-up: study area and rainfall–runoff model

Our study area is the catchment of the upper Wilde Weisseritz (eastern Erzgebirge Mountains near the city of Dresden, Germany) with elevations up to 900 m a.s.l. We focus on the two runoff gauges Rehefeld (15 km^2) and Ammelsdorf (49 km^2). The catchment is characterised by steep slopes and shallow soils. Dominant land use types are forest and grassland, and cropland in the lower parts. The hydrological and the point-based verification consider the entire period from 2004 to 2008.

To simulate the discharge at these gauges, we used the semi-distributed model LARSIM (Ludwig & Bremicker, 2006) which is in operational use at several German flood forecasting centres. The model distinguishes four runoff components: (1) surface runoff, (2) quick subsurface stormflow, (3) interflow, and (4) baseflow. To create 1000 random model parameter sets, we selected nine model parameters whose calibration is recommended by the operational users of LARSIM at German flood forecasting centres. The names and roles of the parameters and their detailed sampling ranges are provided in detail in Kneis & Heistermann (2009).

Case study set-up: rainfall data and benchmarked QPE methods

Thirty raingauges with a temporal resolution of 1 h or less are available in the region, however less than half of them are located close to the study area. Only hourly precipitation sums were considered. The radar data is based on the German Weather Service's hourly C-band composite (1×1 km², 1 h rainfall accumulations) with Dresden-Klotzsche being the nearest radar station (60 km distant). The selected radar product *Borama* (German Weather Service, 2005) is based on an intensity-dependent Z/R-relationship and an automated detection and correction of beam blockage and ground clutter.

Since this is only a case study for demonstration purposes, the selection of benchmarked QPE methods is not intended to be exhaustive. As a first reference method, we selected the Inverse Distance Weighting (IDW) approach in order to interpolate raingauge observations, extended by a quadrant search (meaning that from each 90° sector around the target location only the nearest raingauge was considered for interpolation). As a second reference, the radar product *Borama* (BOR) was used without any further modification. Additionally, four methods were tested which combine radar and raingauge observations with the aim of obtaining a superior QPE product. The mean field bias adjustment, MFB (Goudenhoofdt & Delobbe, 2009), computes a single adjustment factor for the radar image based on a comparison of radar and gauge observations over the entire domain. Two other approaches are based on the assumption that the deviations between gauge and radar observation are spatially correlated and can thus be spatially interpolated in order to correct the radar observation at any desired location. The mentioned deviation can be quantified relatively (as a factor) or absolutely (as a difference). We will refer to these methods as the factor method (FAC) and the difference method (DIF). The factor method is basically equivalent to the approach proposed by Brandes (1975). The forth adjustment method is referred to as the Merging approach, MRG (Ehret *et al.*, 2008). Its philosophy is to preserve the mean field properties as estimated from raingauge observations, but to imprint the spatial variability of the radar image. First, we interpolate the raingauge observations and the radar rainfall *at the gauge locations* on the radar grid using ordinary kriging. We then compute the ratio between radar rainfall and the previously interpolated radar field and use that ratio to multiplicatively correct the field interpolated from the gauge observations.

RESULTS

Applying the ranking approach based on hydrological verification as introduced above, the benchmarked QPE methods fall into one of three groups (see Fig. 1): IDW and MRG perform best in all cases. The unadjusted radar observation (BOR) always performs worst by far. The three alternative methods of radar adjustment obtain intermediate ranks.

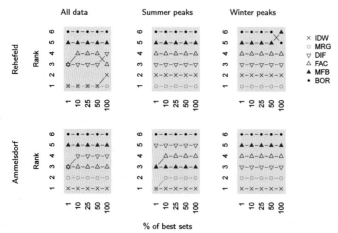

Fig. 1 Ranking of the six competing QPE methods considering different sub-sets of the best-fitting parameter sets. Results are presented separately for the two gauges (rows) and different filters applied to the time series of observed runoff (columns): no filter at all, only summer events, only winter events.

For the smaller sub-catchment (Rehefeld gauge, 15 km^2), the performance of the IDW and MRG method appears to be practically identical when considering the entire simulation period (Fig. 1, top left). However, the MRG approach is superior if we focus on the simulation of flow peaks (Fig. 1, top central and top right). From the competing QPE methods, the difference method (DIF) shows the best performance, which is significantly worse than that of IDW or MRG.

For the larger catchment (Ammelsdorf gauge, 50 km^2), none of the radar-based QPE products is able to outperform the plain interpolated raingauge data (Fig. 1, bottom panel).

In the present case, the ranking of the QPE methods turned out to be rather insensitive to the percentage of selected "best" parameter sets. According to Fig. 1, the ranking based on the best 10–50% of the parameter sets is totally stable. Only for the most rigorous selection (best 10 of 1000 parameter sets) or when no selection is made at all, the ranks of some QPE with similar error do coincide or change.

Table 1 shows the results of the point-based verification. Leave-one-out cross-validation was carried out for all time steps in which at least one raingauge in the entire collective observed a precipitation sum larger than 1 mm (5 mm) per hour. The table contains the 95% confidence intervals of the mean RMSE based on block sub-sampling and subsequent bootstrapping as well as the ranks we assigned based on these confidence intervals.

Table 1 95% confidence intervals (C95) of the mean RMSE (mm/h) for six methods of quantitative precipitation estimation. The ranking is based on these confidence intervals.

Threshold = 1 mm/h			Threshold = 5 mm/h		
Rank	QPE	C95	Rank	QPE	C95
1	MRG	1.00–1.07	1	MRG	2.80–2.86
2	DIF	1.07–1.13	1	DIF	2.85–2.89
2	MFB	1.09–1.16	3	MFB	2.93–3.00
3	IDW	1.13–1.20	4	BOR	3.32–3.38
5	FAC	1.26–1.36	4	IDW	3.31–3.39
5	BOR	1.27–1.36	6	FAC	3.81–3.91

Apparently, MRG, DIF and, to a certain extent, MFB achieve the best results. This statement is independent from the considered intensity threshold. These methods are all radar-based QPE adjusted by raingauge observations. However, a radar-based QPE adjusted by raingauge observations is not necessarily preferable: the method FAC performs poorly for both intensity thresholds. Both the IDW method (which is purely based on raingauge observations) and the BOR method (which is the unadjusted radar observation) show an intermediate or poor performance. The relative ranking of the unadjusted radar (BOR) improves with a higher intensity threshold while the relative ranking of the IDW method is slightly degraded for higher rainfall intensities. This is consistent with our expectation that radar-based QPE is particularly useful for spatially heterogeneous convective events with high rainfall intensities.

However, if the QPE is to force a hydrological model whether it is biased is of particular interest. A biased QPE will most likely introduce bias into the simulated runoff and thus increase the simulation error. However, it is not sufficient to quantify the overall QPE bias because alternating periods of contiguous over- and underestimation of precipitation might also impair the quality of the runoff simulation. Here, the concept of block sub-sampling is particularly useful in order to illustrate the temporal variation of QPE bias. Figure 2 shows the variation of the QPE's mean error (i.e. bias) based on the same block sub-sampling which was used to compute the confidence intervals of the mean RMSE. In this case, however, we are not so much interested in the overall mean, but rather in the variation of the mean error around the line of 0 mm/h. For both intensity thresholds, IDW and MRG show by far the smallest variation around zero. In other words: IDW and MRG are continuously unbiased compared to the other methods. For IDW, the small bias is surprising having the high RMSE in mind: This implies that, for IDW, errors even out more strongly because over- and underestimation of rainfall at different verification points occurs at the same time, or at least, in the same block. The methods MFB, DIF and FAC more often exhibit periods (i.e. blocks) of strong bias in terms of both over- and underestimation. BOR shows by far the strongest variation of the bias around the line of 0 mm/h.

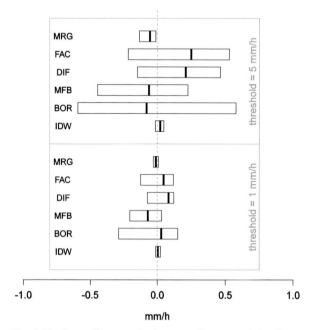

Fig. 2 The boxes illustrate the interquartile range of the distribution of QPE bias in the 1000 blocks generated by the block sub-sampling. Lower panel: The distribution was computed by using only those time steps in which at least one raingauge recorded a minimum precipitation of 1 mm/h. Upper panel: same as lower panel, but for a minimum precipitation of 5 mm/h.

DISCUSSION AND CONCLUSIONS

Based on the ranking established by the mean RMSE from cross-validation (Table 1), the results of runoff simulation are surprising at first sight. This particularly applies to the very good performance of the IDW method. The discrepancy, however, is less pronounced if we focus on QPE bias (Fig. 2) instead of RMSE: both the IDW and MRG method exhibit a low variation around zero bias. Thus, we have to assume that the intermediate rank of the IDW method from cross-validation is largely due to random errors. If we assumed that the transformation of rainfall to runoff acts as a smoothing operation both in space and time, these random errors could be averaged out to a significant extent. In fact, the difference in mean RMSE between the IDW and the MRG method is reduced by over 20% if we use rainfall accumulations of 6 h instead of 1 h rainfall sums (results not shown here).

Despite some discrepancies between the two verification approaches, we can, at least for the two study catchments, already confirm the benefit of adjusting radar observations by raingauge measurements. The unadjusted radar product we used here cannot be recommended for use in hydrological modelling. From the set of considered adjustment procedures, the MRG method appeared to be the most robust and unbiased one. Of course, this verdict is limited to the scope of this case study in terms of both the underlying data and the considered study catchments.

Our analysis shows that the "hydrological verification" approach is able to reveal deficits which remain occluded using a conventional point-based verification. We would have judged the

QPE methods differently if we had based our ranking on the mean RMSE from cross-validation only. Besides, the approach based on runoff simulation is representative exactly for the spatiotemporal scale of interest, the catchment scale. It also allows us to study the variation of the estimation error in space.

On the other hand, the results from the point-based verification helped us to better understand some of the reasons behind the results from hydrological verification. For example, it helped us to understand the reasons for the similar performance of IDW and MRG.

Based on the presented hydrological verification approach we can identify for any (gauged) catchment of interest which QPE method is best suited to represent the precipitation over this catchment. Certainly, this "suitability" is a complex result of potentially interacting factors such as the raingauge network density and configuration, the spatiotemporal error structure of the nearest weather radar and aspects such as topography which might influence the local and regional rainfall generation processes. Hydrological verification allows us to efficiently consider all these sources of error in a scale-sensitive framework. By using Monte Carlo simulations, we minimize the impact of model calibration on our verification results. The subsequent application of block sub-sampling reveals the temporal variability of the estimation error and helps us to assess whether potential differences in the performance of tested QPE methods are significant or not. A pragmatic ranking procedure condenses the voluminous and complex data to an outcome which is easily comprehensible and facilitates the selection of suitable QPE methods.

Future work will elaborate on the potential impact of errors in the structure of the RR model as well as in the discharge observations on the outcome of the hydrological verification procedure.

REFERENCES

Brandes, E.A. (1975) Optimizing rainfall estimates with the aid of radar. *J. Appl. Met.* 14, 1339–1345.

Ehret, U., Götzinger, J., Bardossy, A., Pegram, G. S. (2008) Radar-based flood forecasting in small catchments, exemplified by the Goldersbach catchment, Germany. *Int. J. River Basin Management* 6(4), 323–329.

German Weather Service, DWD (2005) Projekt RADOLAN. Routineverfahren zur Online-Aneichung der Radarnieder-schlagsdaten mit Hilfe von automatischen Bodenniederschlagsstationen. Final Report. DWD, Offenbach, Germany. pdf available at http://www.dwd.de/RADOLAN (in German).

Goudenhoofdt, E. & Delobbe, L. (2009) Evaluation of radar-gauge merging methods for quantitative precipitation estimates. *Hydrol. Earth System Sci.* 13, 195–203.

Gourley, J. J. & Vieux, B. E. (2005) A method for evaluating the accuracy of quantitative precipitation estimates from a hydrologic modeling perspective. *J. Hydromet.* 6(2), 115–133.

Kneis, D. & Heistermann, M. (2009) Quality assessment of radar-based precipitation estimates with the example of a small catchment. *Hydrologie und Wasserwirtschaft* 53(3), 158–169 (in German and English).

Ludwig, K. & Bremicker, M. (2006) The Water Balance Model LARSIM – Design, Content and Application, Freiburger Schriften zur Hydrologie, Vol. 22, University of Freiburg, Institute of Hydrology. pdf available at http://www.hydrology.uni-freiburg.de/publika/band22.html.

Politis, D. N. (2003) The impact of bootstrap methods on time series analysis. *Statistical Science* 18(2), 219–230.

454

Weather Radar and Hydrology
(Proceedings of a symposium held in Exeter, UK, April 2011) (IAHS Publ. 351, 2012).

Study on a real-time flood forecasting method for locally heavy rainfall with high-resolution X-band polarimetric radar information

MAKOTO KIMURA[1], YOSHINOBU KIDO[2] & EIICHI NAKAKITA[2]

1 *Central Research Institute, Nihon Suido Consultants Co., Ltd., PO Box 163-1122, 6-22-1, Nishi-Shinjuku, Shinjuku-ku, Tokyo, Japan*
kimura_m@nissuicon.co.jp
2 *Disaster Prevention Research Institute, Kyoto University, PO Box 611-0011, Gokasyo, Uji Kyoto, Japan*

Abstract In recent times locally heavy rainfall has occurred frequently in Japan and caused serious human accidents; hence the need for flood forecasting systems has increased to reduce inundation damage. However, flood forecasting that secures lead-time for evacuations is extremely difficult because conventional radars cannot adequately measure the rainfall. Under this circumstance, X-band polarimetric radars have been installed, and flood forecasting with a higher accuracy is expected. In order to develop and optimize a real-time flood forecasting method for locally heavy rainfalls in urban drainage areas, we have considered several flood forecasting models with various computational accuracies and loads. Furthermore, we have evaluated these models for comprehensive prediction accuracies through case studies in an actual basin using X-band radar information. As a result, the detailed flood forecasting model might not always have the highest accuracy and the proposed simplified model which can apply the latest rainfall information with lower computational loads was effective.

Key words real-time flood forecasting; X-band polarimetric radar; locally heavy rainfall; urban drainage areas

INTRODUCTION

Recently, locally heavy rainfalls have frequently occurred in highly urbanized areas in Japan and cause serious human accidents in urban rivers and drainage pipes. Moreover, underground spaces in Japanese urban areas are extensively used for subways, underground malls, and other underground facilities. Therefore, the need for flood forecasting systems has increased in order to reduce inundation damage. However, flood forecasting that secures lead-time for evacuations is extremely difficult because raingauges or conventional radars cannot measure the rainfall adequately. Under this circumstance, the X-band polarimetric radars which can observe high-resolution rainfall temporally and spatially have been installed in urban areas of Japan since July 2010; they can be practically applied to flood forecasting with a higher accuracy.

Because for appropriate flood forecasting with locally heavy rainfall it is necessary to finely represent inundation mechanisms of urban drainage areas, combined flow models with one-dimensional (1-D) pipe and 2-D overland surface (physically-based computational modelling; hereinafter referred to as "physical model") are widely applied. However, the use of these detailed models causes heavy computational load and therefore the latest rainfall information is not available because of the limited computational time for forecasting. On the other hand, simplified models with smaller computational loads can utilize the latest rainfall information, but the prediction accuracy of models in itself decreases. Therefore, for adopting the optimal forecasting method, it is necessary to evaluate not only the accuracy of the flood forecasting model, but also the comprehensive prediction accuracy including accuracy of available rainfall information.

In order to develop and optimize a real-time flood forecasting method for reducing damage caused by locally heavy rainfall in urban drainage areas, we have considered several urban flood forecasting models with various computational accuracies and loads. Furthermore we have evaluated these models for comprehensive prediction accuracies by employing case studies in an actual basin using high-resolution X-band polarimetric radar information.

FLOOD FORECASTING MODELS

One approach to flood forecasting is physically-based modelling in which the flow models combine 1-D drainage networks and 2-D overland surfaces. This study has prepared several models of drainage networks and overland surfaces with various resolutions. The other approach is statistically-based modelling in which flooding is predicted as quickly as possible directly from rainfall information without computing the physical process for forecasting.

Physically-based modelling

Detailed model (the foundation of other models) This was developed using the most recently available analysis tool and data in order to finely express the inundation mechanism of urban drainage areas; the standard commercial package "InfoWorks CS" was used as the analysis tool. The drainage network model was constructed considering all its facilities (pipes, manholes, weirs and pumps) with the exception of inlets (see Fig. 1(a)). For the overland surface model, unstructured meshes that reflect the shape of open areas between buildings (minimum width: 2 m) were constructed on the basis of fine-scale grids made from LiDAR data (Light Detection and Ranging data) (see Fig. 1(b) upper). Moreover, in order to express the influence of buildings on the overland flow, we applied a local friction-based approach, whereby areas occupied by buildings were assigned an extremely high roughness coefficient. The exchange of water between the drainage and overland surfaces was modelled using weir equations. In addition, we have confirmed the accuracy of this detailed flood forecasting model by comparing its results with the actual inundation data in a study by Kimura *et al.* (2011).

Simplified model For real-time forecasting, adequate simplifying of models is effective to reduce the computational load. However, the model simplifying should be done with regard to maintaining prediction accuracy because simplifying the drainage pipe and overland surface affects it. Kimura *et al.* (2011) has shown that the prediction accuracy was sufficient with drainage pipe modelling ignoring pipes 300 mm or less in diameter and overland surface modelling using structured 5 m meshes not reflecting building shapes (see Fig. 1(a),(b) lower). Therefore, we applied this model as the simplified model, which can reduce the number of drainage pipes and overland meshes to 38% and 92% for realizing stable calculations and computational load reductions.

More simplified model A more simplified overland surface model is necessary for realizing further reductions in the computational load. Hartnack *et al.* (2009) have shown that overland flow modelling using a fine-scale DEM (digital elevation model) can reduce computational loads substantially with sufficient accuracy. Based on this study we evaluated the method by interpolating the computational flood depth, which was calculated by 25 m coarse meshes and using a 5 m fine-scale DEM, and we confirmed that this method can also obtain the depth with appropriate accuracy. Therefore, we applied this model as a more simplified model. This model can drastically reduce the number of overland meshes to 4% and lead to large reductions in the computational load.

Statistically-based modelling

In a previous study, Kimura *et al.* (2010) has developed a statistically-based model as a real-time and instantaneous flood forecasting method for predicting flooding directly from rainfall information

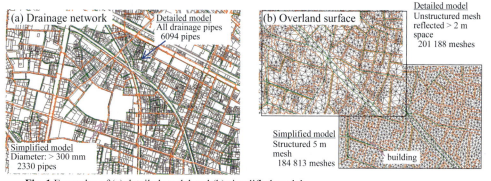

Fig. 1 Examples of (a) detailed model and (b) simplified model.

without computing hydraulic physical processes, and evaluated the method's usefulness through case studies. This statistical model is summarized here.

Modelling concept In order to represent the influences of different scales at each overland mesh point, a statistical model based on "ANN (artificial neural network) model" is applied to each 2-D mesh point. The model expresses relationships between the temporal and spatial rainfall information (as input) and flood depth (as output). However, although much actual inundation data is necessary for the identification of model parameters, typically, actual inundation data have the problem of sufficiency and accuracy. In addition, if the statistical model is expanded to 2-D mesh points, the operations for identifying statistical model parameters become extremely large. Regarding the matter of data sufficiency and accuracy, we have aimed to achieve prediction accuracy as high as the "physical model" and apply its computed flood depths to the training data of the statistical model. On the other hand, in order to reduce the burden of model parameter identification, we apply the "k-means method" to each mesh point as a classification technique and decrease the amount of information to be processed by grouping meshes with similar inundation characteristics (see Fig. 2).

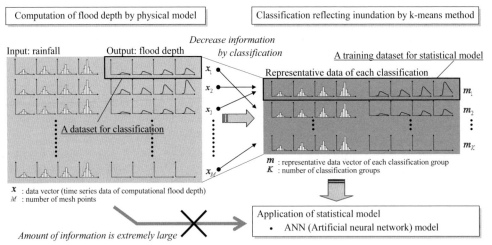

Fig. 2 Concept of statistical model with classification.

Applied methods This study has applied the following conditions on the basis of the conditions proposed by Kimura *et al.* (2010). The physical model for calculating the flood depths at 2-D mesh points was a coupled model comprising of a 2-D flooding model with a 50-m mesh scale and a 1-D drainage pipe model. The computational flood depth of 5-m mesh fine DEM is obtained using the interpolation method as with the "more simplification model". For the classification reflecting the inundation characteristics, this study adopted 36 classifications wherein the inflection point can be defined on the basis of the classification error (see Fig. 3(a)). Subsequently, data of the mesh point where maximum flood depth appears was applied as the representative data of each classification for constructing the ANN model. This model has three layers and formulates relations between rainfall information until the current time and current flood depth of the mesh point. We adopted the following 18 information sources consisting of six temporal cases and three spatial cases as the input data of the ANN model; further, 45 design rainfalls with various temporal and spatial characteristics were adopted as the training data (see Fig. 3(b)).

– Temporal information: 5, 10, 30, 60, 90, 120 min average rainfall intensity.
– Spatial information: rainfall at each mesh point and areal rainfall for two spatial scales; the middle-scale (the study area: 4.75 km^2) and the large-scale (the pump basin: 12.25 km^2).

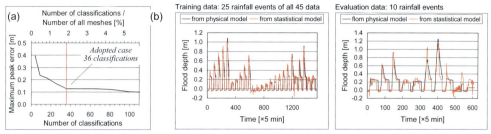

Fig. 3 Applications of (a) model accuracy according to number of classifications, (b) ANN model results for representative group (no. 34).

RESULTS AND DISCUSSION

Study area and rainfall data

For the case study, we adopted a 4.75 km^2 urban drainage area in Osaka, Japan (Fig. 4(b)) representing a completely urbanized area for overland use, underground use, drainage improvements, etc. The stormwater in the entire area is drained away using a combined sewer system and discharged into a river at downstream pumping stations.

The observation and prediction rainfall observed with the X-band polarimetric radars is used as the rainfall data in the case study. This newly installed radar system can observe rainfall with 16 times higher resolution (approx. 250 m mesh scale) and 5 times more frequency (every 1 min) than the conventional C-band radar, and perform 60 min rainfall forecasting in every 5 min, 5-min step using a translation model as a very short-range forecast method. The paper by Kato *et al.* (2011) gives details of the rainfall observation and estimation methods using the X-band and C-band radars. This study applied a typical locally heavy rainfall observed in another region, Tokyo, on 5 July 2010 because such heavy rainfalls have never been previously observed in the case study area (see Fig. 4(a)).

The applied rainfall characteristic is shown in Fig. 4(c). This rainfall was a typical convection rainfall, and its intensity was recorded as >100 mm/h at the rainfall centre. Figure 4(c) indicates that the rainfall observed with the X-band polarimetric radars is similar to the rainfall observed using the raingauge on the ground, and the new radar is superior to the conventional radar, not only in its temporal and spatial resolution, but also in its observation accuracy.

However, Fig. 4(d) shows the prediction rainfall using X-band polarimetric radar observations. This shows that the prediction accuracy deteriorates while its tendency changes according to the base point in forecasting time when the lead-time of the rainfall forecasting is longer. As one example of evaluating the influence of this deterioration tendency, this paper will show the evaluation results of flood forecasting accuracy using predicted rainfall for forecasting to 20:00 h.

Effect of applying X-band polarimetric radar information to flood forecasting

To evaluate the application effect of the X-band polarimetric radar information, Fig. 5(a) compares the flood calculation results from the X-band radar observation rainfall data with those from the conventional radar's data using the "detailed flood forecasting model"; here, the X-band radar's result is used as the standard of comparison. This result suggests that the conventional radar's results are over- or under-estimated as compared to the X-band radar's result, i.e., these results show that the installation of the X-band radar system improves the accuracy of flood forecasting.

On the other hand, in order to evaluate the application influence of the predicted rainfall with the deterioration tendency to flood forecasting, the flood calculation results from the application of the X-band radar predicted rainfall data for each lead-time case to the detailed model are shown in Fig. 5(b); the observation rainfall's result is used as the standard of comparison. This result indicates that the accuracy of flood forecasting deteriorates in the same way as the deterioration tendencies of predicted rainfall; especially when the lead-time for rainfall forecasting is longer (≥25 min), the flooding areas above floor level (flood depth ≥30 cm or more) decrease significantly, and concerns regarding the practical applications of flood forecasting gradually increase.

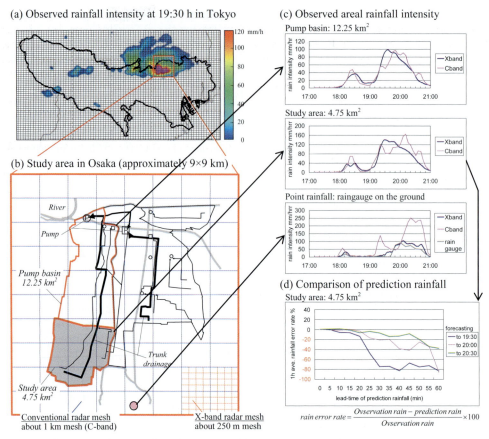

Fig. 4 Characteristics of applied rainfall observed by radar on 5 July 2010.

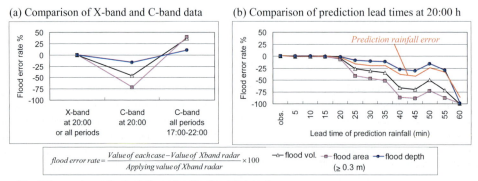

Fig. 5 Comparison of radar information for flood forecasting.

Evaluation of flood forecasting models with prediction rainfall

As a preliminary in evaluating the comprehensive accuracies from the rainfall and flood forecasts, in order to compare each model's accuracy, the flood calculation results from the application of the X-band radar observation rainfall data to each developed flood forecasting model are shown in Fig. 6(a); the detailed model's results are used as the standard of comparison. The statistical model's results have a larger tendency than that of the other models because classification with inundation characteristics is applied and the value of mesh where maximum flood depth appears is applied to the flood depth of each mesh. However, the statistical models can forecast safely, and hence, the applicability of each flood forecasting model including the statistical model is confirmed. Furthermore, the computational time decreases by applying simplified models; therefore, the effectiveness of the simplifying flood forecasting model is indicated.

Then, based on what was found, the flood conditions were computed with the available predicted rainfall for each flood forecasting model from the viewpoint of their computational times. The detailed model's computational time was assumed to be 30 min, and each model's time for 30-min flood forecasting was configured in accordance with the computational time ratios shown in Fig. 6(a). As a result, the available predicted rainfall for each flood forecasting model is listed in Table 1. The result of these cases is shown in Fig. 6(b), and reveals that the detailed model may not always have the highest accuracy for flood forecasting and that the simplified model which can apply the latest rainfall information with lower computational loads is effective.

(a) Comparison of flood forecasting models

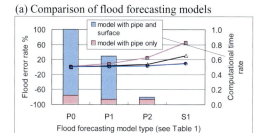

(b) Comparison of comprehensive accuracies

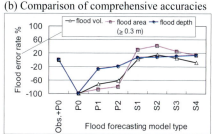

Fig. 6 Comparison of flood forecasting models with predicted rainfall at 20:00 h.

Table 1 Available prediction rainfalls of each flood forecasting model for 30-min forecasting.

Type	Flood forecasting model	Computational time			Lead time of available predicted rainfall
		Rainfall	Flood		
		min	Rate	min	min
P0	Detailed model	5	1.0	30	65
P1	Simplified model	5	0.6	20	55
P2	More simplified model	5	0.1	5	40
S1	Statistical model	5	0.0	0	35

CONCLUSIONS

To develop and optimize a real-time flood forecasting method for locally heavy rainfall in urban drainage areas, this study has considered several urban flood forecasting models with various computational accuracies and loads, and evaluated them for comprehensive prediction accuracies through case studies in an actual basin with X-band polarimetric radar information. As a result, the detailed model might not always have the highest accuracies for flood forecasting and the method which can apply the latest rainfall information with lower computational loads was effective.

REFERENCES

Hartnack, J. H., Enggrob, H. G. & Rungo, M. (2009) 2D overland modelling using fine scale DEM with manageable runtimes. *Flood Risk Management: Research and Practice* 119–124.

Kato, A., Iwanami, K., Maesaka, T., Maki, M. & Misumi, R. (2011) Quantitative precipitation estimate and forecast by the complementary application of X-band polarimetric radar and C-band conventional radar. In: *Proc. 8th Int. Symp. on Weather Radar and Hydrology*.

Kimura, M., Kido, Y. & Nakakita, E. (2010) Development of real-time flood forecasting method in urban drainage areas. In: *Proc. 9th Int. Conf. on Hydroinformatics* 3, 1802–1810.

Kimura, M., Kido, Y. & Nakakita, E. (2011) Fundamental study on real-time flood forecasting method for locally heavy rainfall in urban drainage areas. *J. Hydraul. Engng JICE* 55, 931–936 (in Japanese).

A rainfall–runoff model and a French–Italian X-band radar network for flood forecasting in the southern Alps

D. ORGANDE[1], P. ARNAUD[2], E. MOREAU[3], S. DISS[2], P. JAVELLE[2], J.-A. FINE[1] & J. TESTUD[3]

1 *Hydris hydrologie, 5 Avenue du Grand Chêne, 34270 Saint-Mathieu-de-Tréviers, France*
didier.organde@hydris-hydrologie.fr
2 *IRSTEA, 3275 Route de Cézanne, CS 40061, 13182 Aix en Provence Cedex 5, France*
3 *Novimet, 41 bis Avenue de l'Europe, BP 264, 78140 Vélizy Villacoublay, France*

Abstract The aim of the CRISTAL project (Gestion des CRues par l'Integration des Systèmes Transfrontaliers de prévision et de prévention des bassins versants Alpins) is to develop an operational flood forecasting system for catchments located in the French Southern Alps and Italian Piedmont, based on rainfall data from two dual-polarisation X-band radars. The study deals with the calibration and initialization of the rainfall–runoff model on gauged French catchments (45–461 km² in area) on the Siagne, Paillon and Roya rivers. The GRD conceptual rainfall–runoff model is calibrated in order to reproduce measured flow. The model initialization consists of establishing a calculation rule to define the value of the daily production parameter in relation to known variables (such as previous rainfall or evapotranspiration). Hydrological simulations of recent events measured by X-band radars are presented and compared with raingauge and water-level records.

Key words flood forecasting; X-band radar; rainfall–runoff model; calibration; initialization; French–Italian border

INTRODUCTION

Catchments in the Alpine and Mediterranean regions are prone to flash floods, particularly in the steep mountainous areas and in the coastal areas (Bovo *et al.*, 2008). Therefore, a primary concern is to create an operational flood forecasting system that produces both hydrometeorological information in real-time and employs hydrological models.

Traditional instruments (monitoring networks on the ground) are not effective operational tools to forecast and to prevent floods in these mountainous catchments. The raingauge network is not dense enough to ensure representative data for a real-time hydrological model (Duncan *et al.*, 1993). Currently, only remote sensing instruments can fill in the lack of information.

The aim of the CRISTAL project (Gestion des CRues par l'Integration des Systèmes Transfrontaliers de prévision et de prévention des bassins versants Alpins) is to develop an operational flood forecasting system for catchments located in the Southern French Alps and the Italian Piedmont region. The project draws on experience from the 2006–2008 FRAMEA project (Flood forecasting using Radars in Alpine and Mediterranean Areas,) (Bechini *et al.*, 2008) and uses data from two dual-polarisation X-band radars in the Maritime Alps (the Hydrix® French fixed radar on Mt Vial at an altitude of 1500 m) and at the French–Italian border (an Italian mobile radar sited on the Col de Tende at a height of 1800 m).

This paper deals with the calibration and initialization of an event-based hydrological model. First, the study area and the data are presented; then the rainfall–runoff model is presented and the methodology of its calibration is developed. The third part of the paper treats the study of initialization. Finally, the model is applied in operational conditions with recent events for which radar data were available and that were not used in the calibration and initialization steps.

STUDY AREA AND DATA

Study area

The study area is located in the French Maritimes Alps and the Italian Piedmont region. Four rivers are studied. Three of these are in France (the Siagne, the Paillon and the Roya, which is

located at the French–Italian border) and one is in Italy (the Vermenagna, located on the other side of the border). For each river, one or more catchments are studied:
(a) 4 catchments from 45 to 522 km^2 in area for the Siagne River;
(b) 3 catchments from 69 to 173 km^2 in area for the Paillon River;
(c) 3 catchments from 81 to 445 km^2 in area for the Roya River;
(d) 1 catchment with an area of 133 km^2 for the Vermenagna River.

Rainfall data

Two kinds of hourly rainfall data are available: raingauge data and X-band radar data. Raingauge data are supplied by different organizations; from 1994 through to 2009, between 32 and 51 raingauges are available, depending on the period. The radar data come from X-band radar located at Collobrières (February 2005–June 2007) and at Mont Vial (June 2007). The map in Fig. 1 indicates the locations of the rainfall data sources and the basins.

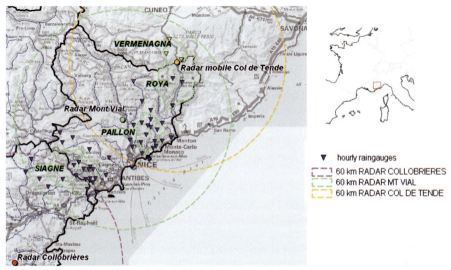

Fig. 1 Location of study area, rainfall data sources and basins.

Flow data

Flow data are available for 10 outlets of the three main French catchments from 1994 to 2010 and for one outlet on the Vermenagna from 2005 on. The flow database includes 25 events from 1994 to 2009, allowing 83 rainfall-discharge pairs to be used in this study.

CALIBRATION OF THE RAINFALL–RUNOFF MODEL GR DISTRIBUTED (GRD)

The rainfall–runoff model

The model is a distributed model that operates in event mode (Javelle *et al.*, 2010) on 1-km^2 cells, using an hourly time-step. The runoff generation in each cell is carried out by a production storage *A* for which a maximum capacity was set in the regionalization study (Javelle *et al.*, 2010). Storage *A* is linked to a storage *B* which ensures nonlinear transfer. All elementary runoffs are aggregated at the catchment outlet simply by adding up the elementary runoffs generated within the catchments cells.

The calibration of the model involves two types of parameters: (1) the production parameter S_0/A, or the initial rate of production store A, and (2) the transfer parameter B, or the maximal capacity of the routing storage.

Calibration methodology

Because of the low amount of radar rainfall data, the model is calibrated using hourly spatialized raingauge data using the squared inverse distance function. The calibration consists in defining two parameters: S_0/A for each rainfall-discharge pair, and B for each catchment. The optimization criterion is the Nash-Sutcliffe Model Efficiency (NSME):

$$\text{NSME} = 1 - \frac{\sum (Q_{\text{cal}} - Q_{\text{meas}})^2}{\sum (Q_{\text{meas}} - \overline{Q}_{\text{meas}})^2} \qquad (1)$$

where Q_{cal} and Q_{meas} are the calculated and measured discharge, respectively.

The B parameter is chosen to be related to the Euclidean distance between the cell and the outlet of the catchment according to:

$$B(ip) = [1 + pB \text{ dist}(ip)] B_{\text{opti}} \qquad (2)$$

with pB varying from 0 to 0.5, $\text{dist}(ip)$ is the Euclidean distance between cell ip and the outlet (km), and B_{opti} is the optimal value of B calculated for each catchment.

Calibration results

The efficiency of the calibration is estimated through the NSME (Fig. 2, left) which indicates the ability of the model to reproduce measured hydrographs and through a comparison of measured and calculated pseudo-specific peak flow ($Q/S^{0.8}$) (Fig. 2, right). For 50% of the events, the NSME exceeds 0.75 and is close to 0.9 between measured and calculated pseudo-specific peak flow.

INITIALIZATION

After the calibration step, each catchment is characterized by a value for parameter B and the optimal value of initial rate of the production store A, S_{opt}, has been calculated for each rainfall–discharge pair.

The objective is to use the hydrological model GRD in a real-time operational mode, so it is necessary to be able to define the S_0/A production parameter. These values are representative of the spatial distribution of the antecedent soil moisture condition, which needs to be known before a flood event.

Continuous daily rainfall and evaporation spatial data are used in a simplified and daily version (without snow module) of the GR_{LOIEAU} model (Folton & Lavabre, 2009) in order to estimate antecedent climatic conditions. The different daily variables from this model are used to establish linear relationships with S_{opt}.

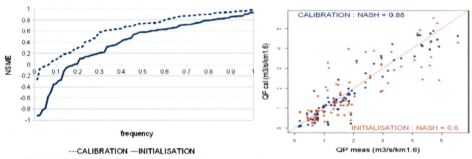

---CALIBRATION —INITIALISATION

Fig. 2 Calibration and initialization results.

One linear relationship is defined for each river and the hourly rainfall–runoff model is run with S_0/A as defined by these linear relationships. The results of these calculations are shown through the NSME statistic and the comparison of measured and calculated pseudo-specific peak flow (Fig. 2).

Simulations using the initialization rules represent an operational situation, and so the results are less efficient than those obtained from the calibration. For 50% of the rainfall–discharge pairs, the NSME exceeds 0.6. The NSME criteria calculated between measured and calculated pseudo-specific peak flow is close to 0.6.

FLOOD WARNINGS

The purpose is to use the hydrological model to activate flood warnings. Two levels of flood warning were defined for several points in the Siagne catchments, and the ability of the model to detect these warnings was tested.

The rainfall data are the same as used in calibration phase. The results are synthesized in a contingency table for flood warnings (Fig. 3). Flow values for warning level 1 and 2 are denoted by Q_1 and Q_2 and the standard peak flow as Q_{pN}. The peak flows (Q_p) are standardized in such a way that for $Q_p = Q_1$, $Q_{pN} = 1$ and for $Q_p = Q_2$, $Q_{pN} = 2$.

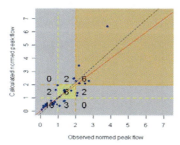

Fig. 3 Contingency table for flood warnings.

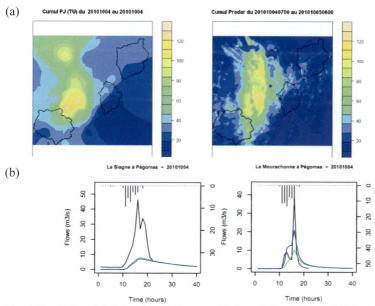

Fig. 4 Simulation of 4–5 October 2010 event. (a) Cumulative daily rainfall data from raingauges (left) and from X-band radar (right). (b) Calculated and "observed" hydrographs: "observed" in black, calculated with corrected radar rainfall in green, and calculated with direct radar rainfall in blue.

(a)

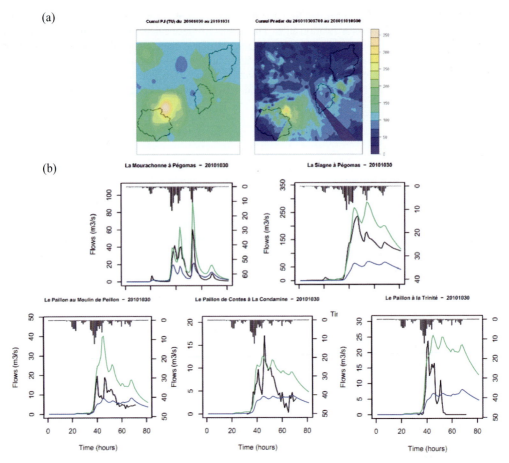

(b)

Fig. 5 Simulation of 30 October to 1 November 2010 event. (a) Cumulative daily rainfall data from raingauges (left) and from X-band radar (right). (b) Calculated and "observed" hydrographs: "observed" in black, calculated with corrected radar rainfall in green, and calculated with direct radar rainfall in blue.

The simulations balance overestimated and underestimated alerts with a good "success rate" (12 "good" warnings for 18 "calculated" warnings).

OPERATIONAL USE WITH RADAR RAINFALL

The aim of this part of the work is to test both the hydrological model and the radar data in a real-time context. Two events are illustrated here (4–5 October and 30 October–1 November 2010) for which there are images from the X-band radar at Mont Vial. The model is initialized and run. Two kinds of rain estimate are tested: a direct measure from radar and a radar rainfall corrected by spatial daily rainfall fields derived from raingauge data. The correction consists in applying a coefficient at each pixel of the study area and for each day (6:00 GMT to 6:00 GMT) in order to equal the daily gauged rainfall and daily radar rainfall; the coefficient is defined as:

$$coef_{\text{pixel}} = \frac{Hr_{\text{pixel}}(d)}{Dr_{\text{pixel}}(d)} \tag{3}$$

where, for day d, $Hr_{pixel}(d)$ is the cumulative hourly radar rainfall for the pixel and $Dr_{pixel}(d)$ is the daily raingauge rainfall at the pixel.

For the first event (Fig. 4), there are few differences between cumulative radar rainfall and raingauge rainfall (Fig. 4(a)). Nevertheless calculated flows underestimate measured flows (Fig. 4(b)) due to the initialization of the model. For the second event (Fig. 5), the cumulative radar rainfall appears to be underestimated (Fig. 5(a)), but thanks to the rainfall correction using equation (3), the hydrological model generates flows close to measured flows (Fig. 5(b)). Here, the model is quite pertinent, but the radar rainfall is underestimated, probably due to bright band. These results highlight the sensitivity of the hydrological model to rainfall data and the uncertainty of the initialization calculation rules. The GRD model performance is highly dependent on the quality of the radar images. A simple correction of radar data, at the daily time-step, increases the relevance of calculated flows.

CONCLUSION

The distributed hydrological model (GRD) has been calibrated and applied on three catchments in the Maritime Alps of Southeastern France and on the Vermenagna basin in Italy. The model was initialized using daily data and allows the simulation of hydrographs in real-time conditions at several outlets of the catchments, using new hourly rainfall data. Flows calculated from Hydrix® X-band radar data are now available on the RAINPOL® platform on the internet, where CRISTAL project partners can visualize simulated hydrographs and flood warnings.

Just how representative the calculated hydrographs and the flood warnings will be depends on calibration and initialization studies and on the relevance of radar data. The operational system is tested and improved as new rainfall–discharge data become available.

REFERENCES

Bechini R., Cremonini R., Campana V., Tomassone L. & Chandrasekar V. (2008) A transportable X-band polarimetric radar in Italy for deployment in complex terrain: results of the first year measurement campaign. ERAD 2008. In: *Fifth European Conference on Radar in Meteorology and Hydrology*. Available at http://erad2008.fmi.fi/proceedings/extended/erad2008-0218-extended.pdf.

Bovo, S., Lavabre, J., Cremonini, R. *et al.* (2008) FRAMEA Flood forecasting using Radar in Alpine and Mediterranean Areas – INTERREG IIIA Italy – France. Relazione finale / Rapport final, available at http://www.arpa.piemonte.it/-pubblicazioni-2/pubblicazioni-anno-2008/pdf-framea.

Duncan, M. R., Austin, B., Fabry, F. & Austin, G. L. (1993) The effect of gauge sampling density on the accuracy of streamflow prediction for rural catchments. *J. Hydrol.* 142(1-4), 445–476.

Folton, N. & Lavabre, J. (2009) Adaptation du modèle GRLOIEAU aux bassins versants à dominante souterraine, Détermination des débits caractéristiques des cours d'eau, Application sur un échantillon de 997 bassins versants, Unité de Recherche Ouvrages Hydraulique et Hydrologie, Cemagref d'AIX-en-Provence.

Javelle, P., Fouchier, C., Arnaud, P. & Lavabre, J. (2010) Flash flood warning at ungauged locations using radar rainfall and antecedent soil moisture estimations. *J. Hydrol.* 394, 267–274.

River flow simulations with polarimetric weather radar

M. A. RICO-RAMIREZ[1], V. N. BRINGI[2] & M. THURAI[2]

1 *Department of Civil Engineering, University of Bristol, Queen's Building, Bristol BS8 1TR, UK*
m.a.rico-ramirez@bristol.ac.uk
2 *Department of Electrical Engineering, Colorado State University, Fort Collins, Colorado 80523-1373, USA*

Abstract Polarimetric weather radars offer advantages over conventional radars such as removal of non-meteorological echoes, attenuation correction, hydrometeor identification and more accurate rainfall estimation in the rain region, all of which lead to an overall improvement in data quality and the subsequent improvement in rainfall estimation. However, how much of that improvement is translated into more accurate river flow simulations given the fact that hydrological models are also subject to uncertainties due to model parameters and model structure? This paper examines the use of radar rainfall estimations from an operational polarimetric C-band weather radar linked directly to a hydrological model for river flow simulations. Several rainfall events from the winter and summer seasons were considered for the analysis. Polarimetric rain-rate algorithms were developed for both seasons, based on several months of disdrometer data. The results are presented in terms of radar and raingauge comparisons (over a large raingauge network) as well as flow simulations.

Key words polarimetric radar; radar errors; attenuation; rainfall estimation; flood forecasting

INTRODUCTION

The hydrological processes in river catchments can be mathematically represented by rainfall–runoff models, which have a variety of applications from reservoir management to flood forecasting. Rainfall–runoff models can be largely classified into two main categories: lumped or distributed. Lumped models consider the whole catchment as a single entity and catchment averages (e.g. rainfall) are used to define its behaviour, whereas distributed models break the catchment into a grid, using the local average for each grid within the catchment as the model basis. Some distributed models use sub-catchments to partition the spatial domain (e.g. REW model, Regianni *et al.*, 1999). The trend for increasingly complex models to model the rainfall–runoff processes more accurately has led to an increase in the number of parameters associated with these models (Todini, 2007). However, the increase in the number of parameters does not necessarily lead to an improvement in model performance. The selection of a particular type of rainfall–runoff model greatly depends upon the application, but it is obvious that precipitation is the main input to any hydrological model and thus, accurate rainfall measurements are required for the model to accurately reproduce the hydrological response of the catchment.

Remote sensing devices such as weather radars are able to produce precipitation estimates over large areas that can be used in real-time hydrological modelling and forecasting. However, there are large uncertainties associated with the estimation of precipitation using single-polarised (SP) weather radar when comparing to raingauge measurements (e.g. Rico-Ramirez *et al.*, 2007; Villarini & Krawjeski, 2010). Dual-polarised (DP) radars provide some advantages over conventional SP radars, for example in identifying non-meteorological echoes (Rico-Ramirez & Cluckie, 2008), in correcting for attenuation in rain (Bringi *et al.*, 2001), and in improving the overall accuracy of rainfall estimation (Bringi & Chandrasekar, 2001). In this paper, we assess some of the advantages of using DP radar rainfall in hydrological modelling.

DATA

The radar data used in this analysis were obtained from the Thurnham C-band DP radar located in southeast England, UK. This radar measures the horizontal reflectivity (Z_h), the differential reflectivity (Z_{dr}), the differential propagation phase (Φ_{dp}), the correlation coefficient (ρ_{hv}) and the radial velocity. The radar measurements have spatial and temporal resolutions of $1° \times 250$ m and 5-min, respectively. A network of 74 tipping-bucket gauges with a resolution of 0.2 mm and river

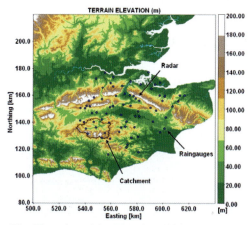

Fig. 1 Location of the C-band DP Thurnham radar, raingauge network and Upper Medway catchment.

flow data with a temporal resolution of 15-min for the Upper Medway catchment were also available within the radar coverage (see Fig. 1). An impact-type disdrometer to measure the Drop Size Distribution (DSD) located in Chilbolton, Hampshire, was also used for this study.

RAINFALL ALGORITHMS

The polarimetric rain-rate algorithms, $R(K_{dp})$ and $R(Z_h, Z_{dr})$, were obtained using measured DSDs from the disdrometer. Three months of disdrometer data during the summer and winter seasons of 2007 were used to calculate the rainfall rate algorithms (see Bringi *et al.*, 2011 for more details of this analysis). The algorithms are summarised in Table 1. In addition, the Marshall and Palmer Z_h-R relationship, $Z_h = 200R^{1.6}$ (Marshall *et al.*, 1955), was also used in this study. This equation is used operationally by the Met Office in the UK and it is suitable for UK rainfall (Harrison *et al.*, 2000). A composite algorithm that uses $Z_h = 200R^{1.6}$, $R(K_{dp})$ and $R(Z_h, Z_{dr})$ was also implemented. This algorithm uses $Z_h = 200R^{1.6}$ in light rain, $R(Z_h, Z_{dr})$ in moderate rain and $R(K_{dp})$ in heavy rain (see Bringi *et al.*, 2011). The reflectivity Z_h was corrected for attenuation using the adaptive algorithm proposed by Bringi *et al.* (2001), which calculates the optimal value of the coefficient α (which is temperature-dependent) in the attenuation correction equation $A = \alpha K_{dp}$ by taking into account the total path-integrated attenuation as a constraint. The correction for differential attenuation in Z_{dr} was performed using the equations shown in Table 1 (i.e. $A_{dp}(K_{dp})$). In all the algorithms, Z_h and Z_{dr} were corrected for attenuation unless otherwise stated. Finally, an additional areal rainfall estimator that uses the "areal Φ_{dp} technique" was also implemented (see Bringi & Chandrasekar, 2001). The areal Φ_{dp} rainfall estimator calculates catchment-averaged precipitation and it is more suitable for heavy rain-rates.

RADAR-RAINGAUGE COMPARISONS

The rainfall events considered for our analysis are summarised in Table 2. Event 1 produced heavy rainfall in different parts of the UK; some of the gauges shown in Fig. 1 recorded total rainfall

Table 1 Summary of relationships derived from DSD data.

Summer	Winter
$R(K_{dp}) = 24.68K_{dp}^{0.81}$	$R(K_{dp}) = 21.5K_{dp}^{0.757}$
$R(Z_h, Z_{dr}) = 0.0112Z_h^{0.822}Z_{dr}^{-1.7486}$	$R(Z_h, Z_{dr}) = 0.01357Z_h^{0.817}Z_{dr}^{-2.387}$
$A_{dp} = 0.0169K_{dp}^{1.345}$	$A_{dp} = 0.0166K_{dp}^{1.277}$

Table 2 Summary of rainfall events.

Event	Dates	Max raingauge accumulation
1	19–24 July 2007	85.4 mm
2	28 June–5 July 2007	67.6 mm
3	7–10 June 2007	36.4 mm
4	20–28 February 2007	49.6 mm
5	6–12 January 2007	36.2 mm

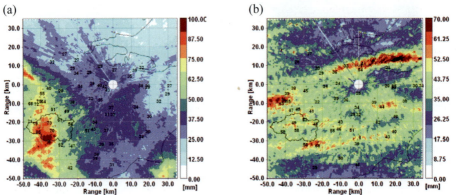

Fig. 2 Total radar and gauge (numbers) rainfall accumulations with the RC algorithm: (a) Event 1 (b) Event 2.

accumulations of up to 55 mm on 20 July 2007 alone. Some DP radar scans for this event showed differential phase shifts of up to 300°, which is equivalent to around 25 dB of attenuation in the horizontal reflectivity at C-band frequencies.

The aforementioned DP algorithms were applied to the events shown in Table 2. The radar rainfall estimations were averaged in polar coordinates over $3° \times 750$ m and using hourly accumulations. In order to perform a quantitative assessment, the composite algorithm (from now on referred to as RC) was compared against the estimated rainfall using the Marshall and Palmer Z_h-R relationship with the radar reflectivity uncorrected for attenuation (from now on referred to as RZ_{Au} algorithm). The Mean Absolute Error (MAE) and the Bias were used as performance indicators. Events 1 and 2 were combined to perform the radar-raingauge comparisons. The RC estimator produced a MAE of 38% and bias of –12%, whereas the algorithm RZ_{Au} produced a MAE of 43% and Bias of –26% when comparing all rain-rates ($R \geq 0.2$ mm/h). This indicates a small improvement when using the composite algorithm RC. However, if the comparisons are performed using high rainfall intensities ($R \geq 6$ mm/h), the composite algorithm produces a significant improvement with a MAE of 24% and Bias of –10%, whereas the rain estimator RZ_{Au} produces a very large underestimation (MAE of 46% and bias of –44%). For event 3, the self-consistency between Z_h and K_{dp} helped to identify a significant calibration problem of 16 dB in Z_h. This can be a very useful way to calibrate Z_h in real time. Events 4 and 5 showed a similar performance between algorithms RC and RZ_{Au} mainly due to the fact that the rainfall intensities were low compared to the summer events. One of the advantages of using DP radar is its ability to measure differential phase measurements, which are immune to attenuation. Events 1 and 2 were affected by attenuation and because Z_h was corrected for attenuation when using the composite rain-rate estimator, it produced an overall improvement in the rainfall estimation when compared to the RZ_{Au} algorithm.

Figure 2 shows the total radar rainfall accumulations using the composite rain-rate algorithm. The total rain recorded by the raingauges is also shown on the same graph for comparisons. As shown, the total rainfall from radar is in broad agreement with the total rainfall measured by the raingauges. Part of the differences between radar and raingauge measurements is attributed to the fact that raingauges provide point measurements, whereas radar provides estimates in a larger volume in space. This results in representativeness errors because of the differences between the sample volumes (Kitchen & Blackall, 1992). Bringi *et al.* (2011) showed that the proportion of the variance of the radar-to-gauge differences that could be explained by the gauge representativeness errors ranged from 20 to 55% when using the composite rain-rate algorithm for events 1 and 2.

RIVER FLOW SIMULATIONS

The hydrological model used in this study was the Probability Distributed Model (PDM), which is a conceptual lumped rainfall–runoff model proposed by Moore (1985). The PDM uses catchment-averaged rainfall and potential evaporation to obtain the flow at the outlet of the catchment.

The PDM has been used in various catchments throughout the UK (Moore, 2007) and it is run operationally in the National Flood Forecasting System (NFFS). The PDM used in this study has been calibrated in previous studies (see Liguori *et al.*, 2009), and the results are shown in Fig. 3(a). The measured (Q_m) and simulated (Q_g) flows and the catchment-averaged rainfall (R) are shown in this figure. The catchment-averaged rainfall was calculated using the Thiessen polygons method from raingauge measurements. The calibration was performed for the whole year of 2006 and the model parameters have been calibrated through manual adjustment followed by automatic calibration to produce a model with the lowest Root Mean Square Error (RMSE) when comparing the model-simulated flow against the measured flow. The model has been calibrated by optimising the objective function on all flows larger than 0 m³/s. However, the model performance decreases if assessed on high flows (see Fig. 3(a)). This model will be referred to as model "A". The validation of the hydrological model is shown in Fig. 3(b) (only 3 months of 2007 are shown). As discussed previously, this year in particular was very wet, with several rainfall events that produced severe flooding in different parts of the UK. The calibration and validation RMSEs were 1.4 m³/s and 1.6 m³/s, respectively.

A second hydrological model was setup to match better the high flows, which are particularly important during severe rainfall events. The objective function chosen to be minimized in the automatic optimisation phase is the same as for model A, but for flows larger than 10 m³/s. This model will be referred to as model "B" (calibration and validation figures not shown). The calibration and validation RMSEs were 2.9 m³/s and 2.5 m³/s, respectively, which indicate that on average, this model produces larger errors than model "A". However, model "B" tends to produce better results in particular for high flows.

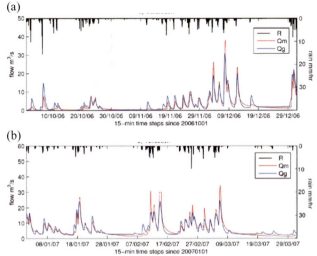

Fig. 3 Calibration and validation of the PDM model "A". R is the measured catchment-averaged rainfall, Qm is the measured flow and Qg is the model-simulated flow using raingauge data. (a) calibration, (b) validation.

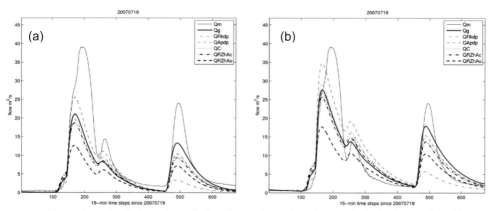

Fig. 4 River flow simulations with polarimetric weather radar using two hydrological models. (a) PDM model "A", (b) PDM model "B".

Figure 4 shows the simulated flows using the different rain estimators. In this figure, Q_m represents the measured discharge in the river, Q_g is the simulated discharge using raingauge data, Q_{RKdp} is the simulated discharge using the $R(K_{dp})$ rain estimator, Q_{Apdp} is the simulated discharge using the areal Φ_{dp} rain estimator, Q_C is the simulated discharge using the composite rain estimator, Q_{RZhAc} is the simulated discharge using the attenuation-corrected Z_h-R rain estimator and Q_{RZhAu} is the simulated discharge using the attenuation-uncorrected Z_h-R estimator. In all cases, except for the areal Φ_{dp} rain estimator, the radar rain-rates were obtained in polar bins and averaged to cover the catchment area shown in Fig. 1.

Figure 4(a) shows the results of using the hydrological model "A" to simulate the flow at the outlet of the catchment for event 1. As shown in Fig. 4(a), the first measured peak flow was around 40 m³/s, which is larger than any of the peak flows used in the calibration of this model. In this case, the model is being applied outside the observed rainfall, which in some cases was >30 mm/h. The simulated flows using raingauge measurements produced the first hydrograph with a peak flow of ~20 m³/s, which is half of the expected peak flow. It is interesting to see that all the DP rain estimators produced peak flows between 18 and 25 m³/s, with the areal Φ_{dp} rain estimator producing the largest peak. The first hydrograph produced by Q_C is very close to the simulated hydrograph produced by the raingauges (Q_g) when computing the RMSE. The only rain estimator that produced poor results was the attenuation-uncorrected Z_h-R algorithm. This is due to the fact that there was a large amount of attenuation of the reflectivity signal. This event clearly indicates one of the advantages of DP radar in correcting for attenuation. However, it also indicates the need for developing hydrological models which are able to cope with more extreme events.

Figure 4(b) shows the results of using the hydrological model "B". This model was calibrated to simulate better the high flows. In this model, the first simulated flow peak using gauge rainfall was around 27 m³/s, and the DP rain estimators produced peak flows in the range of 25–35 m³/s, which are closer, but still below the measured peak flow. Part of this difference could be due to the fact that the calibration of the model was performed using a lower range of flows and consequently the parameters of the model were calibrated accordingly.

Figure 5 shows the simulated flows for the winter event 5. All the rain estimators produced reasonable results except $R(K_{dp})$. This is because this event did not show large differential phase shifts. This is in fact one of the advantages of using a composite rain estimator.

CONCLUDING COMMENTS

The results demonstrate that DP radar offers some advantages over conventional single polarisation radars such as the ability to correct for attenuation, which is particularly important during extreme events. The composite rain estimator offers the advantage of providing better rain-rates by combining the best of different rain-rate estimators. However, more studies on the hydrological applications of polarimetric weather radars are needed in order to fully assess their potential. The hydrological modelling results indicate that any improvement in the radar rainfall estimation will produce an improvement in the simulated flows. However, there is scope to improve the hydrological modelling results by using distributed hydrological models.

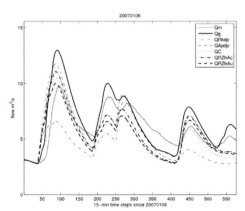

Fig. 5 River flow simulations for event 5 using hydrological model "A".

Acknowledgements The Leverhulme Trust (grant F00182CB) provided support for Prof. Bringi as a visiting professor in the Department of Civil Engineering at the University of Bristol. MT acknowledges support from the National Science Foundation via AGS-0924622. VNB acknowledges partial support from the NASA PMM science program via grant NNX10AJ11G. The authors thank the Met Office, the Environment Agency, the British Atmospheric Data Centre and the Radio Communications Research Unit at the STFC Rutherford Appleton Laboratory for providing the various datasets. We also acknowledge Miss Sara Liguori, who provided the hydrological models for this study.

REFERENCES

Bringi, V. N. & Chandrasekar V. (2001) *Polarimetric Doppler Weather Radar: Principles and Applications.* Cambridge University Press, 636 pp.

Bringi, V. N., Rico-Ramirez, M. A. & Thurai, M. (2011) Rainfall estimation with an operational polarimetric C-band radar in the UK: comparison with a gauge network and error analysis. *J. Hydromet.* 12, 935–954, doi: 10.1175/JHM-D-10-05013.1.

Bringi, V. N., Keenan, T. D. & Chandrasekar, V. (2001) Correcting C-band radar reflectivity and differential reflectivity data for rain attenuation: A self-consistent method with constraints. *IEEE Trans. Geosci. Remote Sens.* 39, 1906-1915.

Harrison, D. L., Driscoll, S. J. & Kitchen, M. (2000) Improving precipitation estimates from weather radar using quality control and correction techniques, *Meteorol. Appl.* 7, 135–144.

Kitchen M. & Blackall, R. M. (1992) Representativeness errors in comparisons between radar and gage measurements of rainfall. *J. Hydrol.* 134, 13–33.

Liguori, S., Rico-Ramirez, M. A. & Cluckie, I. D. (2009) Uncertainty propagation in hydrological forecasting using ensemble rainfall forecasts. In: *Hydroinformatics in Hydrology, Hydrogeology and Water Resources* (ed. by Ian D. Cluckie *et al.*), 3040). IAHS Publ. 331, IAHS Press, Wallingford, UK.

Marshall, J. S., Hitschfeld, W. & Gunn, K. L. S. (1955) Advances in radar weather, *Adv. Geophys.* 2, 1–56.

Moore, R. J. (1985) The probability-distributed principle and runoff production at point and basin scales. *Hydrol. Sci. J.* 30, 273–297.

Moore, R. J. (2007) The PDM rainfall–runoff model. *Hydrol. Earth System Sci.* 11(1), 483–499.

Reggiani P., Sivapalan, M., Hassanizadeh S. M. & Gray W. G. (1999) A unifying framework of watershed thermodynamics: 2. Constitutive relationships. *Adv. Water Resour.* 23, 15–39.

Rico-Ramirez, M. A., Cluckie, I. D., Shepherd, G. & Pallot, A. (2007) A high-resolution radar experiment on the Island of Jersey. *Meteorol. Appl.* 14, 117–129.

Rico-Ramirez, M. A. & Cluckie, I. D. (2008) Classification of ground clutter and anomalous propagation using dual-polarization weather radar. *IEEE Trans. Geosci. Remote Sens.* 46(7), 1892–1904.

Todini, E. (2007) Hydrological catchment modelling: past, present and future. *Hydrol. Earth System Sci.* 11(1), 468–482.

Villarini, G. & Krajewski W. F. (2010) Review of the different sources of uncertainty in single polarization radar-based estimates of rainfall. *Surv. Geophys.* 31, 107–129.

Using combined raingauge and high-resolution radar data in an operational flood forecast system in Flanders

INGE DE JONGH[1], ELS QUINTELIER[2] & KRIS CAUWENBERGHS[2]

1 *VMM, Elfjulistraat 43, B-9000 Gent, Belgium*
i.dejongh@vmm.be
2 *VMM, Koning Albert-II-laan 20, B-1000 Brussels, Belgium*

Abstract Since 2007 the Flemish Environmental Agency has an operational flood forecast system for Flanders. On the website www.overstromingsvoorspeller.be the water manager, civil services and interested citizens can follow in real-time how the situation of the rivers is progressing and what is expected in the next 48 h. For the past 48 h this real-time system uses a pseudo-CAPPI high-resolution radar composite of three radars (Zaventem, Wideumont and Avesnois (France)) combined with real-time raingauge data from more than 40 raingauges spread over Flanders, Belgium. Every 15 minutes, catchment rainfalls are calculated from the updated radar images and the hydrological models are run. Because at some locations a simple data-assimilation technique is used in flood forecast construction, using real-time river flow measurements, the historical catchment rainfall is in fact mainly used to calculate the water balance in the soil. In addition, for the ungauged catchments in the region, the hindcast is built completely with modelling results using raingauge-adjusted radar rainfall as input. Radar rainfall data are known to better resolve the spatial rainfall pattern, in comparison with interpolated raingauge data. Therefore it is a very helpful tool in real-time hydrological forecasting. However, error propagation can make radar rainfall data sometimes spurious. The hydrological forecasts for small catchments are especially more sensitive to the accuracy of the radar rainfall input data. The combination of raingauge and radar data to correct the retrieved radar rainfall is therefore necessary. The influence of merging the raingauge and radar data on the performance of the hydrological forecast system is illustrated for some individual storm events.

Key words flood forecast; raingauge; radar

REAL-TIME FLOOD FORECASTING IN FLANDERS

Since 2007, the Flemish Environmental Agency has an operational flood forecast system for the non-navigable rivers in Flanders (Belgium). The forecasts are published automatically and in real-time on the website www.overstromingsvoorspeller.be. The water manager, civil services and interested citizens can follow in real-time how the situation of the rivers is progressing, and where floods can be expected in the next 48 h by interpreting the modelling results themselves, or following the interpretation of the operator. The forecasts are the results of hydrological and hydrodynamic models which are run with observed rainfall data based on radar rainfall and meteorological forecasts for the subsequent 48 h.

The main source of historical rainfall data besides raingauge data are pseudo-CAPPI (Constant Altitude Plan Position Indicator) high-resolution radar composites from three radars (Wideumont (B), Zaventem (B) and Avesnois (France)). The composite is available every 5 min with a 1 km^2 grid size.

The extra value of radar observations is very clear: a weather radar provides precipitation estimates at very high spatial and temporal resolution over a large area. This resolution is important when the objective is to model the hydrological output of rather small catchments. However, several sources of errors affect the accuracy of the radar rain estimates. The measure of reflectivity itself can suffer from mis-calibration, decrease or increase, ground echoes, parasites (echoes from planes, birds…) and masking effects. Uncertainties also arise with the estimation of rainfall intensity. These are due to the height of measurement (overshooting, condensation and evaporation), bright-band effects due to melting snow and conversion of radar reflectivity Z into rain rate R (the Z-R relationship) (Delobbe, 2007; Goudenhoofdt & Delobbe, 2009). On the other hand, estimates from only a network of raingauges are relatively inaccurate, especially during summer when convective events occur. Raingauges can provide more accurate point-wise measurements, but their spatial representation is limited. So the additional information obtained through a combination of both measurements looks very valuable. This assumption is tested using the PDM hydrological model and the effects on river flow and catchment soil moisture deficit will be shown in the results.

Next, the recalibration method of the radar data is described and a short summary of the hydrological model is given. The results that follow show the advantage of merging the raingauge and radar data for the real-time flood forecasting system in Flanders.

METHODS AND DATA

Radar recalibration

For the purpose of recalibrating the radar data using raingauge observations, the climatological network of the Flemish Environmental Agency has been expanded over the last 5 years. The network now includes 43 raingauges with real-time data access and spread across Flanders (Fig. 1). Every 15 minutes the precipitation totals of all the raingauges are written to the central database. As the forecasting system is used to predict flows even in very small catchments (2–600 km^2) which respond very fast to rainfall, this temporal resolution for the rainfall is necessary to be able to predict rather accurately in time.

In general the merging process applies correction and raingauge-adjustment algorithms to the single radar data and produces catchment- and cell-average rainfall time-series. For this calculation, starting with the raw radar products, the Hyrad Radar Hydrology Kernel is used (CEH, 2007). When no radar data are available, a raingauge surface product can be made in which a surface fitting technique is used to interpolate between raingauge values, to derive a complete rainfall field estimate like that of the grid data.

The output from the merging of radar and raingauges is called the "recalibrated actual". This is an estimate which combines raingauge and radar estimates of rainfall. The form of raingauge radar calibration used is based on fitting a multiquadric surface to calibration factor values defined as a ratio of observed raingauge to coincident radar estimates of rainfall. This "calibration factor surface" is applied to the radar field to obtain the Hyrad rainfall estimate (CEH, 2007). Even for the calculation of the forecast for the next 2 h, the recalibrated images are used. The current recalibrated rainfall field is advected forward, using the calculated velocity from previous radar images.

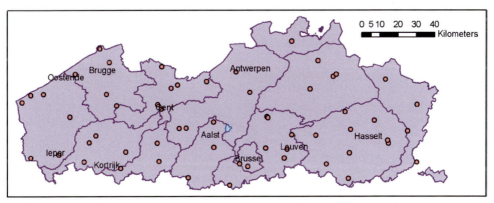

Fig. 1 Map of Flanders where the raingauges used for calibrating the high-resolution radar data are indicated by small circles. The Vondelbeek catchment is indicated in blue.

Hydrological model

Every forecast in the flood forecast system depends on its hydrological input. Indeed, hydrodynamic models are fed with hydrological input at the upstream boundaries. Therefore only hydrological results will be considered here. The hydrological model that is used in the flood forecast system is the Probability-Distributed Model, PDM (Moore, 2007). This model is calibrated for almost 100 monitored catchments in Flanders (Cabus, 2008). A reliable relation between the parameters of the monitored and the ungauged catchments was obtained during the calibration process of the hydraulic models. Catchment precipitation and evaporation are sampled as input for the hydrology in the flood forecast system.

The Vondelbeek catchment

The necessity of merging the radar images and the raingauge data will be shown using model results for an upstream catchment of the Vondelbeek, a small rural catchment located in the central part of Flanders. The catchment area is about 8 km^2. After determination of the Thiessen polygons to determine the catchment rainfall, it could be concluded that data for the Denderbelle raingauge can be used without post-processing. This raingauge is situated near the village of Denderbelle, a few kilometres from the border of the catchment, and has been in operation since November 2006. The Vondelbeek catchment has been gauged since 1996 and the PDM model for this catchment has been calibrated using raingauge data.

RESULTS

Three rainfall sources are used in this research. The precipitation measured by the raingauge at Denderbelle will be compared with the catchment rainfall obtained from the radar images and with the catchment rainfall obtained from the recalibrated radar images. Cumulative rainfall totals are shown over a common period, from 1 May 2008 until 8 November 2010 (Fig. 2). From this figure it is very clear that the cumulated catchment rainfall from the recalibrated radar images differs a lot from the uncalibrated radar rainfall (1760 mm compared to 2300 mm). The cumulated rainfall obtained from the raingauge is about 1830 mm.

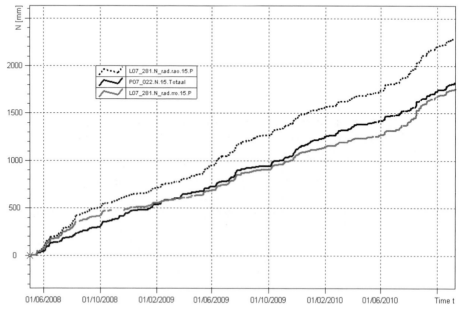

Fig. 2 Cumulative rainfall measured at the raingauge in the catchment (black line), cumulative catchment rainfall derived from the radar (black dotted line) and cumulative recalibrated catchment rainfall (grey line) for the whole data period.

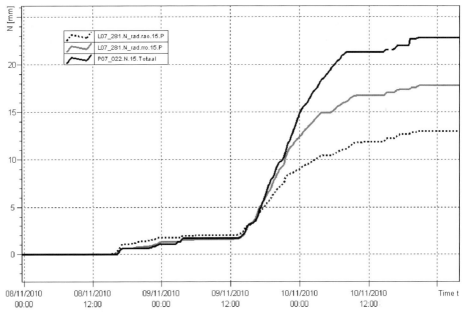

Fig. 3 Cumulative rainfall measured at the raingauge in the catchment (black line), cumulative catchment rainfall derived from the radar (black dotted line) and cumulative recalibrated catchment rainfall (grey line) for the event of November 2010.

Considering the rain data over 3 years, the differences between the rain calculated from the two different radar sources almost equal the annual precipitation of Flanders, which is about 800 mm (De Jongh *et al.*, 2006). When looking at other catchments in Flanders, the same conclusion can be made and the differences between rainfall obtained from raw radar images and rainfall calculated from raingauge-merged radar images are of similar size. One can imagine that these differences will have large effects on the modelled hydrological flows over the long term. This will especially be the case because the PDM has a "hydrological memory", with its soil moisture deficit influencing the modelled flow in the following time-steps.

The event of November 2010 (8 November–11 November at noon) over the Vondelbeek catchment gave between 13 and 23 mm of precipitation, depending on the data source (Fig. 3). According to the uncalibrated radar image the storm event is more intensive at the start, but the total volume of rain is only 56% of the volume that was detected by the raingauge.

Using the PDM model, the evolution of the soil moisture deficit can be plotted for a modelled storm event. Figure 4 shows how the soil moisture deficit decreases while it rains. Note the initial soil moisture deficit calculated with each of the three different rainfall inputs. At the beginning of the storm event the soil is rather wet when warming up the model with uncalibrated radar rainfall. The driest soil is obtained when using raingauge data. This is a direct consequence of the rain data shown in Fig. 2. The difference between the soil moisture obtained with the PDM using raingauge input compared with that calculated using recalibrated radar rain cannot be explained by the difference in total rain volume, as it is almost the same. Further examination of the timing of the rainfall events according to the different sources and the model parameters can help to explain the difference found, but further discussion of this is beyond the scope of this paper.

Although the difference in soil moisture between the three modelled situations is obvious, one can observe that the flow does not differ that much at low flow conditions. The calculated flows at the start, but even more distinct at the end of the event, are almost equal in the three situations (Fig. 5). From Fig. 5 it is obvious that the peak flow, and even the volume, is modelled most accurately using the recalibrated radar rain data. Using the raw radar rain data as model input, the

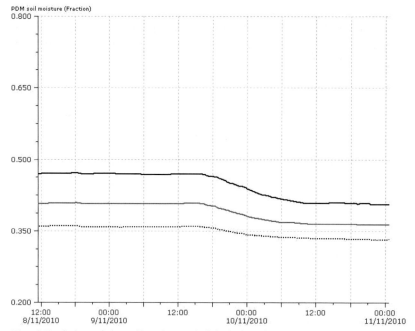

Fig. 4 Evolution of the soil moisture deficit during the event of November 2010 using raingauge rainfall (black line), radar rainfall (black dotted line) and recalibrated radar rainfall (grey line).

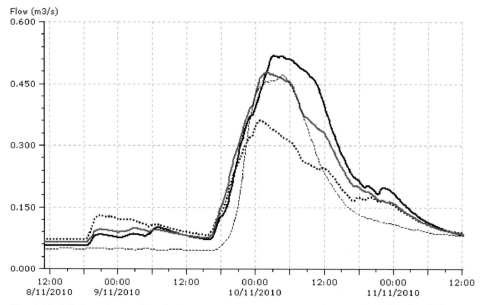

Fig. 5 Modelled catchment flow during the event of November 2010 using raingauge rainfall (black line), radar rainfall (black dotted line) and recalibrated radar rainfall (grey line) compared with the observed flow (dashed line).

flow in this event is underestimated. The hydrograph obtained when using the raingauge data as PDM input tends to overestimate the flow in both peak and volume.

To illustrate that the state of the catchment at the beginning of an event is largely influenced by the type of historical rainfall data used, a *forecast* is calculated at the end of the event discussed earlier. The "forecasted" rainfall is based on the observed recalibrated rainfall over the Vondelbeek catchment. This rainfall is used as input to the PDM which is initialized as described above, using the three different rainfall sources. Figure 6 shows the results for the forecasted period. Starting from an identical flow, the three hydrographs show different peak values and different volumes. Although the rainfall from the uncalibrated radar in the previous event was much less than the other forms of rainfall data, the peak flow is considerably overestimated by about 10%. Despite the overestimation in the previous event, the initialization using raingauge data results in an underestimation.

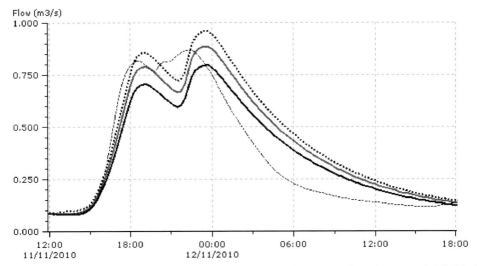

Fig. 6 Forecasted catchment flow after the event of November 2010 using raingauge rainfall (black line), radar rainfall (black dotted line) and recalibrated radar rainfall (grey line) to initialize the model. The observed flow is represented by a dashed line.

CONCLUSION

There is a large difference between the rainfall obtained directly from radar images and the rainfall obtained from raingauge-merged radar images. This has its consequences on hydrological and hydrodynamic modelling and flood forecasting. Flood forecasting in small catchments does really need radar rain data due to the spatial resolution. Without merging the raingauge data and the radar images, the raw radar rain data cannot give enough reliability to be used in a flood forecasting system.

REFERENCES

Cabus, P. (2008) River flow prediction through rainfall–runoff modelling with a probability-distributed model (PDM) in Flanders, Belgium. *Agricult. Water Manage.* 95, 859–868.

CEH (2007) *The Hyrad User Guide*. Technical publication. Centre for Ecology and Hydrology, Wallingford, UK.

De Jongh, I., Verhoest, N. & De Troch, F. (2006) Analysis of a 105-year time series of precipitation observed at Uccle, Belgium. *Int. J. Climatol.* 26(14), 2023–2032.

Delobbe, L. (2007) Schatting van de neerslag met behulp van een meteorologische radar. Scientific and technical publication. Royal Meteorological Institute, Belgium.

Goudenhoofdt, E. & Delobbe, L. (2009) Evaluation of radar-gauge merging methods for quantitative precipitation estimates. *Hydrol. Earth System Sci.* 13, 195–203.

Moore, R. J. (2007) The PDM rainfall–runoff model. *Hydrol. Earth System Sci.* 11(1), 483–499.

478

Weather Radar and Hydrology
(Proceedings of a symposium held in Exeter, UK, April 2011) (IAHS Publ. 351, 2012).

Comparison of raingauge and NEXRAD radar rainfall data for streamflow simulation for a southern Ontario catchment

ROHIT SHARMA[1], RAMESH RUDRA[2], SYED AHMED[2] & BAHRAM GHARABAGHI[2]

1 *Water Resources Analyst, Calder Engineering Ltd, Bolton, Ontario L7E 3B2, Canada*
2 *School of Engineering, University of Guelph, Guelph, Ontario N1G 2W1, Canada*
sahmed@uoguelph.ca

Abstract The aim of this paper is to compare simulated flows using radar rainfall inputs with those obtained using raingauge rainfall. Both versions of simulated flows are also compared with the observed streamflow. The differences in rainfall volume and spatial variability of these datasets were evaluated for 10 storm events at hourly and daily time-scales using the Hydrologic Engineering Center's Hydrologic Modelling System (HEC-HMS). The model was run in event-mode using the SCS Curve Number method and the Green and Ampt Infiltration method for a catchment in southern Ontario, Canada. For most of the events, the runoff hydrographs obtained using raingauge rainfall had better correlation with observed flows than those obtained using radar rainfall, or merged rainfall inputs. However, the merged rainfall gave better runoff simulations than those obtained using only-radar rainfall data. The use of "only" radar rainfall for hydrological modelling resulted in erroneous outputs. Therefore, adjustment of radar rainfall is important prior to its use for runoff simulations. The estimation of antecedent catchment conditions played a dominant role in the event simulations; therefore, the initial parameters should be carefully selected and calibrated. The SCS Curve Number option gave relatively better results in terms of runoff amount, peak flow-rate, and time to peak, than those obtained using the Green-Ampt Infiltration option.

Key words radar hydrology; hydrological modelling; HEC-HMS; heavy rainfall events

INTRODUCTION

A number of radar and raingauge comparison studies were conducted to understand the role of spatial accuracy of radar rainfall and point accuracy of raingauge rainfall, and its role in accurately simulating hydrological processes; however, no specific conclusions have been drawn until now (Skinner *et al.*, 2009). Lopez (2005) used an event-based rainfall–runoff model (Geomorphological Instantaneous Unit Hydrograph – GIUH; Rodriguez-Iturbe & Valdés, 1979) based on the Soil Conservation Service (SCS) method to show that the radar rainfall significantly improved the hydrographs and provided better estimates than those obtained using raingauge rainfall data. Bedient *et al.* (2000) used radar rainfall data with the HEC-1 model for predicting three independent storm events, and observed that the radar rainfall provided more accurate runoff simulations than those obtained using raingauge data. Zhijia *et al.* (2004) showed that corrected-radar rainfall was reasonably accurate for most of the catchments and the distributed modelling results showed that both radar and raingauge rainfall inputs produced runoff hydrographs of similar accuracy. Neary *et al.* (2004) used the HEC-HMS (US Army Corps of Engineers, 1998) model to simulate runoff parameters using radar and raingauge rainfall, and reported that radar rainfall simulations were generally less accurate than raingauge rainfall simulations, in terms of runoff volume, but predicted similar results in terms of magnitude and time-to-peak predictions. Cole & Moore (2008) compared the raingauge, radar, and gauge-adjusted-radar rainfall inputs using lumped and distributed hydrologic models, and concluded that the raingauge rainfall input provided better results than radar input.

The quantitative differences in these two data sources could be due to several factors such as local errors in radar and raingauge data, modelling simplicity or complexity, varying catchment sizes, choice of runoff generation mechanisms, or accuracy of catchment parameters and calibration; however, due to decreasing number of raingauges and spatial variability, there is a profound need for alternative rainfall measuring options (Sharma, 2009). In this study, radar and raingauge rainfall were estimated for 10 storm events, to be used as a different input data to a hydrological model for a southern Ontario catchment in Canada.

METHODOLOGY

Study area

The study was conducted in Upper Welland River Watershed (UWRW) of Niagara Peninsula Conservation Authority (NPCA), Ontario (Fig. 1). The drainage area of approximately 230 km² was discretized into 10 sub-basins with areas ranging from 6.5 to 43.5 km². The average annual precipitation and snow of the sub-basins are around 910 mm and 160 mm, respectively. Brown *et al.* (1980) estimated the regional mean annual actual evapotranspiration as 533 to 559 mm and mean annual water surplus around 279 mm. For the selected study period (2000–2004), the average annual precipitation was recorded as 872 mm, and the mean annual actual evapotranspiration as 559.6 mm. The primary Geographical Information System (GIS) data consists of a 10-m Digital Elevation Model (DEM), stream network, land cover information, and soil distribution layers (prepared by Ontario Ministry of Natural Resources (MNR), Ontario Ministry of Agriculture, Food and Rural Affairs (OMAFRA), and NPCA). The dominant Hydrologic Soil Groups (HSG) are HSG Type C and D, and approximately 85% of land is under crop and agricultural use.

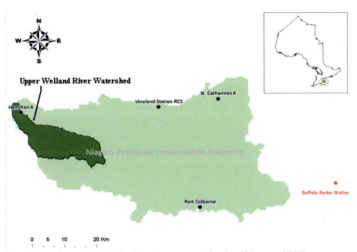

Fig. 1 Location of UWRW with raingauges and radar (Sharma, 2009).

Rainfall data

The raingauge at Hamilton Airport (Climate ID: 6153194) is the only gauge located in the Upper Welland River Watershed. For modelling needs, the Inverse Distance Weighted (IDW) interpolation method was used with data from three additional raingauges in the vicinity, to capture the spatial variability of rainfall events. The radar rainfall data were obtained from the NEXRAD Buffalo Radar Station in Buffalo, New York, USA. The catchment is approximately 150 km from the Buffalo Radar station. The GIS-based Digital Precipitation Array (DPA) product contained the precipitation accumulation in a 131 × 131 array of grid boxes at a grid resolution of approximately 4 km × 4 km. Radar captures the DPA rainfall values every 5 min during rainfall hours, and every 6–10 min during non-rainfall hours.

Selection of runoff events

Ten heavy rainfall–runoff events were selected from a 5-year database (2000–2004) for the comparison of simulated streamflow (excluding the base-flow) with the observed streamflow (Table 1). Main considerations for the selection of these events were: non-winter events, heavy rainfall, spatially covering the entire catchment, availability of both rainfall and observed streamflow information.

Table 1 Summary of rainfall characteristics for the 10 runoff events.

Start time	End time	Rainfall volume (mm)		% Difference
		Raingauge	Radar	
20 April 2000 (02:00)	22 April 2000 (08:00)	36.05	30.82	14.5%
11 May 2000 (17:00)	12 May 2000 (23:00)	18.38	24.60	−33.8%
11 June 2000 (08:00)	14 June 2000 (23:00)	69.90	44.49	36.4%
09 July 2000 (08:00)	09 July 2000 (17:00)	26.45	12.60	52.4%
22 September 2000 (19:00)	23 September 2000 (23:00)	26.79	15.92	40.6%
21 May 2001 (09:00)	22 May 2001 (22:00)	55.32	18.62	66.3%
11 May 2002 (17:00)	14 May 2002 (09:00)	43.90	32.31	26.4%
30 April 2003 (16:00)	02 May 2003 (23:00)	42.27	39.89	5.6%
02 November 2003 (04:00)	04 November 2003 (16:00)	29.62	29.61	0.0%
20 May 2004 (08:00)	25 May 2004 (18:00)	68.76	69.69	−1.4%

Estimation of model parameters

The estimation of model parameters for event-based hydrological models is a critical step due to uncertainty from initial soil-moisture conditions. Some of the parameters were directly estimated based on the catchment physical characteristics (e.g. average slope, sub-basin area, length of stream channels, imperviousness, and hydraulic conductivity). Some other parameters were estimated or calibrated during later stages (e.g. time of concentration, curve number, initial moisture, initial abstraction).

The SCS Curve Number method (SCS, 1972) is a widely used empirical parameter approach and is an efficient method for determining the approximate amount of direct runoff from a rainfall event. The runoff curve numbers are based on the hydrologic soil group, land use, treatment and hydrologic condition of the area. The estimation procedure is discussed in detail in National Engineering Hydrology – NEH-4 handbook (SCS, 1972).

The other modelling option, Green and Ampt Infiltration method (Green & Ampt, 1911) is a physical parameter approach and is a function of the soil suction head, porosity, hydraulic conductivity and time. These parameters were estimated using GIS-based soil and land use information, and previous NPCA reports (NPCA, 2007, 2009). Various related literature sources used for parameter estimation include Rawls et al. (1982, 1998) and Smemoe et al. (2004). The SCS Unit Hydrograph method was selected to simulate the surface runoff hydrographs. It is based on the dimensionless unit hydrograph approach, developed by Victor Mockus in the 1950s and is described in detail in SCS Technical Report (TR-55; SCS, 1986). The Muskingum-Cunge routing method was used for channel routing. The channel geometry was obtained from existing NPCA reports, and the roughness coefficient values were estimated based on Barnes (1967) and McCuen (1998). The channel side-slope and channel length were computed using GIS elevation data and stream network information.

Calibration of model parameters

Both manual and automated calibration options were used to achieve adequate modelling outputs. The manual calibration was done to identify the sensitive input parameters, followed by automatic calibration of the identified parameters to refine the model parameter values. First, the model parameters which affect the runoff volume were calibrated, followed by adjustment of parameters affecting the peak flow, and then time to peak of the events. The calibrated parameters were separately computed using rainfall inputs from raingauge and radar. These estimated parameters were then compared with each other, and in accordance with the antecedent catchment conditions, used to estimate a suitable set of input parameters. The finalized parameters were used to simulate the event streamflow using radar and raingauge rainfall inputs separately and compared with observed streamflow.

Evaluation criteria

ASCE (1993) emphasized the use of both visual and statistical comparisons between simulated and observed flows for a complete and effective comparison of streamflow. Therefore, the following criteria were selected in addition to the visual comparison: Percent Error in Peak (PEP) for the comparison of peak flow rates, Percent Error in Volume (PEV) for volumetric assessments, and Percent Error in Time to peak (PET$_P$) to account for timing errors.

RESULTS AND DISCUSSION

The model was simulated using radar and raingauge rainfall with the best estimated base parameters. Both radar and raingauge input models were calibrated separately, in order to compare the selected parameters with each other and in accordance with the antecedent moisture conditions and catchment information. Finally, the more realistic set of calibrated parameters was selected to run the model with each rainfall input, and the results were compared with each other. The simulated hydrographs were also compared with the observed streamflow. The calibration parameters for 19 out of 20 run-off events (10 SCS and 10 Green-Ampt events) were obtained when raingauge rainfall data were used as an input into the model. The modelling results are divided into two main sections: comparison of the three rainfall products, and comparison of SCS and Green-Ampt modelling options.

Comparison of radar, raingauge and merged radar rainfall

It was seen through the comparison of raingauge and radar rainfall inputs, that radar estimates different rainfall volumes than that obtained using raingauges for most events. In this study, radar underestimated rainfall for 80% of the runoff events. The differences in rainfall inputs are reflected in the simulated hydrographs. In order to take advantage of spatial variability of radar and point accuracy of raingauge rainfall, the two datasets were integrated into a single product, using the Mean Field Bias (MFB) correction in the radar rainfall data. The MFB coefficient was computed for each storm event by dividing the total raingauge rainfall by the total radar rainfall during that event. This coefficient was multiplied with the radar rainfall datasets to obtain the MFB-corrected radar rainfall (or merged radar rainfall). The runoff results estimated with merged rainfall input were significantly better than those obtained using "only" radar rainfall input. Approximately 20.5%, 35.2%, and 10.9% errors were estimated in peak flow, runoff amount, and time to peak computations, respectively, with the SCS modelling option, and 74.3%, 38.1, and 17.6% errors in peak flow, runoff amount, and time to peak, respectively, with the Green and Ampt option. Due to space limitation, scatter plots of only the SCS option are shown here (Fig. 2). The raingauge simulated and observed runoff have a good coefficient of regression of 0.88, while merged radar and "only" radar have R^2 values of 0.75 and 0.78, respectively. Similar results were obtained with the Green-Ampt modelling option, with a R^2 of 0.87 and 0.68 with raingauge and merged-radar rainfall inputs, respectively.

The plot of the observed and simulated runoff hydrographs for 11 May 2000 event is presented in Fig. 3. The results show that the "only" radar input simulation produced approx. 145.7% and 157.4% errors in peak flow and runoff amounts, respectively, while raingauge simulated hydrograph resulted in 5.8% and 2.5% errors in peak flow and runoff amounts, respectively. The MFB corrected radar rainfall produced better results than only-radar rainfall input, with only 18.3% and 9.5% errors

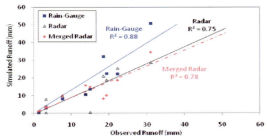

Fig. 2 Simulated *vs* observed runoff amounts with SCS modelling option for raingauge, radar, and merged-radar rainfall inputs.

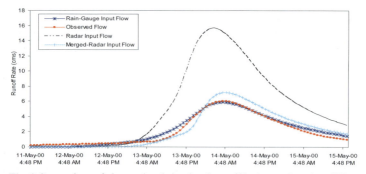

Fig. 3 Comparison of observed and simulated runoff hydrographs using different rainfall inputs for the 11 May 2000 storm event with the SCS modelling option.

Table 2 Percent errors in peak flow rate, runoff amount, and time to peak with SCS and Green-Ampt methods.

Components	SCS curve number method			Green and Ampt method		
	Raingauge	"Only" Radar	Merged-Radar	Raingauge	"Only" Radar	Merged-Radar
Peak flow (m³/s)	14.7	62.9	35.2	22.7	76.7	74.0
Runoff amount (mm)	8.7	59.5	20.5	14.8	154.6	38.1
Time (h)	1.4	20.2	10.9	4.4	15.0	17.6

in peak flow and runoff amounts, respectively. In terms of rainfall input, radar estimated 33.8% more rainfall than the raingauge rainfall total, which was reflected in the output runoff.

Overall, the comparison between the runoff simulations obtained using raingauge and radar rainfall showed mixed trends. For some of the events simulated with the SCS option (11 June 2000, 9 July 2000, 22 September 2000, 21 May 2001, 11 May 2002, and 2 November 2003), the runoff hydrographs computed using merged radar rainfall input showed lower biases in runoff amounts than other rainfall inputs, while for the remaining four events (20 April 2000, 11 May 2000, 30 April 2003, and 20 May 2004), the raingauge simulations showed more accurate results. For most of the events, the merged rainfall resulted in better hydrographs than "only" radar rainfall simulations, but raingauge input simulations were better than merged radar rainfall. In the case of the Green and Ampt modelling option, "only" radar rainfall simulated hydrographs were more accurate than the other rainfall inputs for 20 April 2000, 11 June 2000, 11 May 2002, and 20 May 2004 in terms of runoff volume. The merged radar rainfall simulated better runoff volumes for the 2 November 2003 event. For the remaining five events, the raingauge rainfall gave closer output to observed streamflow. Table 2 summarizes the average percent errors in runoff volume, peak flow rate and time to peak simulations, in comparison to observed flow with both SCS and Green-Ampt options.

Comparison of SCS and Green-Ampt modelling options

To compare the results obtained using the SCS method with the Green-Ampt modelling options, the Sum of Squared Residuals (SSR) and Root Mean Square Error (RMSE) were computed, in addition to visual comparison of the simulated and observed hydrographs. The SCS method gave better results than the Green-Ampt method for the 10 events tested in this study. The SCS method gave lower SSR and RMSE errors than the Green-Ampt option with either rainfall dataset. With the raingauge rainfall input, the SCS method resulted in approximately 773 mm SSR and 1.85 RMSE, while the Green-Ampt method errors are 985 mm SSR and 1.92 RMSE. Similar trends were noticed when radar rainfall was used, but the errors were higher with radar rainfall.

The differences in results are usually attributed to the physical and empirical nature of the SCS and Green-Ampt methods, base model parameters, calibration of the base model parameters, and the accuracy of the model parameters. The SCS method is a simple one-parameter approach, yet robust enough for estimating excess rainfall. However, the parameters required by the Green-Ampt method are physically measured and involve intensive and time consuming soil tests (such as hydraulic conductivity, soil-moisture deficit, and suction at wetting front). The accuracy and authenticity of the physical model parameters is a very critical aspect in event hydrologic modelling. In this study, the simple SCS method provided satisfactory results that validate the strength of the SCS method in hydrological modelling.

CONCLUSIONS

The main conclusions drawn from the study are: (1) Radar under-estimated rainfall for 80% of the events analysed in this study; (2) The SCS Curve Number approach gave relatively good results for the study catchment. Comparable results were obtained with the Green and Ampt infiltration approach. More distinctively, the runoff simulations obtained using the SCS method were better in terms of runoff amount, peak flow rate, and time to peak, than those obtained using the Green and Ampt infiltration model. Since the catchment has dominantly agricultural land and clayey soil type, the variation in SCS Curve Number was minimal. The R^2 value of 0.88, 0.75, and 0.78 were obtained with SCS Curve Number method, with raingauge, "only" radar, and MFB-corrected radar rainfall inputs, while with the Green and Ampt option, the R^2 values of 0.87 and 0.68 were obtained with raingauge and merged-radar rainfall inputs, respectively. (3) For most events, the runoff hydrographs simulated using raingauge rainfall, were better than those obtained using "only" radar rainfall or MFB-corrected radar rainfall. As one might expect, the overall biases in simulated and observed runoff amounts correspond to the biases in input rainfall volumes. (4) Merged radar rainfall input gave reasonable results for most rainfall events. Significant improvements were noticed when compared with radar rainfall simulations. This shows the importance of MFB-correction in radar rainfall data, before its use in hydrologic modelling.

REFERENCES

ASCE (1993) Criteria for evaluation of watershed models. ASCE task committee on definition of criteria for evaluation of watershed models of the watershed management committee. *J. Irrig. Drain. Engng* 119(3), 429–441.

Barnes, H. (1967) Roughness characteristics of natural channels. *US Geol. Survey Water Supply Paper 1849*.

Bedient, P., Hoblit, D., Gladwell, D. C. & Vieux, B. E. (2000) NEXRAD radar for flood prediction in Houston. *J. Hydrol. Engng* 5(3), 269–277.

Brown, D. D., McKay, G. A. & Chapman, L. J. (1980) The Climate of Southern Ontario, Toronto. *Environment Canada, Atmospheric Environment Service*.

Cole, S. J. & Moore, R. J. (2008) Hydrological modelling using raingauge- and radar-based estimators of areal rainfall. *J. Hydrol.* 358(3–4), 159–181.

Green, W. H. & Ampt, G. A. (1911) Studies on soil physics, 1. The flow of air and water through soils. *J. Agric. Sci.* 4(1), 1–24.

Lopez, V., Napolitano, F. & Russo, F. (2005) Calibration of a rainfall–runoff model using radar and raingauge data. *Adv. Geosci.* 2, 41–46.

McCuen, R. (1998) *Hydrologic Analysis and Design*. Prentice Hall, Upper Saddle River, New Jersey, USA.

Neary V. S., Habib, E. & Fleming, M. (2004) Hydrologic modelling with NEXRAD precipitation in Middle Tennessee. *J. Hydrol. Engng ASCE* 9(5), 339–349.

NPCA (2007). Flood Plain Mapping. Niagara Peninsula Conservation Authority, Niagara, Ontario.

NPCA (2009) Water availability study for the Upper Welland River Watershed Plan Area Niagara Peninsula Source Protection Area. *Niagara Peninsula Conservation Authority*, 1–44.

Rawls, W. J., Brakensiek, D. L. & Saxton, K. E. (1982) Estimation of soil water properties. *Trans. ASAE* 25, 1316–1320.

Rawls, W. J., Gimenez, D. & Grossman, R. (1998) Use of soil texture, bulk density and slope of the water retention curve to predict saturated hydraulic conductivity. *Trans. ASAE* 41, 983–988.

SCS (US Soil Conservation Service) (1972) *National Engineering Handbook*, Chapter 4, Hydrology.

Sharma, R. (2009) Simulation of streamflow using rainfall data from gauge and radar. MSc Thesis, University of Guelph, Guelph, Ontario, Canada.

Smemoe, C. M., Nelson, E. J. & Zhao, B. (2004) Spatial averaging of land use and soil properties to develop the physically-based Green and Ampt parameters for HEC-1. *Environ. Modelling and Software* 19, 525–535.

US Army Corps of Engineers, Hydrologic Engineering Center (1998) *HEC-HMS, Hydrologic Modelling System*, Version 1.1. Davis, California, USA.

USDA (1986) Urban hydrology for small watersheds. *Natural Resources Conservation Service, Conservation Engineering Division, Technical Release 55 (TR-55)*.

Zhijia, L., Wenzhong, G., Jintao, L. & Kun, Z. (2004) Coupling between weather radar rainfall data and a distributed hydrological model for real-time flood forecasting. *Hydrol. Sci. J.* 49(6), 945–958.

484

Weather Radar and Hydrology
(Proceedings of a symposium held in Exeter, UK, April 2011) (IAHS Publ. 351, 2012).

Potential of radar data for flood forecasting and warning in lowland catchments in Ireland

M. B. DESTA, F. O'LOUGHLIN & M. BRUEN

School of Architecture, Landscape and Civil Engineering, University College Dublin, and
Centre for Water Resources Research, Belfield, Newstead Building, Dublin 4, Ireland
michael.bruen@ucd.ie

Abstract This paper describes the development of a radar rainfall forecasting method and its use for flood forecasting using a data stream from Met Éireann's radar at Dublin Airport. It is applied to four relatively flat catchments of different sizes on the eastern side of Ireland. The first objective was to determine the value of the radar precipitation information for hydrological applications in general, and the second was to assess if there is added value in applying Quantitative Precipitation Forecasting (QPF). A TREC-type procedure was used to generate QPF. The precipitation estimates are compared to contemporaneous raingauge measurements and the discharge estimates are compared to measured river flows. Preliminary results suggest that, with a 15-min radar cycle, this extends the acceptable performance by only an additional 1 h of lead time. While this is significant for the smaller catchments, it is less so for catchments with longer lag times.

Key words radar; rainfall; flood forecasting; rainfall forecasting; QPF

INTRODUCTION

In Ireland, the main current policy driver on flooding is related to addressing the requirements of the EU Floods Directive. The Office of Public Works (OPW), the government body responsible for the issue of flood forecasting and warning in Ireland, is currently developing a pilot flood forecasting and warning system for the towns of Clonmel, Mallow and Fermony by using hydrological models to study the feasibility of such a system for the whole country. The Irish Meteorological Service (Met Éireann) uses a Numerical Weather Prediction (NWP) model known as HIRLAM (High Resolution Limited Area Model) for short-range weather forecasting (including floods). Up to now, neither the models developed by OPW nor the HIRLAM use radar data quantitatively for flood forecasting. The models developed by OPW rely on gauge rainfall information to run simulations for flood forecasting. Gauge rainfall is very useful for hydrological applications; however, it has limited application for forecasting, particularly for short range, which could be critical in catchments with relatively short response times. In contrast, radar rainfall estimates can improve the quality of advance warnings if suitable rainfall and flood forecasting methods are in place. It is known that radar measurement is also affected by several sources of error that may reduce its potential use for rainfall estimation (Gyuwon, 2006). Notwithstanding this, radar rainfall measurement can be a practical tool for flood forecasting (Mimikou & Baltas, 1996; Moore *et al.*, 2004) owing to its better spatial coverage. The meteorological radar at Dublin airport has good spatial resolution (1 km), but relatively coarse temporal resolution (15 min). The aim of the work reported here is to investigate the potential of precipitation estimates from this radar for flood forecasting and to examine if any added benefit can be obtained by using Quantitative Precipitation Forecasting (QPF) techniques. A previous study concentrated on a hilly area south of Dublin (Parmentier, 2004). In contrast, this study concentrates on four relatively flat catchments north of Dublin.

MATERIALS AND METHODS

Selected catchments

Four relatively flat catchments on the eastern side of Ireland were selected for this study: the Boyne, the Dee, the Glyde and the Nanny (Fig. 1). The Boyne catchment is predominantly pasture, characterised by flat to moderately undulating topography with poorly drained soils that impede drainage to the subsoil and many lakes that slow and attenuate flood runoff. The Boyne River is gauged at a number of locations which allowed division of the catchment into two sub-basins: the southern flat catchment gauged at Ballinter Bridge (area of 1572 km^2) and the northern hilly catchment gauged at Liscarten (area of 714 km^2). The Dee catchment is located northeast of the Boyne catchment. The River Dee is slow-flowing through Whitewood Lake in its upper reaches and fast flowing with pool and riffle areas in its middle reaches. The lower reach has sections of deep pools and slow flowing water to the sea. The River Dee is gauged at Charleville (area of 302 km^2). The Glyde catchment, located north of the Dee Catchment, has almost the same size as the Dee and is characterised by poorly drained soils and a range of grassland-based agricultural land uses. There are two small towns in the catchment: Carrickmacross in Co. Monaghan and Kingscourt in Co. Cavan. The flow data used in this study are gauged at Mansfield Town (area of 323 km^2). The Nanny Catchment lies to the north of Dublin city and east of the Boyne catchment. The catchment is rural and gently sloping with an area of 185 km^2 up to Duleek. There is a history of flooding at Duleek and Julianstown, both of which are on the banks of the Nanny River.

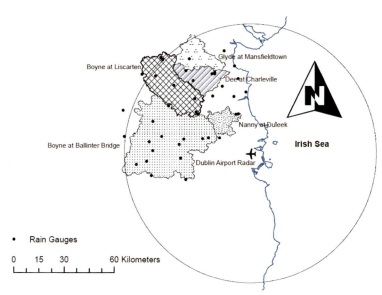

Fig. 1 Map showing location of selected catchments, raingauges and operational range (75 km) of the Dublin Airport Radar.

Characteristics of the data

Flow and rainfall data Daily rainfall data were obtained from Met Éireann for a total of 42 raingauge stations (shown in Fig. 1) all within or close to the study catchments. Two hourly recording stations (Mullingar and Clones) are outside the useful range of the Dublin Airport Radar and could not be included for comparison. Catchment rainfall averages were obtained by the Inverse Distance Weighting (IDW) technique. Fifteen-minute river flow data were obtained from the OPW and the Environmental Protection Agency (EPA).

Radar rainfall data The radar rainfall data used in this study were obtained from Met Éireann's C-band radar at Dublin Airport. The data stream received as a binary file is a pre-processed 15-min Precipitation Accumulation (PAC) over a constant altitude of 1 km above the topographical surface that provides rainfall intensity at 1 km square grid points over a 240-km range. Gauge adjustment factors are applied to the observed catchment average radar to match the total volume of catchment average gauge rainfall values on an annual basis. These factors are assumed to be constant in time (Gjertsen *et al.*, 2004; Germann e*t al.*, 2006), only changing annually. Continuous radar rainfall data were available from 2002 to 2010 with a total of 38 days missing in August 2003 and September 2004 due to breakdown of the radar equipment. Furthermore, there are a number of missing data points ranging from a few (15-min) time steps to a few hours.

Rainfall forecasting method

In this study we assessed the performance of a simple advection-based technique for precipitation nowcasting. This method translates the radar echo using multiple translation vectors determined by a cross-correlation technique (Austin & Bellon, 1974). The basic assumption behind such a nowcasting technique is that features in an image sequence only change shape and not size or intensity over short lead-times (steady state) termed as persistence (Obled *et al.*, 1994; Wilson *et al.*, 1998). The forecast is achieved by advecting the current reflectivity field using a velocity field without any change to the total intensity (Reyniers, 2008). Rectangular echo regions are isolated from the entire field. This method was called Tracking of Radar Echoes with Correlation (TREC) (Rinehart, 1981). To overcome "noisy" velocity vectors resulting from the TREC algorithm, an array of post-processing procedures has been implemented to obtain a reasonably "smooth" motion vector field. This includes filtering the radar image prior to correlation analysis. This filters out small scale patterns which are relatively difficult to forecast, as discussed in Wilson (2004). Further smoothing of the resulting correlation field is carried out by removing vectors that are anomalous when compared with the neighbouring and/or global vectors. Values removed by this procedure are replaced by averaging the neighbouring vectors. By this method, nowcasts were made in 15-min time-steps up to a forecast range of 2 h.

Rainfall–runoff models description

Two different conceptual hydrologic models are used so that the results are not specific to one model. These are the NAM and SMARG models which are described below.

NAM (Nedbor-Afrstromnings-Model) is a lumped conceptual rainfall–runoff model developed at the Technical University of Denmark (Nielson & Hansen, 1973). It has three main storage elements: an upper root zone store, a lower zone store and a groundwater store. The upper root zone represents the soil storage in the upper few centimetres of the surface, water intercepted by vegetation and surface depression storages. The lower zone is the soil moisture storage in the root zone available for vegetative transpiration. Groundwater storage is water stored in the water-bearing stratum capable of affecting surface flow. If a significant amount of snow falls within a catchment for a considerable period, the NAM model incorporates snow storage as a fourth storage unit. In this case temperature data are required as an additional input. The implementation of the model used here has 12 effective parameters.

Calibration of the NAM model involves adjustments of model parameters for the exchange of water between the different storage units to match the observed flow as best as possible. The objective function can be based on either the runoff volume, general shape of the hydrograph, the peak flows or the low flow depending on the problem under consideration (Madsen, 2000).

SMARG is a lumped quasi-physical conceptual rainfall–evaporation–runoff model, with distinct water-balance and routing components. It is a development of the "Layers" rainfall–runoff model introduced by O'Connell *et al.* (1970). Using a number of empirical and assumed relations which are physically plausible, it implements a water balance component over each time-step. The routing simulates the attenuation and diffusive effects of the catchment. This study uses a nine parameter variant of the SMARG model; five parameters control the water balance components, while the remaining four control the various routing components (Tan & O'Connor, 1996).

In this study, these two models were run at a daily time-step and automatically calibrated by minimising the overall runoff volume error. Data from 2002 to 2005 are used for calibration and 2006 and 2007 for validation. First, the models were calibrated using raingauge data and then using observed radar rainfall data. The resulting hydrographs were compared with the observed hydrograph to study the validity of radar rainfall. Next, the parameters obtained after calibrating the model using observed radar rainfall were used to forecast for 1 h and 2 h ranges and the resulting hydrograph was compared with the observed radar derived hydrograph to study the benefit of the utilised QPF technique.

RESULTS

Figure 2 shows comparison of the observed hydrograph with the simulated hydrographs using either radar rainfall or raingauges for the Dee Catchment. The hydrographs resulting from gauge rainfall and observed radar are plotted together. The Nash-Sutcliffe coefficient (NSC) (Nash & Sutcliffe, 1970) values are listed in Table 1. The graph shows that both the raingauge and the radar underestimated the flows and consequently the runoff volume, particularly at the high discharges. These errors are attributed to either the accuracy of the calibration parameters or the reliability of the rating curves used to calculate peak discharges. However, the timings of the peaks were more-or-less accurate. The hydrographs derived from the radar inputs for all the other catchments did not perform as well as those derived from raingauge data as indicated by the Nash-Sutcliffe efficiency. Note that the SMARG model performed slightly better than the NAM model as seen in the Nash-Sutcliffe efficiency.

Figure 3 shows comparison of hydrographs using forecasted rainfall against the observed in the Nanny catchment with the model calibrated using observed radar rainfall. The hydrograph resulting from the forecasted radar rainfall model performs as well as the observed radar-derived

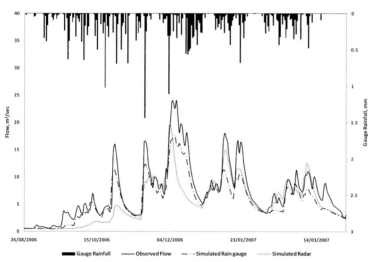

Fig. 2 Hydrographs of observed and simulated hydrographs using gauge and radar rainfall for the Dee catchment.

Table 1 Nash-Sutcliffe-Efficiency for the models using observed radar rainfall.

Model	Stage	Nanny		Dee		Glyde		Boyne at Liscarten		Boyne at Ballinter Br	
		Gauge	Radar	Gauge	Radar	Gauge	Radar	Gauge	Radar	Gauge	Radar
NAM	Calibration	0.74	0.66	0.89	0.72	0.90	0.67	0.93	0.58	0.86	0.71
	Validation	0. 65	0.57	0.77	0.59	0.77	0.66	0.84	0.59	0.76	0.52
SMARG	Calibration	0.84	0.72	0.93	0.82	0.90	0.70	0.88	0.69	0.89	0.78
	Validation	0.57	0.64	0.83	0.72	0.90	0.77	0.86	0.69	0.83	0.72

hydrograph, particularly for low to medium discharges. However, there are instances, such as shown in Fig. 3, where some peak discharges are underestimated. It can be said that the 1 h radar rainfall forecast performed as well as the observed radar rainfall. However, the 2 h forecast did not perform as well as the 1 h forecast. Note that that the 2 h forecast generates below zero efficiency in some of the catchments as shown in Table 2.

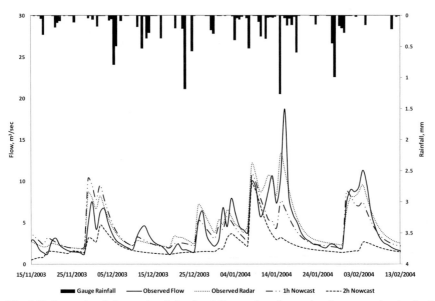

Fig. 3 Hydrographs of observed and simulated flows using observed and nowcasted radar for the Nanny Catchment.

Table 2 Nash-Sutcliffe Efficiency for the models using the observed and forecasted radar rainfall.

Model	Nanny			Dee			Glyde			Boyne at Liscarten			Boyne at Ballinter Br		
	Obs.	1 h	2 h	Obs.	1 h	2 h	Obs.	1 h	2 h	Obs.	1 h	2 h	Obs.	1 h	2 h
NAM	0.66	0.42	0.09	0.72	0.70	−0.40	0.67	0.71	0.55	0.58	0.51	0.09	0.71	0.65	0.31
SMARG	0.72	0.60	0.36	0.82	0.44	0.09	0.70	0.74	0.57	0.69	0.31	0.17	0.65	0.60	0.24

CONCLUSIONS

This study attempted to assess the use of radar rainfall for flood forecasting and if forecasts can be enhanced by forecasted rainfall using QPF techniques. Although the use of radar rainfall did not improve the quality of the forecasted hydrograph compared to a simulation using raingauge data, it can be used in places where there is shortage of rainfall data. Therefore, more work is required to increase the accuracy of radar rainfall measurement so that the resulting hydrograph can be comparable to the raingauge. The Dublin Airport radar is undergoing a refurbishment programme to reduce the sampling time from 15 to 5 min which is expected to increase the accuracy of radar rainfall measurement. The 1 h forecasted radar rainfall performs comparably with the observed radar. However, this same conclusion cannot be made for the 2 h forecast. This agrees with the findings of other researchers (Mecklenburg *et al.*, 2000; Berenguer *et al.*, 2005).

Acknowledgements This research is sponsored by Science Foundation Ireland.

REFERENCES

Austin, G. L. & Bellon, A. (1974) The use of digital weather radar records for short–term precipitation forecasting. *Quart. J. Roy. Met. Soc.* 100(426), 658–664.

Berenguer, M., Corral, C., Sánchez-Diezma, R. & Sempere-Torres, D. (2005) Hydrological validation of a radar-based nowcasting technique. *J. Hydrometeorol.* 6(4), 532–549.

Germann, U., Galli, G., Boscacci, M. & Bolliger, M. (2006) Radar precipitation measurement in a mountainous region. *Quart. J. Roy. Met. Soc.* 132(618), 1669–1692.

Gjertsen, U., Salek, M. & Michelson, D. (2004) Gauge adjustment of radar–based precipitation estimates in Europe. In *3rd European Conference on Radar Meteorology and Cost–717 Final Seminar.* Copernicus GmbH, Visby, Sweden, pp. 7–11.

Gyuwon, L. (2006) Sources of errors in rainfall measurements by polarimetric radar: variability of drop size distributions, observational noise, and variation of relationship between R and polarimetric parameters. *J. Atmos. Oceanic Technol.* 23, 1005–1028.

Madsen, H. (2000) Automatic calibration of a conceptual rainfall–runoff model using multiple objectives. *J. Hydrol.* 235(3–4), 276–288.

Mecklenburg, S., Bell, V. A., Moore, R. J. & Joss, J. (2000) Interfacing an enhanced radar echo tracking algorithm with a rainfall–runoff model for real–time flood forecasting. *Phys. Chem. Earth (B)* 25(10–12), 1329–1333.

Mimikou, M. A. & Baltas, E. A. (1996) Flood forecasting based on radar rainfall measurements. *J. Water Resour. Plann. Manage.* 122(3), 151–156.

Moore, R. J., Jones, A. E., Black, K. B. & Bell, V. A. (2004) Weather radar for flood forecasting: some UK experiences. In: *Sixth International Symposium on Hydrological Applications of Weather Radar*, Melbourne, Australia.

Nash, J. E. & Sutcliffe, J. V. (1970) River flow forecasting through conceptual models. Part 1: a discussion of principles. *J. Hydrol.* 10, 282–290.

Nielson, S. A. & Hansen, E. (1973) Numerical Simulation of the rainfall runoff process on a daily basis. *Nordic Hydrol.* 4, 171–190.

O'Connell, P., Nash, J. E. & Farrell, J. P. (1970) River Flow forecasting through conceptual models. Part 2. The Brosna catchment at Ferbane *J. Hydrol.* 10, 317–329.

Obled, C., Wendling, J. & Beven, K. (1994) The sensitivity of hydrological models to spatial rainfall patterns: an evaluation using observed data. *J. Hydrol.* 159(1–4), 305–333.

Parmentier, B. (2004) Evaluation of radar and rain gauge precipitation estimates and comparison of hydrological catchment models for flood analysis and simulation using a new GIS tool and a new z–transform based hydrological model. PhD Thesis, University College Dublin, Dublin, Ireland.

Reyniers, M. (2008) Quantitative Precipitation Forecasts based on radar observations: principles, algorithms and operational systems. *Royal Meteorological Institute of Belgium. Scientific Report no. 52.*

Rinehart, R. E. (1981) A pattern–recognition technique for use with conventional weather radar to determine internal storm motions. *Atmospheric Technology* 13, 119–134.

Tan, B. Q. & O'Connor, K. M. (1996) Application of an empirical infiltration equation in the SMAR conceptual model. *J. Hydrol.* 185, 275–295.

Wilson, J. W. (2004) Precipitation nowcasting: past, present and future. In: *Sixth International Symposium on Hydrological Applications of Weather Radar*, Melbourne, Australia.

Wilson, J. W., Crook, N. A., Mueller, C. K., Sun, J. & Dixon, M. (1998) Nowcasting thunderstorms: a status report. *Bull. Am. Met. Soc.* 79(10), 2079–2099.

Operational use of nowcasting methods for hydrological forecasting by the Czech Hydrometeorological Institute

LUCIE BŘEZKOVÁ[1], PETR NOVÁK[2], MILAN ŠÁLEK[1], HANA KYZNAROVÁ[2], MARTIN JONOV[3], PETR FROLÍK[2] & ZBYNĚK SOKOL[4]

1 *Czech Hydrometeorological Institute, Kroftova 43, 616 67 Brno, Czech Republic*
lucie.brezkova@chmi.cz
2 *Czech Hydrometeorological Institute, Na Šabatce 17, 143 06 Prague, Czech Republic*
3 *Czech Hydrometeorological Institute, K myslivně 3/2182, 708 Ostrava, Czech Republic*
4 *Institute of Atmospheric Physics ASCR, Bocni II, 1401, 141 31 Prague, Czech Republic*

Abstract The Czech Hydrometeorological Institute is the primary agency responsible for monitoring and forecasting of river stages at national level. In recent years, precipitation estimation and nowcasting tools derived from radar data were established. Together with hydrological models, these tools were tested for use in flash flood forecasting. The high uncertainty of predicting such a type of phenomena leads to using various nowcasting methods for estimation of predicted rainfall totals. This "variant-approach" was tested on case studies and is going to be set up for operational testing for pilot catchments. Detailed case studies of two extreme flash floods, which occurred on 24 June 2006, are presented and serve to demonstrate the possibilities and limitations of this method.
Key words flash flood; rainfall–runoff model; hydrological forecast; weather radar; heavy precipitation

INTRODUCTION

One of the tasks of the Czech Hydrometeorological Institute (CHMI) is the monitoring and forecasting of river stages at national level. Radar-based precipitation data have become a standard input into the hydrological model HYDROG (Starý & Tureček, 2000) for operational flood forecasting in the Morava and Odra river basins, located in the eastern part of the Czech Republic. Quantitative precipitation estimates derived by a radar–raingauge merging algorithm have been used as precipitation input to HYDROG since 2004. Quantitative nowcasting of precipitation based on COTREC (Novák, 2007), with adjustment derived from the merging algorithm, has been used as input precipitation for the first 3 h of the forecasted period since 2007.

Utilization of nowcasting of precipitation in hydrological models also resulted in the efforts of flash flood forecasting, see Šálek *et al.* (2006). The high uncertainty of forecasts associated with these phenomena also leads to employment of various nowcasting methods for estimation of predicted rainfall totals. Based on these precipitation scenarios a "poor man ensemble" of discharge forecasts can be obtained and evaluated. This approach has been motivated by the effort to effectively utilize all the (potentially) available tools and we acknowledge a certain lack of rigour in the proposed scheme.

The procedure was tested for a flash flood that took place on the evening of 24 June 2009. A squall line exhibiting features of a train effect (back-building of new storm cells that were then passing over the same area) moved over the southern part of the Odra catchment to the southwest (Fig. 1). The heavy precipitation caused two extreme flash floods in the subcatchments of the rivers Jičínka and Luha. This event happened at the beginning of a 12-day episode when a baric low, located over the eastern Mediterranean and Balkans, influenced the weather over central and southeastern Europe, causing a long series of severe flash foods. The total damage of this flash flood episode reached approximately 200 million euro and 15 people died.

In this paper we focus on the floods of the Jičínka and Luha catchments. Although the catchments are very close to one another and were hit by the same storm system, the predictability of these flash floods was quite different.

METHOD

The method of flash flood forecasting is based on (semi)continuous analysis of the actual state of the hydrological conditions in the catchment. Every 5 min the set of hydrological forecasts is produced using the quantitative precipitation estimates (QPEs), together with the available quantitative precipitation forecasts (QPFs). The QPEs based on observed precipitation derive from a combination of radar and raingauge measurements, the so called merged product (Šálek *et al.*, 2004; Šálek, 2010). These estimates are calculated for a 1 h time-step and are available 20 min after each hour. For the 5-min analysis, merged precipitation is replaced by precipitation derived from 5-min adjusted radar data with the adjustment coefficients taken from the latest possible calculation. The adjustment coefficients are obtained as a Kriging-interpolated (with nugget effect) array of gauge/radar ratios from the most recent (hourly) observations with a predefined threshold of at least 2 mm of accumulated precipitation. As predicted precipitation, the following nowcasting methods were used:

-   **COTREC** is a radar echo extrapolation technique which uses two consecutive (10- or 5-min step) maximum reflectivity images from the Czech Weather Radar Network (CZRAD) for calculation of motion vector fields; the QPFs are obtained by calculation of the "nowcast" PseudoCAPPI 2 km reflectivity in the area of the CZRAD domain, see Novák (2007).
-   **COTREC ext** is COTREC applied to the CZRAD extended domain that includes radar data from neighbouring countries, see Novák *et al.* (2010).
-   **Celltrack** is an algorithm initially used in the CHMI for identification of cells exhibiting high radar reflectivity, and for their tracking and extrapolation. Cells in the Celltrack algorithm are approximately defined as continuous areas of radar reflectivity equal to 44 dBZ or higher. The extrapolation forecast is made using movement vectors defining cell shifts between the last two radar measurements; see Kyznarová & Novák (2007).
-   **Celltrack ext (history)** This method is the same as Celltrack, but the extrapolation forecast is made using movement vectors derived by identification of the cell movement from a 1-h history with the weight decreasing for older movement vectors (Kyznarová & Novák, 2007).
-   **INCA cz** INCA is a software package developed by ZAMG which combines different inputs such as station data, radar information or NWP outputs, and uses its own algorithms to process the inputs and provide nowcasts and forecasts of meteorological quantities; for a more detail description see Haiden *et al.* (2011). In this case no NWP data were used since the maximum forecast time was 3 h.
-   **INCA ext** is the same as INCA cz, but incorporating measurements from Polish radars.
-   **COSMO1** The COSMO NWP model (Doms & Schattler, 2002), version 4.11, with a horizontal resolution of 2.8 km, is applied, and radar reflectivity data are assimilated using a water vapour correction method. The assimilation uses observed radar reflectivity and extrapolated radar reflectivity. The extrapolation is performed by the COTREC method over a period of 1 h; see Sokol (2011). COSMO1 utilises the standard one-moment microphysics of Lin-Farley-Orville type (Lin *et al.*, 1983) which considers five classes of hydrometeors (rain water, cloud water, snow, ice and graupel).
-   **COSMO2** uses two-moment microphysics developed by Seifert and Beheng (Noppel *et al.*, 2010), and in this case the set of hydrometeors is complemented by hail.
-   **Persistence** repeats the precipitation intensity derived from the last available radar measurement for the following hour. This procedure can simulate the back-building of the storm cell or the quasi-stationarity of the precipitation system.

The performance of the complex forecast system simulates the real operations as much as possible. In Table 1 the time availabilities of the nowcasting data are presented. We assume that the actual forecast is valid until the availability of the next forecast. All the nowcasting products except for the COSMO model are calculated operationally in CHMI; use of the COSMO model for local heavy precipitation events is the subject of research by the Institute of Atmospheric Physics in the Czech Republic.

As an extreme situation, also the variant of **zero precipitation forecast** was assumed as a member of the hydrological ensemble. This member shows the discharge caused by precipitation observed until the given time. The discharge forecast based on such an input precipitation can be considered as the bottom limit of the ongoing estimated discharge assuming complete cessation of the (nowcast) precipitation.

Table 1 Time availability of QPF in minutes after the start of the forecast (nowcasting) procedure.

Nowcasting type	Availability (min)	Nowcasting type	Availability (min)
COTREC	5	INCA cz	15
COTREC ext	10	INCA ext	15
Celltrack	10	COSMO1	15
Celltrack ext (history)	10	COSMO2	15

Table 2 Characteristics of the flash floods which hit the Jičínka and Luha catchments on 24 June 2009.

Catchment	Jičínka River	Luha River
Area (km^2)	95	96
Closing profile (river gauge)	Nový Jičín	Jeseník nad Odrou
Level of highest flood stage (m^3/s)	49	37
Q_{100} (discharge with return period of 100 years) in closing profile (m^3/s)	178	87
Flood extremity	$>>Q_{100}$	$>>Q_{100}$
Time of useful warning	17:30 UTC	19:00 UTC

The sources of error include deficiencies of the particular nowcasting scheme (e.g. nonlinear movement/development of the rainfall pattern *vs* linear extrapolation of the pattern or individual storms) and error of the extrapolated precipitation field; if the nowcasting of rainfall pattern were perfect, the precipitation nowcast would suffer from the same errors as the measured radar-based precipitation estimate. The latter error is partly corrected by the adjustment scheme utilizing the most recent observed estimates (by applying an array of correction factors from the recent hour(s)). Note that these errors are very difficult to quantify and, so far, no steps in this direction have been taken.

HYDROLOGICAL CASE STUDIES

On 24 June 2009 both Jičínka and Luha catchments were hit by extreme flash floods, information on which is summarised in Table 2. The radar-based QPEs were strongly affected by attenuation, especially in the Jičínka catchment (located east of the Luha catchment). Polish radars viewing the squall line from the north showed remarkably better performance, probably due to less pronounced attenuation since the nearest Polish radar at Ramza is approximately at the same range (80 km) as the nearest Czech radar at Skalky (90 km range) (Fig. 1).

Based on the various precipitation forecasts, hydrological simulations of the development of the flood wave at 5-min time-steps were made with the use of the distributed rainfall–runoff model HYDROG (as inputs the time series of rainfall intensities at 5-min time-steps were used). For every term of the forecast the peak discharge exceedance curves were calculated. The passage of the probability of the limit discharge exceedance (related to the full river bed) in time for both catchments is given in Figs 2 and 3.

DISCUSSION

The forecast for the Luha catchment was more successful for two reasons. First, the precipitation was observed better by the radar(s) as is obvious from the radar-(manual) gauge comparison (the factor of the basin-averaged uncorrected radar and the gauge-based (kriging) estimate were 0.40 and 0.34 for the Luha and Jičínka catchments, respectively (see Fig. 4). Second, the automatic raingauge located in the catchment provided reliable data of 15-min precipitation accumulations complying with the basic time-step of the INCA systems.

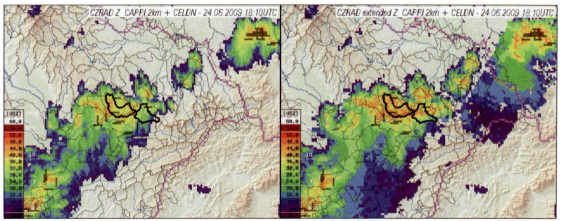

Fig. 1 Comparison of radar reflectivity given by the CZRAD network (Czech radars only – left) and the CZRAD extended network (including data from neighbouring radars, in this case from Poland – right). Jičínka (west) and Luha (east) catchments are depicted by black lines. The Skalky radar is located to the west-southwest of the catchments.

Fig. 2 The estimated probability of limit discharge exceedance in the Jeseník nad Odrou River closing profile in the Luha catchment. As the time proceeds the probability increases and reaches 100%.

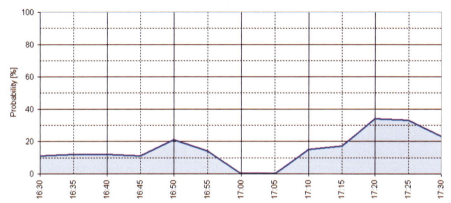

Fig. 3 The probability of limit discharge exceedance in the Nový Jičín River profile in the Jičínka catchment. The probability does not increase in time significantly.

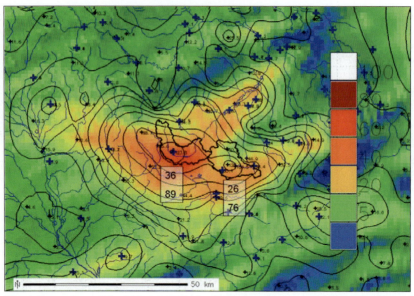

Fig. 4 Comparison of the 24 h uncorrected radar estimate (taken at 06:00 UTC 25 June 2009, in mm) and isolines of (all) raingauge areal estimate computed by Kriging (linear variogram, no nugget effect). The Luha and Jičínka catchments are emphasized in the middle of the picture (Luha catchment is on the left). The numbers in boxes are the mean areal precipitation (MAP) in millimetres from original radar (top) and kriging-interpolated gauge measurements (bottom) for both catchments. The telemetry gauges are marked by blue crosses and the (additional) manual gauges are denoted by smaller black crosses. The isolines (kriging gauge-based estimate) stems from the measurements of all the raingauges.

The forecasts for Jičínka catchment were less successful (lower probability of exceedance of the limit discharge). This is probably due to severe attenuation of the radar signal in the storms, which is not being explicitly corrected, and due to the absence of any telemetry raingauge within the core of the heaviest precipitation.

So far, four catchments have been tested on the proposed algorithm using data from 20 June to 20 July 2009: a season very rich in convective precipitation occurrence. Within this period three flash floods occurred. Each flood was predicted with a probability higher than 25%. However, six false alarms were produced as well. These results proved that, for each catchment, different rules of warning strategy should be applied. For example, concerning the Luha catchment the higher probability of limit discharge exceedance should lead to warning dissemination due to the better location of the catchment (radar visibility, presence of a telemetry raingauge nearby), while the warning for the Jičínka catchment should also be issued in case the low probability of limit discharge exceedance is predicted.

CONCLUSION

In this paper the algorithms for detection of flash flood occurrence have been presented. The first results proved their usefulness in real operation. However, the outlined "poor man ensemble system" should be fully evaluated only after longer (semi)operational testing. We also consider assigning the weights to the particular scenarios based on later tests of the performance.

The case studies presented confirmed the importance of quick exchange of radar data (that is being organized in the framework of the OPERA project) that can help, especially in cases of strong attenuation. Since the radar data are not particularly reliable in terms of absolute accuracy, they have to be complemented by raingauge measurements with a basic time-step not longer than 10 (or 15) min. We have to emphasize that in recent years the availability of these data has increased considerably, but in some areas the density of raingauge data can be considered as insufficient for satisfying results. The average density of the telemetry raingauges over the area of Fig. 4 is about 1 gauge per 154 km^2, and of all gauges 1 gauge per 66 km^2. The simulations also showed that some kind of adjustment is needed for radar-based nowcasting (e.g. COTREC) because it suffers from similar shortcomings to the radar-based QPE. The synthetic QPE from radar and raingauges, so far available operationally only at a 1-h step, is needed to be modified for shorter time intervals (to at least 10 or 15 min) in order to be better utilized for flash flood forecasting. Failure of radar-based (or radar-influenced) QPE can result in total failure of the hydrological forecast. This situation can be improved by installing telemetry raingauges in problematic areas prone to errors of radar-based QPE.

It is necessary to stress that the communication with end-users of the flash flood warnings is an important part of the whole forecasting process: the end users must be well trained and must understand the limitations of the methods used.

The proposed algorithms will be set in operation for selected catchments in summer 2012.

Acknowledgement This work was supported by the Central Europe Programme, INCA-CE project (co-financed by the European Regional Development Fund) and the Czech Republic Ministry of Education, Youth and Sport, Project ME09033.

REFERENCES

Doms, G. & Schattler, U. (2002) A description of the nonhydrostatic regional model LM. Deutscher Wetterdienst.

Haiden, T., Kann, A., Wittmann, C., Pistotnik, G. Bica, B. & Gruber, C. (2011) The Integrated Nowcasting through Comprehensive Analysis (INCA) system and its validation over the Eastern Alpine region. *Weath. Forecasting* 26, 166–183.

Kyznarová, H. & Novák, P. (2009) CELLTRACK – Convective cell tracking algorithm and its use for deriving of life cycle characteristics. *Atmos. Res.* 93, 317–327.

Lin, Y. L., Farley, R. D. & Orville, H. D. (1983) Bulk parameterization of the snow field in a cloud model. *J. Clim. Appl. Met.* 22, 1065–1092.

Noppel, H., Blahak, U., Seifert, A. & Beheng, K. D. (2010) Simulations of a hailstorm and the impact of CCN using an advanced two-moment cloud microphysical scheme. *Atmos. Res.* 96, 286–301.

Novák, P. (2007) The Czech Hydrometeorological Institute's Severe Storm Nowcasting System. *Atmos. Res.* 83, 450–457.

Novák, P., Frolík, P., Březková, L. & Janál, P. (2010) Improvements of Czech Precipitation Nowcasting System. In: *6th European Conf. on Radar in Meteorology and Hydrology (ERAD 2010)* (Sibiu, 6–10 September 2010).

Sokol, Z. (2011) Assimilation of extrapolated radar reflectivity into a NWP model and its impact on a precipitation forecast at high resolution. *Atmos. Res.* 100, 201–212.

Starý, M. & Tureček, B. (2000) Operative control and prediction of floods in the River Odra basin. In: *Flood Issues in Contemporary Water Management*, NATO Science Series, 2. Environmental Security – Vol. 71, Kluwer Academic Publishers, 229–236.

Šálek, M. (2010) Operational application of the precipitation estimate by radar and raingauges using local bias correction and regression kriging. In: *6th European Conf. on Radar in Meteorology and Hydrology (ERAD 2010)* (Sibiu, 6–10 September 2010).

Šálek, M., Březková, L. & Novák, P. (2006) The use of radar in hydrological modelling in the Czech Republic – case studies of flash floods. *Natural Hazards and Earth System Sci.* 6, 229–236.

Šálek, M., Novák, P. & Seo, D.-J. (2004) Operational application of combined radar and raingauges precipitation estimation at the CHMI. In: *Proc. ERAD 2004 2*, 16–20.

496

Weather Radar and Hydrology
(Proceedings of a symposium held in Exeter, UK, April 2011) (IAHS Publ. 351, 2012).

Flood nowcasting in the southern Swiss Alps using radar ensemble

KATHARINA LIECHTI[1], FELIX FUNDEL[1], URS GERMANN[2] & MASSIMILIANO ZAPPA[1]

1 *Swiss Federal Research Institute WSL, Zürcherstrasse 111, CH-8903 Birmensdorf, Switzerland*
kaethi.liechti@wsl.ch
2 *Federal Office for Meteorology and Climatology, MeteoSwiss, Via Monti 146, CH-6605 Locarno, Switzerland*

Abstract Since April 2007 the MeteoSwiss radar ensemble product REAL has been in operation and used for operational flash flood nowcasting by the WSL. REAL is computed for an area in the southern Swiss Alps where orographic and convective precipitation is frequent. These ensemble QPEs are processed by the semi-distributed hydrological model PREVAH. This provides operational ensemble nowcasts for several basins with areas from 44 to 1500 km^2 prone to flash floods and floods, respectively. Performances of discharge nowcasts driven by REAL are compared to performances of nowcasts forced by deterministic radar QPE and by interpolated raingauge data. We show that REAL outperforms deterministic radar QPE over the whole range of discharges, while the intercomparison with interpolated raingauge data is threshold dependent. Further we show that even though REAL nowcasts are underdispersive they have skill and can be a valuable means to produce hydrological nowcasts especially in ungauged catchments.

Key words radar ensemble; nowcasting; flash flood; flood; probabilistic; verification

INTRODUCTION

Mountainous catchments are often prone to flash floods, as their topography favours heavy convective precipitation events (Panziera & Germann, 2010). Due to shallow soils and steep slopes the runoff response time is generally short. Observations covering remote regions are typically scarce. Quantitative radar precipitation estimates (QPEs) are sometimes the only available information about precipitation in a remote area and provide valuable additional information about the spatial distribution of precipitation. However, uncertainties in radar QPEs for Alpine regions are large because of severe shielding of the radar beam by mountains, orographic precipitation mechanisms not fully seen by the radar, and strong mountain returns (clutter) (Germann *et al.*, 2006b). A promising solution to express the residual uncertainty in radar QPEs is to generate an ensemble of precipitation fields (e.g. Krajewski & Georgakakos, 1985). An example can be found in Aghakouchak *et al.* (2010), where three different remotely-sensed rainfall ensemble generators are compared.

The radar ensemble generator used in this study was developed at MeteoSwiss and was coupled to a hydrological runoff model (Germann *et al.*, 2009; Zappa *et al.*, 2011). This resulted in the first operationally running hydro-meteorological model chain using a radar rainfall ensemble and providing hydrological nowcasts at hourly time steps in a mountainous region (Zappa *et al.*, 2008).

During the 4 years of operational experience with this novel model chain, a unique dataset consisting of various types of input data (e.g. radar ensemble, deterministic radar, interpolated raingauge data, numerical weather predictions) and their resulting hydrological simulations could be acquired for validation.

In this study we present a probabilistic assessment of the performance of the radar ensemble (REAL) driven discharge nowcasts in comparison to nowcasts driven by interpolated raingauge data and deterministic radar QPEs. The analysed period ranges from April 2007 to December 2009.

STUDY AREA AND DATA

Test site

Alpine catchments are ideal sites to test a radar ensemble in combination with a hydrological runoff model. Due to their topography, persistent orographic precipitation combined with rapid runoff generation often leads to flash flood events associated with considerable damage (Germann *et al.*, 2009). Our experiments are focused on an area in the southern Swiss Alps including the four test catchments Pincascia, Verzasca, Maggia and Ticino, considered in this study (Fig. 1). The Pincascia with its 44 km^2 is a sub-catchment of the Verzasca. The Verzasca with an area of 186 km^2 down to the gauge in Lavertezzo is little affected by human activities. The Maggia catchment down to Locarno (926 km^2) is influenced by reservoir lakes in the upper part of the catchment for hydropower production. The Ticino catchment down to Bellinzona encompasses an area of 1515 km^2 and is also affected by water management in connection with hydropower production. Major infrastructure (highway, railway) is situated in the main valley.

All the catchments are characterised by snowmelt in spring and early summer and heavy rainfall events in autumn. Elevations for Pincascia, Verzasca, Maggia and Ticino range from 540 to 2500, 490 to 2900, 200 to 3300 and 220 to 3400 m a.m.s.l., respectively.

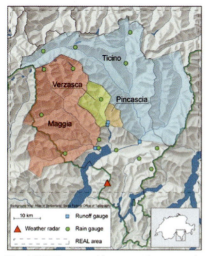

Fig. 1 REAL area with test catchments, raingauges and weather radar location.

REAL and other data

Weather radar offers the possibility to estimate precipitation at a high space and time resolution. But there are many sources of error that have to be dealt with. MeteoSwiss developed and implemented a series of algorithms to correct for several of the many errors inherent in radar reflectivity measurements. Although significant improvements were achieved, the residual uncertainty for hydrological applications is still relatively large (Germann *et al.*, 2006a).

The generation of an ensemble of radar precipitation fields is an elegant way to express the residual uncertainty in radar QPEs (Germann *et al.*, 2009). For the MeteoSwiss radar ensemble (REAL) the radar precipitation field is perturbed with correlated random noise. In this way the residual space–time uncertainty in the radar estimates is taken into account. The perturbation fields are generated by combining stochastic simulation techniques with detailed knowledge on the space-time variance and auto-covariance of radar errors (Germann *et al.*, 2006a). The radar ensemble is generated hourly with a spatial resolution of 2 km. The number of ensemble members is set to 25 (Germann *et al.*, 2009).

The meteorological data to run the hydrological model (air temperature, water vapour pressure, global radiation, wind speed, sunshine duration and precipitation) and the deterministic radar QPEs (1-km^2 spatial resolution) are also provided by MeteoSwiss. The raingauge data are interpolated over the test areas with inverse-distance weighting. The discharge measurements for verification are available in hourly time steps up to December 2009. For analysis, all the resulting nowcasts and the observed discharge time series were aggregated to series of daily maxima. This results in a database of 997 days for evaluation.

METHOD

Model chain

REAL is coupled to the semi-distributed hydrological model PREVAH (Viviroli *et al.*, 2009). Calibration and verification of PREVAH are presented in Wöhling *et al.* (2006) and Ranzi *et al.* (2007). This model chain produces operational nowcasts at hourly time steps ever since April 2007. For the period with available validated discharge data the model chain was re-run using REAL, interpolated raingauge data and deterministic radar QPEs as precipitation input (Fig. 2).

It is our goal to analyse the development of spread of the REAL ensemble QPEs over time. In numerical weather prediction, spread increases with lead time (Bartholmes *et al.*, 2009). By analogy, the spread of radar ensemble products is expected to change with the number of hours the precipitation ensemble is used for the generation of initial states for a subsequent nowcast.

We allow each of the 25 REAL members to build up a 10-day chain of spatially and temporally correlated precipitation values. Thus spread can develop from day to day. During long dry periods the spread can converge. During long wet periods the spread can grow. Our set-up starts from "day minus 10" with identical initial conditions for the hydrological simulations. In case of rainfall the 25 members of weather radar precipitation propagate separately through the hydrological model.

We repeated the 10-day simulations starting them consecutively each day from April 2007 until December 2009. This set-up allows us to create chains of discharge values where the forcing REAL-precipitation input has identical spread development time. These chains are evaluated with standard deterministic and probabilistic verification metrics as generally used for analysing ensemble discharge forecasts for different lead times (e.g. Jaun & Ahrens, 2009; Addor *et al.*, 2011).

Performance measures

For comparison of the performance of the deterministic nowcasts driven by interpolated raingauge data and radar QPEs and the probabilistic nowcasts driven by REAL, the Brier Skill Score (BSS) was chosen. The BSS is the most common measure for the verification of probabilistic forecasts of dichotomous events and allows a direct comparison of deterministic and ensemble predictions (Wilks, 2006). The BSS is based on the Brier Score (BS), which is essentially the mean squared error of the probability forecasts, given a dichotomous event (exceedence of a threshold or not), defined as:

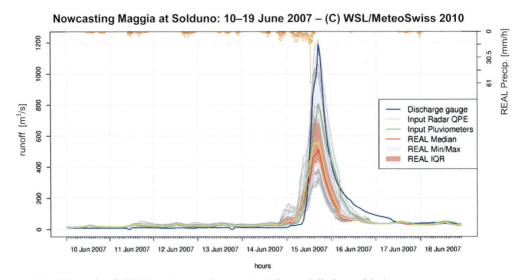

Fig. 2 Example of a REAL nowcast and *a posteriori* observed discharge (blue).

$$BS = \frac{1}{n} \sum_{d=1}^{n} (y_d - o_d)^2 \tag{1}$$

where y_d is the forecast probability of threshold exceedence (between 0 and 1) and o_d is the observed outcome (1 if threshold is exceeded, 0 if not) for forecast d of n assessed. The BSS is the improvement of the forecast in BS over that for a reference forecast BS_{Cl} (Wilks, 2006) such that:

$$BSS = 1 - \frac{BS}{BS_{Cl}}. \tag{2}$$

In this study the reference forecast is the climatological probability of exceedence of the predefined threshold. A perfect forecast has a skill of 1, whereas forecasts worse than the climatological forecast have a skill below 0.

The false alarm ratio (FAR) and the probability of detection (POD) are measures for deterministic predictions, and therefore the REAL ensemble was reduced to its median. The FAR is the fraction of positive forecasts (exceedence of threshold) that turn out to be wrong. The best FAR value is zero and the worst is one.

The POD is the ratio of correctly forecasted threshold exceedences to the number of times it actually happened. The best POD value is one and the worst is zero. The POD can always be improved by forecasting the event more often; however this usually results in higher FAR and for extreme events results in an overforecasting bias (Bartholmes *et al.*, 2009). So both the POD and FAR are shown.

Additionally, rank histograms (RH) provide information about the spread and bias of the REAL ensemble. All performance measures are calculated on the 0.8 and 0.95 quantile of the total available discharge time series for the different catchments (Ticino and Maggia 36 years, Verzasca 21 years and Pincascia 18 years).

RESULTS

The deterministic nowcasts driven by interpolated raingauge data and deterministic radar data are the same for all chains, which implies that the BSS, FAR and POD values are constant over the 10 chains (Fig. 3). The performance of the probabilistic REAL nowcasts vary over the 10 chains. The difference between the chains is however relatively small (Fig. 3).

BSS All nowcasts (except Maggia 0.8 quantile) show positive BSS values. The BSS values for the 0.95 quantile are higher than for the 0.8 quantile for all catchments and data types. From chain 8 to chain 10 BSS values for REAL decrease for all catchments. REAL outperforms deterministic radar over all thresholds, whereas a comparison to the performance of raingauge data is threshold dependent.

FAR For the 0.8 quantile the FAR achieved by raingauge data, deterministic radar and REAL lie very close together in all catchments. Very low FAR values are obtained for the Pincascia catchment, whereas the FAR for the Maggia is comparatively high (0.3–0.4). For Verzasca and Ticino FAR values are slightly higher for the 0.95 quantile than for the 0.8 quantile. For all catchments the REAL FAR increases from chain 8 to 10.

POD The POD for Verzasca, Maggia and Ticino are very similar. For the two threshold quantiles they lie between 0.6–0.7 and 0.7–0.9, respectively. From chain 8 to 10 REAL POD decreases in all catchments. Values for Pincascia are low compared to the other catchments.

Rank histogram The rank histograms show an underdispersion of the REAL ensemble for all catchments, most clearly for the 0.8 quantile. Additionally they show a clear underprediction for the Pincascia for both thresholds. The REAL ensembles for Verzasca and Ticino have a good average spread for the 0.95 quantile (Fig. 4).

DISCUSSION

The positive BSS values show that there is additional skill over a climatological forecast for all data types. The superiority of the REAL performance over the performance of deterministic radar

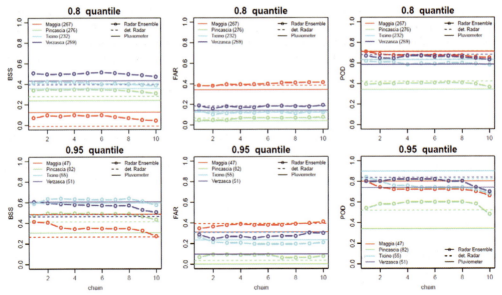

Fig. 3 Brier Skill Score (BSS), False alarm ratio (FAR) and Probability of Detection (POD) for the four test catchments calculated for the different precipitation input and the 0.8 and 0.95 discharge quantiles. The numbers in the brackets are the number of events reaching or exceeding the respective quantile.

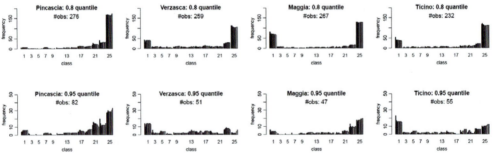

Fig. 4 Rank histograms of the REAL ensemble for the 0.8 and 0.95 quantiles. The thin columns inside each class represent different chains.

QPEs in all catchments is a clear argument for the probabilistic approach. An optimal spread development time for REAL cannot be deduced from the BSS, as the difference between the chains is small. However, the BSS, FAR and POD values all show a decline of performance from chain 8 to 10. This indicates that a spread development time of more than 8 days is not useful.

The comparison of the BSS values for the Verzasca and Pincascia catchments show the value of REAL for ungauged catchments. While in the case of the Verzasca for the 0.95 quantile the raingauge data achieves higher skill than REAL, it is clearly the opposite for the ungauged sub-catchment Pincascia (Figs 1 and 3). For the Pincascia the low FAR and the relatively low POD values are remarkable. This indicates that forecasted events exceeding the thresholds shown in Fig. 3 are reliable and thus will most probably occur. On the other hand, a big part of the events is missed by the forecasts. Due to the lack of a raingauge in the catchment, this is most pronounced for the POD values achieved by interpolated raingauge data. The same signal can be seen in the rank histograms which clearly show an underpredicting bias for the Pincascia for both thresholds.

The results for the Maggia catchment mirror the influence of water management for hydropower production. Precipitation of heavy storm events is often stored in the retention lakes and does not contribute to discharge. This results in low BSS and FAR and high POD values.

It is rather surprising that the scores for the 0.95 quantile are better than for the 0.8 quantile. A seasonal verification or a seasonal varying threshold may change this result. Finally it has to be acknowledged that the presented statistics rely on a series of 33 months only, which strongly limits the number of events exceeding the quantiles relevant for flash flood and flood events in these mountainous catchments.

CONCLUSIONS AND OUTLOOK

For catchments with sparse raingauge networks, REAL is a valuable solution to produce discharge nowcasts. We showed that REAL nowcasts outperform discharge nowcasts forced by deterministic radar QPEs on all thresholds. The REAL ensemble is underdispersive but shows skill nonetheless. We showed that the performance of nowcasts forced by REAL decreases after a spread development time of 8 days.

REAL can be used to generate initial conditions for a subsequent initialisation of the hydrological model with, for example, data from numerical weather prediction models (NWP; Zappa *et al.*, 2011). Such model coupling has already been tested with probabilistic and deterministic NWP products (COSMO-LEPS, COSMO-7, COSMO-2) for forecasts with lead times of a few days. Panziera & Germann (2010) describe a new radar ensemble product tailored for radar QPEs with short lead-times. Such products have the potential to enhance the reliability of nowcasts and short-term forecasts in mountainous areas and will soon be tested for hydrological applications.

Acknowledgements The experiments for the Pincascia and Verzasca and three of the authors are funded by the EU FP7 Project IMPRINTS (grant agreement no. 226555 / FP7-ENV-2008-1-226555). Experiments for the Maggia and Ticino basins are supported by a grant from the administration of Cantone Ticino through the Interreg IV Project FLORA. Discharge data are provided by the Swiss Federal office for the environment.

REFERENCES

Addor, N., Jaun, S. & Zappa, M. (2011) An operational hydrological ensemble prediction system for the city of Zurich (Switzerland): skill, case studies and scenarios. *Hydrol. Earth Syst. Sci.* 15, 2327–2347.

Aghakouchak, A., Habib, E. & Bardossy, A. (2010) A comparison of three remotely sensed rainfall ensemble generators. *Atmos. Res.* 98(2-4), 387–399.

Bartholmes, J. C., Thielen, J., Ramos, M. H. & Gentilini, S. (2009) The European Flood Alert System EFAS – Part 2: Statistical skill assessment of probabilistic and deterministic operational forecasts. *Hydrol. Earth Syst. Sci.* 13(2), 141–153.

Germann, U., Berenguer, M., Sempere-Torres, D. & Salvadè, G. (2006) Ensemble radar precipitation estimation — a new topic on the radar horizon. In: *4th European Conference on Radar in Meteorology and Hydrology* (Barcelona), 559–562.

Germann, U., Berenguer, M., Sempere-Torres, D. & Zappa, M. (2009) REAL – Ensemble radar precipitation estimation for hydrology in a mountainous region. *Quart. J. Roy. Met. Soc.* 135(639), 445–456.

Germann, U., Galli, G., Boscacci, M. & Bolliger, M., 2006. Radar precipitation measurement in a mountainous region. *Quart. J. Roy. Met. Soc.* 132(618), 1669–1692.

Jaun, S. & Ahrens, B. (2009) Evaluation of a probabilistic hydrometeorological forecast system. *Hydrol. Earth Syst. Sci. Discuss.* 6(2), 1843–1877.

Krajewski, W. F. & Georgakakos, K. P. (1985) Synthesis of radar rainfall data. *Water Resour. Res.* 21(5), 764–768.

Panziera, L. & Germann, U. (2010) The relation between airflow and orographic precipitation on the southern side of the Alps as revealed by weather radar. *Q. J. Roy. Meteor. Soc.* 136(646), 222–238.

Ranzi, R., Zappa, M. & Bacchi, B. (2007) Hydrological aspects of the Mesoscale Alpine Programme: Findings from field experiments and simulations. *Q. J. Roy. Meteor. Soc.* 133(625), 867–880.

Viviroli, D., Zappa, M., Gurtz, J. & Weingartner, R. (2009) An introduction to the hydrological modelling system PREVAH and its pre- and post-processing-tools. *Environ. Modell. Softw.* 24(10), 1209–1222.

Wilks, D. S. (2006) *Statistical Methods in the Atmospheric Sciences*. Elsevier, Amsterdam, 627 pp.

Wöhling, T., Lennartz, F. & Zappa, M. (2006) Technical Note: updating procedure for flood forecasting with conceptual HBV-type models. *Hydrol. Earth Syst. Sci.* 10(6), 783–788.

Zappa, M., Jaun, S., Germann, U., Walser, A. & Fundel, F. (2011) Superposition of three sources of uncertainties in operational flood forecasting chains. *Atmos. Res.* 100(2–3), 246–262.

Zappa, M., Rotach, M. W., Arpagaus, M., Dorninger, M., Hegg, C., Montani, A., Ranzi, R., Ament, F., Germann, U., Grossi, G., Jaun, S., Rossa, A., Vogt, S., Walser, A., Wehrhan, J. & Wunram, C. (2008) MAP D-PHASE: real-time demonstration of hydrological ensemble prediction systems. *Atmos. Sci. Lett.* 9(2), 80–87.

502

Weather Radar and Hydrology
(Proceedings of a symposium held in Exeter, UK, April 2011) (IAHS Publ. 351, 2012).

Assessment of typhoon flood forecasting accuracy for various quantitative precipitation estimation methods

TSUNG-YI PAN[1], **YONG-JUN LIN**[1], **TSANG-JUNG CHANG**[1,2,3], **JIHN-SUNG LAI**[1,2,4] & **YIH-CHI TAN**[1,2]

1 *Center for Weather Climate and Disaster Research*, 2 *Department of Bioenvironmental System Engineering*, 3 *Ecological Engineering Research Center*, 4 *Hydrotech Research Institute, National Taiwan University, Taipei, Taiwan, China*
tjchang@ntu.edu.tw

Abstract The main objective of the study is to obtain reliable rainfall estimates using raingauge and radar data. Different quantitative precipitation estimation (QPE) methods are tested and discussed, including (1) using kriging interpolation employing all raingauge data; (2) using radar products based on radar-reflectivity *vs* rain-rate (*Z-R*) formula; (3) using radar products adjusted by all raingauge data; and (4) using radar products adjusted by data from essential raingauges through network optimization with kriging. The estimated rainfalls are employed as the inputs for rainfall–runoff modelling. It is found that the QPE using radar products adjusted by all raingauge data provides superior performance. In addition, we found the estimation method using radar products adjusted by data from essential raingauges through network optimization with kriging can not only provide satisfactory results with efficiency for the spatial heterogeneity of rainfall distributions, but also simplify the raingauge network, reducing maintenance costs.

Key words quantitative precipitation; kriging interpolation; radar; raingauge; flood forecasting accuracy

INTRODUCTION

Taiwan is located in the hot zones of Pacific typhoon tracks. On average, 3.5 typhoons attack Taiwan annually. High rainfall amounts and intensities are induced by the interaction of typhoon rain bands and the mountainous terrains of Taiwan. To monitor rainfall intensities over the entire area of Taiwan, over 400 auto-recorded raingauges and four Doppler radars have been set up. However, the issue of how to obtain reliable rainfall estimates using raingauge and radar data still requires further investigation.

In August 2009, Typhoon Morakot produced a record-breaking rainfall, 2884 mm in 100 h, and it caused serious flood damage in the Gaoping River basin of southern Taiwan. In order to issue a flood warning accurately, rainfall–runoff models with observed rainfall are used to obtain modelled hydrographs. The purpose of this study is to evaluate flood-forecasting accuracy through various quantitative precipitation estimation (QPE) methods. Four scenarios with QPE methods are tested and discussed, including: (1) using kriging interpolation based on data from all raingauges in the basin, (2) using radar products based on a *Z-R* formula calibrated by the Central Weather Bureau (CWB) of Taiwan, (3) using radar products adjusted by data from all raingauges, and (4) using radar products adjusted by data from essential raingauges through network optimization with kriging.

METHODOLOGY

Kriging method

The kriging method is one of the geostatistical approaches applied to analyse the random nature described by a random field $Z(X)$, where X is the location for the random variable, Z, under observation, e.g. rainfalls from radar reflectivity, raingauges, or their difference. By means of the second-order stationarity and intrinsic hypothesis of regionalized variable theory, the spatial variation structure of a random field $Z(X)$ can be represented by its variogram as follows:

$$\gamma(h) = \frac{1}{2}\text{var}\big(Z(x_i + h_{ij}) - Z(x_j)\big) = \frac{1}{2}E\big(Z(x_i + h_{ij}) - Z(x_j)\big)^2 \tag{1}$$

where var(Z) and E(Z) are the variance and the expected value of the random variable Z, respectively. Equation (1) describes that the var(Z) and E(Z) only depend on the relative distance h

between the two points, and are independent of spatial locations. More details of variogram calculation can be found in Journel & Huijbregts (1978).

In this study, an exponential model is used as the theoretical model of the form:

$$\gamma_{ij}(h) = w\left[1 - \exp\left(-\frac{h_{ij}}{a}\right)\right] \tag{2}$$

where w is the sill and $3a$ is the practical range for the relative distance h_{ij} between two locations i and j.

Raingauge network evaluation

Based on the variance of the estimation error derived from cross-validation of ordinary kriging results for each location in a random field, the raingauge with the minimum cross-validation error is regarded as having the minimum influence on the raingauge network. Cheng *et al.* (2008) defined the acceptable accuracy for raingauge network evaluation through the kriging variance. The estimation error of each location in the random field is assumed as a normal distribution with zero mean and kriging variance, and the probability for the estimation error $\tilde{z}(x_0)$ to fall within a given range $(-k\sigma_z, k\sigma_z)$ can be determined through the cumulative probability of the standard normal distribution as follows:

$$\mathrm{P}\left(\left|\hat{z}(x_0) - z(x_0)\right| < k\sigma_z\right) = \mathrm{P}\left(\tilde{z}(x_0) < k\sigma_z\right) = \mathrm{P}\left(\tilde{z}(x_0)/\sigma_K < k\sigma_z/\sigma_K\right) = \mathrm{P}\left(z < k\sigma_z/\sigma_K\right) = \alpha \tag{3}$$

where σ_z is the standard deviation of a random field $Z(X)$, and α is the cumulative probability. The multiplier k in this study is set as 1. For more details of defining the acceptable accuracy, readers are referred to Cheng *et al.* (2008).

A semi-distributed hydrological model

Following on from the principle of the instantaneous unit hydrograph (IUH) with a series of n identical linear reservoirs proposed by Nash (1958), a semi-distributed parallel-type linear reservoir rainfall–runoff model developed by Hsieh & Wang (1999) is used for simulating the typhoon flood. The advantage of the semi-distributed hydrological model is that the entire basin can be separated into several sub-basins with different storage constants calculated from path length, slope, roughness coefficient, and rainfall intensity of each sub-basin. Therefore, the hydrograph simulated by the model is associated with the spatial resolution of rainfall from different sources. For more details of the semi-distributed parallel-type linear reservoir rainfall–runoff model, readers are referred to Hsieh & Wang (1999).

CAST STUDY

Flowchart

For assessing the typhoon flood simulation accuracy for various QPE methods, the study follows the flowchart design shown in Fig. 1. The flowchart circled by a dotted line is the QPE component of a QPESUMS system developed by the CWB of Taiwan. Readers are referred to Chiou *et al.* (2005) and Zhang *et al.* (2007) for a detailed description of the QPESUMS system that integrates data from radar, satellite and raingauge to perform QPE. The QPE adjusted by objective analysis is considered as the true value of precipitation used to calibrate the semi-distributed hydrological model herein. The QPE without adjustment is applied not only to the semi-distributed hydrological model, but also to be adjusted by the kriging of residuals which are added to the radar QPE for raingauge network evaluation. Therefore, four QPE sources mentioned previously are evaluated for typhoon flood simulation through the hydrological model.

Gaoping River basin and Typhoon Morakot

As the largest river in Taiwan, the Gaoping River basin located in southwest Taiwan is 3256.85 km² in area, and the length of the main stream is 171 km. With an area of 2809.1 km², the

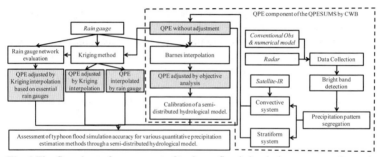

Fig. 1 The flowchart of assessment of typhoon flood simulation accuracy for various QPE methods.

(a)   (b)

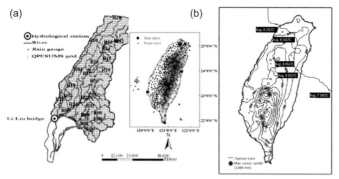

Fig. 2 (a) The upstream sub-basin of Li-Lin Bridge in the Gaoping River basin with the cells of the semi-distributed hydrological model, and the locations of raingauges and radar sites in Taiwan; (b) the track and the cumulative rainfall of Typhoon Morakot from 5 to 10 August 2009.

Table 1 Location and elevation of study raingauges.

Raingauge No. in Fig. 2	Official ID	Name	Location (TM97) X	Y	Elevation (m)
R01	C0R100	Wei-Liao-Shan	216727.14	2526081.87	1018.0
R02	C0R150	San-Ti-Men	212159.94	2512466.10	59.0
R03	C0V250	Chia-Hsien	207224.87	2553420.93	270.0
R04	C0V310	Mei-Nung	199843.72	2533387.82	46.0
R05	C1R090	Li-Kang	197497.29	2522414.32	42.0
R06	C1R110	Ku-Hsia	212746.74	2519047.21	140.0
R07	C1R120	Shang-Te-Wen	218821.12	2518326.06	820.0
R08	C1R130	A-Li	222839.28	2516042.13	1040.0
R09	C1R140	Ma-Chia	217004.73	2509440.49	740.0
R10	C1V160	Min-Sheng	219024.92	2575539.66	1040.0
R11	C1V170	Pai-Yun	244523.00	2595841.44	3340.0
R12	C1V190	Nan-Tien-Chih	240138.72	2574864.54	2700.0
R13	C1V200	Mei-Shan	231100.63	2574104.33	860.0
R14	C1V210	Fu-Hsing	229303.26	2569154.26	700.0
R15	C1V220	Hsiao-Kuan-Shan	230060.73	2561586.04	1781.0
R16	C1V230	Kao-Chung	220186.34	2559480.09	760.0
R17	C1V240	Hsin-Fa	214389.43	2550295.32	470.0
R18	C1V260	Yueh-Mei	201978.22	2541409.37	112.0
R19	C1V270	Hsi-nan	229337.63	2552943.43	1792.0
R20	C1V300	Yu-Yu-Shan	219957.68	2544777.07	1637.0
R21	C1V320	Chi-Tung	203728.66	2527038.75	95.0
R22	C1V330	Chi-Shan	197270.82	2531089.47	63.0
R23	C1V340	Ta-Chin	212864.58	2532181.28	190.0
R24	C1V460	Nan-His	238192.16	2592830.56	1949.0

upstream basin of Li-Lin Bridge in the Gaoping River is selected as the study area as shown in Fig. 2(a). There are 24 raingauges in the study area as listed in Table 1, and one hydrological station is located at Li-Lin Bridge. In order to highlight the spatial resolution of various QPE methods, the study area is separated into 52 cells for building the semi-distributed hydrological model as shown in Fig. 2(a). For analysing the spatial variation in the rainfall structure of the study area, 13 typhoon events listed in Table 2 are used to perform the semivariogram analysis that is then applied in the raingauge network analysis and spatial rainfall interpolation. Finally, Typhoon Morakot is selected as a case study case to evaluate the qualities of various QPEs for hydrological modelling, on account of the record-breaking rainfall, 2884 mm in 100 h, as shown in Fig. 2(b), causing the most serious flood and landslide disaster in southwest Taiwan in 50 years.

Table 2 The profile of typhoons applied to semivariogram analysis.

Year	Typhoon	Warning time	(A)		(B)		Year	Typhoon	Warning time	(A)		(B)	
			(A1)	(A2)	(B1)	(B2)				(A1)	(A2)	(B1)	(B2)
2009	Parma	10/03~10/06	945	43	250	80	2007	Mitag	11/26~11/27	955	35	200	80
2009	Morakot	08/05~08/10	955	40	250	100	2007	Wipha	09/17~09/19	935	48	200	80
2009	Linfa	06/19~06/22	980	28	150	–	2007	Sepat	08/16~08/19	920	53	250	100
2008	Jangmi	09/26~09/29	925	53	280	100	2007	Wutip	08/08~08/09	992	18	100	–
2008	Hagupit	09/21~09/23	940	45	280	100	2006	Shanshan	09/14~09/16	945	48	200	80
2008	Sinlaku	09/11~09/16	925	51	250	100	2006	Kaemi	07/23~07/26	960	38	200	80
2008	Fungwong	07/26~07/29	948	43	220	80							

Column (A): Profile of typhoon centre; (A1): Lowest central pressure (hPa); (A2): Max wind speed (m/s).
Column (B): Radius of wind speed (km); (B1): over 14m/s; (B2): over 25m/s.

RESULTS AND DISCUSSION

Raingauge network evaluation

After fitting the experimental semivariograms via exponential models, the sills for the above three data types are 0.9, 1.2 and 1.1 mm^2, and the influence ranges are 12.9, 33.8 and 20.7 km as shown in Fig. 3. The sills are near one because of the normalization of data in each time-step. The influence ranges show the spatial correlation structures of the three data types. The QPE without adjustment is directly converted from a Z-R relation formula. The shortest influence range implies the largest variability of radar data than that of the other two data types in space. The largest variability implies that the radar data are appropriate to describe the spatial rainfall distribution at high-resolution.

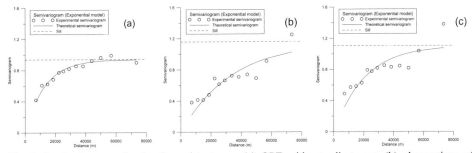

Fig. 3 Semivariograms for the three data types: (a) QPE without adjustment; (b) observations of raingauges; (c) the difference between observations of raingauges and QPE without adjustment.

During the raingauge network evaluation, the acceptable accuracy α is set as 0.9 in this study. The reduction priority of 24 raingauges is listed in Table 3. The average accuracy and the percentage of area passing the acceptable accuracy after removing the raingauge are also calculated. Table 3 shows that the density of raingauges in the upstream sub-basin of Gaoping River is very high because with only six raingauges (* mark in Table 3) the average accuracy remains over 0.9 and with over 98.2% of the area passing the acceptable accuracy. In Fig. 4, the area with accuracy over 0.9 covers most of the upstream basin with a six-raingauge network while a five-raingauge network keeps only about half of the basin with the accuracy over 0.9. This indicates that the six-raingauge network could be used as the essential raingauge network herein.

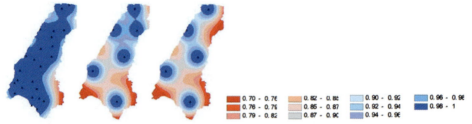

Fig. 4 The acceptable accuracy for raingauge network evaluation: the blue area is higher than 0.9 while the red area is lower than 0.9.

Table 3 The profile of typhoons applied to semivariogram analysis.

Reduction priority	Raingauge name	Accuracy	Percentage of area pass criteria (>0.9)	Reduction priority	Raingauge name	Accuracy	Percentage of area pass criteria (>0.9)
1	Mei-Nung	0.977	98.2%	13	Hsiao-Kuan-Shan	0.962	98.2%
2	San-Ti-Men	0.977	98.2%	14	Chia-Hsien	0.956	98.2%
3	Shang-Te-Wen	0.977	98.2%	15	Yu-Yu-Shan	0.944	97.8%
4	Fu-Hsing	0.976	98.2%	16	Ma-Chia	0.941	97.2%
5	Hsin-Fa	0.976	98.2%	17	Wei-Liao-Shan	0.927	96.0%
6	Chi-Tung	0.976	98.2%	18	Li-Kang	0.909	91.4%
7	Chi-Shan	0.975	98.2%	19*	Nan-Tien-Chih	0.888	87.0%
8	Mei-Shan	0.974	98.2%	20*	Min-Sheng	0.860	72.5%
9	Ku-Hsia	0.973	98.2%	21*	Nan-Hsi	0.793	57.7%
10	Ta-Chin	0.971	98.2%	22*	Hsi-nan	0.684	30.6%
11	Kao-Chung	0.969	98.2%	23*	Yueh-Mei	0.518	10.1%
12	Pai-Yun	0.966	98.2%	24*	A-Li	0.000	0.0%

Rainfall–runoff simulation

With four QPEs from different data sources, interpolation algorithms, and raingauge networks as input to the semi-distributed hydrological model, the simulated hydrographs are illustrated in Fig. 5, and their performance is listed in Table 4 using the following statistics: Coefficient of Efficiency (Nash-Sutcliffe) *CE*, % peak error EQ_p, peak timing error in hours ET_p, % error in discharge volume *VER*, and correlation of observed and simulated discharge *COR*. The analytic results show that QPE E in Table 4 gives underestimated results and with the worst performance among the four cases. It is reasonable because the QPE without adjustment is carried out from the Z-R formula of $Z = 300R^{1.4}$ which is only suitable for stratiform precipitation, not for orographic precipitation. Due to the high density of the raingauge network, the performances of QPE A and QPE D are close. However, the EQ_p of case QPE B and C is larger than that of QPE A because the average precipitations of the six raingauges are greater than that of all raingauges. The comparison between QPE B and C shows the added value of the radar. Although the *COR*s remain the same, the *CE* and EQ_p of QPE B are superior to those of QPE C. Furthermore, the *CE* is improved dramatically from –0.28 to 0.91 after the essential six-raingauge network is applied to adjust the QPE only via a Z-R relation formula. Consequently, the six-raingauge network which is only one quarter of all raingauges can not only provide a satisfactory typhoon rainfall simulation, but also simplify the raingauge network so reducing maintenance costs.

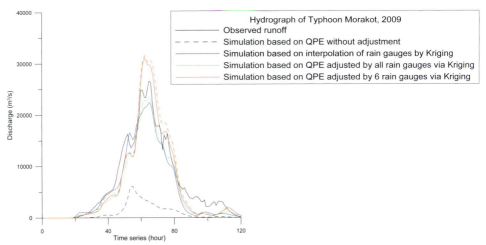

Fig. 5 The typhoon flood simulation using various QPEs.

Table 4 The performance of typhoon flood simulation using various QPEs.

	QPE A	QPE B	QPE C	QPE D	QPE E		QPE A	QPE B	QPE C	QPE D	QPE E
VER	21.13	5.72	−1.76	18.95	85.19	*EQp*	−12.11	16.98	17.60	−15.64	−76.92
CE	0.92	0.91	0.90	0.92	−0.28	*ETp*	0.00	−3.00	−3.00	0.00	−11.00
COR	0.98	0.97	0.97	0.98	0.84						

QPE A, the QPE adjusted by all raingauges via kriging; QPE B, the QPE adjusted by six raingauges via kriging; QPE C, the QPE interpolated by six raingauges via kriging; QPE D, the QPE interpolated by 24 raingauges via kriging; QPE E: the QPE without adjustment.

CONCLUSION

This study evaluates the qualities of four QPEs for flood simulation accuracy of Typhoon Morakot (2009) in the Gaoping River basin of Taiwan. Through semivariogram evaluation, the QPE directly converted from a *Z-R* formula has the largest variability in space than the other two data sources that include less information of spatial variability. Therefore, the radar data are considered to monitor the spatial rainfall distribution better than conventional raingauges. Furthermore, a semi-distributed hydrological model is adopted to simulate rainfall–runoff processes for use in evaluating the qualities of four QPEs for typhoon flood simulation accuracy. It is found that the QPE using radar products adjusted by all raingauges has superior performance. In addition, the QPE using the radar products adjusted by data of the essential six-raingauge network through network optimization with kriging can not only provide satisfactory results for the spatial heterogeneity of rainfall distributions, but also simplify the raingauge network so reducing maintenance costs.

REFERENCES

Cheng, K. S., Lin, Y. C. & Liou, J. J. (2008) Rain-gauge network evaluation and augmentation using geostatistics. *Hydrol. Processes* 22, 2554–2564. Doi: 10.1002/hyp.6851

Chiou, T. K., Chen C. R., Chang, P. L. & Jian, G. J. (2005) Status and outlook of very short-range forecasting system in Central Weather Bureau, Taiwan. In: *Applications with Weather Satellites II* (ed. by W. Paul Menzel & Toshiki Iwasaki), Proc. SPIE 5658, 185–196.

Hsieh, L. S. & Wang, R. Y. (1999) A semi-distributed parallel-type linear reservoir rainfall–runoff model and its application in Taiwan. *Hydrol. Processes* 13, 1247–1268.

Journel, A. G. & Huijbregts, C. J. (1978) *Mining Geostatistics*. Academic Press, London, UK.

Nash, J. E. (1958) The form of the instantaneous unit hydrograph. IAHS Publ. 45, 114–121. www.iahs.info

Zhang, J., Howard, K., Chang, P. L., Chiou, P., Chen, C. R., Langston, C., Xia, W. & Kaney, B. (2007) High-resolution QPE system for Taiwan. In: *Conf. Mesoscale Meteorol. and Typhoon in East Asia* (ICMCSI), 6–9 Nov., Taipei, Taiwan.

508

Weather Radar and Hydrology
(Proceedings of a symposium held in Exeter, UK, April 2011) (IAHS Publ. 351, 2012).

Ensemble nowcasting of river discharge by using radar data: operational issues on small- and medium-size basins

F. SILVESTRO & N. REBORA

CIMA research foundation, Savona, Italy
francesco.silvestro@cimafoundation.org

Abstract Many efforts have been made in order to improve the reliability of quantitative precipitation estimation and to use radar data to forecast future rainfall evolution through nowcasting systems. From this perspective the use of stochastic nowcasting algorithms plays a key role both for taking into account the uncertainty associated with the prediction of rainfall and for generation of possible short-term evolution of the precipitation field. Propagation of the uncertainty to ground effects by using a rainfall–runoff model is a further step to completely exploit the weather radar systems when forecasting the consequences of severe events. We created a nowcasting chain for generating discharge scenarios based on the following procedures: (1) an algorithm for observed rainfall estimation; (2) an algorithm for probabilistic nowcasting (PhaSt); and (3) a distributed hydrological model (DRiFt). Some examples of application in an operational context on small-and medium-sized basins are presented

Key words discharge; flood; nowcasting; ensemble; probabilistic

INTRODUCTION

The phase of monitoring of severe precipitation events is a fundamental part of the work of the hydro-meteorologists involved in Civil Protection activities. The small-scale precipitation structures are still impossible to predict with sufficient precision in terms of localization, time and intensity. Through the forecast process it is only possible to predict what is going to happen on quite a large-scale, but not the occurrence at basin-scale, especially when we deal with small and medium-sized catchments.

The use of rainfall observations is the main starting point to predict the future. Meteorological radars allow capture of the rainfall spatial structure and a great number of nowcasting algorithms have been designed with the aim of forecasting the future rainfall starting from the most recent available observations (Li & Schmid, 1998; Germann & Zawadski, 2002; Seed, 2003).

Various studies investigated the benefits derived by the use of radar rainfall estimation as input to rainfall–runoff models in order to produce simulated hydrographs (Pessoa *et al.*, 1993; Borga, 2002; Kouwen *et al.*, 2004). Recently, some works have been done with the same purpose, but following a probabilistic approach (Germann *et al.*, 2009).

Other authors tried to couple nowcasting techniques and hydrological models with the objective of extending the lead time of hydrological forecasts, by improving the knowledge of future rainfall (Berenguer *et al.*, 2005; Salek *et al.*, 2006; Vivoni *et al.*, 2006).

In this work we try to carry out a similar objective, but following a probabilistic approach. The framework we propose makes a coupling of a stochastic nowcasting algorithm and a rainfall–runoff model: this allows production of an ensemble of discharge scenarios to be used for hydrological nowcasting. An algorithm for optimal observed rainfall estimation is also introduced.

Many works have been devoted to studying the effects of uncertainties in radar rainfall estimation when used as input to a rainfall–runoff model (Carpenter & Georgakakos, 2006a,b; Schroter *et al.*, 2011). The work we present focuses on an attempt to account for the uncertainties related to the forecast rainfall obtained by the nowcasting algorithm; we do not deal with the uncertainties of rainfall–runoff model parameterization and with those associated to the observed rainfall estimation. Detailed investigations about this issue can be found in Zappa *et al.* (2011) and in Carpenter & Georgakakos (2006b).

The paper is organized as follows: first the hydrological nowcasting framework is described, then a description of the application and the results are given, and finally a discussion and conclusions are presented.

FRAMEWORK ELEMENTS

Observed rainfall estimation

The observed rainfall rate is estimated with an algorithm named RIME: it is particularly useful when polarimetric variables are available, and it is described in Silvestro *et al.* (2009).

The rain-rate fields are then accumulated to hourly scale and an algorithm of radar-gauge adjustment is applied. In this way some of the advantages of having rainfall data at high time resolution are lost, but the objective is to try to obtain the best possible quantitative estimation at a still high temporal scale. The employed methodology derives from the algorithms described in Koistinen & Puhakka (1981) and Gabella *et al.* (2001), and gives an estimation of a corrective factor for each point of the radar rainfall fields. Since we suppose to work in a real-time context, the last observed rainfall field (one hour accumulation) is estimated without gauge adjustment because of the incomplete updating of the raingauge database.

The rainfall fields generated by the proposed methodology are affected by uncertainty but they can be considered unbiased if we regard raingauge measurements as true rain.

Probabilistic nowcasting algorithm: Phast

Phast (Metta *et al.*, 2009, Rebora & Silvestro, 2011) is a rainfall nowcasting method based on the combination of an empirical nonlinear transformation of measured precipitation fields and the stochastic evolution in spectral space of the transformed fields. An initial phase-velocity ϖ is obtained from a 2-D FFT of two successive observed rainfall rate fields. This phase velocity is evolved as a Langevin process. In the following we report the main equations of the algorithm:

$$\begin{cases} k_s = \sqrt{k_x^2 + k_y^2} \\ d\phi_{k_s} = \omega_{k_s} dt \\ d\omega_{k_s} = -\dfrac{(\omega_{k_s} - \omega_{k_s}')}{T_s} dt + \sqrt{\dfrac{2\sigma_s^2}{T_s}} k_s dW \end{cases} \tag{1}$$

where ϕ_{kS} is the spectral phase, T_S is the decorrelation-time, σ_s^2 is the variance of the process, k_x and k_y the wave numbers, and dW is a random increment drawn from a normal distribution with zero mean and second-order moment (W is a Wiener process). The time gap between two rainfall fields depends on the radar scan strategy and in this application is 5 or 10 minutes.

After fixing the T and σ parameters, either by direct estimation from a recent sequence of observed fields or choosing them from a library (Metta *et al.*, 2009), possibly conditioned on large-scale synoptic conditions, a certain number of realizations of the precipitation process can be built. The result is an ensemble of forecasted stochastic fields. The complete rainfall scenarios are obtained by joining each forecasted realization with the observed rainfall fields (Fig. 1).

In this application, starting from the results described in Metta *et al.* (2009), we chose the values of the parameters by trying to find a balance between over-representation and under-representation of uncertainty in terms of streamflow. The objective was obtaining a reasonable variability of discharge scenarios to well represent the future possible occurrence, but avoiding production of a spread of predicted streamflows that is too large and therefore not useful. This is a sensitive point and further investigation could be carried out to analyse the effects of parameter variation.

Rainfall–runoff model: DRiFt

DRiFt (Discharge River Forecast) is a linear, semi-distributed, event-scale model based on a geomorphologic approach (Giannoni *et al.*, 2000, 2003; Gabellani *et al.*, 2009). The model is focused on the efficient description of the drainage system in its essential parts: hillslopes and channel networks. The drainage network is delineated by using a Digital Elevation Model (DEM)

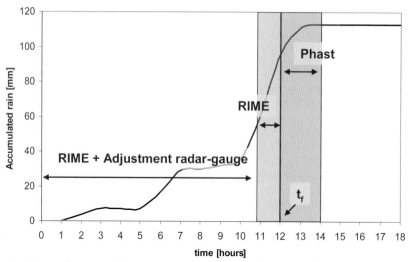

Fig. 1 Scheme of a rainfall scenario represented as cumulative rainfall at basin scale. t_f is the reference time of forecasting. $t < tf$ represent the past part of the rainfall scenario, $t > t_f$ represent the future part of the rainfall scenario.

and each cell is classified as hillslope or channel through a morphologic filter. The propagation of water in the first soil layer is described, so an auto-initialization of the model is reproduced between one event and another. The schematization is valid and applicable when the simulation period is not too long (6–8 days) and the evapotranspiration does not become crucial in the mass balance equation.

The discharge at any location along the drainage network can be mathematically formalized as follows:

$$Q(t) = \int_B M\left(t - \frac{d_0(x)}{v_0} - \frac{d_1(x)}{v_1}, x\right) dx \qquad (2)$$

where B is the drainage basin above the specified location, $M(t,x)$ is the runoff rate at time t and location x, $d_0(x)$ denotes the distances from x to the closest stream channel and $d_1(x)$ denotes the distance from the stream channel closest to x and the outlet, and v_0 and v_1 are the hillslope and channel velocities. The two velocities define a concentration-time for each cell in which the basin is discretized. The runoff estimated at cell scale is routed to the outlet section without accounting for the storage in the channels and without re-infiltration.

APPLICATION

The hydrological nowcasting framework has been applied on two basins of the Liguria Italian region. They are both covered by the C-band polarimetric radar of Settepani mountain and by a dense raingauge network. Two test cases have been considered: an event occurring on 21 October 2009 and another on 16 August 2006. The temporal resolution of the radar data is different for the two events: for the first it is 5 min and for the second 10 min. The spatial resolution of the rain rate fields is 1 km × 1 km.

The two kinematic parameters of the rainfall–runoff model have been set as $v_0 = 0.15$ m/s and $v_1 = 2.5$ m/s. These values well represent the mean behaviour of the basins in the Liguria region environment.

To verify the possible benefit of using the presented system we tried to reproduce two past events. For each event we identified the peak flow with its peak time (t_p) and we carried out the streamflow forecast at a certain forecast time (t_f) defined by the anticipation time τ.

$$t_f = t_p - \tau$$

The maximum value of τ depends on the nowcasting time horizon (t_n) and on the characteristic response time of the basin (t_b). t_n is fixed and is equal to two hours ($t_n = 2$ hours): since Phast uses only observed rainfall as input, we consider two hours as the maximum window of time after which the rainfall forecast no longer makes sense. t_b is estimated with the following equation (McGrawill, 1992):

$$t_b = 0.6 \cdot t_c$$

where t_c is the concentration time of the basin estimated with

$$t_c = 0.27 \cdot \sqrt{A} + 0.25$$

where A is the basin area in km^2 and t_c is in hours.

As a consequence:

$$\tau_{max} = t_b + t_n$$

τ_{max} defines the maximum time window before the peak (or in general before a certain time) for which we can expect a reliable forecast; it can be considered a physical limit. If the nowcasting technique produces a perfect forecast during t_n we can expect to obtain a good discharge forecast for the following t_b time window. Beyond this time the basin response is too influenced by the rainfall that occurs after $t_f + t_n$. We then set $\tau_{min} = t_n$ and we varied the instant of forecast t_f in the range ($t_p - \tau_{max}$, $t_p - \tau_{min}$) with a time-step of 1 h.

In Table 1 are reported the main characteristics of the two study basins (Entella and Scrivia).

Table 1 Main characteristics of the two study basins.

Basin	Area (km^2)	t_c (hours)	t_b (hours)
Entella	364	5.4	3.2
Scrivia	282	4.8	2.9

We decided to use a hydrograph as reference for evaluating the performance of the system, the simulated streamflow obtained using the observed rainfall (estimated as previously described) as input to the rainfall–runoff model (Berenguer *et al.*, 2005). In the panels of Figs 2 and 3 we report the results for the two study events: 21 October 2009 (Entella) and 16 August 2006 (Scrivia). The graphs seem to demonstrate that there is a general improvement of the available information during a hypothetical monitoring phase, derived through the use of the implemented framework. The forecasted discharge scenarios in such cases describe quite well the future occurrence (see Entella, forecast times 21:00 and 22:00; Scrivia, forecast times 6:00 and 7:00) represented by the reference hydrograph; in other situations the contribution of the forecasted rainfall is negligible (see Entella, forecast time 23:00 h).

In some cases the nowcasting system is not able to reproduce the future precipitation volume with sufficient accuracy (see Entella, forecast time 20:00 h; Scrivia, forecast time 05:00 h); in fact the probabilistic approach only partially adjusts this lack of information, and as a consequence the framework fails to forecast the reference hydrograph. Phast can only generate new precipitation structures based on the volume of precipitation observed in the rainfall fields used as input to the algorithm: therefore the process of growth and decay of new cells is not completely described.

CONCLUSIONS

In this work we present the application of a system for the probabilistic nowcasting of river discharge. It is based on three elements: (1) an algorithm for rainfall estimation based on

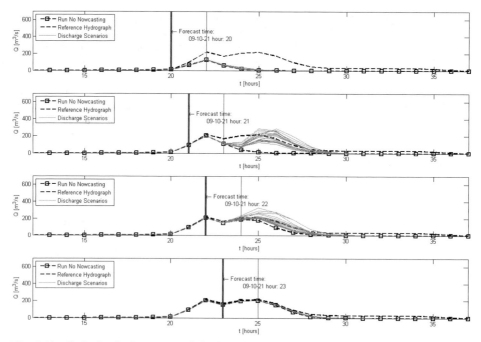

Fig. 2 Entella basin. Performances of the hydrological nowcasting system varying the instant of forecast t_f. The thick grey vertical line represents the time of last available observed field of each nowcasted scenario. The thin grey vertical line represents the end of nowcasting. The Run No Nowcasting is the discharge obtained feeding the rainfall–runoff model with observed rainfall only.

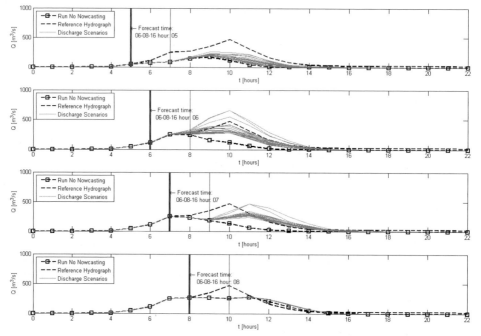

Fig. 3 Scrivia basin. Performances of the hydrological nowcasting system varying the instant of forecast t_f.

polarimetric variables and on radar-raingauge measurements in combination; (2) an algorithm for probabilistic nowcasting (PhaSt); and (3) a semi-distributed hydrological model (DRiFt). The objective is to produce an ensemble of forecasted hydrographs that allows a forecast with time horizons of a few hours (3–5) to be accomplished, accounting for the uncertainty associated with the outputs of a nowcasting technique.

The preliminary results seem to demonstrate that, in general, there is useful information content deriving from the application of the framework. In particular, the probabilistic approach leads to different possible discharge scenarios and their degree of severity can be assessed by the user. The forecaster no longer has a single realization to support his evaluations, but a set of different equi-probable realizations. The system has been applied to two case studies and further analysis should be done.

The use of radar rainfall estimations as unique inputs to the nowcasting algorithm result in this being the major source of uncertainty and of errors, mainly because of the impossibility of correctly and completely reproducing the atmospheric dynamics, the generation of new precipitation cells, and the dissipation of observed structures. The probabilistic approach can only partially deal with these issues that can probably only be addressed through data assimilation in meteorological models, or by introducing more data and information in the nowcasting system.

REFERENCES

Berenguer, M., Corral, C., Sanchez-Diesma, R. & Sempere-Torres, D. (2005) Hydrological validation of a radar-based nowcasting technique, *J. Hydrometeorology* 6, 532–549.

Borga, M. (2002) Accuracy of radar rainfall estimates for streamflow simulation. *J. Hydrol.* 267, 26–39.

Gabella, M., Joss, J., Perona, G. & Galli, G. (2001) Accuracy of rainfall estimates by two radars in the same Alpine environment using gauge adjustment. *J. Geophys. Res.*, 106, 5139–5150.

Gabellani, S., Silvestro, F. Rudari, R. & Boni, G. (2008) General calibration methodology for a combined Horton-SCS infiltration scheme in flash flood modelling. *Natural Hazards* 8, 1317–1327.

Germann, U. & Zawadzki, I. (2002) Scale-dependence of the predictability of precipitation from continental radar images. Part I: Description of the methodology. *Monthly Weather Rev.* 130(12), 2859–2873.

Germann, U., Berenguer, M., Sempere-Torres, D. & Zappa M. (2009) REAL: Ensemble radar precipitation estimation for hydrology in a mountainous region. *Quart. J. Roy. Met. Soc.* 135, 445–456.

Giannoni, F., Roth., G. &. Rudari, R. (2000) A semi-distributed rainfall–runoff model based on a geomorphologic approach. *Physics and Chemistry of the Earth*, 25/7-8, 665–671.

Giannoni, F., Roth., G. & Rudari , R. (2003) Can the behaviour of different basins be described by the same model's parameter set? A geomorphologic framework. *Physics and Chemistry of the Earth* 28, 289–295.

Carpenter, T.M. & Georgakakos, K. P. (2006a) Discretization scale dependencies of the ensemble flow range versus catchment area relationship in distributed hydrologic modeling. *J. Hydrol.* 328, 242–247.

Carpenter, T. M. & Georgakakos, K. P. (2006b) Intercomparison of lumped versus distributed hydrologic model ensemble simulations on operational forecast scales. *J. Hydrol.* 329, 174–185.

Koistinen, J. & Puhakka, T. (1981) An improved spatial gauge-radar adjustment technique. *AMS - Proc. 20th Conf. on Radar Met.*, 179–186.

Kouwen, N., Bingeman, A., Bellon, A. & Zawadzki, I. (2004) Operational issues: Real-time correction and hydrological validation of radar data. Preprints. In: *Sixth Int. Symp. on Hydrological Applications of Weather Radar, Melbourne, Australian Bureau of Meteorology*, CD-ROM.

Li, L. & Schmid, W. (1995) Nowcasting of motion and growth of precipitation with radar over a complex orography. *J. Appl. Met.* 34, 1286–1300.

Metta, S., Rebora, N., Ferraris, L., von Hardernberg, J. & Provenzale, A. (2009) PHAST: a phase-diffusion model for stochastic nowcasting, *J. Hydrometeorology* 10, 1285–1297.

Pessoa, M. L., Raael, L. B. & Earle, R. W. (1993) Use of weather radar for flood forecasting in the Sieve river basin: a sensitivity analysis. *J. Appl. Met.* 32, 462–475.

Rebora, N. & Silvestro F. (2011) PhaSt: stochastic phase-diffusion model for ensemble rainfall nowcasting. In: *WRaH 2011 Proceedings*.

Salek, M., Brezkova, L. & Novak P. (2006) The use of radar in hydrological modeling in the Czech Republic – case studies of flash floods. *Natural Hazards* 6, 229–236.

Schröter, K., Llort, X., Velasco-Forero, C., Ostrowski, M. & Sempere-Torres, D. (2011) Implications of radar rainfall estimates uncertainty on distributed hydrological model predictions. *Atmos. Res.* 100, 237–245.

Seed, A. W. (2003) A dynamic and spatial scaling approach to advection forecasting. *J. Appl. Met.* 42, 381–388.

Silvestro, F, Rebora N. & Ferraris, L. (2009) An algorithm for real-time rainfall rate estimation by using polarimetric radar: RIME. *J. Hydrometeorology* 10, 227–240, ISSN: 1525-755X, doi: 10.1175/2008JHM1015.1

Vivoni, E. R., Entekhabi, D., Bras, R. L., Ivanov, V. Y., Van Horne, M. P., Grassotti, C. & Hoffman, R. N. (2006) Extending the predictability of hydrometeorological flood events using radar rainfall nowcasting. *J. Hydrometeorology* 7, 660–667.

Zappa, M., Jaun, S., Germann, U., Walser, A. & Fundel, F. (2011) Superposition of three sources of uncertainties in operational flood forecasting chains. *Atmos. Res.*, doi:10.1016/j.atmosres.2010.12.005.

514

Weather Radar and Hydrology
(Proceedings of a symposium held in Exeter, UK, April 2011) (IAHS Publ. 351, 2012).

Influence of rainfall spatial variability on hydrological modelling: study by simulations

I. EMMANUEL[1], H. ANDRIEU[2], E. LEBLOIS[3] & N. JANEY[4]

1 *PRES L'UNAM, Ifsttar, Département GER, CS4, 44341 Bouguenais, France*
isabelle.emmanuel@ifsttar.fr
2 *PRES L'UNAM, Ifsttar, Département GER and IRSTV FR CNRS 2488 Bouguenais, France*
3 *CEMAGREF, 3 B Quai Chauveau, 69009 Lyon, France*
4 *LIFC, UFR Sciences et Techniques, 16 route de Gray, 25030 Besançon, France*

Abstract This work presents a simulation chain which enables studying the significance of rainfall spatial variability on flood runoff. A turning-band-method rainfall generator is used to simulate rainfall fields of different space–time variability. Catchments are extracted from Diffusion-Limited Aggregation structures. Three different rainfall–runoff models are implemented and the Hayami function is used to propagate runoff. Two spatial rainfall resolutions are taken into account: 250×250 m^2 and the rainfall average over the catchment. In this context hydrological studies are carried out. The influence of size of the catchment, of its production function and its response time are analysed. Hydrographs are compared in order to determine the interest for hydrology of detailed knowledge of rainfall. This work is currently ongoing.

Key words rainfall generator; catchment simulator; spatial variability; hydrological modelling

INTRODUCTION

The evaluation of the relevance of the spatial distribution of rainfall for hydrological modelling at the catchment outlet remains an open research issue in hydrology. Radar images provide detailed knowledge on the spatial variability of rainfall which was not previously available with raingauge networks. Thus, the use of rainfall radar images for rainfall–flow modelling has been tested in various case studies according to different approaches and several papers have been published on this subject in recent years (see e.g. Segond *et al.* (2007) for a review of the significance of spatial data for flood runoff generation). However, results of such analyses are sometimes contradictory. They highlight that it is not easy to evaluate the contribution of radar data to hydrological modelling since rainfall measurement errors and modelling errors cannot be distinguished from the influence of rainfall variability. In order to overcome errors on rainfall data, to control the rainfall variability and the hydrological behaviour of catchments, we suggest proceeding by simulation. This paper presents the simulation chain (catchment simulator, rainfall field generator and rainfall–runoff model). It gives a first view of the ongoing tests and obtained results, without omitting the difficulties met by this work currently in development.

PRESENTATION OF THE SIMULATION CHAIN

Catchment simulator

The arborescence simulator has been developed by Janey (1992) and adapted to catchments simulation within the framework of this study. It is based on a Diffusion Limited Agregation (DLA) model (Witten & Sander, 1981) which allows simulating irregular arborescences which present characteristics similar to hydrographical networks (Turcotte, 2006). The different simulation steps are as follows:

- Definition of the simulation zone shape and of the catchment outlet position. The outlet represents the seed of the DLA structure.
- Discretisation of the simulation zone by a regular triangular grid. The DLA structure is designed by the aggregation of a set of particles on the apexes of the grid. The particles are moved according to a fractional Brownian motion. A particle becomes fixed as soon as it reaches the immediate neighbourhood of a single fixed particle. This rule prevents cycle creation. The aggregation ends when a stop condition is verified.

- Extraction of the arborescence from the simulated DLA structure. The arborescence represents the channel network of the simulated catchment.
- Determination of the altitude of the particles. The outlet altitude is first arbitrarily fixed. Altitudes of the other particles are then determined step by step. The altitude differences are less important near the outlet than near the sources.
- Generation of the entire relief. The apexes adjacent to at least one fixed particle are considered. Their altitudes are determined such that the created slopes are directed towards the channel network. The outline of the simulated catchment is determined by the external apexes.
- Discretisation of the simulated catchment by a square grid in order to determine the flow direction of each cell. The direction of each cell represents the direction of the steepest descent among the eight permitted choices.

A large sample of 200 catchments of various and random shapes have been simulated. Their bifurcation, length (Horton, 1945) and areas (Schumm, 1956) ratios have been computed. These ratios range in the magnitude of the typical values found in the literature (Rodriguez-Iturbe & Rinaldo, 1997).

For the present study, the shape of the catchment displayed in Fig. 1 has been chosen. Note that the simulator is able to generate catchments of irregular shapes.

Fig. 1 Simulated catchment of 90 km^2 with a spatial resolution of 250×250 m^2. This catchment is of 10.25×12.5 km^2. The represented channel network is defined for a contributing area threshold of 0.5 km^2.

Rainfall generator

A Turning-Band-Method (TBM) (Montoglou & Wilson, 1982) rainfall generator has been used to simulate time series of rainfall fields displaying a defined spatial and rainfall variability (Leblois, 2010). A summary of this geostatistical model can be found in Renard *et al.* (2011). The TBM generates three-dimensional fields, describing two spatial and one time dimension. The final rainfall field is obtained from the product of two independent fields: (1) an indicator field of zero and non-zero rainfall pixels defining the outline of the final field; (2) a field of non-zero rainfall defining the intrinsic variability of the final field. Their simulation requires the prior determination of the rainfall distribution (percentage of zero for the indicator field; mean and variance of an inverse Gaussian distribution for the non-zero field) and the rainfall structure (spatio-temporal variograms) of the observed rainfall to be reproduced. Both fields are simulated by the TBM technique. We can note that the simulation space is distorted in order to take into account the advection of the rainfall field. As the TBM technique simulates Gaussian random fields, a transformation is necessary to obtain the observed distribution. A threshold is applied to the indicator field and an anamorphosis to the non-zero rainfall field.

One main characteristic of this rainfall generator is that it simulates, in continuous form, rainfall fields of given types. An example of a simulated rainfall field is presented in Fig. 2.

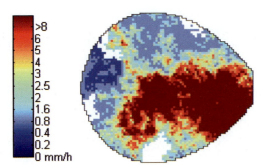

Fig. 2 Example of a simulated rainfall field with a spatial resolution of 250×250 m^2 over a 270 km^2 catchment. This catchment is of 18×21 km^2.

Hydrological model

The hydrological model regroups production and transfer functions which are fully distributed at the pixel scale.

Three simple production functions have been implemented to consider different scenarios of hydrological behaviours:

- a constant runoff coefficient (C in %), applied to all the pixels of the catchment;
- a contributing area model (CA): the C% pixels of the catchment, displaying the highest topographic index, form the contributing areas. Those pixels have a runoff of 100%;
- a Horton model (H): a threshold (S in mm/h) is computed such that, at the time series scale, C% of the raw rainfall is transformed into runoff. For a given pixel and time step, with a raw rainfall R, the net rainfall is equal to 0 if $R < S$, otherwise, its value is $R - S$.

For the sake of simplicity the production function does not change during the entire simulated time series. The three production functions are constrained by the same runoff coefficient, in order to make the scenarios comparable.

The runoff propagation is based on the diffusive wave model. If the two parameters, Celerity (C) and Diffusivity (D) are assumed constant, for the case of a semi-infinite channel the diffusive wave model admits an analytical solution: the Hayami model (Moussa, 1996) whose transfer function (known as the Hayami Kernel function) is defined for each pixel i as follows:

$$K_i(t) = \frac{L}{2(\pi D)^{1/2}} \frac{\exp\left[\frac{CL}{4D}\left(2 - \frac{L}{Ct} - \frac{Ct}{L}\right)\right]}{t^{3/2}} \tag{1}$$

with t the time, C the celerity (in m/s), D the diffusivity (in m^2/s) and L the river path length (in m) from the considered pixel to the catchment outlet.

The outflow at the catchment outlet is then expressed as:

$$Q(t) = \sum_{i=1}^{n} PE_i(t) * K_i(t) \tag{2}$$

with $Q(t)$ the outflow (in m^3/s), $PE_i(t)$ the net rainfall time series at t for the catchment pixel i (in m^3/s) and $K_i(t)$ the Kernel function (in s^{-1}). The symbol * represents the convolution relation.

ILLUSTRATION CASE STUDY

This section presents preliminary results associated with different simulation scenarios. They must not be appreciated as final results but as tests used to define applications conditions of the simulation chain consistent with hydrologists' questions. Let us recall that the goal of this work is to study the influence of rainfall variability on catchment outflows.

Presentation of the different scenarios

A reference scenario has been considered. It consists of a catchment of 90 km^2, of spatial resolution of 250×250 m^2 and of shape defined in Fig. 1. A constant runoff coefficient production function is considered. The celerity (C) is taken equal to 1 m/s with a diffusivity (D) of 500 m^2/s. The pathway is only considered as river travel.

For the other scenarios the reference parameters remain constant, apart from the one whose influence is tested. Two other different catchment surface areas are considered (30 km^2 and 270 km^2, named scenario "Size 30" and "Size 270"). Two celerities are tested (0.5 m/s and 2 m/s named scenario "C05" and "C2"). Finally two production functions are considered (contributing areas and Horton model: "PF-CA" and "PF-H" scenarios). The runoff coefficient constraining the three production functions is taken equal to 30% of the raw rainfall.

The 30, 90 and 270 km^2 catchments have (with a celerity of 1 m/s) a response time of 1, 2 and 4 hours, respectively. With celerities of 0.5 m/s and 2 m/s, the 90 km^2 catchment has a response time of 3.5 and 1.2 h, respectively.

Characteristics of the simulated rainfall fields

The main objective of this first case study being to validate the simulation chain, we have not considered the intermittence of rainfall field; the percentage of zeros has been taken as null. However the intermittence is an important aspect which should be taken into account in further tests, in particular for application to infrequent flash floods. The characteristics of the non-zero rainfall field have been deduced from the analysis of radar images (Emmanuel *et al.*, 2011). The mean and variance have been taken as equal to 5 and 8 mm/h, respectively. The Lagrangian time decorrelation (equivalent to the average life of a rainfall cluster) is equal to 20 min. The advection velocity is equal to 7 m/s. The spatial variogram has an exponential structure. Three spatial ranges (r) have been considered: 4, 8 and 16 km. The range corresponds to the decorrelation distance i.e. the distance from which two points have an independent statistical behaviour.

Methodology

The rainfall generator simulates, in continuous form, rainfall fields of given characteristics. For each rainfall type (i.e. r equal to 4, 8 and 16 km) a time series of 800 hours of rainfall fields has been simulated with a spatial resolution of 250×250 m^2 and a temporal resolution of 5 min over a window of 25×25 km^2. The objective being to study the influence of the spatial rainfall variability, two levels of knowledge of rainfall are taken into account: (1) a 250×250 m^2 spatial resolution, and (2) the rainfall average over the catchment.

For each scenario and for each rainfall type, two time series of catchment outflows are obtained resulting in two levels of knowledge of rainfall (distributed and average). The values of the 50 highest peaks of the first time series (Q_D) and the corresponding values of the second time series (Q_A) are selected. The comparison between the two samples of catchment outflows characterizes the interest of a spatially distributed rainfall forcing. The absolute bias between each couple of associated values Q_D and Q_A are computed [$Bias = (|Q_D - Q_A|/Q_D) \times 100$]. The distribution of those computed biases is studied via four statistical indicators: mean (in %), standard deviation (in %) and quantiles 80% and 90%. The quantile QX means that X% of the studied values are smaller than QX.

Preliminary results

Table 1 regroups the four statistical indicators computed for the different considered scenarios. The higher the average and the quantiles are, the higher the difference between the two time series of outflows (distributed and average) is. Examples of simulated time series are displayed in Figs 3 and 4. They show that the simulated time series depends on the size and response time of the considered catchments, which is not surprising.

Table 1 Values of the four statistical indicators (see Methodology) as percentages computed for each scenario and each rainfall type.

r = 4 km	Reference	Size 30	Size 270	C05	C2	PF- CA	PF - Horton
Mean (%)	6.1	6.3	5.1	4.0	9.5	7.9	12.4
Std (%)	3.3	4.1	2.3	2.1	4.4	4.0	6.6
Q80 (%)	8.8	9.7	7.2	5.8	12.8	11.3	18.3
Q90 (%)	10.6	13.1	8.2	6.5	15.0	12.7	21.2
r = 8 km							
Mean (%)	7.0	6.3	7.7	4.7	10.8	7.4	13.6
Std (%)	4.8	4.3	4.2	3.0	6.4	5.5	8.7
Q80 (%)	11.6	9.6	10.3	7.2	15.0	11.8	22.1
Q90 (%)	13.0	10.7	13.5	8.9	18.4	14.6	25.2
r = 16 km							
Mean (%)	8.5	5.2	8.3	5.3	9.7	9.5	15.1
Std (%)	5.2	3.2	5.5	3.0	6.0	5.6	9.5
Q80 (%)	13.7	8.0	13.1	7.8	14.3	14.9	24.9
Q90 (%)	16.3	9.8	16.6	8.9	18.4	16.7	28.6

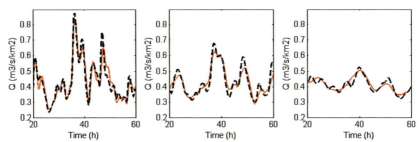

Fig. 3 Example of simulated time series from distributed rainfall (dashed lines) and average rainfall (solid lines) for the Size 30 (left), Reference (centre) and Size 270 (right) scenarios for a range of 8 km.

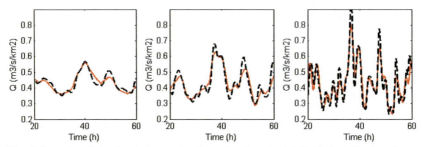

Fig. 4 Example of simulated time series from distributed rainfall (dashed lines) and average rainfall (solid lines) for the C05 (left), Reference (centre) and C2 (right) scenarios for a range of 8 km.

These preliminary results show that, except for the Horton production function scenario, the difference between the 50 highest peaks of the two time series of outflows (distributed and average) is not very significant. The mean of the absolute biases computed between each couple of associated values Q_D and Q_A is less than 10%. It reaches 12–15% for the Horton scenario. Nevertheless, 10% of the computed biases vary between 6.5 and 29% depending on the considered scenario.

In addition, these results confirm that the influence of rainfall type varies according to the catchment size. The computed means and quantiles increase with the range increasing for the

90 km^2 and 270 km^2 catchments, but decrease for the 30 km^2 catchment. For a spatially variable cluster ($r = 4 \text{ km}$) those statistical indicators are higher for the 30 km^2 catchment than for the bigger catchments and this observation is reversed for less variable clusters. These results tend to show that a link exists between catchment size and clusters size.

For the 90 km^2 catchment, whatever the rainfall type, these statistical indicators increase significantly with the celerity increasing.

Finally, these results show that knowledge of the spatial structure of rainfall appears more important for a very nonlinear production function (the Horton production function).

These results have to be taken as first tests in order to assess the good working of the simulation chain.

CONCLUSION

This study deals with the design of a simulation chain able to assess the significance of the spatial variability of rainfall on catchment outflow modelling. It regroups three components: a catchment simulator based on the DLA method, a rainfall fields generator based on an enriched turning-band-method, and a hydrological model able to reproduce different hydrological behaviours. The preliminary tests show that the simulation chain works correctly. They also confirm that the significance of the rainfall spatial variability in hydrological modelling depends on the influence of the catchment size, on the celerity, and on the catchment production function. Discussion with different actors of urban hydrology should allow definition of how to use this simulation chain in accordance with their needs. Before beginning more simulations, it appears necessary to redefine the simulation scenarios more realistically; in particular by e.g. considering the intermittency of the rainfall field which is an important aspect with regard to flash floods.

REFERENCES

Emmanuel, I., Leblois, E. & Andrieu, H. (2011) Temporal and spatial variability of rainfall at urban hydrological scales. In: *Proc. WRaH, 2011*, Exeter, UK. IAHS Publ. 351. IAHS Press, Wallingford, UK (this volume).

Horton, R. E. (1945) Erosional development of streams and their drainage basins; hydrophysical approach to quantitative morphology. *Bull. Geol. Soc. Am.* 56, 275–370.

Janey, N. (1992) Modélisation et synthèse d'images d'arbres et de bassins fluviaux associant méthodes combinatoires et plongement automatique d'arbres et cartes planaires. PhD Thesis, LIFC, Besançon, France.

Leblois, E. (2010) Technical presentation of the turning band method precipitation simulation system. (Personal communication).

Mantoglou, A. & Wilson, J. L. (1982) The turning bands method for simulation of random fields using line generation by a spectral method. *Water Resour. Res.* 18, 1379–1394.

Moussa, R. (1996) Analytical Hayami solution for the diffusive wave flood routing problem with lateral inflow. *Hydrol. Processes* 10, 1209–1227.

Renard, B., Kavetski, D., Leblois, E., Thyer, M. & Kuczera, G. (2011) Towards a reliable decomposition of predictive uncertainty in hydrological modelling: Characterizing rainfall errors using conditional simulation. *Water Resour. Res.* 47, W11516.

Rodriguez-Iturbe, I. & Rinaldo, A. (1997) *Fractal River basins, Chance and Self-Organization*. Press syndicate of the University of Cambridge, Cambridge, UK.

Schumm, S. A. (1956) Evolution of drainage systems and slopes in badlands at Perth Amboy, New Jersey. *Bul. Geol. Soc. Am.* 67, 597–646.

Segond, M.-L., Wheater, H. S. & Onof, C. (2007) The significance of spatial rainfall representation for flood runoff estimation: A numerical evaluation based on the Lee catchment, UK. *J. Hydrol.* 347, 116–131.

Turcotte, D. L. (2006) Self-organized complexity in geomorphology: observations and models. *Geomorphology* 91, 302–310.

Witten, T. A. & Sander, L. M. (1981) Diffusion-limited aggregation, a kinetic critical phenomenon. *Phys. Rev. Let.* 47, 1400–1403.

Quantifying catchment-scale storm motion and its effects on flood response

DAVIDE ZOCCATELLI[1] , MARCO BORGA[1],
EFTHYMIOS I. NIKOLOPOULOS[1] & EMMANOUIL N. ANAGNOSTOU[2]

1 *Department of Land and Agroforest Environments, University of Padova, Legnaro (PD), Italy*
davide.zoccatelli@studenti.unipd.it
2 *Department of Civil and Environmental Engineering, University of Connecticut, Storrs, USA*

Abstract We introduce the concept of catchment-scale storm velocity and illustrate its evaluation for a flash flood case study. The computation of the catchment-scale storm velocity takes into account the overall dynamics of the storm motion over the catchment, reflecting the filtering effect of the catchment morphology with respect to the storm kinematics. Catchment-scale storm velocity is quantified for the 29 August 2003 extreme storm that occurred on the 600 km^2 Fella basin in the eastern Italian Alps. A spatially distributed rainfall–runoff model is used to evaluate the effects of the storm velocity on flood modelling for four sub-basins. It is shown that storm velocity exhibits rather moderate values, in spite of the strong kinematic characteristics of individual storm elements. Consistent with this observation, hydrologic simulations show that storm motion has an almost negligible effect on the flood response modelling.

Key words weather radar; flash flood; space-time rainfall variability; storm motion

INTRODUCTION

The influence of storm movement on flood hydrographs has been widely investigated in hydrology (Niemczynowicz, 1984; Singh, 1998; De Lima & Singh, 2002; among others). Overall, these investigations have shown that the direction of storm movement might augment or reduce flood peaks and modify the hydrograph shape. A general finding is that for the same storm duration, the flood peak is greater for storms moving downstream relative to storms moving upstream (Ogden *et al.*, 1995). Arguably, accounting for storm movement is important for the prediction of floods (Smith *et al.*, 2002). However, in spite of the long-standing research effort, a methodology for the quantification of the storm motion at the catchment scale is still missing. Most of the investigations so far have used idealized storm profiles and motion as an input to catchment models. Attention was mostly focused on the kinematic characteristics of storms, with comparatively less focus on the analysis of the filtering effects exerted by catchment morphology on the storm velocity.

In this work we aim to establish a framework to quantify the concept of "catchment scale storm velocity", defined as the resulting interaction of the observed rainfall space–time storm variability with the catchment morphological properties. The "catchment scale storm velocity" is derived based on the notion of Spatial Moments of Catchment Rainfall (Zoccatelli *et al.*, 2011) and a framework developed by Viglione *et al.* (2010). Moreover, we introduce a methodology to test the impact of neglecting the storm velocity in flood modelling. These concepts are illustrated by using high-resolution rainfall data and flood data from an extreme flash flood that occurred in northeastern Italy in 2003 (Borga *et al.*, 2007).

CATCHMENT SCALE STORM VELOCITY DEFINED

The catchment-scale storm velocity is defined here based on the concept of Spatial Moments of Catchment Rainfall. These moments provide a description of overall spatial rainfall organisation at a certain time *t*, as a function of the rainfall rate $r(x,y,t)$ (L T^{-1}) value at position x,y and of the flow distance $d(x,y)$ (L), which is the distance to the catchment outlet measured along the flow path. The *n*th spatial moment of catchment rainfall p_n (L^{n+1} T^{-1}) is defined as:

$$p_n(t) = |A|^{-1} \int_A r(x,y,t) d(x,y)^n \, dA \qquad (1)$$

where A is the spatial domain of the drainage basin. The *zero*-th order spatial moment $p_0(t)$ yields the average catchment rainfall rate at time t. Analogously, the g_n (L^n) moments of the flow distance are given by:

$$g_n = |A|^{-1} \int_A d(x,y)^n \, dA \tag{2}$$

Non-dimensional (scaled) spatial moments of catchment rainfall can be obtained by taking the ratio between the spatial moments of catchment rainfall and the moments of the flow distance, as follows, for the first two orders:

$$\delta_1(t) = \frac{p_1(t)}{p_0(t)g_1}$$

$$\delta_2(t) = \frac{1}{g_2 - g_1^2}\left[\frac{p_2(t)}{p_0(t)} - \left(\frac{p_1(t)}{p_0(t)}\right)^2\right] \tag{3}$$

Smith *et al.* (2002, 2005) employed a scaled measure of distance from the storm centroid and a scaled measures of rainfall variability, similar to the *Spatial Moments*, to quantify the storm spatial organisation. The first scaled moment δ_1 (-) describes the location of the centre of the mass of catchment rainfall with respect to the average value of the flow distance (i.e. the catchment centre of mass). Values of δ_1 close to 1 reflect a rainfall distribution either concentrated close to the position of the catchment centre of mass or spatially homogeneous, with values <1 indicating that rainfall is distributed near the basin outlet, and values >1 indicating that rainfall is distributed towards the periphery of the drainage basin.

The second scaled moment δ_2(-) relates to the spreading of the rainfall field about its mean position with respect to the spreading of the flow distances. Values of δ_2 close to 1 reflect a uniform-like rainfall distribution, with values <1 indicating that rainfall is characterised by a unimodal distribution along the flow distance. Values >1 are generally rare, and indicate cases of multimodal rainfall distributions. The spatial moments as defined in equation (3) describe the instantaneous spatial rainfall organization at a certain time t. Equations (1)–(3) can also be extended to describe the spatial rainfall organization corresponding to the cumulated rainfall over a certain time period T_s (e.g. a storm event). These statistics, whose meanings are analogous to p_n and δ_n, are termed P_n and Δ_n and are defined as follows:

$$P_n = \frac{1}{T_s} \int_{T_s} p_n(t)dt \tag{4}$$

$$\Delta_1 = \frac{P_1}{P_0 g_1}; \Delta_2 = \frac{1}{g_2 - g_1^2}\left[\frac{P_2}{P_0} - \left(\frac{P_1}{P_0}\right)^2\right] \tag{5}$$

In the following section, Δ_1 and Δ_2 are computed based on the period characterized by the flood-producing rainfall. Zoccatelli *et al.* (2010) showed that the expected effect of a <1 value of Δ_1 is a reduced response lag time. This means that when rainfall is concentrated towards the outlet, the hydrograph responds quicker than for the case of spatially uniform rainfall. The opposite is true for rainfall concentrated towards the periphery of the catchment, with the hydrograph delayed relative to the case of a spatially uniform rainfall. The value of Δ_2 influences the shape of the flood hydrograph and the flood peak. Indeed, in general the effect of decreasing the value of Δ_2 is to increase the flood peak (Zoccatelli *et al.*, 2011). Based on the definition of the first scaled moments, the term "catchment scale storm velocity" V_s is defined as follows:

$$V_s = g_1 \frac{\text{cov}_t[T, \delta_1(t)w(t)]}{\text{var}[T]} - g_1 \frac{\text{cov}_t[T, w(t)]}{\text{var}[T]}\Delta_1 \tag{6}$$

where T is time and weights $w(t)$ are defined as $w(t) = \dfrac{p_0(t)}{P_0}$, $\text{cov}_t[\ ]$ denotes the temporal covariance and $\text{var}[\ ]$ denotes the variance. Equation (6) shows that the storm velocity is defined as the difference between the slope terms of two linear space–time regressions. The first slope term is estimated based on the regression between weighted scaled first moments and time, and the second term is based on the regression between weights and time. For the special case of rainfall constant in time, the storm velocity is defined as the regression over time of the distance of the rainfall centre of mass to the outlet. The sign of the velocity is positive (negative) for the case of upslope (downslope) motion. Zoccatelli *et al.* (2011) showed that the catchment scale storm velocity has an influence only on the shape of the hydrograph, but not on its timing. The type of influence on the flood hydrograph shape depends on the sign of the velocity, with a sharply peaked hydrograph for the case of negative (down-basin) velocity and a more gradually varying hydrograph for the case of positive (up-basin) velocity.

ASSESSMENT OF STORM VELOCITY FOR THE FELLA 2003 FLASH FLOOD

The basin considered in this study is the Fella basin (623 km^2) located in northeastern Italy (Fig. 1(b)). Important sub-basins also analysed in this study include Uque (~24 km^2), Pontebba (~165 km^2) and Dogna (~329 km^2). The flash flood event of the Fella basin during 29 August 2003 has been characterized as one of the most devastating flash flood events in northeastern Italy since the start of systematic observations in the region (Borga *et al.*, 2007). The flood inducing storm started at 10:00 UTC, lasted for approximately 12 h and resulted in loss of lives and damages close to 1 billion euro. The mesoscale convective system responsible for the flooding, exhibited a characteristic persistence of the convective bands over the basin that resulted in very large rainfall accumulations and high spatial variability. Direction and velocity of the storm cells in the bands were strikingly similar during the event, with upstream velocities for these storm elements in the order of 20 m s^{-1} for most of the event. The basin-averaged, half-hourly rainfall accumulations exceeded 80 mm (at Uque sub-basin), while the total rainfall accumulation over the 12 h duration exceeded 400 mm in some parts of the basin (Fig. 1). Weather radar observations and runoff data from streamgauges and a series of post-flood surveys are used to characterise the event. The Triangulated Irregular Network (TIN)-based Real-time Integrated Basin Simulator (tRIBS) (Ivanov *et al.*, 2004) distributed hydrologic model is used to simulate the hydrologic response over a range of sub-basins (Nikolopoulos *et al.*, 2011).

The space–time variability of catchment rainfall is shown in Fig. 2 that reports the basin-averaged rainfall rate at time *t* during the storm, the fractional coverage of the basin by rainfall rates exceeding 20 mm h^{-1}, the first and second scaled moments of catchment rainfall, and the storm velocity. The storm velocity was computed by applying equation (6) over moving time windows. The time windows were related for each basin to the corresponding lag time, ranging from 45 min for the Uque to 2.5 h for Moggio.

Inspection of the basin-averaged rain-rates and of the fractional coverage of heavy rainfall shows that the period of flood-inducing precipitation lasts for 7 h, from 11:00 h to 18:00 h. Despite the large variability in rainfall, the first scaled moment is characterised by a limited variability around 1, showing that the conditional distribution of flow distances, given the spatial rainfall distribution, was close to the distribution in the uniform rainfall case. However, in the last part of the storm and for all basins but Uque, δ_1 takes values significantly >1, indicating a concentration of rainfall towards the headwaters. The temporal evolution of δ_2 reflects the rainfall spatial concentration, with values that are <1 for most of the time. This behaviour is confirmed by the values of Δ_1 and Δ_2, with Δ_1 ranging from 0.916 (Pontebba) to 1.073 (Moggio) and Δ_2 ranging from 0.821 (Dogna) to (0.96) (Uque). Accordingly with the methodology described in this paper, the values of Δ_1 and Δ_2 were computed based on the period of flood producing rainfall.

The values of storm velocity are mainly between –0.7 and +0.8 m s^{-1}, with the exception of the Pontebba catchment, where larger velocity values and a more pronounced temporal variability can be observed. It is likely that the geometry and morphology of this catchment relative to the storm magnifies storm motion effects that are filtered out in the other catchments. Overall, the strong kinematics of the convective cells, with upstream velocity of 20 m s^{-1}, is not reflected in the values of storm velocity. The Triangulated Irregular Network (TIN)-based Real-time Integrated

Basin Simulator (tRIBS) was used to analyse the impact of removing various sources of spatial rainfall variability on simulated flood response for the four study nested basins. We performed numerical experiments in which modelled flash flood response obtained by using spatially variable rainfall forcing ("control" scenario) are contrasted with flood simulations obtained by using two different scenarios. In the first scenario, the model input is given by the spatially uniform rainfall ("uniform" scenario). In the second scenario ("constant variability" scenario), the spatial variability is given by the spatial pattern of the storm cumulated rainfall, which is scaled for each time-step to the corresponding mean areal value. With this scenario, the overall spatial variability of the storm event is preserved. However, any effect related to storm velocity is removed, since the same spatial pattern is applied over time.

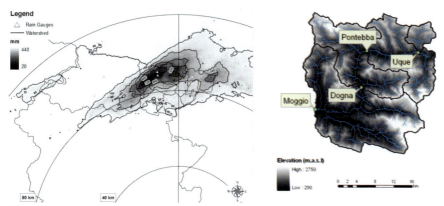

Fig. 1 (a) Storm total rainfall (mm) for 29 August 2003 event. (b) Catchment map of the Fella River basin, with the four study sub-basins.

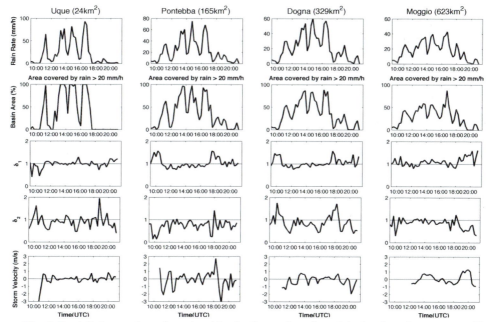

Fig. 2 Precipitation analyses by using 15 min time-series of precipitation intensity, coverage (for precipitation intensity > 20 mm h⁻¹), δ_1 (-), δ_2 (-) and storm velocity for the four study basins.

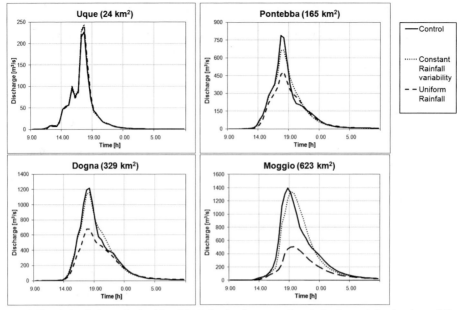

Fig. 3 Flood simulations obtained by tRIBS for the four nested study basins and for the three different rainfall input scenarios ("Control", "Uniform" and "Constant Variability").

The corresponding flood simulations are reported in Fig. 3. Inspection of this figure shows that, as expected, the effect of using a spatially uniform rainfall pattern increases with catchment size. This is consistent with previous findings from Nikolopoulos *et al.* (2011) and Zoccatelli *et al.* (2010). Errors are negligible only for the Uque basin, owing to its relatively small size. Errors are mainly due to differences in runoff volumes and are likely magnified in this event due to the strong nonlinearity introduced by the dry initial wetness conditions. The use of a constant rainfall spatial pattern reduces significantly the flood simulation errors at all scales (with the partial exception of Pontebba).

Using a constant spatial pattern is able to correctly reproduce the spatial rainfall concentration, hence dramatically contributing to reduce the errors in runoff volumes. Contrasting the flood simulations obtained by the "control" and "constant variability" scenarios shows that the effects of the storm velocity are negligible for all basins, except for Pontebba. This is not unexpected, given the outcome of the analysis of the storm velocity, which reported significant values of velocity for that case.

SUMMARY AND CONCLUSIONS

We introduced a methodology for the quantification of storm velocity at the catchment scale and to assess its effect on flood response modelling. The methodology is based on the observation that catchment shape, direction and morphology impose a filtering to the effect of storm motion over the catchment, in spite of the inherent kinematics of the storm elements. This implies that rainfall organisation characterized along the river network using the flow distance coordinate can be a significant property of rainfall spatial variability when considering flood response modelling.

The analysis reported here for the flood of 29 August 2003 on the Fella basin in Italy, illustrates the application of the method for a case where both storm spatial variability and motion were large when viewed from a Euclidean perspective. However, the reported catchment scale storm velocities are within a narrow range of values (generally <0.8 m s^{-1}). This indicates that the magnitude of the storm motion is strongly reduced when examined from the perspective of a distance metric imposed by the drainage network.

Analysis of runoff model responses shows that neglecting the spatial rainfall variability results in a considerable degradation of simulation accuracy, which increases with catchment size. The effect of neglecting the storm motion component is comparatively much less important. This indicates that the near-stationary storm motion of this flood case had a central role in producing large rainfall accumulations and in shaping their spatial distribution. However, it had a less important individual effect on flood response.

An aspect that motivates further research is the results reported here for Pontebba (admittedly limited) indicating that there may be a scale-dependency effect on the impact of storm motion on flood response. Specifically, there may be a range of catchment sizes and positions that maximize the effect of storm motion on flood generation. As we note in our study smaller catchments may be too limited in extension to show sensitivity to storm motion, while larger catchments would filter out the effect of storm motion owing to the variability of flow distances with respect to the storm motion vector.

REFERENCES

Borga, M., Boscolo, P., Zanon, F. & Sangati, M. (2007) Hydrometeorological analysis of the August 29, 2003 flash flood in the eastern Italian Alps. *J. Hydromet.* 8(5), 1049–1067.

De Lima, J. L. & Singh, V. P. (2002) The influence of the pattern of moving rainstorms on overland flow. *Adv. Water Resour.* 25(7), 817–828.

Ivanov, V. Y., Vivoni, E. R., Bras, R. L. & Entekhabi, D. (2004) Catchment hydrologic response with a fully distributed triangulated irregular network model. *Water Resour. Res.* 40(11), W11102, DOI: 10.1029/2004WR003218.

Niemczynowicz, J. (1984) Investigation of the influence of rainfall movement on runoff hydrograph: Part I. Simulation of conceptual catchment. *Nordic Hydrology* 15, 57–70.

Nikolopoulos, E. I., Anagnostou, E. N., Borga, M., Vivoni, E. R. & Papadopoulos, A. (2011) Sensitivity of a mountain basin flash flood to initial wetness conditions and rainfall variability. *J. Hydrol.* 402(3–4), 165–178, doi:10.1016/j.jhydrol.2010.12.020.

Ogden, F. L., Richardson, J. R. & Julien, P. Y. (1995) Similarity in catchment response: 2. Moving rainstorms. *Water Resour. Res.* 31(6), 1543–1547.

Singh, V. P. (1998) Effect of the direction of storm movement on planar flow. *Hydrol. Processes* 12, 147–170.

Smith, J. A., Baeck, M. L., Morrison, J. E., Sturdevant-Rees, P. L., Turner-Gillespie, D. F., & Bates, P. D. (2002) The regional hydrology of extreme floods in an urbanizing drainage basin. *J. Hydromet.* 3, 267–282.

Smith, J. A., Baeck, M. L., Meierdiercks, K. L., Nelson, P. A., Miller, A. J., & Holland, E. J. (2005). Field studies of the storm event hydrologic response in an urbanizing watershed. *Water Resour Res.* 41, W10413, doi:10.1029/2004WR003712.

Viglione, A., Chirico, G. B., Woods, R. & Blöschl, G. (2010) Generalised synthesis of space–time variability in flood response: An analytical framework. *J. Hydrol.* doi:10.1016/j.jhydrol.2010.05.047.

Zoccatelli, D., Borga, M., Zanon, F., Antonescu, B. & Stancalie, G. (2010) Which rainfall spatial information for flash flood response modelling? A numerical investigation based on data from the Carpathian range, Romania. *J. Hydrol.* 394(1–2), 148–161. doi:10.1016/j.jhydrol.2010.07.019.

Zoccatelli, D., Borga, M., Viglione, A., Chirico, G. & Blöschl, G. (2011) Spatial moments of catchment rainfall: rainfall spatial organisation, basin morphology, and flood response. *Hydrol. Earth Syst. Sci.* 15, 3767–3783. www.hydrol-earth-syst-sci.net/15/3767/2011/ doi:10.5194/hess-15-3767-2011.

Improvement of rainfall–runoff modelling with distributed radar rainfall data: a case study in the Lez, French Mediterranean, catchment

M. COUSTAU, V. BORRELL-ESTUPINA & C. BOUVIER

Hydrosciences Montpellier (UMR 5569 CNRS-IRD-UM), 300 avenue du Pr. Emile Jeanbrau, 34 000 Montpellier, France
bouvier@msem.univ-montp2.fr

Abstract The Mediterranean catchments in the south of France are prone to intense rainfall leading to destructive flash floods. These rainfalls mainly occur in autumn and show a high spatial variability. This study aims to assess the quality and impact in hydrological modelling of the radar rainfall data, in the Lez catchment (114 km^2) near Montpellier, France. Comparison of both the raingauges and radar data proved to be satisfactory for events at the beginning of autumn. In contrast, important differences appeared for events occurring at the end of autumn. This can be explained by the weak vertical extension of the clouds and the low altitude of the 0°C isotherm in this period, which could affect the accuracy of radar measurements due to the distance between the basin and the radar (~60 km). To take advantage of the spatial variability of the radar rainfall data, the flood simulations were performed through a distributed event-based rainfall–runoff model. The model was calibrated using a sample of 21 floods observed from 1994 to 2008 where both recording raingauge and radar rainfall data were available. When the radar rainfalls were reliable, they led to: (i) an improvement of the optimal flood simulation at the outlet, and (ii) an improvement of the relationship between the calibrated initial condition of the model and external predictors such as piezometric level, baseflow and Hu2 index from the Meteo-France SIM model. Installation of an X-band radar near the study area could improve rainfall estimation at the end of the autumn for the Lez catchment and the Montpellier agglomeration.
Key words flash flood; distributed rainfall–runoff model; event-based model; radar rainfall

INTRODUCTION

The Mediterranean catchments in the south of France are prone to intense rainfall leading to destructive flash floods. These rainfalls show a high spatial variability which can be captured by weather radars. The radar measurements include a lot of uncertainties in rainfall estimation due to ground clutter, beam blockage (Smith *et al.*, 1998; Bech *et al.*, 2003), variations in Z-R relationships (Chapon *et al.*, 2008; Alfieri *et al.*, 2010) or wind effects (Salles *et al.*, 2010). Thus, radar data have to be checked and improved using raingauge estimates (Borga, 2002; Vieux & Bedient, 2004; Chumchean *et al.*, 2006; Mapiam *et al.*, 2009).

The radar data can then be used in rainfall–runoff models, which are crucial tools for flash flood prediction or real-time forecasting. These models are known to be sensitive to the spatial variability of the rainfall (Obled *et al.*, 1994; Arnaud *et al.*, 2002; Younger *et al.*, 2009). Their efficiency is still limited by uncertainties in the spatial variability of Mediterranean rainstorms (Sangati *et al.*, 2009).

In this study, a comparison is made of the efficiency of a distributed event-based model in a Mediterranean catchment, using either the radar rainfall data or the available raingauge data. Event-based models are found to be efficient in case of data availability problems, but their disadvantage rests on the need to assess the initial wetness condition at the beginning of the event. One of the questions this paper addresses is to show that the radar data are able to make robust the calibration of the initial wetness condition of the model, regarding an external index provided by *in situ* data or operational surface model output. First, the method to check the radar rainfall quality and select reliable radar data is presented. Second, the improvements brought by radar rainfall estimates to the distributed event-based rainfall–runoff model, relating to both the flood simulations and the estimation of the initial condition of the model for each event, are shown.

STUDY AREA

The study site is the 114 km^2 Lez topographic catchment at Lavalette, upstream from Montpellier. Its northern part connects with a 380 km^2 aquifer (Fig. 1) composed of limestones and dolomites with a thickness ranging from 650 to 1100 m (Avias, 1992). The surface relief alternates between limestone plateaus, named "*causses*", and plains. In the plains, limestones are covered by 200 to 800 m thick marls, and soil whose thickness is generally <1 m. Altitudes range from 300 to 700 m in the calcareous *causses* and from 50 to 100 m in the marly plains. Vegetation is primarily garrigue on the *causses* and crops (vineyard, olive trees) on the plains.

Fig. 1 Location of the Lez basin.

Floods occur mainly in autumn during intense rainstorms with up to several hundred millimetres in 24 h (>300 mm at the Matelles in 1976). Lag-times are short (about 5 h) and peak discharges are high (up to 480 m^3 s^{-1} in September 2005). The selected episodes correspond to 21 floods measured from 1994 to 2008. This sample covers a wide range of peak flows from 40 to 480 m^3 s^{-1}. Five events occurred after a long dry period (September 2000, 2002, 2003, 2005 and October 2008) and the 16 others occurred after more or less rainy periods.

Rainfall data from the Nimes radar (Calamar® or Hydram treatment) were available at a 5-minute time-step and a resolution of 1 km^2. They covered a 5000 km^2 area where 20 daily raingauges monitored by Météo-France were available. Raingauge data were recorded at a one-hour time-step at four recording raingauges monitored by Meteo-France, of which only the Prades station is located within the Lez basin, the other three being located at a distance of 5 to 10 km outside the catchment (Fig. 1). The main corrections applied to the radar data consisted of: (i) a deletion of ground clutter and mask effects (beam blockage), (ii) an estimation of the vertical profile of reflectivity and (iii) a reflectivity conversion into a rainfall depth using the Marshall-Palmer Z-R relationship. Discharge time series at the basin outlet were also available at an hourly time-step.

Data concerning catchment wetness were also available. A set of 12 piezometers (Fig. 1) connected to the Lez spring (Karam, 1989) was retained as reference gauges of the aquifer level. These piezometers are evenly distributed throughout the Lez aquifer. A predictor of soil moisture conditions was also calculated based on the soil moisture index given by the Safran-Isba-Modcou (SIM) model developed by Météo-France. This index is available daily at 06:00 UCT for cells of 8 × 8 km^2 at three different levels in the soils (Hu1, Hu2, Hu3 for surface horizon, root horizon, and deep horizon, respectively). In our case, the value of the root layer Hu2 was selected because it is *a priori* the most representative moisture index of the superficial deposits (Marchandise & Viel, 2008). The values used here correspond to the average moisture of all the pixels that comprise the Lez topographic basin.

METHODS

Quality assessment of radar rainfall

To assess radar rainfall quality, linear regressions were established between the raingauge measurements and the corresponding radar pixels. The corresponding radar pixel was derived from the mean value of the central pixel and its eight neighbours. For each radar event, both cumulated rainfall and rainfall intensity influence catchment behaviour on runoff production. To check the total storm depth, a first regression was established between the total storm depth of the 20 daily raingauges and the 20 corresponding radar pixels. Then, a determination coefficient R_e^2 was calculated. To check the rainfall intensity, other regressions were also done using hourly rainfall data at each functioning station. A mean determination coefficient R_h^2 was calculated for each event. Thus, R_e^2 and R_h^2 assessed the closeness of rainfall radar data to ground rainfall "reference" data from raingauges.

For each event, to assess the systematic residual error in radar rainfall estimations, the Mean Field Bias (MFB) coefficient (Vieux & Bedient, 2004), was calculated as:

$$MFB = \left(\frac{1}{n} \sum_i Gi \right) \Big/ \left(\frac{1}{n} \sum_i Ri \right) \tag{1}$$

where n is the number of raingauges taken into account in the calculation (here the 20 Météo-France raingauges); Gi is the rainfall recorded by the raingauge at point i; Ri is the rainfall given by the radar at point i, both for the studied event.

Hydrological model

An event-based model has been selected due to the necessity of prediction or forecasting, even in the case of data scarcity. Modelling only the flood phase also allows reducing the number of parameters of the model, making easier and more robust the calibration of the model. Lastly, distribution of the rainfall was required, to take advantage of the radar data; distribution does not increase the complexity of the model, because no additional parameter was included in the model.

The hydrological model operates over a regular grid mesh of cells, which are not connected. Rainfall is computed for each cell at any time using the Thiessen polygon method.

The runoff from each cell is then calculated using a modified SCS runoff model (Fig. 2). It is based on a runoff coefficient $C(t)$, which depends on both the initial water deficit S of the basin and the cumulated rainfall $P(t)$ since the beginning of the event (Gaume *et al.*, 2004), given by:

$$C(t) = \left(\frac{P(t) - 0.2S}{P(t) + 0.8S} \right) \left(2 - \frac{P(t) - 0.2S}{P(t) + 0.8S} \right). \tag{2}$$

This runoff coefficient divides the rainfall into runoff and infiltration, $i_e(t) = C(t).i_b(t)$, where $i_e(t)$ is the runoff and $i_b(t)$ is the rainfall. The infiltration then fills a reservoir whose level, $stoc(t)$, is computed by :

$$\frac{dstoc(t)}{dt} = i_b(t) - i_e(t) - ds.stoc(t) \tag{3}$$

where ds is the coefficient of discharge of the reservoir. In order to represent slow discharges, an additional runoff was considered, as a part of the cumulated infiltration since the beginning of the event. The additional delayed runoff $i_d(t)$ is expressed as a proportion of the release of this reservoir:

$$i_d(t) = \min(1, \frac{w}{S}).ds.stoc(t) \tag{4}$$

where w is a value assumed to be constant for a given catchment. The total runoff $i_t(t)$ at the time t is thus given by :

$$i_t(t) = i_e(t) + i_d(t) \tag{5}$$

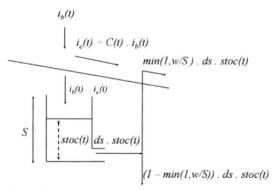

$i_b(t)$

$i_e(t) = C(t) \cdot i_b(t)$

$min(1, w/S) \cdot ds \cdot stoc(t)$

$i_b(t) - i_e(t)$

S

$stoc(t)$ $ds \cdot stoc(t)$

$(1 - min(1, w/S)) \cdot ds \cdot stoc(t)$

Fig. 2 Runoff model for each mesh of the basin. Here, S represents the initial water deficit of the catchment for an event.

Each cell produces an elementary hydrograph, routed to the outlet using a Lag and Route model. This routing model has two parameters: V_0, a velocity of travel which controls the runoff routing time to the outlet, and K_0 which controls the runoff diffusion (see Lhomme *et al.*, 2006). The complete hydrograph of the flood is finally obtained after addition of the elementary hydrographs.

The complete model finally deals with five parameters: S, w, ds, V_0, K_0, which are uniform in space. Thus the model accounts for the spatial variability of the rainfall, but not for the spatial variability of the runoff conditions. The routing conditions are also distributed by using the distance from each cell to the outlet. The model was implemented on the ATHYS modelling platform (www.athys-soft.org).

RESULTS

Quality assessment of radar rainfall

Whatever was the radar treatment (Calamar® or Hydram), it was observed that the MFB was always greater than 1 (see Table 1), indicating that radar systematically underestimates rainfall. This error could come from the Z-R relation and radar calibration effects (Borga, 2002). Except in 2002, it was also noted that both determination coefficients R_e^2 and R_h^2 of the linear regressions were >0.7 for storms at the beginning of autumn (September and October). For the other storms (November, December and January), at least one of the two determination coefficients was <0.5. The second result highlights the poor quality of radar measurements at the end of autumn, which is likely the result of both the weak vertical extension of the clouds and the low altitude of the 0°C isotherm, because of the distance between the basin and radar (~60 km). For each event, and for each pixel, radar rainfall intensities were corrected by the MFB coefficient.

Improvement for the hydrological model

The model parameters were first calibrated for each event, optimizing the Nash-Sutcliffe criterion. Then, a mean value was set for all the parameters which remained constant for all the flood events, except S which represents the initial water deficit of the basin; this parameter has to be optimized for each event. Simulations were performed on the sample of 21 events. Firstly, the simulations were performed using only raingauge data (21 floods). Secondly, when the radar rainfall measurements were reliable and corrected by the MFB (for the five floods in Table 2), they were used to replace the raingauge measurements.

As an event-based model, the efficiency of the model was not only assessed for the accuracy of the flood simulations (in the sense of the Nash-Sutcliffe criterion), but also for the quality of the relationship between S and external predictors of the initial water content in the catchment (in the sense of the R^2 determination coefficient).

Table 1 Determination coefficients of the linear regressions between raingauge and radar rainfall depths and MFB values for each rainfall radar event. The reliable radar rainfall data are in bold.

		Dec 97	Nov 99	Sep 00	Dec 00	Jan 01	Oct 01	Sep 02	Oct 02	Dec 02	Sep 03	Nov 03	Dec 03	Sep 05	Jan 06	Oct 08
Hydram	R_e^2	0.13	0.26	**0.91**	0.11	0.01		0.95	0.31	0.11	**0.87**	0.48	0.10	**0.80**	0.00	**0.76**
	R_h^2	0.28	0.47	**0.93**	0.31	0.64		0.51	0.72	0.54	**0.87**	0.85	0.64	**0.70**	0.45	**0.80**
	MFB	1.74	1.09	**1.79**	1.50	1.53		1.80	1.74	1.69	**1.27**	1.58	1.05	**1.29**	1.24	**1.07**
Calamar	R_e^2						0.90			0.12	**0.83**			0.12	**0.87**	
	R_h^2						0.99			0.54	**0.85**			0.25	**0.75**	
	MFB						1.03			1.62	**1.17**			1.29	**1.00**	

Table 2 Initial condition (*S* in mm) and Nash-Sutcliffe ("Nash") criterion obtained with recording raingauge and radar rainfall inputs with the following calibrated parameters $ds = 0.28$ d^{-1}; $w = 101$ mm; $V_0 = 1.3$ m.s^{-1}; $K_0 = 0.3$ (constant values for all the floods).

	Recording Raingauges		HYDRAM radar treatment		CALAMAR radar treatment	
Event	*S*	Nash	*S*	Nash	*S*	Nash
September 2000	143	0.94	267	0.91	No data	
October 2001	164	0.81	No data		139	0.94
September 2003	481	0.81	254	0.90	276	0.89
September 2005	33	0.72	246	0.81	240	0.90
October 2008	386	0.81	392	0.88	No data	

Generally speaking, for 18 out of 21 floods, the model provides satisfactory simulations of the flood events, with Nash-Sutcliffe criterion values ranging between 0.66 and 0.94 (with an average value of 0.80). When radar rainfall data inputs are used, they greatly improve the simulations: this is the case for September 2003, September 2005, October 2001 and October 2008. For the September 2003 and September 2005 floods (Table 2) the improvement is essentially due to the dysfunction of the Prades recording raingauge, located in the topographic basin. In addition, radar rainfall can lead to a significant change in the estimation of *S*, like in September 2000 (which occurs after a long dry period) because the Prades recording raingauge data did not represent the total storm depth falling in the basin.

The improvement brought by radar rainfall inputs also appeared in the correlation between the initial condition of the model and different predictors. Using reliable radar rainfall measurements instead of only the raingauge measurements leads to an improvement in the determination coefficients for the linear regression: between *S* and the Hu2 predictor from $R^2 = 0.29$ to $R^2 = 0.69$ (Fig. 3), and between *S* and the different piezometric levels: from a median $R^2 = 0.27$ (between 0.17 and 0.64) to $R^2 = 0.63$ (between 0.48 and 0.81).

Piezometric levels supply the best correlation ($R^2 = 0.81$) with the initial condition of the model, but the limited number of events available did not allow a significant hierarchy of the performances of these predictors to be established.

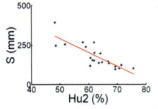

Fig. 3 Linear regression established between S (mm) and the Hu2 predictor at the beginning of the event using reliable radar rainfall measurements.

Use of reliable rainfall radar data allows for a significant improvement in rainfall–runoff modelling. This improvement is mainly due to the low density of raingauges in the basin, which leads to a poor spatial description of the rainfall, a problem at times exacerbated by the malfunction of a raingauge. A study of the Gardon d'Anduze catchment (Tramblay *et al.*, 2011) demonstrated that a higher density of raingauges in the basin can provide results similar to those of the radar measurements.

CONCLUSION

The quality-control showed that the radar rainfall at the beginning of autumn was reliable whereas the quality of the radar rainfall at the end of autumn was poor. This result can be explained by the weak vertical extension of the clouds and the low altitude of the 0°C isotherm in this period, which can affect the efficiency of radar measurements due to the distance between the basin and the radar (~60 km).

Furthermore, the study showed that use of reliable radar rainfall in the hydrological model allows for improvements in discharge simulations and estimation of initial condition. Thus, the model proved to be able to satisfactorily reproduce the Lez basin flash floods. The initial condition is also correlated with water content predictors such as the Hu2 index of the SIM model or the piezometric level, allowing for the use of this distributed event-based model for future events.

REFERENCES

Alfieri, L., Claps, P. & Laio, F. (2010) Time-dependent Z–R relationships for estimating rainfall fields from radar measurements. *Nat. Hazards Earth Syst. Sci.* 10, 149–158.

Arnaud, P., Bouvier, C., Cisneros, L. & Dominguez, R. (2002) Influence of rainfall spatial variability on flood prediction. *J. Hydrol.* 260, 216–230.

Avias, J. (1992) Karstic aquifer of Mediterranean type, geological controls: Lez spring (North Montpellieran karsts) example. *IAH, Hydrogeology of Selected Karst Regions* 13, 89–113.

Bech, J., Codina, B, Lorente, J. & Bebbington, D. (2003) The sensitivity of single polarization weather radar beam blockage correction to variability in the vertical refractivity gradient. *J. Atmos. Oceanic Tech.* 20, 845–855.

Borga, M. (2002) Accuracy of radar rainfall estimates for streamflow simulation. *J. Hydrol.* 267, 26–39.

Chapon, B., Delrieu, G., Gosset, M. & Boudevillain, B. (2008) Variability of rain drop size distribution and its effect on the Z–R relationship: A case study for intense Mediterranean rainfall. *Atmospheric Research* 87, 52–65.

Chumchean,S., Seed, A., Sharma, A. (2006) Correcting of real-time radar rainfall bias using a Kalman filtering approach. *J. Hydrol.* 317, 123–137.

Gaume, E., Livet, M., Desbordes, M. & Villeneuve, J. P. (2004) Hydrological analysis of the river Aude, France, flash flood on 12 and 13 November 1999. *J. Hydrol.* 286, 135–154.

Karam, Y. (1989) Essais de modélisation des écoulements dans un aquifère karstique. Exemple de la source du Lez (Hérault France). Thèse. Montpellier, Université des sciences et techniques du Languedoc.

Lhomme, J., Bouvier, C., Perrin, J. L. (2004) Applying a GIS-based geomorphological routing model in urban catchment. *J. Hydrol.* 299, 203–216.

Mapiam, P. P., Sharma, A., Chumchean, S. & Sriwongsitanon, N. (2009) Runoff estimation using radar and rain gage data. In: *18th World IMACS/MODSIM Congress* (Cairns, Australia).

Marchandise, A. & Viel, C. (2008) Utilisation des indices d'humidité de la chaîne Safran-Isba-Modcou de Météo-France pour la vigilance et la prévision opérationnelle des crues. *Colloque SHF, 191ᵉ CST "Prévision hydrométéorologiques".*

Obled, Ch., Wendling, J. & Beven, K. (1994) The sensitivity of hydrological models to spatial rainfall patterns: an evaluation using observed data. *J. Hydrol.* 159(1–4), 305–333.

Salles, C., Cres, F.-N., Tournoud, M.-G. & Ibrahim, H. (2010) Assessment of rainfall fields in a small Mediterranean basin: radar versus rain gauge data. In: *ERAD 2010 The sixth European conference on radar in meteorology and hydrology.*

Sangati, M., Borga, M., Rabuffetti, D. & Bechini, R. (2009) Influence of rainfall and soil properties spatial aggregation on extreme flash flood response modelling: an evaluation based on the Sesia river basin, North Western Italy. *Adv. Water Resour.* 32, 1090–1106.

Smith, P. L. (1998) On the minimum useful elevation angle for weather surveillance radar scans. *J. Atmos. Oceanic Tech.* 18(3), 841–843.

Tramblay, Y., Bouvier C., Martin C., Didon-Lescot J. F., Todorovik D. & Domergue J. M. (2010) Assessment of initial soil moisture conditions for event-based rainfall–runoff modelling. *J. Hydrol.* 387(3–4), 176–187.

Vieux, B. E. & Bedient, P. B. (2004) Assessing urban hydrologic prediction accuracy through event reconstitution. *J. Hydrol.* 299, 217–236.

Younger, P. M., Freer, J. E. & Beven, K. J. (2009) Detecting the effects of spatial variability of rainfall on hydrological modelling within an uncertainty analysis framework. *Hydrol. Processes* 23(14), 1988–2003.

Representing the spatial variability of rainfall for input to the G2G distributed flood forecasting model: operational experience from the Flood Forecasting Centre

DAVID PRICE[1], CHARLIE PILLING[1], GAVIN ROBBINS[1], ANDY LANE[1], GRAEME BOYCE[1], KEITH FENWICK[1], ROBERT J. MOORE[2], JOANNE COLES[3], TIM HARRISON[3] & MARC VAN DIJK[4]

1 *Flood Forecasting Centre, Met Office, FitzRoy Road, Exeter, Devon EX1 3PB, UK*
david.a.price@environment-agency.gov.uk
2 *Centre for Ecology & Hydrology, Wallingford OX10 8BB, UK*
3 *Environment Agency, Horizon House, Deanery Road, Bristol BS1 5AH, UK*
4 *Deltares, Inland Water Systems, Deltares, Delft, The Netherlands*

Abstract Over the year 2010 the Flood Forecasting Centre (FFC) calibrated and implemented a distributed flood forecasting model to support the FFC's remit to provide flood risk forecasts across England and Wales, UK. The distributed nature of the model, designed to run at 15-min time-steps on a 1 km^2 grid, enables the spatial variability of rainfall measurements and forecasts, rather than lumped catchment averages, to be captured. Such a distributed model should therefore benefit greatly from the spatial and temporal resolution afforded by radar observations. Initial results have highlighted the importance of the quality of the gridded rainfall fields and in a number of cases erroneous radar rainfall data have been shown to contribute to poor model performance. It is suggested that gridded datasets of sufficient quality will be best provided by capturing the spatial variability inherent in radar data together with raingauge data in a merged product.

Key words flood forecasting; distributed flood forecasting model; Flood Forecasting Centre; radar; Grid-to-Grid model

INTRODUCTION

In April 2009 the Flood Forecasting Centre (FFC) was established as a joint venture between the Environment Agency and the Met Office as a direct response to the Pitt Review (Pitt, 2008). The FFC is uniquely placed to deliver some of the key technical and operational recommendations set out in the Pitt Review in England and Wales.

In this paper we describe the introduction of a grid-based flood forecasting model for England and Wales (UK) and describe the ways in which radar data are employed within the modelling system. The model is referred to as Grid-to-Grid (G2G). The use of a grid-based model allows the spatial nature of rainfall, as represented by radar data, to be captured in a way that is not possible using lumped hydrological catchment models, which have traditionally been used for flood forecasting in the UK. Thus, capturing the spatial nature of rainfall using the G2G offers a significant advantage over lumped models, but it also raises challenges. Examples are given that highlight a number of issues arising from using radar rainfall data as an input to the G2G model, raising the challenge for the provision of more reliable data for use in an operational environment.

This paper first introduces the G2G model and its environment. The ways in which radar rainfall data are used within the modelling system are then explored, first with regard to the rainfall input data hierarchy used to generate a complete time-series prior to each model run, and then in relation to short-term forecasts (nowcasts). The impact of erroneous rainfall input data on model output is considered and finally, the paper considers opportunities for future developments in relation to radar rainfall.

THE G2G MODEL AND IMPLEMENTATION AT THE FFC

Overview of the G2G Model

The G2G (Moore *et al.*, 2006; Bell *et al.*, 2009; Environment Agency, 2010) developed at the Centre for Ecology & Hydrology (CEH) is able to provide a forecast of river flow across England

and Wales at a high temporal and spatial resolution (down to 1 km^2). As such, it has the capacity to forecast "everywhere" and lends itself to large-scale applications, including at national level. Figure 1 presents a schematic of the G2G model.

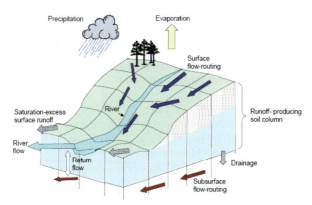

Fig. 1 Schematic of the G2G model.

The G2G is a physical-conceptual distributed, grid-based runoff production and routing model which contrasts with current hydrological model networks that generally comprise a connected set of catchment-based (lumped) rainfall–runoff models feeding into river routing reach models, providing forecasts at specific locations. Thus, the G2G employs a "grid-to-grid" formulation rather than the often used "source-to-sink" approach. It is designed to work with gridded rainfall estimates and can be used to forecast river flows at both gauged and ungauged sites.

The model has been calibrated using flow records from gauging stations on rivers throughout England and Wales. The model implemented at the FFC uses soil, geology and land cover spatial datasets, as well as terrain slope. This formulation makes explicit use of soil properties, including soil depth. These spatial datasets reduce the number of model parameters that require calibration.

Probability-distributed model theory is applied when representing surface runoff production. Water accounting principles applied to each grid square provides gridded surface and subsurface runoff for input to the G2G routing scheme. This scheme employs kinematic wave principles applied to channel and lateral inflows, and sub-surface runoff. The G2G is hosted on a platform based on Delft-FEWS (Werner *et al.*, 2004) and the forecasting system is referred to hereafter as the National Flood Forecasting System–Flood Forecasting Centre (NFFS–FFC).

Model input data: the data hierarchy

Prior to a G2G Model run, a complete time-series of gridded 15-min data at a 1 km^2 resolution must be generated for the entire domain of the G2G model. For precipitation, the following data sources are available to NFFS-FFC:

– Observed raingauge rainfall, from a network of 974 0.2-mm tipping-bucket raingauges across England and Wales.
– Observed radar rainfall rates, 1 km, 5 min resolution.
– Radar-based forecast rainfall accumulation, 2 km, 15-min time-steps out to 6 h – the STEPS (Short Term Ensemble Prediction System) control forecast – run every 15 minutes (96 forecasts per day).
– NWP forecast rainfall accumulation from the 4 km, NAE (North Atlantic and European, 12 km) and Global (25 km) Met Office Unified Models.
– Met Office Global and Regional Ensemble Prediction System (MOGREPS) ensemble forecast, 25 km, 24 members.

An integral part of this process requires HyradK (Moore *et al.*, 1994, 2005; Moore, 1999; Cole & Moore, 2009) to generate gridded rainfall data based on one of three options: raingauge-only, radar-only or raingauge-adjusted radar. These products are generated using a simple multiquadric surface-fitting technique.

Based on this set of available data, a hierarchy is used to prepare the gridded rainfall input to the G2G model (see Fig. 2). For the period up to the time of forecast, T0, the raingauge-only grid estimate from HyradK is the first data source, then the raingauge-adjusted radar from HyradK, and then the radar-only product. Operation rules are currently set for the raingauge-only option if up-to-date data are available from >75% of raingauges. The last-resort backup in this period, expected to be activated only rarely, is zero precipitation; note that for the last 2 h up to T0, radar-based forecasts or even NWP forecasts are used as a backup source in preference. For the first 6 h of the forecast the radar-based rainfall product (STEPS control forecast) will be used supplemented with (up to 36 h) the 4 km NWP product; this also serves as a backup for the radar-based forecast in the first 6 h of the forecast. For the longer lead-times, up to 5 days ahead, the 12 km NAE model and 25 km Global model products are used (12 km out to 54 h, then 25 km out to 5 days). If all fails, the zero precipitation backup option is used. A separate procedure is set up for the forecast using the MOGREPS ensemble product. In that case MOGREPS is the only source of precipitation in the forecast period (up to 54 h).

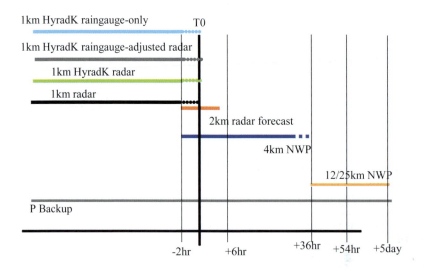

Fig. 2 Diagram showing the available precipitation sources relative to the time of forecast (T0) and their position in the precipitation data hierarchy (top line is first priority) that is applied when constructing input for the G2G.

SYSTEM EVALUATION

The G2G model is currently under a period of system evaluation within the FFC. As well as system and reliability performance, not considered here, particular attention was paid to the model's response to erroneous input data. As already stated, radar rainfall data coupled with the G2G enables the spatial nature of rainfall to be captured and exploited. However, the problems of erroneous radar rainfalls are exacerbated when operating at high spatial resolutions. During the period of system evaluation (since October 2010) the following examples highlighted the effects of such erroneous data.

Example 1 shows how clutter in radar data caused unusually high flows to be predicted at the Environment Agency's Manchester Racecourse flood warning site when little or no rainfall was actually observed or forecast (Fig. 3(a)). The radar observations, shown in Fig. 3(b), show an area of intense rainfall above the catchment feeding the Manchester Racecourse site. Investigation has shown these rainfall intensities not to be genuine, but due to radar clutter caused by Scout Moor wind farm. Clutter is an intermittent but common problem associated with radar rainfall and radar-based rainfall forecasts.

Example 2 shows how a radar spike from High Moorsley Radar (Fig. 4(a)) has resulted in G2G grid squares showing high return period flows where there are no actual river channels (Fig. 4(b)).

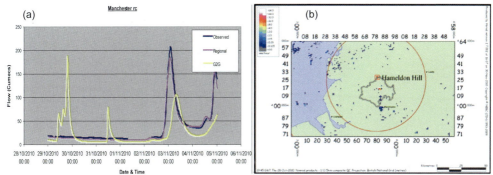

Fig. 3 (a) G2G and Regional model simulations of river flow and observations for Manchester Racecourse (the Regional model assumes knowledge of measured flows upstream) (b) radar actual showing anomalous echoes over the Manchester Racecourse catchment (both courtesy of Lindsay Ness, Environment Agency North West Region).

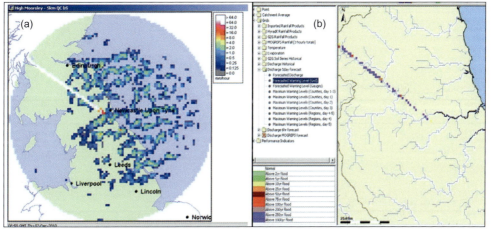

Fig. 4 (a) A radar spike in High Moorsley radar, 2 December 2010, (b) Grid-to-Grid output showing unrealistically high flow predictions away from river channels as a result of a radar spike.

Example 3 shows a test case where HyradK was used to produce raingauge-only and raingauge-adjusted radar estimates of 1-km gridded rainfall (the radar data product used was the UK 1/2/5 km radar composite) to simulate likely G2G performance during the notable Cumbria floods of 2009. Figure 5 maps the 1-km gridded rainfalls accumulated over a 3-day period obtained using three rainfall estimators: HyradK raingauge-only, UK 1/2/5 km radar and HyradK raingauge-adjusted radar. The underestimation of rainfall over Cumbria by the weather radar when compared to the interpolated raingauge image is apparent from an inspection of these maps.

Figure 6 shows the resulting G2G and observed hydrographs for the Cocker and Derwent catchments, which experienced some of the most severe flooding during this event. During this period it is clear that the hydrograph generated using the raingauge-only input compares best to the observations. The raingauge-adjusted radar data also provides a good match although the unadjusted radar data underestimates significantly. These results are consistent with those of Cole & Moore (2009), where further discussion can be found.

(a)   (b)   (c)

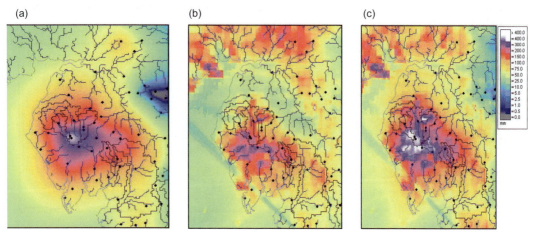

Fig. 5 Map of rainfall accumulations (mm) over Cumbria for the 3-day period ending 09:00 h 20 November 2009 obtained using the following rainfall estimators: (a) HyradK raingauge-only, (b) UK 1/2/5 km radar, and (c) HyradK raingauge-adjusted radar; dots indicate raingauges.

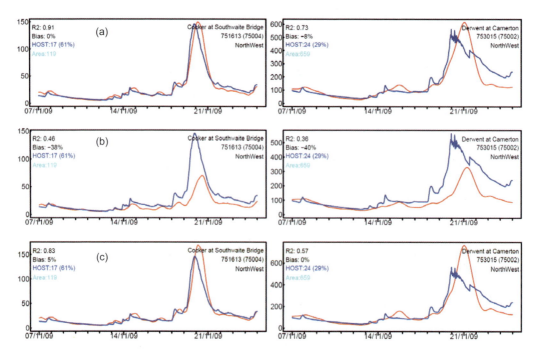

Fig. 6 Simulated hydrographs obtained using the G2G Model for catchments in Cumbria affected by the November 2009 floods. (a) HyradK raingauge-only rainfall, (b) Unadjusted radar rainfall and (c) HyradK raingauge-adjusted radar rainfall. Blue: observed flow; red: simulated flow.

DISCUSSION

For the benefits of a distributed modelling approach to flood forecasting to be fully realised the input data needs to accurately reflect the spatial variability of the rainfall. Such data are uniquely available from the UK radar network which has full coverage for England and Wales, with varying resolution, at a 5-minute interval. The examples given here show that these radar rainfall data are not always of a sufficient standard: with incidences of clutter, spurious radar spikes and radar under-estimating rainfall at the ground, resulting in poor model performance. Poor or erroneous radar rainfall data will impact on operational flood forecasting since radar data are used both in the data hierarchy invoked to generate a complete gridded rainfall time-series for the G2G model and as a basis for producing nowcasts, through the use of STEPS.

In a way that a lumped catchment approach may not, the G2G challenges the consistent quality of the radar rainfall data. As shown in the examples above, small areas of anomalous radar rainfalls can clearly adversely affect subsequent G2G forecasts presented at a 1 km scale, the effects of which would not be so clearly highlighted within forecasts from lumped catchment models. The requirement for good high-resolution rainfall accumulations, reflecting the spatial variability that only radar can provide, will continue as we are challenged to test our capability to forecast for rapidly responding catchments that can be of the order of 25 km^2 or smaller. It is evident that to generate accurate flood forecasts from a distributed hydrological model, such as G2G, requires a reliable and complete time-series of gridded precipitation data which must capture the spatial and temporal variability in observed rainfall. For real-time forecasting, timeliness is as important as capturing spatial and temporal variability and therefore radar still provides the best source for complete and timely rainfall accumulations over England and Wales. It is proposed that a reliable and accurate merged rainfall dataset – incorporating the strengths of both raingauge and radar sources – will play a critical role in meeting future challenges, including the provision of forecasts for rapidly responding catchments.

REFERENCES

Bell, V. A., Kay, A. L., Jones, R. G., Moore, R. J. & Reynard, N. S. (2009). Use of soil data in a grid-based hydrological model to estimate spatial variation in changing flood risk across the UK. *J. Hydrol.* 377, 335–350.

Cole, S. J. & Moore, R. J. (2009) Distributed hydrological modelling using weather radar in gauged and ungauged basins. *Adv. Water Resour.* 32, 1107–1120.

Environment Agency (2010) *Hydrological Modelling using Convective Scale Rainfall Modelling.* Science Report – SC060087. Authors: J. Schellekens, A. R. J. Minett, P. Reggiani, A. H. Weerts (Deltares); R. J. Moore, S. J. Cole, A. J. Robson, V. A. Bell (CEH Wallingford). Environment Agency, Bristol, UK, 240 pp.

Moore, R. J. (1999) Real-time flood forecasting systems: Perspectives and prospects. In: *Floods and Landslides: Integrated Risk Assessment* (ed. by R. Casale & C. Margottini), Chapter 11, 147–189. Springer.

Moore, R. .J., May, B. C., Jones, D. A. & Black, K. B. (1994) Local calibration of weather radar over London. In: *Advances in Radar Hydrology* (ed. by M. E. Almeida-Teixeira, R. Fantechi, R. Moore & V. M. Silva) (Proc. Int. Workshop, Lisbon, Portugal, 11–13 November 1991), Report EUR 14334 EN, European Commission, 186–195.

Moore, R. J., Jones, D. A., Black, K. B., Austin, R. M., Carrington, D. S., Tinnion, M. & Akhondi, A. (1994) RFFS and HYRAD: Integrated systems for rainfall and river flow forecasting in real-time and their application in Yorkshire. In: *Analytical Techniques for the Development and Operations Planning of Water Resource and Supply Systems.* BHS National Meeting, University of Newcastle, 16 November, BHS Occasional Paper no. 4, 12 pp.

Moore, R. J., Bell, V. A. & Jones, D. A. (2005) Forecasting for flood warning. *C. R. Geoscience* 337, 203–217.

Moore, R. J., Cole, S. J., Bell, V. A. & Jones, D. A.(2006) Issues in flood forecasting: ungauged basins, extreme floods and uncertainty. In: *Frontiers in Flood Research* (ed. by I. Tchiguirinskaia, K. N. N. Thein & P. Hubert), (8th Kovacs Colloquium, UNESCO, Paris, June/July 2006), 103–122. IAHS Publ. 305. IAHS Press, Wallingford, UK.

Pitt, M. (2008) Learning lessons from the 2007 floods. An independent review by Sir Michael Pitt. Cabinet Office, London, UK http://webarchive.nationalarchives.gov.uk/20100807034701/http://archive.cabinetoffice.gov.uk/pittreview/thepittreview/final_report.html.

Werner, M., van Dijk, M. & Schellekens, J. (2004) DELFT-FEWS: an open shell flood forecasting system. In: *Proc. 6th Int. Conf. on Hydroinformatics* (World 10 Scientific Publishing Company), 1205–1212.

Countrywide flood forecasting in Scotland: challenges for hydrometeorological model uncertainty and prediction

**MICHAEL CRANSTON[1], RICHARD MAXEY[1], AMY TAVENDALE[1],
PETER BUCHANAN[2], ALAN MOTION[2], STEVEN COLE[3], ALICE ROBSON[3],
ROBERT J. MOORE[3] & ALEX MINETT[4]**

1 *Scottish Environment Protection Agency, Flood Forecasting and Warning Section, 7 Whitefriars Crescent, Perth, UK*
michael.cranston@sepa.org.uk
2 *Met Office, Operations Centre, Davidson House, Aberdeen Science and Technology Park, Aberdeen, UK*
3 *Centre for Ecology & Hydrology, Wallingford, UK*
4 *Deltares, Rotterdamsewag 185, Delft, The Netherlands*

Abstract The Scottish Flood Forecasting Service, a new partnership between the Met Office and the Scottish Environment Protection Agency, aims to make best use of weather and river forecasting expertise in providing improved flood resilience and vigilance for emergency responders across Scotland. Flood guidance employs a blend of experience, professional assessment and input from meteorological and hydrological models. For countrywide forecasts, the CEH-developed Grid-to-Grid model is planned to be the key forecasting tool: it employs rainfall estimates from raingauges, radar and weather models to produce forecast river flows up to 5 days ahead on a 1-km grid across the Scottish mainland. Probabilistic flood forecasts, using ensemble rainfalls as input, are planned in a future phase. Use of rainfall as input to hydrological models is a challenge in Scotland, especially given the terrain and sparse radar and raingauge network coverage, and makes forecasting uncertain. However, the merged hydrological and meteorological capabilities of the new service bring tangible benefits for improved flood forecasting.

Key words flood; forecasting; hydrometeorology

INTRODUCTION

The Flood Risk Management (Scotland) Act 2009 is the framework for managing flood risk in Scotland and gives the Scottish Environment Protection Agency (SEPA) strengthened and formalised duties for flood forecasting and warning. Under the Act, SEPA are committed to deliver a number of flood warning service developments: these include developing methods of working more closely with the Met Office, aiming to improve its technical capability to forecast, model and warn against all sources of flooding. A model for much closer working with the Met Office was developed through consultation with emergency responders, and based on international best practice such as the Flood Forecasting Centre (FFC) in England and Wales and the Service Central d'Hydrometeorologie et d'Appui a la Prevision des Inondations (SCHAPI) in France. The service aims include a combined flood forecasting service for Scotland, with fully integrated meteorological and hydrological aspects providing knowledge-transfer between the meteorological and hydrological services and the provision of regular, consistent information on flood threat to emergency responders (Cranston & Tavendale, 2012). As a result, the Scottish Flood Forecasting Service (SFFS) formally became operational on 8 March 2011.

Routinely, the service consists of collaboration between a SEPA Flood Forecasting Hydrologist generally based at SEPA's Perth office, working on a call-out basis; and the Met Office Public Weather Services desk, based at Met Office Aberdeen, which is a 24-h a day shift-working operation. Although working virtually, capability has been developed for SEPA and Met Office staff to work on their own separate corporate networks at the other organisation's location and co-locate as required. This will facilitate closer co-working during flood events and encourage interchange of knowledge and the build-up of shared expertise in weather and flood forecasting. On a daily basis there is routine dialogue, supported by model outputs, between the Flood Forecasting Hydrologist and the Duty Meteorologist to discuss the upcoming weather and potential flooding situation. The Flood Guidance Statement (FGS) is then compiled, with additional input from SEPA's Flood Warning Duty Officers to assess regional scale impact. The FGS displays information on flood risk from all sources, whether fluvial, tidal or surface water, using colour-coded maps to represent the next five days. The level of detail is greater for days 1 and 2. The flood risk for each area is calculated from an impact-likelihood matrix and is allocated a status of Very Low, Low, Medium or High. The FGS is distributed to emergency responders in Scotland with the aim of improving flood vigilance and resilience to potential flooding. The FGS complements the respective severe weather and flood warning services currently offered by the Met Office and SEPA.

This paper reviews the background to flood forecasting in Scotland, outlines the ongoing implementation of the G2G model for countrywide forecasting especially in relation to its use of rainfall data from radar and weather models, and future plans for probabilistic flood forecasting; the conclusion recognises the critical value of radar rainfall to the new flood forecasting service.

BACKGROUND TO FLOOD FORECASTING IN SCOTLAND

Real-time and forecast weather

The SFFS aims to bring together access to respective meteorological and hydrological forecasting data and tools. This includes sea- and land-based observations and data from rain, river and tide gauges. These are rapidly accessible and monitored via SEPA's Flood Early Warning System (FEWS) Scotland and the Met Office SWIFT system. High-resolution remotely-sensed data – including satellite; lightning detection and radar – contribute to monitoring and validating the rainfall pattern. These data are employed to initialise, calibrate and validate the dynamic weather prediction and hydrological forecast models and diagnostic tools used by the SFFS. Weather prediction models (both deterministic and probabilistic) and post-processing systems cover a variety of temporal and spatial resolutions from the next few hours to beyond 5 days. Deterministic weather model data and visualisation includes a "nowcast" from the UK Post Processing System (UKPP), UKV (1.5 km model to replace the UK 4 km) (Fig. 1), the North Atlantic and European model (NAE) and the Global Model (GM). These data as well as outputs in ensemble form are to be utilised in hydrological modelling, discussed further below.

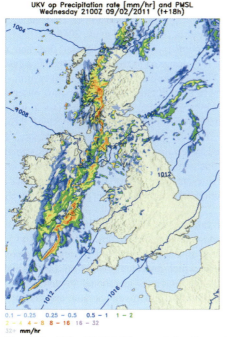

Fig. 1 Met Office UKV 1.5 km resolution forecast rainfall.

Hydrological forecasting

The drive for improved flood forecasting and warning in recent years fits with the move towards a more sustainable flood risk management approach (Tavendale, 2009). Flood warnings provide benefit to communities by allowing them the opportunity to take preventative action to mitigate flood impacts. To ensure confidence in the service, it is essential that sound science underpins the warnings: forecasts should aim to be timely and accurate whilst including an appreciation of their uncertainty. The drivers for an improved flood warning service have led to significant investment in flood forecasting capabilities in recent years. The introduction of major new flood warning schemes for the Strathclyde Region resulted in the development of the FEWS Scotland system – a national flood forecasting system bringing together hydrological and meteorological observations in a real-time environment (Cranston *et al.*, 2007). The best practice developed as part of this work has subsequently led to a rapid expansion of forecasting capabilities (Cranston & Tavendale, 2012).

HYDROLOGICAL MODEL AND FORECAST SYSTEM DEVELOPMENT

Flood forecasting systems have conventionally evolved in a targeted way with regard to cost–benefit appraisals of specific locations at risk, with deployment normally at a catchment, river basin or regional level: e.g. see Moore *et al.* (2009). The development of FEWS Scotland, along with related developments for England and Wales, provided the opportunity to develop integrated systems with consistency of forecasting software infrastructure at a national scale (Werner *et al.*, 2009). The model networks now configured reflect past and ongoing developments at catchment, area and regional scales and are targeted to make forecasts at specific locations.

It became apparent that a complementary, countrywide vision of flood risk in space and time was required. Research had progressed on distributed flood forecasting models for practical use, with particular focus on forecasting for ungauged areas and with the capability to forecast river flow "everywhere" over a continuous gridded domain (Moore *et al.*, 2006). A further impetus came from the Pitt Review of the summer 2007 floods, which recognised the need for a consistent countrywide early-alert of flood risk. The operational requirement and research activity combined to accelerate the development and assessment of CEH's Grid-to-Grid (G2G) model for countrywide operational deployment (Environment Agency, 2010; Cranston & Tavendale, 2012), together with improvements in its formulation (Moore *et al.*, 2007; Bell *et al.*, 2009).

The G2G model is a physical-conceptual distributed hydrological model that has runoff production and runoff routing components (Fig. 2). It is designed to make use of spatial datasets on terrain, soil, geology and land-cover properties. These underpin the spatial configuration of the model, leaving only a modest number of parameters to be calibrated against river flow observations across the country. Runoff production is controlled by a saturation excess mechanism in which the capacity of the soil to absorb water is controlled by soil properties and terrain slope through a probability-distributed formulation (Moore, 1985; Bell & Moore, 1998; Cole & Moore, 2009). Lateral transport of water through the soil, controlled by terrain slope and soil properties, can also be simulated. Routing of surface runoff through hillslope and channel pathways, under terrain control, employs kinematic wave approximations; similar representations are used to route subsurface flows in the groundwater component of the model. Interaction between water in the ground and the channel is allowed for through "return flow" functions.

In 2010, the Scottish Government approved funding to implement the G2G model across Scotland in support of the new flood forecasting service to enable river flow forecasts out to 5 days. Delivery of a configuration across Scotland was made in early March 2011 for testing purposes. A component of the system is HyradK (discussed later) which provides gridded rainfall estimates as input to the G2G from radar rainfall and/or raingauge sources. Operational implementation and evaluation was planned to start in June 2011, and model improvement and assessment using historical datasets is ongoing. Key ongoing developments will include the integration of a snowmelt component to the gridded model and production of probability-based forecasts.

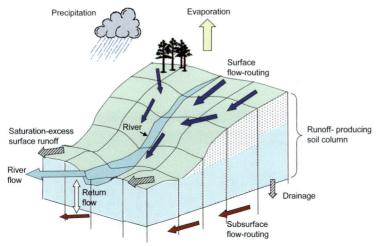

Fig. 2 Conceptual schematic of the G2G model.

Scotland's current flood forecasting system, FEWS Scotland, employs the Delft-FEWS operational flood forecasting platform (Werner *et al.*, 2004; Cranston *et al.*, 2007). This brings together meteorological (radar and forecast grids), hydrological (raingauges and river levels), reservoir (levels and outflows) and tidal (observed and astronomical) information. These data are imported, processed and displayed and are available to drive external hydrological and hydrodynamic river models whose outputs can be post-processed and evaluated.

To enable countrywide flood forecasting, both the HyradK and G2G models have been integrated as external modules into Delft-FEWS. HyradK brings together observed precipitation from SEPA's network of 185 raingauges and the Met Office composite radar product to produce an adjusted precipitation grid. The adjusted precipitation is then merged in Delft-FEWS with the forecasted NWP products to produce a continuous estimate of precipitation for input to the G2G model. River flow data from 214 gauges is also input to the G2G model to enable state-correction.

In a post-processing module, the G2G modelled river flows on a 1-km grid are compared with "flows of a given return period" grids for Scotland to give an estimate of the exceedence return period for each grid square. These are then transformed to give the warnings at gauge locations and maximum warning levels for larger flood alert areas. This enables the forecasting hydrologist to quickly identify where floods may occur for further verification and analysis.

UNCERTAINTY AND PREDICTION

Precipitation source hierarchy

The G2G model uses as input estimates of 15 min rainfall accumulations on a 1 km grid across Scotland. The observation sources available to make these estimates, up to the time the flood forecast is made, are: (i) rainfall values from a network of 182 tipping-bucket raingauges, and (ii) radar rain-rate values at 5 min intervals on a 1 km grid, derived as a composite primarily from the four radar installations in Scotland.

HyradK, the hydrological radar processing kernel of Hyrad developed by CEH, is used to form 15-min 1-km gridded rainfall accumulations of the three forms shown in Fig. 3. The left image shows the raingauge-only estimate, for an example time-frame, obtained by applying a multiquadric surface-fitting technique (Moore *et al.*, 2001, 2004; Cole & Moore, 2009) to the raingauge values alone (there are 25 gauges not reporting values, indicated in red). The right image shows the radar-only rainfall estimate formed using the 5 min rain-rates. A raingauge-adjusted radar estimate is shown in the central image, obtained by surface fitting to adjustment factors at each raingauge, formed as a modified ratio of gauge to coincident radar pixel rainfall values. Raingauge-adjusted estimates are seen in this example to be modified most in the central area of the image, where raingauges report larger values than indicated by the radar, but little change in the south where there is reasonable agreement.

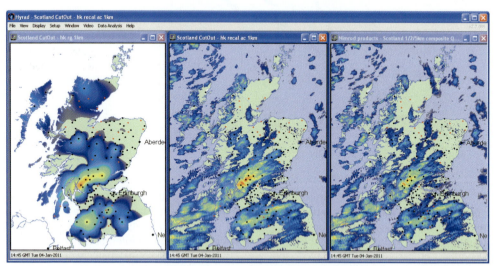

Fig. 3 HyradK 1 km gridded rainfall estimates over Scotland for 15 minute totals ending 14:45 h 4 January 2011. Left: raingauge-only; Middle: raingauge-adjusted radar; Right: radar-only. Raingauge locations indicated by black dots (red when missing observation).

A configurable hierarchy of gridded rainfall estimates is used within FEWS Scotland to decide on the rainfall input G2G will use. One possibility, in review as part of G2G calibration, is for the raingauge-adjusted radar estimate to have priority followed by radar-only, and then raingauge-only if radar data become unavailable. For future times, to obtain extended lead time forecasts out to 5 days, the system will use a 6 h 2 km deterministic "nowcast" (merging of advected radar data with short-period NWP) along with a 5-day NWP forecast based, as appropriate, on the UKV, NAE and Global atmospheric models. Provision is also being made to use 5-day temperature forecasts in the G2G snowmelt hydrology scheme which produces gridded estimates of snowmelt.

Weather radar and precipitation forecasting

Although the Central Lowlands of Scotland has high-resolution radar rainfall coverage, many parts suffer from poor (e.g. Highlands) or no coverage (Shetland Islands). However, the use of radar for flood forecasting in Scotland has demonstrable benefits (Cranston & Black, 2006), with raingauges unable to provide as effective spatial analysis of rainfall. Also, despite advances in numerical modelling, there remains uncertainty in the predictions of rainfall, especially at the resolutions and timescales associated with extremes storms of convective origin. Short-range rainfall forecasting tools such as STEPS (Bowler *et al.*, 2006), used as the basis of the deterministic nowcast, are heavily influenced in the first few hours by the quality of radar rainfall data. Poor coverage and radar anomalies can inevitably affect the skill of the rainfall forecasts and modelled river flows using them e.g. radar "anaprop" can lead to false alarms. However, with knowledge of these uncertainties the benefit of using rainfall predictions in flood forecasting is still greater than using none (Werner & Cranston, 2009). At larger synoptic scales and dealing with organised dynamic rainfall, tools like the Met Office's ensemble rainfall forecasts can be used to support probabilistic rainfall and flood forecasting. The "likelihood" of a flooding event occurring at a particular time and place can be predicted. Ensemble rainfall prediction in combination with G2G area-wide flood forecasting are seen as key tools helping address the operational challenge of dealing with uncertain forecasts.

CONCLUSION

The Scottish Flood Forecasting Service, although in relative infancy, is developing methods of close collaborative working so as to improve its capability to model, forecast and warn against all sources of flooding in Scotland. Uncertainties associated with both meteorological and hydrological forecasting techniques may be compounded when merged into a unified flood forecasting procedure for the whole of Scotland. However, methods such as HyradK and the precipitation source hierarchy presented here, alongside expert evaluation, aim to mitigate inaccuracies with radar rainfall, raingauge and weather model sources. Weather radar is a critical data source in this information chain leading to improved flood guidance. The forecasting service, by combining tools and expertise of meteorologists and hydrologists, issues a Flood Guidance Statement underpinned by a careful analysis of likelihood against impacts. This daily assessment of risk forms a valuable strategic and sustainable enhancement to flood management in Scotland.

REFERENCES

Bell, V. A. & Moore, R. J. (1998) A grid-based distributed flood forecasting model for use with weather radar data: Part 1. Formulation. *Hydrol. Earth System Sci.* 2(2–3), 265–281.

Bell, V. A., Kay, A. L., Jones, R. G., Moore, R. J. & Reynard, N. S. (2009) Use of soil data in a grid-based hydrological model to estimate spatial variation in changing flood risk across the UK. *J. Hydrol.* 377, 335–350.

Bowler, N. E., Pierce, C. E. & Seed, A. W. (2006) STEPS: a probabilistic precipitation forecasting scheme which merges an extrapolation nowcast with downscaled NWP. *Quart. J. Roy. Met. Soc.* 132, 2127–2155.

Cole, S. J. & Moore, R. J. (2009) Distributed hydrological modelling using weather radar in gauged and ungauged basins. *Adv. Water Resour.* 32, 1107–1120.

Cranston, M. & Black, A. (2006) Flood warning and the use of weather radar in Scotland: a study of flood events in the Ruchill Water catchment. *Met. Appl.* 13, 43–52.

Cranston, M., Werner, M., Janssen, A., Hollebrandse, F., Lardet, P., Oxbrow, J. & Piedra, M. (2007) Flood Early Warning System (FEWS) Scotland: an example of real time system and forecasting model development and delivery best practice. In: *Defra Conference on Flood and Coastal Management*, Paper 02-3, York, UK.

Cranston, M. & Tavendale, A. (2012) Advances in operational flood forecasting in Scotland. *Water Manage.* 165(WM2), 79–87.

Environment Agency (2010) Hydrological Modelling using Convective Scale Rainfall Modelling. *Science Report – SC060087*. Authors: J. Schellekens, A. R. J. Minett, P. Reggiani, A. H. Weerts (Deltares); R. J. Moore, S. J. Cole, A. J. Robson, V. A. Bell (CEH Wallingford). Environment Agency, Bristol, UK, 240pp.

Moore, R. J., Bell, V. A., Cole, S. J. & Jones, D. A. (2007) Rainfall-runoff and other modelling for ungauged/low-benefit locations. *Science Report – SC030227/SR1*, Research Contractor: CEH Wallingford, Environment Agency, Bristol, UK, 249 pp.

Moore, R. J., Bell, V. A. & Jones, D. A. (2005) Forecasting for flood warning. *C. R. Geoscience* 337, 203–217.

Moore, R. J., Cole, S. J., Bell, V. A. & Jones, D. A. (2006) Issues in flood forecasting: ungauged basins, extreme floods and uncertainty. In: *Frontiers in Flood Research* (ed. by I. Tchiguirinskaia, K. N. N. Thein & P. Hubert), 103–122. 8th Kovacs Colloquium, UNESCO, Paris, June/July 2006. IAHS Publ. 305. IAHS Press, Wallingford, UK.

Moore, R. J., Jones, A. E., Jones, D. A., Black, K. B. & Bell, V. A. (2004) Weather radar for flood forecasting: some UK experiences. In: *6th Int. Symp. on Hydrological Applications of Weather Radar* (2–4 February 2004, Melbourne, Australia).

Moore, R. J., Watson, B. C., Jones, D. A. & Black, K. B. (1991) Local recalibration of weather radar. In *Hydrological Applications of Weather Radar* (ed. by I. D. Cluckie & C. G. Collier), 65–73. Ellis Horwood, Chichester, UK.

Tavendale, A. C. W. (2009) Grant-aided flood management strategies in Scotland and England 1994–2004: Drivers, policy and practice. PhD thesis, University of Dundee (unpublished).

Werner, M., van Dijk, M. & Schellekens, J. (2004) DELFT_FEWS: an open shell flood forecasting system. In: *6th Int. Conf. on Hydroinformatics* (ed. by S. Y. Liong, K. Phoon & V. Babovic), 1205–1212. World Scientific Publishing Company, Singapore.

Werner, M. & Cranston, M. (2009) Understanding the value of radar rainfall nowcasts in flood forecasting and warning in flashy catchments. *Met. Appl.* 16, 41–55.

Werner, M., Cranston, M., Harrison, T., Whitfield, D. & Schellekens, J. (2009) Recent developments in operational flood forecasting in England, Wales and Scotland. *Met. Appl.* 16, 13–22.

Distributed flood forecasting for the management of the road network in the Gard Region (France)

J.-P. NAULIN, E. GAUME & O. PAYRASTRE

French Institute of Science and Technology for Transport, Development and Networks, Centre de Nantes, France
jean-philippe.naulin@ifsttar.fr

Abstract A prototype of a road submersion warning system, providing a rating of road submersion risks every 15 minutes during a storm event, at about 2000 points where roads and rivers intersect, has been developed for the Gard region (French Mediterranean area). The computed risks result from the confrontation between discharges produced by a distributed rainfall–runoff model and the estimated susceptibility to flooding of the intersection points. The warning system is validated against road inundations reported by the local road management service. The comparison of the performances of this framework fed with various quantitative precipitation estimates (1-km^2 grid interpolation of point rainfall measurements or radar products) is presented herein. This case study, based on highly distributed rainfall–runoff modelling and on a rich set of indirect observations of the flood magnitudes, provides *a priori* an ideal framework to evaluate the usefulness of weather radar products for hydrological applications.

Key words flash flood; radar; road network; rainfall–runoff; ungauged

INTRODUCTION

Operational flash flood forecasting represents a major issue for the definition of efficient management strategies for flood disasters. Since damage often occurs in headwater catchments (Ruin *et al.*, 2008), flash flood forecasts should ideally be spatially distributed (i.e. based on high-resolution quantitative precipitation estimates and distributed rainfall–runoff models). Most of these catchments are ungauged. The validation of distributed flash flood forecasting tools has therefore to rely partly on indirect observations of the flood magnitudes.

A prototype of a road inundation warning system (RIWS) has been developed in a small test area covering about 100 km^2 in the Gard region (French Mediterranean area) which is frequently affected by severe flash floods inducing disruptions of the road network (Versini *et al.*, 2010a,b). This prototype has been recently extended to the entire Gard region, providing warnings at a 15-min time-step at about 2000 intersection points between roads and rivers over a 5000 km^2 area. The warning system combines two components: a highly distributed rainfall–runoff model fed with 1-km^2 grid QPEs producing discharge estimates at each considered intersection point that are compared to discharge thresholds evaluated through a model of susceptibility to flooding of the intersections.

Since 2002 and more systematically since 2007, the local road management service records the reported inundated roads. In 2007 and 2008, four floods induced inundations of roads and 77 sections have been reported to be flooded. During the 2002 extreme flood event, 367 flooded intersections were recorded. Even if this inventory is probably not exhaustive, it constitutes a unique dataset providing indirect information on the magnitude of the floods affecting the ungauged headwater catchments for the evaluation of the warning system itself, and indirectly for the assessment of each of its elements. It will be used here, along with more standard stream gauge measurements available on the main streams of the region and post-event peak discharge estimates on intermediate range catchments, to evaluate the added value of radar QPEs for flood simulation and forecasting.

DATA AND METHODOLOGY

The various tested QPEs

The Gard region is covered by a dense observation network including 37 automatic raingauges and two S-band weather radars. On the basis of this available dataset, four quantitative precipitation estimates at a 1 km^2 grid have been produce (the various QPE products have been computed by our colleagues of the LTHE, Laboratoire d'étude des Transferts en Hydrologie et Environnement, Université de Grenoble (CNRS, UJF, IRD, INPG), France):

– CLIM: interpolated average hourly rainfall fields obtained through ordinary Kriging with a "climatologic" isotropic variogram (Lebel *et al.*, 1987).
– KED: interpolated average hourly rainfall fields obtained through Kriging with external drift. Contrary to the ordinary Kriging, the expectancy of the locally interpolated intensity E[Z(x)] is not supposed to be constant, equal to the average of the measured rainfall intensity field, but is a linear function of the estimated radar intensity s(x) at the same location: E[Z(x)]=a+b s(x).
– TRADHYhora: hourly radar QPEs produced with the approach described in Delrieu *et al.* (2009) combining radar signal analysis, rainfall structure decomposition to separate areas with dominant convective or stratiform rain having contrasted vertical reflectivity profiles, and radar visibility diagnosis based on the digital elevation model. It includes three steps. First, the ground clutters are identified and eliminated. The vertical reflectivity profiles are then identified and different error sources are corrected (attenuation, partial beam blockage). Finally, the Z-R relationship is adjusted using the raingauge measurements.
– TRADHY15min: same as TRADHYhora but at a 15-minute time-step.

According to the density of the raingauge network (about 1 gauge for 150 km²) and to the decorrelation distances (about 25 km for hourly rainfall intensities), it did not appear reasonable to interpolate spatially average intensities over time-steps shorter than one hour. For the purpose of the rainfall–runoff simulation at a 15-minute time resolution, each interpolated intensity value has been repeated four times.

The road inundation warning system

The RIWS relies on a comparison between simulated stream discharges and discharge thresholds that depend on the estimated susceptibility to flooding of each intersection and on the estimated discharge quantiles of its upstream watershed. The CINECAR rainfall–runoff model is used to simulate the discharges (Gaume *et al.*, 2004). It is a simple distributed model in which the watershed is described as a network of stream reaches to which one or two hillslopes are connected. For this specific application, the Gard region has been divided into 3288 hillslopes with a mean surface of 1.8 km². At each time-step, the SCS-CN model is used to compute the effective rainfall on each hillslope. The resulting runoff is then propagated over the hillslopes and into the river network using the kinematic wave model or the Hayami diffusive wave model for river reaches with slopes lower the 1%. The "Curve Number" values proposed by the USDA-Soil Conservation Service (1985) depending on the land use, soil type and 5-days antecedent rainfall appeared to produce satisfactory results for the Gard and were therefore selected without further calibration. The celerity and diffusivity of the flood wave propagation model were adjusted based on the available streamgauge measurements for major flash floods occurring in 2002, 2003, 2005 and 2006.

In parallel, the analysis of the reported past inundations of roads, led to the definition of four classes of susceptibility to flooding (Versini *et al.*, 2010a). The allocation of each intersection between a road and a river to a class depends on its location (flood plain, mountainous area), on the local slope, altitude and the upstream watershed area. The high susceptibility sections are located in the downstream flood plains (Fig. 1). A contingency table (Table 1) defines the inundation risk level at each section and computation time-step, depending on the simulated discharge, on the local estimated discharge quantiles and on the susceptibility level of the considered section.

Methodology for evaluation of RIWS performances

The comparison between the four QPEs has been conducted in three steps extending progressively the analysis to the headwater catchments and hence necessitating *a priori* higher spatial resolution rainfall inputs.

The hydrograph simulated on the main streams were first compared to the hydrographs measured at the 32 available stream gauges (upstream watershed areas ranging from 50 to 2000 km²) with the Nash-Sutcliffe criterion.

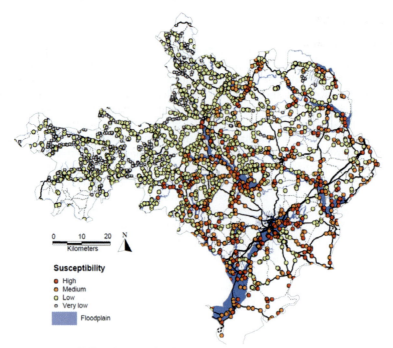

Fig. 1 Susceptibility classes defined at each intersection between roads and stream networks in the Gard Department.

Table 1 Contingency table for the determination of the inundation risk level at each road/river intersection.

Susceptibility level	Discharge thresholds			
	$Q_2/2$	Q_2	Q_{10}	Q_{50}
High	Level 2	Level 3	Level 3	Level 3
Medium	Level 1	Level 2	Level 3	Level 3
Low	No risk	Level 1	Level 2	Level 3
Very low	No risk	No risk	No risk	Level 1

Q_2: 2-year return period discharge, Q_{10}: 10-year return period discharge, Q_{50}: 50-year return period discharge.

Post-event surveys conducted after two major flash floods occurring in 2002 and 2008 provided estimates of peak discharges for 79 cross-sections with upstream watershed areas ranging from 1 to 200 km². The relative absolute differences between the simulated (Q_s) and estimated (Q_e) peak discharges were compared, i.e.

$$A_d = \frac{|Q_e - Q_s|}{Q_e}$$

Finally, the efficiency of the RIWS fed with the various QPEs in detecting the reported inundated road intersections have been compared for five recent floods events for which inundations were reported. The upstream watershed area of the considered intersections ranges from 3.10^{-4} to 2500 km², with a median of 1 km². Two criteria were therefore computed. The probability of detection (POD) is the ratio between the number of roads affected at least by a risk level 1 during the event and which have been inundated (correct warnings) and the total number of road sections reported as inundated. The probability of false detections (POFD) is the ratio between the number of roads with no inundation reported where at least a risk level 1 has been computed (false alarm) and the total number of roads with no inundation report. A perfect warning would provide a POD of 1 and a POFD of 0.

RESULTS AND DISCUSSION

Comparison of flood hydrographs

The synthesis of results for 26 flood hydrographs that were not considered for the adjustment of the rainfall–runoff model is presented as boxplots in Fig. 2. The Nash-Sutcliffe criterion are modest, but in the range of other similar studies focused on the same issue: flood hydrograph simulations at short time-steps, on Mediterranean catchments in autumn. As in previous studies, the possible added value of the radar data is hardly detectable. The simulations based on the radar QPEs seem to produce a lower proportion of poor criterion values, but this does not appear to be statistically significant according to a Friedman test.

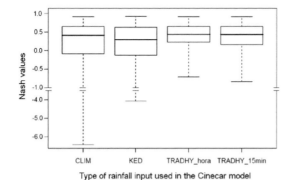

Fig. 2 Boxplot showing the distribution of the Cinecar performances based on Nash-Sutcliffe criterion and obtained with four rainfall inputs. Percentiles shown on the boxplot are 0, 0.25, 0.50, 0.75 and 1.

The information on the spatio-temporal repartition of the rainfall is probably partly filtered out by the rainfall–runoff response for such scales of watersheds. The spatially variable accuracy of the radar QPEs can also be put forward to explain this result: local gains may be compensated by errors in other areas.

Comparison of flood peak discharges

The overall comparison between simulated discharges and post-event estimates is satisfactory (relative absolute difference A_d generally lower than 50%), validating both: the rainfall–runoff model and the post-event field estimates. The synthesis of results in Table 2 seems to indicate that radar-based QPEs (KED and TRADHY) lead to more fluctuating A_d than interpolated rainfall (i.e. larger standard deviations). Figure 3 shows that this fluctuation is due to a limited number of very large A_d values. As a consequence, if the median values of A_d are smaller for the radar-based QPEs, it is not the case for the mean values. Radar-based QPEs seem to lead generally to better discharge estimates, but appear to be locally affected by large errors that reduce their overall performance.

The large deviations from the estimated discharges are systematically larger with TRADHY15min than with TRADHYhora for small watersheds. They may correspond to large over-estimations of rainfall intensities that can be observed on radar-based QPEs at small time resolutions. The differences between TRADHY15min and TRADHYhora is less pronounced for watershed areas exceeding 20 km^2, indicating that large deviations are rather due in this case to systematic local QPEs errors that could not be completely reduced by the radar data processing algorithms.

Comparison of the RIWS performances

The RIWS performance over the whole Gard region measured through the POD and POFD are similar to those obtained previously for limited tested areas and other flood events (Versini,

Table 2 Synthesis of absolute differences obtained for CLIM, KED and TRADHY QPEs.

	8 September 2002 (40 values)			19 October 2008 (39 values)		
	Mean	Median	Standard-Deviation	Mean	Median	Standard-Deviation
CLIM	0.45	0.43	0.29	0.48	0.43	0.25
KED	0.44	0.29	0.58	0.43	0.39	0.42
TRADHY hora	0.53	0.35	0.65	0.49	0.38	0.51
Tradhy15min	0.56	0.32	0.80	0.49	0.42	0.54

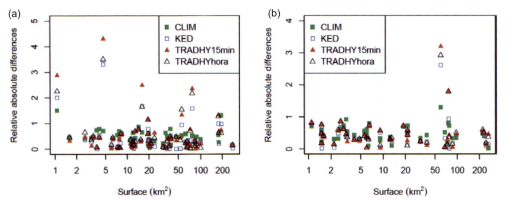

Fig. 3 Comparison between CLIM, KED and TRADHY absolute differences as a function of the surface for the events of (a) 8 September 2002 and (b) 19 October 2008.

2010b). This validates the extension of the system to the entire region. Considering the numerous sources of uncertainties, such results can be considered as satisfactory from a practical point of view: POD between 60 and 99% (larger for the most intense events) and POFD lower than 25%, except for the 2002 extraordinary and generalized flood event.

Contrary to what could have been expected according to the limited median upstream watershed areas of the considered intersections, the POD and POFD of the RIWS are not sensitive to the QPEs (Fig. 4) except for the case of the November 2007 flood where only 11 inundated roads have been reported: the differences in POD corresponding to ±1 detection are not significant. These global scores mask nevertheless significant differences between the RIWS results obtained with the various QPEs as illustrated in Fig. 5.

Depending on the event, the risk levels computed using CLIM and TRADHY15min differ for 30–60% of the intersections for which an alarm is issued. The various QPEs lead to clearly contrasted inundation risk maps. The RIWS results are sensitive to the QPEs used. The fact that

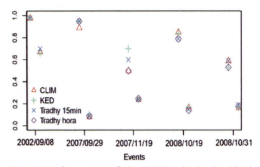

Fig. 4 Performances of the RIWS obtained with the four tested QPEs. For each event, POD are

represented on the left and POFD on the right.

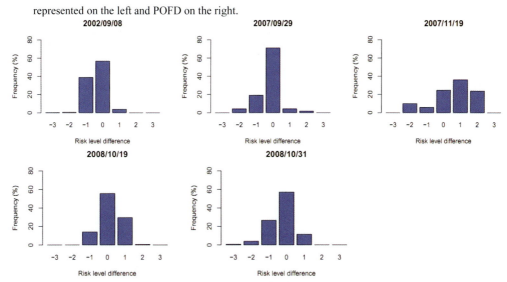

Fig. 5 Distributions of the differences between risk levels computed with TRADHY15min and CLIM for the intersections affected by a warning in the case of the five considered floods.

this is not reproduced in the global scores indicates that local improvements are compensated by worsening in other areas. This is in accordance with the conclusion based on the comparison with the post-event discharge estimates. The spatial variability of the accuracy of the radar-based QPEs used herein limits their added value for the tested applications if compared to the interpolated raingauge data.

CONCLUSION

The dense dataset collected in the Gard region and the considered highly distributed flood forecasting exercise provided a new and *a priori* ideal framework to evaluate radar-based QPEs for hydrological applications. The tested radar QPEs lead to forecasting and simulation efficiencies that are at least as good as the efficiencies obtained with interpolated raingauge measurements, but not significantly better. According to the presented results, the added value of the radar-based QPEs used herein for the tested applications when compared to the interpolated raingauge data, seems to be mainly limited by the remaining spatial variability of their accuracy.

REFERENCES

Delrieu, G., Boudevillain, B., Nicol, J., Chapon, B., Kirstetter, P. E., Andrieu, H. & Faure, D. (2009) Bollène-2002 Experiment: radar quantitative precipitation estimation in the Cévennes-Vivarais region (France). *J. Appl. Meteor.* 48(7), 1422–1447.

Gaume, E., Livet, M., Desbordes, M. & Villeneuve, J. P. (2004) Hydrological analysis of the river Aude, France, flash flood on 12 and 13 November 1999. *J. Hydrol.* 286, 135–154.

Lebel, T., Bastin, G., Obled, C. & Creutin, J.-D. (1987) On the accuracy of areal rainfall estimation: a case study. *Wat. Resour. Res.* 23, 2123–2138.

Ruin, I., Creutin, J. D., Anquetin, S. & Lutoff, C. (2008) Human exposure to flash floods – Relation between flood parameters and human vulnerability during a storm of September 2002 in Southern France. *J. Hydrol.* 361, 199–213.

USDA-SCS (1985) *National Engineering Handbook, Section 4 – Hydrology.* Washington, DC, USA.

Versini, P.-A., Gaume, E. & Andrieu, H. (2010a) Assessment of the susceptibility of roads to flooding based on geographical information – test in a flash flood prone area (the Gard region, France). *Nat. Hazards Earth Syst.Sci.* 10, 793–803.

Versini, P.-A., Gaume, E. & Andrieu, H. (2010b) Application of a distributed hydrological model to the design of a road inundation warning system for flash flood prone areas. *Nat. Hazards Earth Syst. Sci.* 10, 805–817.

550

Weather Radar and Hydrology
(Proceedings of a symposium held in Exeter, UK, April 2011) (IAHS Publ. 351, 2012).

The AIGA method: an operational method using radar rainfall for flood warning in the south of France

PIERRE JAVELLE[1], JEAN PANSU[2], PATRICK ARNAUD[1], YVES BIDET[2] & BRUNO JANET[3]

1 *Cemagref, Centre Régional d'Aix-en-Provence, CS 40061, 13182 AIX EN PROVENCE Cedex 5, France*
pierre.javelle@cemagref.fr
2 *Météo-France, Direction Interrégionale Sud-Est, 2, Bd Chateau-Double, 13098 Aix-en-Provence cedex 02, France*
3 *Ministère de l'Ecologie, du développement durable des transports et du logement (MEDDTL), Direction gérérale de la prévention des risques (DGPR) Service des risques naturels et hydrauliques (SRNH), Service Central d'Hydrométéorologie et d'Appui à la Prévision des Inondations (SCHAPI), 42, avenue Gaspard Coriolis 31057, Toulouse Cedex 01, France*

Abstract This paper aims to present the operation of the AIGA flood warning system. Developed by Cemagref and Météo-France, this method combines radar rainfall and a simple distributed hydrological model taking into account antecedent soil moisture conditions. Discharges are calculated at ungauged points on the river network and compared to statistical reference values. Depending on the occurrence level of the on-going event, different warnings are emitted in real-time. The case study presented focuses on the dramatic event of 15 June 2010 in the area surrounding the town of Draguignan, south of France.

Key words flood warning system; ungauged catchments; post event analysis

INTRODUCTION

Flash flood forecasting is one of the most difficult tasks in operational hydrology: by definition, affected catchments have a very short time response and, very often, no data are available to calibrate the models. In Europe, as mentioned by Gaume *et al.* (2009), particularly Mediterranean regions are subject to such events, but other inner continental countries can be affected as well, with dramatic consequences. Thus, there is a real need for operational tools, which make it possible to better anticipate this kind of event. The classical approaches for flood forecasting generally combine conceptual rainfall–runoff models that generate upstream discharges and hydraulic models that propagate this through the monitored river network (Moore *et al.*, 2005; Rabuffetti & Barbero, 2005). In such systems, real-time streamflow information can be assimilated in order to improve the forecasts (Berthet *et al.,* 2009). But this approach cannot be adapted to small ungauged catchments located outside the monitored network. The flash flood guidance (FFG) method, used routinely in the USA, aims at providing flood warnings at ungauged locations. However, as mentioned by Reed *et al.* (2007), this method is based on a lumped model (the Sacramento model), using parameters transferred from bigger gauged catchments, and therefore can lead to scale problems. As shown by Reed *et al.* (2007), Blöschl *et al.* (2008) and Cole & Moore (2009), simple distributed models can now be used at an operational level, and provide realistic forecasts at ungauged locations.

In this context, the aim of this paper is to present a flood warning system currently used in the south of France for ungauged catchments. The case study analysed concerns a dramatic event (causing 25 casualties) which affected the region of Draguignan on 15 June 2010. This approach, called AIGA, combines radar rainfall and a simple distributed hydrological model taking into account antecedent soil moisture conditions. The next section describes the method used in the AIGA approach. Then a case study application is presented followed by a few concluding remarks concerning future needs.

THE AIGA APPROACH

Operational since 2005, the AIGA method has been developed by Cemagref and Météo-France with financial support from the French Ministry in charge of ecology. The method is still subject to research and improvement. Run by Météo-France for the French Mediterranean region, the system produces, every 15 minutes, a map of the river network with a colour chart indicating the return period of the ongoing flood event. The system is based on a simple distributed hydrological model using radar rainfall as input. The model and the methodology used for parameter calibration have been described by Javelle *et al.* (2010).

Soil moisture accounting scheme

As shown by Merz & Blöschl (2009), the response of catchments to a rainfall event is greatly influenced by antecedent soil moisture conditions which can be correctly estimated using soil moisture accounting (SMA) models. The AIGA method uses daily soil moistures calculated in each cell by a SMA scheme derived from the lumped model proposed by Perrin *et al.* (2003).

Rainfall–runoff model

The distributed rainfall–runoff model is run in an event-based mode, at an hourly time-step. For each 1-km^2 cell, elementary discharges are generated using a simple conceptual scheme (production and transfer store). These discharges are then aggregated to the catchments outlets. Initialisation of the production store is done using information given by the SMA model, as described by Javelle *et al.* (2010).

Warning levels determination

Real-time calculated discharges are compared with reference annual peak discharge values. In order to give a simple scale of the gravity of the event, a colour code is adopted: yellow if the discharge is ranging from the 2-year to the 10-year flood, orange if the discharge is ranging from the 10-year to the 50-year flood, and red if the discharge exceeds the 50-year flood. Thus, 2-year, 10-year and 50-year floods should be defined at any point of the river network. To do that, a rainfall generator was regionalised (Arnaud *et al.*, 2008) and coupled to the rainfall–runoff model using the method described by Arnaud & Lavabre (2002).

Real-time data used

In real-time, the system needs only precipitation data. They are of two types: daily interpolated raingauges (around one gauge for 100 km^2) and a 15-min multi-sensor radar-gauge product called "PANTHERE". Both types of data are provided on 1 km^2-grids. The daily interpolated rainfall is used by the SMA model to estimate initial soil moisture. The PANTHERE grids are used by the rainfall–runoff model. This Météo-France product combines data provided by the weather radar network and raingauge data (Parent du Chatelet *et al.*, 2003; Tabary, 2007; Tabary *et al.*, 2007). Since the hydrological model runs at an hourly time-step, the 15-min grids are cumulated to 1 h, and four models are run independently with a 15-min shift, in order to always give the most updated information. It should be mentioned that the system does not need other real-time data such as discharges or air temperature. Concerning the evaporation needed by the SMA model, it is simply estimated using mean air temperature derived from long-term averages, as described by Oudin *et al.* (2005a,b).

Results dissemination

AIGA real-time products are delivered every 15 minutes. These results are used as an input for a web site, dedicated to French local authorities. They are also sent to a network of specific computers, used by flood forecasting services and allowing forecasters to visualize and manipulate various calculated data. AIGA may also be re-run on past events, enabling more accurate studies.

Comparison with rainfall threshold warnings

Often, warnings at ungauged catchments are simply issued using predefined rainfall thresholds. Javelle *et al.* (2010) compared this approach with the AIGA method. The data that was used for this comparison covered the 2005–2009 period and involved 176 French catchments located in the south of France. Results showed that the AIGA method detected more alerts than the rainfall method, with fewer false alarms. This can be explained by the fact that the hydrological modelling takes into account the initial soil moisture, while the rainfall method emitted warnings for the same rainfall threshold independently of the initial moisture conditions.

THE 15 JUNE 2010 FLOOD AROUND DRAGUIGNAN

Description of the rainstorm event

On 15–16 June, the region of Draguignan in the south of France was affected by a rainstorm event of an exceptional intensity caused by stationary thunderstorms (Fig. 1). Precipitation started on the morning of 15 June and lasted almost 24 h, with a maximal intensity in the afternoon of the same day. During this time, the maximal amount recorded by a raingauge located in the most exposed area reached 461 mm (at Lorgue), with a maximal intensity of almost 80 mm/h at 15:00 h (local time). Recorded values at raingauges exceeded by far the highest historically known values (the longest time series in the area are around 80 years). The spatial extent of the event was considerable, with, for example, a 100 km^2 area experiencing at least 300 mm.

Consequences

The local authorities reported 26 casualties. Emergency services saved 2450 people, including 1350 rescued by helicopters. The area around the town of Draguignan was the most damaged. It is within the Nartuby River catchment (234 km^2), a tributary of the Argens River which is a small coastal river (2500 km^2). The evacuated areas concerned some sensitive zones, such as a hospital, a retirement home, the prison of Draguignan, camping sites, as well as residential areas.

Warning issued by the AIGA approach

The river network in this area was not monitored in real-time (now it is). So the only information concerning river response to the rainfall was given by the AIGA approach. Figure 2 presents the

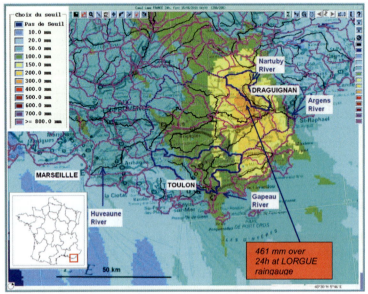

Fig. 1 Total rainfall during the event (ANTILOPE radar product) and localisation of the highest value recorded by the raingauge at Lorgue (461 mm).

calculated situation at 14:45 h, 16:15 h, 16:30 h and 17:15 h (local time), on 15 June. At 16:15 h, it indicates that the discharge of the Nartuby River (crossing the town of Draguignan) has just exceeded its 50-year flood level. According to witnesses, this also corresponds to the moment when the situation became problematic. The onset of flooding was reported to start around 16:00 h. A large water level increase was observed in Draguignan at 17:00 h. At Trans-en-Provence, a village located 5 km downstream, the central street was submerged by 17:30 h and the first cars were swept away at 19:30 h (local time). Downstream, on the Argens River, the flooding occurred later during the night.

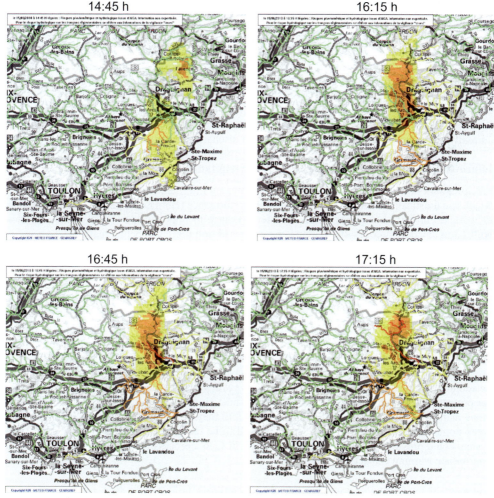

Fig. 2 Situation according to the AIGA method between 14:45 h and 15:15 h (local time): yellow, orange and red coloured rivers correspond to estimated discharges exceeding the 2, 10 and 50-year flood.

According to the emergency services, the situation reported by AIGA was one of the elements taken into account when deciding to put the alert at its maximum level and to ask for more intervention resources. It helped the emergency services have a "synthesized" view of the situation while they, in the meantime, had to deal with a multitude of partial and local information emerging from the field, sometimes with difficulty due to communication problems (for example caused by the mobile network crashing down). Even if the anticipation given by the approach seems small (around 1 hour), it seems that the information was worthwhile providing.

WHAT DO WE NEED TO GO FURTHER?

Even if the case study presented here showed the usefulness of the AIGA approach, further efforts need to be made to better anticipate this kind of event and reduce their consequences.

Improving and validating the method

The presented method still needs to be improved and validated, in particular concerning the modelling approach. Research is currently being done to better take into account local catchment characteristics, both for the real-time modelling and estimation of the regional flood quantiles. But the validation is not straightforward since the method is supposed to be helpful at ungauged locations, where by definition no data are available (gauged catchments can be taken as if they were ungauged, but they are often larger than most of the potential target catchments). For this reason, we try to collect the maximum amount of information from post-event analysis. For example, for the case study presented here, peak discharge estimations at different locations are currently being done, and they will be compared to discharges calculated by the model.

Improving the observation and forecasting of precipitations

Precipitation estimated by weather radars give significant information, but are still subject to errors which can drastically reduce hydrological model performance (Borga *et al.*, 2000). A large amount of work is currently being carried out by Météo-France to improve operational radar estimates, with some encouraging results. Among the proposed solutions are better real-time corrections using telemetry raingauge data or the use of polarimetric radars (Friedrich *et al.*, 2009; Gourley *et al.*, 2009; Tabary *et al.*, 2009). Furthermore, anticipation could be improved by integrating rainfall forecasts. This is not yet the case in the AIGA method, but ongoing studies are investigating this.

Be prepared

Some improvements can also be achieved through better preparation of the population and local authorities. Often, the memory of the last damaging events is lost, and properties are built in potential flooding areas with the residents unaware of the encountered risks. In France, the "risk prevention plans" (plans de prévention des risques) are supposed to address this issue. They must be adopted at a local level, but many towns are not yet covered. Furthermore, the "communal plans for protection" (plan communaux de sauvegarde) must also be generalised as they enable a better crisis management. For each kind of risk (and before they occur), these plans analyse where the weak points are, and which defined actions should to be taken in the case of a crisis. The AIGA system could provide some of the pre-alert signals in such a process that can be used to trigger the appropriate warning and action.

Acknowledgement The authors thank the French ministry in charge of Ecology for its financial support.

REFERENCES

Arnaud, P. & Lavabre, J. (2002) Coupled rainfall model and discharge model for flood frequency estimation. *Water Resour. Res.* 38(6), 1–11.

Arnaud, P., Lavabre, J., Sol B. & Desouches, Ch. (2008) Régionalisation d'un générateur de pluies horaires sur la France métropolitaine pour la connaissance de l'aléa pluviographique. *Hydrol. Sci. J.* 53(1), 34–46.

Berthet, L., Andréassian, V., Perrin, C. & Javelle, P. (2009) How crucial is it to account for the antecedent moisture conditions in flood forecasting? Comparison of event-based and continuous approaches on 178 catchments. *Hydrol. Earth Syst. Sci.* 13(6), 819–831.

Blöschl, G., Reszler, C. & Komma, J. (2008) A spatially distributed flash flood forecasting model. *Environmental Modelling & Software* 23(4), 464–478.

Borga, M., Anagnostou, E. N. & Frank, E. (2000) On the use of real-time radar rainfall estimates for flood prediction in mountainous basins. *J. Geophys. Res.* 105, 2269–2280.

Cole, S. J. & Moore, R. J. (2009) Distributed hydrological modelling using weather radar in gauged and ungauged basins. *Adv. Water Resour.* 32(7), 1107–1120.

Friedrich, K., Germann, U. & Tabary, P. (2009) Influence of ground clutter contamination on polarimetric radar parameters. *J. Atmos. Oceanic Technol.* 26(2), 251–269.

Gaume, E. *et al.* (2009) A compilation of data on European flash floods. *J. Hydrol.* 367(1–2), 70–78.

Gourley, J.J., Illingworth, A.J. & Tabary, P. (2009) Absolute calibration of radar reflectivity using redundancy of the polarization observations and implied constraints on drop shapes. *J. Atmos. Oceanic Technol.* 26(4), 689–703.

Javelle, P., Fouchier, C., Arnaud, P. & Lavabre, J. (2010) Flash flood warning at ungauged locations using radar rainfall and antecedent soil moisture estimations. *J. Hydrol.* 394(1–2), 267–274.

Merz, R. & Blöschl, G. (2009) A regional analysis of event runoff coefficients with respect to climate and catchment characteristics in Austria. *Water Resour. Res.* 45(1).

Moore, R.J., Bell, V.A. & Jones, D.A. (2005) Forecasting for flood warning. *C. R. Geoscience* 337(1-2), 203–217.

Oudin, L., Hervieu, F., Michel, C., Perrin, C., Andréassian, V., Anctil, F. & Loumagne, C. (2005a) Which potential evapotranspiration input for a lumped rainfall-runoff model? Part 2 – Towards a simple and efficient potential evapotranspiration model for rainfall-runoff modelling. *J. Hydrol.* 303(1–4), 290–306.

Oudin, L., Michel, C. & Anctil, F. (2005b) Which potential evapotranspiration input for a lumped rainfall-runoff model? Part 1 – Can rainfall–runoff models effectively handle detailed potential evapotranspiration inputs? *J. Hydrol.* 303(1–4), 275–289.

Parent du Chatelet, J., Guimera, M. & Tabary, P. (2003) The Panthere Project of Meteo-France: extension and upgrade of the French Radar Network. In: *Proc. 31st Conf. Radar Meteor.* 2, 802–804.

Perrin, C., Michel, C. & Andréassian, V. (2003) Improvement of a parsimonious model for streamflow simulation. *J. Hydrol.* 279(1–4), 275–289.

Rabuffetti, D. & Barbero, S. (2005) Operational hydro-meteorological warning and real-time flood forecasting: the Piemonte Region case study. *Hydrol. Earth Syst. Sci.* 9(4), 457–466.

Reed, S., Schaake, J. & Zhang, Z. (2007) A distributed hydrologic model and threshold frequency-based method for flash flood forecasting at ungauged locations. *J. Hydrol.* 337(3–4), 402–420.

Tabary, P. (2007) The New French operational radar rainfall product. Part I: Methodology. *Weather and Forecasting* 22(3), 393–408.

Tabary, P., Desplats, J., Do Khac, K., Eideliman, F., Gueguen, C. & Heinrich, J. C. (2007) The new French operational radar rainfall product. Part II: validation. *Weather and Forecasting* 22(3), 409–427.

Tabary, P., Vulpiani, G., Gourley, J. J., Illingworth, A. J., Thompson, R. J. & Bousquet, O. (2009) Unusually high differential attenuation at C Band: results from a two-year analysis of the French Trappes polarimetric radar data. *J. Appl. Met. Clim.* 48(10), 2037–2053.

Uncertainty estimation of deterministic river basin response simulations at gauged locations

ZACHARY L. FLAMIG[1], EMMANOUIL ANAGNOSTOU[2], JONATHAN GOURLEY[3] & YANG HONG[1]

1 *Atmospheric Radar Research Center, University of Oklahoma, 120 David L. Boren Blvd. Norman, Oklahoma 73072, USA*
zac.flamig@noaa.gov

2 *Civil and Environmental Engineering, University of Connecticut, 261 Glenbrook Rd, UNIT-2037, Storrs, Connecticut 06269, USA*

3 *NOAA/National Severe Storms Laboratory, 120 David L. Boren Blvd. Norman, Oklahoma 73072, USA*

Abstract This study presents a method to supply uncertainty estimates to flood predictions based on deterministic river basin response simulations from an uncalibrated, distributed hydrological model. A 15-year radar rainfall archive was used to run a hydrological model, thus providing a time series of simulated flows at every model grid cell. At grid cells corresponding to streamgauge locations, the time periods at which observed streamflow exceeded pre-computed observed flow frequency thresholds (e.g. 2-, 5-, 10-year return period flows) were identified. The distributions of simulated flows within (i.e. flooding at the respective frequency threshold) and outside (non-flooding at the respective frequency) these time intervals were then computed. The accuracy of the method is evaluated during an independent validation period where probabilities of flood >0.9 during flood cases are predicted more than 90% of the time, while probabilities of flood equal to zero occurred 75% of the time during non-flood cases.

Key words flood; distributed hydrological model; uncertainty estimation; probabilistic forecasting

INTRODUCTION

The American Meteorological Society's (AMS) 2000 policy statement on the prediction and mitigation of flash floods identifies flash flooding as one of nature's worst killers (AMS, 2000). The same policy statement acknowledges several important research challenges that lie ahead, including quantifying forecast uncertainty for flash flood forecasts.

Recent focus has been on how to deterministically forecast flash flooding given a streamflow forecast from a hydrological model. The challenge here for distributed hydrological models is that there is little specific information on discharge thresholds known *a priori* regarding the flooding flows that exceed bankfull conditions at all grid cells and result in impacts to life and property. Reed *et al.* (2007) demonstrated the threshold frequency (TF) technique where historical simulations from a hydrological model were used to derive the simulated streamflow thresholds for different return periods, including the two-year return period, which is generally assumed to represent bankfull conditions (Carpenter *et al.*, 1999). In gauged catchments, the same datasets used to derive threshold frequencies with simulations can be used to produce probabilistic information (i.e. uncertainty estimates) associated to the deterministic forecasts.

This study will explore a novel method by which a deterministic streamflow simulation from a hydrological model can be used to estimate conditional probabilities useful for predicting the occurrence of flooding. This technique is demonstrated with a case study on the Illinois River basin in Oklahoma, USA. The validity and skill of the probabilistic forecasts will be examined based on an independent validation time period for the same streamflow gauge. Finally, general conclusions from the study and suggestions for future work directions will be shared.

HYDROLOGICAL MODEL AND STUDY DOMAIN

For the purposes of this study the Hydrology Laboratory Research Distributed Model (HL-RDHM; Koren *et al.*, 2004) was set up over the Arkansas-Red River basin located in south-central USA (Fig. 1). HL-RDHM was developed by the US National Weather Service (NWS) for operational flood forecasting. HL-RDHM is a gridded version of the Sacramento Soil Moisture Accounting

Model with kinematic wave routing between grid cells. The model has a native grid cell resolution of 4 km at mid-latitudes and was forced using the StageIV precipitation product from the NWS Arkansas-Red River Basin River Forecast Center. The StageIV precipitation product is a radar-based quantitative precipitation estimate with automatic and human adjustment using quality-controlled raingauge data. The hydrological model was run at an hourly time-step in an uncalibrated manner using *a priori* parameter estimates provided with the model by the developers. This is important because model calibration can be a challenge in ungauged basins so there is a desire to be able to predict flooding without the requirement to calibrate the hydrological model.

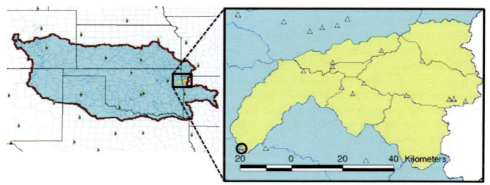

Fig. 1 The study domain located in the south-central USA. The Illinois River Basin (yellow area) is contained within the larger Arkansas-Red River Basin that is outlined in red. The circled marker is the location of the streamflow gauge on Illinois River near Tahlequah, Oklahoma.

To demonstrate the method for deriving the probability of flooding, a single streamgauge was identified. The gauge providing hourly streamflow measurements used for this study is on the Illinois River near Tahlequah, Oklahoma (United States Geological Survey ID 07196500). This gauge has a contributing drainage area of approximately 2484 km^2 with a streamflow of 549 and 130 m^3/s corresponding to the two-year and one-year return frequency, respectively. The HL-DHM model was run for this basin for the period of record for the precipitation product to produce a simulation of streamflow. This period runs from 1 January 1996 to 30 September 2009. Approximately the first half of this time interval from 1 January 1996 to 1 January 2003 is being used as the calibration period, while the probability of flooding method is then validated on the second half of the period from 1 January 2003 to 30 September 2009.

DERIVING PROBABILTY OF FLOODING

The first step in deriving the probability of flooding is to identify time intervals during a calibration period in which flooding is occurring. For this study, these time intervals are defined as when the observed streamflow exceeded the two-year return period flow, although any return period could be used. These time intervals will be referred to as the flood events. The second step identifies time intervals where the observed streamflow exceeds the one-year return period flow but is less than the two-year return period flow. A one-year return period reference value was chosen here so that the data analysed were associated with rainfall-driven events rather than the significant amount of time at which flows are at or below baseflow. This second set of time intervals will be referred to as the non-flood events hereafter. Because we are working explicitly with time periods for this portion there is a concern that with an uncalibrated model the observed peaks in streamflow may not be properly aligned with the simulated peaks in streamflow. The next step is to perform a simple "calibration" on the simulated time-series of streamflow by shifting it either backwards or forwards in time in order to maximise the correlation coefficient between the simulated and observed streamflow. Future work may show that this step is either unnecessary or dependent on the basin's response time. The fourth step is to use the time intervals identified in the first step to make histograms of the simulated streamflow for both the non-flood and flood cases. Gaussian distributions are fitted to both the non-flood and flood histograms to derive conditional probabilities defined as follows:

Flood Conditional Probability: $P(Q^{sim} \mid Q^{obs} > Q_t^{obs})$

Non-flood Conditional Probability: $P(Q^{sim} \mid Q_r^{obs} < Q^{obs} < Q_t^{obs})$

where Q^{obs} is the observed streamflow, Q^{sim} is simulated streamflow, and the subscripts correspond to return periods of time t in years. Q_r is the lower reference threshold. We chose to use the one-year return period flow for Q_r and two-year return period flow for Q_t. Figure 2 illustrates how the time intervals are identified, as well as the resulting histograms and modelled PDFs. As shown in the figure this process produces modelled PDFs, which are well behaved. A point to note about the two PDFs is that the overlap area defines the degree of uncertainty in the simulated flows to predict the flood occurrences. Ideally, if flow simulations were identical to the observed flows (i.e. accurate flow simulation) then the two distributions would have no overlap. In that case, the Q^{sim} values that belong in the flood conditional distribution would accurately represent the flood occurrence for the corresponding Q_t, and *vice versa* for the non-flood PDF. However, the simulated flows are not perfectly accurate, which is apparent by the overlap area of the two conditional distributions (Fig. 2(b)).

The next step is to use the conditional PDFs in the Bayesian framework to predict the conditional probability of flooding for any given value of simulated streamflow (Q^{sim}):

$$P\left(Q^{obs} > Q_t^{obs} \mid Q^{sim}\right) = \frac{P\left(Q^{sim} \mid Q^{obs} > Q_t^{obs}\right)P\left(Q^{obs} > Q_t^{obs}\right)}{P\left(Q^{sim} \mid Q^{obs} > Q_t^{obs}\right)P\left(Q^{obs} > Q_t^{obs}\right) + P\left(Q^{sim} \mid Q_r^{obs} < Q^{obs} < Q_t^{obs}\right)P\left(Q_r^{obs} < Q^{obs} < Q_t^{obs}\right)} \tag{1}$$

This is also shown in Fig. 2(b) (black solid line) in a cumulative probability form representing the probability of flooding for any simulated streamflow.

AN EXAMINATION OF THE PROBABILITY OF FLOODING

Using the method described above, the probability of flooding was computed for the entire validation period of the study basin. Figure 3 shows the observed and simulated streamflows along with the computed probability of flooding for the validation period. The probability of flooding is

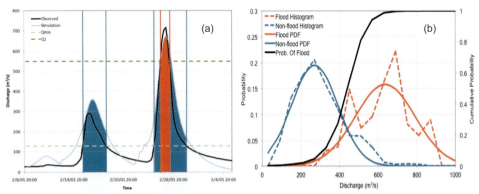

Fig. 2 Panel (a) with observed and simulated streamflow hydrographs for an example non-flood and flood case along with one- and two-year return period flows (horizontal dashed green lines). The red portion of the simulation contributes to the flood event histogram in panel (b) while the teal portion contributes to the non-flood event histogram. Panel (b) shows the flood and non-flood normalised histograms with the normal distribution fits and the derived cumulative probability of flood for the Illinois River near Tahlequah, Oklahoma. There are 48 non-flood events and eight flood events during the calibration period from 1 January 1996 to 1 January 2003, comprised of 1543 and 135 hourly data points in the non-flood and flood histograms, respectively.

well matched with observed events that exceed the observed two-year return period flow. Furthermore, the event during January 2009 that is clearly a false alarm occurred during a winter storm event where the precipitation falling was frozen and thus not contributing to direct runoff. Currently the hydrological model is not configured to adequately deal with frozen precipitation but steps to mitigate the impact of snow, sleet, freezing rain, etc. such as using a surface temperature threshold, or the surface precipitation type product available with the National Severe Storms Laboratory's National Mosaic and QPE system (http://nmq.ou.edu), will be considered in future work. Figure 3 illustrates a single flooding event to show the characteristics of the modelled probability of flooding. The probability of flooding quickly ramps up in value from 0 to 1 as the hydrological model responds to the precipitation forcing. In this event the probability of flood remains at 1 for nearly the duration of the event indicating that a flood is very likely taking place. The observed streamflow for this event verifies that there was a flood as the streamflow easily exceeds the two-year return period flow value. The probability of flood is colour-coded with green for probabilities between 0 and 0.5, yellow for probabilities between 0.5 and 0.75 and red for probabilities greater than 0.75. This is an example of how the probability of flood can be used to provide a more visual indication that a flood is likely to occur.

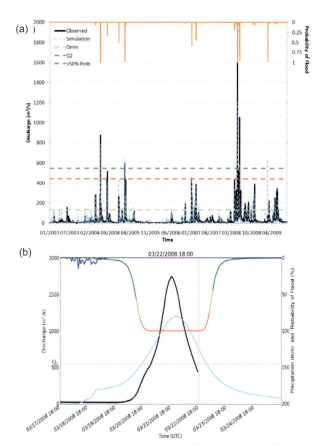

Fig. 3 (a) Observed and simulated flows for the validation time period from 1 January 2003 to 30 September 2009 along with the probability of flooding shown on the upper *x*-axis. (b) Example event illustrating how the probability of flood is visualised in a format for end-users. The gauge observation is plotted in black with the deterministic model simulation plotted in light blue. The probability of flood is colour-coded in traffic-light fashion with green being used for probabilities <50%, yellow for probabilities >50% but <75% and red for probabilities >75%. The figure is also designed to look like a theoretical forecast with the current time being 18:00 UTC 22 March 2008; thus, the gauge observations are limited to only the time period before this cut-off time.

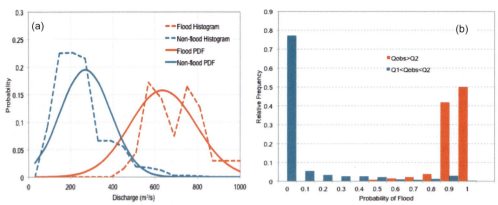

Fig. 4 (a) Same as in Fig. 2(b) but for flood and non-flood normalised histograms during the validation period from 1 January 2003 to 30 September 2009. The Gaussian curves are the modelled PDFs derived during the calibration period. (b) Relative frequency of derived probability of flood during the validation period for events in which $Q^{obs} > Q_2$ (red) and for non-flood events where $Q_1 < Q^{obs} < Q_2$ (in teal).

VALIDITY AND SKILL OF PROBABILITY OF FLOODING

In order to evaluate the computed probability of flooding, histograms of simulated flows for the non-flood and flood events during an independent validation period are produced and shown in Fig. 4(a). There is an apparent shift in the histogram of non-flood simulations toward lower flows in comparison to its modelled PDF, derived during the calibration period. However, this disagreement at low flows where there is little overlap between the flood and non-flood PDFs has negligible impact on the estimated conditional probability of flooding (see Fig. 2(b)). Overall, there is good effective agreement between the histograms and their respective modelled PDFs that were generated during the calibration period. From this analysis we can conclude that: (a) simulated flows for non-flood and flood cases are well separated using the developed methodology, (b) the normal approximation of the data distributions assumed for the modelled PDFs is appropriate for this basin, and c) eight years of data is sufficient to estimate the sample PDFs.

Figure 4(b) shows the relative frequencies of the derived probability of flood during the validation period for flood events (defined when $Q^{obs} > Q_2$) and non-flood events ($Q_1 < Q^{obs} < Q_2$). The mode of probabilities is 1 for flood events with the second highest frequency being the 0.9 probability bin. For non-flood events, over 75% of the probabilities fall in the 0 probability bin, indicating the method has excellent capability in separately predicting flood *vs* non-flood events. We note a slight increase in the 0.9 probability bin for non-flood events, suggesting a tendency for the method to have a non-zero false alarm rate. Specific examples were examined and the hydrological model was noted to produce hydrographs that were not as peaked as the observed flows. As shown in Fig. 3(b), instances can be seen immediately before and after the observed peakflow occurred when the probability was between 0.9 and 1, yet the observed flow was less than the 2-year return period flow.

CONCLUSIONS AND FUTURE WORK

This study demonstrates that it is possible to use a long-term simulation from a hydrological model to estimate a sample conditional probability of flooding. The method does not require detailed parameter estimation and it readily corrects for model bias. The probabilities were derived during a calibration period and the time series was graphically enhanced to show how the probabilities could be visualised to provide forecasters and emergency managers with an easy-to-understand product. The method was evaluated during an independent validation period and was shown to reproduce results that agreed well with the modelled PDFs. Probabilities of flood during flood cases were > 0.9 more than 90% of the time and were predominantly zero during non-flood cases. Note the non-flood cases had observed flows that exceeded the 1-year return period flows, but were less than the 2-year return flow. Model-simulated hydrographs had peak flows that were broader than and not as peaked as the observed flows, which caused a non-zero false alarm rate in predicting flooding.

While this research study employs a split calibration and validation period for examining the skill and validity of the derived probabilities, an operational deployment should utilise the entire period of data availability to increase the sample size and achieve the most reliable probabilities of flooding. Furthermore, the time-shift derived from the simple calibration should be assumed to be a constant and applied to real-time operations and forecasts. In an operational scenario it is also likely that the hydrological model of choice will be run utilising both data assimilation and quantitative precipitation forecasts. Given either or both of these cases this method could still be utilised assuming the quantitative precipitation forecasts are made available in hindcast mode for the hydrological model calibration period. Data assimilation will reduce state uncertainty of model variables, which will lead to better deterministic model simulations and subsequent probabilistic partitioning between flood and non-flood events. This is similar to a scenario where the model parameters are more accurately estimated to yield better performance for deterministic simulations. While neither data assimilation nor traditional model calibration is required for this method to yield reasonable results as demonstrated in this study, both approaches will reduce state and parametric uncertainty. Furthermore, different precipitation forecast lead-times can be used to generate differing sets of probabilities allowing for the incorporation of precipitation uncertainty, treated as a function of lead-time, into the framework to yield the overall probability of flood. This can be conducted by repeating the experiment detailed in this study but with forcing from precipitation forecasts with 1 h of lead-time, then 2 h, 3 h, etc.

This study lends itself to many possible future research topics including how to expand the method to ungauged basins. The authors hypothesise that PDFs derived on gauged basins can be extrapolated to ungauged basins using relationships between basin area or geomorphological parameters and streamflow. Furthermore it is possible to expand the source of information on observed floods to other datasets such as human reports of flooding and satellite remotely-sensed inundation maps. As this method is applied to more basins in the region, the ability to better validate the accuracy and reliability of the probabilities will be established. Data from additional basins will also be used to evaluate the normal approximation currently used for the flood and non-flood PDFs. These results from an uncalibrated hydrological model can be compared against those from a calibrated one to see if calibration greatly improves the model performance in predicting flooding events using the probability of flooding method. Lastly, we will expand the thresholds presently used for "nuisance flooding" to more rare, catastrophic flooding cases.

REFERENCES

AMS (2000) Policy statement: Prediction and mitigation of flash floods. *Bull. Amer. Meteor. Soc.* 81 1338–1340.

Carpenter, T. M., Sperfslage, J. A., Georgakakos, K. P., Sweeney, T. & Fread, D. L. (1999) National threshold runoff estimation utilizing GIS in support of operational flash flood warning systems. *J. Hydrol.* 224, 21–44.

Koren, V. I., Reed, S., Smith, M., Zhang, Z. & Seo, D. J. (2004) Hydrology Laboratory Research Modeling System (HL-RMS) of the US National Weather Service. *Hydrol.* 291, 297–318.

Reed, S., Schaake, J. & Zhang, Z. (2007) A distributed hydrologic model and threshold frequency-based method for flash flood forecasting at ungauged locations. *J. Hydrol.* 337, 402–420.

Flash flood forecasting using Data-Based Mechanistic models and radar rainfall forecasts

PAUL J. SMITH[1], KEITH BEVEN[1], LUCA PANZIERA[2] & URS GERMANN[2]

1 *Lancaster Environment Centre, Lancaster University, UK*
p.j.smith@lancs.ac.uk
2 *MeteoSwiss, Locarno Monti, Switzerland*

Abstract The parsimonious time series models used within the Data-Based Mechanistic (DBM) modelling framework have been shown to provide reliable accurate forecasts in many hydrological situations. In this work the DBM methodology is applied to forecast discharges during a flash flood in a small Alpine catchment. In comparison to previous work this catchment responds rapidly to rainfall. It is demonstrated, by example, that the use of a radar-derived ensemble quantitative precipitation forecast coupled to a DBM model allows the forecast horizon to be increased to a level useful for emergency response. A treatment of the predictive uncertainty in the resulting hydrological forecasts is discussed and illustrated.

Key words DBM; NORA; flash flood; IMPRINTS; Verzasca

INTRODUCTION

A significant step in the issuing of timely flood alerts is the ability to predict future values of water level or discharge. This paper focuses on the provision of short-range (up to 6 h lead time) predictions of discharge in small catchments whose natural response time is rapid: typically less than the lead time required for a useful forecast. An example analysis based upon the Verzasca catchment (Ticino, Switzerland) is presented. In producing such forecasts the hydrological forecasting problem (predicting the water level or discharge at points of interest) is exacerbated by the meteorological forecasting problem (predicting the precipitation) which must be addressed to achieve useful lead times.

Four types of quantitative precipitation forecast can be identified: those based on extrapolation of radar fields (e.g. Germann & Zawadski, 2002), those from numerical weather prediction (NWP) models (e.g. Marsigli *et al.*, 2005, 2008), those from stochastic models of rainfall evolution (e.g. Sirangelo *et al.*, 2007) and those derived from historical analogues (e.g. Obled *et al.*, 2002). Forecasts based upon the temporal extrapolation of a weather radar field offer the potential for a level of spatial resolution that is not currently generally achievable through NWP models; but the level of accuracy can decay rapidly (Germann *et al.*, 2006). This in part is due to limitations in the techniques used to evolve the rainfall field forward in time and predicting the formation of new rainfall cells.

The ensemble precipitation forecasts utilised in this study are provided by the Nowcasting ORographic precipitation in the Alps (NORA) product outlined in more detail elsewhere in these proceedings (Panziera *et al.*, 2010, 2011). NORA precipitation forecasts are derived from historical analogues based on the radar field and upper atmospheric conditions. As such, they avoid the need to explicitly model the evolution of the rainfall field through for example Lagrangian diffusion. The technique does depend upon the availability of an historic archive of suitable length and accuracy to allow analogues to be developed.

The hydrological model is formulated within the inductive Data Based Mechanistic (DBM) modelling approach. This calls for the identification and estimation of a numerical representation of a catchment to be carried out with reference to available data reflecting the catchment dynamics. The parameterisation of this cause-effect relationship is achieved by applying the minimum degree of model complexity required to fit the observed data to some acceptable level of accuracy. Importantly however, the DBM process imposes a further requirement: the resulting model structure and parameterisation must provide a suitable mechanistic interpretation. A successful mechanistic interpretation should provide some degree of protection against erratic model behaviour when extrapolating the model outside the range of data used for model identification and estimation. Further details and a fuller discussion of DBM modelling along with further hydrological applications can be found in Young (2001, 2002, 2003, 2006), Young *et al.* (2004) and Lees *et al.* (1994).

The use of ensemble precipitation forecasts in hydrology has attracted much attention (see for example Schaake *et al.*, 2007; Cloke & Pappenberger, 2009; Dietrich *et al.*, 2009; Thielen *et al.*, 2009). This paper addresses one area of ongoing research which is the construction of a predictive distribution for the hydrological variable from an ensemble of hydrological forecasts.

THE VERZASCA CATCHMENT AND DATA AVAILABILITY

The Verzasca basin with an area of 186 km^2 is located in the Ticino region of southern Switzerland. It is an alpine basin with an elevation range of 490–2870 a.m.s.l. Forests (30%), shrub (25%), rocks (20%) and alpine pastures (20%) are the predominant land cover. The discharge regime is governed by snowmelt in spring and early summer and by heavy rainfall events in the autumn (Wöhling *et al.*, 2006; Ranzi *et al.*, 2007). The steep valley sides and shallow soils (generally <30 cm) make the Verzasca River prone to flash floods with peak discharges often exceeding 400 m^3/s. The lower part of the catchment is dominated by a reservoir controlled by the Versasca dam. Here, discharge data from a gauging site above the reservoir have been used. These data are available from 1990, with corresponding meteorological records available from up to six gauges in the surrounding area. Precipitation forecasts are available for 71 events in the period 2004–2009. An ensemble of precipitation fields (at a 1 km × 1 km resolution) comprising 12 members was generated hourly in the NORA system, for 8 h ahead on an hourly time step.

A DATA BASED MECHANISTIC MODEL OF THE VERZASCA

This section outlines a deterministic DBM forecasting model for the Verzasca catchment. Initially a parsimonious simulation model of the catchment is identified, and its parameters estimated. The simulation model is then cast in a state space form and embedded in a data assimilation algorithm to produce the forecasting model. Observations of discharge are assimilated as they become available to improve the deterministic forecast of discharge. Both the forecasting and simulation models are constructed offline using historical data from 1990 to 1995. In calibration, future observations of the precipitation are presumed to be available where required in predicting to longer lead times. As such, the model is estimated as though the predictions of precipitation exactly matched what was later observed.

For compatibility with the meteorological forecasts the DBM model is evaluated on a 1 h time step. The variable being predicted, y_t, is the discharge volume observed within the *t*th time step. Two input series are available on a similar time step. These are catchment averaged precipitation totals $\mathbf{u} = (u_1, \ldots, u_T)$ and temperatures $\mathbf{k} = (k_1, \ldots, k_T)$.

Simulation model identification and estimation

Initial analysis of the system proceeded by fitting a linear discrete time transfer function driven by precipitation to the observe data. Analysis of the residuals of this initial model suggested that the relationship between the input and the observed water level was non-linear. The non-linearity is heavily influenced by precipitation falling as snow and its subsequent melting. A past DBM modelling exercise covering the Ticino region (Young *et al.*, 2007) used a transformed temperature as a second input series to capture the dynamics of this behaviour. In this more detailed study such an approximation proved inadequate.

Exploration using state dependant parameter estimation (Young *et al.*, 2001) indicates that the following simple state space model evolving the snow store s_t may be used to compute the snow melt v_t and an effective rainfall input r_t :

(1) If $k_t \geq k_{rain}$ then $r_t = y_t^\phi u_t$ and $s_+ = s_t$ else $r_t = 0$ and $s_+ = s_t + u_t$;

(2) If $k_t \geq k_{melt}$ compute the snowmelt $v_t = \min\left(s_+, \gamma\left(1+\theta_t\right)\left(k_t - k_{melt}\right)\right)$ else set $v_t = 0$;

(3) Evaluate $s_{t+1} = s_+ - v_t$.

The parameters (ϕ, γ) require estimation from the observed data. The value of θ_t is given by a sinusoid function. The snowmelt model is therefore closely related to the Positive Degree-Day Index (PDDI) methodology for computing snowmelt. The annual maxima of the sinusoid function was set for 21 June and minima for 21 December, thus providing a crude representation of seasonal variation in solar radiation (Braun, 1985). The values of k_{melt} and k_{rain} were set to 0.

The two effective input series were related to the observed discharge by a multiple input single output linear transfer function and stochastic noise series υ_t, such that:

$$y_t - \min_{t=1,\ldots,T}(y_t) = \frac{b_{0,r} + \ldots + b_{m_r,r} z^{-m_r}}{1 + a_1 z^{-1} + \ldots + a_n z^{-n}} r_{t-d_r} + \frac{b_{0,r} + \ldots + b_{m_v,r} z^{-m_v}}{1 + a_1 z^{-1} + \ldots + a_n z^{-n}} v_{t-d_v} + \upsilon_t . \tag{1}$$

The parameters (ϕ, γ) were optimised by a grid search with the aim of minimising $\sum_t \upsilon_t^2$. The structures of the linear transfer function determined by (n, m_r, m_v, d_r, d_v) was identified and estimated for each point on the grid search. The resulting values (ϕ, γ) are $(0.4, 0.025)$ with associated model structure $(n, m_r, m_v, d_r, d_v) = (2, 2, 2, 2, 7)$. When simulated, this model accounts for 85% of the variance of the observed calibration series.

Forecasting model

As time passes additional data become available which can be used to condition the DBM model and improve the forecasts given. As in past studies (e.g. Young, 2002; Romanowicz et al., 2006, 2008) data assimilation is performed by casting the linear transfer function in a state space form and embedding it in the Kalman filter.

A state space form representing two parallel paths of response is selected. The state vector $x_t \sim \left(\overline{x}_t, X_t\right)$ is summarised by its mean $\overline{x}_t$ and variance X_t and evolves according to:

$$x_{t+1} = Ax_t + B\begin{bmatrix} r_{t+1-d_r} \\ v_{t+1-d_v} \end{bmatrix} + \zeta_{t+1} \quad \zeta_{t+1} \sim \left(0, Q\omega_t^2\right) \tag{2}$$

$$y_{t+1} = hx_{t+1} + \xi_{t+1} \quad \xi_{t+1} \sim \left(0, \omega_t^2\right) \tag{3}$$

$$\omega_t^2 = \lambda_0\left(1 + \lambda_1 y_t^{\lambda_2}\right) . \tag{4}$$

The matrices A and B are derived from the roots and poles of the linear transfer function in equation (1) and h is a suitably sized vector of ones. The parameters $(Q, \lambda_0, \lambda_1, \lambda_2)$ describe the stochastic noise terms denoted by Greek letters in equations (2) and (3). For each lead time $(Q, \lambda_1, \lambda_2)$ are optimised to minimise the sum of squared errors of the expected value of the forecasts. Since in later sections only the expected values of the forecast are utilised, λ_0 is left unspecified, though a Maximum Likelihood estimate can always be computed (Schweppe, 1965). The optimised 6 hour ahead forecasts (using future observed rainfall) explains 90% of the variance of observed discharge series in the calibration period with a mean bias of –0.6 m³/s.

Coupling the DBM model and precipitation forecasts

The structure of the DBM model indicates that the expected value of the $f = 1,2$ step ahead predictions, denoted $\overline{y}_{t+f|t}$, can be computed without using predicted inputs. For $f > 2$ forecast the ensemble precipitation predictions generated by NORA at the current time step are utilised to drive the model. The future temperature series is presumed known since the events presented are not significantly affected by snowfall or melt. The *j*th member of the NORA ensemble results in an f step ahead predicted value of $\overline{y}_{t+f|t,j}$.

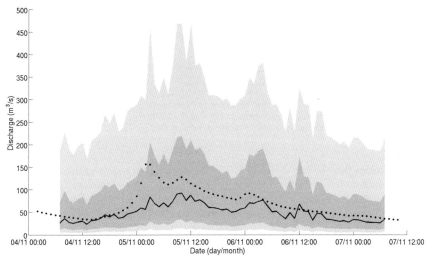

Fig. 1 Six hour ahead forecasts of discharge for an event in November 2008 generated using the NORA precipitation forecasts. Shaded areas represent 75% and 90% prediction confidence intervals, the line the median of the predictive distribution and points observed data.

PREDICTIVE UNCERTAINTY

A number of works (e.g. Krzysztofowicz, 1999; Todini, 2008) have proposed methodologies for computing predictive uncertainty by constructing the joint distribution of the *f* step ahead model forecasts and observed data (possibly along with other variables such as past prediction errors). For $f = 1,2$ the predictive uncertainty is summarised by the conditional distribution $\Pr\left(y_{t+f} \middle| \overline{y}_{t+f|t}\right)$. Since the NORA ensemble forecasts are not exchangeable the predictive distribution for $f > 2$ is a conditional distribution of $\Pr\left(y_{t+f}, \overline{y}_{t+f|t,j} : j = 1,\ldots,12\right)$. For the purposes of this study, and due in part to the limited data available, this is constructed in a simplistic fashion based on the bivariate case presented in Todini (2008). In this approach each of the variables undergoes a normal quantile transform (NQT) to produce thirteen standard normal variables. The joint distribution of these variables is modelled by a multivariate Gaussian distribution. Figure 1 shows the 6-h ahead forecasts for one of the 71 events generated using the NORA ensemble precipitation predictions. This event is not influenced by snowmelt or fall. Figure 2 shows the corresponding forecasts generated assuming the future observations of precipitation were known. For this idealised situation (and for $f = 1,2$) the joint distribution of the modelled and observed data is bivariate as in Todini (2008).

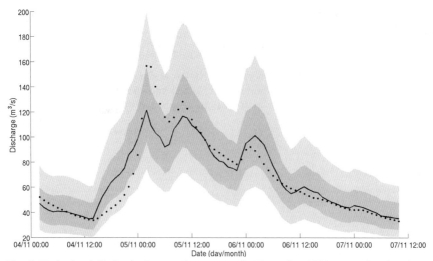

Fig. 2 Six-h ahead discharge forecast for an event in November 2008 presuming the observed future precipitation was known. Shaded areas represent 75% and 90% prediction confidence intervals, the line the median of the predictive distribution and points observed data.

Figure 2 indicates that the DBM model of the Verzasca River is able to produce efficient 6-h ahead predictions when driven by the (unknown) future observed precipitation values. Comparing Fig. 1 to Fig. 2 indicates that the forecasts derived from the NORA input have a much wider predictive distribution. This is in part due to the difference in sample sizes. The joint distribution utilised in Fig. 1 has approximately 1/25th of the sample size of that used in Fig. 2. Further NORA forecast periods, particularly those corresponding to periods of high discharge, are required to refine the predicted confidence intervals but were not available to this study.

A second cause of the additional forecast uncertainty is the difficulty in capturing timing of the precipitation and hence the timing of the rising limb of the hydrograph. The prediction error model outlined should, if enough data are available, capture these timing errors as part of the joint distribution. Improving on the simple model of the joint distribution utilised may be key to achieving this.

CONCLUSIONS

The work presented shows a DBM model aimed at capturing the dynamics of snowmelt in small Alpine basins. Coupled with be predictions of future precipitation, such as NORA, forecasts of future discharge can be derived. A methodology for representing the predictive distribution of the discharge is outlined. The provisional results show some promise. Improving the characterisation of the joint distribution by incorporating further hydrological information (such as differentiating between rising and falling limbs and snowmelt and non-snowmelt periods) and more appropriate statistical models is a topic of current research. The robustness of the statistical models to different temporal and spatial scales may be of particular interest if this approach is to be used broadly in an operational environment.

REFERENCES

Braun, L. (1985) Simulation of Snowmelt-Runoff in Lowland and Lower Alpine Regions of Switzerland. In: *Zürcher Geographische Schriften, Heft 21, ETH Zürich* [available from the Institute for Atmospheric and Climate Science ETH, Winterthurerstr. 190, Zürich, Switzerland].

Cloke, H. L. & Pappenberger, F. (2009) Ensemble flood forecasting: a review. *J. Hydrol.* 375, 613–626.

Dietrich, J., Schumann, A. H., Redetzky, M., Walther, J., Denhard, M., Wang, Y., Pfuetzner, B. & Buettner, U. (2009) Assessing uncertainties in flood forecasts for decision making: prototype of an operational flood management system integrating ensemble predictions. *NHESS* 9, 1529–1540.

Germann, U. & Zawadski, I. (2002) Scale Dependence of the predictability if precipitation from continental radar images. Part I: Description of the methodology. *Monthly Weather Rev.* 130, 2859–2873.

Germann, U., Zawadzki, I. & Turner, B. (2006) Predictability of precipitation from continental radar images. Part IV: Limits to prediction. *J. Atmos. Sci.* 63, 2092–2108.

Krzysztofowicz, R. (1999) Bayesian theory of probabilistic forecasting via deterministic hydrologic model. *Water Resour. Res.* 35, 2739–2750.

Lees, M. J., Young, P. C., Ferguson, S., Beven, K. J. & Burns, J. (1994) An adaptive flood warning scheme for the River Nith at Dumfries. In: *2nd Inter-national Conference on River Flood Hydraulics* (ed. by W. Watts & J. Watts), Wiley, UK.

Marsigli, C., Boccanera, F., Montani, A. & Paccagnella, T. (2005) The COSMO-LEPS mesoscale ensemble system: Validation of the methodology and verification. *Nonlinear Proc. Geoph.* 12, 527–536.

Marsigli, C., Montani, A. & Paccangnella, T. (2008) A spatial verification method applied to the evaluation of high-resolution ensemble forecasts. *Meteorol. App.* 15, 125–143.

Martinec, J., Rango, A. & Major, E. (1983) The snowmelt runoff model users manual. *NASA Tech. Report 1100.*

Obled, C., Bontron, G. & Garcon, R. (2002) Quantitative precipitation forecasts: a statistical adaptation of model outputs through an analogues sorting approach. *Atmos. Res.* 63, 303-324.

Panziera, L. & Germann, U. (2010) The relationship between airflow and orograohic precipitation on the southern side of the Alps as revealed by weather radar *Quart. J. Roy. Met. Soc.* 136, 222–238

Panziera, L., Germann, U., Hering, A. & Mandapaka, P. (2011) Nowcasting of orographic rainfall by using Doppler weather radar. In: *Weather Radar and Hydrology* (Proceedings of a symposium held in Exeter, UK, April 2011). IAHS Publ. 351. IAHS Press, Wallingford, UK.

Ranzi, R., Zappa, M. & Bacchi, B. (2007) Hydrological aspects of the Mesoscale Alpine Programme: Findings from field experiments and simulations. *Quart. J. Roy. Met. Soc.* 133, 867–880.

Romanowicz, R. J., Young, P. C. & Beven, K. J. (2006) Data assimilation and adaptive forecasting of water levels in the River Severn catchment, United Kingdom. *Water Resour. Res.* 42, W06407.

Romanowicz, R. J., Young, P. C., Beven, K. J. & Pappenberger, F. (2008) A data based mechanistic approach to nonlinear flood routing and adaptive flood level forecasting. *Adv. Water Resour.* 31, 1048–1056.

Schaake, J. C., Hamill, T. M., Buizza, R. & Clark, M. (2007) The hydrological ensemble prediction experiment. *B. Am. Meteorol. Soc.* 88, 1541–1547.

Schweppe F. C. (1965) Evaluation of likelihood functions for gaussian signals. *IEEE Transactions on Information Theory* 11(1), 61–70.

Sirangelo B., Versace, P. & De Luca, D. L. (2007) Rainfall nowcasting by at site stochastic model PRAISE. *Hydrol. Earth System Sci.* 11, 1341–1351.

Thielen, J., Bogner, K., Pappenberger, F., Kalas, M., del Medico, M. & de Roo, A. (2009) Monthly-, medium-, and short-range flood warning: testing the limits of predictability. *Meteorol. App.* 16, 77–90.

Todini, E. (2008) A model conditional processor to assess predictive uncertainty in flood forecasting. *Int. J. River Basin Manage.* 6(2), 123–137.

Wöhling, T., Lennartz F. & Zappa, M. (2006) Updating Procedure for Flood Forecasting with conceptual HBV-Type Models. *Hydrol. Earth System Sci.* 10, 783–788.

Young, P. C. (2001) Data-based mechanistic modelling and validation of rainfall-flow processes. In: *Model Validation: Perspectives in Hydrological Science* (ed. by M. G. Anderson & P. D. Bates), 117–161. Wiley, Chichester, UK.

Young, P. C. (2002) Advances in real-time flood forecasting. *Philos. T. Roy. Soc. A.* 360(1796), 1433–1450.

Young, P. C. (2003) Top-down and data-based mechanistic modelling of rainfall-flow dynamics at the catchment scale. *Hydrol. Process.* 17(11), 2195–2217.

Young, P. C. (2006) The data-based mechanistic approach to the modelling, forecasting and control of environmental systems. *Annual Reviews in Control* 30(2), 169–182.

Young, P. C., McKenna, P. & Bruun, J. (2001) Identification of non-linear stochastic systems by state dependent parameter estimation. *Int. J. Control* 74(18), 1837–1857.

Young, P. C., Chotai, A. & Beven, K. J. (2004) Data-based mechanistic modelling and the simplification of environmental systems. In: *Environmental Modelling: Finding Simplicity in Complexity* (ed. by J. Wainwright & M. Mullgan), 371–388.

Young, P. C., Castelletti, A. & Pianosi, F. (2007) The data-based mechanistic approach in hydrological modelling. In: *Topics on System Analysis and Integrated Water Resource Management* (ed. by A. Castelletti & R. S. Sessa), 27–48. Elsevier, Amsterdam, The Netherlands.

Urban flood prediction in real-time from weather radar and rainfall data using artificial neural networks

ANDREW P. DUNCAN, ALBERT S. CHEN, EDWARD C. KEEDWELL,
SLOBODAN DJORDJEVIĆ & DRAGAN A. SAVIĆ

Centre for Water Systems, University of Exeter, Harrison Building, North Park Road, Exeter EX4 4QF, UK
apd209@exeter.ac.uk

Abstract This paper describes the application of Artificial Neural Networks (ANNs) as Data Driven Models (DDMs) to predict urban flooding in real-time based on weather radar and/or raingauge rainfall data. A 123-manhole combined sewer sub-network from Keighley, West Yorkshire, UK is used to demonstrate the methodology. An ANN is configured for prediction of flooding at manholes based on rainfall input. In the absence of actual flood data, the 3DNet / SIPSON simulator, which uses a conventional hydrodynamic approach to predict flooding surcharge levels in sewer networks, is employed to provide the target data for training the ANN. The ANN model, once trained, acts as a rapid surrogate for the hydrodynamic simulator. Artificial rainfall profiles derived from observed data provide the input. Both flood-level analogue and flood-severity classification schemes are implemented. We also investigate the use of an ANN for nowcasting of rainfall based on the relationship between radar data and recorded rainfall history. This allows the two ANNs to be cascaded to predict flooding in real-time based on weather radar.

Key words artificial neural network; manhole; multi-layer perceptron; nowcasting; prediction; rainfall; urban flood; weather radar

BACKGROUND

Recent studies (Min *et al.*, 2011; Pall *et al.*, 2011) have documented the increased frequency and likelihood of extreme precipitation events. At the same time, the complete redesign and construction of urban drainage networks to prevent flooding during such events in every case would be prohibitively expensive with increasing urbanisation further exacerbating this problem. Therefore models are required, which can provide predictions of location, severity and/or risk of flooding. In order to be operationally useful, these need to provide at least a 2-h lead-time (Einfalt *et al.*, 2004).

Conventional hydraulic simulators have been used to model the response of Urban Drainage Networks (UDNs) to rainfall events. However, for large networks, these can be slow and computationally expensive. A faster surrogate method is sought, which would permit modelling of very large networks in real-time, without unacceptable degradation of accuracy. Also, in the worst case, the predictive ability of such models is limited by the "time of entry" for the sewer network, with the possibility of flooding commencing from this time onwards, following the start of precipitation. In practice, this would normally be of the order of minutes, rather than hours.

Therefore prediction of rainfall is a requirement to achieve lead-times sought. Many papers have been written on rainfall nowcasting methods from radar rainfall images (Schellart *et al.*, 2009; Wang *et al.*, 2009). In this study, rainfall intensity predictions are made for a 3 × 3 km catchment, using Met Office Nimrod UK-1km composite radar images with 5-minute temporal resolution.

As part of University of Exeter's research under Work Package 3.6 of the Flood Risk Management Research Consortium Phase 2 (FRMRC2) Project, we developed the 'RAdar Pluvial flooding Identification for Drainage System' (RAPIDS) using ANN's to predict flooding in sewer systems. The RAPIDS project includes two phases: RAPIDS1, which addresses the need for a faster surrogate for hydraulic simulators, and RAPIDS2, which provides nowcasting for rainfall over the catchment containing the modelled UDN. It is hoped to be able to demonstrate the cascading of these two systems to provide the required urban flood predictive model.

METHODOLOGY

RAPIDS1

The ANN framework is based on a 2-layer, feedforward MLP (Multi-Layer Perceptron), used to relate incoming rainstorm data to the extent of flooding present at each manhole in the UDN. It has the same number of output neurons as manholes. The number of neurons in the hidden layer and number of input nodes are varied to establish an optimum. The supervised training regime uses a backpropagation of error quasi-Newton gradient-descent method. A moving time-window approach is implemented whereby three time-series traces (rainfall intensity, cumulative rainfall and elapsed time) are provided as inputs to the ANN. The number of input nodes is therefore three times the number of 3-min time-steps in the input time-window (e.g. for a 30-min input time window, 30 input nodes are used). Output target signals for training and evaluation of ANN performance are provided from the flood-level hydrographs generated by the SIPSON (University of Belgrade, 2010) hydrodynamic simulator outputs for each manhole. The trained ANN thus aims to generate the same hydrographs, based on learning the relationship between the provided input signals and the SIPSON-generated targets. Figure 1(a) illustrates the architecture of the ANN system to predict SIPSON outputs. The target signals selected are the flood levels at each manhole at a time-step that corresponds to the desired prediction lead-time (i.e. up to 60 min). Storm profile data arrays of the three input-signals are prepared for use as the time-series input to the ANN as illustrated in Fig. 1(b). Input data are normalised.

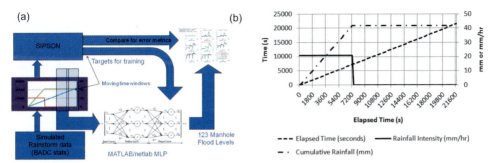

Fig. 1 (a) Architecture of RAPIDS1, (b) ANN Input signals for a typical design storm.

A constant 6-h simulated period for each rainstorm is used throughout, with design storms of 0.5, 1, 2 and 3-h duration and return periods of 1, 10, 50 and 100 years. A sampling period of 3-min applies in all cases.

RAPIDS2

Treatment of radar rainfall images directly by an ANN is still computationally prohibitive since, for example, for a 3-h prediction there would be 36-images, each with at least 360^2-pixels (allowing for a maximum storm velocity of 60 km/h). This would potentially require $\sim 5 \times 10^6$ neurons (at 1-neuron per pixel). Therefore features are extracted from the rain echoes in each time-step and associated with features from the previous time-step using a 1-nearest neighbour approach. These can then be applied to the inputs of an ANN as time-series signals.

Rain echoes are first distinguished and labelled by thresholding the image, smoothing and pre-filtering to remove clutter. A low threshold (e.g. 0.25 mm/h) is used to ensure rejected rainfall has very low probability of contributing to flooding. Features extracted for each echo include: positions of geometric centroid and centre of rainmass, area, total rainmass, north, south, east and west extremities and peak intensity. It is proposed to use principal component analysis to rank these features in terms of predictive skill.

The same time-windowed ANN framework as for RAPIDS1 can then be implemented. Target rainfall for training and evaluating the ANN is derived from the individual rainfall intensities for the radar image pixels covering the required catchment containing the UDN to be modelled.

The final stage of the RAPIDS project will be to cascade the two stages together, RAPIDS2 providing predicted rainfall, which can be applied to RAPIDS1 inputs to provide flood-severity predictions for each manhole in the UDN.

CASE STUDY

An ANN with 123-outputs is used to model the Stockbridge sub-section of the combined rain/wastewater drainage system for the town of Keighley, West Yorkshire, UK (Fig. 2), containing 123 manholes and one combined sewer overflow. This implements a surrogate DDM for the 3DNet / SIPSON simulator, by using its output hydrographs as target data for training the ANN. The neural network will output a floating-point estimate of the level of flooding at each manhole. However, this level of accuracy is unlikely to be required for flood-warnings. Therefore we use a classification scheme for flood severity shown in Table 1. This is used by a wrapper function around the ANN to convert flood levels to classes. The flood classification threshold edges are deliberately nonlinear to demonstrate flexibility of the approach.

   A full 16-storm, leave-one-out cross validation (LOOCV standard method) (Cawley & Talbot, 2003) is conducted, using each of the 16 design storms in turn to test the ANN and measure errors. The mean of the results then provides a summary of overall performance. During ANN training, for each test storm, SIPSON data from a second storm are used to validate and terminate training (early stopping). The remaining 14 storms are used as target signals to train the ANN. Both target and ANN output are then post-processed to classify flood severity for each manhole at each time-step. ANN setup parameters (number of input time-steps (Nin), number of hidden units (Nhu), weight decay coefficient (α)) are varied in combination to establish an optimum setup. The results presented below are for the optimum setup (Nin = 10, Nhu = 10 and α = 10.0). Both the analogue flood level and classified flood severity data are analysed for error. Timing both for training and running the trained ANN are compared to both SIPSON simulation time and to real-time (assuming the sampling period of 3-minutes used throughout).

RESULTS AND DISCUSSIONS

RAPIDS1 – timing analysis and benchmarking

Figure 3 presents mean timings in seconds for ANN training and test for all storms. This is against a mean simulation run time of 195 seconds for SIPSON. Results are shown for a mean of samples taken at $T_{TSAdvance} = \{0,10,20\} \times 3$-min time-steps. Overall, mean results for the trained ANN

Table 1 Flood severity classification scheme.

Flood class	Description	Flood depth
3	Severe	Above 5.00
2	Moderate	Between 1.00 and 5.00
1	Slightly	Between 0.00 and 1.00
0	None	Less than 0.00

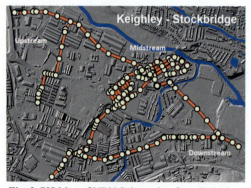

Fig. 2 GIS Map of UDN Sub-section from Keighley, West Yorkshire, UK.

were as follows. Training was typically achieved in 0.58 × mean duration of a SIPSON simulation run (i.e. 1.7 × faster). This provides the possibility of regular re-calibration of the live system, during periods of system quiescence (e.g. during periods of observed baseflows). Run times for the trained ANN for the 123-manhole UDN over a period of 6-h, with 3-min sampling rate was better than 0.12 seconds on an Intel quad core i5-960 2.7GHz processor, running 64-bit MS Windows 7 and MATLAB 2010b.

TYPICAL FLOOD DEPTH TRACE AND FLOOD CLASSIFICATION TRACE RESULTS

Figure 4 illustrates typical ANN performance for individual manholes (ANN outputs) *vs* time-steps. Each plot has four traces: solid black: target flood level (m); solid green: ANN output flood level; dotted red: target flood severity class (0–3); dashed blue: ANN output flood severity class (0–3). The plot on the left illustrates manhole 1898's training results over all 14 storms. On the right are the corresponding test run results for a 30-minute prediction advance for manholes 1898 and 1931 for a 50-year return-period, 2-hour, design storm. Trials were conducted to analyse performance for this setup, for all 16-storms and all values of prediction $T_{TsAdvance}$.

Figure 5(a) illustrates variation of flood level percentage errors, with an overall mean of 16.0%. Figure 5(b) similarly shows variations for flood severity percentage classification errors, with an overall mean of 2.65%. Results for the 10-year RP, 3-hour storm are significantly worse than the mean for all storms. Analysis shows this is due to the UDN being at the threshold between recharge and surcharge under the catchment conditions created by this storm. Flood severity class errors for storms of 1-year RP are low because they do not lead to surcharge for the UDN studied.

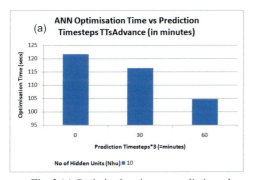
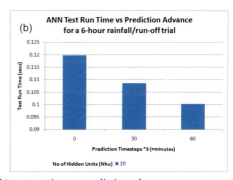

Fig. 3 (a) Optimisation time *vs* prediction advance, (b) test run time *vs* prediction advance.

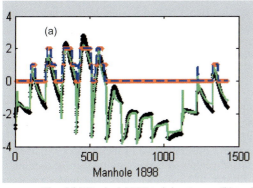
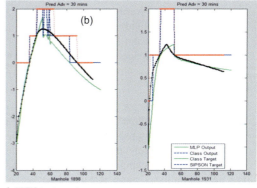

Fig. 4 (a) Typical ANN training traces; (b) typical ANN test run traces.

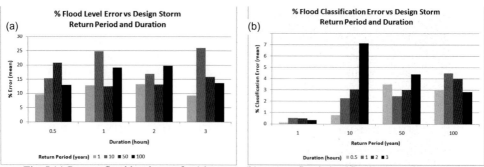

Fig. 5 (a) Percent flood level error for 16 storms, (b) percent flood classification error for 16 storms.

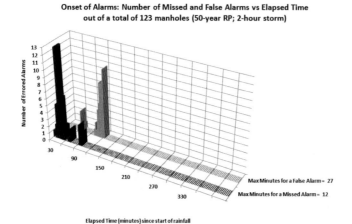

Fig. 6 Missed and false alarms *vs* time for single storm.

Figure 6 illustrates the onset of alarms across all 123 manholes (y-axis) *vs* time in min (x-axis). The two ranks show the number of missed or false alarms for a 50-year RP, 2-hour typical storm.

In this analysis, the alarm classes 1 to 3 are merged together. The maximum time during which a false alarm occurs is 27 min and for a missed alarm is 12 min. The zero-level section to the right of each rank shows that no missed or false alarm is sustained beyond the time of peak flood in this case. Calibration of the system to eliminate either false or missed alarms or achieve a trade-off between the two (as in the case illustrated) would be possible by adjusting threshold offsets for the classification wrapper function. Further analysis of the more severe (2–3) alarm states would also be worthwhile. Alternatively, a Bayesian Belief Network could be used here, not only to predict the flooding class, but also the probability of prediction. As the time goes on and the analysis is repeated, the probability would be expected to increase (or decrease). This would be invaluable to practitioners to see and assess the evolution of the storm and flooding.

RAPIDS2

The radar rainfall images used for the Keighley catchment rainfall nowcasting study contain 361 × 361 pixels, so as to allow up to 3-h storm travel time in any direction at a maximum of 60 km/h. In pattern-recognition terms, this corresponds with 130 k dimensions, since the rainfall intensity for each pixel is effectively an independent variable. Figure 7 illustrates two of the features extracted from the images: tracking of centroids (white traces) and centres-of-mass (grey traces) for each of two echoes during a storm that occurred between 28 and 30 November 2009. The longer tracks represent an elapsed time of 8.5 h. Keighley is located at the centre of the image (white square).

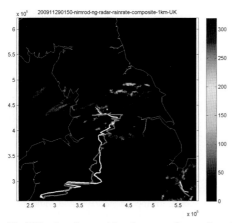

Fig.7 Tracks of centroid and centre of mass for echoes.

CONCLUSIONS

Results for RAPIDS1 show that ANNs can provide a very significant speed improvement over conventional hydraulic simulators without excessive degradation in performance. This is particularly so for flood severity classification. The method presents opportunities for automated generation of flood alarms / warnings right down to the individual manhole, including potentially for UDNs of considerable size, without being computationally expensive. However, flood prediction based on rainfall alone cannot provide operationally useful lead-times. Instead, prediction is limited in the worst case by the Time of Entry of the UDN (typically <30 min).

Possibilities for extending prediction time to operationally useful values of 2+ hours are being explored through a process of radar rainfall echo feature extraction and feature time-series prediction using ANNs. More work is needed to determine the value of this approach.

Assuming that RAPIDS2 achieves satisfactory results, the possibility of cascading the two systems to provide flood-level prediction at manholes based on live radar rainfall images will be tested.

Acknowledgement The research reported in this paper was conducted as part of the Flood Risk Management Research Consortium, with support from the Engineering and Physical Sciences Research Council, the Department of Environment, Food and Rural Affairs/Environment Agency Joint Research Programme, UK Water Industry Research, Office of Public Works Dublin, and Northern Ireland Rivers Agency. Data were provided by the British Atmospheric Data Centre, Environment Agency, Met Office, Ordnance Survey and Yorkshire Water. Our thanks go to all the above organisations for their support.

REFERENCES

Cawley, G. C. & Talbot, N. L. C. (2003) Efficient leave-one-out cross-validation of kernel Fisher discriminant classifiers. *Pattern Recognition* 36, 2585–2592.
Einfalt, T., Arnbjerg-Nielsen, K., Golz, C., Jensen, N.-E., Quirmbach, M., Vaes, G., *et al.* (2004) Towards a roadmap for use of radar rainfall data in urban drainage. *J. Hydrol.* 299, 186–202.
FRMRC (2005–2011) Flood Risk Management Research Consortium 2. http://www.floodrisk.org.uk/ (accessed 23 February 2011).
Min, S.-K., Zhang, X., Zwiers, F. W. & Hegerl, G. C. (2011) Human contribution to more-intense precipitation extremes. *Nature* 470, 378–381.
Pall, P., Aina, T., Stone, D. A., Stott, P. A., Nozawa, T., Hilberts, A. G. *et al.* (2011) Anthropogenic greenhouse gas contribution to flood risk in England and Wales in autumn 2000. *Nature* 470, 382–386.
University of Belgrade (2010) 3DNet Users' Manual. Belgrade, Serbia: University of Belgrade, Faculty of Civil Engineering, Institute for Hydraulic and Environmental Engineering.
Wang, P., Smeaton, A., Lao, S., O'Connor, E., Ling, Y. & O'Connor, N. (2009) Short-term rainfall nowcasting: using rainfall radar imaging. In: *Eurographics Ireland 2009, 9th Irish Workshop on Computer Graphics* (Dublin, Ireland), 9 pp.

Contribution of weather radar data to hydropower generation optimization for the Rhone River (France)

BENJAMIN GRAFF[1], DOMINIQUE FAURE[2], GUILLAUME BONTRON[1] & SEBASTIEN LEGRAND[1]

1 *Compagnie Nationale du Rhône, 2, rue André Bonin, 69316 Lyon Cedex 04, France*
b.graff@cnr.tm.fr
2 *Alicime, 14 rue du Docteur Bordier, 38100 Grenoble, France*

Abstract CNR operates 19 hydropower plants all along the Rhone River in France. To forecast hydropower generation and to ensure the security of people and plants, CNR has developed hydrometeorological forecasting tools, rainfall–runoff modelling and propagation models, for the Rhone River and some of its main tributaries. With the assistance of Alicime, CNR achieved an application named AEGIR to integrate radar QPE into its hydrometeorological forecasting process. AEGIR allows CNR: to assess spatial rainfall pattern and rainfall variability in time and space over the Rhone River basin; to compare radar QPE and observed raingauge measurements; to forecast rainfall hyetograph for each sub-basin of the Rhone River. The paper focuses on the contribution of radar QPE regarding CNR's purposes, by analysing the quality of observed and forecasted radar quantitative estimates over the Rhone River basin, and finally assessing the operational use of radar QPE as an input into pre-existent rainfall–runoff models.
Key words weather radar; QPE; rainfall–runoff forecasting; Rhone River; hydropower generation

INTRODUCTION

Founded in 1933, the Compagnie Nationale du Rhone (CNR) is France's second largest electricity producer. CNR operates 19 run-of-river hydropower development schemes all along the Rhone River (97 000 km^2), from the border with Switzerland to the Mediterranean Sea. Since the year 2000 and the deregulation of the French energy market, CNR has become an independent electricity producer and has been trading its energy. Thus, anticipating its energy generation and therefore discharges all along the Rhone River and its main tributaries is a key point for CNR. In order to perform its activities and to provide safety control, CNR has been developing a complete operational hydrometeorological forecasting system for 10 years (Bompart *et al.*, 2008). This forecasting system consists of several modelling tools linked by Hydromet, which is the CNR hydrometeorological database. The more recent tool is the AEGIR application developed by CNR with the collaboration of Alicime, which is a small development company, in order to integrate weather radar data into CNR hydrometeorological forecasting process. AEGIR helps CNR operators to monitor hydrometeorological events on the Rhone River basin and to improve the discharge forecasting. In this paper, we will present the CNR hydrometeorological forecasting system and the integration of AEGIR into this system. Then, we will analyse the quality of the radar quantitative precipitation estimates (QPE) over the Rhone River basin before presenting the first results of the integration of radar QPE as an input to operational rainfall–runoff modelling.

AEGIR AS PART OF THE CNR HYDROMETEOROLOGICAL FORECASTING SYSTEM

The hydrometeorological forecasting system of CNR is composed of several applications that are interconnected via their databases. Yet, to ensure the robustness of the results, this system requires a human operator to validate each step of the forecasting process. The tools are built to allow the operator to change the output of an application before its use as input into another application: the models and algorithms just propose results that operators can adjust.

The process of forecasting discharges on the Rhone River and its main tributaries has two main steps. The first one is the generation of a rainfall input dataset containing both observed and forecasted QPE at the sub-basin scale up to 4 days ahead. This dataset is based on:
– mean raingauge QPE for each sub-basins, estimated by Météo-France using a kriging method from the measurement of the French real time raingauge network;

- precipitation forecasts at sub-basin scale from several independent data sources: (i) commercial products from the weather forecasting model ARPEGE of Météo-France, (ii) an analogue sorting method developed and run by CNR (Obled *et al.*, 2002; Bontron, 2004), which is based on the outputs of the Global Forecast System (GFS) of NOAA and which links meteorological forecasts to an historical weather archive and to the associated observed rainfall, (iii) public forecasts available on the Internet;
- estimation of snowmelt on alpine sub-basins provided by a forecasting model based on a degree-day factor methodology.

The second step generates hydrological forecasts over the main tributaries of the Rhone River, based on the sub-basin rainfall scenario forecasts. This step relies on:
- discharge data at different time scales (from instantaneous to hourly data) provided by a network of around 220 staff gauges covering 88 000 km^2 (Grimaldi & Haond, 2008);
- statistical rainfall–runoff models calibrated on each sub-basin of importance.

The first step of the hydrometeorological forecasting process runs with a 6 h time step, which corresponds to the time step of the input used by CNR (0–6 h, 6–12 h, 12–18 h, 18–24 h). Besides, observed raingauge data are made available with a mean delay of 90 min after the end of each 6-h time step. The second step of the hydrometeorological process runs on a 1-h time step since discharge data are available in real time. This difference between the time steps implies that the 6-h rainfall values are disaggregated into six equal 1-h rainfall values before being used as input into rainfall–runoff models. This approach is appropriate for a catchment the size of the whole Rhone basin and for the general purpose of CNR. However, a 6 h time step can be too long for smaller sub-basins because of the short response time and the nonlinearity of the basin response.

In order to have more real time and spatial information on rainfall, CNR decided to integrate weather radar data into this hydrometeorological forecasting system, by developing the AEGIR application. Using geographic information system (GIS) facilities, AEGIR allows CNR to easily access, in real time, the rainfall variability over the whole Rhone River basin (Fig. 1). AEGIR facilities make it easier to monitor the rainfall events, and to validate the CNR forecasted rainfall scenarios. AEGIR can also display radar data for past events for post event analysis.

But, operated on a 15-min time step basis that can be aggregated to hourly time step, AEGIR provides equally useful quantitative data to the CNR operators during flood events, particularly in

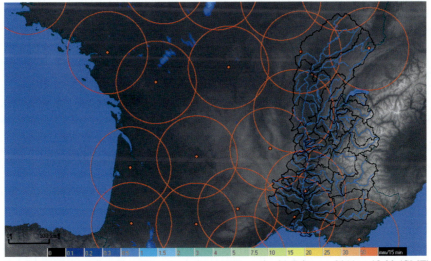

Fig. 1 Overview of the Rhone River basin in AEGIR on the 23 February 2011 at 12:30 (GMT) showing all the Rhone River sub-basins in the southeast of France, the weather radar locations with 100 km circles, and some rainy areas.

the areas of the Rhone River basin where there are few raingauges. These data could complement the previous 6 h data and forecasts, not only with observed radar data but also with a limited forecast for the next hour. This radar data assimilation should improve the first 6 h time step of the forecasting process.

RADAR INPUT AND DATA PROCESSING FOR AEGIR

The quantitative precipitation estimation

The weather radar data used by AEGIR are provided by Météo-France. Every 5 min, Météo-France updates a map of QPE named PANTHERE over the whole French Territory using the latest measurements of the ARAMIS network of 24 weather radars. Each radar is operated with a volumetric scanning protocol, and data are processed specifically for each radar to reduce the effects of the problems of measurement: ground clutter, natural and anthropogenic partial beam blockage, vertical variation of the reflectivity (VPR) integrating a correction function of the distance from the radar, motion of the rainy areas between two radar images, and a limited correction of the resulting QPE at ground level with the hourly measurements of the national real time raingauge network (about 1000 units). Concurrently, a quality code is estimated for each radar pixel, and the QPE estimated from the 24 radars are merged using a weighting procedure as a function of these quality codes. The final PANTHERE QPE is composed of two maps of 1 km^2 pixel size: the map of the best QPE at ground level for the last time step chosen (every 5 or 15 minutes), and a map of the estimated quality code for each pixel. This process has been upgraded several times during the last years in order to improve the quality of the quantitative estimations (Tabary 2007; Tabary *et al.*, 2007). In several studies, this quality has been simulated for different weather situations (Faure, 2006; Delrieu *et al.* 2009) or validated using operational real time data (Faure, 2008), and the spatial or temporal variation of this quality is relatively well known in the southeast of France and the Rhone Valley.

Data processing in AEGIR

Every 15 min, AEGIR receives the latest 15 min PANTHERE QPE from Météo-France, and uses several specialised routines provided by Alicime for automatic data processing. The first one changes the Geodetic Datum and projection of the maps so that the human-machine interface (HMI) of AEGIR presents the PANTHERE QPE like a GIS layer. Another routine estimates automatically several variables over a few pre-selected geographic domains, in order to monitor the rain events, to select the QPE maps with significant rainfall for recording in the database, and to provide alerts for hydrological risks depending on intensity, surface area of rainfall and identified type of rainfall.

Then mean radar QPE are estimated over 46 basins or sub-basins of the Rhone valley, for time durations of 15 min, 1 h and 6 h. A forecasting program provides quantitative precipitation forecasts (QPF) on the same areas for the next 60 min. This program uses a cross-correlation approach on a number of fixed sub-areas to estimate a set of independent displacement vectors with associated confidence indexes. These vectors are used to interpolate a global displacement field taking into account the confidence index of each vector and the spatial and temporal consistency of the vectors values. This field is then used to extrapolate the motion of the rainfall areas with two algorithms. The first one forecasts the best spatial distribution of cumulated rainfall. The second one forecasts the instantaneous spatial structure of rain fields. All the observed and forecasted values are saved in Hydromet.

FIRST USE OF RADAR QPE AS AN INPUT IN THE CNR HYDROMETEOROLOGICAL FORECASTING PROCESS

Comparison between radar and raingauge QPE over the Rhone River basin

Before testing the use of radar QPE as an input into rainfall–runoff models, CNR has compared radar QPE and previously used 6 h raingauge QPE. This comparison aimed at assessing radar QPE quality over all the 46 sub-basins of the Rhone River, and in all ranges of values, to use them as an input in rainfall–runoff modelling. Data used covered a one year time period starting in February 2010 and ending in January 2011. A minimum threshold of 0.2 mm has been used for the QPE estimated for the 6 h time steps, and the results presented do not distinguish between intensities of rainfall.

Two criteria are presented in this paper: the Nash efficiency criterion (equation (1)) and the global bias criterion (equation (2)):

$$Nash = 1 - \frac{\sum_{i=1}^{n}(QPE_{RG}(i) - QPE_{Radar}(i))^2}{\sum_{i=1}^{n}(QPE_{RG}(i) - \overline{QPE_{RG}})^2} \tag{1}$$

$$Bias = \frac{\sum_{i=1}^{n} QPE_{Radar}(i)}{\sum_{i=1}^{n} QPE_{RG}(i)} \tag{2}$$

where QPE_{RG} the 6 h raingauge QPE; QPE_{Radar} the 6 h QPE based on radar data; n, the number of time steps with both QPE over 0.2 mm.

The results are globally consistent with the previous knowledge of the PANTHERE radar data on the Rhone valley: the quality is better in the main Rhone valley, and decreases dramatically in the Alps area (Fig. 2). Note that the radar QPE integrates a limited correction with the hourly measurements of the same real time raingauge network as the one used in the raingauge QPE.

The global bias criterion indicates that in the Rhone valley, the radar QPE could slightly overestimate the raingauge QPE near the radar location, which is in accordance with the results of previous studies (Faure, 2006, 2008). The global bias criterion also shows an important underestimation in the Alps area. Faure (2006, 2008) attributed this underestimation principally to few sources of error in the radar measurement: partial radar beam blockage by the relief, high altitude of the radar measurement, great distance to the radars. If one assumes that raingauge QPE represent the ground reference, the Nash criterion indicates the variation of the radar QPE quality between the different sub-basins. This information is very important for CNR operators.

Comparison between radar QPF and radar QPE over the Rhone River basin

A comparison between hourly radar QPF and hourly radar QPE is carried out to inform forecasters whether the radar predictions made by AEGIR are consistent with radar observations 1 h later. The

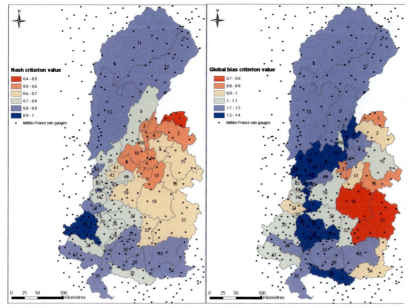

Fig. 2 Comparison between radar and raingauge QPE over the Rhone River basin (time step 6 h).

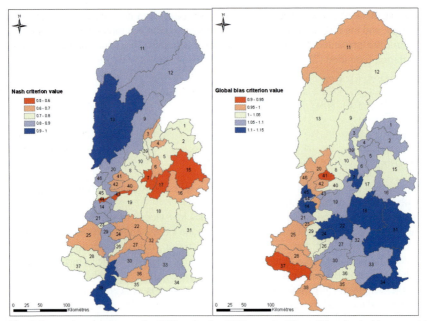

Fig. 3 Comparison between radar QPF and radar QPE over the Rhone River basin (time step 1 h).

same criteria as in the previous paragraph are used here. However, one may replace QPE_{RG} and QPE_{Radar} by QPE_{Radar} and QPF_{Radar}, respectively, in equations (1) and (2).

The results indicate that hourly radar QPF are consistent with hourly radar QPE for a great number of sub-basins: the bias criterion values vary from 0.9 to 1.1 for 38 of the 46 sub-basins (Fig. 3). Nevertheless, the bias criterion shows a tendency to overestimate the QPF in the mountain areas where the radar QPE are underestimated. It will be necessary to verify whether this phenomenon is linked to the rain motion coming from less overestimate areas or not. The Nash criterion indicates that the quality of the radar forecasting varies with the size and the location of the sub-basins. The 7 sub-basins having the largest areas (above 3000 km^2) correspond to Nash criterion values above 0.75.

These results indicate that the QPF provided by the AEGIR data processing could be used in the hydrometeorological forecasting process and would become very useful information for CNR operators.

First test of use of radar QPE as an input to rainfall–runoff model

Tests are still in progress and mainly concern the radar QPE assimilation in the CNR hydrometeorological forecasting process, in order to estimate radar QPE impacts on the forecasted discharges of the Rhone River and its main tributaries. The Doux sub-basin has been selected (sub-basin no. 14 in previous figures). This sub-basin has an area of 630 km^2, is sensitive to heavy rainfall observed during a short period of time, and is located in the north of the Cevennes area affected by heavy rain events. The Doux sub-basin is also located in an area of medium quality regarding radar QPE.

Two different rainfall–runoff models inputs were tested from February 2010 to November 2010: (i) hourly QPE disaggregated from observed 6 h raingauge QPE and (ii) hourly radar QPE, without any forecasted radar values. Figure 4(a) shows the evolution of the Nash efficiency criterion computed for the two inputs and for different forecast lead-time. Results of the persistence model are also addressed in order to illustrate the mean response-time of the basin. This basic model assumes that discharge will remain constant for the next coming hours, from +1 h to +48 h ahead.

Figure 4(a) highlights that both QPE input lead to very good and similar performances from +1 h to +6 h forecast lead-time. After this time horizon, using radar QPE leads to better performance than using raingauge QPE. For example, at +24 h lead-time, radar QPE leads to an increase of 0.2 regarding the Nash criterion. Even if these are only preliminary results, it can be suggested that they are due to two main reasons. Firstly, radar QPE provides a fine-detail realistic rainfall time chronicle, whereas hourly raingauge QPE is calculated assuming that rainfall intensity is stationary during the 6 h time step of raingauge QPE. This is a major difference between radar QPE and raingauge QPE. Secondly, the raingauge network has a mean density of 1 raingauge per 210 km^2 in this area. Thus, raingauge QPE may miss some specific rainfall patterns that are observed by weather radar, mostly during convective events.

This second point is illustrated by Fig. 4(b), which presents the results achieved during the flood event of the 1 November 2010 at +24 h forecast lead-time. For this particular flood event, Nash criterion values estimated between observed discharges and calculated discharges by the rainfall–runoff model, are 0.07 and 0.79 for raingauge QPE and radar QPE inputs, respectively. For both inputs, the shapes of forecasted hydrograms fit well with the observed one. However, the raingauge QPE overestimates the runoff volume. This overestimation is surely not due to a bias of the rainfall–runoff model, as it was calibrated considering a raingauge QPE archive. It is more likely due to an overestimation of raingauge QPE during this event (145 mm *versus* 91.5 mm for radar QPE).

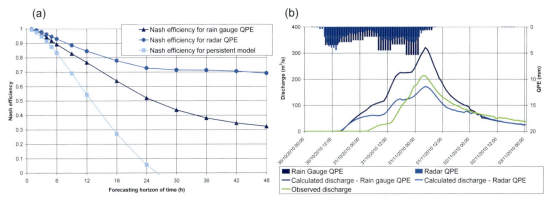

Fig. 4 Use of radar or raingauge QPE as an input to rainfall–runoff model: results on the Doux sub-basin. (a) Nash efficiency criterion for the whole 10 month period, (b) example of flood event: 1 November, 2010.

CONCLUSION

This study highlights that weather radar data are very useful for CNR purposes, not only in terms of qualitative information, but also in terms of quantitative information. In the case of the Doux sub-basin, radar QPE improves rainfall–runoff modelling performances. This encouraging result seems due to the high temporal and spatial resolution of the radar QPE, even using a rainfall–runoff model that is not a distributed one.

In this paper, we mainly focused on radar QPE quality over the Rhone River basin and the use of radar QPE as input to CNR rainfall–runoff models. CNR is not currently using radar QPE as an input into its rainfall–runoff models for operational purposes. Nevertheless, CNR is already using radar data as a qualitative input thanks to the expertise of its forecasters and as quantitative data since forecasters monitor in real time radar QPE values to check for flood events.

Besides, every single tool of the CNR hydrometeorological forecasting process is able to use radar QPE values made available by the AEGIR application, since both observed and forecasted radar QPE are displayed in Hydromet. In order to validate these preliminary results, CNR will now test more largely the integration of radar QPE in its hydrometeorological forecasting process for a selection of sub-basins. The use of radar QPE and QPF should be a major benefit in order to enhance hydrological forecast, which is a major issue for CNR regarding both energy production forecasting and flood forecasting.

Acknowledgements The authors wish to thank Météo-France for providing PANTHERE data and raingauge QPE.

REFERENCES

Bompart, P., Bontron, G., Celie, S. & Haond, M. (2008) Une chaine opérationnelle de prévision hydrométéorologique pour les besoins de la production hydroélectrique de la CNR. In: *Prévisions hydrométéorologiques* (191st SHF workshop, Lyon, France, November 2008), 59–66. SHF Publ., 191 SHF edition, France.

Bontron, G. (2004) Prévision quantitative des précipitations: adaptation probabiliste par recherche d'analogues. Utilisation des réanalyses NCEP-NCAR et application aux précipitations du Sud-Est de la France. Thesis, Institut National Polytechnique de Grenoble, France.

Delrieu, G., Boudevillain, B., Nicol, J., Chapon, B., Kirstetter, P.-E., Andrieu, H. & Faure, D. (2009) Bollène 2002 experiment: radar rainfall estimation in the Cévennes-Vivarais region, France. *J. Appl. Met. & Climatology* 48(7), 1422–1447.

Faure, D. (2006) Cartographie globale de la visibilité hydrologique du réseau radar métropolitain ARAMIS à l'horizon 2006. *Alicime study reports for Météo-France RE05001, RE05002, RE05003, RE06001.*

Faure, D. (2008) Évaluation spatialisée des lames d'eau radar PANTHERE par rapport aux pluviomètres aux pas de temps journalier et horaire sur la région PACA. *Alicime study report for Météo-France RE08122.*

Grimaldi, L. & Haond, M. (2008) Le réseau de mesure en continu de la Compagnie Nationale du Rhône. *TSM* 2, 93–108.

Obled, Ch., Bontron, G. & Garcon, R. (2002) Quantitative precipitation forecasts: a statistical adaptation of model outputs through an analogues sorting approach. *J. Atmos. Res.* 63, 303–324.

Tabary, P. (2007) The new French radar rainfall product. Part I: methodology, *Weather Forecasting* 22(3), 393–408.

Tabary, P., Desplats, J., Do Khac, K., Eideliman, F., Gueguen, C. & Heinrich, J.-C. (2007) The new French radar rainfall product. Part II: Validation. *Weather Forecasting* 22(3), 409.

Weather Radar and Hydrology
(Proceedings of a symposium held in Exeter, UK, April 2011) (IAHS Publ. 351, 2012).

581

Relations between streamflow indices, rainfall characteristics and catchment physical descriptors for flash flood events

P. A. GARAMBOIS[1,2], H. ROUX[1,2], K. LARNIER[1,2] & D. DARTUS[1,2]

1 *Université de Toulouse, INPT, UPS, IMFT (Institut de Mécanique des Fluides de Toulouse), F-31400 Toulouse, France*
pierre-andre.garambois@imft.fr
2 *CNRS; IMFT; F-31400 Toulouse, France*

Abstract Flash flood is a very intense and quick hydrologic response of a catchment to rainfall. This phenomenon has a high spatial-temporal variability as the generating storm often hits small catchments (few km²). Given the small spatial-temporal scales and high variability of flash floods, their prediction remains a hard exercise as the necessary data are often scarce. This study investigates the potential of hydrologic indices at different scales to improve understanding of flash floods dynamics and characterize catchment response in a model independent approach. These hydrologic indices gather information on hydrograph shape or catchment dynamic for instance and are useful to examine catchment signature in function of their size. Results show that for middle-size (>100 km²) catchments response shape can be correlated to storm cell position within the catchment contrarily to smaller catchments. In a multi-scale point of view, regional characteristics about catchment geomorphology or rainfall field statistics should provide useful insight to find pertinent hydrologic response indices. The combined use of these indices with a physically-based distributed modelling could facilitate calibration on ungauged catchments.

Key words flash flood; hydrologic indice; ungauged catchment

INTRODUCTION

The stream hydrograph is a spatial and temporal integration of all the water input, storage and transfer processes within a catchment. Thus the shape and magnitude of the hydrograph potentially yields a wealth of information about catchment hydrological process.

Streamflow indices are calculated from the streamflow hydrograph data of a catchment (Shamir *et al.*, 2005); thus such indices as runoff ratio or time to peak flow are catchment specific: they contain the unique signature of the catchment behaviour. Indices derived from hydrograph shape are also called dynamic response characteristics.

Numerous streamflow hydrograph indices are introduced in the literature. Olden & Poff (2003) address a review of 171 currently available hydrologic indices using stream flow records from 420 sites from across the continental USA. They examine patterns of redundancy in hydrologic indices with principal component analysis (PCA) and conclude that the statistical framework provided can be helpful for the selection of hydrologic indices in hydro-ecological studies. Chinnayakanahalli *et al.* (2005) examined possible links between various hydrologic indices from Olden & Poff (2003) and physical characteristics of catchments to predict hydrologic flow regimes for biological assessment in ungauged basins.

Of particular interest for hydrological modellers are studies like Farmer *et al.* (2003) who use "water balance signatures", which are derived from streamflow records at three different temporal time scales to evaluate the model complexity that is required to reproduce these signatures. Morin *et al.* (2001) propose a peak density measure and a conceptual basin response time scale that is defined as the time to aggregate the precipitation so that the hyetograph and hydrograph are of comparable shape. An objective measure of shape similarity is provided by Morin & Konstantine (2002) with the Rising Limb Density and the Declining Limb Density (RLD, DLD) index calculated for the aggregated hyetograph and the hydrograph.

This paper presents the study at the regional scale of model independent dynamic response characteristics in the case of flash floods. Several physical descriptors are derived from available catchment physiographic data and rainfall field maps. We calculate dynamic response characteristic correlations to physical characteristics of catchments and rainfall fields. The objective here is to improve understanding of the flash flood generating mechanisms by linking dynamic flow indices to physical descriptors in a model independent approach.

The study site is first presented, and then the chosen hydrologic indices are listed: descriptors of physiographic catchment characteristics, descriptors of dynamic response characteristics and descriptors of rainfall characteristics. Finally, correlations between dynamic response characteristics and the other indices are calculated.

STUDY SITE AND DATASET

A set of seven small to medium size (45–619 km^2) gauged catchments located in the Cévennes-Vivarais region was selected for this study (Fig. 1). These catchments are characterized by a strong topographic gradient, and high spatial variability in their litho-pedology and soil cover and occupation. None of the catchment streamflows used here are significantly affected by abstractions or other alterations. All of them are hit by highly variable thunderstorms generating flash floods.

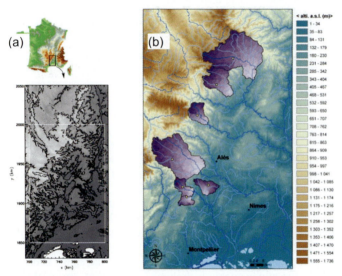

Fig. 1 (a) OHMCV (Observatoire Hydrométéorologique Cévennes Vivarais) pilot site and topography. The white lines delineate rainfall data spatial extend (source (Berne *et al.*, 2009)). (b) Flowpath distances for the seven catchments of interest, yellow dots are the basin centroids in terms of flowpaths. Main rivers and towns are plotted.

We constituted a set of 51 events, occurring during the last decade, by selecting more than eight events for each of the seven catchments. Event hydrograph data, at five minute time-steps, are considered from the beginning of the storm to the end of the rapid recession.

Synoptic events can show some spatial coherence (Merz & Blöschl, 2003); this is the case of few events of the dataset used in the current study, and consequently few points can be statistically dependent. One of the main obstacles to flash flood prediction is the lack of data (Gaume & Borga, 2008; Gaume *et al.*, 2009) so these points have still been considered in this study. However, a better dataset in terms of statistical independence, location and more contrasting catchment characteristics is being collected for a larger area of the French Mediterranean region. Note that most of the events considered here are convective and data are mainly statistically independent, so the method and results presented here are useful at the regional scale.

INDICES DESCRIPTION

Description of catchment characteristics

The variables for the catchments considered here are listed in Table 1.

Table 1 Descriptors of physical characteristics.

Name	Units	Description	Ref
Slope	[–]	Global slope index	–
Kcomp	[–]	Compacity index (Gravelius constant)	–
Rcirc	[–]	Circularity index	–
Dmoy	m	Mean flowpath distance	–
elance	[–]	Horton index (S / Lmax^2)	(Wagener *et al.*, 2004)
DPSBAR	m/km	Index of watershed steepness	(Wagener *et al.*, 2004)
DPLBAR	km	Index describing watershed size and drainage path	(Wagener *et al.*, 2004)
APSBAR	[–]	Index representing dominant watershed slopes	(Wagener *et al.*, 2004)
APSVAR	[–]	Index of invariability of aspect of watershed slopes	(Wagener *et al.*, 2004)
tan(α)m	m/m	Mean topographic slope index	(Hjerdt *et al.*, 2004)
tan(α)s	m/m	Topographic slope index standard deviation	(Hjerdt *et al.*, 2004)

Table 2 Descriptors of dynamic response characteristics.

Name	Units	Description	Ref
RLD	h^{-1}	Rising limb density	(Morin *et al.*, 2002)
DLD	h^{-1}	Declining limb density	(Morin *et al.*, 2002)
HFDisch	[–]	High flow discharge (mean of the 99th percentile)	(Clausen & Bigs, 2000)
Hpcount	[–]	High pulse count (3 times median)	(Clausen & Bigs, 2000)
HPdur	h^{-1}	High pulse duration	(Clausen & Bigs, 2000)
Skew	[–]	Skewness (mean flows divided by median flows)	(Clausen & Bigs, 2000)
Cvar	[–]	Streamflow variability	(Clausen & Bigs, 2000)
Grise	m^3/s/km^2/jour	Mean rising limb gradient	(Clausen & Bigs, 2000)
Gdec	m^3/s/km^2/jour	Mean declining limb gradient	(Clausen & Bigs, 2000)
Vruiss	m^3	Coefficient of variation in streamflow	–
Vruispe	m^3/km^2	Flow volume / catchment area	–

Description of streamflow shape

Studies about flash floods and their generating storms (Le Lay & Saulnier, 2007; Castaings *et al.*, 2009), and technical breakthroughs lead to improved understanding of this physical phenomenon. Information might be extracted from the statistical characterization of catchment response behaviour during a flash flood where most of the surface and subsurface flow paths are active. Several shape descriptors derived from event hydrographs (Table 2) have been used in this study. The idea is to find simple relevant shape descriptors of the flash flood hydrographs.

Description of rainfall characteristics

We follow the intent of (Jakeman & Hornberger, 1993) and attempt to determine: "What reliable information may reside in concurrent observed precipitation-streamflow measurements for assessing the dynamic characteristic of catchment response?"

The area of interest for this study is the Cévennes-Vivarais region (Fig. 1) particularly hit by thunderstorms generating flash floods in the Cévennes foothills. We limit this preliminary study to the OHMCV hourly Kriged rainfall data where raingauge density is high. The OHMCV hourly rainfall data are gridded at 1-km resolution and hourly time-steps.

In the Mediterranean climatic zone, precipitation is highly variable, both in time and space, and this variability increases with elevation in mountainous regions (Moussa *et al.*, 2007). Spatial variability of rainfall was measured by Smith *et al.* (2004) who developed a general variability index and a location index. We calculate this variability index I_σ and the rainfall field location I_L with the Kriged rainfalls maps according to the methodology proposed by Smith *et al.* (2004). The flowpath distance for each cell and topographic characteristics are derived from a 50 m resolution DEM (IGN).

CORRELATIONS BETWEEN STREAMFLOW INDICES, CATCHMENT PHYSIOGRAPHIC DATA AND RAINFALL CHARACTERISTICS

In this section we relate hydrologic response characteristics both to rainfall characteristics and to observable physical catchment characteristics in order to build regional regression relationships. The behavioural information contained in these relations is model independent.

Single correlations are calculated for the 51 event set, i.e. over the seven different catchments (Table 3). For reasons of clarity, only correlation coefficients greater than 0.5 are indicated. Rainfall field characteristics such as intrastorm variability (Iσ) are correlated to dynamic response characteristics such as rising limb gradient (Grise) or declining limb gradient (Gdec). No significant correlations were found with Hu2 the daily wetness index from SIM platform (Habets *et al.*, 2008), I$_L$, Pente, Kcmp, Rcirc: so these do not appear in the table.

On the basis of these correlations, some streamflow indices can be explained in multiple regressions by physical descriptors (Fig. 2). For the seven catchments dataset, results show that the mean rising limb gradient (Grise) is correlated with intrastorm variability (Iσ) and the mean topographic index (tan(α)) with a r^2 of 0.61; the maximum specific discharge (Qpspe) is correlated with the total accumulated rainfall (Cumul) and intrastorm variability (Iσ) (r^2 = 0.72). Further investigation will include some other shape and frequency characteristics, peak descriptors or wavelet built indices.

Table 3 Single correlations between streamflow indices and physical descriptors for the full dataset.

	Cumul	Iσ	Vpluie	Dmoy	elance	Dpsbar	tan(α)	tan(α)s
Mean	0.73	0.77	0	0	0	0	0	0
Med	0.63	0	0	0	0	0	0	0
Std.dv	0.66	0.79	0	0	0	0	0	0
RLD	0	0	0	0	0.54	0	0	0.52
DLD	0	0	0	0	0	0	0	0
HFDis	0	0	0	0	0	−0.53	0	0
Hpcnt	0.70	0	0	0	0	0	0	0
HPdur	0	0	0	0	−0.53	0	0	0
Skew	0	0.57	0	0	0	−0.56	−0.55	0
Cvar	0	0	0	0	0.58	−0.56	0	0.51
Grise	0	0.59	0	0	0.52	0	0	0.55
Gdec	0.59	−0.70	0	0	0	0	0	0
Vruisp	0	0	0.63	0.56	−0.51	0	0	0
Qpoint	0.74	0.52	0	0	0	0	0	0
Qpspe	0.50	0.65	0.70	0	0	0	0	0

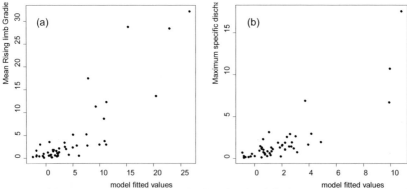

Fig. 2 Multiple regressions (gaussian family) for the full dataset. (a) Mean rising limb gradient estimated from intrastorm variability (Pvalue = 5.04e-05) and mean topographic index (Pvalue = 0.00034); r^2 = 0.61. (b) Maximum specific discharge estimated from total accumulated rainfall (Pvalue = 0.0047) and intrastorm variability (Pvalue = 7.98e09); r^2 = 0.72.

Table 4 Single correlations between streamflow indices and physical descriptors. (left) Set of four catchments (area < 100 km^2) (right) Set of three catchments (area > 100 km^2).

	area < 100 km^2							area > 100 km^2						
	Cumul	I$_L$	Iσ	Vpluie	Pente	Kcmp	Rcirc	Cumul	I$_L$	Iσ	Vpluie	Pente	Kcmp	Rcirc
Mean	0.81	0	0.83	0.61	0	0	0	0.62	0	0.73	0.53	0	0	0
Med	0.66	0	0.57	0.51	0	0	0	0.58	0	0	0	0	0	0
Std.dv	0.76	0	0.84	0.55	0	0	0	0.57	0	0.91	0.59	0	0	0
RLD	0	0	0	0	0	0	0	0.69	0	0	0	0	0	0
DLD	0	0	0	0	0	0	0	0.51	0	0	0	0	0	0
HFDis	0	0	0	0	0	0	0	0	−0.60	0.61	0	0	−0.56	0.56
Hpcnt	0.85	0	0.54	0.75	0	0	0	0.57	0	0	0	0	0	0
HPdur	0	0	0	0.51	0	0	0	0	0	0	0	0	0	0
Skew	0	0	0.58	0	0	0	0	0	−0.60	0.68	0	0	−0.54	0.54
Cvar	0	0	0	0	0	0	0	0	−0.50	0.61	0	−0.61	−0.66	0.66
Grise	0	0	0.63	0	0	0	0	0	0	0.94	0.51	0	0	0
Gdec	−0.71	0	−0.78	−0.53	0	0	0	0	0	−0.88	−0.53	0	0	0
Vruisp	0.86	0	0.53	0.86	0	0	0	0	0	0	0	0	0	0
Qpoint	0.88	0	0.71	0.77	0	0	0	0	0	0	0	0	0	0
Qpspe	0.80	0	0.73	0.68	0	0	0	0.51	0	0.89	0.70	0	0	0

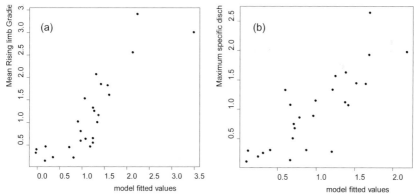

Fig. 3 Multiple regressions for set of 3 catchments (area > 100 km^2) (a) Mean rising limb gradient estimated from intrastorm variability (Pvalue= 1.29e-07) and rainfall location index (Pvalue= 0.023); r^2 = 0.72. (b) Maximum specific discharge estimated from total accumulated rainfall (Pvalue= 0.00097) and intrastorm variability (Pvalue= 0.0024); r^2 = 0.73.

Considering the previous correlations, the dataset has been split into two subsets: catchments with area smaller than 100 km^2 and the others (Table 4). We can see in the case of small catchments that cumulative rainfall intensity or rainfall volume are correlated with several streamflow indices, whereas in the case of middle-size catchments, location index (Iσ) and storm variability (I$_L$) play a greater role, as do hydrograph skewness and variability.

Indeed, for middle-size catchments, we can now correlate the mean rising limb gradient (Grise) with intrastorm variability (Iσ) and storm location index (I$_L$) with a r^2 = 0.72 (Fig. 3); whereas this regression gives no significant correlation (r^2 < 0.4 and bad P-values) for the small catchment set. This is due to a scale problem: indeed storm cells can be larger than small catchments and so location within a catchment loses importance. Moreover these small basins can have very contrasting responses depending on forcing variability and physiographic characteristic variability and often show a small dampening effect. That is why a large dataset is needed to study small catchments in the statistical framework we propose.

CONCLUSIONS

This study, for the case of flash floods events, tackles the problem of physical process description and catchment signature analysis through streamflow indices, physical descriptors and rainfall characteristics. The correlation between mean rising limb gradient, rainfall location and variability index, for middle-size catchments, is quite good contrary to the one for small catchments (<100 km^2). Knowing this threshold could be useful to study a catchment's dampening effect and response. We can also wonder which descriptors or parameters and which spatial averaging are representative of the processes and can explain catchment response dynamics. Moreover this approach gives a statistical framework to quantify catchment response dispersivity/similarity, and could allow scale analysis.

The study will be carried on with radar data at 5-min time-steps on more catchments and events to improve the hydrograph shape description and statistical independence. We hope to obtain strong regression relations between streamflow indices, physical descriptors and rainfall characteristics. Another explanatory variable may be the soil storage capacity derived from high resolution pedologic data. This could enable the elaboration of a model independent calibration methodology at the regional scale, and its application to ungauged catchments (Yadav *et al.*, 2007; Zhang *et al.*, 2008).

REFERENCES

Berne, A., Delrieu, G. & Boudevillain, B. (2009) Variability of the spatial structure of intense Mediterranean precipitation. *Adv. Water Resources* 32(7), 1031–1042.

Castaings, W., Dartus, D., Le Dimet, F.-X. & Saulnier, G.-M. (2009) Sensitivity analysis and parameter estimation for distributed hydrological modeling: potential of variational methods. *Hydrol. Earth System Sci.* 13(4), 503–517.

Chinnayakanahalli, K. J., Tarboton, D. G. & Hawkins, C. P. (2005) Predicting hydrologic flow regime for biological assessment at ungauged basins in the western United States. AGU, Fall Meeting, abstract #H54C-04.

Clausen, B. & Biggs, B. J. F. (2000) Flow variables for ecological studies in temperate streams: groupings based on covariance. *J. Hydrol.* 237(3–4), 184–197.

Farmer, D., Sivapalan, M. & Jothityangkoon, C. (2003) Climate, soil, and vegetation controls upon the variability of water balance in temperate and semiarid landscapes: Downward approach to water balance analysis. Water Resour. Res. 39(2), 1035.

Gaume, E., V. Bain, P. Bernardara, O. Newinger, M. Barbuc, A. Bateman, L. Blaskovicová, G. Blöschl, M. Borga, A. Dumitrescu, I. Daliakopoulos, J. Garcia, A. Irimescu, S. Kohnova, A. Koutroulis, L. Marchi, S. Matreata, V. Medina, E. Preciso, D. Sempere-Torres, G. Stancalie, J. Szolgay, I. Tsanis, D. Velasco & A. Viglione, (2009). A compilation of data on European flash floods. *J. Hydrol.* 367(1–2), 70–78.

Gaume, E. & Borga, M. (2008) Post-flood field investigations in upland catchments after major flash floods: proposal of a methodology and illustrations. *J. Flood Risk Manage.* 1(4), 175–189.

Habets, F., Boone, A., Champeaux, J. L., Etchevers, P., Franchistéguy, L., Leblois, E., Ledoux, E., Le Moigne, P., Martin, E., Morel, S., Noilhan, J., Quintana Seguí, P., Rousset-Regimbeau, F. & Viennot, P. (2008) The SAFRAN-ISBA-MODCOU hydrometeorological model applied over France. *J. Geophys. Res.* 113, D06113.

Hjerdt, K. N., McDonnell, J. J., Seibert, J. & Rodhe, A. (2004) A new topographic index to quantify downslope controls on local drainage. *Water Resour. Res.* 40, W05602.

Jakeman, A. J. & Hornberger, G. M. (1993) How much complexity is warranted in a rainfall-runoff model? *Water Resour. Res.* 29(8), 2637–2649.

Le Lay, M. & Saulnier, G.-M. (2007) Exploring the signature of climate and landscape spatial variabilities in flash flood events: Case of the 8–9 September 2002 Cévennes-Vivarais catastrophic event. *Geophys. Res. Lett.* 34(L13401), doi:10.1029/2007GL029746.

Merz, R. & Blöschl, G. (2003) A process typology of regional floods. *Water Resour. Res.* 39(12), 1340.

Morin, E., Enzel, Y., Shamir, U. & Garti, R. (2001) The characteristic time scale for basin hydrological response using radar data. *J. Hydrol.* 252(1–4), 85–99.

Morin, E. & Konstantine, P. (2002) Objective, observations-based, automatic estimation of the catchment response timescale. *Water Resour.* 38(10), 1212.

Moussa, R., Chahinian, N. & Bocquillon, C. (2007) Distributed hydrological modelling of a Mediterranean mountainous catchment – Model construction and multi-site validation. *J. Hydrol.* 337, 35–51.

Norbiato, D., Borga, M. Sangati, M. & Zanon, F. (2007) Regional frequency analysis of extreme precipitation in the eastern Italian Alps and the August 29, 2003 flash flood. *J. Hydrol.* 345, 149–166.

Olden, J. D. & Poff, N. L. (2003) Redundancy and the choice of hydrologic indices for characterizing streamflow regimes. *River Research and Applications* 19(2), 101–121.

Shamir, E., Imam, B., Gupta, H. V. & Sorooshian, S. (2005) Application of temporal streamflow descriptors in hydrologic model parameter estimation. *Water Resour. Res.* 41(6), W06021.

Smith, M. B., Koren, V. I., Zhang, Z., Reed, S. M., Pan, J.-J. & Moreda, F. (2004) Runoff response to spatial variability in precipitation: an analysis of observed data. *J. Hydrol.* 298(1–4), 267–286.

Wagener, T., Wheater, H. & Gupta, H. (2004) *Rainfall–Runoff Modelling in Gauged and Ungauged Catchments*. Imperial College Press, UK.

Yadav, M., Wagener, T. & Gupta, H. (2007) Regionalization of constraints on expected watershed response behavior for improved predictions in ungauged basins. *Adv. Water Resour.* 30(8), 1756–1774.

Zhang, Z., Wagener, T., Reed, P. & Bhushan, R. (2008) Reducing uncertainty in predictions in ungauged basins by combining hydrologic indices regionalization and multiobjective optimization. *Water Resour. Res.* 44, W00B04.

Radar for hydrological modelling: new challenges in water quality and environment

M. BRUEN

Centre for Water Resources Research, School of Civil, Structural and Environmental Engineering, Newstead Building, University College Dublin, Belfield, Dublin 4, Ireland
michael.bruen@ucd.ie

Abstract Any exploration of actual and potential uses of radar information in hydrological models must start with a survey of actual and potential uses of hydrological models. Their early history is closely associated with the increasing availability of computing power while the more recent past shows an increasing range of applications. Early developments were for water resources management, floods and droughts, but the range of potential uses has expanded considerably. For instance, in Europe, the requirements of the EU Water Framework Directive will increase the use of either distributed or semi-distributed models in "design" mode to identify critical source areas contributing contaminants and sediment to rivers and lakes. The critical sources areas can be a small percentage of the total catchment area, but contribute most of the sediment and a lot of the associated particulate contamination. The rainfall climatology at such small scales may be developed from radar records and may be important for the management of risk if there is significant spatial variability at such scales. In the past, radar-based rainfall forecasts would only be used for river flow forecasting; however their use can be extended to such water quality applications as forecasting bathing water quality on beaches as a public information service.

Key words hydrological modelling; radar; Water Framework Directive; Floods Directive

INTRODUCTION

Hydrological models require numerical values for their parameters and good input data and both cause considerable difficulty for hydrologists. Radar is a source of spatially-distributed and remotely-sensed input data for flood forecasting applications, but, as the range of applications for hydrological models expands, the potential uses of radar information also increase. In particular, concerns about deteriorating water quality and efforts to deal with water-borne contaminants require more complex hydrological models. Now, as much, if not more, hydrological modelling is directed at water quality and ecological issues, with an increasing focus on better modelling of individual physical processes and pathways. The development of physically-based distributed models, not often required for flood forecasting, increased with the need to simulate water quality and, along with measures to manage it, required more detailed information. Another dimension is added by concerns about the impacts of anticipated climate change which encouraged the development of coupled hydrological atmospheric models. The spatial coverage given by radar precipitation information is important in all these applications. This paper starts with a short survey of trends in the radar/hydrology literature and then focuses discussion on two key drivers of the future uses for hydrological models and radar data.

WHITHER HYDROLOGICAL MODELLING?

One way of attempting to predict the future for hydrological models is to analyse past trends and project them forward into the future in a manner analogous to Quantitative Precipitation Forecasting (QPF). Models are simplifications of reality and different models are developed for different applications with the model structure and its data requirements determined by the purpose of the model. Therefore any broad exploration of the actual and potential use of radar information in hydrological models must start with a survey of the current uses of hydrological models. A search of a widely used bibliographic technical database (COMPENDEX) was undertaken to determine the frequency of other keywords used in papers that also had the keywords "radar" and "hydrology" or related forms. A separate search was conducted for each decade, starting from the period 1970–1979 to the period 2000–2009. A number of very general terms were removed from the analysis, e.g. hydrodynamics, probability, mathematical and some terms were combined, e.g. rain and precipitation. The number of papers increased exponentially in each decade, as seen in Fig. 1. The frequencies of specific keywords in each decade have been normalised by dividing by the total number of papers in the decade.

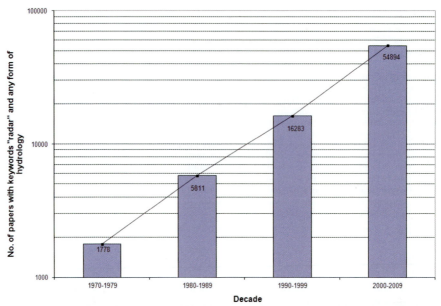

Fig. 1 Exponential growth of numbers of papers with radar and hydrology themes from 1970 to 2009.

For this paper, the first two decades were combined (i.e. 1970–1989) and the second two were also combined (i.e. 1990–2009). Table 1 shows the ranks of each keyword within each 20-year group and also shows the change in rank. Key words with a negative change in rank have reduced in relative frequency (or ceased to occur) and those with positive changes have increased in relative frequency (or did not occur in the earlier grouping). Table 2 shows the top five keywords that have dropped in rank, gained in rank, and those that have maintained a high rank.

There is a trend from the general to the more specific, for example, from atmosphere to climate change and from data processing to GIS and simulation. In both cases a very broad topic has been replaced by a specific aspect of the topic. This finer focus is to be expected with the exponential growth in numbers of papers. However, there is also a significant change from a major focus on water resources applications (e.g. agriculture, forestry, irrigation) to a concentration on specific physical processes, particularly evapotranspiration and infiltration and on modern technology (GIS and remote sensing). The increased focus on the two most nonlinear, and difficult to model, components of the hydrological cycle is driven by a broadening of the use of hydrological models in environmental, ecological and decision-support applications. From the hydrologist's point of view, much of this theory and modelling is not new, but the broader range of potential uses of hydrological models provides the catalyst to seek further improvements in modelling these physical processes.

The necessary theoretical relationships have been known and fully- and semi-distributed hydrological models have been developed and have been used for some time (Abbott *et al.*, 1986); (Garrote & Bras, 1995). While, initially, their use has been limited to specific areas and applications, their potential for use in river basin management has long been recognised by hydrologists (Bathurst & &O'Connell 1992). What is different now is: (a) the driving factor of

Table 1 Twenty year changes in keyword frequency rankings (additional keywords in papers with "radar" and "hydrology" and their variants).

Additional keyword	Rank in 1970–1989	Rank in 1990–2009	Change in rank
Agriculture	14	19	−5
Aquifer(s)	4	1	3
Atmosphere	5	6	−1
Climate change	26	17	9
Computer simulation	13	7	6
Data processing	19	26	−7
Environment	22	24	−2
Evapotranspiration	21	15	6
Floods/flow modelling	2	3	−1
Forestry	18	25	−7
Geology/GIS	16	14	2
Hydraulic conductivity	26	21	5
Hydraulics	12	13	−1
Infiltration	26	18	8
Irrigation	15	20	−5
Land use	26	22	4
Rain	7	5	2
Remote sensing	17	16	1
Reservoirs/lakes	11	12	−1
Rivers/Streams	3	4	−1
Runoff	1	2	−1
Sediment transport	20	23	−3
Soils	8	8	0
Water pollution	10	10	0
Water resources	9	11	−2
Watershed(s)	6	9	−3

Table 2 Prevalence of the top keywords that have changed in relative frequency.

Reducing frequency	Ranks lost	Increasing frequency	Ranks gained	Consistently high rank	Mean rank
Data Processing	−7	Climate change	9	Runoff	1.5
Forestry	−7	Infiltration	8	Aquifer(s)	2.5
Agriculture	−5	Evapotranspiration	6	Floods	2.5
Irrigation	−5	Computer simulation	6	Rivers/streams	3.5
Sediment Transport	−3	Hydraulic conductivity	5	Atmosphere	5.5
Watershed(s)	−3	Land use	4	Rain	6

European legislation and the legal necessity to respond to the EU Water Framework and Floods Directives by producing specific and achievable management plans, and (b) the improved availability of the required spatial datasets with national coverage. Together, these factors open the real possibility of hydrological models being deployed, Europe-wide, for flood warning and environmental/ecological management of river basins, with radar precipitation as a potential source of input data, as explained below.

WATER QUALITY AND ECOLOGY

The Water Framework Directive (2000/60/EC) is one of the main drivers of current research in Europe related to water quality. It requires member states to achieve "good" chemical and ecological water quality in all its water bodies. Hydrologists have become closely involved in such work, together with ecologists and catchment managers, because of the recognition of the need to quantify water flows and to understand how contaminants are mobilised and transported by water through the landscape, (Bruen, 2009). The concept of "critical source area" (CSA) is widely used and phosphorus, nitrogen and microbial (bacteria and viruses) contaminants are the main concerns. Depending on conditions, a relatively small part of a catchment can contribute most of the contamination, especially sediment, sediment-bound and other particulate contaminants. For instance, >80% of particulate pollution can originate in <10% of a catchment's area. Also, much of the pollution is delivered in a small number of extreme storm events: e.g. Pionke *et al.* (2000) report 90% of phosphorus exported in the seven largest storms in a year. Once the contaminant pressures have been quantified and the water bodies at risk identified, measures to manage the problem must be implemented by EU member states. Some measures, such as changes to legislation, regulations or enforcement practices are appropriate at national scales. Others, mainly related to agricultural practices (land-use change, nutrient and wastes management, e.g. land-spreading) and diffuse source pollution, would be too costly and too wasteful of resources to apply at a national scale and are likely to be applied at catchment or sub-catchment scales in an approach that targets critical source areas, e.g. Cho *et al.* (2010). This requires an understanding of the factors that create a critical source area and an ability to simulate the mobilisation and transport of contaminants from these areas. Although physically-based fully distributed catchment models would seem to be the most appropriate choice of model, practical considerations favour semi-distributed models, such as SWAT (Busteed *et al.*, 2009; Jha *et al.*, 2009; McDowell & Srinivasan, 2009) and the related AGNPS (Sarangi *et al.*, 2007). The main factors determining critical source areas for contaminants transported by surface and near-surface runoff are land-use, soil type, ground slope and rain-rate. Information on the spatial distribution of all these factors is required by either semi-distributed or distributed models. Some measure of slope can be derived from topographic information and considerable progress has been made in the last decades in mapping land-use and soil types to greater resolutions so that, together with radar, all the required distributed data sources are available. For instance, NEXRAD (4 km) radar precipitation was used by White *et al.* (2009) to study critical source areas in Oklahoma for a catchment that had large variations in topography and spatial rainfall distribution. Radar has the advantage of determining the spatial and temporal distribution of rainfall at scales useful for determining CSAs.

In addition to producing critical source area maps for management purposes, hydrological models may be used in a forensic context to investigate possible causes of significant "events" in the aquatic environment. For instance, if a significant incident of pollution or "fish-kill" occurred in a river or lake, a hydrological model might be used to investigate if it could have been related to weather, i.e. whether there could have been sufficient contaminated runoff to cause the event. The model might, for instance, be able to rule out diffuse source pollution. For this purpose, rainfall and flow data for the period surrounding the "event" is required. If the catchment has a number of critical source areas for the pollutant implicated in the "fish-kill" then a key question is whether there was intense rainfall on the CSA. Radar, with its good spatial resolution may be able to address this type of issue.

FLOODS

The European Floods Directive (2007/60/EU) requires EC member states to assess and map flood risk by December 2011 and to develop flood management plans by December 2015. The implementation of the Water Framework Directive and the Floods Directive should take into account the objectives of both. Article 7 of the Floods Directive specifically mentions that flood forecasts and warning are a component of "preparedness". What role can radar play here? In large river basins, dominated by fluvial flooding issues, the most useful sources of information for flood forecasting are real-time measurements of water levels and estimation of the corresponding discharges in upstream river sections and tributaries. In many cases decisions on warnings, mobilisation of prevention measures and/or evacuation may be based on predictions based on these data. In some Irish towns, a warning of the main river flood with sufficient lead-time to erect demountable barriers may be possible. However, there are also towns for which the most serious flood damage is caused by: (a) "flash" floods from a nearby small sub-catchment "piggy-backing" on the slower main river flood, or by (b) localised flooding from surcharged storm-drains within the urbanised area. In addition, there is some section of the upstream river reach from which floods reach the town within the desired forecast lead time. In all such cases upstream hydrograph information is not sufficient and real-time precipitation estimation and forecasting is required and mainly for the critical sub-basin(s) or over the urban area itself. Thus the main requirement for radar can vary between extensive or complete coverage of the catchment and relatively small scale coverage of a number of small but critical, often urban, areas. In the latter case, a number of strategically placed short-range and, hopefully, cheap radars may meet flood forecasting requirements. This could be located to give good coverage of critical sub-catchments or vulnerable urban areas and complement the information produced by national radar networks. They are particularly suited to urban areas at risk of pluvial flooding for which the smaller space and time scales are more important (Villarini *et al.*, 2010), even though they may not yet be capable of delivering the desired combination of spatial resolution and accuracy (Berne *et al.*, 2004).

The quantification of uncertainty is receiving considerable attention, although it did not feature as one of the increasing rank keywords in the literature survey above. The EU Concerted Action COST731 (cost731.bafg.de/) addressed uncertainty in the flood forecast chain from numerical weather prediction, through radar measurements of atmospheric variables (winds as well as precipitation) (Rossa *et al.*, 2010) to hydrological models and flood forecasts (Zappa *et al.*, 2010).

Warnings must be communicated effectively to end-users, both the general public and the people who make decisions relating to the emergency response and there is considerable variation in how this can be done effectively (Bruen *et al.*, 2010).

Acknowledgements The ideas in this paper come from the author's involvement in a research project supported by Science Foundation Ireland (07/RFP/ENMF292) and with the work of COST731.

REFERENCES

Abbott, M. B., Bathurst, J. C., Cunge, J. A., O'Connell, P. E. & Rasmussen, J. (1986) An introduction to the European Hydrological System – Systeme Hydrologique Europeen, "SHE", 1: History and philosophy of a physically-based, distributed modelling system. *J. Hydrol.* 87(1–2), 45–59.

Bathurst, J. C. & O'Connell, P. E. (1992) Future of distributed modelling: The Systeme Hydrologique Europeen. *Hydrol. Processes* 6(3), 265–277.

Berne, A., Delrieu, G., Creutin, J. D. & Obled, C. (2004) Temporal and spatial resolution of rainfall measurements required for urban hydrology. *J. Hydrol.* 299, 166–179.

Bruen, M. (2009) Hydrology and the Water Framework Directive in Ireland. *Biology & Environment, Proc. Royal Irish Academy* 109B(3), 207–220.

Bruen, M., Krahe, P., Zappa, M., Olsson, J., Vehvilainen, B., Kok, K. & Daamen, K. (2010) Visualizing flood forecasting uncertainty: some current European EPS platforms—COST731 working group 3. *Atmos. Science Letters* 11(2), 92–99.

Busteed, P. R., Storm, D. E., White, M. J. & Stoodley, S. H. (2009) Using SWAT to target critical source sediment and phosphorus areas in the Wister Lake Basin, USA. *American J. Environmental Sciences* 5(2), 156–163.

Cho, J., Lowrance, R. R., Bosch, D. D., Strickland, T. C., Her, Y. & Vellidis, G. (2010) Effect of watershed subdivision and filter width on SWAT simulation of a coastal plain watershed. *J. American Water Resources Association* 46(3), 586–602.

Garrote, L. & Bras, R. L. (1995) An integrated software environment for real-time use of a distributed hydrologic model. *J. Hydrol.* 167, 307–326.

Jha, M. K., Schilling, K. E., Gassman, P. W. & Wolter, C. F. (2009) Targeting land-use change for nitrate-nitrogen load reductions in an agricultural watershed. *J. Soil and Water Conservation* 65(6), 342–352.

McDowell, R. W. & Srinivasan, M. S. (2009) Identifying critical sources areas for water quality: 2. Validating the approach for phosphorus and sediment losses in grazed headwater catchments. *J. Hydrol.* 379, 68–80.

Pionke, H. B., Gburek, W. J. & Sharpley, A. N. (2000) Critical source area controls on water quality in an agricultural watershed located in the Chesapeake Basin. *Ecological Engineering* 14(4), 325–335.

Rossa, A., Haase, G., Keil, C., Alberoni, P., Ballard, S., Bech, J., Germann, U., Pfeifer, M. & Salonen, K. (2010) Propagation of uncertainty from observing systems into NWP: COST-731 Working Group 1. *Atmos. Science Letters* 11(2), 145–152.

Sarangi, A., Cox, C. A. & Madramootoo, C. A. (2007) Evaluation of the AnnAGNPS Model for prediction of runoff and sediment yields in St Lucia watersheds. *Biosystems Engineering* 97(2), 241–256.

White, M. J., Storm, D. E., Busteed, P. R., Stoodley, S. H. & Phillips, S. J. (2009) Evaluating nonpoint source critical source area contributions at the watershed scale. *J. Environmental Quality* 38, 1654–1663.

Villarini, G., Smith, J. A., Baeck, M. L., Sturdevant-Rees, P. & Krajewski, W. F. (2010) Radar analyses of extreme rainfall and flooding in urban drainage basins.. *J. Hydrol.* 381(3–4), 266–286.

Zappa, M., Beven, K. J., Bruen, M., Cofiño, A. S., Kok, K., Martin, E., Nurmi, P., Orfila, B., Roulin, E., Schröter, K., Seed, A., Szturc, J., Vehviläinen, B., Germann, U. & Rossa, A. (2010) Propagation of uncertainty from observing systems and NWP into hydrological models: COST-731 Working Group 2. *Atmos. Science Letters* 11(2), 83–91.

7 Urban hydrology and water management applications

Weather Radar and Hydrology
(Proceedings of a symposium held in Exeter, UK, April 2011) (IAHS Publ. 351, 2012).

595

Advances in the application of radar data to urban hydrology

HANS-REINHARD VERWORN

Institute for Water Resources Management, Leibniz University, Appelstr. 9A, 30167 Hannover, Germany
verworn@iww.uni-hannover.de

Abstract When radar data are to be used in urban hydrology a few special aspects have to be considered. The smaller scales in time and space require a higher resolution and rapid availability if utilised for real-time applications, and have consequences for the processing of data. This review focuses on some of the problems and the developments and improvements in recent years.

Key words urban hydrology; urban drainage; radar rainfall data; nowcasting

INTRODUCTION

The past decades have seen rapid improvements to radar-derived rainfall information. New radar technology as well as improved data processing algorithms and procedures – in part made possible by still increasing computing power – are the basis for the currently available rainfall information.

The end-user, nevertheless, still seems to be sceptical, and the inclusion of radar data into the rainfall–runoff calculations is not as high as would be expected. This may still be due to the belief that radar data have to be largely consistent with ground-measured point data, still widely believed to be the ground truth. Amongst others, the representativeness errors arise from the very different sample volumes (Kitchen & Blackall, 1992). A methodology to estimate the point-to-area variance was proposed by Ciach & Krajewski (1999) and applied by Bringi *et al.* (2011), resulting in point-to-area differences of 20% and 40% for rainfall accumulations larger than 1 mm/h and larger than 6 mm/h, respectively. With all comparisons, however, one has to bear in mind that radar rainfall largely depends on the Z (reflectivity) to R (rainfall) conversion which, due to its dependency on the drop spectrum, is one of the main error sources.

Another reason for reluctant acceptance is the fact that the standard processing procedures to produce a good quantitative precipitation estimate (QPE) are only available with a certain delay. The assimilation of data from radars and other sources needs time-steps in hours rather than in minutes, and the data collection from various sources and the calculations need some time as well. So in general, the processed rainfall data are available too late for real-time applications in urban drainage. End-users are therefore often confronted with the need to use the more-or-less instantly available raw data and develop and apply their own processing procedures in order to improve the quality of the derived rainfall data.

In the following sections the requirements, restrictions and procedures related to applications in urban drainage modelling and operation are dealt with.

TYPICAL APPLICATIONS

Urban applications of weather radar-derived rainfall data fall into two categories: real-time and non-real-time. With the increasing interest in real-time control of urban drainage systems the conceptual planning and off-line simulations made it clear that an effective and successful control strategy depends largely on reliable rainfall input data, not only up to the actual moment but for some time into the future as well, to pre-estimate the expected load and the consequences of potential control measures. Short-term forecasts – generally referred to as nowcasts – require the rainfall patterns and their development which cannot be provided from point data on the ground.

In contrast to the large interest in real-time control there are still not many cases in which an operational system with automatic or supervised real-time control is implemented (Pleau *et al.*, 2004; Fuchs *et al.*, 2005; Maeda *et al.*, 2005; Schroeder *et al.*, 2005). Apart from the Vienna project, where the rain input was derived from raingauges and radar data (Kraemer *et al.*, 2007), the use of radar data is mainly restricted to qualitatively visual information and not directly incorporated in the simulation and decision-finding processes. Nevertheless, interest in reliable forecasts exists and various investigations have shown that nowcasting up to 90 minutes ahead lies within tolerable error margins and may improve predictive real-time control significantly (Golding, 2000; Pierce *et al.*, 2004; Verworn & Krämer, 2008; Achleitner *et al.*, 2009).

Non real-time applications include calibration and verification of drainage system parameters, recalculation of past events, system performance assessment runs, and design and planning scenarios. In many cases, post-processed or even original radar datasets are used or at least considered: for example, see Bouilloud *et al.* (2010) and Villarini *et al.* (2010).

When models are calibrated with point rainfall data, re-calibration seems to be necessary when distributed rainfall from various gauges or radar is used. Differences in system output, however, have been found to be insignificant when calibrations with point rainfall and distributed rainfall from several gauges are compared (Kleidorfer *et al.*, 2006). This is only transferrable to radar rainfall input if there is no bias between radar and ground data.

REQUIREMENTS FOR URBAN APPLICATIONS

In comparison to the use of weather radar data for river catchments, or other large-area related applications, the scales in time and space are much smaller for urban drainage modelling. In general, the sub-catchments are in the size range of m^2 to hectares and the response times of these small sub-catchments to storm runoff are in minutes or even less. Consequently, detailed rainfall–runoff calculations with distributed hydrological or hydrodynamic models not only require short calculation time-steps in seconds or minutes, but also an adequate high resolution of the rainfall data in time and space.

Theoretically, radars are able to provide such data, but in practice there are some problems and physical restrictions that have to be tackled or accounted for (Met Office, 2009). Apart from installing advanced polarimetric radar systems, many efforts have been made to develop more sophisticated algorithms and procedures for processing radar data and to incorporate information from other sources like ground gauges, disdrometers and microwave links to produce better QPEs and forecasts.

In the following sections some aspects of the various steps towards that aim are described.

RESOLUTION IN TIME AND SPACE

The resolution in time and space clearly depends on the kind of application. For urban drainage modelling the resolution has to be definitely higher than for large river catchments. The highly distributed modelling and the fast reaction times require an appropriate resolution of the input data. While end-users would like a resolution as high as possible – 1 min and 100×100 m have been stated – the sensible resolution depends on the radar technology as well as on interests of other users (Berne *et al.*, 2004). The polar format of the radar data and the range are the main restrictions. Nowadays, operational radars generally have a resolution of 1° in azimuth and 1000 m in range. Though higher resolutions are possible, the restrictions come from the need to use the weather radars not only for precipitation detection but for volumetric information for assimilation into Numerical Weather Prediction models as well. Higher resolutions would demand longer scanning times or a reduction in the number of pulses to be integrated for each azimuthal step. Generally, a complete volumetric scan takes about 15 min, and to produce more frequent data the precipitation scans with the lowest possible elevation are inserted in shorter time-steps.

The final rainfall product is generally provided in Cartesian coordinates. With the conversion from polar to Cartesian data the true resolution of the polar data has to be accounted for. While there are many polar data elements that fall in one square bin near the radar, the spread and pulse volume of the radar beam expands with range. At a distance of 60 km and an azimuthal step of 1° the centres of the polar data elements are already about 1 km apart. For a Cartesian resolution of 1 km not all elements beyond that distance get a radar data hit. To overcome this problem, sophisticated grid conversion techniques have been developed (Henja & Michelson, 1999) as an alternative to simple interpolation. Nevertheless, the true variance within the pulse volume that increases with range remains unknown, so that the Cartesian resolution does not necessarily resemble the rainfall patterns. Additionally, one has to bear in mind that the main beam width is generally larger than the azimuthal data resolution.

To increase the resolution in time it would be best to increase the scanning frequency. The more frequent the capture of the momentary situation the less the danger that rainfall from rapidly moving convective cells is not "delivered" to all pixels on its way. But more frequent scanning may be avoided by tracking and accumulation methods. In the French operational network two-dimensional advection fields are used that are derived using standard cross-correlation methods (Tabary, 2007). The resulting interpolated rain-rate fields at 1 min intervals may then be directly used or accumulated to the required coarser resolution.

QUANTITATIVE PRECIPITATION ESTIMATE

To derive sufficiently accurate estimates of quantitative precipitation a number of quality control and correction procedures are necessary. The potential radar errors fall into two categories: errors relating to the basic measurement of reflectivity and those resulting from the conversion of measured reflectivity to precipitation falling at the ground (Harrison *et al.*, 2009; Met Office, 2009). The focus here is only on three of the error sources.

Clutter

Statistical filters in combination with advanced technology generally enable successful clutter detection. A successful clutter reduction, however, means that the clutter affected pixels are not only erased or marked but that the resulting "holes" in the radar matrix have to be filled. This may be done by cross-interpolation techniques if the "holes" are not too big.

Occasionally-appearing clutter echoes may be reduced by simple structure-detecting algorithms. In this way all structures with e.g. <8 coherent pixels can be erased, since rain structures generally consist of a larger number of coherent pixels.

Attenuation

Attenuation leads to under-estimation of precipitation and this increases with frequency. Standard attenuation correction procedures are not always applied. The German Weather Service (DWD) does not include any attenuation correction in its data processing. For the UK network the correction is done by a cumulative gate-by-gate algorithm equivalent to attenuation $A = 0.0044\,R^{1.17}$ (with $Z = 200\,R^{1.6}$ used to convert between reflectivity Z and rain-rate R). This is capped at a maximum of a factor of two increase in rain-rate to avoid instability (Harrison *et al.*, 2009).

The generally applied cumulative gate-by-gate algorithm has the disadvantage that small errors may lead to instabilities by producing either implausibly high corrections or too low ones. Investigations to improve the attenuation correction have come up with a backward correction algorithm where the total attenuation is derived from radial microwave link data to find the optimal parameters for the actual situation. The transfer to the radar field together with capping implausibly high reflectivities showed a significant increase in compliance with the reference data from the ground (Krämer & Verworn, 2009).

All attenuation correction algorithms suffer from the fact that the total attenuation along the radar beam is not known. Additional information to be used in attenuation correction may be derived from disdrometers and especially from radial microwave links to obtain path-integrated attenuation and rainfall rates. The advantages of the integration of microwave link and disdrometer data have been demonstrated by Krämer & Verworn (2008), although only one link was installed. Polarimetric radars will definitely reduce the uncertainties concerning attenuation but will still need some references from other sources.

Reflectivity to rainfall conversion

In most radar networks the final conversion from reflectivities to radar intensities is done by the Marshall-Palmer relation of $Z = 200R^{1.6}$. Though it is well known that the Z-R relation varies significantly in its dependence on the drop spectrum, in most cases there is no information available on which way the relation would change. In some cases a parallel displacement of the relation which is linear in a logarithmic grid is applied. This displacement, however, may lead to implausible results when extrapolation to the whole range of Z values is necessary. Utilizing measurements from a disdrometer, the effects of the derived drop-size distributions on the parameters of the Z-R power-law model were investigated by Chapon *et al.* (2008). It was shown that the linear regression techniques for the parameters are significantly better than event-fitted Z-R relationships due to intra-event changes in the drop-size distributions.

An alternative approach has been developed making use of disdrometer measurements of the drop spectrum from which R and Z values are derived every minute (Krämer & Verworn, 2008). The mean of the last 10–30 minutes of R-Z combinations is used as the third point, in addition to fixed upper and lower points, to calculate the parameters of the R-Z relation which is restricted to the area of potential R-Z combinations (Fig. 1). This procedure is currently being tested in comparison to a fixed Z-R relationship.

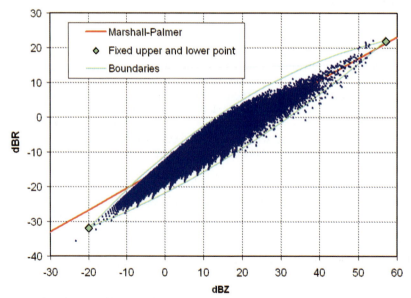

Fig. 1 dBZ-dBR combinations from 10 months of minutely disdrometer drop spectra (blue), Marshall-Palmer relation (red), fixed upper and lower points and the boundaries for valid variations of the quadratic Z-R relation.

General adjustment

The final stage in processing radar data to derive quantitative reliable rainfall fields often is a general bias adjustment either for the whole radar area or differently for parts of it. The main reference source for this still is the information from ground gauges, though more advanced methods are increasingly being developed and applied. Microwave links provide path-integrated rainfall data which may be combined with gauge data (Cummings *et al.*, 2009). The deployment of existing telecommunication networks for deriving attenuation and rainfall data may in future yield additional information about rain-rates (Goldshtein *et al.*, 2009).

QUANTITATIVE PRECIPITATION FORECAST (QPF)

According to Golding (2000), all short-term forecasting up to 6 h ahead may be called nowcasting. Generally, nowcasting is based on extrapolation techniques of the dynamics of the rainfall field in a short time before the nowcasting is made. Simple procedures to detect pattern and displacement of rain cells and their extrapolation over the nowcasting horizon generally do not perform well in convective situations, since the emergence, growth and decay of convective cells are generally not accounted for. The simple techniques have been improved by scale separation techniques using low pass filters, multifractal methods and wavelet analysis in addition to the correlation-based algorithms (Seed *et al.*, 2003). For small urban drainage systems a nowcasting horizon of up to 2 hours is often sufficient. Consequently, the focus so far has been mainly on tracking procedures. Various investigations (Verworn & Krämer, 2008; Achleitner *et al.*, 2009; Fabry & Seed, 2010) have shown that nowcasts within acceptable uncertainty margins are possible for horizons of up to 30 minutes for convective situations and up to 90 minutes for stratiform rainfall. In larger urban areas, however, the need is for nowcasts beyond that horizon, for which the ever improving performance of NWPs is a prospect, especially when combined with the tracking procedures. An example of this approach is STEPS (Short-Term Ensemble Prediction System) which merges precipitation forecasts from a nowcasting system with downscaled deterministic NWP forecasts (Bowler *et al.*, 2006). The results of an investigation of STEPS performance in comparison with a simple tracking model (HyRaTrac) are presented in this symposium (Schellart *et al.*, 2011).

OUTLOOK

Performance assessment with event-based or continuous long-term simulation requires long time series of recorded rainfall which up to now have been available only from ground measurements, and in most cases even from only one point resulting in uniform rainfall distribution even for large areas. The availability of radar data time series of now nearly 20 years duration opens up new perspectives for a more detailed assessment, as was shown by Krämer *et al.* (2005). The data further in the past, however, are bound to be of lower quality than the newer ones, and a general and thorough quality control and processing of these data is necessary if they are to be used.

In spite of all efforts to increase the reliability of rainfall data, it is a fact that error sources remain and will not be completely overcome, mainly due to the fact that with radar the rainfall is not directly measured, but estimated from other measurable data. Nevertheless, radar rainfall for the past, present and nowcasting provides additional knowledge for which the uncertainty margins have to be known and incorporated in any derived result or decision.

The problem that it is not the rainfall measurement alone, but the whole rainfall–runoff-forecast-decision chain that incorporates uncertainties, is increasingly addressed and made aware of and accepted by the end-user (Collier, 2009; Rossa *et al.*, 2011).

REFERENCES

Achleitner, S., Fach, S., Einfalt, T. & Rauch, W. (2009) Nowcasting of rainfall and of combined sewer flow in urban drainage systems. *Water Sci. & Technology* 59(6), 1145–1151.

Berne, A., Delrieu, G., Creutin, J.-D. & Obled, C. (2004) Temporal and spatial resolution of rainfall measurements required for urban hydrology. *J. Hydrol.* 299(3–4), 166–179.

Bouilloud, L., Delrieu, G., Boudevillain, B. & Kirstetter, P-E. (2010) Radar rainfall estimation in the context of post-event analysis of flash flood events. *J. Hydrol.* 394, 17–27.

Bowler, N. E. H., Pierce, C.E. & Seed, A (2006) STEPS: A probabilistic precipitation forecasting scheme which merges an extrapolation nowcast with downscaled NWP. *Quart. J. Roy. Met. Soc.* 132, 2127–2155.

Bringi, V. N, Rico-Ramirez, M. A. & Thurai, M. (2011) Rainfall estimation with an operational polarimetric C-band radar in the UK: Comparison with a gage network and error analysis. *J. Hydromet.* ISSN: 1525-755X, DOI: 10.1175/JHM-D-10-05013 (in press).

Chapon, B., Delrieu, G., Gosset, M. & Boudevillain, B. (2008) Variability of rain drop size distribution and its effect on the Z-R relationship: A case study for intense Mediterranean rainfall. *Atmos. Res.* 87(1), 52–65.

Ciach, G. J. & Krajewski, W. F. (1999) On the estimation of radar rainfall error variance, *Adv. Water Resour.* 22, 585–595.

Collier, C. G. (2009) On the propagation of uncertainty in weather radar estimates of rainfall through hydrological models. *Met. Appl.* 16(1), 35–40.

Cummings, R. J, Upton, G. J. G., Holt, A. & Kitchen. M. (2009) Using microwave links to adjust the radar rainfall field. *Adv. Water Resour.* 32(7), 1003–1010.

Fabry, F. & Seed, A. W. (2010) Quantifying and predicting the accuracy of radar-based quantitative precipitation forecasts. In: *Proc. 6th European Conf. on Radar in Meteorology and Hydrology: Adv. in Radar Technology* (Sibiu, Romania).

Fuchs, L. & Beeneken, T. (2005) Development and implementation of a real-time control strategy for the sewer system of the city of Vienna. *Water Sci. & Technol.* 52(5), 187–194.

Golding, B. W. (2000) Quantitative precipitation forecasting in the UK. *J. Hydrol.* 239, 286–305.

Goldshtein, O., Messer, H. & Zinevich, A. (2009) Rain rate estimation using measurements from commercial telecommunication links. *IEEE Trans. Signal Processing* 57(4), 1616–1625.

Harrison, D. L., Scovell, R. W. & Kitchen, M. (2009) High-resolution precipitation estimates for hydrological uses. Proc. Inst. of Civil Engineers, *Water Management* 162(2), 125–135.

Henja, A. & Michelson, D. (1999) Improved polar to Cartesian radar data transformation. In: *Proc. 29th American Meteorol. Soc. Conf. on Radar Meteorology*, Montreal, 252–255.

Kitchen, M. & Blackall, R. M. (1992) Representativeness errors in comparisons between radar and gauge measurements of rainfall. *J. Hydrol.* 134, 13–33.

Kleidorfer, M., Meyer, S. & Rauch, W. (2006) Aspects of calibration of hydrological models for the estimation of CSO performance. In: Ertl, T.; Pressl, A.; Kretschmer, F.; Haberl, R. In: *Proc. 2nd Int. IWA Conf. on Sewer Operation and Maintenance SOM 06*. London: IWA Publishing, ISBN 978-3-900962-67-8, S. 353–360.

Krämer, S., Verworn, H.-R. & Ziegler, J. (2005) Radar rainfall time series for the performance assessment of sewer systems. In: *Proc. 10th Int. Conf. on Urban Drainage* (21–26 August 2005. Copenhagen, Denmark).

Krämer, S., Fuchs, L. & Verworn, H.-R. (2007) Aspects of radar rainfall forecasts and their effectiveness for real time control - the example of the sewer system of the city of Vienna. *Water Practice and Technology* 2(2), doi10.2166/wpt.2007.042.

Krämer, S. & Verworn, H.-R. (2008a) Integration of radar microwave link and disdrometer data for improved quantitative rainfall estimation – part I: Attenuation analysis and correction methodology. In: *Int. Symp. on Weather Radar and Hydrology* (Grenoble, France, 10–12 March 2008). Paper P1-027.

Krämer, S. & Verworn, H.-R. (2008b) Integration of radar, microwave link and disdrometer data for improved quantitative rainfall estimation – Part II: advanced attenuation correction and R-Z-procedures. In: *Int. Symp. on Weather Radar and Hydrology* (Grenoble France, 10–12 March 2008). Paper P1-028.

Krämer, S. & Verworn, H.-R. (2009) Improved radar data processing algorithms for quantitative rainfall estimation in real time. *Water Sci. & Technol.* 60(1), 175–184

Maeda, M., Mizushima, H. & Ito, K. (2005) Development of the real-time control (RTC) system for Tokyo sewage system. *Water Sci. & Technol.* 51(2), 213–220.

Met Office (2009) Fact Sheet 15, Weather Radar. http://www.metoffice.gov.uk/media/pdf/j/h/Fact_sheet_No._15.pdf.

Pierce, C. E., Seed, A., Ebert, E. E., Fox, N., Sleigh, M., Collier, C. G., Donaldson, N., Wilson, J., Roberts, R. & Mueller, C. (2004) The Nowcasting of precipitation during Sydney 2000: an appraisal of the QPF algorithms. *Weather & Forecasting* 19(1), 7–21.

Pleau, M., Colas, H., Lavallèe, P., Pelletier, G. & Bonin, R. (2004) Global optimal real-time control of the Quebec urban drainage system. *Environmental Modelling & Software* 20(4), 401–413.

Rossa, A., Liechti, K., Zappa, M., Bruen, M., Germann, U., Haase, G. G., Keil, Ch. & Krahe, P. (2011) The COST 731 Action: A review on uncertainty propagation in advanced hydro-meteorological forecast systems. *Atmos. Res.* doi:10.1016/j.atmosres.2010.11.016.

Schellart, A., Ligouri, S., Krämer, S., Saul, A. & Rico-Ramirez, M. (2011) Analysis of different quantitative precipitation forecast methods for run off and flow prediction in a small urban area. In: *Proc. Int. Symp. on Weather Radar and Hydrology* (April 2011, Exeter, UK).

Schroeder, K. & Pawlowsky-Reusing, E. (2005) Current state and development of the real-time control of the Berlin sewage system. *Water Sci. & Technol.* 52(12), 181–187.

Tabaray, P. (2007) The new French operational radar rainfall product. Part I: Methodology. *Weather & Forecasting* 22(3), 393–408.

Verworn, H. R. & Krämer, S. (2008) Radar based nowcasting of rainfall events- analysis and assessment of a one year continuum. In: *Flood Risk Management: Research and Practice* (ed. by P. Samuels). CRC Press, Leiden, The Netherlands.

Villarini, G., Smith, J. A., Baeck, M. L., Sturdevant-Rees, P. & Krajewski, W. F. (2010) Radar analyses of extreme rainfall and flooding in urban drainage basins. *J. Hydrol.* 381, 266–286.

Weather Radar and Hydrology
(Proceedings of a symposium held in Exeter, UK, April 2011) (IAHS Publ. 351, 2012).

601

What is a proper resolution of weather radar precipitation estimates for urban drainage modelling?

JESPER E. NIELSEN, MICHAEL R. RASMUSSEN & SØREN THORNDAHL

Aalborg University, Department of Civil Engineering, Sohngaardsholmsvej 57, DK-9000 Aalborg, Denmark
jen@civil.aau.dk

Abstract The resolution of distributed rainfall input for drainage models is the topic of this paper. The study is based on data from high-resolution X-band weather radar used together with an urban drainage model of a medium-sized Danish village. The flow, total runoff volume and CSO volume are evaluated. The results show that the model to some extent is dependent on the rainfall input resolution and recommendations for the resolution are given. However, none of the investigated resolutions can be characterized as "unusable".

Key words weather radar measurements; urban drainage modelling; MOUSE; radar data resolution: LAWR

INTRODUCTION

Urban drainage modelling is widely used to analyse and evaluate the rainfall response of urban sewer systems in order to determine if the system in question is well functioning and meets the require-ments set by the authorities. Failure of the system is typically defined as limits for the frequency and time periods, e.g. flooding, surcharge, or combined sewer overflow to receiving waters.

When failure occurs in an urban drainage system, it is in most cases due to heavy rainfall which makes the precipitation input one of the most important boundary conditions for the model, if the failure situation of the system is investigated. Traditionally, the precipitation input is derived from raingauge measurements located either within the catchment or the closest available raingauge in the nearby area. Using raingauge measurements as the input for the drainage model has limitations, as the raingauge only covers an area of a few hundred square centimetres and it is therefore not able to describe the spatial variation and distribution of the precipitation. This means that unless a high density network of raingauges is installed within the catchment (which is rarely the case), it is assumed that the raingauge is representing the rainfall for the whole catchment without any spatial description of the rainfall event.

The importance of using distributed rainfall as input for urban drainage models instead of raingauge point measurements has been investigated by Pedersen *et al.* (2005), who found that the spatial variability of the precipitation could be so extensive, that it affected the model result significantly.

Unlike the raingauge, weather radars have the benefits of describing the spatial variability of the precipitation and cover large areas, which in principle make the weather radar measurements ideal as precipitation inputs for urban drainage models. Even though the use of weather radars in urban drainage is still fairly limited, and mainly performed in relation to research, it has become more and more common to use weather radars for supplementing raingauges in urban drainage. The radars are included in almost all aspects of the field, e.g. in system design, operation, real-time surveillance and real-time control of the sewer system, e.g. Einfalt *et al.* (2004), Pedersen *et al.* (2006) and Rasmussen *et al.* (2010).

Different types of weather radars are in operation today, ranging from massive long range S- and C-band radars to relatively low cost X-band radars. The radars are operated with different configura-tions and thereby different properties of the radar measurement regarding both range and resolution in time and space. Denmark is covered by five meteorological C-band weather radars with resolutions of 5–10 min and 500–2000 m. Furthermore, seven cities have installed their own high resolution X-band radars, i.e. Local Area Weather Radars (LAWR) with resolutions of 1–5 min and 100–500 m.

When the radar estimates are used as input for urban drainage models, one can question the requirements for the radar data. In principle data from all kinds of radars can be used as input for the drainage model as long as the radar covers the catchment of interest, but how dependent is the model result of the radars' capability of describing the variability of the precipitation? These are the topics of this paper, and they are investigated by using high-resolution X-band radar (LAWR) together with a drainage model simulating an urban sewer system for a medium-size Danish village.

The catchment and urban drainage model

Frejlev is a medium-size village in northern Denmark. The village is located on a hillside facing north with approx. 2000 inhabitants and a total catchment area of 87 ha. The drainage system is partly separated and partly combined with a combined sewer overflow (CSO) to the receiving stream, Hasseris å. The catchment and drainage system are illustrated in Fig. 1.

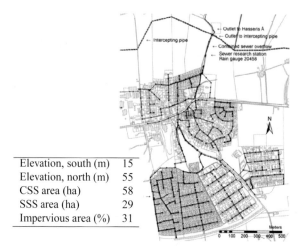

Elevation, south (m)	15
Elevation, north (m)	55
CSS area (ha)	58
SSS area (ha)	29
Impervious area (%)	31

Fig. 1 The catchment and drainage system of Frejlev. The hatched areas mark the separate catchments. (Thorndahl *et al.*, 2008).

Table 1 Hardware specifications and data products for the radar (Jensen, 2004).

LAWR X-band (Furono1525)		Data products
Frequency	9.41 Ghz	500 × 500 m
Wavelength	3.2 cm	Maximum range: 60 km
Emission power	25 kW	Quantitative range: 15–20 km
Temporal resolution	5 min	300 × 300 m
Angular resolution	0.95°Azimuth	Maximum range: 15 km
Vertical resolution	±10°	Quantitative range: 15 km
Data resolution	255 classes	100 × 100 m
Rotation	24rpm	Maximum range: 15 km
Scanning elevation	0°	Quantitative range: 15 km

An existing MOUSE model of Frejlev drainage system (Thorndahl *et al.*, 2008) is used to simulate the response of the radar estimated rainfall. In order to use the radar measurement as the rainfall boundary, minor adjustments have been made. Each of the 560 manholes is addressed to their corresponding radar pixel, and a rainfall boundary time series is created and connected to the model, from each of the radar pixels that cover the catchment. The MOUSE model is configured with default settings and parameters for both the surface runoff and the pipe flow computation.

The radar data

The radar used for the experiments is a high-resolution LAWR weather radar (Table 1). This type of radar is based on a marine X-band radar and is adapted for the weather radar application by DHI (Jensen & Overgaard, 2002).

The radar data used for the experiments is from a three month period starting 1 June and ending 1 September 2009. The calibration method is based on Thorndahl & Rasmussen (2012) and performed as a static calibration by radar-raingauge comparison of accumulated rainfall in seven raingauges within the range of the radar, yielding the following values for the calibration parameters: $C_1 = 2.65\text{e-}4$ mm/DRO and $C_2 = 0.13$ km^{-1}.

The radar is located close to Frejlev at a distance of approximately 2.5 km. Unfortunately, this short distance is not optimal due to the saturation of the radar signal which can occur even at low rain intensities. It has therefore been necessary to relocate the catchment fictitiously, so that the precipitation input to the drainage model is conducted further away from the radar.

Investigated resolutions and events

Even though the data products from the radar are Cartesian and come in three different resolutions, as mentioned in Table 1, the products originate from the same polar data. The polar radar data are transformed individually into each Cartesian resolution to make the radar data more convenient for the applications. However, as a result the data products are not fully consistent, meaning that a 500 m pixel is not necessarily the mean of the corresponding 5 × 5 100 m pixels.

For this reason, the investigated resolutions are conducted from the 100 m product by spatial averaging into the coarser resolution of interest. In this way, differences in the drainage model results can be directly addressed to the change of resolution. A side benefit of this approach is that more resolutions can be analysed than preset by the radar software. Six different resolutions ranging from 100 × 100 m to 2 × 2 km are examined (Fig. 2), matching the finest spatial resolution of the LAWR and the spatial resolution of conventional meteorological C-band radar (Gill *et al.*, 2006).

In order to evaluate the model results, two points in the drainage system are selected, i.e. the total discharge from the whole catchment and CSO to the receiving Hasseris Å. Therefore, only events that cause CSO are used. Table 2 shows the time and dates for the seven examined events.

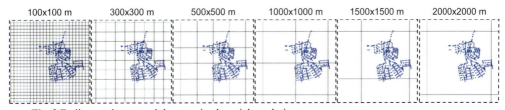

Fig. 2 Frejlev catchment and the examined spatial resolutions.

Table 2 Simulation time period for the seven events.

	Event 1	Event 2	Event 3	Event 4	Event 5	Event 6	Event 7
Start time of simulation	09.06.09 15:00	10.06.09 03:00	11.06.09 15:00	18.07.09 05:00	24.07.09 14:00	27.07.09 18:00	20.08.09 20:00
End time of simulation	10.06.09 02:00	10.06.09 14:00	12.06.09 22:00	18.07.09 22:00	24.07.09 20:00	28.07.09 04:00	21.08.09 06:00
Rain depth	9.0 mm	10.2 mm	31.5 mm	13.1 mm	4.0 mm	6.8 mm	7.2 mm
Type of event	Widespread rain	Widespread clusters of rain	Widespread	Frontal passage with clusters	Clustered rain	Clustered rain	Clustered rain

RESULTS

As a representative example of the model results, the runoff hydrographs for the largest event (Event 3) are plotted in Fig. 3. The runoff hydrographs (Fig. 3) illustrate that the flow progression through the event is so similar that it is difficult to distinguish between the model results. Only around some of the peaks are the differences visual. If the total runoff volume is considered, as presented in Table 3,

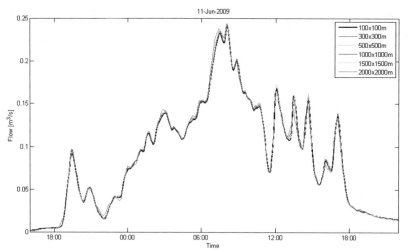

Fig. 3 Runoff hydrograph for the whole Frejlev catchment from Event 3.

Table 3 Total runoff from the Fejlev catchment.

Resolution (m)	Event 1 (m³)	Event 2 (m³)	Event 3 (m³)	Event 4 (m³)	Event 5 (m³)	Event 6 (m³)	Event 7 (m³)
100 × 100	2664	2984	9508	3925	1141	1986	2060
300 × 300	2665	2992	9517	3931	1142	1992	2061
500 × 500	2694	3043	9631	3991	1159	2019	2088
1000 × 1000	2707	3062	9667	4034	1177	2034	2104
1500 × 1500	2698	3119	9643	3993	1145	2031	2100
2000 × 2000	2651	3064	9578	3964	1143	1940	2084
Summary, resolution with maximum deviation from the 100 × 100 m resolution							
Resolution	1000 × 1000	1500 × 1500	1000 × 1000	1000 × 1000	1000 × 1000	1000 × 1000	1000 × 1000
Deviation (m³)	43	135	159	109	36	48	44
Deviation (%)	1.6	4.5	1.7	2.8	3.2	2.4	2.1

Table 4 CSO volume to the recipient.

Resolution (m)	Event 1 (m³)	Event 2 (m³)	Event 3 (m³)	Event 4 (m³)	Event 5 (m³)	Event 6 (m³)	Event 7 (m³)
100 × 100	369	99	1061	681	71	57	226
300 × 300	368	99	1066	677	72	58	226
500 × 500	385	114	1136	713	82	69	244
1000 × 1000	394	124	1149	741	90	76	254
1500 × 1500	393	120	1148	702	67	71	257
2000 × 2000	381	153	1133	716	55	57	249
Summary, resolution with maximum deviation from the 100 × 100 m resolution							
Resolution	1000 × 1000	2000 × 2000	1000 × 1000	1000 × 1000	1000 × 1000	1000 × 1000	1500 × 1500
Deviation (m³)	25	54	88	60	18	19	31
Deviation (%)	6.9	54.7	8.3	8.9	25.9	33.7	13.7

all the differences are accumulated through the event and the results show variation dependent on the input resolution of the radar data.

The same impression is given if the CSO volume is considered, as presented in Table 4.

For the majority of the events, the models yield fairly similar results in both CSO and total runoff volume for the input resolutions of 100 m, 300 m and 500 m. For the coarser resolutions, the results deviate mostly with larger volumes compared to the 100 m resolution, and the 1000 m resolution is in most events causing the largest volumes and deviation.

DISCUSSION AND CONCLUSION

As the results show, the urban drainage model computation is somewhat dependent on the spatial resolution of the radar precipitation input, although the coarser resolutions are generated from the same high-resolution radar data.

If the coarser resolutions (1000–2000 m) are related to the spatial extent of the catchment area as illustrated in Fig. 2, it makes good sense that the model result is deviating, because precipitation outside the catchment area is averaged over the catchment and thereby introduced into the model. This could consequently lead to the similar conclusion that the small deviation for the finer resolutions (100–500 m) is due to the lack of this effect because the finer grid fits the edges of catchment sufficiently. However, this is only correct, if temporal resolutions of the data were high enough.

The temporal resolution of the radar data is a 5 minute average. In general, it is beneficial for the urban drainage model approach that the measurement is conducted as an average over time instead of a snapshot image of the precipitation, as most conventional radars do. Although, the average measurement reduces the peak rain intensities, as illustrated in Fig. 4, the average measurement also ensures that the mass balance of the precipitation is correct and thereby, the total mass balance of the urban drainage model, as shown in Table 5.

A side effect of the temporal average is that temporal and spatial resolution becomes connected, because the precipitation can be detected several times at different locations within the temporal average. If a fixed pattern precipitation field is considered, as illustrated in Fig. 5, it is clear that the effective spatial scale of the measurement is dependent on the velocity of the precipitation in the principle relationship: $\Delta s = v\ \Delta t$, where v is the velocity of the precipitation, Δt is time period for the average and Δs is "effective spatial resolution".

The movement of the precipitation is determined by the wind: thus the "effective spatial resolution" will vary from event to event. At high wind speeds e.g. 20 m/s, the "principle effective spatial scale" for a five minute average is 6.000 m, while at low wind speeds, e.g. 1 m/s, it is 300 m. This "effective scale" cannot be directly translated into the pixel size of the radar measurement and the effect will be different perpendicular to the movement, as illustrated in Fig. 5.

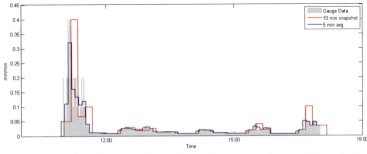

Fig. 4 Resampling of a one minute time series from a raingauge into a 5 min average and a 10 min snapshot time series.

Table 5 Total rain depth of the gauge measurement, five and ten minute average resample, and five and 10 minute snapshot resample.

	Gauge	5 min avg.	10 min avg.	5 min snap.	10 min snap.
Rain depth (mm)	10.4	10.4	10.4	11.5	12

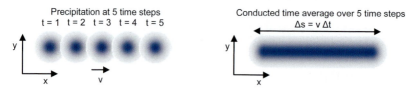

Fig. 5 Sketch showing the principle relationship between temporal and spatial resolution.

The actual movement of the precipitation for the seven events is not investigated further within this context. However, the runoff results indicate that the "effective" spatial resolution could be lower than specified, due to the lack of deviation between the finer resolutions. The spatial information, which are lost by the averaging are limited for all the events, or at least so small that they vanish in the integration by the surface-model and the pipe-flow computation.

A short and simple answer to the title of this paper is difficult to state. But when weather radar data are applied as input for urban drainage modelling, it is shown that it is important to keep the origin of the radar data in mind. In general, it is recommended that measurements averaged in time are used for this type of application, because this ensures mass-balance of the input, unless discrete data are available with a high temporal resolution (>1 min). Consequently, time-averaged measurements also result in a relation between the temporal and spatial resolution, which can cause coarser spatial properties of the measurement than the specified data product.

The model itself also converges the model output, because of the transformation by the surface runoff and pipe flow. Peaks in the input boundary are of course detectable in the model output, but the averaging nature of the system reduces the response. It is therefore recommended that the spatial resolution used represents the catchment of interest and fits the edges of the catchment area sufficiently. In this particular case, the results show that a threshold for the investigated catchment is in the region of the 500 m resolution, if the CSO volumes are considered.

It can be suggested, that the deviation would be more significant for catchments larger than Frejlev, but if the catchment is considered as a sub-catchment for a larger urban catchment, the results will still be relevant. Other result parameters, e.g. the water depth in manholes within the catchment, could also (most properly) be more sensitive to spatial description of the precipitation input.

That being said, it is important to emphasise that the similarities in runoff hydrographs and the total runoff volume from the whole catchment suggest that if the point of interest is far downstream in the system, the deviations are limited. All the investigated resolutions give reasonable results with similar hydrographs and <5% deviation in the total runoff volume compared with 100 m resolution. Therefore, none of the investigated resolutions can be characterized as "completely unusable" based on this study.

REFERENCES

Einfalt, T., Arnbjerg-Nielsen, K., Golz, C., Jensen, N. E., Quirmbach, M., Vaes, G. & Vieux, B. (2004) Towards a roadmap for use of radar rainfall data in urban drainage. *J. Hydrol.* 299, 186–202.

Gill, R. S., Overgaard, S. & Bøvith, T. (2006) The Danish weather radar network. In: *4th European Conference on Radar in Meteorology and Hydrology* (Barcelona 18–22 September).

Jensen, N. E. (2004) Local Area Weather Radar Documentation v. 3.0 (2004), DHI, Institute for the Water Environment.

Pedersen, L, Jensen N. E., Rasmussen, M. R. & Nicolajsen, M. G. (2005) Urban run-off volumes dependency on rainfall measurement method – Scaling properties of precipitation within a 2 × 2 km radar pixel. In: *Proc. 10th International Conference on Urban Drainage* (Copenhagen, Denmark).

Pedersen, L., Jensen, N. E. & Madsen, H. (2006) Application of local area weather radar (LAWR) in relation to hydrological modelling–identification of the pitfalls of using high resolution radar rainfall data. In: *Proc. 4th European Conference on Radar in Meteorology and Hydrology* (Barcelona), 414–417.

Rasmussen, M. R., Thorndahl, S., Grum, M., Neve, S. & Poulsen, T. S. (2010) Vejrradarbaseret styring af spildevandsanlæg, *By- og landskabsstyrelsen* (in Danish).

Thorndahl, S, Beven, K. J., Jensen, J. B. & Schaarup-Jensen, K. (2008) Event based uncertainty assessment in urban drainage modelling, applying the GLUE methodology. *J. Hydrol.* 357, 421–437.

Thorndahl, S. & Rasmussen, M. R. (2012) Marine X-band weather radar data calibration. *Atmospheric Res.* 103, 33–44.

Weather Radar and Hydrology
(Proceedings of a symposium held in Exeter, UK, April 2011) (IAHS Publ. 351, 2012).

607

The flooding potential of convective rain cells

EFRAT MORIN & HAGIT YAKIR

Geography Department, Hebrew University of Jerusalem, Jerusalem 91905, Israel
msmorin@mscc.huji.ac.il

Abstract Flash floods caused by convective rain storms are highly sensitive to the space–time characteristics of rain cells. In this study we exploit the high space–time resolution of the radar data to study the characteristics of the rain cells and their impact on flash flood magnitudes. A rain cell model is applied to the radar data of an actual storm and the rain fields represented by the model further serve as input into a hydrological model. Global sensitivity analysis is applied to identify the most important factors affecting the flash flood peak discharge. As a case study we tested an extreme storm event over a semi-arid catchment in southern Israel. The rain cell model was found to simulate the rain storm adequately. We found that relatively small changes in the rain cell's location, speed and direction could cause a three-fold increase in flash flood peak discharge at the catchment outlet.

Key words convective rain cells; flash floods; weather radar

INTRODUCTION

Extreme flash-floods occur under rare hydrometeorological conditions in which intense rainstorms interact with catchments such that the generated flow magnitude is maximized. The presented study focuses on the "flooding potential" of convective rainstorms. Flooding potential is assessed utilizing a unique approach where the storm convective rain cells are explicitly represented using a rain cell model and they are then used as an input into a hydrological model with a range of configurations (Morin *et al.*, 2006; Yakir & Morin, 2011).

Rain cell models are designed to represent the basic elements of the convective rain storm, the rain cells, and describe their spatial and temporal evolution. These models are commonly applied to rainfall data from meteorological radar systems, which provide detailed space–time rain rate information. Several studies focused on rain cell modelling describing circular or elliptical cell shapes (Cox & Isham, 1988; Northrop, 1998; Willems, 2001; Feral *et al.*, 2003; von Hardenberg *et al.*, 2003; Karklinsky & Morin, 2006), with the rain rate spatial distribution within the cell represented by a uniform function or by a Gaussian or exponential decay of rain from the cell centre. The Hycell model (Feral *et al.*, 2003) combines exponential and Gaussian equations to describe both the high rain rates with the fast decay at the cell core and the lower rain rates and gradients at the margins. The Hycell rain cell model is used in the current study. Direction and velocity of rain cells can be derived by applying tracking algorithms to observed rain cells (e.g. Dixon & Wiener, 1993; Johnson *et al.*, 1998).

The main objective here is to study the hydrologic response of a semi-arid catchment to convective rain cell characteristics. This is achieved using the following three stages:

(a) Applying a model to describe the rain cells and their characteristics in time and space;
(b) Using the model generated rain cells as an input to a calibrated hydrological model;
(c) Inspecting the relations between the rain cell characteristics and the catchment hydrological response, and testing the catchment sensitivity to changes in the rain cells characteristics.

CASE STUDY DESCRIPTION

The 94 km^2 semi-arid Beqa catchment is located in southwest Israel (Fig. 1). The catchment height spans from 260 to 460 m a.s.l. and most of the area is rural, partly covered by cultivated fields and sparsely inhabited territories. The mean annual rainfall in the region is 196 mm with a standard deviation of 83 mm. On average there are 40 rainy days in a rainy season that typically spans between November and March, while summer is hot and dry. Between 1947 and 2006, 290 flows

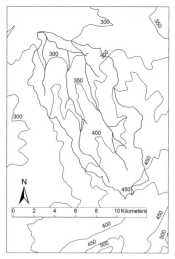

Fig. 1 The Beqa catchment, the channel network and elevation contours in m.

were measured in the Beqa catchment with a mean annual runoff of 0.32×10^6 m³. The maximal observed peak discharge occurred on December 1951 and was 240 m³/s.

The rain storm selected for the study is an extreme storm event that occurred in December, 1993, and which caused major floods in the Negev desert area (Ziv *et al.*, 2005). Over the Beqa catchment the storm occurred on the 22–23 December. Rain was mostly in the form of afternoon convective showers, with rain totals that in some cases exceeded the long-term December averages. This rainstorm resulted in extreme flash floods with return periods of 35–100 years for several catchments. The raingauge near the Beqa catchment measured 24 mm for this storm but the meteorological radar data indicate > 70 mm over some parts of the catchment. The resulting flash-flood had a peak discharge of 81.4 m³/s, which is ordered fifth in the catchment record (1951–2006), and has a return period of about 10 years. The flash-flood event was the largest one with sufficient hydrological and meteorological data and was therefore selected for this study.

Radar data for the storm are obtained from the Shacham meteorological radar system located at Ben-Gurion airport, 90 km from the studied catchment. The radar data resolution is 5 min in time and $1.4° \times 1$ km in space. Radar maps at polar coordinates are transformed into 1×1 km² Cartesian maps of 34×34 km² around the catchment area. Rain rate data were computed from radar reflectivity data using the methodology described in Morin & Gabella (2007).

CONVECTIVE RAIN CELL IDENTIFICATION AND MODELLING

Convective rain cells are identified and modelled from storm radar data applying the following steps: segmentation, spatial model fitting, rain cell tracking, and spatio-temporal model fitting.

The first step of the rain cell modelling process is segmentation. A segment including a rain cell is defined as the area around a local maximum, contoured by a threshold rain rate or a neighbour segment. In order to exclude noise from the analysis, segments that are smaller than 9 km², or have a maximum rain rate of less than 30 mm/h, are removed. Adjacent segments where the difference between the peak and the pixel bordering the segments <25 mm/h are united into a single segment. It should be taken into account that the removal and merging of small and low intensity segments may affect the rain cell distribution described in the next section.

The spatial rain cell model Hycell, proposed first by Feral *et al.* (2003), is fitted to each of the segments. The model is combined from a Gaussian decay function (cell inner part) and an exponential decay function (cell outer part) and it includes several parameters: R_1 (mm/h) is the lower threshold rain rate for the Gaussian function, R_2 is a lower threshold for the exponential function, R_E and R_G (mm/h) are peak rain rate for the exponential and Gaussian functions, respectively, and, a_E, a_G, b_E, b_G (km) represent the decay rate along the major (a) and minor (b) of the exponential (E) and Gaussian (G) radii. These parameters are found by an optimization algorithm applied for each segments. More details can be found in Yakir & Morin (2011).

Rain cells were tracked through observed radar maps in order to examine their dynamic properties, i.e. their starting point location, speed and direction of movement, life span and the change of spatial model parameters with time. In the current study a manual tracking procedure was applied where every cell in each time step was examined and a decision was made whether this cell was new or a cell from the former time step that moved.

The spatio-temporal model assumes that the rain cells' parameters remain constant in time and that the cells move at a constant speed and direction during their life span. The cell spatial parameters were taken as the median value throughout the cell's life span. While the simplified assumptions above allow the present model being parsimonious they imply that changes of rain cell properties during its passage are not accounted for.

The above procedure was applied to the studied storm and 56 rain cells were identified and modelled. Figure 2 presents histograms of the maximal rain rate (R_G), rain cell area, cell life span, and cell speed (for rain cells with life span of 10 minutes or more). The weighted average of the rain cell maximum rain rate (weighted by cell life span) is 78 mm/h, and the weighted average cell area is 100 km^2. Twelve out of the 56 cells are larger than the 94 km^2 catchment size and have the potential to completely cover it. The average life span is 13.6 minutes and the average cell speed for rain cells with life span of 10 minutes or more is 6.7 (m/s). The most intense cell, referred later as the "flooding rain cell" has a maximum rain rate (R_G) of 167 mm/h, area of 283 km^2, duration of 26 minutes, and a speed of 4.2 m/s.

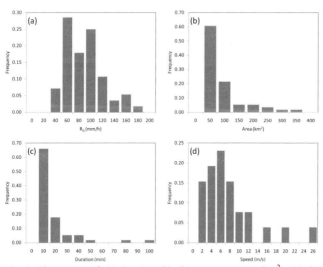

Fig. 2 Histograms of: (a) R_G (mm/h), (b) segment area (km^2), (c) duration (min), and, (d) speed (m/s) for the 56 rain cells derived for the analysed storm.

HYDROLOGICAL MODELLING

The hydrological model used in this research is an event-based distributed hydrological model describing the generation of rainfall excess, routing of surface water over hillslopes and in channels toward the outlet, with infiltration into the channel alluvium. The model was used in previous studies to simulate catchment runoff for arid and semi-arid catchments (Bahat *et al.*, 2009; Morin *et al.*,

2009) and was calibrated for the studied storm event. The catchment was divided into 17 sub-catchments. Rain rate is assumed to be uniform over each sub-catchment and is computed as the spatial average of the rain rate over the sub-catchment. When the accumulated rainfall depth is larger than the initial loss parameter value (20 mm) and the rain rate is higher than the constant infiltration capacity (10 mm/h), rainfall excess is generated as the difference between the rain rate and the infiltration capacity. The Kinematic Wave equation is used to compute water routing over the hillslopes and in the channels (Bahat *et al.*, 2009). Manning parameters was taken as 0.08 for hillslopes and 0.025 for channels, and the constant alluvium infiltration rate was 100 mm/h.

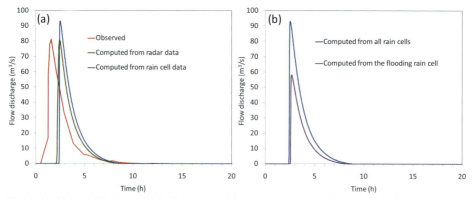

Fig. 3 (a) Observed hydrograph (red), computed hydrograph using radar data (green), and computed hydrograph using rain cell model (blue), (b) Computed hydrographs using all rain cells (blue line) and the computed hydrograph using only the flooding rain cell (purple line).

Two rainfall inputs were fed into the hydrological model: the original radar-based rain rate data and the rain rates as obtained by applying the rain cell model. The computed outlet runoff hydrographs are similar for the two inputs (Fig. 3(a)) suggesting that the rain cell model represents the important elements of the storm. There is, however, a time shift between the observed and the modelled hydrographs, which is suspected to be a result of inaccuracies in the observed flow timing caused by the mechanical recorders of the hydrometric stations.

By running the model with one rain cell at a time it was found that only one major cell, the most intense rain cell, the "flooding rain cell", produced flow at the catchment outlet while the rest of the cells did not generate any outlet flow individually. No outlet flow was generated, even if all cells except to the flooding cell were input to the model. The outlet hydrograph generated from the flooding cell alone, as computed by the model, is presented in Fig. 3(b).

FLOOD SENSITIVITY TO RAIN CELL CHARACTERISTICS

Sensitivity analysis was conducted by varying characteristics of the flooding rain cell and examining the effect on total rain and outlet runoff (peak discharge and runoff volume). The flooding cell starting location was changed over the analysed area, leaving all the other cell parameters unchanged. The total rain over the catchment (mm), peak discharge (m³/s) and total runoff volume (m³) computed for each starting location are shown in Fig. 4. These results show that if the rain cell starting location was about 4 km northwest of the original point, the peak discharge and runoff volume could be doubled.

The effect of cell movement direction on the catchment hydrological response is shown in Fig. 5. Three different starting locations are considered: the original starting point, the catchment outlet and the upstream edge of the catchment. The direction that produced the highest peak discharge was the direction in which the cell spent the majority of its lifespan over the catchment.

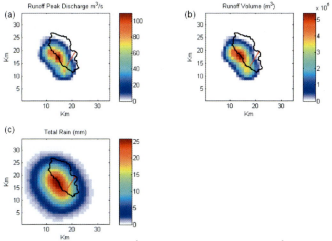

Fig. 4 (a) Peak discharge (m³/s), (b) total runoff volume (m³), and, (c) total rain over the catchment (mm), for each starting point of the flooding cell. The original direction of the flooding cell over the basin (black) is indicated by the red line.

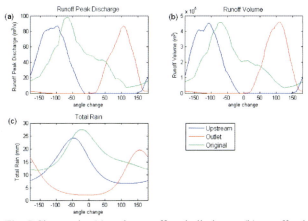

Fig. 5 Changes in: (a) outlet runoff peak discharge, (b) runoff volume, and, (c) total rain as a function of cell direction of movement for three starting locations: original position (green), outlet (red) and upstream (blue). The x axis is the angle difference relative to the cell's original direction.

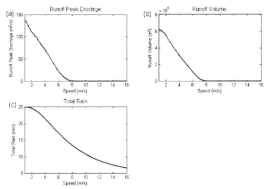

Fig. 6 Changes in: (a) peak discharge, (b) runoff volume, and, (c) total rain as a function of the flooding cell's speed.

For a cell starting at the original starting point, a 65° rotation counter-clockwise (down slope the catchment topography, see Fig. 1) would double the peak discharge. For a cell starting at the catchment outlet a 110° rotation clockwise from the original cell direction would produce the highest peak discharge while movement in the original direction would not produce any runoff. As for a cell starting upstream, a rotation of 95° counter-clockwise from the original direction would produce the highest peak discharge.

The rain cell speed was changed from 0 to 16 m/s while keeping the rest of the characteristics unchanged. A decrease in peak discharge as the speed increases can be seen (Fig. 6) due to the fact that higher velocities cause the cell to pass over the catchment faster and with less rain.

A global sensitivity analysis of runoff peak discharge to rain cell location, direction and speed was conducted and the sensitivity indexes are presented in Table 1. The most influencing factor is the cell location, both as a main effect (caused by this factor only) and total effect (caused by the factor and all its interactions with other factors). Rain cell speed is the second most important of the three factors examined here.

RAIN CELL FLOODING POTENTIAL

What is the largest flash flood the analysed convective cell can generate without changing its magnitude? A range of cell configurations were examined in terms of starting location, direction and speed. The highest peak discharge obtained from the flooding rain cell as a single cell was 175 m³/s. Taking into account the whole storm conditions, all other rain cells (55 cells) were input to the model prior to the convective rain cell to maximize initial soil moisture conditions. The flash flood peak discharge has increased to 260 m³/s. The computed flash flood hydrograph is presented in Fig. 7.

Table 1 Global sensitivity indexes

Factor	Main effect	Total effect
Speed	0.08	0.69
Movement direction	0.01	0.31
Starting point	0.30	0.91

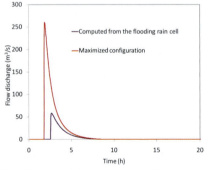

Fig. 7 Computed hydrograph with the original rain cell properties (blue) and computed hydrograph under maximizing conditions

The above results indicate that a potential flash flood with a peak discharge more than three times higher than the observed flash flood could potentially be generated by the analysed storm without changing cell magnitude but by inducing changes related to location, direction and cell speed and changes related to the order of rain cells in the storm. It is interesting to put this enhancement in relation to the envelope curve of the region. The maximum peak discharge for the Negev catchments of the same area as the Beqa catchment is about 700 m^3/s. It can be shown that the obtained peak discharge falls inside the envelope curves for the Negev area (Meirovich *et al.*, 1998) and that the highest flash flood peak measured in the Beqa catchment was 240 m^3/s (December 1951), close to the value obtained in the current analysis, but still lower than the maximal possible.

We can postulate that the maximal flood in the Beqa catchment requires storms with more intense rain cells or larger rain depth to increase initial soil moisture conditions.

SUMMARY

The rainstorm of the 22–23 December 1993 over the Beqa catchment was analysed. Rain maps obtained from radar data were divided into segments and for each segment rain cell model (Hycell) was applied. The rain cells were tracked in time yielding 56 rain cells. The modelled cells were used as an input to the hydrological model and one of the cells was found to be the most significant in generating the flash flood. Sensitivity analysis was conducted based on this major cell and it was found that the catchment runoff was very sensitive to the convective rain cell characteristics. In particular, a small change of cell location, direction and speed, could cause a three-fold higher flood peak discharge. The flooding potential of the analysed storm was estimated here as 260 m^3/s, the maximal observed discharge in the catchment.

Acknowledgements The research project was funded by the Israel Science Foundation (grant no. 880/04). Radar data were provided by E. M. S. Mekorot, raingauge data by the Israel Meteorological Service and runoff flow data by the Israel Hydrological Service.

REFERENCES

Bahat, Y., Grodek, T., Lekach, J. & Morin, E. (2009) Rainfall-runoff modeling in a small hyper-arid catchment. *J. Hydrol.* 373, 204–217.

Cox, D. R. & Isham, V. (1988) A simple spatial-temporal model of rainfall. *Proc. R. Soc. London Ser. A-Math. Phys. Eng. Sci.* 415, 317–328.

Dixon, M. & Wiener, G. (1993) TITAN – Thunderstorm identification, tracking, analysis, and nowcasting – a radar-based methodology. *J. Atmos. Ocean. Tech.* 10, 785–797.

Feral, L., Sauvageot, H., Castanet, L. & Lemorton, J. (2003) HYCELL - A new hybrid model of the rain horizontal distribution for propagation studies: 1. Modeling of the rain cell. *Radio Sci.* 38, doi:10.1029/2002rs002802.

Johnson, J., MacKeen, P., Witt, A., Mitchell, E., Stumpf, G., Eilts, M. & Thomas, K. (1998) The storm cell identification and tracking algorithm: An enhanced WSR-88D algorithm. *Weather Forecast.* 13, 263–276.

Karklinsky, M. & Morin, E. (2006) Spatial characteristics of radar-derived convective rain cells over southern Israel. *Meteorol. Z.* 15, 513-520, doi:10.1127/0941-2948/2006/0153.

Meirovich, L., Ben-Zvi, A., Shentsis, I. & Yanovich, E. (1998) Frequency and magnitude of runoff events in the arid Negev of Israel. *J. Hydrol.* 207, 204–219.

Morin, E., Goodrich, D. C., Maddox, R. A., Gao, X. G., Gupta, H. V. & Sorooshian, S. (2006) Spatial patterns in thunderstorm rainfall events and their coupling with watershed hydrological response. *Adv. Water Resour.* 29, 843–860, doi:10.1016/j.advwatres.2005.07.014.

Morin, E. & Gabella, M. (2007) Radar-based quantitative precipitation estimation over Mediterranean and dry climate regimes. *J. Geophys. Res.* 112, doi:10.1029/2006JD008206.

Morin, E., Jacoby, Y., Navon, S. & Bet-Halachmi, E. (2009) Towards flash-flood prediction in the dry Dead Sea region utilizing radar rainfall information. *Adv. Water Resour.* 32, 1066–1076.

Northrop, P. (1998) A clustered spatial-temporal model of rainfall. *Proc. R. Soc. A-Math. Phys. Eng. Sci.* 454, 1875–1888.

von Hardenberg, J., Ferraris, L. & Provenzale, A. (2003) The shape of convective rain cells. *Geophys. Res. Lett.* 30, 2280, doi:10.1029/2003gl018539.

Willems, P. (2001) A spatial rainfall generator for small spatial scales. *J. Hydrol.* 252, 126–144.

Yakir H. & Morin E. (2011) Hydrologic response of a semi-arid watershed to spatial and temporal characteristics of convective rain cells. *Hydrol. Earth Syst. Sci.* 15, 393–404, doi:10.5194/hess-15-393-2011.

Ziv, B., Dayan, U. & Sharon, D. (2005) A mid-winter, tropical extreme flood-producing storm in southern Israel: Synoptic scale analysis. *Meteorol. Atmos. Phys.* 88, 53–63, doi:10.1007/s00703-003-0054-7.

Analysis of different quantitative precipitation forecast methods for runoff and flow prediction in a small urban area

ALMA SCHELLART[1], SARA LIGUORI[2], STEFAN KRÄMER[3], ADRIAN SAUL[4] & MIGUEL RICO-RAMIREZ[2]

1 *School of Engineering, Design and Technology, University of Bradford, Richmond Road, Bradford BD7 1DP, UK*
  a.schellart@bradford.ac.uk
2 *Department of Civil Engineering, University of Bristol, Bristol BS8 1TR, UK*
3 *Institute for Technical and Scientific Hydrology (itwh) Ltd., Engelbosteler Damm 22, 30167 Hanover, Germany*
4 *Dept. of Civil and Structural Engineering, University of Sheffield, Mappin Street, Sheffield S1 3JD, UK*

Abstract Due to the relatively small spatial scale as well as rapid response of urban drainage systems, the use of quantitative rainfall forecasts for providing quantitative flow forecasts is a challenging task. Due to urban pluvial flooding and receiving water quality concerns it is, however, worthwhile to investigate the potential. In this paper, two radar nowcast models have been compared and used to create quantitative forecasts of sewer flows in the centre of a small town in the north of England.

Key words urban; rainfall–runoff; radar nowcasting; (NWP) flow forecasting; urban drainage

INTRODUCTION

Increasingly strict legislative drivers for water quality and urban flood risk are driving a growing need to predict the performance of urban drainage systems with a lead time of a few hours. Due to the dynamic rainfall–runoff response in urbanized areas the combination of Quantitative Precipitation Forecast (QPF) and flow prediction is a challenging task. A limited amount of research on the field of quantitative rainfall forecasting for the prediction of flows in urban drainage systems has been carried out. The application of radar-based rainfall forecasts to model runoff on a sub-catchment of the urban drainage system of Vienna (Austria) was discussed by Krämer *et al.* (2006) and the implementation of a real-time control system for Vienna was described by Fuchs & Beeneken (2005). Other examples of the use of radar nowcasting techniques for predictive real-time control of urban drainage systems are described in, e.g. Yuan *et al.* (1999) for Bolton, UK and Achleitner *et al.* (2009) for Linz, Austria. As described by Seed *et al.* (2003) the lifetime of a feature in a rainfall field is dependent on the spatial scale of this rainfall feature. This is of particular interest for relatively small urban areas, where small intense rainfall features can have a considerable impact. Specific requirements of temporal and spatial scale of rainfall data and rainfall forecasts for urban areas have been described by, e.g. Schilling (1991) and Berne (2004). The case study described in this paper comprises the sewer system serving the centre of a small town in the north of England. Two radar nowcasting models used for simulating sewer flow forecasts have been compared.

DESCRIPTION OF CASE STUDY AND DATA

This paper is based on a small urban case study area in the UK, with a population just over 13 000 and approximately 60 km of sewers, the majority of which are combined. The catchment benefits from having 1 km^2 resolution and 5-min frequency coverage of composite radar rainfall data from the Met Office network of 15 C-band radars (Harrison *et al.*, 2009). The radar data have been quality-controlled by the Met Office (Harrison *et al.*, 2000) to account for all the errors inherent in radar rainfall measurements (see Rico-Ramirez *et al.*, 2007). Radar data over the catchment are a composite of Hamildon Hill and Ingham radar stations (Met Office, 2010), located 30 km and 90 km from the catchment, respectively. Rainfall events where both radars were operational have been considered for this study, which has focused only on the sewer system of the central area of the town. The flows entering this part of the sewer system are influenced mainly by rainfall recorded in seven 1 × 1 km radar pixels. Raingauge data and sewer flow and depth data have been collected in the case study catchment since 2007, the research described in this paper, however, only focuses on the application of radar rainfall data. Using information from a depth level monitor in the main Combined Sewer Overflow (CSO) chamber, a number of rainfall events that caused a CSO spill event have been selected.

RADAR FORECASTS

Method 1 – Short-Term Ensemble Prediction System (STEPS)

A recent and major advance to combine radar-based forecasts (nowcasts) and Numerical Weather Prediction (NWP) forecasts was developed at the Met Office in collaboration with the Australian Bureau of Meteorology. They developed a new stochastic precipitation forecasting system known as STEPS (Short-Term Ensemble Prediction System) which merges precipitation forecasts from a nowcasting system with high-resolution deterministic NWP forecasts (Bowler *et al.*, 2006). The blending incorporates stochastic components to account for the inherent uncertainties in the forecasts. The nowcasting system employed in the STEPS is based on the spectral decomposition proposed by Seed (2003) with the incorporation of the optical flow equation proposed by Bowler *et al.* (2004). The deterministic nowcasting system from STEPS is used in this study (i.e. there is no blending with NWP forecasts and noise). The deterministic nowcasting system was set up to produce precipitation forecasts with spatial and temporal resolutions of 2 km and 15 min, respectively over the whole UK, and covering an area of 1000×1000 km.

Method 2 – Hydrological Radar Tracking (HyRaTrac)

The HyRaTrac (Hydrological Radar Tracking) model was developed to predict rainfall as a basis for decision support for real-time control of urban drainage systems (Krämer *et al.*, 2007). Since rainfall–runoff processes in urban areas are highly dynamic, a rapid and frequent update of the forecast within <5 min as well as the inter process communication with different hydrodynamic sewer network models is essential (Verworn & Krämer, 2005). To meet these requirements the HyRaTrac model is simply based on localized tracking of individual storm cells by comparing rainfall structures in two subsequent radar images. The workflow is subdivided into three processes:

(1) Definition of individual storm structures using a defined number of contiguous radar pixels exceeding a certain intensity threshold.
(2) Recognition of storm cells and analysis of local displacement vectors using cross-correlation techniques.
(3) Forecast by linear extrapolation of rainfall structures.

Since storm structures are treated as objects, detailed information about the storm evolvement is generated at run time to support decision finding. Depending on the parameters settings for storm cell definition (8 pixels in this study) and the amount of storm cells detected, the run time was from several seconds up to around 2 min.

SEWER MODEL

A hydrodynamic sewer network model built using InfoWorks CS, v 10.0 was applied to model the rainfall–runoff from the urban area, as well as the flow through the sewer network conduits. The hydrodynamic sewer model obtained from the sewer operator had been calibrated following current industry standards, Wastewater Planner User Group WaPUG (2002). The model consists of several hundreds of nodes, conduits and subcatchment areas, of which 48 ha is impermeable, 238 ha is permeable, park and garden areas in the town centre and 695 ha of steep moorland immediately surrounding the town centre. Sewer flow predictions have been generated by importing QPFs from the nowcasting models described above. Model simulations were also created using the observed radar data as input, in order to focus the analysis on the difference in results between observed and forecast radar data and not on the uncertainty involved in the calibration of the sewer network model or differences between radar and raingauge data.

RESULTS AND DISCUSSION

Four rainfall events have been studied, the average cumulative rainfall depth and rainfall intensity is provided in Table 1. Figures 1 and 2 show the cumulative rainfall as recorded by the radar and forecasted by HyRaTrac and STEPS, respectively. Figure 3 shows the Root Mean Square Error (RMSE) between observed radar rainfall and forecasts calculated for the areal average cumulative rainfall over 7 radar pixels, comparing HyRaTrac and STEPS. For both models a maximum forecast horizon (lead-time L_t) of 180 min was set with an update frequency for the STEPS nowcasts of 15 minute intervals and the HyRaTrac nowcasts updated every 5 min. The time resolution of rainfall nowcasts over lead-time was 5 min for both models. Figure 2 therefore shows the lead-times over 15-minute intervals, as cumulative values are provided at 15-min frequency in contrast to Fig. 1 where selected cumulative values are given for 5-min integration times.

Table 1 Event characterization and HyRaTrac forecast results.

Event	Character	Observed intensity (mm/h)	Observed radar rainfall (mm)	Advection (m/s)		Forecast results Correlation r (L_t) ≥ 0.5		
				v_x	v_y	L_t (min)	r (–)	Bias (%)
5/6 July 2007	Conv. with advection	15.9	19.8	11.4	–2.8	30	0.547	–24.2
21 Sep. 2007	Stratiform	9.5	10.7	20.2	–8.3	65	0.579	15.3
21 Jan. 2008	Stratiform	14.5	40.8	21.6	–3.4	40	0.546	2.0
7 July 2008	Local conv.	30.2	11.6	5.8	1.1	10	0.725	–15.0

In general it can be seen that the forecast models do not differ significantly: the RMSE is generally below 0.8 mm for lead times up to 60 min for both models. During the stratiform rainfall events on 21 January and 21 September, a distinct advection was observed (Table 1); the STEPS model performs slightly better than the HyRaTrac model. For these events the speed of the advection in the x-direction as analysed by HyRaTrac (approx. 75 km/h west to east) was such that after a lead time of 100 min the forecast is limited by the size of the Nimrod composite radar grid and after this lead-time the cells are not extrapolated anymore. The event on 7 July contained very small localised high intensity rainfall and is therefore difficult to forecast; the STEPS model especially shows a considerably error at L_t 15 to 30 min. Both models show comparable results for the event on 5–6 July. Note that the event on 21 September 2007 commenced at 03:00 h but the radar data were not available until 07:20 h; note that Figs 1 and 2 show two different methods of dealing with this issue.

The flow forecasts were provided by importing the nowcasts into the InfoWorks CS model of the sewer system. In order to compare both the deterministic STEPS nowcasts and HyRaTrac nowcasts, a new nowcast was imported every 15 min. The flow forecasts were all created offline and the observed radar data were used to provide a preliminary state of the sewer system at the start for each forecast. For example, for a nowcast provided at 13:15 h, an InfoWorks CS model simulation had been created using observed radar data until 13:15 h to provide the preliminary state of the flows and water depths in the sewer system. The simulated flow forecasts are provided at 5-minute frequency. Figure 4 shows the cumulative flow for the simulation made using observed radar rainfall data as well as the simulations made using the nowcasts, at the most downstream conduit of the sewer system. Due to limited space only two events using the deterministic STEPS nowcasts as input, are shown. The flow forecast results for 7 July 2008 (Fig. 4 – right) shows how the forecasts with lead times between 15 and 45 min provide the same total cumulative flow at the end of the event, due to an underestimation of rainfall followed by an over-estimation. It has to be noted

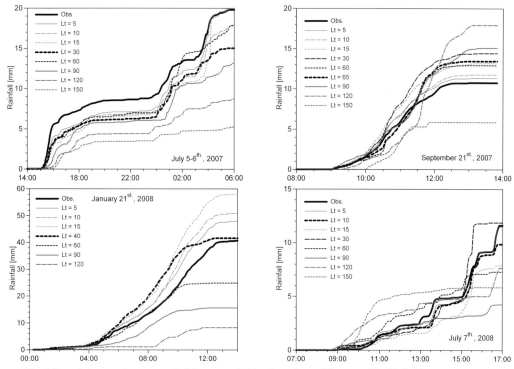

Fig. 1 Cumulative areal average rainfall as recorded by the radar and forecasted by HyRaTrac.

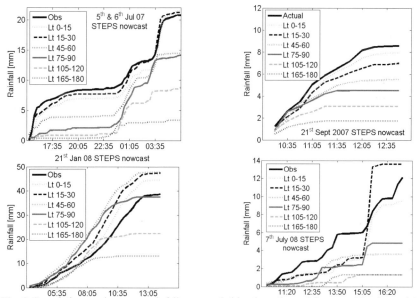

Fig. 2 Cumulative areal average rainfall as recorded by the radar and forecasted by deterministic STEPS nowcast.

that combined sewer overflows occurred during all selected events, and hence differences between forecasted peak flow intensity will not be picked up in these graphs because the "tops" of the peak flow will have been spilled over the CSO weir. Research on the forecasted CSO spill volumes is currently underway, as well as research into the use of ensemble forecasts for this catchment, as described by Liguori *et al.* (2012).

Separate sewer flow simulations (not shown here) indicated a lag-time of approximately 15 min between rainfall peaks and sewer flow peaks. In order to see the effect or errors in rainfall forecasts on sewer flow forecasts, a direct comparison between RMSE values has been made (Fig. 5). Because of the lag time, sewer flow forecasts with lead-time of 30 min have been plotted against rainfall forecast with a lead time of 15 min, etc. Figure 5 shows an almost linear relationship between rainfall and flow forecast errors for the event on 21 January 2008, whereas the results for the 7 July 2008 event are more scattered.

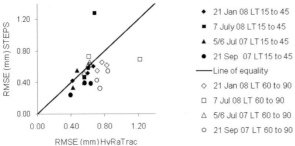

Fig. 3 Intercomparison of HyRaTrac and STEPS nowcasting models. RMSE between observed data and forecasts calculated for the areal average rainfall over 7 1 × 1 km radar pixels covering the town centre.

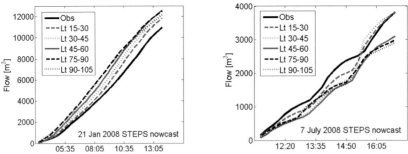

Fig. 4 Examples of cumulative flow at the downstream end of the system, derived using observed radar rainfall data and rainfall nowcasts as input to the hydrodynamic sewer model.

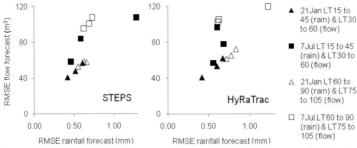

Fig. 5 Intercomparison of HyRaTrac and STEPS nowcasting models when used to forecast sewer flows.

CONCLUSIONS

The paper compared two radar nowcast models and illustrates the potentially complex interaction between rainfall nowcasts and flows in a relatively small urban drainage system for four rainfall events. In general both nowcast models provided comparable results, with the deterministic STEPS nowcast providing slightly better results for the advective events. The rainfall forecasts were deemed to be acceptable for lead times up to 15–90 min, depending on the event type, and for flow forecasts this can be increased by 15 min due to the lag-time. This lag-time is, however, dependent on the size and steepness of the catchment. Further upstream in the catchment this lag time will be less, and for larger urban areas this lag time will increase. Care needs to be taken when analysing cumulative flow forecasts, as subsequent over- and under-estimation in the rainfall forecast can lead to a closely matched final cumulative flow, hiding considerable errors in flow peaks. The timing and occurrence of flow peaks is of concern for urban pluvial flooding and negative impacts of CSO spills on receiving water; research on this is currently ongoing. Plotting the errors in flow forecasts directly against errors in rainfall forecasts can provide a useful indication of the required accuracy of a nowcasting model for different applications. For example for 21 January 2008 a RMSE of 1 mm in rainfall in the STEPS nowcast forecast over a 15-min period results in approximately 100 m^3 error in the flow forecast over a 15-min period.

Acknowledgements The authors would like to thank the Met Office for providing the radar data and the STEPS model, and Yorkshire Water Ltd. for providing the sewer model. We thank Dr Alan Seed and Clive Pierce for providing valuable advice to run the STEPS model. The first author has been funded through the Flood Risk Management Research Consortium 2 (FRMRC2). The FRMRC is supported by Grant EP/F020511/1 from the Engineering and Physical Sciences Research Council (EPSRC), in partnership with the Department of Environment, Food and Rural Affairs/Environment Agency (Defra/EA) Joint Research Programme on Flood and Coastal Defence, United Kingdom Water Industry Research (UKWIR), the Office of Public Works (OPW) Dublin and the Northern Ireland Rivers Agency (DARDNI). The second and fifth authors acknowledge the support of EPSRC grant EP/I012222/1.

REFERENCES

Achleitner, S., Fach, S., Einfalt, T. & Rauch, W. (2009) Nowcasting of rainfall and of combined sewer flow in urban drainage systems. *Wat. Sc. Tech.* 59(6), 1145–1151.

Berne, A., Delrieu, G., Creutin, J. & Obled, C. (2004) Temporal and spatial resolution of rainfall measurements required for urban hydrology. *J. Hydrol.* 299, 166–179.

Bowler, N. E. H., Pierce, C. E. & Seed, A (2004) Development of a precipitation nowcasting algorithm based upon optical flow techniques, *J. Hydrol.* 288, 74–91.

Bowler, N. E. H., Pierce, C. E. & Seed, A (2006) STEPS: a probabilistic precipitation forecasting scheme which merges an extrapolation nowcast with downscaled NWP, *Q. J. R. Meteorol. Soc.*, 132, 2127–2155.

Fuchs, L. & Beeneken, T. (2005) Development and implementation of a real time control strategy for the sewer system of the city of Vienna. *Wat. Sc. Tech.* 52(5), 187–194.

Harrison, D. L., Driscoll, S. J. & Kitchen, M. (2000) Improving precipitation estimates from weather radar using quality control and correction techniques. *Meteorol. App.* 7, 135–144.

Harrison, D. L., Scovell, R. W. & Kitchen, M. (2009) High-resolution precipitation estimates for hydrological uses. In: *Proc. of the ICE - Wat. Man.* 162, 125–135.

Krämer, S., Fuchs, L. & Verworn, H.-R. (2007) Aspects of radar rainfall forecasts and their effectiveness for real time control – the example of the sewer system of the city of Vienna. *Wat. Pract. Tech.* 2(2) doi:10.2166/wpt.2007.042.

Liguori, S., Rico-Ramirez, M. A., Schellart, A. N. A. & Saul, A. J. (2012) Using probabilistic radar rainfall nowcasts and NWP forecasts for flow prediction in urban catchments. *Atmos. Res.* 103, 80–95. doi: 10.1016/j.atmosres.2011.05.004 .

Met Office (2010) Fact Sheet 15, Weather Radar. http://www.metoffice.gov.uk/media/pdf/j/h/Fact_sheet_No._15.pdf.

Rico-Ramirez, M. A., Cluckie, I. D., Shepherd, G. & Pallot, A. (2007) A high-resolution radar experiment on the island of Jersey. *Meteorol. App.* 14, 117–129.

Seed, A. W. (2003) A dynamic and spatial scaling approach to advection forecasting, *J. Appl. Meteorol.* 42, 381–388.

Schilling, W. (1991) Rainfall data for urban hydrology: what do we need? *J. Atmos. Res.* 27, 5–21.

Verworn, H. R. & Krämer S. (2005) Aspects and effectiveness of real time control in urban drainage systems combining radar rainfall forecasts, linear optimization and hydrodynamic modelling. In: *Proc. 8th Int. Conf. on Computing and Control for the Water Industry* (5–7 September 2005, University of Exeter, UK), 307–312.

Verworn, H. R. & Krämer, S. (2008) Radar based nowcasting of rainfall events- analysis and assessment of a one year continuum. In: *Flood Risk Management: Research and Practice* (ed by P. Samuels). CRC Press, Leiden.

WaPUG (2002) Code of practice for the hydraulic modelling of sewer systems. http://www.ciwem.org/media/44426/-Modelling_COP_Ver_03.pdf (accessed 10 February 2011).

Yuan, J. M., Tilford, K. A., Jiang, H. Y. & Cluckie, I. D. (1999). Real-time urban drainage system modelling using weather radar rainfall data. *Phys. Chem. Earth (B)*, 24(8), 915–919.

On comparing NWP and radar nowcast models for forecasting of urban runoff

S. THORNDAHL[1], T. BØVITH[2], M. R. RASMUSSEN[1] & R. S. GILL[2]

1 *Aalborg University, Department of Civil Engineering, Sohngaardsholmsvej 57, DK-9000 Aalborg, Denmark*
st@civil.aau.dk
2 *Danish Meteorological Institute, Lyngbyvej 100, DK-2100 Copenhagen Ø, Denmark*

Abstract The paper compares quantitative precipitation forecasts using weather radars and numerical weather prediction models. In order to test forecasts under different conditions, point-comparisons with quantitative radar precipitation estimates and raingauges are presented. Furthermore, spatial comparisons of forecasts and observations have shown good results during stratiform conditions, but more scattered results during convective conditions. Finally, the potential for applying forecasts as input to urban drainage models is investigated. Results prove promising.
Key words numerical weather prediction; radar nowcasting; QPE; QPF; urban flow forecasting

INTRODUCTION

Short-term forecasting of urban storm water runoff is a rapidly evolving product which is in high demand by municipalities, water utility companies, waste water treatment plants, etc., as there are large economic, environmental and climate adaptation potentials in real-time control of drainage and treatment systems. The use of extrapolation-based quantitative precipitation forecasts (QPF) using weather radar data (subsequently named radar nowcast) for predicting urban runoff has previously shown great potential in forecasting within a lead time of 1–2 h, using a single radar, e.g. Germann & Zawadzki (2002), Thorndahl *et al.* (2010), and somewhat larger lead times using radar composites during quasi-stationary conditions. The quality of the radar extrapolation is highly dependent on the atmospheric conditions, and is thus larger during stratiform conditions compared to convective conditions. With regards to the latter, the numerical weather prediction (NWP) models have an advantage compared to the radar, as the atmospheric advection is modelled directly. This means that the precipitation in the model is generated by the advection, humidity, temperature, pressure, etc., and transient conditions are thus possible to predict with lead-times up to several days. The quality of the precipitation estimates however decreases significantly for lead-times larger than 24 h.

This paper investigates the potential for combining NWP forecast and radar nowcast models in order to increase forecast lead-times on urban drainage systems. Initially, the precipitation field generated by the NWP forecast is compared to both radar observations, i.e. quantitative precipitation estimates (QPE) as well as raingauge observed precipitation in different points. Secondly, a spatial comparison of NWP precipitation, radar QPE and radar nowcast is conducted. How differences in spatial and temporal resolution between the two model types are affecting the prognoses and how quality of the urban runoff forecast varies as a function of the lead-time were studied. Furthermore, if a temporal optimum for shifting between the two types of models can be estimated was investigated. Finally, NWP forecasts, radar QPE and nowcasts, as well as raingauge rain rates are applied as input to an urban drainage model of a small Danish urban catchment in order to study the potential for short-term flow and overflow forecasting. Urban short-term flow forecasting has previously been studied, by e.g. Achleitner *et al.* (2009) using radar nowcasts, and by Rico-Ramirez *et al.* (2009) using NWP forecasts.

DATA

In order to test both the NWP forecast model and the radar nowcast model during different conditions, the analysis is based on three different events. The first event, Event 1, is on 18 November 2009 and is a clearly stratiform event with a uniform advection from the southwest during the initial phase. The event evolves to a cyclonic low-pressure rotation during the day. Event 2, on 28 May 2010, is a convective event with small scattered rain cells, and Event 3, on 7 June 2010, is a partly convective partly stratiform event with one relatively small, but very high intensive rain cell. This event caused a minor urban flooding in a town in the northern part of Denmark.

Numerical weather prediction forecast model

HIRLAM (HIgh Resolution Limited Area Model, http://hirlam.org) is a hydrostatic model for short-range weather forecasting. The Danish Meteorological Institute runs various set-ups of HIRLAM and for this study the DMI-HIRLAM-S05 is used. The model is run every 6 h at 00:00, 06:00, 12:00, 18:00 UTC with a horizontal spatial resolution of 0.05° (approx 5 km) over 40 vertical levels. The model predicts accumulated stratiform and convective precipitation as well as snow with a temporal resolution of 1 h. Einfalt *et al.* (2004) recommends a minimum resolution of 1 km and 5 min. for urban hydrology. This paper will however investigate coarser spatial and temporal resolutions. For this study the total accumulated precipitation was used, i.e. the sum of stratiform and convective precipitation. The outer boundary for the model are prognoses from ECMWF (European Centre for Medium range Weather Forecasts).

Radar and radar nowcast model

The radar applied in the study is a 250 kW single polarization C-band radar located in Sindal in the northern part of Denmark. The radar is owned and operated by the Danish Meteorological Institute (DMI, Gill *et al.*, 2006) and has a range of 240 km, with a temporal resolution of 10 min. The data are an extraction of the lowest cappi layer with a Cartesian grid of 2000 × 2000 m pixels. DMI recommends quantitative precipitation estimates in the range of 0 to 75–100 km. The radar data are calibrated against raingauges using techniques similar to those presented in Thorndahl & Rasmussen (2012).

The radar nowcast algorithm has been developed at Aalborg University, Denmark, and is currently running in real time operation. It is based on the TREC-algorithm (TRacking of Echoes with Correlation, Rinehart & Garvey, 1978) and the CO-TREC (Li *et al.*, 1995; Continuous-TREC, Mecklenburg *et al.*, 2000). Van Horne (2002) has inspired some of the spatial smoothing algorithms which are implemented in the model. The model is described in detail in Thorndahl *et al.* (2009, 2010). It does not include growth/decay terms and is solely an extrapolation of the observed rain.

Raingauges, runoff model, urban catchment

The raingauges applied in this study are all part of the national Danish network of 20 cm diameter tipping-bucket raingauges with a tip resolution of 0.2 mm. Data are monitored and verified by the Danish Meteorological Institute. The raingauge time series has a temporal resolution of 1 min. Two of the gauges are located in the urban catchment described below. The extracted area of the NWP model covers 80 gauges, and the quantitative range of the radar covers 10 raingauges.

In recent years, the urban catchment of Frejlev has been studied intensively, as Aalborg University has a flow monitoring station located in the catchment (Schaarup-Jensen *et al.*, 1998). Consequently, different models and model calibrations have been performed using the catchment as a case-study, e.g. Thorndahl *et al.* (2006, 2008). The runoff model applied in this study is the MOUSE model DHI (2009). The catchment covers an area 0.8 km^2. The sewer system is partly combined and partly separated with a local overflow to a small river.

POINT COMPARISON OF NWP FORECAST, RADAR, AND RAINGAUGES

In order to compare how the NWP-model performs in predicting rainfall volumes compared to raingauges and radar, and also to investigate differences in timing, Fig. 1 show examples of accumulated precipitation as a function of time/lead-time. Obviously this is a comparison of

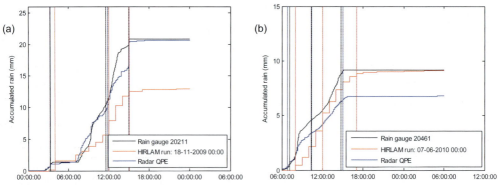

Fig. 1 (a) Example of point comparison of NWP forecast, radar and raingauge, Event 1: 18 November 2009. (b) Example of point comparison of NWP forecast, radar and raingauge, Event 3: 7 June 2010.

different scales as the sampling volumes are different in raingauges (point measurement) and radar (area measurement).

The vertical dotted lines indicate the time of the 5% quantile of the accumulated precipitation and the vertical dashed and dash-dotted lines indicate the 50 and 95% ditto. In the first example, Fig. 1(a), quite a large difference in the accumulated precipitation between radar/raingauge and the NWP model is observed, and the NWP-model clearly underestimates the precipitation. However, if the vertical lines are studied, it is obvious that almost total timing convergence is present; for example the 50% quantile is located around 12:00 h for all three precipitation estimates, and even though the volumes are slightly different, it is an important result that timing errors are negligible in this event. In the second example, Fig. 1(b), almost perfect agreement between raingauge and NWP is apparent, but the radar precipitation estimates are smaller. Analysis of a whole range of these types of point comparisons gives the following results. As expected, the deviations between NWP-model and raingauges on volumes and timing take on the lowest values with regards to the first part of the stratiform event (Event 1) and a bit larger deviations as the event transforms into the cyclonic rotation. The largest deviations (on both volumes and timing) are, as expected, observed with regards to Event 2, as this consists of scattered convective rain cells.

It is not possible analysing the point comparison plots to conclude significant tendencies on how the errors in volumes and timing evolve as function of the lead time. For these types of analyses, spatial compositions using radar observations are much more appropriate. These are presented in the next section.

SPATIAL COMPARISON OF NWP FORECAST, RADAR QPE AND NOWCAST

In this section the NWP forecast and the radar nowcast are compared to the radar observations (quantitative precipitation estimates, QPE) in order to investigate the quality of the two QPF models as a function of the lead time. Apart from a visual comparison the observations and models are compared using two skill measures, namely the well known 2-D correlation coefficient (r^2) and the critical success index (CSI, Li *et al.*, 1995). CSI is a measure of the model's ability to predict rain in the correct pixels. A rain/no-rain criterion was used and the score therefore does not take the intensity levels into account. The intensity levels are however covered by r^2. Both quality indices range from zero to one. The highest quality is given the value of one. In Fig. 2, the radar QPE is assumed to be the reference for both the NWP forecast and the radar nowcast.

In event 1 the forecast at 00:00 h shows very low quality measures on the nowcast model. This is due to the fact that it is not raining prior to 00:00 h. Hence, it is not possible to generate an extrapolation of the rain. However, prior to 12:00 h it is raining and both CSI and r^2 for the radar nowcast show a higher quality index, than the same measures of the NWP forecast, until a lead time of 3–4 h. The quality of both NWP and nowcast are higher during the stratiform event

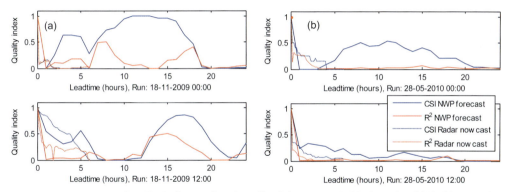

Fig. 2 (a) Example of quality indices as function of lead-time, event 1, 18 November 2009. (b) Example of quality indices as function of lead-time, Event 2, 28 May 2010.

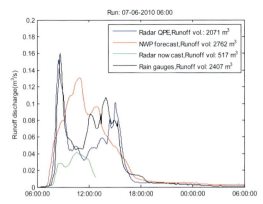

Fig. 3 Forecast runoff with radar QPE, raingauges as well as NWP forecast and radar nowcast run at 06:00 h, Event 3, 7 June 2010.

(Fig. 2(a)) compared to the convective conditions, Event 2 (Fig. 2(b)) and Event 3 (not shown). In general, the NWP forecast shows a good performance at a lead time of 6 to approx. 20 h, contrary to the radar nowcast which only gives reliable results from 0 to 3–4 h during stable conditions and 0 to 1–2 h during convective conditions.

FLOW FORECASTING

The different runs of the NWP forecast and the radar nowcast are applied as input to the urban drainage model described above. For comparison both observed radar data and the two raingauges which are located in the catchment are included as input to the runoff model as well. Figure 3 presents the runoff from the whole catchment applying the run from Event 3, 06:00 h. It is obvious that radar QPE and raingauges show somewhat the same runoff both with regards to temporal variations as well as runoff volume. The NWP forecast shows a runoff volume of the same order of magnitude as radar nowcast and raingauge input, but a different temporal variation. This is partly due to the coarser temporal resolution of the NWP forecast compared to the other inputs (1 h *vs* 10 and 1 min). As the drainage system has a concentration time of approx. 20–30 min, the modelled flow has smaller fluctuations. Furthermore, some timing errors are present as the peak flow using the NWP forecast arrives about 1.5 h later than the peak modelled with radar QPE and raingauges, respectively. This timing error is also present in the raw rainfall intensity data, and is thus not related to the prediction of the flow, but can be related to some of the issues discussed in the previous section. In this example, the flow forecast using the radar nowcast is seen to under-predict both volume and peak. This is due to the short lead time of this model. The peak arrives more than 3 h after the initiation of the model and at this time the quality of the radar nowcast is so low that it does not make sense to apply the results. It is however possible to run the radar nowcast model every 10 min and not, as done here, at the same time as the NWP forecast. If the radar nowcast is initiated 1 or 1.5 h later, results are similar to the radar QPE simulation.

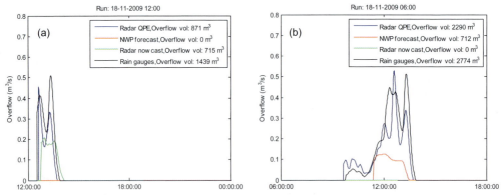

Fig. 4 (a) Forecasted overflow discharge with radar QPE, raingauges as well as NWP forecast and radar nowcast. Run: 12:00 h, Event 1, 18 November 2009. (b) Forecasted overflow discharge with radar QPE, raingauges as well as NWP forecast and radar nowcast. Run: 12:00 h, Event 1, 18 November 2009.

In Fig. 4 the modelled overflow discharge and volume are presented for two different runs at 12:00 and 06:00 h 18 November 2009. In the 12:00 h-simulation, there is a good agreement between radar QPE, raingauges and radar nowcast with regards to the overflow volumes; the NWP forecast, however, fails. If the NWP forecast is initiated 6 h earlier (Fig. 4(b)) the modelled overflow volumes become of the same order of magnitude as modelled with radar QPE and raingauges, respectively. This is a typical example of the NWP-model performing with low quality when forecasting the nearest future (0–6 h) compared to forecasting the future from 6 h and beyond. The low performance of the NWP is probably due to spin-up effects in the first hours after the model run due to inconsistencies between the interpolated field from the boundary model which lead to a drop-off in precipitation in the first hours of the forecast and a subsequent stabilization and better estimate of precipitation. If the temporal flow variations are compared, again the discharge peak values using the NWP is much smaller compared to the runoff modelled with the other rainfall inputs. This is due to the coarser temporal resolution of the NWP data.

DISCUSSION

In the paper different examples of comparisons between the numerical weather prediction forecast of rainfall, extrapolated radar rainfall, and the observed rainfall with radar and raingauges is presented. Generally, good correlation of all the rainfall estimates on both timing and rainfall volumes are observed during stratiform conditions, and more scattered results for the convective conditions. However, this paper only contains selected examples. In order to conclude significant trends, more events during different conditions have to be analysed. Generally, the NWP forecast shows the best results with regards to rainfall volumes, whereas the rainfall intensity estimates deviates more from observed data. This is primarily due to a coarse temporal resolution of 1 h. In the cases presented here, the location of the NWP rainfall has been equal to the radar observations; however, NWP forecasts can contain location errors which might be large compared to the urban area.

The radar nowcast model performs better compared to the NWP forecast when estimating intensity levels, but as the lead time of this model is rather short, the volume prediction becomes rather uncertain. Depending on the meteorological conditions, the radar nowcast model has produced an acceptable quality forecast with a lead time of 0 to 2–4 h. A corresponding optimum has been observed using the NWP forecast with a lead time of 6 to 12–20 h. Evidently, there is a period of time between lead times of 2–6 h in which the quality of both model performances are low. As a consequence, it is difficult to recommend when to shift from one forecast model to the other in terms of obtaining the best possible forecast. This recommendation is omitted from the paper as more events have to be analysed in order to conclude significant trends. However, blending techniques as proposed by Atencia *et al.* (2010) and STEPS by Bowler *et al.* (2006) might be the solution in order to gradually shift from the radar nowcast to the NWP forecast. Continuous assimilation of the NWP model with radar QPE data might be a solution to improve quality of short-term NWP forecasts but this is outside the range of this paper. The radar nowcast model might be improved by using the advection field from the NWP model to extrapolate radar data. As the atmospheric conditions are included in the NWP model the advection field is more precise than the field that can be obtained by correlation analysis of radar images (Bowler *et al.*, 2004).

In terms of forecasting urban runoff, the results obtained in this paper definitely show potential in prediction of discharge or overflow volumes, even though the discharge modelled with observed data, at some points differs from the discharges modelled with forecast data. This is partly due to differences in spatial and temporal resolution between the different types of input data. Since the catchment applied in this study is relatively small, with a short time of concentration, the coarse temporal resolution of the NWP model propagates through the runoff model with a coarse hydrograph as output. However, if a larger catchment with larger time of concentration is used, these types of errors would be smaller. This system shows great potential for modelling the inflow to waste water treatment plants or inflow to large storage basins since they often have a runoff time of up to several hours. In the case of modelling local surcharge or flooding, the radar nowcast would be preferable because of short lead times and fine temporal resolutions.

REFERENCES

Achleitner, S., Fach, S., Einfalt, T. & Rauch, W. (2009) Nowcasting of rainfall and of combined sewage flow in urban drainage systems *Water Science and Technology* 59(6).
Atencia, A., Rigo, T., Sairouni, A., Moré, J., Bech, J., Vilaclara, E., Cunillera, J., Llasat, M. C. & Garrote, L. (2010), Improving QPF by blending techniques at the meteorological service of Catalonia, Natural Hazards and Earth System Science, v. 10.
Bowler N. E., Pierce, C. E. & Seed, A. (2004) Development of a precipitation nowcasting algorithm based upon optical flow techniques. *J. Hydrol.* 288.
Bowler, N. E., Pierce, C. E. & Seed, A. W. (2006) STEPS: A probabilistic precipitation forecasting scheme which merges and extrapolation nowcast with downscaled NWP. *Quart. J. Roy. Met. Soc.* 132, 2127–2155.
DHI (2009) MOUSE Pipe Flow reference manual, MIKE by DHI.
Einfalt, T., Arnbjerg-Nielsen, K., Golz, C., Jensen, N.-E., Quirmbach, M., Vaes, G. & Vieux, B. (2004) Towards a roadmap for use of radar rainfall data in urban drainage. *J. Hydrol.* 299, 186–202.
Germann, U., Zawadzki, I. (2002) Scale-dependence of the predictability of precipitation from continental radar images. Part I: Methodology. *Mon. Weath. Rev.* 130, 2859–2873.
Gill, R. S., Overgaard, S. & Bøvith, T. (2006) The Danish weather radar network. In: *Proc. Fourth European Conf. on Radar in Meteorology and Hydrology 2006*.
Li, L., Schmid, W. & Joss, J. (1995) Nowcasting of motion and growth of precipitation with radar over a complex orography. *J. Appl. Met.* 34.
Mecklenburg, S., Joss, J. & Schmid, W. (2000) Improving the nowcasting of precipitation in an Alpine region with an enhanced radar echo tracking algorithm. *J. Hydrol.* 239.
Rico-Ramirez, M. A., Schellart, A. N. A., Liguori, S & Saul, A. J. (2009) Quantitative precipitation forecasting for a small urban area: use of a high-resolution numerical weather prediction model. In: *8th International Workshop On Precipitation In Urban Areas* (St Moritz, Switzerland).
Rinehart, R. E. & Garvey, E. T. (1978) Three-dimensional storm motion detection by conventional weather radar. *Nature* 273.
Schaarup-Jensen, K., Hvitved-Jacobsen, T., Jütte, B., Jensen, B. & Pedersen, T. (1998) A Danish sewer research and monitoring station. *Water Sci. Technol.* 37(1).
Thorndahl, S., Beven, K. J., Jensen, J. B. & Schaarup-Jensen, K. (2008) Event based uncertainty assessment in urban drainage modelling, applying the GLUE methodology. *J. Hydrol.* 357(3–4).
Thorndahl, S., Johansen, C. & Schaarup-Jensen, K. (2006) Assessment of runoff contributing catchment areas in rainfall runoff modelling. *Water Sci. Technol.* 54(6–7).
Thorndahl, S. & Rasmussen, M. R. (2012) Marine X-band weather radar data calibration. *Atmos. Res.* 103, 33–44.
Thorndahl, S., Rasmussen, M. R., Grum, M. & Neve, S. L. (2009) Radar based flow and water level forecasting in sewer systems – a Danish case study. In: *Proc. 8th Int. Workshop on Precipitation in Urban Areas* (10–13 December, 2009, St Moritz, Switzerland).
Thorndahl, S. L., Rasmussen, M. R., Nielsen, J. E. & Larsen, J. B. (2010) Uncertainty in nowcasting of radar rainfall: a case study of the GLUE methodology. In: *Proc. 6th European Conference on Radar in Meteorology and Hydrology* (Sibiu, Romania).
Van Horne, M. P. (2002) Short-term precipitation nowcasting for composite radar rainfall fields. Master's thesis, Massachusetts Institute of Technology, Massachusetts, USA.

626

Weather Radar and Hydrology
(Proceedings of a symposium held in Exeter, UK, April 2011) (IAHS Publ. 351, 2012).

Decision support for urban drainage using radar data of HydroNET-SCOUT

ARNOLD LOBBRECHT[1,2], THOMAS EINFALT[3], LEANNE REICHARD[1] & IRENE POORTINGA[1]

1 *PO Box 2177, 3800 CD Amersfoort, the Netherlands*
info@hydrologic.com

2 *UNESCO-IHE, PO Box 3015, 2601 DA Delft, the Netherlands*

3 *Hydro & meteo GmbH&Co.KG, Breite Str. 6-8, 23552 Lübeck, Germany*

Abstract Users of hydro-meteorological data often face problems with collecting, handling and quality control of data from radar and raingauges. Current web technologies allow centralised storage, data management and integration of software tools. HydroNET and SCOUT tools have been integrated to produce accurate precipitation information and to present easy-to-understand interfaces to practitioners. The SCOUT software has been developed by hydro&meteo for obtaining calibrated precipitation information from raw radar data. HydroNET has been developed by HydroLogic with the aim of bringing meteorological data to the desktop of water managers and to support their daily work. Co-creation with users has led to HydroNET portal (www.hydronet.eu). This portal integrates the functionalities and supports water managers in assessing historical, current and forecasted precipitation events. The portal has been built using the Software as a Service (SaaS) paradigm. It is highly customisable and permits the user to configure its own interface, tools and warning levels.

Key words precipitation; raingauge; radar; RTC; DSS; web-service; SaaS; HydroNET; SCOUT

NEED AND AVAILABILITY OF URBAN PRECIPITATION DATA

Current needs

The frequent occurrence of excessive rainfall in urbanised areas is an important reason for water managers to look for opportunities to have easy online access to real-time and historical precipitation information. During the past few years HydroLogic has developed, together with several Dutch municipalities and water boards, the HydroNET portal which gives access to a variety of precipitation information up to a spatial resolution of 1×1 km and time series with 5 min intervals. The main reason for municipalities to look for such a tool was:

(a) The need for high quality historical precipitation datasets for calibration of urban drainage models.
(b) Accurate data for technical analysis of sewer networks, e.g. detecting clogging of sewers, evaluation of the operation of storage tanks and of pumping stations.
(c) Accurate data for handling of water-damage claims of the general public and companies.
(d) Cooperation between various water authorities such as municipalities, water boards, river management organisations and sewer departments.
(e) Operational use of rainfall data for flood prevention and managing flood situations: optimising the use of storage in sewer systems; applying spatial rainfall distribution in real-time control (RTC) algorithms and; assessing the risks of flooding.
(f) Availability of short term forecasts, preferably ensembles, up to 3 h.
(g) To have a set of tools for visualisation of information, specifically devoted to hydrological use.

A lot of meteorological information is presently available at national weather services, for free or at delivery costs. This recent development is stimulated by the European and national policies of free meteorological data availability. This availability only is not enough for hydrological use. Often barriers exist to the use of these data, in particular because of the difficult formats, the timing of delivery, issues of completeness and the varying quality of the data. In general the user wants the best available dataset, not being bothered by all these issues.

In a study with several municipalities and water boards in the Netherlands a definition of necessary data and tools for decision support was made and implemented in HydroNET. This has resulted in the use of the system by more than 90 municipalities in their daily work of analysis, monitoring and forecasting of high water levels, overflows and floods. We found a solution to maintain only one dataset for all users, store it centrally and giving access to it for local use by means of a web portal. This solution also matches the current policy to have less software installed and maintained in government organisations. The single dataset is maintained, quality-controlled and kept up to date for all users by specialised service providers.

Availability of data

For a well balanced management of the urban water system, high quality precipitation data with a high resolution in time and space are necessary. This is a very difficult task to accomplish by the use of raingauges alone. On average, one raingauge per e.g. 4 km^2 (proposed approach in the Netherlands) would be necessary and this presents enormous investments to municipalities. Many municipalities have few raingauges and collect these data by telemetry networks, which makes the monitored data available together with other monitored data, e.g. pump operation, discharges and water levels. In practice the positioning of raingauges is a problem in urbanised areas. Following international standards of placement, the majority of these raingauges are in inappropriate locations such as on tops of buildings, next to buildings, under trees, etc. As a result, the associated errors in recording may rise up to 40%, which presents very inaccurate results. The general conclusion when reviewing the precipitation data collected by municipalities is that the coverage of the raingauge monitoring, as well as its quality, are poor.

On the other hand, a variety of hydro-meteorological data from the national weather services exists. The most important one for urban water management is the (uncalibrated) radar data, made available every 5 min by the national weather service KNMI. Calibrated composites exist in the Netherlands: 3-h sums and 24-h sums. However, these time resolutions are inappropriate for use in urban water management.

Apart from the radar, the national weather service makes available precipitation data from raingauges: 33 automatic gauges and approx. 330 manually read gauges, of which daily sums become available after half a day. These data appear to be very useful for calibration of radar data.

HydroNET-SCOUT

HydroLogic and hydro&meteo joined forces to fill the gap between information needs of urban water managers and available data by combining the tools which were developed and improved during many years. The SCOUT tools are used to create the best datasets possible on the basis of radar data and raingauge monitoring, including quality labelling, correction, calibration and updating as soon as new data become available. The HydroNET tools are used for storage, visualisation (time series, GIS) and decision support to water managers, using web-interfaces. The joint activity has resulted in a product which is welcomed by end-users in both the Netherlands and Germany.

A SAAS SOLUTION FOR HYDROLOGY

In hydro-meteorological information services for urban water management, the availability of abundant computing power is needed when multiple computations are requested, e.g. for running models in ensemble, but more importantly, to permit many users to simultaneously access the services, e.g. during heavy rainstorms, without loss of performance.

One of the most important recent developments in information and communication technology (ICT) for hydrology is cloud computing. Clouds present networks of servers which are connected to share loads of traffic and computations. In cloud computing often the Software as a Service (SaaS) paradigm is used, which is basically software that is fully running on servers, presenting interfaces to clients in the form of interactive websites.

Hydro-meteorological applications of interest use both private and public clouds. Private clouds use local clusters of servers, which are these days at the premises of the supplier. Public clouds are made available by providers of cloud services, e.g. Amazon, Microsoft, IBM (these days there are around 4000 cloud providers). The interesting feature of public cloud is that it permits replication of services, using virtualisation technologies. Good water-engineering examples have been presented by Xu *et al.* (2010). The approach creates the possibility of having any number of servers available to the user on demand. Private and public clouds can be combined, where private clouds function as the primary computing environment, managed by a provider of information. In such a configuration a public cloud can be used for expansion of computing power when needed, also referred to as "cloudburst" (not to be confused with rainstorms).

HydroNET has been developed as a SaaS cloud application, presenting interesting advantages for users and the organisations providing the services:

– All information is presented via a customisable web-portal.
– No desktop software needs to be installed.
– No server software is needed on premises of the municipalities and water boards.
– Updates to the software are available to all, right after installation in the private cloud.
– Costs of operation and maintenance can be reduced and the level of service enhanced.
– All data are maintained and made available to all users in one central place.
– Virtually unlimited computing power is available for calibration and modelling tools, solving performance problems.

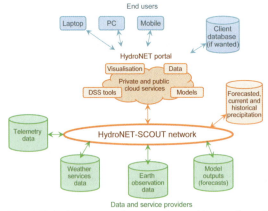

Fig. 1 HydroNET network connecting and interacting with various data sources and providing services through a web-portal.

The service network designed and developed is presented in Fig. 1. The picture shows the concept of a secure network in which services of several parties such as added-service providers and research organisations can be joined. These parties can make their data available through the same portal, having the advantage of the flexibility in configuration as well as the scalability and associated high level of service. Examples of these data are: radar data, raingauge data, evaporation data, and Earth Observation data (in sparsely measured areas). Because of the use of web applications (Fig. 2), users can run them from desktops, laptops and mobiles. The availability of real-time bench marking tools permit the water managers of municipalities and water boards to access the current situation and prepare for flood preventive measures, warn operational services and inform the public.

Some authorities, for administrative purposes or further integration purposes, also want the data from HydroNET in their own databases. For that purpose the solution provides a specific web service which permits these organisations to abstract the data and store it in corporate databases or telemetry systems.

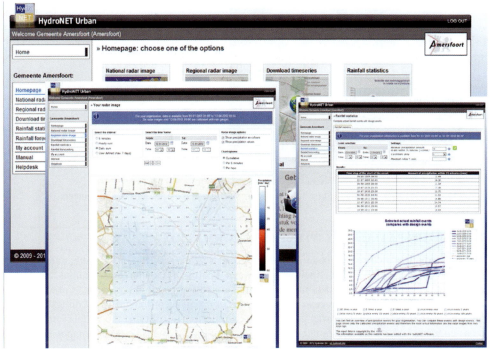

Fig. 2 Web-portal interface to the hydro-meteorological information showing map with precipitation sums and the associated forecast made available one day before the event.

CO-CREATION IN DEVELOPMENT

One of the key factors for a successful implementation of software is the involvement of users in the process of definition of functionalities, user interfaces and testing (Fig. 3). In HydroLogic the DSDM approach is successfully followed for this purpose (www.dsdm.org). Parts of the approach are early interactive sessions with users to define what they exactly want with the application. Some generally applicable conclusions of these co-creation sessions were:

(a) Keep interfaces very simple, so that also infrequent users can easily find their way.
(b) Use web-technology, so that all users always have the latest functionalities.
(c) Make interfaces user-customisable, e.g. by allowing them to make their own front page or dashboard with information in the portal.
(d) Make the application fast, so that users don't have to wait more than a few seconds for information.

More specific user requirements which followed from the sessions were:

(a) Include tools for fast assessment of recent and historic rainfall events, e.g. by comparing actual events with design events.
(b) Make series downloadable for use in other packages for statistical analysis or modelling, supporting different formats of these packages.
(c) Connect to existing telemetry systems and deliver the best precipitation data as soon as available.
(d) Add simple summation tools and GIS tools for rapid analysis, visualisation and presentation of spatial rainfall.
(e) Allow local raingauge data to be included in the radar calibration process.

Fig. 3 Co-creation with end-users: brainstorming, sketches and discussions on the contents.

TOOLS FOR SPATIAL PRECIPITATION INFORMATION

The data which are currently made available by the weather services in the Netherlands do not meet the requirements of municipalities and water boards. The latter organisations have clear requests for inclusion of local data into the entire service chain and in particular to use their own raingauges in the calibration process. To permit this, an analysis of the quality of raingauges to be included and their positioning was performed. Also the functionality of the SCOUT software to cross-compare radar data and raingauge data was enhanced. Other important components in the development of the integrated solution were:

(a) Fast online compositing and data processing by SCOUT.
(b) Data quality assessment and labelling.
(c) Radar data correction (Golz *et al.*, 2006).
(d) Radar/raingauge cross-comparison.
(e) Radar statistics: maximum hourly rainfall, daily sum, etc.
(f) Spatial rainfall information generation by calibration.
(g) Several methods of radar-based nowcasting, including ensemble approaches.

The calibration of radar data forms an important point in the development of accurate spatial precipitation information. This process has been developed for real-time applications as a work flow in which the continuous flow of data from its sources within weather services and water authorities is streamlined (Fig. 4). On the basis of new data which becomes available continuously, further calibration and nowcasting is performed using data of radar and raingauges. Finally, on the basis of a set of approx. 330 raingauges the dataset is fully updated and made available for offline analysis to the clients being water boards and municipalities.

The original cell-tracking based nowcasting approach in SCOUT was extended with three more nowcasting approaches, using vector fields for cell movement description. Additionally, ensembles can now be created based on the observed variability of cell speed, direction, growth and decay. The ensembles are a mix of nowcasting approaches and the use of observed uncertainties and developments in the data (Tessendorf & Einfalt, 2011).

The coverage of the two radars of the Netherlands in De Bilt and Den Helder is limited, resulting in inaccurate data in the northeastern and southwestern parts of the country. For the northeastern part this problem has been solved by using the German radar of Emden, which is right across the border. The entire procedure of integrating the data streams of the Netherlands and Germany has been investigated, tested and included in the procedure of data processing. The experiment performed, using raw 5-min radar data from the KNMI and the German weather service DWD, is presented in Einfalt & Lobbrecht (2011).

Users can define warning levels for recent, current and forecasted rainfall intensity. By doing so, the system can support their operations, warning them when thresholds are passed. This supports decisions on when operators should take action such as extra pumping, operation of reservoirs for temporal storage, etc. Also, this approach allows water managers to warn inhabitants in cases of a high probability of high water levels, floods and calamities.

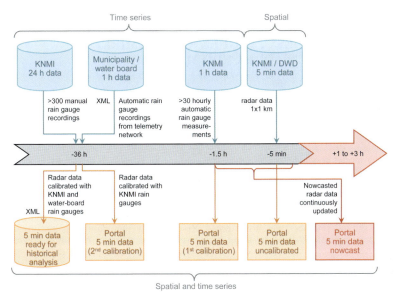

Fig. 4 Process-flow diagram for calibration of radar data to create the best composition of information in real-time and post processing of data to generate up-to-date and calibrated spatial and time series.

CONCLUSIONS

The integration of HydroNET and SCOUT into a combined solution for hydrological practitioners of municipalities and water boards, has been performed and tested extensively. Users have been involved by means of co-creation of the tools and are very satisfied with the portal solution. They do not have to worry anymore about installed software and databases; all services are outsourced, maintained and available to them whenever needed. Municipalities use the portal in operational management of excessive rainfall and floods and for post-flood analysis. The high quality spatial information provided, supports them in discussions with other water authorities and in cases of claims by third parties.

The current Software as a Service paradigm has been fully integrated in the decision-support system. It allows various providers of data, information and services to work together and provide added-value to end-users. The new HydroNET-SCOUT solution provides an excellent basis for cooperation of industry partners, research institutions and practitioners all over the world. End-users have access to high quality data and online customisable tools; the basis for decision-support in strategic and operational urban water management.

REFERENCES

Einfalt, T. & Lobbrecht, A. (2011) Compositing international radar data using a weight-based scheme. *Weather Radar and Hydrology*. (Proc. Symp., Exeter, UK). IAHS Publ. 351. IAHS Press, Wallingford, UK (this volume).

Golz, C., Einfalt, T. & Galli. G. (2006) Radar data quality control methods in VOLTAIRE. *Meteorologische Zeitschrift*, 15(5), 497–504.

Tessendorf, A. & Einfalt, T. (2011) Ensemble radar nowcasts - a multi-method approach. *Weather Radar and Hydrology*. (Proc. Symp., Exeter, UK). IAHS Publ. 351. IAHS Press, Wallingford, UK (this volume).

Xu, Z., Vélez, C., Solomatine, D. & Lobbrecht, A. (2010) Use of cloud computing for optimal design of urban wastewater systems. In: *9th Int. Conf. on Hydroinformatics 2010* (ed. by J. Tao, Q. Chen *et al.*), 1, 930–938. Chemical Industry Press, Tiajin, China.

Radar-based pluvial flood forecasting over urban areas: Redbridge case study

LI-PEN WANG[1], NUNO SIMÕES[1,2], MIGUEL RICO-RAMIREZ[3], SUSANA OCHOA[1], JOAO LEITÃO[4] & ČEDO MAKSIMOVIĆ[1]

1 *Department of Civil and Environmental Engineering, Imperial College London, Skempton Building, South Kensington Campus, London SW7 2AZ, UK*
li-pen.wang08@imperial.ac.uk

2 *Department of Civil Engineering, University of Coimbra, Coimbra, Rua Luís Reis Santos, 3030-788 Coimbra, Portugal*

3 *Department of Civil Engineering, University of Bristol, UK*

4 *Laboratório Nacional de Engenharia Civil, Av. do Brasil 101, 1700-066 Lisboa, Portugal*

Abstract A nowcasting model coupled with an urban drainage model is used in this study to assess the forecasting of pluvial floods in urban areas. The deterministic nowcasting model used in this paper is part of the Met Office STEPS (Short-Term Ensemble Prediction System) system, and the hydraulic model is run based on the 1D/1D dual drainage simulation scheme. A highly-urbanised catchment, Cranbrook (located in the London Borough of Redbridge), is employed for this case study to analyse the associated uncertainties. The aim of this work is to assess the impact of using rainfall forecasts with different spatial and temporal resolutions to forecast pluvial flooding over urban areas. Results show that promising performance in hydraulic forecasting is in general observed by using higher spatial and temporal resolution nowcasts as inputs; this implies the necessity of using advanced radar-based nowcasting techniques to improve the state-of-the-art pluvial flood forecasting over urban areas.

Key words nowcasting; flood forecasting; radar; rainfall; urban drainage

INTRODUCTION

In the last decades, urban pluvial (surface) flooding has caused enormous economic losses all over the world, and has affected thousands of people. For this reason, it has been pointed out as an important issue that urgently needs to be appropriately tackled (Pitt, 2008). In order to accurately predict floods over urban areas, high-resolution rainfall measurements and forecasts, as well as efficient hydraulic models, are required.

Many studies have thus been conducted on the use of weather radar in urban hydrology (Einfalt *et al.*, 2004), as these devices are able to provide rainfall measurements with high spatial and temporal resolutions. Furthermore, the use of radar to produce a rainfall forecast and its subsequent combination with runoff models has been recently studied (Krämer *et al.* 2007; Liguori *et al.* 2012) and constitutes the focus of several current studies, like the one presented here.

Regarding the runoff models used for urban pluvial flood forecasting, the dual-drainage concept, which entails integrating the overland and sewer networks, has been widely accepted as a feasible physically-based representation of pluvial flooding (Leitão *et al.*, 2009; Maksimović *et al.*, 2009; Simões *et al.*, 2010).

A radar-based nowcasting model coupled with an urban drainage model is therefore employed in this study to simulate urban (pluvial) floods. The main aim of this work is to assess the impact of using rainfall nowcasts with different spatial and temporal resolutions to forecast pluvial flooding over urban areas.

EXPERIMENTAL SITE AND DATASET

Cranbrook catchment

The case study used for testing our pluvial flood forecasting methodology is the Cranbrook catchment, which is located within the London Borough of Redbridge (situated in the northeast part of Greater London) and is drained by the River Roding (Fig. 1(a)). According to the Environment Agency (2006), the River Roding has a rapid response to rainfall, which is typical of densely-urbanised catchments overlying London Clay. Flood events have been recorded in the Roding catchment since 1926, with the most recent events being in 2000 and 2009, when several properties were flooded. These events are relatively well documented and can be used in the development of advanced flood prediction methodologies.

The drainage area of the Cranbrook catchment is approximately 910 ha; the main water course is about 5.75 km long, of which 5.69 km are piped or culverted.

Radar data

The Cranbrook catchment is in the coverage of two radars: Chenies and Thurnham (Fig. 1(c)). The radar data are provided by the Met Office through the British Atmospheric Data Centre (BADC) with spatial and temporal resolutions of 1 km and 5 min, respectively. This dataset has been averaged to obtain a spatial resolution of 2 km. The radar data have been quality-controlled by the Met Office following the correction techniques proposed by Harrison *et al.* (2000) to account for all the errors inherent to radar rainfall measurements (see Rico-Ramirez *et al.*, 2007).

Monitoring system

A real-time accessible monitoring system is installed in this catchment (Fig. 1(d)), including three tipping-bucket raingauges, one pressure sensor for Roding River level monitoring, two sensors for water depth measurement in sewers, and one sensor for water depth measurement in open channels. The collected measurements however are not used to verify the forecasting results but for the calibration of the hydraulic model because the focus of this study lies in assessing the forecast errors rather than the forecast plus hydraulic errors.

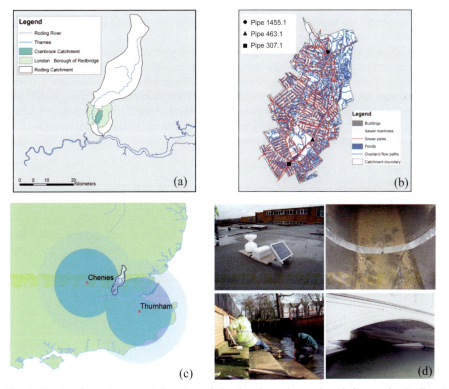

Fig. 1 Cranbrook catchment: (a) location of Cranbrook catchment in relation to the Roding River catchment; (b) dual drainage networks of the Cranbrook catchment; (c) radars which cover the study area; (d) monitoring system installed in the study area.

NOWCASTING MODEL

The deterministic nowcasting system employed in this study is the Short-Term Ensemble Prediction System (STEPS) developed by the Met Office and the Australian Bureau of Meteorology. The deterministic nowcasting system in STEPS is based on the spectral decomposition proposed by Seed (2003) with the incorporation of the optical flow equation proposed by Bowler *et al.* (2004). The deterministic nowcasts produced in this study are purely based on radar.

The nowcasting model has been setup to run at 1 km and 2 km spatial resolutions and 5 min, 10 min and 15 min temporal resolutions. At both spatial resolutions, the domain size was fixed to 500 km × 500 km covering the window with lower left coordinates of 200 km (easting), –100 km (northing) and top right coordinates of 700 km (easting) and 400 km (northing). The radar data have been pre-processed to cover this domain size and also to remove any spurious echoes. The nowcasting model has been setup to produce 3 h forecasts initialised every 10 min for all the events.

HYDRAULIC MODEL

Based upon the dual-drainage concept, the hydraulic model of the Cranbrook catchment was implemented in InfoWorks CS by coupling the overland network generated by the AOFD (Automatic Overland Flow Delineation) (Maksimovic *et al.*, 2009) and the sewer system. The overland network model is implemented using a LiDAR (Light Detection And Ranging) DEM (Digital Elevation Model) with 1 m resolution and vertical accuracy of approximately 0.15 m. The sewer network model was obtained from the water utility of the study area. The original simulation parameters were unchanged for all simulations.

The AOFD, an in-house tool developed by the Urban Water Research Group (UWRG) at Imperial College London, was employed in this study to generate the overland network. This tool automatically creates the overland flow network to enable its interaction with the sewer drainage system. The AOFD analyses several GIS layers such as DEM, buildings, manhole location, etc. to generate a 1-D overland flow network model consisting of ponds and flow pathways. The resulting model can be further coupled with the sewer network model in order to simulate and forecast pluvial flooding (Leitão, 2009; Maksimovic *et al.*, 2009). The advantage of having a 1-D model of the overland network (instead of a 2-D model) is its short computational time, which makes it suitable for real-time forecasting of pluvial urban flooding. Analyses conducted by Leitão *et al.* (2010) justify the fact that a 1-D model produces reasonable results when compared to 2-D models, while requiring significantly less computational time.

CASE STUDY

Three events occurring since August 2010 (22–23 August 2010, 1 October 2010 and 17–18 January 2011) are selected in this analysis. The event 22–23 August 2010 was associated with a warm front and the rainfall falling within the Redbridge catchment was approximately 30 mm in around 18 h, with more than 20 mm falling in a period of 5 h. The event 1 October 2010 produced around 35 mm of rain in approximately 72 h. The event 17–18 January 2011 was associated with an occluded front passing over southeast England, producing heavy rain in the Redbridge catchment with total accumulations of around 30 mm in 24 h.

RESULTS

Precipitation forecasting

Figure 2 shows the performance of the radar-based forecasts versus lead time at different spatial and temporal scales for the event on 22–23 August 2010. Figure 2(a)–(c) show the results of the simulations performed at 1 km spatial resolution and 5 min, 10 min and 15 min temporal resolutions.

In the same figures, the performances of the forecasts are also shown when averaging the rainfall forecasts at 1 km to produce larger spatial scales (e.g. 2 km, 5 km and 10 km). As shown in these figures (Fig. 2(a)–(c)), the performance of the rainfall forecasts is very similar for the different temporal resolutions (i.e. 5 min, 10 min and 15 min). On the other hand, Fig. 2(d)–(e) show the results of the simulations performed at 2 km spatial resolution and different temporal resolutions. These results indicate that the performance of the forecasts shown in Fig. 2(d)–(e) is slightly higher than the performance of the forecasts shown in Fig. 2(a)–(c) for the same spatial scale of 2 km. Similar results are obtained for the other two events (not shown in this paper). These results have important implications for urban flood forecasting, where forecasts with small spatial scales are required. The results shown in Fig. 2 also indicate that the performance of the precipitation forecast depends on the spatial scale, with lower performance obtained at small scales but higher performance at larger scales, which is consistent with the results shown by Liguori *et al.* (2011).

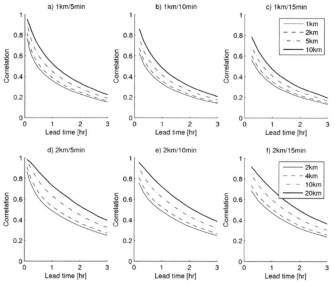

Fig. 2 Performance of the radar-based forecasts versus lead time at different spatial and temporal scales for event 22–23 August 2010.

Hydraulic modelling

Three pipes with significantly different drainage area (located in the upper, mid and lower part of the Cranbrook catchment) have been selected to demonstrate the impact of the spatial and temporal resolution of rainfall forecasts on the forecast of pluvial flooding over the Cranbrook catchment. The location of these pipes is shown in Fig. 1(b).

Figure 3 shows the performance of different flood forecasts for the event on 22–23 August 2010 against lead time for each pipe considered in the analysis. The different flood forecasts are obtained by feeding rainfall forecasts with different spatial and temporal resolution into the hydraulic model. The benchmark results are the hydraulic model results obtained using the real Nimrod data with 1 km and 5 min resolution. In the figure "*1-Relative Error*" is used as a surrogate measure of performance. The *Relative Error (RE)* is defined by:

$$1 - Relative\ Error = 1 - \left\langle \frac{\left| Y_{Nowcast} - Y_{Nimrod} \right|}{Y_{Nimrod}} \right\rangle \tag{1}$$

where <·> means the average, and the Y_{Nimrod} and $Y_{Nowcast}$ represent the flow depths in a specific pipe estimated, respectively, using the Nimrod data and Nowcasting results. When

1-Relative Error is equal to unity, it represents a perfect forecast. It is expected that this quantity will decrease with increasing forecasting lead-time. Furthermore, Fig. 4 shows the rain-rate profiles over the Cranbrook catchment on 23 August 2010 (00:00–08:00 h), as well as the water levels estimated in pipe 1455.1 with Nimrod data and with nowcasting results at different spatial and temporal resolutions.

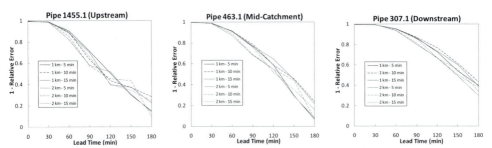

Fig. 3 Three pipes (1455.1, 463.1 and 307.1 located at the upstream, mid-catchment and downstream areas of the Cranbrook catchment, respectively) are used here to show the uncertainties in using rainfall nowcasts over different spatial and temporal scales for event 22–23 August 2010.

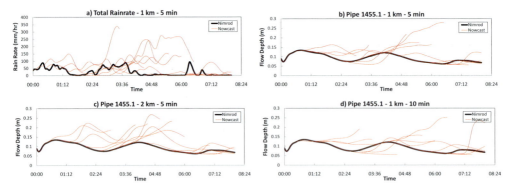

Fig. 4 Profiles of (a) the total rain-rate over the Cranbrook catchment on 23 August 2010 (00:00 h – 08:00 h) and the associated water levels of pipe 1455.1: (b) 1 km – 5 min, (c) 2 km – 5 min, and (d) 1 km – 10 min scales generated from Nimrod rainfall (dark lines) data and 3 h nowcasting results, respectively (red lines).

The results show that both the spatial and temporal resolutions of the rainfall forecast have an impact on the flood forecast. Regarding the spatial resolution, it can be seen that, in general, the flood forecasts obtained with 1 km resolution rainfall forecast are better than those obtained with the 2 km resolution rainfall forecast. However, for lead-times longer than 120 min the difference in the flood forecast obtained for the two spatial resolutions decreases, and in some cases better performance is achieved with the 2 km rainfall forecasts as input. This indicates that although 2 km nowcasts have higher correlation with Nimrod data, in this case the 1 km nowcasts, which exhibit higher spatial variability, are better inputs for the hydraulic model with lead-times up to 120 min.

In terms of temporal resolution, the results show that for lead-times up to 90 min the rainfall forecast at 5 min resolution generates the best flood forecast. After 90 min lead-time the performance of the 5 min resolution rainfall forecast decreases significantly and better flood forecasts are achieved with 10 min and 15 min rainfall forecasts.

In addition, it can be noticed that the magnitude of the impact of the spatial and temporal resolution of the rainfall forecast on the flood forecast also depends on the drainage area of the pipe or conduit that is being analysed. As shown in Fig. 3, the impact of the temporal and spatial resolution decreases as the area drained by the pipe increases. This can be explained by the fact that in a greater drainage area the differences in the spatial and temporal distribution of the rainfall are smoothed.

CONCLUSIONS

In this study, a radar-based nowcasting model coupled with an advanced 1D/1D hydraulic model is used to simulate flood forecasting over urban areas. The Cranbrook catchment in London is employed for this case study to demonstrate the feasibility of using high-resolution radar rainfall data as inputs to forecast floods. An uncertainty study has been carried out by applying various spatial (1 and 2 km) and temporal (5, 10 and 15 min) rainfall nowcasts to a hydraulic model. The results show that better performance in flood simulation is in general obtained using higher spatial resolution rainfall nowcasts as inputs. This implies that the reliability of operational urban flood models may be improved by employing finer resolution rainfall information which has a better ability to reproduce the high spatial and temporal variability of urban rainfall. The perspective research will be focused on generating street-scale rainfall nowcasts using radar-based nowcasting techniques in synergy with statistically-based downscaling methods and studying the associated uncertainties.

Acknowledgement The research was conducted as part of the Flood Risk Management Research Consortium (FRMRC2, SWP3). The first author acknowledges the full financial support of the Ministry of Education of Taiwan for the postgraduate research programme. The second author acknowledges the financial support from the Fundação para a Ciência e Tecnologia – Ministério para a Ciência, Tecnologia e Ensino Superior, Portugal [SFRH/BD/37797/2007]. The third author acknowledges the support of the Engineering and Physical Sciences Research Council (EPSRC) grant EP/I012222/1. The authors thank the Met Office for providing the radar data and the STEPS model, thank the British Atmospheric Data Centre (BADC) for providing some of the datasets, thank MWH Soft for providing the InfoWorks CS software and thank Thames Water for providing the information of the sewer network over the Cranbrook catchment.

REFERENCES

Bowler, N. E. H., Pierce, C. E. & Seed, A. W. (2004) Development of a precipitation nowcasting algorithm based upon optical flow techniques. *J. Hydrol.* 288(1–2), 74–91.

Bowler, N. E. H., Pierce, C. E. & Seed, A. W. (2006) STEPS: A probabilistic precipitation forecasting scheme which merges an extrapolation nowcast with downscaled NWP. *Quart. J. Roy. Met. Soc.* 132(620), 2127–2155.

Einfalt, T., Arnbjerg-Nielsen, K., Golz, C., Jensen, N.-E., Quirmbach, M., Vaes, G. & Vieux, B. (2004) Towards a roadmap for use of radar rainfall data in urban drainage. *J. Hydrol.* 299(3–4), 186–202.

Environment Agency (2006) River Roding flood risk management strategy. *Strategic Environmental Assessment Environmental Report, Non technical summary*.

Harrison, D. L., Driscoll, S. J. & Kitchen, M. (2000). Improving precipitation estimates from weather radar using quality control and correction techniques. *Meteorol. Appl.* 7(2), 135–144.

Krämer, S., Fuchs, L., Verworn, H. (2007) Aspects of radar rainfall forecasts and their effectiveness for real time control – the example of the sewer system of the city of Vienna. *Water Practice & Technology* 2(2), 0042-0049.

Leitão, J. P. (2009) Enhancement of digital elevation models and overland flow path deliniation methods for advanced urban flood modelling. PhD Thesis, Imperial College London, London, UK.

Leitão, J. P., Simões, N. E., Ferreira, F., Prodanović, D., Maksimović, Č., Matos, J. & Sá Marques A. (2010) Real-time forecasting urban drainage models: full or simplified networks? *Water Sci. Technol.* 62(9), 2106–2114.

Liguori, S., Rico-Ramirez, M. A., Schellart, A. N. A. & Saul, A. J. (2012) Using probabilistic radar rainfall nowcasts and NWP forecasts for flow prediction in urban catchments. *Atmos. Res.* 103, 80-95. DOI: 10.1016/j.atmosres.2011.05.004.

Maksimović, Č., Prodanović, D., Boonya-aroonnet, S., Leitão, J. P., Djordjević, S. & Allitt, R. (2009) Overland flow and pathway analysis for modelling of urban pluvial flooding. *J. Hydraul. Res.* 47(4), 512–523.

Pitt, M. (2008) The Pitt review: learning lessons from the 2007 floods. *Technical Report*, UK Cabinet Office, London, UK.

Rico-Ramirez, M. A., Cluckie, I. D., Shepherd, G. & Pallot, A. (2007) A high-resolution radar experiment on the island of Jersey. *Meteorol. Appl.* 14(2), 117–129.

Seed, A. W. (2003) A dynamic and spatial scaling approach to advection forecasting. *J. Appl. Met.* 42(3), 381–388.

Simões, N. E., Leitão, J. P., Maksimović, Č., Sá Marques, A. & Pina, R. (2010) Sensitivity analysis of surface runoff generation in urban floods forecasting. *Water Sci. Technol.* 61(10), 2595–2601.

A new FEH rainfall depth-duration-frequency model for hydrological applications

ELIZABETH J. STEWART, DAVID G. MORRIS, DAVID A. JONES &
CECILIA SVENSSON

Centre for Ecology & Hydrology, Wallingford, Oxfordshire OX10 8BB, UK
ejs@ceh.ac.uk

Abstract Recent research funded by the Joint Environment Agency/Defra Flood and Coastal Risk
Management R&D Programme has developed a new statistical model of point rainfall depth-duration-
frequency (DDF) for the UK. The analysis made use of an extensive set of annual maximum rainfall depths
for daily and recording raingauges across the UK. The new model will eventually replace the *Flood
Estimation Handbook* (FEH) rainfall DDF model to provide estimates of rainfall depth for storm durations
ranging from under 1 h to 8 days and return periods from 2 years to >10 000 years. The paper reports on
current progress to generalise the new model so that it can be applied at any point, catchment or user-defined
area, and potential links between the new model and hydrological applications of weather radar are
highlighted.

Key words rainfall; depth-duration-frequency; Flood Estimation Handbook; radar rainfall; urban drainage

INTRODUCTION

Recent research funded by the Joint Environment Agency/Defra Flood and Coastal Risk
Management R&D Programme has developed a new statistical model of point rainfall depth-
duration-frequency (DDF) for the UK (Stewart *et al.*, 2010a). The model was developed for
rainfall durations from 1 hour to 8 days. Although it was originally envisaged that it would be
applicable primarily to the long return periods which are typically used in hydrological analyses
for reservoir flood risk assessment, the new model has been developed to provide estimates of
rainfall frequency for a wide range of return periods from 2 to >10 000 years. Therefore, it is
proposed that the new DDF model should eventually replace that published in Volume 2 of the
Flood Estimation Handbook (FEH) (Faulkner, 1999).

This paper describes the main results obtained from applying the new FEH DDF model at
over 70 sites throughout the UK, and discusses recent progress in generalising the model to
provide both gridded point and catchment average rainfall frequency estimates. Although work is
ongoing to develop a new rainfall model utility which will be delivered within the next version of
the FEH CD-ROM (CEH, 2009), other ways of providing access to the new rainfall frequency
estimates are currently being explored, which include replacing the use of the existing FEH model
within the Hyrad system (Moore *et al.*, 2005).

APPLICATIONS OF THE EXISTING FEH DDF MODEL

Rainfall frequency estimates from the existing FEH DDF model are used in various approaches to
hydrological design studies using rainfall–runoff techniques, for example in application of the ReFH
design methodology (Kjeldsen, 2007), and for assessing the rarity of particular rainfall events in the
UK, and it is the latter application that is most relevant to the hydrological use of weather radar.
Until recently, most notable rainfall events have been measured by individual raingauges or gauge
networks, but increasingly weather radar is capturing information about the spatial and temporal
characteristics of extreme storms, for example the Boscastle event of 2004 (Fenn *et al.*, 2005). The
rarity of individual radar-derived catchment average rainfall estimates can be assessed directly using
the stand-alone implementation of the FEH DDF model on the FEH CD-ROM 3 (CEH, 2009), but
the software does not allow estimates to be derived off-line or for user-defined areas rather than river
catchments. Cole *et al.* (2011) describe a recent development of the Hyrad Weather Radar System
(Moore *et al.*, 2005) that utilises the existing FEH DDF model to allow post-event analysis of storm

events captured by weather radar for Scottish Water. Following flood events, the estimates are used to monitor whether urban drainage systems performed within design specifications or if remedial action is required to comply with the regulatory framework.

DEVELOPMENT OF THE NEW DDF MODEL

Motivation

The FEH (Institute of Hydrology, 1999) introduced a new set of procedures for the estimation of rainfall and flood frequency in the UK. For rainfall frequency, it superseded the previous UK design standard of the Flood Studies Report (FSR) (NERC, 1975). Both methods derive the rainfall frequency curve by multiplying a (local) index variable by a regionally derived growth curve. However, while the FSR used just two regions for the UK, the FEH pools data from circular regions centred on the point of interest, resulting in growth curves that vary relatively smoothly in space. A further innovation of the FEH was to employ annual values of the largest rainfall observed within a region, together with a spatial dependence model, to estimate the regional growth curve at higher return periods than would be available from single gauge records.

The FEH procedures and subsequent updates and refinements have been widely adopted for flood risk management. However, the FEH rainfall DDF model was developed for return periods of up to 2000 years and, especially amongst reservoir engineers, concern has been voiced about the results it produces for the very long return periods used in reservoir flood safety assessment. Babtie Group (2000) compared the FEH DDF model with the FSR, and found that the FEH often gave significantly higher rainfall depth estimates at the highest return periods. MacDonald & Scott (2001) found that in some cases the FEH 10 000-year return period rainfall exceeded the estimate of probable maximum precipitation (PMP) derived from the FSR. On the basis of these concerns, the project was commissioned to consider some aspects of the FEH DDF model at return periods above 100 years, although the final analysis has considered the entire range of return periods from 2 years upwards.

Key components of the new model

The new DDF model was developed using an extensive dataset of annual maximum rainfall depths from raingauges throughout the UK. The possibility of making use of archives of radar data was explored in the early stages of the project, but the task of reconciling raingauge and radar estimates was considered to be too complex to be practicable. However, radar data were utilised in the examination of selected extreme events that were used to validate the final results of the DDF analysis (Dempsey & Dent, 2009).

The development of the new model was based on an extensive statistical analysis of the annual maximum dataset. The basic approach taken mirrored that of the FEH rainfall analysis, but with key revisions: (i) the standardisation of the rainfall maxima is now more complex, making the rainfalls at the different sites more similar prior to data pooling; (ii) the model of spatial dependence now allows the dependence to decrease with increasing return period, rather than being the same for both large and small events; and (iii) changes were made to the pooling methodology to overcome anomalous behaviour observed across a wide range of test cases. In addition, the new model makes use of an extended dataset.

Rainfall frequency curves were produced by the revised methodology for durations from 1 hour to 8 days at 71 test sites and a new DDF model was then fitted to the results. Full details of the analysis are given by Stewart *et al.* (2010a).

Model results

Rainfall frequency estimates from the new DDF model were compared with those derived from the published FEH model at 71 sites throughout the UK. The test sites were selected primarily on the basis of raingauge record length and/or proximity to large reservoirs, and also to give good coverage of the UK. Estimates were derived for 11 key durations from 1 to 192 hours (i.e. 8 days) and for return periods from 100 to 10 000 years. To illustrate the results, Fig. 1(a) shows a map comparing estimates from the new model with the FEH model for a 2-h duration and a return period of 100 years. It can be seen that the estimates from the new model are generally similar to or slightly lower than those from the FEH for most of England, Wales and Northern Ireland. However, in northwest England, at one site in Northern Ireland and for much of Scotland, the new rainfall estimates exceed those from the FEH. This is mainly due to the new, larger dataset available, which comes from a denser network.

From the comparisons for all durations and return periods studied, several notable features emerge. Firstly, the estimates from the new model are higher over most of Scotland at the shortest durations (<6 h). Secondly, the estimates from the new model tend to be lower than the FEH at higher return periods (>200 years) and this is thought to be due mainly to the improved model of spatial dependence (see Fig. 1(b) for an example). At extremely high return periods, estimated rainfalls from the new DDF model are often considerably lower than the FEH model because the extrapolation of the new model is an approximate straight line on the Gumbel scale whereas the FEH model curves upwards (an exponential extrapolation). Finally, whilst FEH 10 000-year rainfall estimates commonly exceeded FSR PMP, this is rarely the case with estimates from the new model.

Areal rainfall frequency

The results discussed thus far were derived by applying the new DDF model at individual sites. These can be used to estimate the rarity of particular rainfall events, usually measured at a single

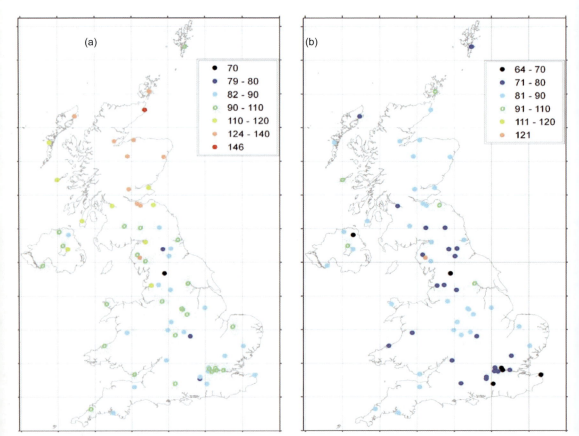

Fig. 1 New rainfall estimates as a percentage of estimates from the FEH model: (a) duration 2 h and return period 100 years, (b) duration 24 h and return period 1000 years.

raingauge. However, many applications require estimates of areal average rainfall frequency over river catchments or storm sewer networks as inputs to hydrological models, or for an ungauged site, and for this reason work is now under way to estimate the parameters of the new DDF model on a gridded basis throughout the UK. As was the case with the FEH model, construction of the new DDF model requires the prior estimation for each of the key durations of the variable termed RMED, the median annual maximum rainfall, over a 1-km grid of the UK. A new methodology for estimating this is being adopted.

In a pilot study, the new DDF model has been applied to two catchments in west Cumbria, the Derwent at Camerton (area 661.9 km^2) and the Leven at Newby Bridge (area 247.8 km^2), to estimate the frequency of the extreme event that caused widespread flooding in November 2009 (Stewart *et al.*, 2011). The variable *RMED* for the key durations was mapped using a new interpolation method that incorporates gridded (1-km resolution) values of the standard average annual rainfall (*SAAR*) as a predictor, as well as the *RMED* calculated from observations at gauged sites. The new DDF model was fitted at every point on a 1-km grid covering the two catchments, and used to estimate rainfall depths for the key durations and for return periods from 2 to 10 000 years at every grid point. To estimate the rarity of the catchment *areal* rainfall, a catchment-representative *point* rainfall of a particular duration and return period was first derived for each catchment: for each combination of duration and return period, the point rainfall depths were averaged across each catchment. These catchment-representative point rainfalls were then multiplied by the areal reduction factors presented in the FEH (Keers & Wescott, 1977) to give the catchment average rainfall of the appropriate return period and duration.

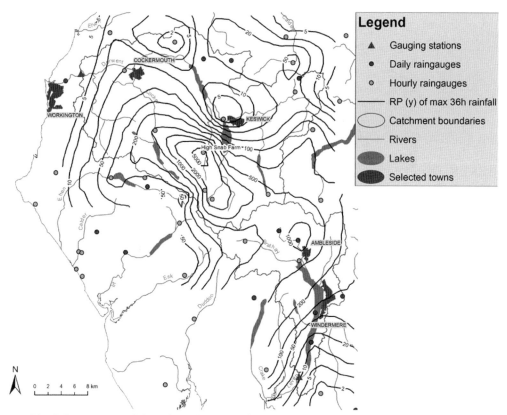

Fig. 2 Contour map of the estimated return period of the maximum 36-h rainfall depth recorded over west Cumbria during the event in November 2009.

The most extreme (rarest) rainfall recorded at raingauges at Seathwaite Farm and Honister Pass during the November 2009 event occurred over a 36-h period (Sibley, 2010). Figure 2 illustrates the estimated return period from the new model of the maximum 36-h rainfall recorded over the two catchments. The contours indicate that the highest return periods occur in the vicinity of the High Snab Farm raingauge, just to the north of the raingauges that recorded the highest rainfall depths, and show how extreme the event was over the Derwent catchment. The maximum 36-h rainfall over the Derwent catchment during the event was estimated at 155.7 mm from the available raingauges, and the associated return period for the catchment rainfall was assessed at 193 years by the new model.

Figure 3 shows a comparison of the frequency curves for catchment average rainfall for one of the catchments derived from the new DDF model and from the FEH model for the 36-h duration. It can be seen that the new model gives higher rainfall estimates than the FEH for return periods between 2 and 50 years, and lower estimates for return periods in excess of 50 years.

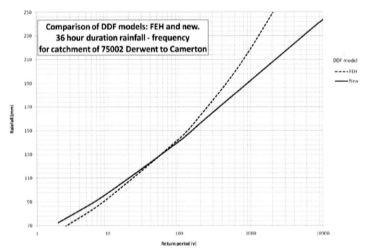

Fig. 3 Comparison of rainfall frequency curves derived from the new and FEH rainfall DDF models for the catchment of the Derwent to Camerton for a duration of 36 h.

NEXT STEPS

Work is continuing to explore the behaviour of the new rainfall DDF model in different parts of the UK and to compare the results with the FEH for the full range of durations and return periods to which the model is applicable. Until now, the focus of the research has been on return periods of over 100 years, which are relevant to fluvial flood risk management, but further exploration of the model results for shorter return periods is planned, particularly in southeast England where rainfall estimates from the FEH model for shorter durations have been questioned.

Other aspects of the model are also being considered, including the spatial resolution of the outputs. It is likely that a 1-km grid will be used as it was in the FEH rainfall model implementation, although the possibility of using a finer grid spacing in upland areas such as the Lake District will be evaluated.

Finally, as in the FEH and FSR models, areal reduction factors (ARFs) are used to convert point rainfall estimates to areal estimates. There is a general need to update the values of ARF since the current methodology dates back to 1975 and takes no account of possible variation with return period or geographical location.

Future applications of the new model

For hydrological design studies using rainfall–runoff modelling, the new DDF model will be incorporated into a revised software utility which will be released on an upgrade to the FEH CD-

ROM in the near future. Applications such as urban drainage design generally require rainfall frequency estimates for shorter durations and, although the finest temporal resolution of the new DDF model is currently 1 h, reflecting the data that were used in the analysis, further work will look at the feasibility of extrapolating to sub-hourly durations. This will allow the new model to be incorporated into an upgrade of the Hyrad system to allow post-event analysis of rainfall events identified by weather radar in urban areas. In the longer term, it would be preferable to develop a model specifically for shorter durations and return periods should the necessary data be available from recording raingauges and possibly weather radar.

CONCLUSIONS

The development of a new rainfall depth-duration-frequency (DDF) model has been described and examples of its application at individual points and over catchment areas have been presented. For other applications such as urban drainage modelling and compliance monitoring using weather radar, new software solutions are being explored. Possible future delivery mechanisms include the development of a web service to "plug in" to other software systems to provide rainfall frequency estimates for user-defined points or areas, or a more interactive "pay-per-view" web system.

Acknowledgements Development of the new rainfall DDF model was funded by Defra under the Joint Environment Agency/Defra Flood and Coastal Risk Management R&D Programme (contract WS 194/2/39). Rainfall data for the model development were supplied by the Environment Agency, SEPA, the Met Office and Met Éireann. Data for the study of the rainfall event in Cumbria in November 2009 were supplied by the Environment Agency. The assistance of Harry Gibson (CEH) in analysis and mapping is gratefully acknowledged.

REFERENCES

Babtie Group (2000) Reservoir safety – floods and reservoir safety: clarification on the use of FEH and FSR design rainfalls. Final report to DETR. Babtie Group, Glasgow, UK, 36 pp.

Centre for Ecology & Hydrology (2009) FEH CD-ROM 3. CEH Wallingford, UK.

Cole, S. J., McBennett, D., Black, K. B. & Moore, R. J. (2011) Use of weather radar by the water industry in Scotland. In: *Weather Radar and Hydrology* (Proc. Symposium, Exeter, UK, April 2011). IAHS Publ. 351. IAHS Press, Wallingford, UK (this volume).

Dempsey, P. & Dent, J. (2009) Report on the extreme rainfall event database. Report no. 8 to Defra, Contract WS 194/2/39 Met Office, Exeter, UK, 87 pp.

Faulkner, D. S. (1999) *Rainfall Frequency Estimation*. Flood Estimation Handbook, Volume 2. Institute of Hydrology, Wallingford, UK.

Fenn, C.R., Bettess, R., Golding, B., Farquharson, F.A. & Wood, T. (2005) The Boscastle flood of 16 August 2004: Characteristics, causes and consequences. In: *40th Defra Flood and Coastal Management Conference* (5–7 July 2005, York, UK).

Institute of Hydrology (1999) *Flood Estimation Handbook* (five volumes). Institute of Hydrology, Wallingford, UK.

Keers, J. F. & Wescott, P. (1977) A computer-based model for design rainfall in the United Kingdom. Met Office Scientific Paper no. 36, HMSO, London, UK.

Kjeldsen, T. R. (2007) The revitalised FSR/FEH rainfall–runoff method. FEH Supplementary Report no. 1, Centre for Ecology & Hydrology, Wallingford, UK.

MacDonald, D. E. & Scott, C. W. (2001) FEH vs FSR rainfall estimates: an explanation for the discrepancies identified for very rare events. *Dams & Reservoirs* 11(2), 28–31.

Moore, R. J., Bell, V. A. & Jones, D. A. (2005) Forecasting for flood warning. *C. R. Geosciences* 337(1–2), 203–217.

NERC (1975) *Flood Studies Report* (five volumes). Natural Environment Research Council, London, UK.

Sibley, A. (2010) Analysis of extreme rainfall and flooding in Cumbria 18–20 November 2009. *Weather* 65(11), 287–292.

Stewart, E. J., Jones, D. A., Svensson, C., Morris, D. G., Dempsey, P., Dent, J. E., Collier, C. G. & Anderson, C. W. (2010a) *Reservoir Safety – Long Return Period Rainfall* (two volumes). R&D Technical Report WS 194/2/39/TR, Joint Defra/EA Flood and Coastal Erosion Risk Management R&D Programme.

Stewart, E. J., Morris, D. G., Jones, D. A. & Spencer, P. S. (2010b) Extreme rainfall in Cumbria, November 2009 – an assessment of storm rarity. In: *Proc. BHS Third International Symposium, Managing Consequences of a Changing Global Environment* (Newcastle, UK).

Stewart, E. J., Morris, D. G., Jones, D. A. & Gibson, H. S. (2011) Frequency analysis of extreme rainfall in Cumbria 16–20 November 2009. *Hydrology Res.* (submitted).

Use of weather radar by the water industry in Scotland

STEVEN J. COLE[1], DOMINIC MCBENNETT[2], KEVIN B. BLACK[1] & ROBERT J. MOORE[1]

1 *Centre for Ecology & Hydrology, Wallingford OX10 8BB, UK*
  scole@ceh.ac.uk
2 *Scottish Water, Invergowrie, Dundee DD2 5BB, UK*

Abstract Rainfall data are a key source of information used by the UK water industry to perform its diverse regulatory functions. Raingauges have traditionally been used, but radar rainfall data are increasingly being utilised. Within Scotland, the public body Scottish Water has the responsibility for supplying drinking water and the collection and treatment of wastewater. An outline of Scottish Water's requirements and use of weather radar data is presented along with a brief description of the Hyrad Weather Radar System. A case study illustrates a novel method for post-event analyses of storm events associated with surface water flooding incidents. These analyses combine the analytical capabilities of Hyrad with the Flood Estimation Handbook depth-duration-frequency rainfall model to obtain estimates of rainfall return periods. The estimates are used to assess whether urban drainage systems performed within design specifications or if remedial action is required to comply with the regulatory framework. Finally a look forward is given of future planned applications of weather radar within the water industry in Scotland.

Key words pluvial; flood; radar rainfall; rainfall return period; FEH; urban drainage

INTRODUCTION

Rainfall data are a key source of information used by the UK water industry to perform its diverse regulatory functions. Raingauges have traditionally been used for this purpose, but national networks of permanent gauges rarely meet the water industry's spatio-temporal user requirement for rainfall data (Han, 2009), particularly in urban areas. In addition, the number of raingauges in these networks is generally in decline (Eden, 2009) and only a subset of them may be accessible to the water industry.

UK water utility companies may operate their own raingauge networks and can temporarily deploy local dense networks for particular activities, including for flow surveys and research studies such as the City RainNet project (Collier *et al.*, 2010). Taken on their own, the water company networks do not permanently meet their user requirement for spatio-temporal rainfall estimation, tend to be expensive to deploy and/or maintain, and are difficult to site in urban areas. In part, this has led to radar rainfall data increasingly being used by water utilities as an additional source of rainfall information. Another major attraction is the national coverage of the UK radar network (Harrison *et al.*, 2009) which provides real-time, frequent (5 min) and high-resolution (1 km) estimates of rainfall and meets many of the water industry's spatio-temporal user requirements for rainfall data.

REQUIREMENTS FOR USING WEATHER RADAR WITHIN SCOTTISH WATER

Within Scotland, the public body Scottish Water has the responsibility for supplying drinking water and the collection and treatment of wastewater. A recent review of the Scottish raingauge network (McGregor & MacDougall, 2009) recommended that it should be expanded in urban, remote and upland areas. It highlighted that the density of tipping-bucket raingauges (TBRs) needed to measure sub-daily rainfall is generally low and did not meet the user requirement, particularly within Glasgow, Dundee and Aberdeen. In 2007, the construction of a large wind farm resulted in an existing radar at Corse Hill being replaced by two new C-band radars at Munduff Hill and Holehead (Fig. 1). The improved radar coverage over the Midland Valley of Scotland, encompassing the major cities of Glasgow and Edinburgh, presented an important opportunity for Scottish Water to consider the operational use of radar rainfall data (McLachlan *et al.*, 2008).

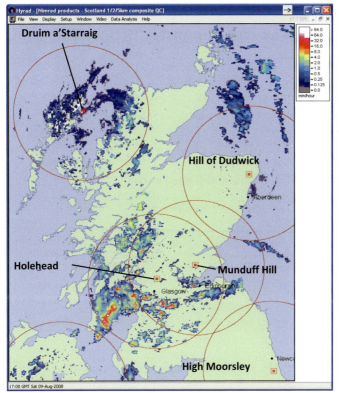

Fig. 1 Hyrad display of the UK composite radar data over Scotland. Also shown are the radar sites (red squares) which provide coverage over Scotland, along with their respective 100 km radar range circles.

The following issues were considered:
- Could radar data be used in rainfall return period analyses for sewer flood compliance monitoring?
- What other radar-based applications could support business processes?
- What platform and infrastructure is required to retrieve, store and analyse radar data?

A particular requirement was the post-event analysis of storm events associated with surface water (pluvial) flooding incidents, including reliable estimates of rainfall rarity or return period. Such estimates can then be used to assess whether urban drainage systems performed within design specifications if remedial action is required to comply with the regulatory framework. An ability to perform these post-event analyses "in-house" was desirable as previously such reports were obtained from outside organisations, incurring expense and delays of several weeks.

During this process a wide range of other potential weather radar applications and spatio-temporal rainfall requirements have been identified, here summarised in Table 1. Whilst there are many applications that can take advantage of historical records of radar rainfall, such as hydraulic model verification, there are also several real-time or near real-time applications (e.g. predictive maintenance) that require appropriate automated systems to be in place. The review concluded that radar data should be used, in conjunction with raingauge data from all possible sources, and in 2009 resulted in Scottish Water commissioning the Hyrad Weather Radar System.

THE HYRAD WEATHER RADAR SYSTEM

Spatial rainfall information from weather radar and/or raingauge networks can be of limited use for routine and real-time water management applications unless tools for automated reception,

Table 1 Summary of current and planned (grey) use of radar rainfall data by Scottish Water.

Team	Use	Resolution Time resolution	Space resolution	Reporting frequency
Asset Intelligence	Capital scheme review - assessment of rainfall events relative to design standards	5 min	1 km	Sporadic
Asset Intelligence	Hydraulic models (model verification)	5 min	1 km	Sporadic
Flood Investigation	Event analysis – real-time and historical	5 min	1 km to catchment	Daily
Pollution Incident	Event analysis – real-time and historical	5 min	1 km	Sporadic
Environmental	Event analysis for bathing water quality	5 min	Catchment	Sporadic
Distribution, Operation and Maintenance Strategy	Network investigation - linking rainfall to major pipe bursts	Daily	1 km	Sporadic
Leakage Planning	Consumption analysis - relationship between rainfall and water use	Daily	1 km	Monthly
Regulatory Reporting	Flow return for regulatory body (WICS)	Daily	Catchment	6 monthly
Annual Return Reporting	Yearly rainfall by region & wastewater treatment work areas for SEPA	Daily	1 km	Yearly
Water Resources	Water resource planning	Daily	Catchment	Monthly
Predictive Maintenance	Use of rainfall for predictive maintenance	Daily	1 km	Daily
Control Centre	Forecast data – extreme weather	5/15 min	1/5 km	24/48 h

processing and visualisation are available. The Hyrad Weather Radar System (Moore *et al.*, 2005) meets this need by supporting real-time receipt of radar rainfall and other hydro-meteorological products. Hyrad is the standard system for weather radar display, processing and analysis used by government agencies in support of flood warning across England, Wales, Scotland and Belgium.

A schematic of the Hyrad client-server system and its functionality is summarised in Fig. 2. It incorporates a hydrological processing kernel which improves rainfall estimation and forecasting by merging raingauge and radar data and removing anomalies. The analytical capabilities of Hyrad can be used either interactively through the display client or via an automated interface to flow forecasting and modelling systems (e.g. Delft-FEWS in the UK and FloodWorks in Belgium).

Significant recent developments include the "Statistics Analysis" module within the display client and the "Merging Tool" server utility. The statistics module allows a user to interactively interrogate a sequence of spatial data and derive a statistical summary for a selected rectangular region, a set of points or a catchment. This is particularly useful for post-event analysis as it can identify the maximum pixel rainfall total and location within a rainfall event and/or catchment which can then be used for deriving rainfall return period estimates.

The Merging Tool can merge different sources of actual and forecast data, allow for different spatio-temporal resolutions, convert between projections, create averages or totals (e.g. hourly or daily), cut out smaller regions and join images together. Of particular interest for the water industry are automatically derived products that support daily monitoring and reporting of storm events. These make everyday analytical work more efficient by avoiding the need for users to repeatedly form daily products from the base data (usually available at a 5- or 15-min interval).

RETURN PERIOD ANALYSIS METHODOLOGY USING RADAR DATA

The return period analyses use the industry standard Flood Estimation Handbook (FEH) CD-ROM (Centre for Ecology & Hydrology, 2009) depth-duration-frequency rainfall model (Faulkner, 1999) to obtain estimates of rainfall return periods. For a given event, time-series of radar rainfall data are extracted via Hyrad for locations and catchments of interest at the base temporal resolution (5 min here). Rainfall accumulations of these time-series are made using rolling windows of varying length (e.g. 30 min, 1, 6 or 24 h) and the maximum accumulation extracted for each window length. Different durations are required to identify the "maximum" return period of the rainfall event as the return period varies as a function of both duration and depth. When radar rainfall totals for an individual 1 km^2 radar pixel are considered, an appropriate Areal Reduction Factor is inversely applied to obtain the equivalent point rainfall for use with FEH.

Furthermore, when attempting to derive an estimate of the rainfall return period at a given point (e.g. a pluvial flooding location), all radar pixels within a given distance of the point are considered (e.g. a 3 × 3 km box). This attempts to mitigate the impact of the positional uncertainties of the radar data (e.g. wind drift and antenna pointing accuracy (Harrison *et al.*, 2009)), issues that may affect an individual pixel (e.g. clutter) and the fact that the rain causing a pluvial flood may have fallen a small distance away from the actual flood location. For pluvial events in larger catchments, the maximum pixel rainfall within the catchment and the catchment average rainfall are also considered.

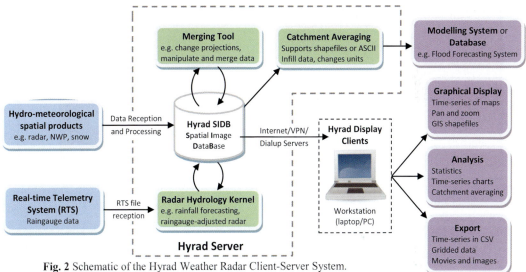

Fig. 2 Schematic of the Hyrad Weather Radar Client-Server System.

CASE STUDY: POST-EVENT RETURN PERIOD ANALYSIS

At about 09:30 GMT 14 August 2008, localised showers began to develop south of Glasgow. Some individual storm cells were very intense and did not move very quickly or very far but the region of storm initiation spread northwards over Glasgow. Figure 3 shows the daily radar rainfall total for 14 August and reveals the high spatial variability of the rainfall totals due to the passage of individual showers. Subsequently, several reports of pluvial flooding were received with the most serious incidents at locations 1 and 6 in Fig. 3 which are associated with areas of relatively high rainfall totals.

Hyetographs for the pixels with the largest rainfall totals within a 3 × 3 km box centred on locations 1 and 6 are presented in Fig. 4: these show that most of the rain fell in a short period of time and, for location 1, was very intense at times. Due to the nature of the event, rolling windows of 3, 2, 1 and 0.5 h were deemed appropriate for calculating the maximum rainfall totals and these are presented in Table 2 along with their associated FEH return periods.

According to this radar data analysis, the location that experienced the most extreme rainfall was location 1 with the four nearest radar grid-cells having daily totals ranging between 39.2 mm and 43.1 mm. This was mostly due to a particularly intense and localised shower which passed over between 13:30 and 14:30 GMT; see Fig. 4. The resulting short duration totals had very high return periods with 132 years estimated for the 30 min total of 28.5 mm.

In this case, had an asset designed to withstand a 1 in 20 year, 1-h rainfall event failed at location 1, the analysis would have given strong evidence to the regulator that the failure was due to extreme rainfall beyond the design specification of the asset. In general, the preference would be to include raingauge data but none were available for this case study. However, the study still serves to highlight the potential utility of radar data for return period analysis, especially for localised events which are a challenge for conventional permanent raingauge networks to capture.

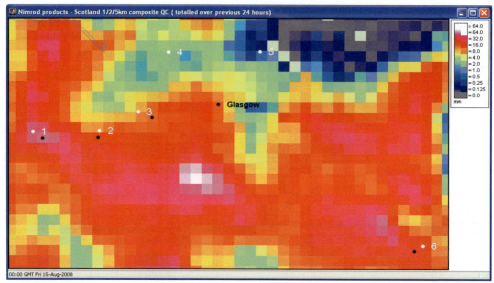

Fig. 3 24-h 1 km radar rainfall accumulations for the period ending 00:00 GMT 15 August 2008. Black circles are the "max" pixels identified within a 3 × 3 km box for locations 1, 2, 3 and 6.

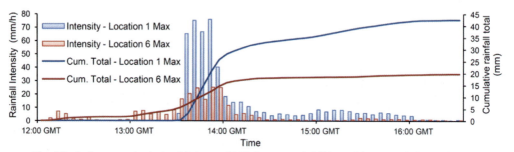

Fig. 4 Radar hyetographs during 14 August 2008. Bars are rainfall intensities at 5-min intervals.

Table 2 Radar rainfall return period (peak-over-threshold) analyses during 14 August 2008 for locations 1 and 6.

Duration	Location 1 Max (242500 661500)			Location 6 Max (276500 650500)		
	End time	Rainfall	Return period	End time	Rainfall	Return period
30 min	14:00 GMT	28.5 mm	132 years	13:55 h GMT	10.6 mm	5.9 years
1 h	14:25 h GMT	33.3 mm	86 years	14:00 h GMT	14.3 mm	7.3 years
2 h	15:30 h GMT	39.4 mm	58 years	14:05 h GMT	16.8 mm	5.3 years
3 h	16:25 h GMT	42.6 mm	41 years	15:00 h GMT	18.5 mm	4.2 years

A LOOK FORWARD

Over the last few years, Scottish Water have moved to using radar data routinely in several aspects of their business and plan to apply it more widely, as indicated in Table 1. The Hyrad Weather Radar System has played a key role in this uptake and an active development plan is in place to refine existing applications and to facilitate new ones (e.g. interfacing to IBM SPSS to run predictive maintenance models in real-time). In particular, automation of the FEH return period analysis of radar rainfalls is planned for both the Hyrad server and display client. This will save significant time and remove many of the manual steps that can introduce human-errors.

Another area of active operational research is how to combine radar and raingauge data. One method being considered is the multiquadric surface fitting scheme that forms part of the Hyrad Hydrology Kernel: this has been shown to have benefits for hydrological modelling when compared to using radar rainfall alone (Cole & Moore, 2008). The Probability Matching Method is another approach being considered under the City RainNet project (Collier *et al.*, 2010).

Scottish Water is also planning significant investment in their observing network, in stark contrast to recent reductions in their raingauge network. Having already invested in a dense raingauge network for the City RainNet project, there are plans to strategically increase their permanent network, possibly use "fill-in" X-band radars and to exchange and share rainfall data with others such as the Scottish Environment Protection Agency (SEPA). The combination of investing in observations, systems and operational research means that the use of radar rainfall by the water industry has a bright future in Scotland.

REFERENCES

Centre for Ecology & Hydrology (2009) *FEH (Flood Estimation Handbook) CD-ROM 3*. Centre for Ecology & Hydrology, Wallingford, UK.

Cole, S. J. & Moore, R. J. (2008) Hydrological modelling using raingauge-and radar-based estimators of areal rainfall. *J. Hydrol.* 358(3–4), 159–181.

Collier, C. G., Hawnt, R. & Powell, J. (2010) Real-time adjustment of radar data for water management systems using a PDF technique: The City RainNet Project. In: *Proc. 6th European Conf. on Radar in Meteorology and Hydrology: Adv. in Radar Technology, Sibiu, Romania*, 6 pp.

Eden, P. (2009) Traditional weather observing in the UK: An historical overview. *Weather* 64(9), 239–245.

Faulkner, D. S. (1999) *Rainfall frequency estimation. Flood Estimation Handbook, Volume 2*. Institute of Hydrology, Wallingford, UK.

Han, D. (2009) Editorial: Weather radar for water management. *Water Management* 162(2), 63–64.

Harrison, D. L., Kitchen, M. & Scovell, R. W. (2009) High-resolution precipitation estimates for hydrological uses. *Water Management* 162(2), 125–135.

McLachlan, I., Lang, I. & McBennett, D. (2008) The use of radar rainfall data in Scotland. *WaPUG Autumn Conference 2008*, Paper 14, 11 pp.

McGregor, P. & MacDougall, K. (2009) A review of the Scottish rain-gauge network. *Water Management* 162(2), 137–146.

Moore, R. J. (1999) Real-time flood forecasting systems: perspectives and prospects. *Floods and landslides: integrated risk assessment* (ed. by R. Casale & C. Margottini), Chap. 11, 147–189. Springer, Berlin, Germany.

Moore, R. J., Bell, V. A. & Jones, D. A. (2005) Forecasting for flood warning. *C. R. Geosciences* 337(1–2), 203–217.

Impact of Z-R relationship on flow estimates in central São Paulo

ROBERTO V. CALHEIROS & ANA M. GOMES

Instituto de Pesquisas Meteorológicas – UNESP, Brazil
calheiros@ipmet.unesp.br

Abstract Mean areal radar rainfall over catchments in the State of São Paulo is an operational product under development by the Meteorological Research Institute – IPMet. A pilot project is being carried out which focuses on the important Corumbataí River basin, under surveillance by the IPMet-operated Bauru radar. Previous work on the project explored the relative impact of factors like time resolution of radar data and reflectivity to rain-rate conversion relationships, when the relevance of the latter was verified. This paper deals with the stratification of those relationships by daily intervals and its impact on flow estimates. Daily values of radar mean rainfall using gauges and different conversion relationships are plotted against the corresponding flow at the basin outlet. Flow estimates derived by applying the rainfall from the different relationships to a previously obtained rainfall–runoff curve for the basin is compared to the historical hydrograph. Preliminary results suggest stratification has hydrological significance.

Key words areal radar rainfall; Z-R relationship; Corumbataí River basin; rainfall–runoff relationship

INTRODUCTION

The Meteorological Research Institute – IPMet/UNESP started radar operations back in 1974, when a non-coherent C-band system was installed at Bauru, in the central region of the State of São Paulo, Brazil. Hydrological applications were one of the major operational objectives of the radar project (Calheiros, 1982) and were carried out systematically during the 1980s when radar observations were used in synergy with hydrological prediction models, pioneering radar-based flow forecasts in the region.

Radar facilities operated by IPMet evolved to the present two-radar set of S-band Doppler radars, providing quantitative coverage for a substantial area of the State, and much enhancing the potential for hydrological use. Basic to this use is the mean areal radar rainfall product, whose development is presently underway. A pilot project was deployed that focuses on the Corumbataí River basin, an important catchment situated at an approximate 140 km range of the Bauru (BRU) radar. The first efforts concentrated on the characterization of the radar rainfall field (Calheiros *et al*, 2008) and in the selection of *Z (radar reflectivity)-R (rain-rate)* relationships (Calheiros & Gomes, 2010).

This paper deals with a refinement of the conversion relationship represented by its stratification as a function of daily intervals. Relationships were derived for different daily intervals based on the evolution of the hourly rainfall along the day for the Bauru radar coverage area. Series of daily mean areal radar rainfall were then derived with different relationships, e.g. conventional, single locally-adjusted (Calheiros & Gomes, 2010) and stratified composite. Flow estimates were obtained inputting the rainfall, derived in one case from the single locally-adjusted relationship and, in the other, from the stratified composite relationship, to a rainfall–runoff curve previously obtained (Calheiros *et al*, 2008). Resulting flow estimates were then compared to the corresponding historical hydrograph.

METHODOLOGY AND DATA

The Corumbataí basin selected for the pilot project covers an approximate area of 1600 km^2 and is situated about 140 km from BRU, where distance effects are still less pronounced. In fact, one study in progress based on BRU accumulated precipitation for one full rainy season indicates range effects of 2.0–2.5 dB at ~140 km range. Yet, in another work Calheiros & D'Oliveira (2007) show an average range impact of about 3.7 dB at that range, but referring only to peak reflectivity.

Due to its importance, this basin has been the subject of many studies, including that of Rennó (2004) in which it was used to test a new system of analysis of hydrological simulation, called SASHI. Figure 1 depicts the basin boundaries and the raingauge network used in the study.

Stratified Z-R relationships were derived for four daily intervals, e.g. 0–6 h, 6–14 h, 14–19 h and 19–24 h local time, which were defined based on the variation of the hourly rainfall within the 240 km of the quantified Bauru range, as seen in Fig. 2. Radar reflectivity data are filtered for ground clutter in the operational routine, using the SIGMET proprietary G-Map product (SIGMET, 2007). IPMet's radar calibration variations have been kept within ±1 dB, as checked routinely every two weeks, on the average. Also cumulative probability curves indicative of calibration shifts are generated for the rainy periods.

Using a probability equating technique (Calheiros & Zawadzki, 1987), which takes into account radar calibration as well as the other factors mentioned below, three relations were generated, e.g. $Z = 16R^{1.7}$, $Z = 25R^{1.4}$ and $Z = 40R^{1.65}$. In fact, the values obtained for the multiplying coefficients in these relationships reflect the calibration status for BRU. Calibration differences among radars, such as those verified for the NEXRAD radars (Anagnostou *et al.*, 2001) and BRU (Anagnostou *et al.*, 2000) impact the respective Z-R relationships. Contributing to this, also, are issues like the polar-to-Cartesian conversions (Einfalt *et al.*, 2004), reflectivity

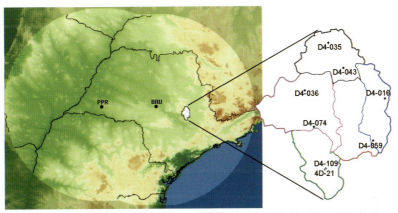

Fig. 1 Radar network operated by IPMet: BRU= Bauru, PPR=Presidente Prudente, range 450 km. In the detail: Corumbataí River basin and its five sub-catchments with 7-raingauge network and flow gauge station 4D-21 (Calheiros & Gomes, 2010).

February 1993 - 2005

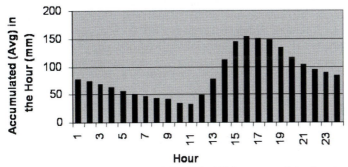

Fig. 2 Hourly distribution of rainfall within 240 km range of the Bauru radar, as an average for the month of February, spanning the period 1993–2005 (Calheiros & Tepedino, 2006).

averaging in the process of proper comparison to gauges (Calheiros, 1982), and radar product generation involving averaging as is the case with the CAPPI upon which are based the stratified relationships. The relations are for the daily intervals 0–14 h, 14–19 h and 19–24 h. A single relation was used for both 0–6 h and 6–14 h intervals, since the individual curves were too close together. An indication of the degree of stratification can be seen in Fig. 3.

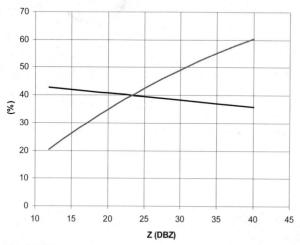

Fig. 3 Differences in rain-rate (%) for each given reflectivity, with reference to the relationship for the daily interval 14–19 h, when using the relationships for 0–14 h (blue line) and for 19–24 h (pink line).

Mean areal rainfall values were computed for each day for the month of November 1997 from: (a) raingauges, (b) a single Z-R relationship previously derived for the local conditions, e.g. radar calibration, and (c) the stratified conversion relationships. Radar data were CAPPIs at approximately 3.5 km above ground level, to the 240 km range, generated every 7.5 min, composed of reflectivity values represented by unit cells of 1×1 km^2 resolution and structured in a 480×480 matrix. Conventional rainfall data were daily accumulations from the raingauge network shown in Fig. 1, and river flow data (DAEE, 2002) were daily maxima values derived from measurements of river level at a key cross-section at the basin outlet (station 4D-21).

RESULTS AND ANALYSIS

Radar areal mean rainfall was computed with the SIGMET proprietary module Hydromet, structured with the SRI (Surface Rain Intensity) task and the CATCH (rain accumulation for a pre-defined time interval, over selected catchments (SIGMET, 2007)) task. The SRI product includes a VPR correction for stratiform rainfall. Hourly-accumulated CAPPI rainfall is used as input to CATCH providing the daily rainfall. Computations were made for each one of the Z-R conversion relationships, i.e. conventional (Marshall-Palmer), single locally-adjusted, and stratified composite. In the case of stratification, rainfall was computed for each daily interval using the respective relation, and the resulting three values were added to provide the daily rainfall. The resulting curves are shown in Fig. 4. Also plotted are the curve for the raingauge network and the corresponding maximum flow curve.

The time series of rainfall for the single locally-derived and the stratified composite relationships were then converted to the corresponding estimated flow time series through the rainfall–runoff relationship previously developed in the pilot project (Calheiros *et al.*, 2008). This relationship was from 30 events selected from the wet-to-dry transition and summer seasons, for

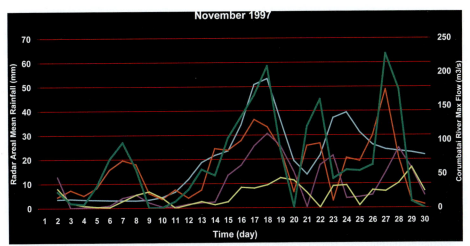

Fig. 4 Time evolution of values of: max flow (blue), mean areal raingauge rainfall (pink) and mean areal radar rainfall for the following Z-R relationships: Marshall-Palmer (yellow), RC_32 (Calheiros & Gomes, 2010) (orange) and stratified composite (green), for the Corumbataí River basin.

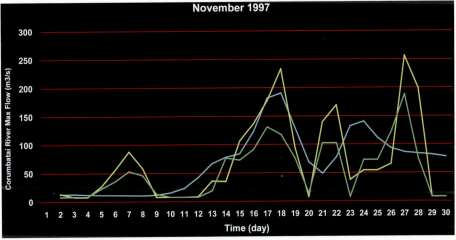

Fig. 5 Daily maximum flow curves for November 1997: historical (blue), with rainfall input from RC_32 single locally-derived *Z-R* relationship (green), and with rainfall input from stratified composed conversion relationships (yellow).

which flow data were available within the 1994–2004 BRU 10-year observation series. Daily radar areal mean rainfall obtained with the locally-adjusted Z-R relationship RC_32 (Calheiros & Gomes, 2010) was plotted against the maximum daily flow, and the fitted line provided the Q = 4.60 P – 36.7 [(Q = flow, m³ s⁻¹, P = rainfall (mm)] equation used in this paper. The resulting curves are presented in Fig. 5.

Curves in Fig. 4 show that, when the daily rainfall is computed with the stratified composite relationships, a better agreement in shape occurs for the main hydrological period from approximately 14 to 20 November. The maximum value of rainfall in this period increases by about 60% when the accumulated rain is computed with the stratified composite relationships as compared to the single locally-adjusted relationship.

In Fig. 5, for the same period of November, the flow using the stratified composite relationship – coincident with the historical peak – overestimates the historical peak by about 21%, while the use of the single locally-adjusted relationship results in an underestimation of about 32%. For the latter, peak time differs from that of the historical curve. It is also noted that for the period a rough estimate of the water volume from the stratified composite derived curve is comparable to that from the historical curve. All curves in Figs 4 and 5 are moving averages on the original data.

COMMENTS AND CONCLUSIONS

While results attained to date in the pilot project are encouraging, caution must be exercised when considering, in particular, the implementation of the mean areal radar rainfall product for an experimental phase, in the operational scenario at IPMet. Validation, which must be carried out, is not an easy task, due to major factors like the relatively scarce raingauge network with long time series. In this first project the raingauge network used to date features a density of only about 1 gauge per 230 km^2, and the measurement points are not evenly distributed throughout the catchments, as seen in Fig. 1. This is a priority problem to be tackled in the continuation of the work.

Notwithstanding, the Bauru radar has already been shown to be a substantial asset for hydrological applications in central São Paulo, as reported in previous works. The preliminary results obtained in this paper suggest that the stratification of the Z-R relationship by daily intervals may prove to be a requirement for many hydrological applications envisaged, and clearly justify the continuation of the studies. Although the radars operate at a practically non-attenuating wavelength (S-band) suited to observing strong convective storms and this results in less pronounced range effects, the use of range-dependent Z-R relationships is now being worked on by IPMet in relation to vertical profile effects.

Acknowledgements Thanks are due to Paulo Borges for support with data processing and figure editing, and to Hermes França and José Bassan for data handling.

REFERENCES

Anagnostou, E. N, Morales, C. A. & Tufa, D. (2001) The use of TRMM precipitation radar observations in determining ground radar calibration biases. *J. Atmos. Oceanic Technology* 18, 616–628.

Anagnostou, E. N., Morales, C. A. & Calheiros, R. V. (2000) Calibration of ground weather radar systems from TRMM precipitation radar observations: application to the S-Band Radar in Bauru, Brazil. In: *Proc. XI Brazilian Congress of Meteorology (CBMet2000)*. Available at http:// www.cbmet.com/.

Calheiros, R. V. (1982) Spatial resolution of radar rainfall estimates with hydrological radar. PhD Thesis, University of São Paulo, São Carlos, São Paulo, Brazil (in Portuguese).

Calheiros, R. V., Gomes, A. M. & D'Oliveira, A. B. (2008) Flow forecasting in the Corumbataí River Basin: radar rainfall characterization. In: *Int. Symp. on Weather Radar and Hydrology – WRaH2008* (Grenoble, France) (on CD-ROM).

Calheiros, R. V. & Gomes, A. M. (2010) Flow forecasting in the Corumbataí River Basin: radar rainfall stratification and runoff-rainfall relations. In: *Proc. 6th European Conference on Radar in Meteorology and Hydrology ERAD2010*, 429–433.

Calheiros, R. V. & Tepedino, P. (2006) Daily interval and range stratifications in rainfall measurements with the Bauru Radar, In: *Proc. 4th European Conference on Radar in Meteorology and Hydrology ERAD2006*, 202–205.

Calheiros, R. V. & Zawadzki, I. I. (1987) Reflectivity-rain rate relationships for hydrology in Brazil. *J. Climate Appl. Meteor.* 33, 682–693.

Calheiros, R. V. & D'Oliveira, A. B. (2007) Retrieval of the gross structure of reflectivity: correcting for degraded peak values. In: *33rd Conf. on Radar Meteorology*, Paper P6B.3, 7pp. Available at: https://ams.confex.com/ams/33Radar/ techprogram/paper_123112.htm.

DAEE (2002) Departamento de Aguas e Energia Elétrica (São Paulo State Water and Energy Survey). Available at: http://www.sighr.sp.gov.br.

Einfalt, T., Jessen, M. & Goltz, C. (2004) Searching for rainfall truth: multisensor thunderstorm analysis. In: *Proc. 3rd European Conference on Radar in Meteorology and Hydrology, ERAD2004*, 230–232.

Rennó, C. D. (2004) Construction of a hydrological analysis and simulation system: application to hydrographical basins. PhD Thesis. National Institute for Space Research – INPE. São José dos Campos, São Paulo, Brazil (in Portuguese).

SIGMET (2007) Interactive radar information system IRIS, IRIS products & display manual. Available at: http://www.sigmet.com/products/iris.

Derivation of seasonally-specific *Z-R* relationships for NEXRAD radar for a sparse raingauge network

SAMUEL H. RENDON[1], BAXTER E. VIEUX[1] & CHANDRA S. PATHAK[2]

1 *School of Civil Engineering and Environmental Science, University of Oklahoma, National Weather Center, 120 David L. Boren Blvd. Suite 5340, Norman, Oklahoma 73071, USA*
bvieux@ou.edu

2 *Operations and Hydro Data Management Division, SCADA and Hydro Data Management Department, South Florida Water Management District, 3301 Gun Club Road, West Palm Beach, Florida 33416-4680, USA*

Abstract Radar-based hydrological prediction relies on available raingauges to correct for bias in rainfall estimates. Standard *Z-R* (radar reflectivity factor against rain-rate) relationships have been developed which are characteristic of storm types, e.g. convective or tropical storms. However, the evolution of storm drop-size distribution and radar-specific factors can affect the accuracy of these standard relationships. Deriving *Z-R* relationships from raingauge observations for specific radars offers the potential for improved rainfall estimation. The derived *Z-R* relationship would be more representative of local climatology and radar characteristics, and can be used when raingauges are not available in real-time for bias correction. The purpose of this project is to derive and evaluate regionally- and seasonally-specific *Z-R* relationships for use in the South Florida Water Management District (SFWMD). These regionally specific relationships are expected to reduce bias in rainfall estimates found when using standard *Z-R* relationships, and lead to improved rainfall estimation for operational decisions. Validation of the derived *Z-R* relationships for dry, intermediate, and wet seasons revealed significant bias reduction to essentially 1:1 agreement during the respective seasons. While such relationships are not expected to replace bias adjustment using raingauges on a storm total or real-time basis, they do represent a better starting point for gauge adjustment of the *Z-R* relationship.

Key words radar; NEXRAD; *Z-R* relationships; hydrological forecasting; rainfall estimation

INTRODUCTION

The radar reflectivity factor, *Z*, is directly related to the drop-size distribution and thus can be used in the estimation of precipitation amounts as:

$$Z = \sum\nolimits_{i=1}^{n} N_i D_i^6 \tag{1}$$

$$Z = AR^b \tag{2}$$

where *Z* is the radar reflectivity factor in mm^6 m^{-3}, *n* is the number of drop size categories, N_i is the number of drops of diameter D_i to $D_i + \delta D$, *R* is the rainfall rate (mm h^{-1}) and *A* and *b* are fitting coefficients related to the drop-size distribution (Battan, 1973; Doviak & Zrnic, 1993; Bedient *et al.*, 2008). The most widely used forms of the *Z-R* relationship are the convective form where *A* = 300 and *b* = 1.4 and the tropical from where *A* = 250 and *b* = 1.2. These two forms of the *Z-R* relationship are referred to as the standard or default *Z-R* relationships herein. However, these standard *Z-R* relationships assume drop-size distributions that may not be valid for the South Florida area. Radar-based factors such as antenna, transmitter and receiver characteristics may also affect radar rainfall estimates. Of significance is the large number of *Z-R* relationships reported by Battan (1973), even for the same type of storm. Doviak & Zrnic (1993) suggest that radar equipment and the power transmitted/received by specific installations used in deriving these relationships, if not the variation in micro-physics and resulting drop-size distributions, determine the storm specific *Z-R* relationship. Methods of adjusting the *Z-R* coefficients *A* and *b* have been researched by Steiner & Smith (2000) where the multiplicative coefficient, *A*, is found to be far more sensitive than the exponent, *b*, when comparing storm-to-storm variations. Sensitivity of the *A* and *b* coefficients were demonstrated through application to streamflow prediction by Vieux & Bedient (2004) and Habib *et al.* (2008). Operational radar applications often rely on using *Z-R* relationships for a given type of storm, i.e. stratiform, convective, or tropical.

Several methods can be applied in deriving and adjusting the coefficients, *A* and *b*, in *Z-R* relationships. These calibration techniques are referred to here as the FIX method where only *A* is adjusted and *b* is fixed as a constant; the least square fitting (LSF) method that optimizes both *A* and *b*; and the BIAS_RMSE method used by Habib *et al.* (2008) that is based on minimizing root-mean-square errors to remove bias. Previous studies have shown that the BIAS_RMSE method and the FIX methods are superior estimation methods, especially at coarse time scales (Malakpet *et al.*, 2007). It has also been found that the multiplicative constant in a *Z-R* relationship, *A*, has a large effect on hydrological models and is especially sensitive to storm-by-storm and spatial variation (Steiner *et al.*, 2000; Bedient *et al.*, 2008; Habib *et al.*, 2008). Real-time adjustment of the *Z-R* relationships with gauge measured accumulations has been found to improve the accuracy of radar rainfall estimates, and so offline tests are needed to identify whether derived *Z-R* relationships can lead to more accurate rainfall measurements with radar even in the absence of raingauge bias adjustment, and whether seasonally-specific relationships are advantageous (Vieux & Rendon, 2009).

STUDY AREA

There are two distinct seasons in the South Florida area: the wet season and the dry season. In the wet season, from June to October, rainfall events can be characterized as either convective or tropical, with two thirds of the area's rainfall occurring in this season. The dry season is from December to April with either stratiform or frontal events. During the intermediate months of November and April, events with characteristics of both seasons may occur. The South Florida area is serviced by five NEXRAD stations KBYX (Key West), KAMX (Miami), KMLB (Melbourne), KBTW (Tampa) and KJAX (Jacksonville) radars. Although there are many other radar sites in the South Florida area, SFWMD uses only National Weather Service (NWS) owned NEXRAD stations as its radar input for reliability and longevity reasons. For this study KAMX will be used as the radar data input. There are 243 raingauges contained by the KAMX radar umbrella, which extends out to a radius of 230 km.

To derive the appropriate *Z-R* relationship for the KAMX to be used by SFWMD for radar-estimated rainfall, a series of events from each of the seasonal periods were selected. The study period spans 2007–2008 and is broken into seasonal periods corresponding to the natural seasons for the South Florida area: wet (June–September), dry (November–April) and intermediate transition periods. To gain a representative distribution, a minimum number of storm events to be analysed per month in each season were established, as presented in Table 1. From the radar and gauge data assembled for the candidate events, 20 events covered over 400 hours, with 12 events in the wet, five in the intermediate, and three in the dry season.

Table 1 Selection criteria for minimum number of storm events to be analysed.

Year	Season	Months	Number per month	Number selected
2007–2008	Dry	Nov–Apr	1	3
2008	Wet	Jun–Sep	2	12
2008	Transitions	May and Oct	2	5

METHODOLOGY

It is known that when using linear regression trend lines the coefficient of determination r^2 value only depends on the chosen *b* coefficient. It was then assumed that it was more efficient to calibrate the *b* coefficient first and then calculate an event-specific *A* coefficient. Varying the *b* coefficient and using the r^2 value, an optimum *b* coefficient may be found. Each event is then also analysed using the FIX method at the chosen *b* coefficient values. Both *G* (gauge accumulation) and R_{uncal} (radar estimated accumulation) rainfall hyetographs were converted to consistent units of mm h^{-1} and then the event totals were accumulated for each value of the *G* and R_{uncal} pairs for each gauge location. Each event was then analysed using *b* coefficients equal to 1.2, 1.3 and 1.4. The

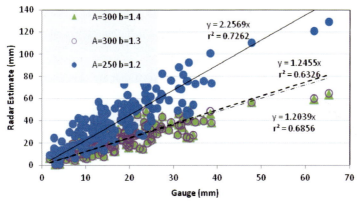

Fig. 1 Scatterplot of differing *Z-R* relationships for 15 June 2008 event.

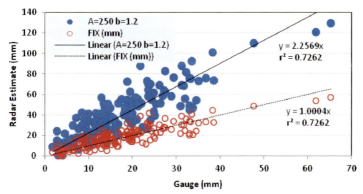

Fig. 2 Rainfall event totals for radar and gauge accumulations; filled circles are uncorrected and open circles are corrected radar rainfall.

scatter plot of radar estimate *versus* gauge for each of these *b* coefficient values for the 15 June 2008 event is shown in Fig. 1 where each point corresponds to a given raingauge.

As shown in Fig. 1, each of the individual *Z-R* relationships yield a different r^2 value as well as a different slope of the trend line (bias). In the MFM the first step is to choose the *b* coefficient value with the highest r^2 value. The highest r^2 value for this event is 0.72 for *b* = 1.2. Once the value of the *b* coefficient is chosen, the bias is corrected. The scatterplot of R_{uncal} *versus* G is shown in Fig. 2.

Using the slope of the trend line (R_{uncal} *versus* G) a simple bias correction was made while holding the *b* exponent constant at a set value of 1.2. The result is a storm-specific multiplicative coefficient, *A*, for the set *b* in the *Z-R* relationship.

RESULTS

The goal of this study was to find a *Z-R* relationship for each radar and seasonal period in the south Florida area. After evaluation of the bias *versus* the number of gauges used in the analysis, dependence was found in terms of the *Z-R* coefficients when fewer gauges were used. Therefore, a weighted average that gives more weight based on the number of gauges was used to calculate the bias, and then the *A* and *b* coefficients for each of the seasonal periods. The weighted average took the sum of the biases for each event multiplied by the number of gauges used in that event, and then divided by the total number of gauges used in that seasonal period. Using the MFM, event-specific *Z-R* relationships are found and compared with using seasonal *Z-R* relationships.

Seasonal evaluation

Seasonal evaluation was performed to identify the agreement in terms of r^2 and to determine appropriate A and b coefficients. These values are shown in Table 2 with the derived seasonally-specific Z-R coefficient values shown in bold.

As shown in Table 2, derived Z-R relationships are obtained that are characteristic of each seasonal period: $Z = 518R^{1.3}$, $591R^{1.2}$, and $615R^{1.2}$ for Dry, Intermediate, and Wet seasons, respectively. Investigation of the seasonal bias obtained with using derived Z-R relationships throughout, reduces the bias in rainfall estimation. The average biases before and after correction by seasonal period are shown in Table 3. The column indicated as the "Derived" results are obtained using the derived Z-R relationships, which presents a considerable improvement over those obtained with the default Z-R relationships (Standard). It should be emphasized that these are seasonal biases that can be significantly reduced using the derived relationships. However, individual storm biases can still be large even when the derived Z-R relationships are used. Thus, improved seasonal bias does not obviate the need for real-time or storm event bias correction.

Table 2 Derived Z-R coefficients by season.

	$b = 1.4$		$b = 1.3$		$b = 1.2$	
Season	A	r^2 est.	A	r^2 est.	A	r^2 est.
Dry	406.7	0.56	518.1	0.58	686.3	0.58
Intermediate	299.5	0.72	420.4	0.73	591.8	0.74
Wet	311.2	0.58	436.7	0.60	615.9	0.61

Table 3 Biases by seasonal period before (standard) and after correction (derived).

Season	Standard	Derived
Dry	1.24	1.00
Intermediate	2.05	1.00
Wet	2.11	0.99

Storm total evaluation

Storm total maps were generated for selected events to illustrate the difference between using standard and derived Z-R relationships. Sample storm total maps for both Z-R relationships are shown in Figs 3 and 4. These maps are for the 15 June 2008 event using $A = 250$ and $b = 1.2$ for the standard and $A = 615$ and $b = 1.2$ for the defined Z-R relationships.

As shown in Figs 3 and 4, the use of a different Z-R relationship does not change the spatial or temporal distribution of the rainfall, but does increase/decrease the storm total. The use of a derived Z-R relationship changes and improves the magnitude of the rainfall estimated over any given area, especially where raingauges are absent or sparsely located. Thus, the new A and b coefficients lead to a much closer match to the rainfall gauged accumulated over a season. The difference in magnitude is further demonstrated in Fig. 5, where the use of the standard Z-R relationship yields a bias as illustrated by the slope of the linear trend line of 3.06, which equates to the radar overestimating the rainfall rate by approximately 200%. Some over- or under-estimation is expected for any particular event. Thus, when using the seasonally-derived Z-R relationship with $A = 615$ and $b = 1.2$, rainfall is better estimated for this event than with the standard relationship, i.e. closer to a slope of 1.0 (filled blue circles in Fig. 5). For rainfall less than about 12 mm, the agreement is quite close to the event-specific relationship $A = 547$ and $b = 1.4$ when using the seasonally-specific derived Z-R relationship (Wet Season).

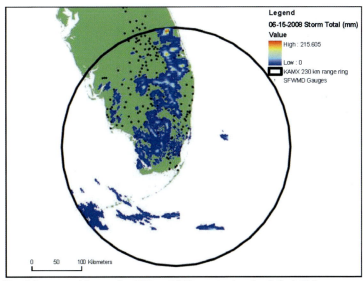

Fig. 3 Storm total in mm for 15 June 2008 event using the default *Z-R*.

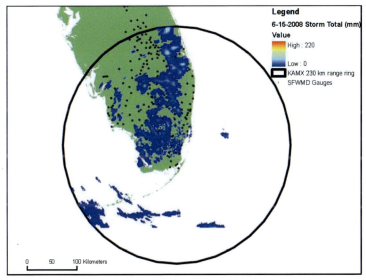

Fig. 4 Storm total in mm for 15 June 2008 event using the derived *Z-R* relationship.

SUMMARY

Seasonally derived *Z-R* relationships from raingauge observations for specific radars offer the potential for improved rainfall estimation. The derived *Z-R* relationship can be used when raingauges are not available in real-time for bias correction or before the storm has encountered gauges in a sparse network. In the South Florida region, use of the standard relationship $Z - 300R^{1.4}$ can lead to large biases in radar estimated rainfall when compared to raingauge data. Based on the 243 gauges intersecting the KAMX radar located in Miami, Florida, a set of storm events were selected that are representative of each seasonal period. From the radar and gauge data assembled for the

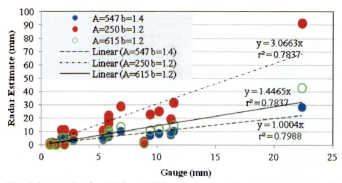

Fig. 5 Scatterplot for 15 June 2008 event using Standard, Event Specific and Seasonal *Z-R* relationships.

candidate events, 20 events covered over 400 h, with 12 events in the wet, five in the intermediate, and three in the dry season.

From comparison of the radar and gauge hyetographs, event-specific *Z-R* relationships are derived. The two-step method first identifies the *b* coefficient within the range 1.2 to 1.4 that maximizes the coefficient of determination r^2, and then the *A* coefficient is found that minimizes bias. Optimum *Z-R* coefficients were found for each event, and the mean *A* coefficient obtained for the three seasons based on the gauge-weighting scheme. Because the micro-physics of any specific event can deviate from the seasonal *Z-R* relationships, the seasonally-adjusted *Z-R* relationships correct for seasonal bias in rainfall estimation. Validation of the derived *Z-R* relationships for dry, intermediate, and wet seasons revealed significant bias reduction from 1.24, 2.05 and 2.11 to essentially 1.0, i.e. 1:1 agreement during the respective seasons. While such relationships are not expected to replace bias adjustment using raingauges on a storm total or real-time basis, they do represent a better starting point for gauge adjustment of the *Z-R* relationship.

REFERENCES

Battan, L. J. (1973) *Radar Observation of the Atmosphere*. Univ. of Chicago Press, Chicago, Illinois, USA.

Bedient, P. B., Huber, W. C & Vieux, B. E. (2008) *Hydrology and Floodplain Analysis*, 4th edn. Prentice Hall, New Jersey, USA.

Doviak, R. J. & Zrnic, D. S. (1993) *Doppler Radar and Weather Observations*. Academic Press.

Habib, E., Malakpet, C. G., Tokay, A. & Kucera, P. A. (2008) Sensitivity of streamflow simulations to temporal variability and estimation of *Z-R* relationships. *J. Hydrol. Engng* 13(12), 1177–1186.

Malakpet, A. G., Habib, E. & Meselhe, E. A. (2007) Sensitivity analysis of variability in reflectivity-rainfall relationships on runoff prediction. In: *AMS 21st Conference on Hydrology*, Session 3.

Marshall, J. & Palmer, W. M. (1948) The distribution of raindrops with size. *J. Atmos. Sci.* 5, 165–166.

Steiner, M. & Smith, J. A. (2000) Reflectivity-rain rate and kinetic energy flux relationships based on raindrop spectra. *J. Appl. Met.* 39(11), 1923–1940.

Vieux, B. E. & Bedient, P. B. (2004) Assessing urban hydrologic prediction accuracy through event reconstruction. *J. Hydrol.* 299(3–4), 217–236.

Vieux, B. E. & Rendon, S. (2009) Derivation and evaluation of seasonally specific *Z-R* relationships. *Final Project Report, South Florida Water Management District, West Palm Beach, Florida, USA. 23 September 2009.*

Weather Radar and Hydrology
(Proceedings of a symposium held in Exeter, UK, April 2011) (IAHS Publ. 351, 2012).

661

Weather radar to predict bathing water quality

MURRAY DALE[1] & RUTH STIDSON[2]

1 *Halcrow Group Ltd, Ash House, Falcon Road, Exeter EX2 7LB, UK*
dalem@halcrow.com

2 *Scottish Environment Protection Agency, Clearwater House, Heriot Watt Research Park, Avenue North, Riccarton,
Edinburgh EH14 4AP, UK*

Abstract Weather radar has significant theoretical advantages over raingauges when used for predicting episodes of poor bathing water quality in UK beaches: radar measures rainfall over areas, rather than at a point; radar data are available in real-time and do not require telemetry links; and the detail within a spatial radar image can isolate suspected pollution sources. Poor bathing water quality, characterised by high faecal coliform concentrations, is primarily caused by pollutants mobilised during wet weather in river and urban drainage catchments discharging close to beaches. With a revised Bathing Water Directive (2006/7/EC, repealing current Directive 76/160/EEC), which came into force on 24 March 2006, interest is increasing throughout the UK in developing techniques to predict faecal coliform exceedences. This paper describes the findings of a recent research project in which radar data were used to develop a methodology to improve real-time predictions of faecal coliform concentrations in bathing waters.

Key words radar; rainfall; bathing water quality; prediction

BACKGROUND

Rainfall is acknowledged as having a primary influence in causing episodes of high faecal coliform (FC) concentration in bathing water, a principal indicator of poor water quality (Crawther *et al.*, 2001). This pollution occurs through two key pathways; increasing runoff from agricultural land and combined sewer overflows (CSOs) spilling during times of heavy rainfall.

With a revised Bathing Water Directive (2006/7/EC, repealing current directive 76/160/EEC), which came into force on 24 March 2006, interest is increasing throughout the UK in the development of techniques to predict when FC exceedences will occur – both to enable information to be provided to the public and to enhance compliance with the Directive.

As a means of managing bathing beaches that are at risk of failing to meet the European Union (EU) Bathing Water Directive mandatory water quality standards, the Scottish Environment Protection Agency (SEPA) piloted a Scottish Government project to predict bathing water quality based on prior rainfall. These predictions were then displayed using electronic messaging signs at selected sites. Six sites in southwest Scotland were targeted in the 2003 bathing water season and 10 from 2004, including two outside the southwest area.

SEPA bathing waters signage currently uses point source rainfall measured by telemetered raingauges to make predictions of FC concentration to inform bathing water quality on a daily basis during the bathing water season. River gauge measurements are also used operationally by SEPA as an additional indicator of rainfall. The method is based on a correlation between rainfall data and coliform levels. The electronic messaging signs are updated following a poll of available telemetry raingauges and rivergauges at 09:00 h each morning during the bathing season. An example of a beach sign is shown in Fig. 1.

The Scotland and Northern Ireland Forum for Environmental Research (Sniffer) on behalf of SEPA and the Environmental Heritage Service of Northern Ireland (EHS) initiated this project on the premises that:

(a) existing prediction methods could potentially be improved by using rainfall radar data as an alternative or in addition to rain and river gauge data;

(b) the revised directive includes tighter FC standards to be met by 2015, requiring increased accuracy of predictions;

(c) using weather radar may be more practical than installing telemetered rain and river gauges at new sites;

(d) using weather radar may be better for obtaining midday updates and projections than raingauge and rivergauge networks.

Fig. 1 Example of bathing signage at Prestwick beach, Scotland (image courtesy of SEPA).

Since completion of the project, SEPA has been considering the potential for using radar data operationally in real-time for bathing water quality predictions.

PROJECT AIM

The aim of the project was to examine whether radar data can improve on existing methods of prediction of exceedences of FC concentrations to inform of poor water quality events. The project did not set out to compare and contrast radar and raingauge data directly, but to examine whether radar-derived thresholds could improve the prediction of exceedences of FC concentrations in bathing water.

The project aimed to develop techniques for use in real-time signage provision at bathing water sites in the UK, to inform the public of likely bathing water quality, and to aid future compliance with the EU Bathing Water Directive.

The study analysed two sites on the southwest coast of Scotland (Saltcoats and Irvine) and two sites in England (Fleetwood in Lancashire and Paignton-Preston Sands in Devon). Early results of this study that report on the findings of the Scottish sites only were published by Dale & Stidson (2009).

SUITABILITY OF RADAR DATA FOR QUANTITATIVE ANALYSIS AND OPERATIONAL USE

Radar data over the UK produced by the Met Office have increased in quality over recent years, such that they can now be used quantitatively with confidence in a range of hydrological applications (Dale, 2004; Lau & Sharpe, 2004). Quantitative applications of weather radar to urban hydrology are becoming more mature with better understanding of limitations and achievable accuracy. Research in the USA and Europe has led to operational applications of radar to site-specific forecasting of floods and management of sewer overflow during wet weather (Einfalt *et al.*, 2004; Vieux *et al.*, 2005)). However, there are no known examples of weather radar being used in real-time to predict water quality for bathing beach quality compliance purposes.

Raingauge data have the disadvantage of being point source data, requiring a high density of gauges to measure spatially variable rainfall, and they need to be established with telemetry to be used in real-time. This is most noticeable during localised, convective rainfall, which can result in very high rainfall intensities over a limited area and zero rainfall in adjacent areas. As well as ensuring that such localised rainfall is measured, radar can also identify peak intensities within a wet period, 24 h for example. Radar data can be received and analysed centrally without need for telemetry from the field.

Radar data can be used operationally in the same way as telemetered raingauges to allow a prediction of likely water quality to be made for the day ahead at 09:00 h. This prediction can then be displayed on electronic bathing water signs at beaches. A further advantage of radar data is that the data are available in real-time for any antecedent period at 5 min temporal resolution. In contrast, real-time raingauge data are available at 09:00 h for the previous 24 h period, with any alternative periods requiring further downloading of the data. Therefore, radar data can allow for bathing water sign updates to occur more easily throughout the day, should rain continue through the morning after the initial message has been issued to the public. Owing to this envisaged difference in operational use, the end time for all radar rainfall totals used in this project was the time of sampling, generally between 10:00 and 14:00 h.

Radar data for this study were from radar sites at Corse Hill (for the Scottish sites), Hameldon Hill (for Fleetwood) and Cobbacombe Cross (for Paignton-Preston Sands). The Scottish sites had 1 km resolution radar data coverage; the English sites have 2 km resolution coverage. For all events analysed in this study, a check was made of the time-series to identify missing radar data. This was not found to be a significant issue, with only a small number of events having missing data.

The radar data used in this project were measurements of actual rainfall (observations) – they were not forecast data. Products are available from the Met Office that use a mixture of sources, including weather radar, to provide quantitative precipitation forecasts that could be used in providing forecasts of likely water quality, probably 24 h or possibly 36 h in advance. Hence the potential exists to extend the lead-time of warning messages issued, but this must be balanced against the reduced accuracy of quantitative forecast data.

METHODOLOGY

Radar rainfall data were obtained from three different radars to cover the four study areas. These data were provided in the form of rainfall intensities at a temporal resolution of 5 min and 1-km or 2-km spatial resolution. A GIS software utility was developed for the project to construct the time-series of instantaneous rainfall values for each event and each catchment. This was used to evaluate the average rainfall over specified areas at 5-min intervals during multiple events. The end output was then a calculation of average rainfall across a number of grid cells within the catchment areas specified in a shape file. Using this function, grid squares with centres located within the boundaries of a polygon area were included in the calculations.

The radar data were processed to derive rainfall averages over a variety of potential pollution source areas. In total, 11 suspected pollution source areas were analysed for the four sites.

A method was derived to estimate the threshold rainfall depth as a predictor of FC exceedence. The method involved the assessment of as many rainfall-driven events as possible for the four catchments. An "event" in this context refers to a microbiology sample date that follows a period of wet weather that resulted in an exceedence of the pre-established rainfall depth thresholds in any one of the catchments' raingauges. Some events were attributable to suspected wet weather not measured by the raingauges but implied by an exceedence of the FC value of 500 cfu/100 ml.

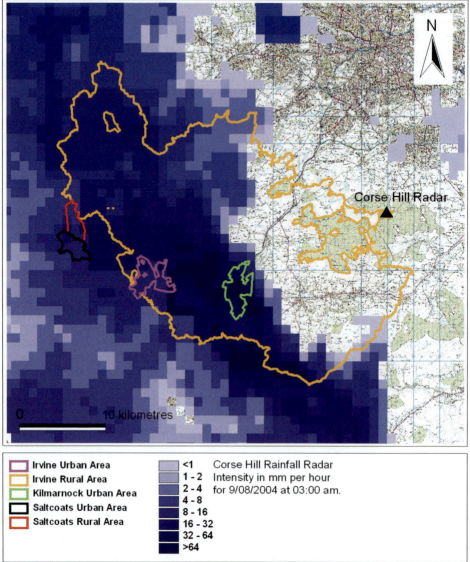

Fig. 2 Radar rainfall intensity data covering the Scottish study sites. (This map is reproduced with the permission of the Ordnance Survey, © Crown Copyright, SEPA, 100020538, September 2008.)

The threshold was determined by fitting a best-fit line through the event data (rainfall depth against FC concentration) plotted on a graph. At the point where the best-fit line intersected the FC500 level, the rainfall depth was identified from the graph and this formed the threshold value for a specific event duration (e.g. 1-h, 6-h, 24-h, 72-h). The fitting method was improved during the project and is described in detail in the project final report (SNIFFER, 2008).

RESULTS

In order to best compare the raingauge results with the radar results a like-for-like comparison was made by comparing the data from both sources using the thresholds determined by the radar data (optimising) method. An overview of results is shown in Table 1. This shows a general improvement in the ability of radar to correctly predict poor water quality events over raingauges at the sites with smaller upstream catchments (Saltcoats and Paignton). This increase in the overall percentage of correct predictions includes both an increase in the number of poor water quality events correctly predicted (termed "hits") and also a reduction in the number of incorrect predictions of poor water quality, occasions where the water quality is actually good but poor water quality is incorrectly predicted (termed "false alarms"). For the larger catchments, at Fleetwood and Irvine, the radar-derived results are similar for Fleetwood but poorer for Irvine than using raingauge data.

Table 1 Summary of project results, comparing raingauge against radar-derived predictions of poor water quality at 4 locations around the UK.

Site	Data source	Number of correct predictions (raingauge predictions use radar derived thresholds)	Percentage correct (%)
Saltcoats	Ashgrove raingauge	13 / 23	57%
	Saltcoats Rural Radar	16 / 23	70%
	Saltcoats Urban Radar	15 / 23	65%
	Saltcoats Radar aggregated	15 / 23	65%
Irvine	Raingauges (aggregated)	17 / 26	74%
	Irvine Urban Radar	9 / 26	35%
	Kilmarnock Urban Area	9 / 26	35%
	Irvine Rural Area	12 / 26	46%
	Irvine Radar aggregated	10 / 26	38%
Fleetwood	Raingauges (aggregated)	25 / 40	63%
	Wyre Urban Radar	23 / 40	58%
	Wyre Rural Radar	27 / 40	68%
	Lune Rural Radar	22 / 40	68%
	Fleetwood radar aggregated	27 / 40	68%
Paignton	Raingauges (aggregated)	6 / 11	55%
	Torbay Radar	9 / 11	82%
	Okham Radar	9 / 11	82%
	Paignton radar aggregated	9 / 11	82%

CONCLUSIONS

The project found that the radar performs well overall, particularly for smaller catchments in which it is shown to improve on the SEPA and Environment Agency prediction methods using raingauges. The radar performs slightly less well for larger catchments. This is likely to be due to factors including peak intensities in radar data being "smoothed" due to spatial averaging and additional complexities in catchment processes affecting the results rather than the quality of the radar data.

The fact that use of radar rainfall data has been shown to be at least as good as raingauge data in predicting exceedences of faecal coliform concentrations means that a radar-based system could operate where no raingauges exist and may be preferable for cost and practicality reasons (a cost–benefit analysis is a recommendation of the project). Furthermore, the ability to use forecast rainfall products from the Met Office up to 6-h ahead mean that increased lead-times can be achieved, provided the forecast quantities are reasonable predictions of actual quantities.

A further benefit of weather radar is that it can measure localised, convective rainfall which may occur in individual events or within frontal systems as embedded convection. At sites at which localised, high intensity rainfall is known to result in FC exceedences, the ability to analyse the peak intensity using radar, and have data available throughout the day, is an advantage. Other key conclusions from the study are as follows:

– Radar data can excel during highly localised, convective rainfall events (e.g. thunderstorms), or frontal events with embedded convective rainfall, which are missed by raingauges unless these are installed in a very dense network. At the Paignton site there was evidence that radar had measured a localised event that raingauges had missed.

– For the Scottish sites, it appeared that that the rainfall events were predominantly widespread frontal systems crossing the western coast of the UK from the southwest or west resulting in relatively uniform rainfall, with the exception of orographic enhancement (rainfall totals increasing with ground altitude). For this reason the radar data did not manage to explain the raingauge method's "misses" at the Scottish sites.

– Spatial averaging of radar rainfall data, where the radar pixels are averaged over polygon pollution-source areas, may result in "smoothing" of localised intense rainfall events (for example, from convective rainfall).

Acknowledgements The authors wish to thank SNIFFER (Scotland and Northern Ireland Forum for Environmental Research) as the body commissioning the research project described in this paper, and the project funders: the Scottish Environment Protection Agency (SEPA), Northern Ireland Environment Agency (NIEA) and the Environment Agency of England and Wales. The authors would also like to thank the members of the project Steering Group (SNIFFER, SEPA, Scottish Government, Scottish Water, Environment Agency and NIEA) for their support and assistance with this project. The Met Office is also gratefully acknowledged for commenting on the research and the radar data used. The views expressed by the authors of the paper are their own and do not necessarily reflect the views and policies of SEPA.

REFERENCES

Crawther, J., Kay, D. & Wyer, M. D. (2001) Relationships between microbial water quality and environmental conditions in coastal recreational water: the Fylde coast, UK. *Water Research* 35, 4029–4038.

European Parliament and Council (2006) Directive 2006/7/EC of the European Parliament and of the Council of 15 February 2006 Concerning the Management of Bathing Water Quality and Repealing Directive 76/160/EEC. Council Directive 76/160/EEC concerning the quality of bathing water. Official Journal of European Communities, 197-6, L31 (5.2.1976), 1–7.

Dale, M. (2004) Modelling with radar data – how and why? In: *Proc. Wastewater Planning Users Group (WaPUG) Autumn Meeting*, November 2004.

Dale, M. & Stidson, R. (2009) Weather radar to predict exceedences of faecal coliforms *Water Management* 162(2), 65–72.

Einfalt, T., Arnbjerg-Nielsen, K., Golz, C. & Jensen, N. E. (2004) Towards a roadmap for use of radar rainfall data in urban drainage. *J. Hydrol.* 299, 186–202.

Lau, K. T. & Sharpe, A. W. (2004) Sprint – spatial radar rainfall integrating tool for hydraulic models. In: *Proc. 6th Int. Conf. on Urban Drainage Modelling* (Dresden, Germany), 174–176.

SNIFFER (2008) Methods for estimating impacts of rainfall on bathing water quality – report on results for Saltcoats, Irvine, Fleetwood and Paignton, Project UKLQ07.

Vieux, B. E., Bedient, P. B. & Mazori, E. (2005) Real-time urban runoff simulation using radar rainfall and physics-based distribution modelling for site-specific forecasts. In: *Proc. 10th Int. Conf. on Urban Drainage* (Copenhagen, Denmark).

Key word index

Changes in Flood Risk in Europe

Edited by

Zbigniew W. Kundzewicz
Institute for Agricultural and Forest Environment, Polish Academy of Sciences,
Poznań, Poland, and
Potsdam Institute for Climate Impact Research (PIK), Potsdam, Germany

Floods are the most prevalent natural hazard in Europe. But, has flood risk increased in the continent? How, where, and why? Are climate change impacts apparent? How do socio-economic trends and associated land-use change impact flood risk?

Changes in Flood Risk in Europe

Edited by Z. W. Kundzewicz

This interdisciplinary book, authored by an international team, offers:

- A comprehensive overview of flood risk in Europe, past and present, and future
- National/regional chapters covering Central Europe, Western Europe, Southern Europe and Northern Europe, the Alpine region and the Iberian Peninsula.
- A focus on detection and attribution of change with respect to climate change and its impacts, water resources and flood risk, the re-insurer's view point, and future projections of flood risk
- Rectification of common-place judgements, e.g.: "climate is warming so floods should become more frequent and intense"; observations do not always confirm this expectation

The book will be of interest to those interested in floods and flood risk, including research scientists and educators, students, engineers, planners, risk reduction specialists, staff of specialized national and international agencies, and the media.

IAHS Special Publication 10

(*April 2012*) ISBN 978-1-907161-28-5 (Paperback); 516 + xvi pages
Price £85.00

Order **online** at *www.iahsmembers.info/shop.php* or contact:

Mrs Jill Gash
IAHS Press, Centre for Ecology and Hydrology
Wallingford, Oxfordshire OX10 8BB, UK

jilly@iahs.demon.co.uk
tel.: + 44 1491 692442
fax: + 44 1491 692448/692424

Book prices include postage worldwide. IAHS Members receive discounts on IAHS publications.
See *www.IAHS.info* for information about membership, publications, meetings and other activities.

Revisiting Experimental Catchment Studies in Forest Hydrology

Edited by

Ashley A. Webb, Mike Bonell, Leon Bren, Patrick N. J. Lane, Don McGuire, Daniel G. Neary, Jami Nettles, David F. Scott, John D. Stednick & Yanhui Wang

IAHS Publ. 353 (*2012*) ISBN 978-1-907161-31-5, 240 + viii pp. Price £56.00

Most of what we know about the hydrological role of forests is based on paired catchment experiments whereby two neighbouring forested catchments are jointly monitored during a calibration period of several years, after which one of the catchments is kept untouched as a reference (control), while the second is submitted to a forest treatment (impact). This volume, generated from a workshop that gathered forest hydrologists from around the world, with the aim of revisiting results and promoting a renewal of international collaboration on this topic, is divided into four sections:

- Addressing new questions using historical data sets
- Impacts of fires
- Water quality and sediment loads
- Ecosystem services

Abstracts of the papers in this volume can be seen at:

www.iahs.info